1st European Conference on Gas Micro Flows

(GasMems 2012)

Journal of Physics: Conference Series Volume 362

Skiathos, Greece
6-8 June 2012

ISBN: 978-1-62276-779-3
ISSN: 1742-6588

Printed from e-media with permission by:

Curran Associates, Inc.
57 Morehouse Lane
Red Hook, NY 12571

Some format issues inherent in the e-media version may also appear in this print version.

Printed by Curran Associates, Inc. (2013)

For permission requests, please contact the Institute of Physics
at the address below.

Institute of Physics
Dirac House, Temple Back
Bristol BS1 6BE UK

Phone: 44 1 17 929 7481
Fax: 44 1 17 920 0979

techtracking@iop.org

Additional copies of this publication are available from:

Curran Associates, Inc.
57 Morehouse Lane
Red Hook, NY 12571 USA
Phone: 845-758-0400
Fax: 845-758-2634
Email: curran@proceedings.com
Web: www.proceedings.com

Papers

Preface

The aim of the 1st European Conference on Gas Micro Flows is to advance research in Europe and worldwide in the field of gas micro flows as well as to improve global fundamental knowledge and to enable technological applications.

Gas flows in microsystems are of great importance and touch almost every industrial field (e.g. fluidic microactuators for active control of aerodynamic flows, vacuum generators for extracting biological samples, mass flow and temperature micro-sensors, pressure gauges, micro heat-exchangers for the cooling of electronic components or for chemical applications, and micro gas analyzers or separators). The main characteristic of gas microflows is their rarefaction, which for device design often requires modelling and simulation both by continuous and molecular approaches. In such flows various non-equilibrium transport phenomena appear, while the role played by the interaction between the gas and the solid device surfaces becomes essential. The proposed models of boundary conditions often need an empirical adjustment strongly dependent on the micro manufacturing technique.

The 1st European Conference on Gas Micro Flows is organized under the umbrella of the recently established GASMEMS network (www.gasmems.eu) consisting of 13 participants and six associate members. The main objectives of the network are to structure research and train researchers in the fields of micro gas dynamics, measurement techniques for gaseous flows in micro experimental setups, microstructure design and micro manufacturing with applications in lab and industry.

The conference takes place on June 6–8 2012, at the Skiathos Palace Hotel, on the beautiful island of Skiathos, Greece.

The conference has received funding from the European Community's Seventh Framework Programme FP7/2007–2013 under grant agreement ITN GASMEMS no. 215504. It owes its success to many people. We would like to acknowledge the support of all members of the Scientific Committee and of all referees for their thorough reviews and evaluation of the full papers. Above all, we would like to sincerely thank all authors for their valuable contributions to these proceedings as well as all the participants for creating a stimulating atmosphere through their presentations and discussions and making this conference a great success.

Dr Arjan Frijns
Editor and Event Coordinator

Prof. Dimitris Valougeorgis
Local Organizer

Prof. Stéphane Colin
Network Coordinator

Dr Lucien Baldas
Assistant Network Coordinator

Conference Organization

This 1st European Conference on Gas Micro Flows was organized by:

Local Organization
Prof. Dimitris Valougeorgis (University of Thessaly, Greece)
Secretarial support: Mrs Nikoleta Sachinidou (University of Thessaly, Greece)
Technical support: Mr Ioannis Lihnaropoulos; Mr Nikos Chiotoglou (University of Thessaly, Greece)

Overall Coordination
Dr Arjan Frijns (TU Eindhoven, The Netherlands)
Secretarial support: Mrs Linda Essink (TU Eindhoven, The Netherlands)
Prof. Stéphane Colin (INSA Toulouse, France)
Dr Lucien Baldas (INSA Toulouse, France)
Secretarial support: Mrs Chloé Prado (INSA Toulouse, France)
Prof. Dimitris Valougeorgis (University of Thessaly, Greece)

Scientific Committee
Dr Lucien Baldas (INSA Toulouse, France)
Dr Robert Barber (STFC, UK)
Dr Tobias Bauer (TU Dresden, Germany)
Prof. Jürgen J. Brandner (KIT, Germany)
Prof. Stéphane Colin (INSA Toulouse, France)
Prof. Giulio Croce (Universita degli Studi di Udine, Italy)
Dr Arjan Frijns (TU Eindhoven, The Netherlands)
Dr Silvia Gaastra-Nedea (TU Eindhoven, The Netherlands)
Prof. Irina Graur (Université de Provence, France)
Dr Anne-Marie Gué (LAAS, France)
Dr Gennady Markelov (AOES, The Netherlands)
Prof. Gian Luca Morini (Universita di Bologna, Italy)
Dr David Newport (University of Limerick, Ireland)
Dr Pierre Perrier (Université de Provence, France)
Prof. Jason Reese (University of Strathclyde, UK)
Prof. Stefan Stefanov (ImechBas, Bulgaria)
Prof. Dimitris Valougeorgis (University of Thessaly, Greece)
Dr Martin Wüest (INFICON, Liechtenstein)
Dr Yonghao Zhang (University of Strathclyde, UK)

Peer review statement

All papers published in this volume of *Journal of Physics: Conference Series* have been peer reviewed through processes administered by the proceedings Editors. Reviews were conducted by expert referees to the professional and scientific standards expected of a proceedings journal published by IOP Publishing.

Direct Simulation of Transport Properties from Three-Dimensional (3D) Reconstructed Solid-Oxide Fuel-Cell (SOFC) Electrode Microstructures

Naga Siva Kumar Gunda[1] and Sushanta K. Mitra[2]

Micro and Nano-scale Transport Laboratory, Department of Mechanical Engineering, University of Alberta, Edmonton, Canada, T6G 2G8

E-mail: [1]nagasiva@ualberta.ca and [2]sushanta.mitra@ualberta.ca

Abstract. A well-known approach to develop a high efficiency solid-oxide fuel-cell (SOFC) consists of extracting the microstructure and transport properties such as volume fractions, internal surface area, geometric connectivity, effective gas diffusivity, effective electronic conductivity and geometric tortuosities from three-dimensional (3D) microstructure of the SOFC electrodes; thereafter, performing the SOFC efficiency calculations using previously mentioned quantities. In the present work, dual-beam focused ion beam - scanning electron microscopy (FIB-SEM) is applied on one of the SOFC cathodes, a lanthanum strontium manganite (LSM) electrode, to estimate the aforementioned properties. A framework for calculating transport properties is presented in this work. 3D microstructures of LSM electrode are reconstructed from a series of two-dimensional (2D) cross-sectional FIB-SEM images. Volume percentages of connected, isolated and dead-ends networks of pore and LSM phases are estimated. Different networks of pore and LSM phases are discretized with tetrahedral elements. Finally, the finite element method (FEM) is applied to calculate effective gas diffusivity and electronic conductivity of pore and LSM phases, respectively. Geometric tortuosities are estimated from the porosity and effective transport properties. The results obtained using FEM are compared with the finite volume method (FVM) results obtained by Gunda et al. [J. Power Sources, 196(7), 35929(2011)] and other numerical results obtained on randomly generated porous medium. Effect of consideration of dead-ends and isolated-ends networks on calculation of effective transport properties is studied.

1. Introduction

A solid-oxide fuel-cell (SOFC) is an electrochemical conversion device that produces electricity by oxidizing a fuel directly [1]. Fuel cells are typically characterized by their electrolyte material; the SOFC having a solid-oxide or ceramic electrolyte.

Understanding the microstructure properties [2, 3, 4, 5, 6, 7, 8, 9, 10, 11, 12] and computing the transport properties [3, 4, 5, 13] in such solid-oxide electrolytes provides the information to improve the SOFC performance. Typically, these solid-oxide electrolytes are porous in nature and composite in material. Scanning electron microscopy (SEM) is a powerful technique for extracting two-dimensional (2D) images of the microstructures but it does not provide the third component (i.e depth) of the sample which is important to find interconnected regions and pore volumes, shapes and sizes [13]. In the last decade, extensive use of tomography [14] techniques like X-ray micro or nano computed tomography [11] and focused ion beam - scanning electron microscope (FIB-SEM)[6, 7, 13] tomography has been applied to extract series of cross-sectional images of porous samples in order to generate 3D structures. These 3D structures are used to understand the microstructural parameters (porosity, surface area, etc.) to calculate the effective transport properties (effective gas diffusivity, electronic conductivity, permeability, etc.).

The main focus of this paper is to provide a framework of calculating the effective transport properties of SOFC electrodes from the point of preparing a sample to solving the diffusion equation in 3D reconstructed structures using the finite element method (FEM). Significant amount of qualitative and quantitative information for measuring the microstructural parameters as well as the transport properties from the porous materials is provided. In this work, a SOFC cathode electrode, lanthanum strontium manganite (LSM), is investigated using a dual-beam FIB-SEM system and the transport properties are computed using the FEM. Section 2 explains the materials and methods used to prepare the LSM sample, collecting the cross-sectional images, 3D reconstruction from the collected images, and the solution methodology to calculate the transport properties. In section 3, results are discussed. Finally, in section 4, important points are pointed out to conclude the paper.

2. Materials and Methods

2.1. Sample Preparation

The SOFC cathode, LSM, was prepared by pulse spraying $(La_{0.8}Sr_{0.2})_{0.95}MnO_{3\pm\delta}$ powder (LSM, Praxair) onto a thick electrolyte disk $((Y_2O_3)_{0.08}(ZrO_2)_{0.92}$ (YSZ, Tosoh)) using an automated air-brush. Interested readers can refer Gunda et al. [13] for further details of sample preparation. Fig. 1 explains the step-by-step procedure of preparing the LSM sample.

2.2. Collection of 2D cross-sectional images

The methodology involves taking a 2D cross-section SEM image of the LSM cathode and then etching a thin layer from the LSM sample using the FIB to take subsequent SEM cross-section images. A series of cross-sectional images of LSM sample was collected using a dual beam FIB-SEM system (Zeiss N-Vision 40 crossbeam workstation). Figure 2(a) shows the FIB-SEM system (Zeiss N-Vision 40 crossbeam workstation) and Fig. 2(b) shows the set of cross-section images of the LSM sample. The detailed procedure of collecting the LSM sample using the FIB-SEM system can be obtained in Gunda et al. [13]. The collected images are $10.94nm \times 10.94nm \times 40nm$ each in voxel size. LSM has two phases, one is solid phase and other is pore phase.

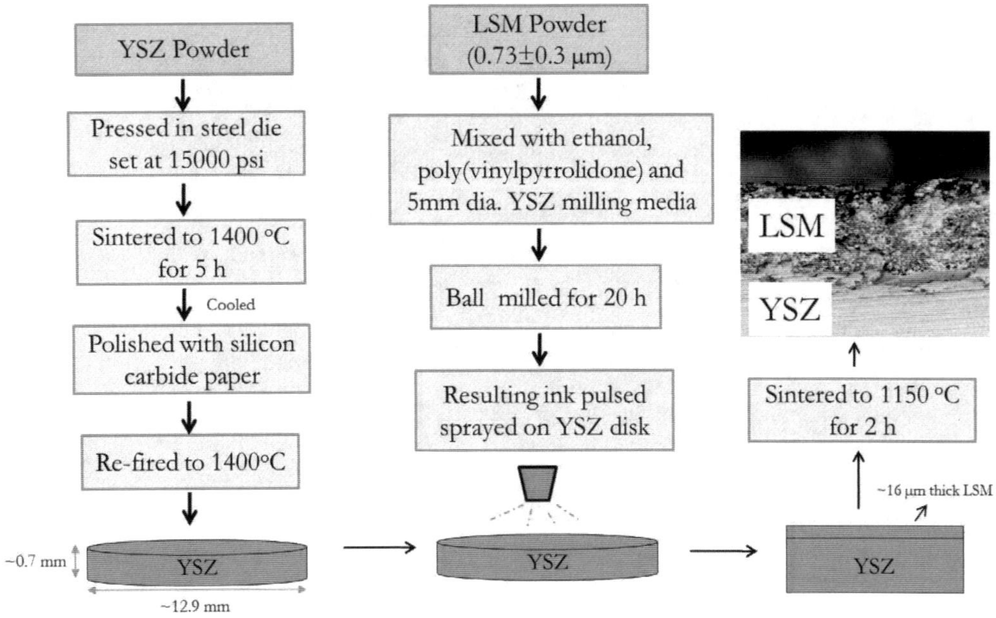

Figure 1. Step by step procedure of LSM sample preparation

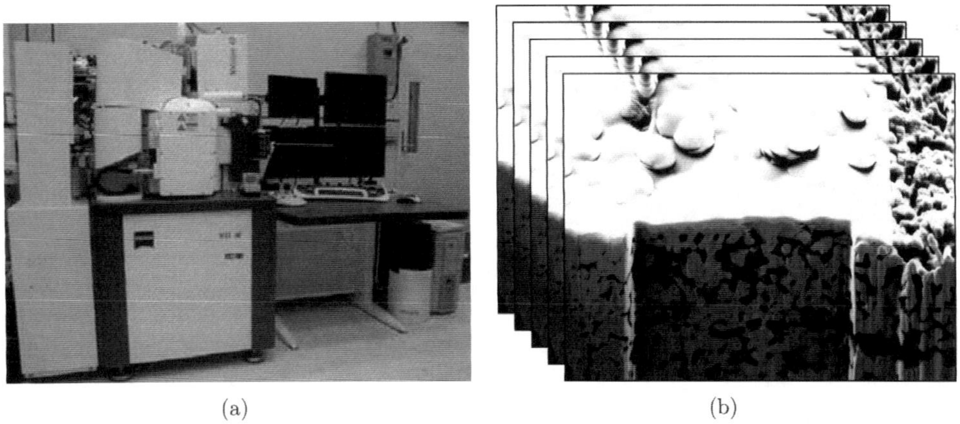

Figure 2. Collection of 2D cross-sectional images using FIB-SEM; (a) FIB-SEM system Zeiss N-Vision 40 crossbeam workstation; (b) set of cross-section images of the LSM sample

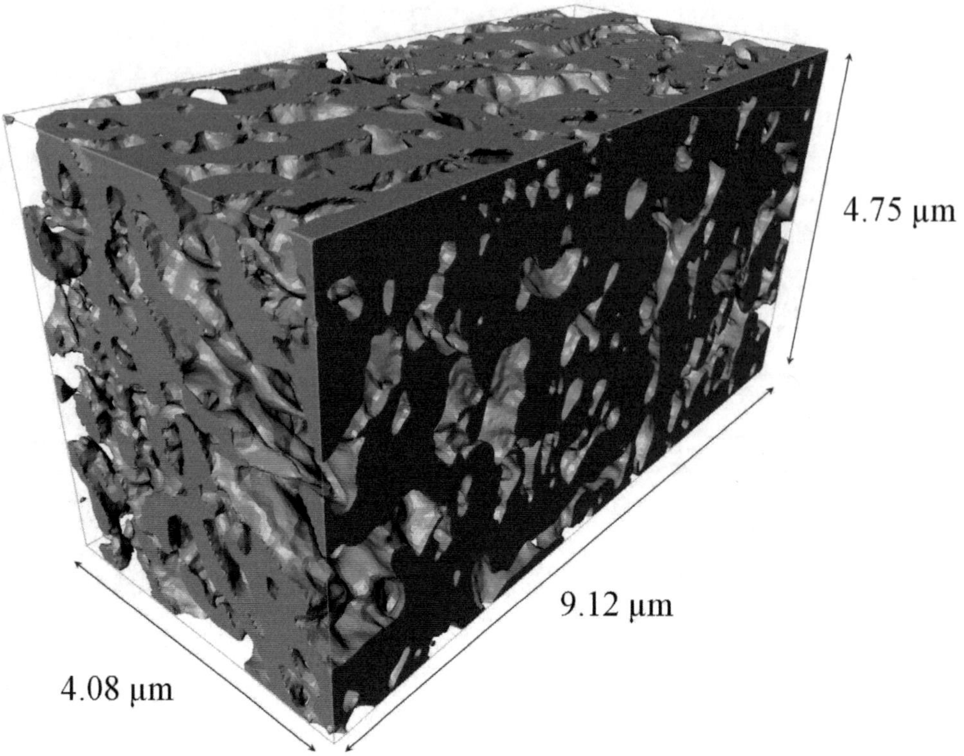

Figure 3. 3D microstructure of LSM sample

2.3. 3D reconstruction of LSM sample
Image processing of the collected images was performed using Avizo Fire edition version 7 (VSG, Visualization Sciences Group, Inc., Burlington, MA), before reconstructing the images. The details of different image processing steps and the reconstruction method applied to such series of images for 3D reconstruction can be found in Gunda et al [13]. A volume of 9.12 μm $\times$ 4.75 μm $\times$ 4.08 μm was selected from the 3D data which is shown in Fig. 3.

2.4. Computational Domain and Solution Methodology
The reconstructed 3D microstructures were used for computing the transport properties of the LSM sample. A computational domain of 2.74 μm $\times$ 1.22 μm $\times$ 1.43 μm was selected for calculating these transport properties. Based on computational source availability, we have restricted the computational domain for these dimensions. There is no thumb rule to take these dimensions. We have taken the domain from the middle of the reconstructed sample where there is no crater and no interface of electrode and electrolyte. Since we have taken the domain from the middle of the electrode, we assumed the transport process is taking place due to diffusion [15]. The concentration gradient is

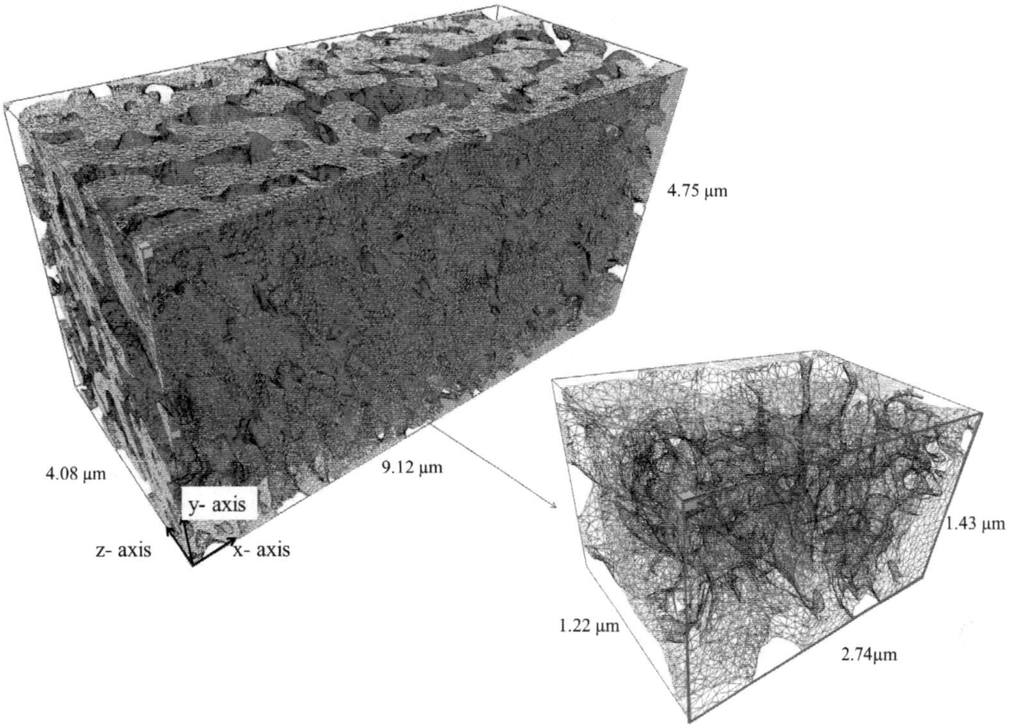

Figure 4. Computational domain used for calculating transport properties

the driving force to the diffusion. Dead-ends are the pores cannot be flushed, but they can cause fluid movement by release of pressure like gas expansion. Isolated-ends are the excluded closed pores (or non-connected cavities). Avizo Fire has the capability of removing these dead-ends and isolated-ends from the 3D reconstructed microstructures.

For each phase, the relevant potential field ϕ is governed by the diffusive process as follows:

$$\nabla \cdot (\alpha \nabla \phi) = 0, \quad \phi \in \Omega_{\text{phase}}, \tag{1}$$

where α denotes the bulk diffusion coefficient and Ω_{phase} is the computational volume of each phase. Note that each phase volume (Ω_{phase}) is embedded in the electrode sample volume (Ω) such that $\Omega = \Omega_{\text{LSM}} \bigcup \Omega_{\text{pore}}$. Dirichlet (fixed value) boundary conditions are imposed on the top and bottom boundaries of each phase, i.e., $\phi_{\text{top}} = 1$ and $\phi_{\text{bot}} = 0$. Symmetry (zero gradient) boundary conditions, i.e., $\frac{\partial \phi}{\partial n} = 0$, are presumed on the sides of the boundaries of each phase as well as the solid and pore phase interfaces.

In the present work, FEM based commercial software COMSOL Multiphysics Version 4.2a (COMSOL, Inc., Burlington, MA) was used to compute the effective gas diffusivity (D^{eff}) and the electronic conductivity (κ^{eff}) by solving the diffusion equation, shown in Eq.(1). Dead-ends and isolated-ends were removed from the 3D structures before discretizing the computational domain into tetrahedral elements. Computational domain considered in this work is depicted in Fig. 4.

The local transport flux q for each phase can be related to the generalized local concentration gradient such that:

$$q = \alpha \nabla \phi, \tag{2}$$

where the potential field ϕ is calculated from Eq.(1). The average (surface integrated) local flux across the surface boundary of each phase should be satisfied such that:

$$\int_{\partial\Omega} \alpha^{\text{eff}} \frac{\partial \phi}{\partial n}\, dS = \int_{\partial\Omega_{\text{phase}}} \alpha \frac{\partial \phi}{\partial n}\, dS. \tag{3}$$

Hence, the normalized effective transport coefficient can be evaluated by:

$$\frac{\alpha^{\text{eff}}}{\alpha} = \frac{\int_{\partial\Omega_{\text{phase}}} \frac{\partial \phi}{\partial n}\, dS}{\frac{\Delta\phi}{L} S}, \tag{4}$$

where $\Delta\phi$ represents the potential difference, L is the selected height of the computational domain and S denotes the selected boundary surface area of each phase. Note that $\frac{\alpha^{\text{eff}}}{\alpha} = \frac{D^{\text{eff}}}{D}$ represents molecular diffusion ratio for the pore phase and $\frac{\alpha^{\text{eff}}}{\alpha} = \frac{k^{\text{eff}}}{k}$ represents electrical conductivity ratio for the LSM phase.

3. Results and Discussions

Figure 3 represents the 3D reconstructed microstructure obtained after stacking the FIB-SEM 2D cross-sectional images. The coloured part in the Fig. 3 is the LSM solid phase and the transparent part is the pore phase. In this 3D structure, volume fraction of the pore phase is 49.4 % and volume fraction of the LSM phase is 50.6 %. The volume fractions of pore phase after removing the dead-ends and isolated-ends, is found to be 47.8 %(approx.) which tells that around 5.53 % of total pore phase contains the dead-ends and isolated-ends. Dead-ends contribute some error in the calculating effective tranpsort properties whereas isolated ends do not change the effective transport values. But, the dead-ends and isolated-ends contribute to convergence problems while calculating the effective transport properties. Here in this work, we have removed the dead-ends and isolated-ends to compute the exact effective transport properties.

A computational domain of 2.74 μm × 1.22 μm × 1.43 μm was selected from the volume of 9.12 μm × 4.75 μm × 4.08 μm for calculating the transport properties. Figure 4 shows the computational domain considered. The 3D computational domain shown in Fig. 4 was discretized by the finite element tetrahedral elements (108,376 elements). The normalized effective gas diffusivity and electronic conductivity are computed by Eq.(4) for this 3D computational domain. Table 1 summarizes the normalized effective properties in x, y, and z-directions calculated for this computational domain as well as the values obtained by the finite volume method (FVM)[13]. Gunda et al. computed normalized effective transport properties on computational domains reconstructed by the FIB-SEM as well as by a numerical model. Numerical models are generated based on the assumption of the same sample size and porosity of FIB-SEM reconstructed microstructure. The values for effective transport properties obtained in the present work using FEM after removing dead-ends and isolated-ends are 10% lesser than the values obtained by Gunda et al.[13]. Lesser size of the sample might be the reason for this low values. The present method is more efficient for smaller microstructures but not for bigger ones.

	FVM of Gunda et al. [13]				Present Work
Sample Volume	$9.12\mu m \times 4.75\mu m \times 4.08\mu m$		$4.98\mu m \times 3.04\mu m \times 4.08\mu m$		$2.74\mu m \times 1.22\mu m \times 1.43\mu m$
Technique	FIB-SEM	Numerical	FIB-SEM	Numerical	FIB-SEM
D_x^{eff}/D	0.304	0.298	0.276	0.265	0.243
D_y^{eff}/D	0.297	0.305	0.284	0.270	0.212
D_z^{eff}/D	0.225	0.312	0.211	0.290	0.189
k_x^{eff}/k	0.242	0.159	0.265	0.184	0.211
k_y^{eff}/k	0.225	0.176	0.294	0.207	0.197
k_z^{eff}/k	0.146	0.174	0.165	0.214	0.122

Table 1. Normalized effective properties in x, y, and z-directions based on the present work and Gunda et al.[13].

4. Conclusion

This paper presented a framework of calculating the effective transport properties of solid-oxide fuel-cell porous electrodes. The framework explained in this work has significant potential to extend into computation of other transport properties like permeability, hydraulic conductivity, etc. It can be used to compute transport properties of reservoir core samples. The information obtained in this work will be helpful to elucidate the microstructure/performance relationship of SOFC cathodes such as the influence of internal surface area on oxygen gas adsorption and electrochemical performance of the cathode. In addition, it will give a better understanding of transport through porous media as it applies to sintered SOFC cathodes with non-uniform particle sizes and perhaps, non-spherical shapes.

5. Acknowledgments

The authors would like to acknowledge Douglas Vick (National Research Council-National Institute for Nanotechnology) for performing the FIB-SEM on the LSM sample. Important discussions and support from John Pharoah's and Kunal Karan Group's (Queens University) are gratefully acknowledged. The authors gratefully acknowledge the financial support of Alberta Innovates - Technology Futures in the form of the scholarship provided to NSKG.

6. References

[1] Huang K and Goodenough J B 2009 *Solid oxide fuel cell technology: principles, performance and operations* (Woodhead Publishing)
[2] Giannuzzi L A and Stevie F A 2005 *Introduction to Focused ion beams: Instrumentation, Theory, Techniques and Practice* (Springer)

[3] Choi H W, Berson A, Kenney B, Pharoah J G, Beale S and Karan K 2009 *ECS Transactions* vol 25 pp 1341–1350
[4] Wilson J R, Cronin J S, Rukes S, Duong A, Mumm D and Barnett S 2009 *ECS Transactions* vol 25 pp 2283–2292
[5] Wilson J R, Duong A T, Gameiro M, Chen H Y, Thornton K, Mumm D R and Barnett S A 2009 *Electrochemistry Communications* **11** 1052–1056
[6] Wilson J R, Kobsiriphat W, Mendoza R, Chen H Y, Hines T, Hiller J M, Miller D J, Thornton K, Voorhees P W, Adler S B, Mumm D and Barnett S A 2007 *ECS Transactions* vol 7 pp 1879–1887
[7] Wilson J R, Kobsiriphat W, Mendoza R, Chen H Y, Hiller J M, Miller D J, Thornton K, Voorhees P W, Adler S B and Barnett S A 2006 *Nature Materials* **5** 541–544
[8] Iwai H, Shikazono N, Matsui T, Teshima H, Kishimoto M, Kishida R, Hayashi D, Matsuzaki K, Kanno D, Saito M, Muroyama H, Eguchi K, Kasagi N and Yoshida H 2010 *Journal of Power Sources* **195** 955–961
[9] Kenney B, Valdmanis M, Baker C, Pharoah J G and Karan K 2009 *Journal of Power Sources* **189** 1051–1059
[10] Shearing P, Golbert J, Chater R and Brandon N 2009 *Chemical Engineering Science* **64** 3928–3933
[11] Shearing P R, Gelb J and Brandon N P 2010 *Journal of the European Ceramic Society* **30** 1809–1814
[12] Ostadi H, Rama P, Liu Y, Chen R, Zhang X X and Jiang K 2010 *Journal of Membrane Science* **351** 69–74
[13] Gunda N S K, Choi H W, Berson A, Kenney B, Karan K, Pharoah J G and Mitra S K 2011 *Journal of Power Sources* **196** 3592 – 3603
[14] Ritman E L 2004 *Annual Review of Biomedical Engineering* **6** 185–208
[15] Bove R and Ubertini S 2008 *Modeling solid oxide fuel cells: methods, procedures and techniques* vol 1 (Springer Verlag)

Design of micro distribution systems consisting of long channels with arbitrary cross sections

S Misdanitis, D Valougeorgis

Department of Mechanical Engineering, University of Thessaly, 38334 Volos, Greece

E-mail: semisdan@mie.uth.gr, diva@mie.uth.gr

Abstract. Gas flows through long micro-channels of various cross sections have been extensively investigated over the years both numerically and experimentally. In various technological applications including microfluidics, these micro-channels are combined together in order to form a micro-channel network. Computational algorithms for solving gas pipe networks in the hydrodynamic regime are well developed. However, corresponding tools for solving networks consisting of micro-channels under any degree of gas rarefaction is very limited. Recently a kinetic algorithm has been developed to simulate gas distribution systems consisting of long circular channels under any vacuum conditions. In the present work this algorithm is generalized and extended into micro-channels of arbitrary cross-section etched by KOH in silicon (triangular and trapezoidal channels with acute angle of 54.74^o). Since a kinetic approach is implemented, the analysis is valid and the results are accurate in the whole range of the Knudsen number, while the involved computational effort is very small. This is achieved by successfully integrating the kinetic results for the corresponding single channels into the general solver for designing the gas pipe network. To demonstrate the feasibility of the approach two typical systems consisting of long rectangular and trapezoidal micro-channels are solved.

1. Introduction

Rarefied gas flows through long channels of various cross sections have been extensively investigated over the years based on linear kinetic theory [1–3]. It has been clearly shown that this approach is the most reliable one in the case of low speed flows providing very accurate results in the whole range of the Knudsen number with minimal computational effort. Very good agreement with corresponding experimental work developed in micro or either in vacuum experimental facilities has been obtained [4–6].

However, in many microfluidics applications, the gaseous distribution system consists of single channels accordingly combined in order to form a pipe network. Despite the fact that computational algorithms dedicated to the design of gas pipe networks in the continuum regime are well developed, corresponding tools for the design of gaseous pipe networks operating under rarefied conditions have not been developed so far. Recently, a first attempt to deal with this problem has been performed for the design of pipe networks consisting of long tubes [7, 8]. This is achieved by successfully integrating the kinetic results obtained for the rarefied flow through each tube of the network into a typical network algorithm solving the whole distribution system. In the present work this methodology is generalized for gas distribution systems consisting of long channels with arbitrary cross sections. Results are provided for pipe networks consisting of rectangular and trapezoidal channels in a wide range of the Knudsen number.

It is emphasized that the present work is limited to long channels, i.e. when the ratio of the length over the hydraulic diameter is larger than about one hundred. Based on this assumption, the pressure drop due to the end effects as well as the junction elements such as bends, T junctions, cross junctions etc. may be considered negligibly small compared to the pressure drop through the channel and is not taken into account.

2. Formulation and Algorithm

A typical pipe network may be considered as a directed linear graph consisting of a finite number of channel elements interconnected in a specified configuration. Each element is characterized of its specific geometrical characteristics such as diameter D for cylindrical channels, width W and height H for rectangular channels, size of big base B, small base b, height h and the acute angle ϕ for trapezoidal channels or side size a and acute angle ϕ for triangular channels. All channels have a specific length L, a cross section area A' and a perimeter Γ'. At this point we introduce the hydraulic diameter D_h, defined as

$$D_h = 4\frac{A'}{\Gamma'} \tag{1}$$

and it is used as the characteristic length of each piping element. In Table 1 the expressions for the hydraulic diameter, the cross section area and the perimeter for channels of various cross sections are shown. A point where two or more of the piping elements consisting a network are joined is known as a junction node or simply as a node. The closed path uniquely formed by adjacent pipes is a loop, while the open path connecting two fixed-grade nodes is a pseudo-loop. A fixed grade node is a node where a consistent energy grade is maintained (e.g. a constant pressure reservoir). Usually the geometry of the network is specified and the objective is to compute the flow quantities, i.e., the mass flow rates $\dot{M}$ through each tube and the pressure head P at each node.

Independent of the flow regime, the system of equations describing such a network consists of the pressure drop equations along each pipe element and the mass conservation equations at each node of the network. The pressure drop equations may be reduced to a set of the energy balance equations along the closed loops of the network. The latter ones, along with the mass conservation equations, form a closed set which may be solved for the unknown mass flow rates. Then, the pressure heads at the nodes are estimated through the pressure drop equations. In the case when the Knudsen number characterizing the flow through the network is very small and the flow is in the continuum or slip regimes, the pressure drop equations along each channel are given by explicit algebraic expressions and their integration in the whole algorithm is straightforward. In contrary, when the flow is in the transition regime, such expressions are not available. Here, this information is obtained from a data base, which has been developed for this purpose by solving a linearized kinetic equation in a wide range of the Knudsen number and obtaining the corresponding data.

This is a well described methodology recently applied to the design of networks consisting of circular elements [7, 8] and in the present work it is extended into channels of various cross sections. As a result, the corresponding equations for the kinetic solution for rarefied flow through a single tube are generalized and presented in the next section, followed by the description of the integrated algorithm for the flow solution of the whole network.

2.1. Generalized formulation for single pipes with arbitrary cross sections

The solution of a pressure driven rarefied gas flow through long channels based on linear kinetic theory is a very well known problem [1–4, 9–11] and therefore here, only information which is directly connected and needed to the solution of the whole network is reviewed. Since $D_h << L$, the flow is considered as fully developed and therefore the pressure (or density) varies only in

Table 1. Expressions of D_h, A' and Γ' for channels of various cross sections.

Cross section	D_h	A'	Γ'
Cylindrical	D	$\frac{\pi D^2}{4}$	πD
Rectangular	$2\frac{W\times H}{W+H}$	$W \times H$	$2\,(W+H)$
Equilateral triangular	$\frac{a}{\sqrt{3}}$	$\frac{\sqrt{3}}{4}a^2$	$3a$
Equilateral trapezoidal	$\frac{2(B+b)h}{B+b+\frac{2h}{\sin\phi}}$	$\frac{B+b}{2}h$	$B+b+\frac{2h}{\sin\phi}$

the flow direction being constant at each cross section, while end effects at the connecting nodes of the channel are ignored. The mass flow rate at each cross section through the piping element is given by

$$\dot{M} = \int \rho u' dA', \tag{2}$$

where ρ is the mass density and u' is the bulk velocity. The equation of state is given by $\rho = 2P/\upsilon_0^2$, where P is the local pressure along the piping element and υ_0 the most probable molecular velocity. By introducing into Equation 2 the equation of state as well as the corresponding dimensionless space variables along with the dimensional quantity for the bulk velocity $u = u'\,(\upsilon_0 X_P)$, where $X_P = (D_h/P)\,(dP/dz')$ is the local pressure gradient and following a straightforward procedure [4, 11, 12] it is found that

$$\dot{M} = G^* \frac{A' D_h}{\upsilon_0} \frac{P_1 - P_2}{L}. \tag{3}$$

Here, P_1 and P_2 denote the pressure head at the two ends of the tube and

$$G^* = \frac{1}{\delta_1 - \delta_2} \int_{\delta_1}^{\delta_2} G\,(\delta) d\delta. \tag{4}$$

The local rarefaction parameter δ is proportional to the inverse Knudsen number and it is defined by

$$\delta = \frac{P D_h}{\mu \upsilon_0} = \frac{\sqrt{\pi}}{2} \frac{1}{Kn}, \tag{5}$$

with μ denoting the gas viscosity at some reference temperature. The values δ_1, δ_2 correspond to P_1, P_2 and denote the rarefaction parameters at the two ends of the tube. Finally, the quantity $G\,(\delta)$ is the dimensionless flow rate obtained by solving a suitably chosen linearized kinetic model equation subject to some boundary condition. For the purposes of the present work, the dimensionless flow rates are estimated based on the linearized BGK equation with Maxwell purely diffuse boundary conditions. It is noted that the latter choice is related to the assumed roughness of the tube. The quantity $G\,(\delta)$ may be easily found in the literature (or computed) in the whole range of δ. These results are kept in a data base and are used in the solution of the network. In particular, in the process of solving the whole network, the pressures P_1, P_2 and therefore δ_1, δ_2 are estimated and then the values of $G\,(\delta)$, obtained from the data base, are used in the integration according to Equation 4 to yield G^*, which is substituted next into Equation 3 to deduce the mass flow rate for each piping section of the network.

2.2. Generalized network equations

As mentioned above, the initial system of equations describing the network consists of the pressure drop equations along each piping element and the mass conservation equations at each node of the network. The former ones are given by solving Equation 3 for the pressure difference to yield

$$(P_1 - P_2)_j = \Delta P_j = \frac{\dot{M}_j}{G_j^*} \frac{L_j \upsilon_0}{A'_j D_{h_j}}, \tag{6}$$

where the index $1 \leq j \leq p$ denotes each of the p pipes of the network. The latter ones may be expressed as

$$\left[\sum_j (\pm) \dot{M}_j - Q \right]_i = 0, \tag{7}$$

where the index $1 \leq i \leq n$ denotes each of the n junction nodes of the network, while the summation index j refers to the pipes connected to the node i and Q_i is the external demand (if any) at node i. The plus and minus signs are used for flow into and out of the node i respectively. Equations 6 and 7 are coupled and may be solved for the unknown pressure heads and mass flow rates.

However, it is convenient to reduce the number of equations by combing the pressure drop equations (Equation 6) along each uniquely determined closed loop of the network to derive the so-called energy balance equations:

$$\left[\sum_j (\pm) (\Delta P_j) \right]_k = 0. \tag{8}$$

Here the summation index j pertains to the pipes that make up a loop, while the index $1 \leq k \leq l$, denotes each of the l loops. The plus sign is used if the flow in the element is positive in the clockwise sense; otherwise the minus sign is employed. By substituting Equations 6 into 8 the energy balance equations become

$$\left[\sum_j (\pm) \left(\frac{\dot{M}_j}{G_j^*} \frac{L_j \upsilon_0}{A'_j D_{h_j}} \right) \right]_k = 0. \tag{9}$$

Following this procedure, the pressure heads have been eliminated from the set of Equations 9, which along with the set of Equations 7 form a well defined system having as unknowns only the mass flow rates. Once this system is solved for the mass flow rates, then the pressure heads are obtained from Equations 6. In the case when there are fixed-grade nodes in the network, the system of equations for the mass flow rates is amplified by the energy balance equation around each pseudo-loop connecting two fixed grade nodes:

$$\left[\sum_j (\pm) \left(\frac{\dot{M}_j}{G_j^*} \frac{L_j \upsilon_0}{A'_j D_{h_j}} \right) + \Delta H \right]_m = 0. \tag{10}$$

Here, the summation index j pertains to the pipes that make up a pseudo-loop, the index $1 \leq m \leq f - 1$, denotes each of the $f - 1$ pseudo-loops (f is the number of fixed grade nodes) and ΔH is the difference in magnitude of the fixed-grade nodes in the path ordered in a clockwise fashion across the imaginary pipe in the pseudo-loop. The plus and minus signs follow the same arguments given for Equations 8 and 9.

Based on the above, the final system of equations consists of $n + l + f - 1$ equations to be solved for the p unknown mass flow rates $\dot{M}_j$. However, it is important to note that since in the system consisting of Equations 7, 9 and 10 the quantities G_j^* are not known a priori, an overall iterative algorithm incorporating the set of Equations 6 is needed. The detailed description of this algorithm is presented in the next section.

2.3. Numerical algorithm

The developed code includes the drawing of the network in a graphical environment and the formulation and solution of the governing equations describing the flow conditions of the micro distribution system. At the present stage the code is able of providing the input data and solving pipe networks of various complexity.

The drawing of the network and the input of the data are prepared on a graphical user interface (GUI). The development of the graphic interface is based on available GNU General Public License (GPL) libraries including some new libraries written in javascript to match the needs of the application. As a result, the user is able to draw the desired network by adding nodes and edges and the corresponding data, i.e. the coordinates of the nodes in a 3D space, the specific geometrical characteristics of each piping elements, the pressure heads of the fixed-grade nodes and information for the type of the gas and its properties (viscosity, most probable molecular velocity, etc.). The demands (if any) at the nodes are also provided. Therefore, a connectivity matrix for each node and element of the network is formed providing all necessary information as input data.

Once the network is drawn, the resulting input file is introduced into a Fortran code. The code is scanning to find all possible closed loops, keeping finally the l uniquely determined loops consisting of the least number of piping elements and the $f-1$ pseudo-loops between the f fixed-grade nodes. In addition, the code is based on an iterative process between the pressure drop equations (Equation 6) and the system of mass and energy conservation equations consisting of Equations 7, 9 and 10, which may be summarized as follows:

1. At all nodes of the network, where the pressure head is unknown it is initially assumed and the pressure differences ΔP_j along each tube are stored.
2. The rarefaction parameter at each node is estimated by Equation 5.
3. The quantity G_j^* is estimated by Equation 4 for each tube using the available data bank for the dimensionless flow rate $G(\delta)$. Cubic splines are used to interpolate if needed between the values provided in the data bank.
4. The system of mass conservation and energy balance equations (Equations 7, 9 and 10) is solved by applying Gauss elimination with full pivoting to compute the mass flow rates $\dot{M}_j$ through each pipe section.
5. Equations 6 are solved to estimate the updated pressure drops ΔP_j.
6. The updated values of ΔP_j are compared with the ones in Step 1 and the whole process is repeated upon convergence.

3. Results

To demonstrate the feasibility and the effectiveness of the proposed methodology the sample network shown in Figure 1 is simulated. The network consists of $p = 14$ tubes, $n = 9$ junction nodes, $f = 2$ fixed-grade nodes and $l = 4$ loops. Nodes 1 and 11 refer to two reservoirs, where the pressure is held constant. The micro distribution system is characterized by the availability of introducing demands at nodes 4 and 8, however, for comparison reasons these two values are set to 0. The reference temperature is set to $T_0 = 293.7$ K. The conveying gas is argon (Ar), with molar mass $m = 0.039948$ kg/mol, gas constant $R = 208$ J/(kg $\cdot$ K), most probable molecular velocity $\upsilon_0 = 347.74$ m/s and viscosity $\mu = 22.7985$ μPa $\cdot$ s.

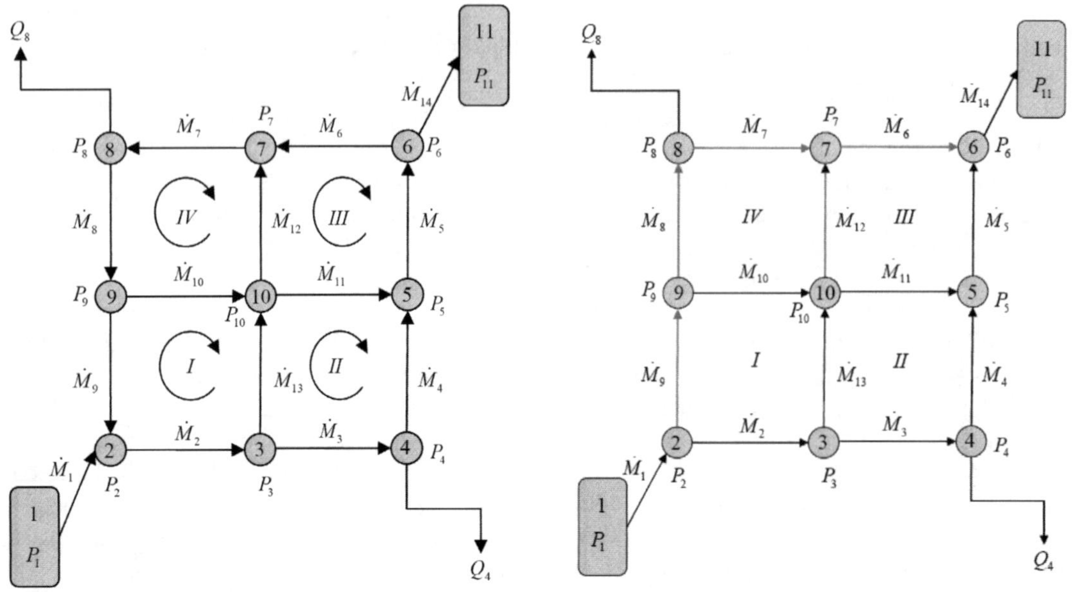

Figure 1. Schematic representation of pipe network showing a) the initially assumed flow directions (left) and b) the flow directions derived after the completion of the code (right).

The system of governing equations includes nine mass conservation equations at nodes 2,3,...,10, four energy balance equations along the closed loops I,II,III,IV and one energy balance equation along the open pseudo-loop formed along the nodes 1,2,3,4,5,6,11. The total number of equations of the system is 14 and its solution returns the 14 unknown mass flow rates $\dot{M}_1, ..., \dot{M}_{14}$. Then, from the pressure drop equations, the pressure heads $P_2, ..., P_{10}$ are found.

Two typical simulations are performed. The first one by simulating a network which consists of rectangular micro-channels with $H = 2.00$ μm and $W = 20.00$ μm, leading to an aspect ratio of $H/W = 0.1$ and a hydraulic diameter $D_h = 3.63$ μm and the second one where the micro-distribution system consists of equilateral trapezoidal elements with size of big base $B = 6.87$ μm, small base $b = 1.80$ μm, height $h = 3.60$ μm and acute angle $\phi = 54.74^o$, leading to a hydraulic diameter $D_h = 3.56$ μm. In all cases considered, the length of the piping elements is kept constant and equal to $L = 5.00$ mm, while the pressures at both the upstream and downstream reservoirs are fixed to $P_1 = 0.125$ MPa and $P_{11} = 140$ Pa, respectively. Both networks simulated are compared, while keeping the same configuration, to a network consisting of cylindrical tubes of the equivalent hydraulic diameter D_h.

The results are shown in Figure 2 for the rarefaction parameter, the pressure heads [Pa] and the mass flow rates [kg/s] for both networks simulated. The corresponding results using the hydraulic diameter concept are also included. The rarefaction parameter covers the range $0.0642 \leq 1/Kn \leq 57.33$ for the rectangular case and $0.0630 \leq 1/Kn \leq 56.28$ for the trapezoidal one. The flow in the network covers the slip, transition and free molecular regimes. The negative values at some of the mass flow rates indicate that the final direction of the flow in this piping element is opposite to the one initially assumed. The red arrows shown in Fig. 1 indicate the corrected flow direction obtained after convergence of the code.

Next, we comment on the comparison of the simulated networks with the corresponding ones consisting of cylindrical channels with the equivalent hydraulic diameter. In the case where all channels have the same hydraulic diameter, the friction factor is the same and therefore the pressure drop is similar. As a result, the rarefaction parameter is close for every node of the

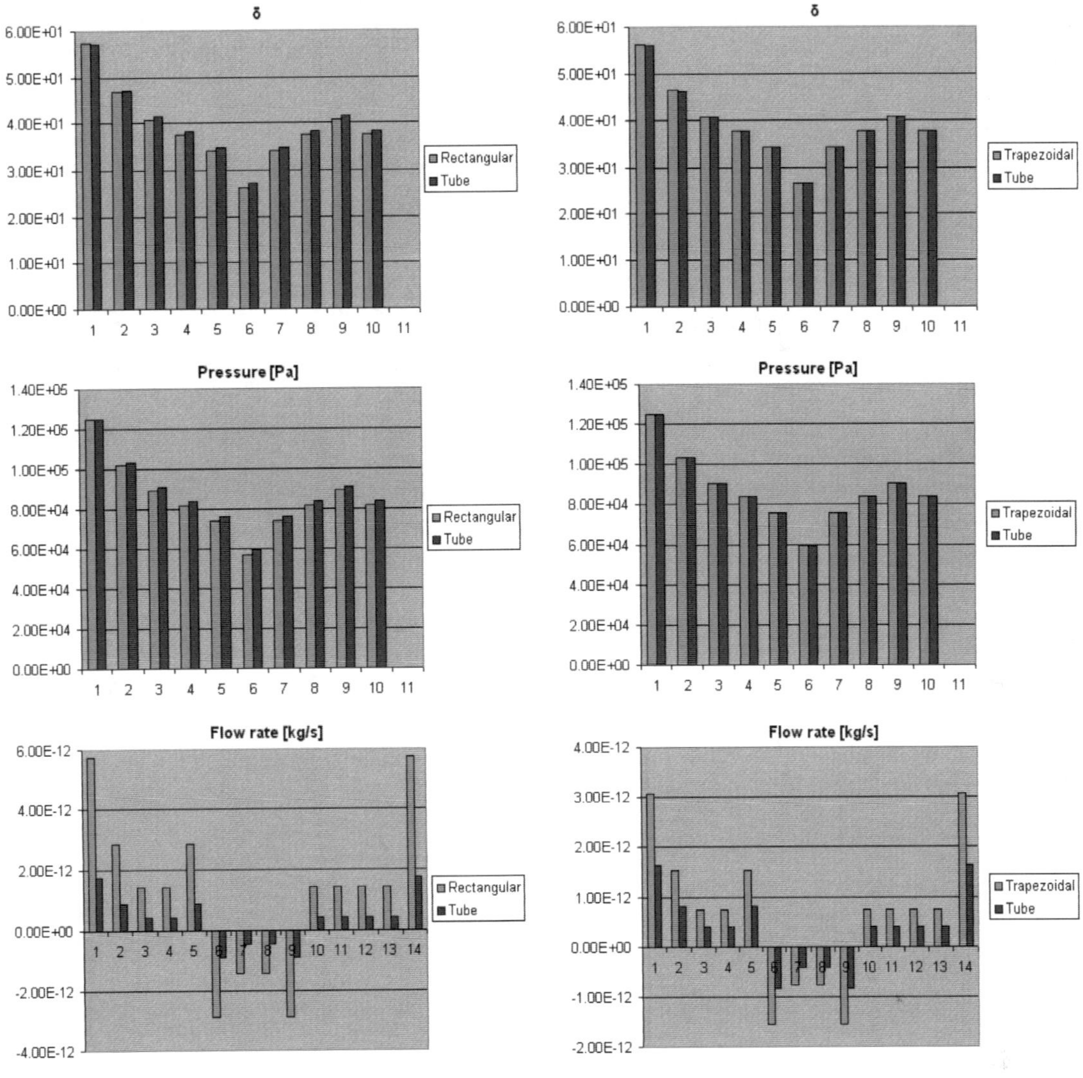

Figure 2. Rarefaction parameter (top), pressure heads (middle) and flow rates (bottom) for a network consisting of rectangular (left) and trapezoidal (right) channels. The corresponding results with the hydraulic diameter concept are also shown.

network in both geometries simulated. The relative error between the rectangular channels and the cylindrical ones is less than 4.11% for the pressure heads. Similar results can be seen for the comparison between the trapezoidal and the cylindrical channels, where the the relative error is less than 0.09% for the pressure heads. However, in both comparisons, despite the good agreement between the rarefaction at the nodes and the pressure heads, the comparison with regard to the mass flow rate through each of the piping elements of the networks shows that there are significan differences. For the case of the rectangular channels there is a dicrepancy around 220% for the flowrates and for the case of the trapezoidal channels around 85%.

In order to provide an estimation of the involved computational effort it is noted that the solution of the sample network requires the CPU time of seconds on a 3 Ghz dual core system. It is obvious that the involved computational effort is negligible. For more complex networks consisting of hundreds of tubes the computational effort it will be increased but not significantly.

4. Concluding remarks

An existing "in house" algorithm for the design of gaseous distribution systems consisting of long channels has been generalized by taking into account channels with arbitrary cross sections. A graphical interface is developed in order to draw the network and create the output file which is directly linked to the main iterative algorithm. The algorithm is based on the integration of a data base created for this purpose by solving the linearized BGK equation with Maxwell purely diffuse boundary conditions in the whole range of the Knudsen number. As a result the integrated algorithm may successfully handle gas pipe networks of any complexity operating under any vacuum conditions from the free molecular, through the transition up to the slip and hydrodynamic regimes yielding the mass flow rate through the piping elements as well as the pressure heads and the Knudsen number at the nodes of the network. The effectiveness of the methodology has been demonstrated by solving a network consisting of rectangular channels and one of trapezoidal channels and comparing them with the equivelant network consisting of cylindrical channels of the same hydraulic diameter.

The developed algorithm may be further extended to pipe networks consisting not only of long channels but also of channels of finite length. This goal is going to be achieved once the available kinetic data base is enhanced with the corresponding results. It is hoped that present algorithm will constitute a significant engineering tool in the design and optimization of gaseous micro distribution networks operating under any rarefied conditions.

Acknowledgments

This work has been supported by the European Communities under the contract of Association EURATOM / Hellenic Republic. Registration and accommodation expenses to the 1st European Conference on Gas Micro Flows have been covered from the European Community's Seventh Framework programme (FP7/2007-2013) under grant agreement no 215504. The views and opinions expressed herein do not necessarily reflect those of the European Commission.

References

[1] Valougeorgis D, Thomas J R 1986 *Phys. Fluids* **29** 423–429
[2] Sharipov F 1999 *J. Vac. Sci. Technol. A* **17** 3062–3066
[3] Naris S, Valougeorgis D 2008 *Eur. J. Mech. B/Fluids* **27** 810–822
[4] Varoutis S, Naris S, Hauer V, Day C, Valougeorgis D 2009 *J. Vac. Sci. Technol. A* **27** 89–100
[5] Pitakarnnop J, Varoutis S, Valougeorgis D, Geoffroy S, Baldas L, Colin S 2010 *Microfluid & Nanofluid* **8** 57–72
[6] Szalmas L, Pitakarnnop J, Geoffroy S, Colin S, Valougeorgis D 2010 *Microfluid & Nanofluid* **9** 1103–1114
[7] Misdanitis S, Valougeorgis D 2011 *Proc. of the 3rd GASMEMS Workshop* (Bertinoro, Italy)
[8] Misdanitis S, Valougeorgis D 2011 *J. Vac. Sci. Technol. A* **29** 061602:1–7
[9] Breyiannis G, Varoutis S, Valougeorgis D 2008 *Eur. J. Mech. B/Fluids* **27** 609–622
[10] Graur I, Sharipov F 2008 *Eur. J. Mech. B/Fluids* **27** 335–345
[11] Sharipov F, Seleznev V 1998 *J. Phys. Chem. Ref. Data* **27** 675–706
[12] Valougeorgis D 2007 *Phys. Fluids* **19** 091702:1–4

Flow rate measurements of binary gas mixtures through long trapezoidal microchannels

Lajos Szalmas[1,2], Stéphane Colin[1] and Dimitris Valougeorgis[2]

[1] Université de Toulouse; INSA, UPS, Mines Albi, ISAE; ICA (Institut Clément Ader); 135 avenue de Rangueil, F-31077 Toulouse, France

[2] Department of Mechanical Engineering, University of Thessaly; 38334 Volos, Greece

E-mail: lszalmas@gmail.com, stephane.colin@insa-toulouse.fr, diva@mie.uth.gr

Abstract. The flow rate of two noble gas mixtures, namely He/Ar and He/Kr, is measured through a microsystem containing 400 long trapezoidal microchannels placed in parallel configuration. Each microchannel has a trapezoidal cross section with long base 5.38 micrometers and height 1.90 micrometers, while its length is 5000 micrometers. The experiment is based on the constant volume method. The flow is driven by pressure gradient. The flow rate measurements refer to downstream pressures of 15.1 kPa and 8.05 kPa. The pressure ratio is in the range of 3-7 and 4-7 for the larger and smaller downstream pressures, respectively. The investigated rarefaction range is in the slip and early transition regions. The concentration of He varies from zero to one. The measured flow rates are compared to the corresponding computational ones obtained by the numerical solution of the McCormack kinetic model. Very good agreement between the experimental and computational results is reached. The difference between the corresponding results is less than the experimental uncertainty. Typical pressure and concentration profiles along the axis and the velocity profiles in the center of the channel obtained from the numerical solution are also presented.

1. Introduction

Over the last years, gaseous micro-flows have captured an increasing attention in the scientific community. This interest is well-justified by the appearance of new technologies at the micro-scale, such as micro-electromechanical systems or vacuum technology. The description of the gaseous flow in small scale devices requires the consideration of the molecular distribution function and the Boltzmann or other kinetic equations. By comparing the predictions of different theoretical models with experiments, the validity of these models can be evaluated.

Previous experimental works on gaseous micro-flows have mainly focused on the measurement of the flow rate of single gases through various micro-channels [1, 4, 5, 7, 22]. For micro-flows, the constant volume or the bubble tracking methods may be used for these purposes. Slip models have been developed or tested for gaseous micro-flows [2, 3, 10]. The slip coefficients have also been determined in several studies. The properties of single gases are generally well-known; however, there are only a few works on gaseous mixtures. Recently, the flow rate of gas mixtures through rectangular micro-channels has been measured, and the results have been compared to the solution of the McCormack kinetic model [13, 17, 18].

Much work has been devoted to the numerical solution of linearized kinetic models for single gas flows through various channels with rectangular [14], elliptical [6], triangular [11, 20] or

Published under licence by IOP Publishing Ltd

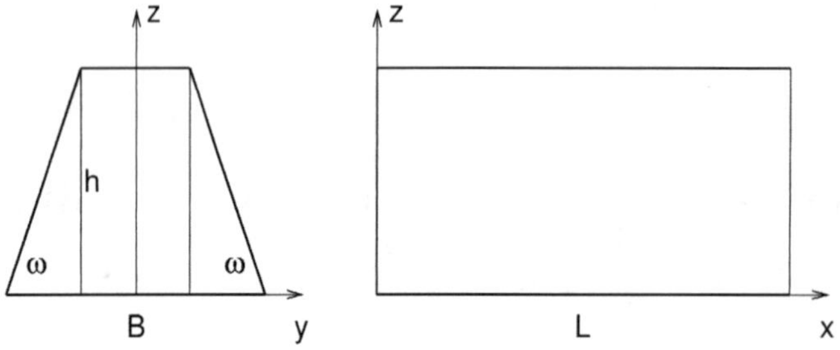

Figure 1. Geometry of the trapezoidal channel.

trapezoidal [21] cross sections. The solution of these models has been carried out by the discrete velocity method. The McCormack kinetic model has been used to model the flow of gaseous mixtures in channels [12, 15, 19]. The McCormack kinetic model has the advantageous property that the hydrodynamic behavior of the mixture can be properly captured. In Ref. [19], a flow rate formula has been derived for deducing the flow rate in long channels. It has been used for comparison between computational and experimental flow rates through rectangular channels [17, 18].

The flow of gas mixtures is affected by diffusion effects. At finite rarefactions, the different components of the gas tend to travel with different velocities in the channel due to their different molecular speeds. Hence, the species tend to separate. The separation phenomenon caused by the gas diffusion has been investigated in Refs. [8, 19, 18, 16].

In the present work, the flow rate of the noble gas mixtures He/Ar and He/Kr is measured through long trapezoidal micro-channels and compared to the numerical solution of the McCormack model. The flow in micro-channels with this cross section has not been analyzed earlier for gaseous mixtures. The measurements are carried out for a wide range of concentrations for the He/Ar mixture, while the concentration of 50% is used for the He/Kr mixture. The downstream pressure is $15.1kPa$ and $8.05kPa$. In addition to the comparative study in terms of the flow rates, the representative axial distributions of the pressure and the concentration and the typical velocity profiles in the center of the channel are analyzed.

2. Definition of the problem

The isothermal fully-developed flow of binary gaseous mixtures through long trapezoidal channels is considered. The channel lies along the x coordinate, whereas the cross section is located in the (y, z) coordinate sheet. The longer base and the height of the cross section are along the y and z coordinates. The acute angle of the cross section is $\omega = 54.74^o$, while the height of the channel is $h = 1.90 \pm 0.013\mu m$, its longer base is $B = 5.38 \pm 0.019\mu m$ and its length is $L = 5000\mu m$. The geometry of the channel is presented in Figure 1. The origin of the Descartes coordinate system is in $(x, y, z) = (0, 0, 0)$.

The mixture has two components $\alpha = 1, 2$. The molar densities and masses of the species are denoted by n_α and m_α. The molar masses of the specific gases are given by $m_{He} = 4.003g/mol$, $m_{Ar} = 39.95g/mol$ and $m_{Kr} = 83.80g/mol$. The concentration of the first component is defined by $C = n_1/(n_1 + n_2)$. The flow is driven by the pressure difference between the inlet and the outlet of the channel. The downstream and upstream pressures and concentrations are denoted by P_A, P_B and C_A, C_B. In the present work, $C_A = C_B$. The temperature of the gas T is

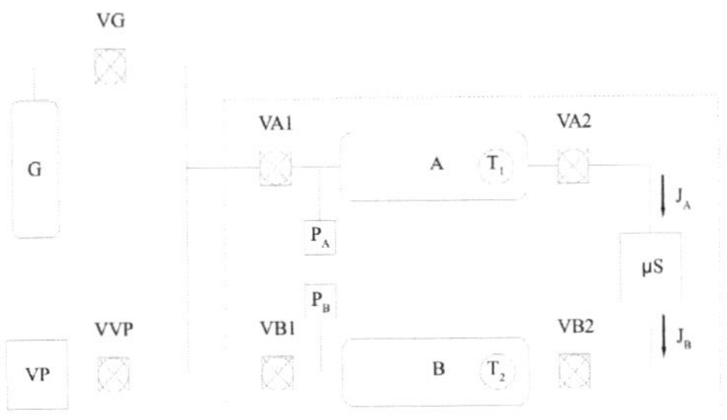

Figure 2. Layout of the experimental setup.

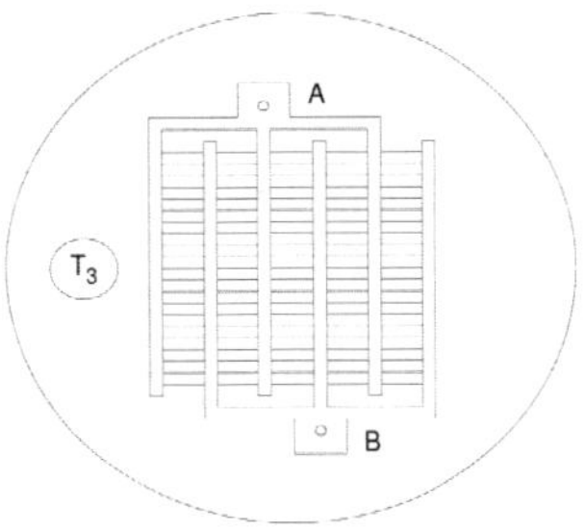

Figure 3. Layout of the microsystem.

constant. The flow is characterized by the rarefaction parameter

$$\delta(x) = \frac{P(x)D_h}{\mu(x)v_0(x)}, \tag{1}$$

where P is the pressure, μ is the viscosity, D_h is the hydraulic diameter of the channel and $v = (2R_gT/m)^{1/2}$ is the characteristic molecular speed. In addition, R_g is the global gas constant and $m = Cm_1 + (1-C)m_2$ is the average molar mass. The average Knudsen number in the channel is defined as the inverse of the average rarefaction parameter $Kn_0 = 2/(\delta_A + \delta_B)$, where $\delta_{A,B}$ is the inlet or outlet rarefaction parameter.

In this study, one of the important quantities is the molar flow rate of the components

$$J_\alpha = \int_A n_\alpha u_{\alpha x} dA, \tag{2}$$

where $u_{\alpha x}$ is the axial velocity of the components and A is the area of the cross section. The total molar flow rate is given by $J = J_1 + J_2$.

3. Experimental approach

The measurement of the flow rates is based on the constant volume method. The experimental setup, which was described in detail in Ref. [13], is located in a thermally insulated plastic chamber, Figure 2. The circuit consists of upstream A and downstream B reservoirs and the microsystem μS, which is a collection of microchannels etched into a silicon plate, Figure 3. The inlet A and the outlet B of the microchannels are connected to reservoirs A and B via valves VA2 and VB2. The reservoirs can be vacuumed out by the vacuum pump VP or filled up with the gas mixture from the tank G. The pressure in the reservoirs is measured by two Inficon capacitance gauges (CDG025D). The capacitance gauges are denoted as P_A and P_B in Figure 2. The accuracy of the capacitance gauges is $\pm 0.2\%$ of reading. The temperature is regulated by Peltier modules and accurately measured by three PT100 temperature sensors. The accuracy of the temperature sensors is $0.15K$. Two sensors are attached to the inlet and outlet reservoirs as denoted by T_1 and T_1 in Figure 2, while the third senor T_3 is connected to the microsystem, Figure 3. These sensors provide the same results within their accuracy. The third sensor is used to define the temperature in the system.

The measurement is carried out in a three-step procedure. First, the whole circuit is filled up with the gas at the downstream pressure. Then, valve VA2 is closed, and the possible leakage or outgassing flow rate J_{Bo} is measured in circuit B. Secondly, reservoir A is filled up with the mixture at the upstream pressure. After thermal regulation is reached, VA2 is opened, and the gas flows through the microsystem. In this step, the inlet and outlet flow rates J_A and J_B are measured on the basis of the pressure variations in the two containers. Finally, the whole circuit is filled up with the mixture at the actual upstream pressure. Valve VB2 is closed, and the possible leakage or outgassing flow rate J_{Ao} is measured in circuit A. The inlet and outlet experimental flow rates, J_A^e and J_B^e, are deduced by correcting J_A and J_B, with J_{Ao} and J_{Bo}. The measured flow rate is defined as the average $J^e = (J_A^e + J_B^e)/2$. The duration of the outgassing/leakage measurement is between 1.5 and 2 hours, while the typical time of the flow rate measurement is around 40 minutes. The allowed pressure variation in the flow rate measurement is less than 2% in the downstream reservoir. The temperature of the gas is $T = 299.5K$ and the volumes of the upstream and downstream containers are $V_A = 1.708e-4m^3$ and $V_B = 1.658e-4m^3$. In the experiments, the pressure variation in the reservoirs is less than 2%, which indicates that the amount of the gas flowing through the microsystem is less than $0.02\times P_B V_B/(R_g T) = [2e-4mol, 1.1e-4mol]$ for $P_B = 15.1kPa$ and $P_B = 8.05kPa$, respectively.

The flow rates are determined from the pressure variations in the reservoirs in accordance with the equation of state

$$J_{A,B} = \mp\frac{dN_{A,B}}{dt} = \mp\frac{V_{A,B}}{R_g}\frac{d}{dt}\left(\frac{P_{A,B}}{T_{A,B}}\right), \tag{3}$$

where $N_{A,B}$ is the molar amount of the gas in the upstream and downstream containers. The above equation can be written by

$$J_{A,B} = \mp\frac{dN_{A,B}}{dt} = \mp\frac{V_{A,B}}{R_g T_{A,B}}\frac{dP_{A,B}}{dt}\left(1 - \frac{dT_{A,B}/T_{A,B}}{dP_{A,B}/P_{A,B}}\right). \tag{4}$$

In the experiments, the relative temperature and pressure variations are in the order of $dT_{A,B}/T_{A,B} = 4\text{x}10^{-4}$ and $dP_{A,B}/P_{A,B} = 2\text{x}10^{-2}$. As a result, the flow rate can be determined as

$$J_{A,B} = \mp\frac{V_{A,B}}{R_g T_{A,B}} a_{A,B} c_{A,B}, \tag{5}$$

where the time derivative of the pressure variation is denoted by $a_{A,B}$ with $P_{A,B}(t) = a_{A,B}t + b_{A,B}$, and $c_{A,B} = (1 - dT_{A,B}/T_{A,B})/(dP_{A,B}/P_{A,B}) = 1 \pm 0.02$. Eq. (5) is used to calculate the flow rates, $J_{A,B}$, after the flow rate measurement. The time-averaged value of the temperatures $T_{A,B}$ is taken from sensor T_3, the parameter $a_{A,B}$ is determined from a least square fit of the pressure profiles and $c_{A,B}$ is unity. The uncertainty of the measurement can be estimated as follows. The relative standard deviation of $a_{A,B}$ is less than $\pm 0.5\%$. This value is determined in the way as given in Ref. [13]. The standard deviation defines confidence interval with 68.2% acceptance. The relative standard deviation is used to estimate the uncertainty of $a_{A,B}$. The total uncertainty of the experiment consists of the uncertainty of the volume $V_{A,B}$, the temperature $T_{A,B}$ and the coefficients $a_{A,B}$ and $c_{A,B}$. The volume was determined by using the method given in Ref. [13], and its uncertainty was estimated as $\pm 1.3\%$. The overall uncertainty of the temperature is estimated by the deviation of the temperature inside the chamber in a test measurement lasting six hours. The deviation of the temperature is less than $\pm 0.2\%$. The uncertainty of $c_{A,B} = (1 - dT_{A,B}/T_{A,B})/(dP_{A,B}/P_{A,B})$ is in the order of $\pm 2\%$. Combining the these factors, the total experimental uncertainty is estimated as

$$\frac{\Delta J_{A,B}}{J_{A,B}} = \frac{\Delta V_{A,B}}{V_{A,B}} + \frac{\Delta T_{A,B}}{T_{A,B}} + \frac{\Delta a_{A,B}}{a_{A,B}} + \frac{\Delta c_{A,B}}{c_{A,B}} = \pm(1.3\% + 0.2\% + 0.5\% + 2\%) = \pm 4\%. \tag{6}$$

4. Computational approach

The McCormack linearized kinetic model is utilized to describe the flow. Since the channel is long, the speed of the gas is small compared to the characteristic molecular speed and the linearized description can be used. The end effects can also be neglected. The calculation is divided into two steps. First, the local flow problem at a particular cross section is solved. Secondly, the global behavior of the flow including the flow rates and the distributions of the pressure and the concentration is deduced.

The flow is described by the J_P, J_C thermodynamic fluxes [18, 19] defined by

$$J_P = -\int_A (n_1 u_{1x} + n_2 u_{2x}) dA, \quad J_C = -n_1 \int_A (u_{1x} - u_{2x}) dA. \tag{7}$$

These fluxes are the linear functions of the local pressure and concentration gradients

$$X_P = \frac{\partial P}{\partial x}\frac{D_h}{P}, \quad X_C = \frac{\partial C}{\partial x}\frac{D_h}{C} \tag{8}$$

such that

$$J_P = \frac{n v_0 A}{2}(\Lambda_{PP} X_P + \Lambda_{PC} X_C), \tag{9}$$

$$J_C = \frac{n v_0 A}{2}(\Lambda_{CP} X_P + \Lambda_{CC} X_C), \tag{10}$$

where $n = n_1 + n_2$ and Λ_{PP}, Λ_{PC}, Λ_{CP}, Λ_{CC} are the so-called kinetic coefficients. These coefficients are obtained by the numerical solution of the McCormack kinetic model, which is achieved by an accelerated discrete velocity method [19, 20]. The realistic potential of Kestin et al. is used in the McCormack collision frequencies [9]. The coefficients are the functions of the local rarefaction parameter and the concentration. By solving the McCormack model in a wide range of the rarefaction and the whole range of the concentration a database is created for the kinetic coefficients. The discrete velocity algorithm has been used with the computational grid having a number of nodes $M \times N = 16 \times 300$ for $\delta < 1$ and $M \times N = 16 \times 72$ for $\delta \geq 1$ in the velocity space, where M and N denote the number of magnitudes and polar angles. In the spatial space, the number of nodes along the large base is 1500 for all δ. The process of iteration for the calculation of the kinetic coefficients is terminated when the relative convergence error is less than $1e-6$.

The flow rates of the components can be expressed by the thermodynamic fluxes according to

$$J_1 = -C J_P - (1-C) J_C, \quad J_2 = -(1-C)(J_P - J_C). \tag{11}$$

The flow rates of the species are constant along the axis of the channel due to the conservation of mass. By using the ideal gas law, the definition of the rarefaction parameter and Eqs. (7)-(10), Eq. (11) can be written by

$$J_1 = -\frac{PAD_h}{m v_0 L}\left[(C\Lambda_{PP} + (1-C)\Lambda_{CP})\frac{\partial P}{\partial x'}\frac{1}{P} + (C\Lambda_{PC} + (1-C)\Lambda_{CC})\frac{\partial C}{\partial x'}\frac{1}{C}\right], \tag{12}$$

$$J_2 = -\frac{PAD_h}{m v_0 L}(1-C)\left[(\Lambda_{PP} - \Lambda_{CP})\frac{\partial P}{\partial x'}\frac{1}{P} + (\Lambda_{PC} - \Lambda_{CC})\frac{\partial C}{\partial x'}\frac{1}{C}\right], \tag{13}$$

where $x' = x/L$ is the dimensionless coordinate along the axis of the channel. These equations are supplemented with the boundary conditions for the pressure and the concentration at the inlet and the outlet of the channel such that

$$P(0) = P_A, \quad P(1) = P_B, \quad C(0) = C_A, \quad C(1) = C_B. \tag{14}$$

This boundary condition is justified by the facts that relatively small pressure variation is allowed in the two reservoirs during the measurement and the concentration variation due to the gaseous separation is also negligible. Eqs. (12)-(13) constitute a system of two ordinary differential equations for the primitive functions $P(x')$ and $C(x')$. For deducing the kinetic coefficients, the rarefaction parameter is determined from the local value of the pressure and the concentration and the viscosity function of the gaseous mixtures, see Eq. (1). The actual value of the kinetic coefficients are deduced from the pre-calculated database mentioned above. Eqs. (12)-(13) together with Eq. (14) are numerically solved by discretizating the spatial coordinate and using the Euler method to compute the spatial derivatives. The spatial spacing is $\Delta x' = 1/500$ and the convergence criterion on the downstream pressure and concentration is $1e-5$ [19]. The solution yields the flow rates of the components J_1, J_2 and the distributions of the pressure $P(x')$ and the concentration $C(x')$.

5. Results and discussion

The flow rate of He/Ar and He/Kr gas mixtures are measured and compared to the kinetic calculation at downstream pressures $15.1kPa$ and $8.05kPa$. The microsystem consists of 400 parallel channels. The channels are etched in a circular silicon plate and covered with glass plates via anodic bonding. The parallel channels are connected to the inlet A and the outlet B of the microsystem, Figure 3. The flow rate through an individual channel is considered in the following.

Table 1 shows the flow rates for downstream pressure $15.1kPa$. For the He/Ar mixture, the concentrations $C_A = [0.0, 0.1017, 0.3012, 0.5010, 0.7019, 0.9014, 1.0]$, which include the pure single gas cases, is investigated, while the concentration $C_A = 0.5010$ is studied for the He/Kr mixture. The flow rates are measured in the range of the pressure ratio $3-7$. In the table, the columns from left to right in order show the concentration of the gas, the upstream pressure, the pressure ratio, the average Knudsen number, the experimental flow rate, the difference between the inlet and outlet experimental flow rates, the kinetic flow rates of the species J_1, J_2, the total kinetic flow rate J and the discrepancy between the experimental and the total kinetic flow rates, defined by $\Delta = 100(1 - J^e/J)$, respectively. The experimental flow rate J^e is determined as described in section 3. The difference between the inlet and outlet experimental flow rates are defined by $\Delta_e = 100(1 - J_A^e/J_B^e)$. The J_A^e and J_B^e experimental flow rates are determined as defined by Eq. (5) in section 3. The kinetic flow rates J_1, J_2, J are obtained by the solution of Eqs. (12)-(14) in section 4. The inlet and outlet values of the pressures P_A, P_B measured in the experiment and the constant concentrations $C_A = C_B$ in Table 1 are used as input parameters for the calculation. In the table, it can be seen that the flow rates increase with increasing pressure drop. In addition, they also increase with decreasing concentration. This latter behavior is caused by the smaller mass of the first species, which results into a larger molecular speed. This correlation remains valid in the whole range of the gas rarefaction. This has been confirmed by us both numerically and experimentally. The flow rates for the He/Kr mixture are smaller than the corresponding ones for the He/Ar mixture at the same concentration, which is caused by the larger mass of the Kr than that of the Ar. The difference between the inlet and outlet experimental flow rates is always less than the experimental uncertainty, which justifies that possible leakage or outgassing effects are well taken into account. It is found that the discrepancy between the experiment and kinetic flow rates are always less than the experimental uncertainty; hence, a very good agreement is reached between the two different approaches.

Table 2 shows the corresponding results at downstream pressure $8.05kPa$. For the He/Ar mixture, the same concentrations excluding the single gas cases are examined as for the larger downstream pressure. For the He/Kr mixture, the flow with the same concentration is studied as in Table 1. The pressure ratio is in the range of $4-7$. The columns of Table 2 presents the data in the same scenario as in Table 1. The flow rates are deduced in a similar manner

Table 1. Flow rates through an individual microchannel for He/Ar and He/Kr mixtures at downstream pressure $P_B = 15.1kPa$.

C_A	$P_A(Pa)$	P_A/P_B	Kn_0	$J^e(mol/s)$	Δ_e	$J_1(mol/s)$	$J_2(mol/s)$	$J(mol/s)$	Δ
He/Ar									
0.0000	45355	3.00	0.11	1.02e-11	1.22	0	1.01e-11	1.01e-11	-0.35
	59422	3.94	0.09	1.66e-11	-1.41	0	1.64e-11	1.64e-11	-1.35
	74977	4.97	0.07	2.49e-11	-2.45	0	2.45e-11	2.45e-11	-1.75
	90110	5.97	0.06	3.39e-11	-2.63	0	3.36e-11	3.36e-11	-0.90
	105010	6.95	0.05	4.41e-11	1.66	0	4.38e-11	4.38e-11	-0.67
0.1017	45328	3.00	0.12	1.04e-11	-2.16	1.30e-12	9.30e-12	1.06e-11	2.37
	59944	3.97	0.09	1.76e-11	1.64	2.00e-12	1.54e-11	1.74e-11	-1.58
	74994	4.97	0.08	2.56e-11	2.59	2.80e-12	2.26e-11	2.54e-11	-0.73
	90439	5.99	0.07	3.51e-11	1.61	3.74e-12	3.11e-11	3.48e-11	-0.93
	105938	7.02	0.06	4.69e-11	-1.50	4.80e-12	4.05e-11	4.53e-11	-3.47
0.3012	45340	3.00	0.13	1.15e-11	0.76	4.08e-12	7.82e-12	1.19e-11	2.88
	59885	3.97	0.11	1.86e-11	0.52	6.28e-12	1.27e-11	1.90e-11	2.05
	74925	4.96	0.09	2.81e-11	-1.38	8.83e-12	1.87e-11	2.75e-11	-2.41
	90251	5.98	0.08	3.79e-11	-2.96	1.17e-11	2.56e-11	3.73e-11	-1.64
	105761	7.00	0.07	4.91e-11	-1.07	1.51e-11	3.34e-11	4.85e-11	-1.27
0.5010	45200	2.99	0.15	1.40e-11	-2.52	7.61e-12	5.99e-12	1.36e-11	-2.49
	59840	3.96	0.12	2.11e-11	-3.79	1.16e-11	9.88e-12	2.15e-11	1.89
	74731	4.95	0.10	3.10e-11	-3.23	1.62e-11	1.44e-11	3.06e-11	-1.12
	90368	5.98	0.09	4.09e-11	-0.39	2.15e-11	1.98e-11	4.13e-11	1.06
	105126	6.96	0.08	5.21e-11	-0.89	2.72e-11	2.54e-11	5.26e-11	0.78
0.7019	45082	2.99	0.19	1.55e-11	0.37	1.21e-11	4.03e-12	1.61e-11	3.35
	60203	3.99	0.15	2.51e-11	1.41	1.87e-11	6.75e-12	2.54e-11	1.24
	75204	4.98	0.12	3.52e-11	-3.85	2.60e-11	9.74e-12	3.57e-11	1.41
	90316	5.98	0.11	4.69e-11	-3.84	3.41e-11	1.31e-11	4.72e-11	0.72
	105449	6.98	0.09	5.99e-11	-3.49	4.29e-11	1.70e-11	5.99e-11	-0.02
0.9014	45255	3.00	0.24	1.95e-11	1.82	1.85e-11	1.61e-12	2.01e-11	2.57
	59883	3.97	0.19	3.02e-11	-2.80	2.86e-11	2.63e-12	3.12e-11	3.08
	74922	4.96	0.16	4.30e-11	-0.59	3.98e-11	3.81e-12	4.36e-11	1.37
	90239	5.98	0.14	5.64e-11	-3.63	5.24e-11	5.14e-12	5.75e-11	1.95
	105660	7.00	0.12	7.17e-11	-1.96	6.61e-11	6.69e-12	7.28e-11	1.50
1.0000	45482	3.01	0.30	2.29e-11	-3.07	2.32e-11	0	2.32e-11	1.35
	60343	4.00	0.24	3.57e-11	3.70	3.63e-11	0	3.63e-11	1.70
	75293	4.99	0.20	4.94e-11	-0.90	5.07e-11	0	5.07e-11	2.66
	89976	5.96	0.17	6.54e-11	2.01	6.61e-11	0	6.61e-11	1.05
	105450	6.98	0.15	8.23e-11	-1.93	8.37e-11	0	8.37e-11	1.77
He/Kr									
0.5010	45275	3.00	0.12	1.11e-11	-3.32	6.47e-12	4.83e-12	1.13e-11	1.87
	59875	3.97	0.10	1.80e-11	-0.89	9.83e-12	7.97e-12	1.78e-11	-1.24
	74947	4.96	0.08	2.53e-11	-2.84	1.36e-11	1.18e-11	2.54e-11	0.18
	90320	5.98	0.07	3.50e-11	-3.29	1.80e-11	1.61e-11	3.41e-11	-2.62
	105832	7.01	0.06	4.43e-11	-1.50	2.29e-11	2.11e-11	4.40e-11	-0.56

Table 2. Flow rates through an individual microchannel for He/Ar and He/Kr mixtures at downstream pressure $P_B = 8.05kPa$.

C_A	$P_A(Pa)$	P_A/P_B	Kn_0	$J^e(mol/s)$	Δ_e	$J_1(mol/s)$	$J_2(mol/s)$	$J(mol/s)$	Δ
He/Ar									
0.1017	31974	3.97	0.17	7.50e-12	-3.50	9.92e-13	6.30e-12	7.29e-12	-2.88
	40237	5.00	0.14	1.03e-11	-1.97	1.32e-12	9.08e-12	1.04e-11	1.32
	47965	5.96	0.12	1.33e-11	-0.77	1.62e-12	1.18e-11	1.34e-11	0.74
	56065	6.96	0.11	1.68e-11	0.46	1.95e-12	1.50e-11	1.70e-11	1.03
0.3012	32038	3.98	0.20	8.58e-12	1.48	3.12e-12	5.31e-12	8.43e-12	-1.82
	40155	4.99	0.16	1.16e-11	-0.11	4.15e-12	7.65e-12	1.18e-11	1.14
	47940	5.96	0.14	1.48e-11	1.22	5.20e-12	1.00e-11	1.52e-11	2.80
	56086	6.97	0.12	1.92e-11	-0.66	6.39e-12	1.27e-11	1.91e-11	-0.56
0.5010	32063	3.98	0.23	9.77e-12	-1.02	5.88e-12	4.12e-12	1.00e-11	2.50
	40143	4.99	0.19	1.35e-11	2.62	7.85e-12	5.95e-12	1.38e-11	2.14
	47955	5.96	0.16	1.72e-11	-0.07	9.79e-12	7.91e-12	1.77e-11	2.53
	55789	6.93	0.14	2.23e-11	-1.54	1.19e-11	9.92e-12	2.18e-11	-1.96
0.7019	31879	3.96	0.28	1.18e-11	-2.76	9.24e-12	2.76e-12	1.20e-11	1.24
	40248	5.00	0.23	1.61e-11	-0.54	1.26e-11	4.05e-12	1.66e-11	2.57
	47842	5.94	0.20	2.05e-11	-2.04	1.57e-11	5.32e-12	2.10e-11	2.03
	55793	6.93	0.17	2.55e-11	-3.54	1.91e-11	6.74e-12	2.58e-11	1.10
0.9014	31945	3.97	0.36	1.48e-11	0.26	1.40e-11	1.02e-12	1.50e-11	1.26
	40188	4.99	0.30	1.99e-11	-2.62	1.91e-11	1.54e-12	2.06e-11	3.35
	47947	5.96	0.26	2.60e-11	-0.29	2.41e-11	2.09e-12	2.62e-11	0.62
	55821	6.93	0.23	3.18e-11	0.91	2.95e-11	2.61e-12	3.21e-11	0.85
He/Kr									
0.5010	31971	3.97	0.19	7.99e-12	0.64	5.10e-12	3.20e-12	8.30e-12	3.70
	40264	5.00	0.15	1.11e-11	-2.13	6.70e-12	4.70e-12	1.14e-11	2.34
	47974	5.96	0.13	1.43e-11	-3.37	8.31e-12	6.29e-12	1.46e-11	2.03
	55979	6.95	0.12	1.76e-11	-2.14	9.98e-12	7.52e-12	1.75e-11	-0.21

as described in the previous paragraph. It can be seen that the flow rates are always smaller than the corresponding values at the larger downstream pressure because of the smaller value of the density at smaller pressure. In terms of the pressure ratio and the concentration, a similar observation can be made as for Table 1. The difference between the inlet and outlet flow rates is also always smaller than the experimental uncertainty, and the discrepancy between the experimental and kinetic results are also smaller than the experimental uncertainty. A very good agreement is reached between the two approaches again. The agreement between experimental and computational flow rates is similar to the one observed in [17] and much better than the one given in [18]. This improvement has been caused by the improved experimental device, where the plastic connections have been replaced by metal ones, which provides better tightness. The good agreement in both situations justifies that the McCormack model is a viable tool for the description of the present flow problem.

In Table 3, a comparison of single gas, approximated and total kinetic flow rates, J_s, J^* and J, are shown for $P_B = 8.05kPa$ and $C_A = 0.5010$. The single gas results are deduced by assuming a single gas with the average mass and the proper viscosity of the mixture. For the He/Ar and He/Kr mixtures, the average masses and the viscosities are given by $m = [21.9g/mol, 43.8g/mol]$

Table 3. Comparison of single gas, approximated and exact flow rates for the two mixtures at downstream pressure $P_B = 8.05kPa$ and $C_A = 0.5010$.

P_A/P_B	$J_s(mol/s)$	$J^*(mol/s)$	$J(mol/s)$	Δ_s	Δ^*
He/Ar					
3.98	8.33e-12	1.02e-11	1.00e-11	16.7	-1.9
4.99	1.17e-11	1.41e-11	1.38e-11	15.0	-2.0
5.96	1.53e-11	1.81e-11	1.77e-11	13.6	-2.1
6.93	1.92e-11	2.24e-11	2.18e-11	12.0	-2.6
He/Kr					
3.97	6.24e-12	8.53e-12	8.30e-12	24.9	-2.7
5.00	8.95e-12	1.18e-11	1.14e-11	21.5	-3.7
5.96	1.17e-11	1.51e-11	1.46e-11	19.6	-3.4
6.95	1.49e-11	1.88e-11	1.75e-11	14.9	-7.2

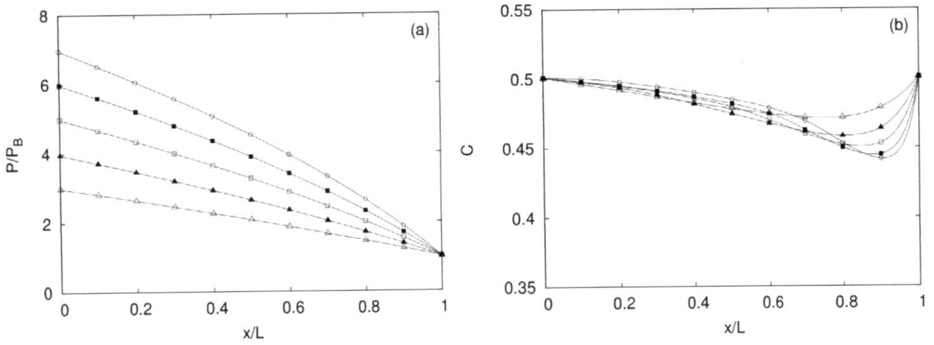

Figure 4. The distributions of the pressure (a) and the concentration (b) for He/Ar mixture at $C_A = 0.5010$ and $P_B = 15.1kPa$. The symbols, empty and solid triangles, empty and solid squares and empty circles, represent the cases with pressure ratios 2.99, 3.96, 4.95, 5.98, 6.96, respectively.

and $\mu = [2.38e-5Pas, 2.70e-5Pas]$. The single gas flow rate is obtained from the solution of Eqs. (12)-(14) by setting $C = 1$ and using the average mass and the viscosity as input quantities to determine the inlet and outlet rarefaction parameters. The approximated flow rates J^* are deduced by using the methodology neglecting the concentration gradient for pressure driven flow and developed in Ref. [19]. The differences $\Delta_s = 100(1 - J_s/J)$ and $\Delta^* = 100(1 - J^*/J)$ are tabulated in the last columns of the table. It can be seen that the single gas assumption does not provide reliable results. The Δ_s difference can be as large as 16.7% and 24.9% for the He/Ar and He/Kr mixtures. The approximated formula provides an improvement with the largest absolute deviation of 2.6% and 7.2% for the two cases.

In addition to the flow rates, Figures 4 and 5 show the typical pressure and concentration profiles along the axis of the channel as deduced from the numerical calculation for He/Ar and He/Kr mixtures at concentration of 0.5010 and downstream pressure of $15.1kPa$. As it is described in section 4, the solution of Eqs. (12)-(14) provide the flow rates J_1, J_2 and the distribution of the pressure and the concentration along the axis of the channel. The input parameters for the calculation of the pressure and concentration distributions are exactly the

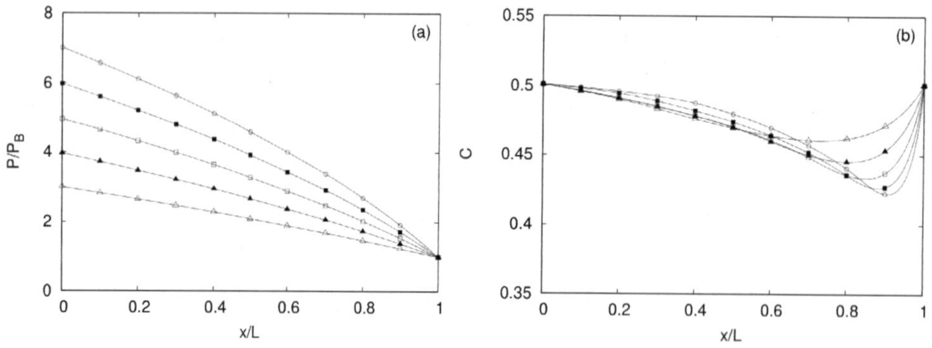

Figure 5. The distributions of the pressure (a) and the concentration (b) for He/Kr mixture at $C_A = 0.5010$ and $P_B = 15.1kPa$. The symbols, empty and solid triangles, empty and solid squares and empty circles, represent the cases with pressure ratios $3.00, 3.97, 4.96, 5.98, 7.01$, respectively.

same as those in Table 1 at $C_A = 0.5010$. By examining the pressure profiles, it can be concluded that the distributions are nearly non-linear at this downstream pressure. The non-linearity is caused by the compressibility and reduced by the rarefaction of the gas. The dimensionless flow rate depends on the rarefaction degree at intermediate values of the rarefaction. The concentration profiles exhibit the separation phenomenon. The concentration starts at the upstream value at the entrance of the channel, it starts to decrease, taking its minimum at the second half of the channel, and, finally, it reaches the downstream value at the exit. The non-uniformity of the concentration depends on the pressure ratio. With increasing pressure drop, the gaseous separation becomes larger. The corresponding distributions for the two mixtures are nearly the same, but it can be seen, on the concentration profiles, that the separation is stronger for the He/Kr mixture, which has larger mass ratio compared to the He/Ar mixture.

Figure 6 shows the typical dimensional velocity profiles of the species of the gas along the center line of the trapezoidal cross section, $y = 0$, at the center of the channel, $x = L/2$, for the two types of mixture at $P_B = 8.05kPa$, $C_A = 0.5010$ and $P_B/P_A \sim [4,6]$. The exact pressure ratios are given in Table 2. The dimensional velocity is calculated according to $u_\alpha = v(u_\alpha^{(P)} X_P + u_\alpha^{(C)} X_C)$, where $u_\alpha^{(P)}$ and $u_\alpha^{(C)}$ are the dimensionless gas velocities for uniform pressure or concentration driven flows and X_P and X_C are the local pressure or concentration gradients. This calculation is based on the methodology of Ref. [19] for the solution of the McCormack model. For the examined cases, the local rarefaction and concentration values and X_P, X_C at the center of the channel are deduced from the distribution of the pressure and the concentration. In the figure, the separation of the components is clearly visible. The lighter species have larger velocity than that of the heavier one due to their larger molecular speed. The velocities are larger if the pressure ratio becomes larger. In addition, the velocities of the components of the He/Kr mixture are smaller those of the He/Ar mixture. This can be explained by the fact that the He/Kr gas has a smaller characteristic molecular speed than that of the He/Ar mixture.

6. Conclusions

In the present work, the flow rates of He/Ar and He/Kr mixtures through long trapezoidal micro-channels are measured by using the constant volume method. The downstream pressures $15.1kPa$ and $8.05kPa$ are investigated. The pressure ratio is in the range of $3-7$ and $4-7$ for the larger and smaller outlet pressures. The flow rates are compared to the results of the McCormack

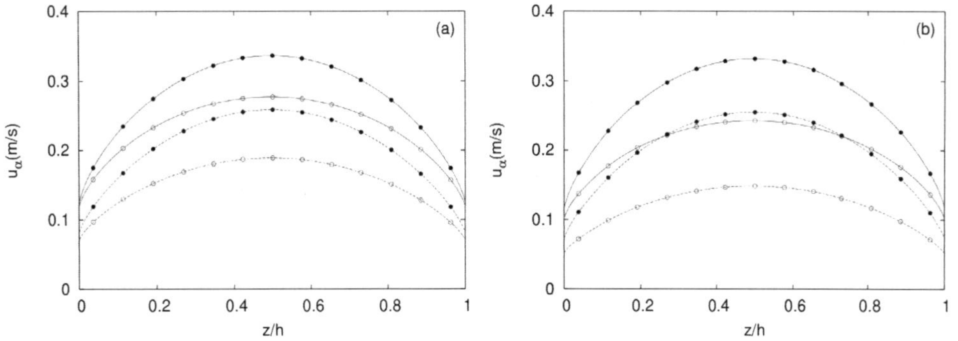

Figure 6. The distribution of the velocity along the center line of the cross section at the center of the channel at $P_B = 8.05kPa$, $C_A = 0.5010$ for the He/Ar (a) and He/Kr (b) mixtures at $P_B/P_A \sim [4,6]$. Solid and dashed lines stand for the lighter and heavier species, while empty and solid circles denote the velocity profile for $P_B/P_A \sim [4,6]$, respectively.

kinetic model. The relative difference between the experimental and kinetic results is less than the experimental uncertainty, which is $\pm 4\%$. This indicates a complete agreement between the two approaches within the range of the precision (uncertainty) of the experiment. This agreement can be considered as very good. The McCormack model reproduces the experimental results within their uncertainty. It is shown that the single gas assumption does not provide reliable results for this type of flow. Hence, the kinetic modelling of the gas mixture, taking into account the gas diffusion, is necessary. The typical axial distributions of the pressure and the concentration are also analyzed. The typical dimensional velocity profiles are also studied. The gaseous mixture exhibits the separation phenomenon. The good agreement between the theoretical and experimental flow rates suggests that the McCormack model is a viable tool for modeling pressure driven isothermal flows of gas mixtures in micro-channels at the considered parameters.

Acknowledgments

The research leading to these results has received funding from the European Community's Seventh Framework Programme (FP7/2007-2013) under grant agreement no 215504. Financial supports from the Région Midi-Pyrénées and the Fédération de Recherche Fermat are also acknowledged.

References

[1] Arkilic E B, Breuer K S and Schmidt M A 2001 Mass flow and tangential momentum accommodation in silicon micromachined channels *J. Fluid Mech.* **437** 29-43
[2] Aubert C and Colin S 2001 High-order boundary conditions for gaseous flows in rectangular microchannels *Microscale Thermophys. Eng.* **5** 41-54
[3] Colin S, Lalonde P and Caen R 2004 Validation of a second-order slip flow model in rectangular microchannels *Heat Transfer Eng.* **25** 23-30
[4] Ewart T, Perrier P, Graur I and Meolans J G 2006 Mass flow rate measurements in gas micro flows *Exp. Fluids* **41** 487-98
[5] Ewart T, Perrier P, Graur I and Meolans J G 2007 Mass flow rate measurements in a microchannel from hydrodynamic to near free molecular regimes *J. Fluid Mech.* **584** 337-56
[6] Graur I and Sharipov F 2007 Gas flow through an elliptical tube over the whole range of the gas rarefaction *Eur. J. Mech. B/Fluids* **27** 335-45
[7] Harley J C, Huang Y Y, Bau H H and Zemel J N 1995 Gas flow in micro-channels *J. Fluid Mech.* **284** 257-74

[8] Kalempa D and Sharipov F 2011 Flows of rarefied gaseous mixtures with a low mole fraction. Separation phenomenon *Eur. J. Mech. B/Fluids* **30** 466-73

[9] Kestin J, Knierim K, Mason E A, Najafi B, Ro S T and Waldman M 1984 Equilibrium and transport properties of the noble gases and their mixture at low densities *J. Phys. Chem. Ref. Data* **13** 229303

[10] Maurer J, Tabeling P, Joseph P and Willaime H 2003 Second-order slip laws in microchannels for helium and nitrogen *Phys. Fluids* **15** 2613-21

[11] Naris S and Valougeorgis D 2008 Rarefied gas flow in a triangular duct based on a boundary fitted lattice *Eur. J. Mech. B/Fluids* **27** 810-22

[12] Naris S, Valougeorgis D, Sharipov F and Kalempa D 2005 Flow of gaseous mixtures through rectangular microchannels driven by pressure temperature and concentration gradients *Phys. Fluids* **17** 100607.1-12

[13] Pitakarnnop J, Varoutis S, Valougeorgis D, Geoffroy S, Baldas L and Colin S 2010 A novel experimental setup for gas microflows *Microfluid. Nanofluid.* **8** 57-72

[14] Sharipov F 2002 Rarefied gas flow through a long rectangular channel *J. Vac. Sci. Technol. A* **17** 3062-66

[15] Sharipov F and Kalempa D 2002 Gaseous mixture flow through a long tube at arbitrary Knudsen numbers *J. Vac. Sci. Technol. A* **20** 814-22

[16] Sharipov F and Kalempa D 2005 Separation phenomena for gaseous mixture flowing through a long tube into vacuum *Phys. Fluids* **17** 127102

[17] Szalmas L, Colin S and Valougeorgis D 2011 Flow rate measurement of rarefied binary gases in long rectangular microchannels *Proc. 3rd Micro and Nano Flows Conf. (Thessaloniki, Greece)* pp 5.1-9

[18] Szalmas L, Pitakarnnop J, Geoffroy S, Colin S and Valougeorgis D 2010 Comparative study between computational and experimental results for binary rarefied gas flows through long microchannels *Microfluid. Nanofluid.* **9** 1103-14

[19] Szalmas L and Valougeorgis D 2010 Rarefied gas flow of binary mixtures through long channels with triangular and trapezoidal cross sections *Microfluid. Nanofluid.* **9** 471-87

[20] Szalmas L and Valougeorgis D 2010 A fast iterative model for discrete velocity calculations on triangular grids *J. Comp. Phys.* **229** 4315-26

[21] Varoutis S, Naris S, Hauer V, Day C and Valougeorgis D 2009 Experimental and computational investigation of gas flows through long channels of various cross sections in the whole range of the Knudsen number *J. Vac. Sci. Technol.* **27** 89-100

[22] Zohar Y, Lee S Y K, Lee W Y, Jiang L and Tong P 2002 Subsonic gas flow in a straight and uniform microchannel *J. Fluid Mech.* **472** 125-51

Numerical analysis of thermal creep flow in curved channels for designing a prototype of Knudsen micropump

V Leontidis[1], J J Brandner[2], L Baldas[1] and S Colin[1]

[1] Université de Toulouse; INSA, UPS, Mines Albi, ISAE; ICA (Institut Clément Ader) 135 avenue de Rangueil, F-31077 Toulouse, France
[2] Karlsruhe Institute of Technology, Institute of Micro Process Engineering, Campus North, Hermann-von-Helmholtz-Platz 1, D-76344 Eggenstein-Leopoldshafen, Germany

E-mail: stephane.colin@insa-toulouse.fr

Abstract. The possibility to generate a gas flow inside a channel just by imposing a tangential temperature gradient along the walls without the existence of an initial pressure difference is well known. The gas must be under rarefied conditions, meaning that the system must operate between the slip and the free molecular flow regimes, either at low pressure or/and at micro/nano-scale dimensions. This phenomenon is at the basis of the operation principle of Knudsen pumps, which are actually compressors without any moving parts. Nowadays, gas flows in the slip flow regime through microchannels can be modeled using commercial Computational Fluid Dynamics softwares, because in this regime the compressible Navier-Stokes equations with appropriate boundary conditions are still valid. A simulation procedure has been developed for the modeling of thermal creep flow using ANSYS Fluent®. The implementation of the boundary conditions is achieved by developing User Defined Functions (UDFs) by means of C++ routines. The complete first order velocity slip boundary condition, including the thermal creep effects due to the axial temperature gradient and the effect of the wall curvature, and the temperature jump boundary condition are applied. The developed simulation tool is used for the preliminary design of Knudsen micropumps consisting of a sequence of curved and straight channels.

1. Introduction

The increasing demand on microdevices over the last years led to the development of new fabrication techniques for manufacturing micro-electro-mechanical systems (MEMS), such as micro-heat exchangers, pressure gauges, micro-sensors for mass flow and temperature measurements and micro-pumps. The need to manipulate very small fluid volumes for example for medical, biological, or electronics cooling purposes led to the development of various micropumps, but even today in most microfluidic applications the pumping system remains the largest in size and the most expensive (mainly due to power consumption) part of a fluidic microsystem [1-4]. Miniaturization of these pumps would lead to minimization of the total size and cost and enable the portability (for example in chemical analysis systems). At the same time the necessity of simulation tools is equally important and crucial, especially in the case of gaseous fluids. For that reason various analytical and numerical approaches have been developed for the simulation of rarefied gas flows (continuum and kinetic models, Direct Simulation Monte Carlo method…) [5-6].

Rarefied gas flows can be encountered under low pressure and/or in micro/nano-scale dimensions. The degree of rarefaction of a gas is determined by the Knudsen number:

$$Kn = \frac{\lambda}{D_h} \tag{1}$$

defined as the ratio of the mean free path λ of the gas molecules over a characteristic length, for example the hydraulic diameter D_h of the microchannel. The Knudsen number encountered in classic microsystems (micropumps, microvalves, micronozzles, microflow sensors) is frequently between 10^{-3} and 10^{-1}, which is the typical range of the well-known slip flow regime [7]. In this regime, even if the Navier-Stokes equations are still valid in the bulk flow, the gas is no longer in local thermodynamic equilibrium near the wall (Knudsen layer). In this region, a variety of rarefaction effects are exhibited, including the presence of non-negligible velocity slip and temperature jump at the walls [8, 9]. The slip flow regime is quite convenient from a theoretical point of view because it generally leads to analytical or semi-analytical solutions. In the literature, however, various forms of the boundary conditions can be found for the velocity slip and the temperature jump boundary conditions [10].

In rarefied gas flows, it is possible to generate a flow just by imposing a tangential temperature gradient along the walls without the existence of an initial pressure difference. This phenomenon is called thermal transpiration or thermal creep. In figure 1 a schematic representation of thermal driven gas flow in a narrow channel connecting two reservoirs at different temperatures is given. The phenomenon is well known since 1878 when Reynolds [11] experimentally observed it for the first time, namely in a tube connected to two reservoirs at different temperatures. Actually, the gas molecules move from the lower to the higher temperature zones, which eventually will cause a pressure difference and a reversed Poiseuille flow due to this pressure gradient. In addition, Reynolds showed that at the equilibrium the pressures are related to the temperatures according to the law:

$$\frac{P_2}{P_1} = \sqrt{\frac{T_2}{T_1}} \tag{2}$$

where P and T are the pressure and the temperature, in the case of molecular flow regime in the microchannel. The subscript 1 corresponds to the cold side and 2 to the hot reservoir.

Figure 1. Schematic representation of thermal transpiration flow.

After Reynolds, several studies have been devoted both to experimental [12-14] and computational analyses [15-19] of thermal creep flows. The modeling of this kind of flows, however, goes back to 1879 when Maxwell [20] proposed a mathematical expression, based on the kinetic theory, of a flow induced by a temperature gradient. A few years later (1909), Knudsen proved Reynolds law and mentioned the existence of two counter-current flows, due to the temperature gradient in one direction and due to the developed pressure difference in the opposite direction. After a transient phase during which the flow engaged by the temperature difference initially dominates, the effect of the Poiseuille counter-flow increases and finally leads to an equilibrium state with zero net flow.

The above flow is at the origin of the basic operation principle of the Knudsen pumps. A Knudsen pump is actually a compressor without moving parts that works for a gas under rarefied conditions. The basic design of such a pump consists of a thin channel, in which the gas is at least in the slip flow regime, connecting a cold and a hot region. In order to increase the efficiency of the pump it is possible to use an array of parallel thin channels and/or a cascade system consisting of several units connected in series by intermediate thick channels or reservoirs (figure 2a). It is obvious that in order to operate a Knudsen pump at atmospheric pressure, the dimensions of the narrow channel must be in sub-micro or nano-scale. Until recently, it was difficult to fabricate such a channel, but this limitation was overcoming by operating the system at low pressures and with channels of some ten to hundreds of micrometers. Today, it is possible to manufacture channels with very small height and in addition to work in vacuum conditions. Since the work of Knudsen, several researchers have tried to developed micromachined pumps [21-25] or pumps using alternative materials for the narrow channels [26-29].

In parallel, there are also some theoretical approaches to describe the operation of a Knudsen pump, either with the Direct Simulation Monte Carlo method [15] or with the BGK model [18, 19, 30, 31]. Recently, an alternative configuration has been proposed and numerically studied [18, 19], in which the sequence of narrow-wide segments could be replaced with a sequence of curved-straight segments. The curved (semicircular) segments have the same width as the straight segments (figure 2b). In this case, there is a uniform temperature gradient along the walls of the straight part (the temperature decreases linearly from T_H to T_L along the channel), while an opposed temperature gradient is applied along the walls of the curved part. As a consequence, two opposed thermal creep flows will be generated in the two parts. Due to the different geometry, however, the transpiration effect is stronger in the curved segments and a global net flow is observed. In addition, it has been shown [18, 19] that a multi-stage Knudsen compressor consisting of curved and straight channels can actually operate between the slip and the free molecular flow regimes. More precisely, there is an optimum Knudsen number, depending on the curvature of the bend part and also on the number of the stages, for achieving the highest compression efficiency. Under certain circumstances, it is possible to have maximum compression ratio in the slip regime or in the early transition regime.

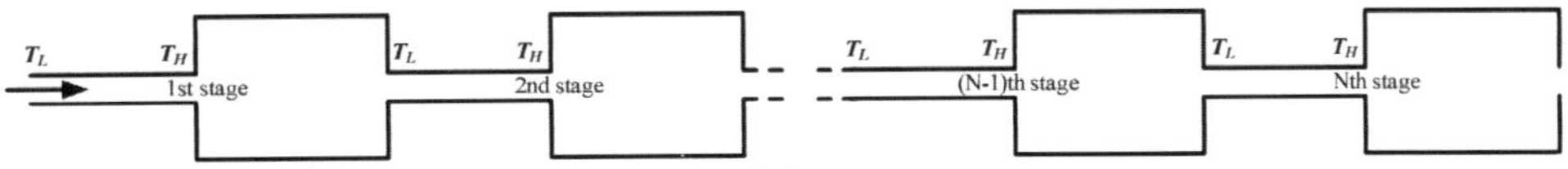

(a)

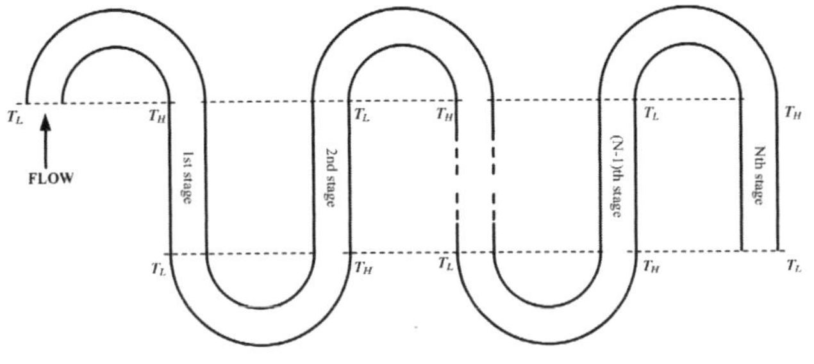

(b)

Figure 2. Configurations of a) classic Knudsen pump and b) curved-channel Knudsen pump.

Motivation of the present work is the development of a simulation tool which will help in the preliminary design of a Knudsen micro-pump with curved channels. A two-dimensional numerical model has been developed for implementing the correct first order boundary conditions in the case of the slip flow regime, taking into consideration the thermal creep flow due to the axial temperature gradient and the effect of the wall curvature. The modeling of the thermally driven flow of rarefied gases in microchannels is achieved using the commercial CFD software ANSYS Fluent®, while C++ routines have been developed in order to apply the boundary conditions at the walls of the channels.

2. Boundary conditions

As already pointed out, the rarefied gas flow in the slip flow regime can be modeled using the Navier-Stokes equations, but together with the appropriate boundary conditions describing the slip velocity and the temperature jump at the solid walls. For the case of a 2-D flow with a curved wall or with significant roughness, the expression of slip velocity at the walls can be written as [10, 32, 33]:

$$u_{fluid} - u_{wall} = \alpha \frac{(2-\sigma_v)}{\sigma_v} \lambda \left(\frac{\partial u}{\partial n} + \frac{\partial v}{\partial s} \right) \bigg|_{wall} + \frac{3}{4} \frac{\mu R}{P} \frac{\partial T}{\partial s} \bigg|_{wall} \tag{3}$$

where α and σ_v are a corrective coefficient and the tangential momentum accommodation coefficient, u and v are the tangential and the normal velocities, n and s the normal and the tangential coordinates, μ and R the dynamic viscosity and the specific gas constant, P and T the pressure and the temperature. The first term of the right hand side is the first order isothermal slip velocity which derives from the tangential shear stress at the wall and the second term is the thermal creep due to the tangential temperature gradient. It is important to note that the additional derivative $\partial v/\partial s$ must be taken into account in the case of curved channels, although in most published works this term is neglected or has been sometimes erroneously replaced with the derivative $\partial u/\partial s$.

In a similar way, the first order temperature jump boundary condition is:

$$T_{fluid} - T_{wall} = \beta \frac{(2-\sigma_T)}{\sigma_T} \frac{2\gamma}{\gamma+1} \frac{\lambda}{Pr} \frac{\partial T}{\partial n} \bigg|_{wall} \tag{4}$$

where β and σ_T are a corrective coefficient and the thermal accommodation coefficient, γ and Pr the ratio of the specific heats and the Prandtl number. Both corrective coefficients in equations (3) and (4) have been added to the initial Maxwell-Smolukowski boundary conditions, derived from numerical simulations [34].

3. Numerical Model

The numerical modeling has been carried out using the commercial CFD code ANSYS Fluent® 12.1, which implements the finite element method. The velocity slip boundary condition is imposed as the velocity of a moving wall, and the temperature jump as a modified temperature of the wall, both of them applied cell by cell at the wall. For that reason User Defined Functions (UDFs) have been developed by means of C++ routines and compiled in the solver. Through the UDFs the derivatives of the velocities and the temperature are extracted from the solver and the slip velocity and temperature jump are calculated based on equations (3) and (4). In addition, the temperature of the walls is imposed with the UDFs, either as a constant temperature or as a temperature gradient when needed.

Due to the surface curvature, the local wall coordinate system is in general not aligned with the solver's coordinate system. This makes a transformation of coordinate system necessary. The velocity vectors and all derivatives can be extracted from the solver, but are expressed in the solver coordinate system (x, y). In the case of straight channels, the normal (y direction) and tangential (x direction) derivatives of the velocity and the temperature, which are included in equations (3) and (4), are aligned with the solver coordinate system and can be replaced with the terms $\partial u/\partial y$, $\partial v/\partial x$, $\partial T/\partial y$ and $\partial T/\partial x$, respectively. In curved channels, the derivatives $\partial u/\partial n$, $\partial v/\partial s$, $\partial T/\partial n$ and $\partial T/\partial s$ must be expressed as functions of the known $\partial u/\partial x$, $\partial u/\partial y$, $\partial v/\partial x$, $\partial v/\partial y$, $\partial T/\partial x$ and $\partial T/\partial y$.

All calculations are performed with the two-dimensional segregated solver assuming steady state laminar flows. A second order upwind spatial discretization scheme is applied. The dilute gas is considered as an ideal compressible gas, while the temperature-dependent viscosity is calculated from the power law:

$$\mu = \mu_{ref} \left(\frac{T}{T_{ref}} \right)^{\omega} \tag{5}$$

where μ_{ref} is the dynamic viscosity of the gas at the reference temperature T_{ref} and ω is a power law coefficient. The values of these coefficients are available in the literature for various gases [35]. The corrective coefficients α and β and the accommodation coefficients σ_v and σ_T of equations (3) and (4) have been kept equal to unity throughout the numerical simulations presented here. Assuming a dilute gas, the Prandtl number is set to 2/3 for a monatomic and to 0.72 for a diatomic gas, whereas the ratio

γ is equal to 5/3 for a monatomic and to 7/5 for a diatomic gas. The specific gas constant R is the ratio of the gas constant $\mathscr{R}$ = 8.314 J mol^{-1} K^{-1} to the molar mass M of the gas.

It is worth mentioning that the commercial CFD code ANSYS Fluent® gives the possibility of implementing the first-order slip and temperature jump boundary conditions, with the so-called Low Pressure Boundary Slip (LPBS) method. The choice of velocity slip boundary condition is, however, limited to Maxwell's first-order condition, without taking into account the effect of the curvature of the wall and the thermal creep. In addition, the mean free path can be adjusted only through the Lennard-Jones length. The validity of the LPBS method was tested for the simplest case of pressure driven gas flows in straight microchannels, and comparison with analytical solutions, which can be found in literature [35], showed a very good agreement. It has been shown [36], however, that the obtained results are not always very accurate, especially in three-dimensional flows with cross-sections involving acute angles. To overcome these drawbacks and also to implement, if needed, higher-order boundary conditions, specific UDFs have been developed in the present work.

4. Validation of the numerical model

Several benchmark cases for gas flows in the slip flow regime are simulated in order to test the validity of the developed numerical procedure used for implementing the boundary conditions, e.g. pressure and/or temperature driven gas flows in straight and curved microchannels of rectangular and circular cross sections. The results of the model are compared either with analytical or numerical results when possible.

Initially, the numerical model (UDF method) has been validated for the simplest case of pressure driven gas flows for the below cases:

- isothermal flow in a plane microchannel limited by parallel plates compared with an analytical solution [35],
- isothermal and non-isothermal flows (without taking under consideration the thermal creep and the effect of the curved wall) in rectangular and circular straight channels compared with the numerical results of ANSYS Fluent® using the LPBS boundary condition method and
- isothermal and non-isothermal flows (without taking under consideration the thermal creep and the effect of the curved wall) in a sequence of straight and curved rectangular microchannels compared with the LPBS method.

Next, the UDF method is validated for the case of temperature driven flows in a straight rectangular microchannel and a sequence of straight and curved rectangular channel, taking into account the curvature of the bent channels and thermal creep effect.

4.1. Pressure driven flows in bent microchannels

The isothermal and non-isothermal flow in a straight and curved rectangular microchannel connecting two reservoirs maintained at different pressures is considered (see figure 6a). In the isothermal case the whole system is at constant temperature (T = 300 K) and neither the temperature jump nor the thermal creep terms are taken into account. On the other hand, in the non-isothermal case the wall temperature T_L = 300 K is imposed at the inlet and outlet sections A and C, and, the wall temperature T_H = 900 K is imposed at the intermediate section B; between these sections, the temperature of the wall varies linearly. In that case both the velocity slip and temperature jump at the walls are taken into account, while the thermal creep effect is still not considered. In both cases, the effect of the wall curvature expressed by the additional derivative of the normal velocity ($\partial v/\partial s$) in equation (3) is not taken into account. The results of the simulations using the UDF Method are compared with those obtained imposing the LPBSC provided by ANSYS Fluent®.

The width of the microchannel is D = 100 µm, the length of the straight part is L_s = 314 µm, while the radius of the centerline of the curved part is R_c = 100 µm. A pressure drop $\Delta P = P_A - P_C$ = 100 Pa is imposed between the inlet and the outlet of the whole microchannel. The gas is argon, considered as an ideal gas. Under these conditions the gas is in rarefied conditions and the Knudsen number at the inlet is $Kn_A = 0.012$. The grid used in the simulations contains 200×50 cells. It has been checked that a finer grid leads to identical results, which means that the mesh is fine enough.

In figure 3 the pressure distributions along the centerline are presented (s represents the curvilinear abscissa, with $s = 0$ corresponding to the inlet A of the curved channel) for both the isothermal (in black) and non-isothermal flow (in red). The pressure distribution along a microchannel should be a 2nd order polynomial, as shown in [35, 37-39], with opposite effects of rarefaction and compressibility on the curvature of the polynomial distribution. The main effect is due to compressibility and rarefaction effects (characterized by Kn) tend to reduce the curvature of the pressure distribution. In the isothermal case of figure 3, however, the pressure distribution is very close to a linear one due to the small imposed pressure drop which results in negligible compressibility effects. On the other hand, for the non-isothermal case, compressibility effects induced by temperature variations –density is strongly temperature dependent– cause the curvature on the pressure distribution between the two cold ends of the microchannel and the hot junction of the straight and bent part.

Moreover, it is observed in figure 3 that the agreement between the two simulation procedures is good and the results are very close. For the isothermal flow the deviation between the mass flow rates per unit of depth calculated with both methods is about 1.2 %, the LPBS method giving 4.74×10^{-5} kg s^{-1} m^{-1} and the UDF 4.68×10^{-5} kg s^{-1} m^{-1}. For the non-isothermal flow, the mass flow rate calculated with the LPBS method is equal to 1.43×10^{-5} kg s^{-1}m^{-1} and to 1.41×10^{-5} kg s^{-1}m^{-1} with the UDF method.

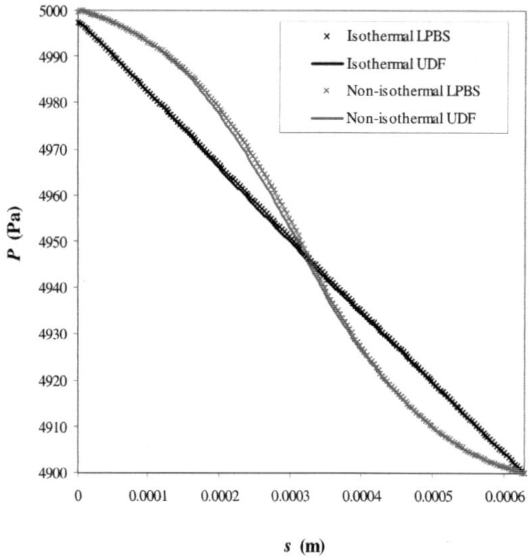

Figure 3. Pressure profiles along the centerline for pressure driven isothermal and non-isothermal gas flows - Comparison between UDF and LPBS methods.

4.2. Thermal creep flows in straight microchannels

The temperature driven fully developed flow of rarefied gas between parallel plates is simulated taking under consideration the thermal creep in this test case. The computations are carried out for a two-dimensional flow limited by the symmetry plane and the upper plate (figure 4). The temperatures at the inlet and the outlet are noted T_L and T_H, respectively, with $T_L < T_H$ and the wall temperature is varying as a linear function of the streamwise coordinate x. The pressures at the inlet and outlet are always the same. The channel has a width of D = 10 μm and a length of L = 1 cm. The grid has 100x25 cells. The gases are helium or nitrogen, considered as ideal monatomic or diatomic gases. The data used in the simulations are presented in table 1.

The results are compared with the numerical solution of the Navier-Stokes equations with the appropriate boundary conditions calculated by Méolans & Graur [17]. The model developed by these authors is based on the exact numerical solution of the compressible Navier-Stokes equations associated with similar boundary conditions as those described by equations (3) and (4). In addition, Méolans & Graur derived a simplified analytical model using a perturbation method according to a small geometric parameter. The boundary conditions used in the UDF model have been modified for the present comparisons, in order to be applied to the equations proposed by the above authors.

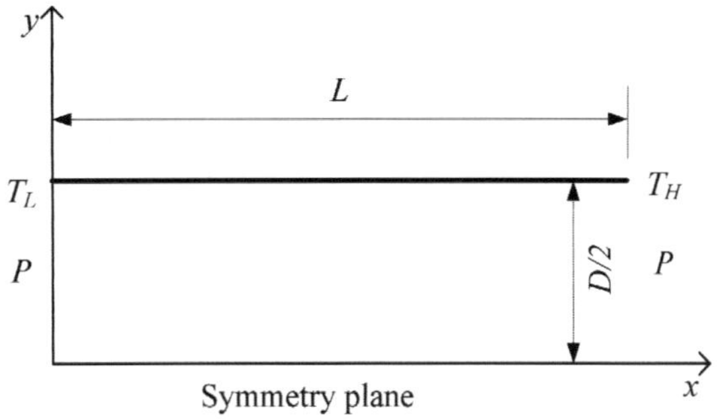

Figure 4. Straight channel parameters for thermal creep simulation.

Table 1. Simulation parameters of the thermal creep flow in a straight microchannel and mass flow rates calculated by the present UDF method and the numerical simulation of Méolans & Graur [17].

Case	1	2	3	4	5	6
Gas	He	N_2	N_2	N_2	N_2	N_2
T_{in} **(K)**	295.6	295	295	295	295	295
T_{out} **(K)**	573	435	595	895	595	595
***P* (Pa)**	101325	101325	101325	101325	10132.5	5066.25
Kn_{out}	0.036	0.009	0.0126	0.021	0.126	0.253
	Mass flow rate (10^{-8} kg s^{-1} m^{-1})					
[17]	1.884	0.725	1.465	2.686	1.470	1.427
UDF	1.879	0.724	1.462	2.678	1.463	1.466
Deviation (%)	**0.263**	**0.143**	**0.161**	**0.281**	**0.462**	**0.342**

The mass flow rates are calculated per unit of depth of the channel and are compared in table 1. It is obvious from this table that the constant pressure level in the reservoirs connected to the inlet and outlet of the microchannel has little influence on the mass flow rate (see cases 3, 4 and 6). This result is expected, as the flow is not pressure driven and the relative pressure variation along the microchannel is small (see figure 5b), and the thermal creep term in equation (3) is inversely proportional to the pressure. This term is the major contribution of the velocity slip, as can been proved by comparing the order of magnitude of the two terms of the RHS of equation (3): the ratio of the first over the second term of the RHS is of the order of 10^{-4}. As a result, the velocity slip and the local velocity are also roughly inversely proportional to the pressure and the mass flow rate defined as:

$$\dot{m} = \iint_S \rho \, u \, dS = \frac{P}{R} \iint_S T \, u \, dS \qquad (6)$$

is almost not pressure dependent. On the other hand, the temperature gradient across the microchannel strongly affects the flow rate (see cases 2, 3 and 4). An increase in the temperature of the hot reservoir causes an increase in the flow rate. The deviation in the calculated mass flow rate between the two specific approaches is very small, less than 0.5 %, for all the studied cases. Méolans & Graur [17] compared their results with a numerical solution of the Boltzmann equations (S-model) and they found that for small Knudsen numbers there is a very good agreement, but for large Knudsen numbers there is a significant deviation (15 %). This deviation, however, is expected as the regime deviates from the slip flow regime.

The velocity profiles in various cross-sections of the channel are presented in figure 5a. The streamwise velocity u has been normalized with the velocity:

$$u_R = \frac{R\mu_{out} T_{w,out}}{2P_{out}\, Pr\, L} \tag{7}$$

and the normal coordinate y with the width D of the channel. In the first half of the channel, where the pressure increases, the maximum of the velocity is observed at the wall, while on the other half the maximum is on the centerline.

Figure 5b shows the pressure profiles along the streamwise direction on the centerline for two values of the Knudsen number. In that case the pressure is normalized with the outlet pressure and the streamwise position x with the length L of the microchannel. It can be seen that the pressure distribution along the channel is non-linear with a non-symmetric distribution and a maximum value downstream from to the mid-section of the channel.

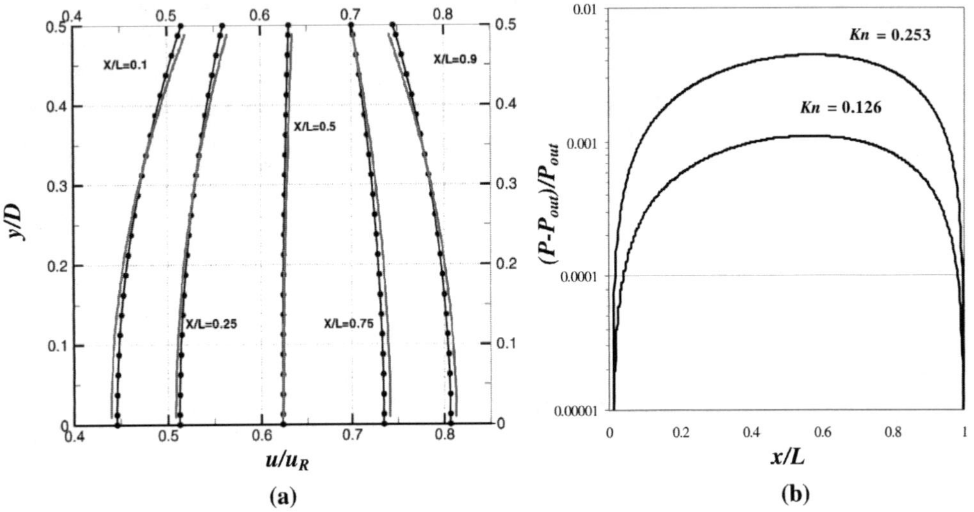

Figure 5. a) Velocity profiles in various cross-sections (He, Kn = 0.036). The red curves correspond to the present UDF model and the black ones to the numerical solution of Méolans & Graur [17]. b) Pressure profiles calculated by the UDF model for two Knudsen numbers. Flow of nitrogen.

4.3. Temperature driven flows in ringed shaped channels

In this section, the layout proposed by Aoki et al. [18, 19] is considered. The complete expressions of the boundary conditions, as described by the equations (3) and (4), are used. The basic unit of this device consists of a straight channel with length L_s connected to a semi-circular part with radius of the centerline R_c (figure 6a). The height of the straight and bent channel is D. A linear temperature gradient is applied in the straight part and the temperature is increasing from T_L to T_H. An opposed temperature gradient is applied to the curved part. By completing this configuration with a symmetrical part, a close ring can be formed (figure 6b). It has been proved with theoretical approaches by Aoki and co-workers that a circulating flow is generated inside the ring. Using the Boltzmann-BGK model as the governing equation for the gas flow, Aoki et al. have developed numerical methods based either on the asymptotic technique of the diffusion approximation [18] or on the deterministic finite volume scheme [19].

The geometry of the ring is symmetric with reference to the centre of the ring and so the configuration of figure 6a is simulated with the present method. Periodic boundary conditions are applied at the inlet and outlet sections A and C, expressed as $T_A = T_C$ and $P_A = P_C$. The steady state 2-

D compressible flow of an ideal gas (argon) is simulated and the results are compared with the numerical simulation of Aoki et al. [19].

Three different layouts are tested, the parameters of which are presented in table 2. The pressure in the inlet is P_A = 0.15 Pa and the hot and cold temperatures are T_H = 900 K and T_L = 300 K, respectively. The Knudsen number defined for the global average density in the channel and given by Aoki et al. [19] is Kn = 0.1, but when calculated for the entrance conditions, its value is Kn_A = 0.04. Due to the pressure level, the system is not in micro-scale, but a device with macro dimensions working under very low pressures, which ensures that the gas is in rarefied conditions.

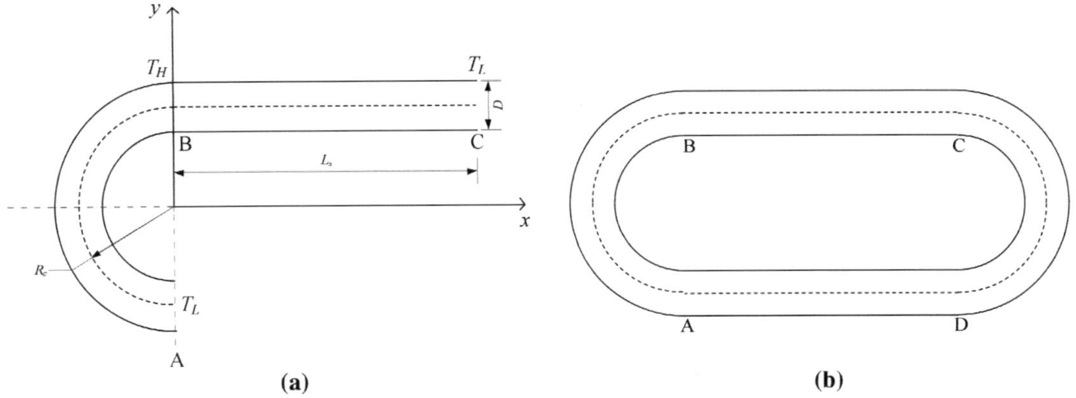

Figure 6. a) Curved and straight microchannel geometry, b) ring-shaped configuration.

The channel is discretized with a number of cells depending on its size: the mesh contains 200×200 cells for case 1, 200×100 cells for case 2 and 400×100 cells for case 3. In all cases the grid is non-uniform in the spanwise direction with an increased density near the walls. Coarse and fine meshes either uniform or non-uniform have been tested and the specific grids were chosen by comparing the results obtained with the different meshes and also with the numerical simulations of Aoki et al. [19].

Table 2. Simulation parameters of thermal creep flow in a curved channel and mass flow rates calculated by the present UDF method and the numerical simulation of Aoki et al. [19].

Case	**1**	**2**	**3**
***D* (m)**	1	1	1
L_s (m)	3.14	6.28	15.7
R_c (m)	1	2	5
	Mass flow rate (10^{-8} kg s^{-1} m^{-1})		
[19]	79.645	8.833	0.511
UDF	78.867	8.734	0.513
Deviation (%)	**0.977**	**1.126**	**0.314**

In figure 7 the distribution of the dimensionless temperature, density and pressure along the centerline of the whole channel are shown, for all cases described in table 2. In this figure the normalization has been done with the cold temperature T_L, the average density ρ_{av}, and the term $R\rho_{av}T_L$, respectively, and S represents the dimensionless curvilinear abscissa normalized by the length of the straight part L_s with S = 0 corresponding to the inlet A of the curved channel. In addition, in table 2 the mass flow rate per unit of depth is included for both methods. In every case the UDF method shows a very good agreement with the numerical analysis and especially for case 3.

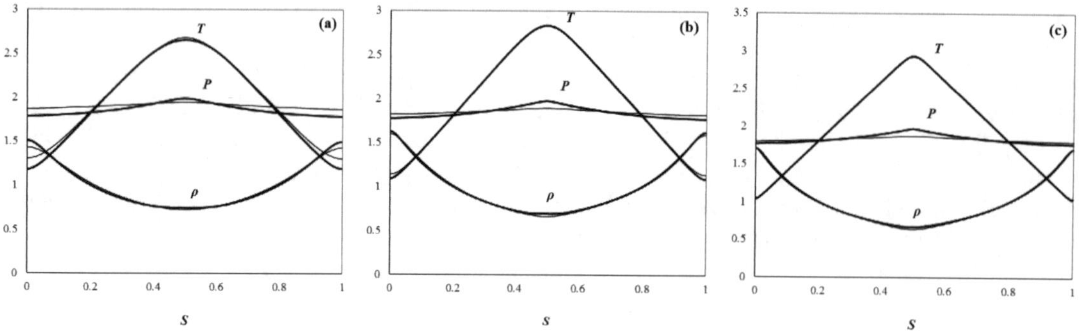

Figure 7. Distributions of the dimensionless temperature T, density ρ and pressure P along the centerline of the channel for a) case 1, b) case 2 and c) case 3. The thick curves correspond to the results of the present UDF model and the thin ones to the numerical solution of Aoki et al. [19].

5. Application to the design of a multistage Knudsen micropump

The device presented in figure 2b, which actually is a Knudsen pump (compressor) consisting of N stages, and has been proposed by Aoki and co-workers [18, 19] is now simulated. The cascade system consists of two basic units joined alternately; the unit ABC of figure 6a and the mirror image of part ADC of figure 6b. It is considered that the end of the first stage is closed by a wall kept at temperature T_L, while the other end of the N stage is open and kept at pressure P_o and temperature T_L. The slip and temperature jump conditions are also imposed at the wall of the closed end of the pump. Again, S is the dimensionless curvilinear abscissa across the centerline, with $S = 0$ referring to the closed end and $S = N$ to the open end of the cascade system.

The size of each unit is defined by $D = 100$ μm, $L_s = 314$ μm and $R_c = 100$ μm. The hot and cold temperatures are $T_H = 900$ K and $T_L = 300$ K, respectively, and the pressure at the open end of the device is $P_o = 1500$ Pa. The Knudsen number calculated for this conditions is $Kn = 0.04$ (or equal to 0.1 if calculated based on the average density in the whole device). The mesh of one unit (stage) contains 200×100 cells and the grid is non-uniform in the spanwise direction with higher density near the walls. The simulations are made for N equal to 1, 2, 4, 8 and 25 stages.

In figure 8a the dimensionless pressure P along the centerline of the whole channel for the different cascade systems is shown as a function of the dimensionless distance S. As expected, the pressure profiles oscillate with the same frequency as the imposed wall temperature and the cumulative effect of the number of stages is evidenced. Calculating the dimensionless average pressure $P_{average}$ for every stage (figure 8b) it can be concluded that the dependence on the number of stages is rather close to a linear function which depends on each system. In figure 8c the pressure gain P_{gain}, defined as the ratio of the pressure difference (between the two ends of the cascade system) to the outlet pressure P_o, and the density gain ρ_{gain}, defined as the ratio of the density difference between the two ends of the cascade system to the density on the open end, are plotted in percent. Both pressure and density gains are linearly dependent on the number of stages. Additional simulations with a larger number of stages and under different conditions, however, are required in order to confirm this observation. For the present simulated device the pressure and the density gains are almost 45% for the 25 stage device. Extrapolating the line of figure 8c, it is found that 100% of gain could be achieved with around 60 units.

In their work Aoki et al. [19] have simulated a different multistage Knudsen pump ($D = 1$ m, $L_s = 6.28$ m, $R_c = 2$ m, $T_H = 900$ K, $T_L = 300$ K, $Kn = 0.5$, $N = 2$-16). In their system a gain of 100% is achieved with around 32-36 units. The results of figure 8, however, present the same trend and the observations are almost identical. The difference in the required number of stages for achieving 100% of gain is most likely due to the different Knudsen number which is higher in the case simulated by Aoki et al. It is expected that the operation of the Knudsen pump in this early transition regime is a bit more efficient. Nevertheless, the similarities between the results from the present simulation and the numerical simulation using the Boltzmann-BGK model support our methodology.

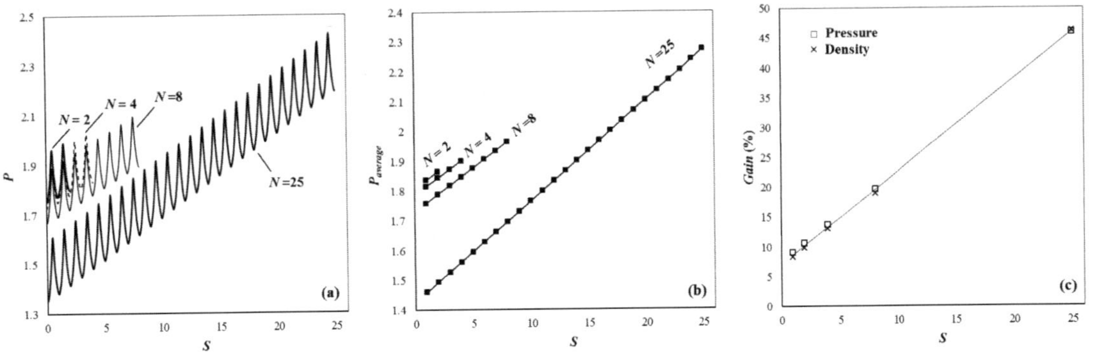

Figure 8. a) Distributions of the dimensionless pressure P along the centerline of the channel, b) distributions of the dimensionless average pressure $P_{average}$, and c) pressure and density gain due to the pumping of the device.

6. Conclusions

It has been proven that thermal transpiration in the slip flow regime can be numerically simulated using commercial CFD codes, if the correct boundary conditions at the walls are used and handled in an appropriate way. The numerical model that has been developed showed a very good agreement with analytical and other numerical approaches when compared for benchmark cases.

The present numerical model proved that Knudsen pumps based on curved channels can be efficient in the slip flow regime. The simulation results are close to the numerical data obtained from a kinetic approach [18, 19].

Motivation of the present project is the development of a simulation tool which would be useful in the design of Knudsen pumps. At the moment, a multistage Knudsen pump with curved segments (see figure 2b) at millimetric scale has been designed, taking into account manufacturing limitations and based on the numerical results, and is being fabricated at the Karlsruhe Institute of Technology (Germany). Its performance will be experimentally analyzed and compared with the numerical results and its structure and operation will be optimized.

Acknowledgments

The research leading to these results has received funding from the European Community's Seventh Framework Programme (FP7/2007-2013) under grant agreement GASMEMS n° 215504. The authors thank Prof. Luc Mieussens and Prof. Kazuo Aoki for the provided comments and tabulated data from their published works.

References

[1] Hobson J P and Salzman D B 1999 *J. Vac. Sci. Technol.* A **18** 1758
[2] Nguyen N-T, Huang X and Chuan T K 2002 *T ASME* **124** 384
[3] Laser D J and Santiago J G 2004 *J. Micromech. Microeng.* **14** R35
[4] Tsai N-C and Sue C-Y 2007 *Sensor. Actuat.* A **134** 555
[5] Bird G A 1994 *Molecular Gas Dynamics and the Direct Simulation of Gas Flows* (Oxford University Press)
[6] Karniadakis G, Beskok A and Aluru N 2005 *Microflows and nanoflows: fundamentals and simulation* (Springer)
[7] Colin S 2005 *Microfluidic. Nanofluidic.* **1** 268
[8] O'Hare L, Lockerby D A, Reese J M and Emerson D R 2007 *Int. J. Heat Fluid Flow* **28** 37

[9] Dongari N, Zhang Y, and Reese J M 2011 *J. Fluids En.g* **133** 071101
[10] Colin S 2012 *J. Heat Transf.* **134** 020908
[11] Reynolds O 1878 *Proceedings of the Royal Society of London* **28** 303
[12] York D C, Chambers A, Chew A D and Troup A P 1999 *Vacuum* **55** 133
[13] Han Y-L, Muntz E P, Alexeenko A and Young M 2007 *Nanosc. Microsc. Therm. Eng.* **11** 151
[14] Cardenas M R, Graur I, Perrier P and Meolans J G 2011 *Phys. Fluids* **23** 031702
[15] Sone Y, Waniguchi Y and Aoki K 1996 *Phys. Fluids* **8** 2227
[16] Alexeenko A A, Gimelshein S F, Muntz E P and Ketsdever A D 2006 *Int. J. Therm. Sci.* **45** 1045
[17] Méolans J G and Graur I A 2007 *Eur. J. Mech.* B **27** 785
[18] Aoki K, Degond P, Mieussens L, Takata S and Yoshida H 2008 *Multiscale Model. Sim.* **6** 1281
[19] Aoki K, Degond P and Mieussens L 2009 *Commun. Comput. Phys.* **6** 919
[20] Maxwell J. C 1879 *Philos. T. R. Soc. Lond.* **170** 231
[21] Phan-Van-Diep G, Keely P, Muntz E P and Weaver D P 1995 *Proceedings of the 19th International Symposium on Rarefied Gas Dynamics* **I** 715 (London UK)
[22] Vargo S E and Muntz E P 1996 *Proceedings of the 34th Aerospace Sciences Meeting & Exhibit* (Reno)
[23] Young R M 1999 *J. Vac. Sci. Technol.* B **17** 280
[24] MacNamara S and Gianchandani Y B 2005 *J. Microelectromech. S.* **14** 741
[25] Sugimoto H 2009 *Proceedings of the 26th International Symposium on Rarefied Gas Dynamics* **1123** (Kyoto Japan)
[26] Han Y-L and Muntz E P 2007 *J. Vac. Sci. Technol.* B **25** 703
[27] Gupta N K and Gianchandani Y B 2008 *Appl. Phys. Lett.* **93** 193511
[28] Pharas K and MacNamara S 2010 *J. Micromech. Microeng.* **20** 125032
[29] Gupta N K and Gianchandani Y B 2011 *J. Micromech. Microeng.* **21** 095029
[30] Aoki K, Takata S, Tatsumi E and Yoshida H 2010 *Phys. Fluids* **22** 112001
[31] Takata S and Umetsu H 2011 *Proceedings of the 27th International Symposium on Rarefied Gas Dynamics* **1333** 998 (California USA)
[32] Lockerby D A, Reese J M, Emerson D R and Barber R W 2004 *Phys. Rev.* E **70** 017303
[33] Barber R W, Sun Y, Gu X J and Emerson D R 2004 *Vacuum* **76** 73
[34] Sharipov F 2011 *J. Phys. Chem. Ref. Data* **40** 023101
[35] Kandlikar S G, Garimella S, Li D, Colin S and King M R 2006 *Heat Transfer and Fluid Flow in Minichannels and Microchannels* (Elsevier Chap 2)
[36] Pitakarnnop J, Geoffroy S, Colin S and Baldas L 2008 *Int. J. Heat Technol.* **26** 167
[37] Jie D, Diao X, Cheong K B and Yong L K 2000 *J. Micromech. Microeng.* **10** 372
[38] Graur I A, Méolans J G and Zeitoun D E 2006 *Microfluid. Nanofluid.* **2** 64
[39] Verma N, Mewes D and Luke A 2010 *Int. J. Heat Mass Transfer* **53** 3175

Rarefied gas mixture flow between plates of arbitrary length due to small pressure difference

C Tantos[1], **S Naris, D Valougeorgis**
University of Thessaly, Dept. of Mechanical Engineering, Volos 38334, GR

E-mail: christantos@mie.uth.gr

Abstract. The linearized flow of a binary gas mixture between two parallel plates of any length driven by a small pressure difference is numerically solved based on the McCormack kinetic model. In this case the flow is not fully developed and the simulated flow field includes, in addition to the channel, efficiently large upstream and downstream regions significantly increasing the required computational effort. In the present work results are provided for one gas mixture, namely He-Ne with a specific concentration of 0.5 in a wide range of gas rarefaction. The ratio of the length over the height of the channel is taken to be equal to 0, 1 and 5 corresponding to flow through channels of very short (slit), short and moderate lengths respectively. Results are provided for the two kinetic coefficients related to Poiseuille and barodiffusion flow rates and the velocity profiles of each species. The effect of the length to height ratio and of the gas rarefaction on the flow quantities is investigated deducing that as the ratio in increased the separation phenomenon becomes more dominant, while a Knudsen minimum is observed at the ratio equal to 5.

1. Introduction

Most of the work of gas mixture flows through micro channels is based on the assumption that the channel is long enough to introduce a fully developed flow approach [1-4]. The corresponding work through short micro channels is very limited mainly due to the involved computational cost. When the flow is fully developed the pressure varies only in the flow direction and it is possible to simulate the flow field in several cross sections along the channel. Then, following a well known integrating procedure based on the mass conservation principal yields the flow rate through the channel [4]. Contrary, when the fully developed assumption is not valid as in the case of channels of short and moderate lengths the whole flow field including the channel with adequate large inlet and outlet regions must be simulated.

In the latter case when the flow field is driven by large pressure differences the DSMC method may be implemented in an efficient manner [5]. However, when small pressure differences are imposed the flow velocity is rather small and the DSMC method becomes computationally inefficient due to the introduced statistical noise. It has been shown that in the case of low-speed flows linearized kinetic theory is the most suitable computational approach.

In the present work a system of linearized Boltzmann equations is solved via a fully deterministic approach to simulate the flow of a binary gas mixture between parallel plates of arbitrary length. The McCormack kinetic model [6] is introduced to replace the complicated collision term of the Boltzmann equation. Purely diffuse boundary conditions at the wall boundaries are assumed. The

[1] To whom any correspondence should be addressed.

proposed algorithm is initially validated by comparing the present results for the flow through a slit with corresponding ones available in the literature [7] and then it is used to simulate binary gas flow between parallel plates of finite length. Preliminary results are provided for one mixture and a limited number of channel lengths in a wide range of the Knudsen number. A more complete investigation of the effect of all involved geometric and flow parameters will be performed in the short future.

2. Formulation

Consider two large reservoirs connected by a channel of length L. The reservoirs contain a binary gas mixture having a reference concentration C_0 and they are maintained at pressures $P_0+\Delta P$ and P_0 with $\Delta P/P_0 \ll 1$, while the temperature in both reservoir is equal to T_0. Due to the pressure difference there is a gas mixture flow through the channel. The height of the channel is denoted by H, while the width of the channel is much larger than its height in order to ignore the end effects in that direction. However, the end effects in the flow direction are taken into account with the present formulation and channels of arbitrary length may be studied provided that the pressure difference is adequately small to justify a linear treatment. In Figure 1 the flow configuration is shown. The coordinate system is (x, y), with x denoting the flow direction and y the direction vertical to the plates.

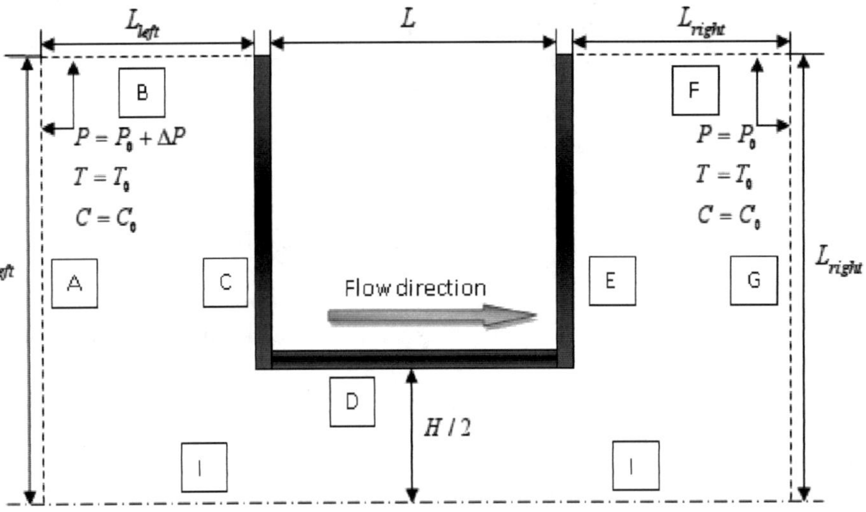

Figure 1. Flow configuration.

The concentration of the mixture is defined as

$$C(x,y) = n_1(x,y)/n(x,y) \tag{1}$$

where $n_a(x,y)$, $a=1,2$ is the number density of the species and $n=n_1+n_2$ is the number density of the mixture. Always the number density of light species is denoted by "1". The reference concentration C_0 is defined as

$$C_0 = n_{B1}/n_B \tag{2}$$

where n_{1B} and n_B are the number density of the light species and of the mixture respectively at the downstream container. In the present configuration the concentrations of the mixture in the downstream and upstream containers are equal. The molecular mass of the mixture is

$$m = C\,m_1 + (1-C)\,m_2 \tag{3}$$

where m_a, $a=1,2$ is the molecular mass of each species.

The formulation of the McCormack kinetic equation is identical to the one described in [6,7] and therefore only the final projected kinetic equations, as implemented in the present work, are provided. In particular starting from the second order McCormack model and introducing the well known projection procedure yields the following set of four coupled kinetic equations:

$$
\begin{aligned}
& c_p \sin\theta \frac{\partial R_\alpha}{\partial y} + c_p \cos\theta \frac{\partial R_\alpha}{\partial x} = \frac{\delta_\alpha}{\gamma_\alpha} \sum_{\beta=1}^{2} \Big\{ -\gamma_{\alpha\beta} R_\alpha + \gamma_{\alpha\beta} \nu_\alpha \\
& +2\left[\gamma_{\alpha\beta} u_{\alpha x} - \left(u_{\alpha x} - \sqrt{\frac{m_\alpha}{m_\beta}} u_{\beta x} \right) \nu_{\alpha\beta}^{(1)} \right] c_p \cos\theta + 2\left[\gamma_{\alpha\beta} u_{\alpha y} - \left(u_{\alpha y} - \sqrt{\frac{m_\alpha}{m_\beta}} u_{\beta y} \right) \nu_{\alpha\beta}^{(1)} c_p \right] c_p \sin\theta \\
& +2\left[(\gamma_{\alpha\beta} - \nu_{\alpha\beta}^{(2)}) \Pi_{\alpha xx} + \nu_{\alpha\beta}^{(3)} \Pi_{\beta xx} \right] c_p^2 \cos^2\theta + 2\left[(\gamma_{\alpha\beta} - \nu_{\alpha\beta}^{(2)}) \Pi_{\alpha yy} + \nu_{\alpha\beta}^{(3)} \Pi_{\beta yy} \right] c_p^2 \sin^2\theta \\
& +\left[(\gamma_{\alpha\beta} - \nu_{\alpha\beta}^{(2)}) \Pi_{\alpha zz} + \nu_{\alpha\beta}^{(3)} \Pi_{\beta zz} \right] + 4\left[(\gamma_{\alpha\beta} - \nu_{\alpha\beta}^{(2)}) \Pi_{\alpha xy} + \nu_{\alpha\beta}^{(3)} \Pi_{\beta xy} \right] c_p^2 \cos\theta \sin\theta \Big\}
\end{aligned}
\tag{4}
$$

$$
c_p \sin\theta \frac{\partial S_\alpha}{\partial y} + c_p \cos\theta \frac{\partial S_\alpha}{\partial x} = \frac{\delta_\alpha}{\gamma_\alpha} \sum_{\beta=1}^{2} \left\{ -\gamma_{\alpha\beta} S_\alpha + 2\left[(\gamma_{\alpha\beta} - \nu_{\alpha\beta}^{(2)}) \Pi_{\alpha zz} + \nu_{\alpha\beta}^{(3)} \Pi_{\beta zz} \right] \right\} \tag{5}
$$

All quantities are in dimensionless form. Here $R_\alpha(x, y, c_p, \theta)$ and $S_\alpha(x, y, c_p, \theta)$, $\alpha = 1,2$ are the unknown projected distribution functions, (x, y) the two spatial independent variables and (c_p, θ) the magnitude and the polar angle of the molecular velocity. The quantities γ_α, $\gamma_{\alpha\beta}$ and $\nu_{\alpha\beta}$ are model parameters described in [6,7]. At the right hand side the dimensionless perturbed macroscopic quantities of number density ν_α, bulk velocity $\boldsymbol{u}_\alpha$ and shear stress tensor $\Pi_{\alpha ij}$ of each species are given by the moments of R_α and S_α according to

$$
\nu_\alpha = \frac{1}{\pi} \int_0^{2\pi} \int_0^{\infty} c_p R_\alpha e^{-c_p^2} dc_p d\theta \tag{6.1}
$$

$$
u_{\alpha x} = \frac{1}{\pi} \int_0^{2\pi} \int_0^{\infty} c_p^2 \cos\theta R_\alpha e^{-c_p^2} dc_p d\theta \tag{6.2}
$$

$$
u_{\alpha y} = \frac{1}{\pi} \int_0^{2\pi} \int_0^{\infty} c_p^2 \sin\theta R_\alpha e^{-c_p^2} dc_p d\theta \tag{6.3}
$$

$$
\Pi_{\alpha xx} = \frac{1}{3\pi} \int_0^{2\pi} \int_0^{\infty} c_p \left[\left(c_p^2 \left(2\cos^2\theta - \sin^2\theta \right) - \frac{1}{2} \right) R_\alpha - S_\alpha \right] e^{-c_p^2} dc_p d\theta \tag{6.4}
$$

$$
\Pi_{\alpha yy} = \frac{1}{3\pi} \int_0^{2\pi} \int_0^{\infty} c_p \left[\left(c_p^2 \left(2\sin^2\theta - \cos^2\theta \right) - \frac{1}{2} \right) R_\alpha - S_\alpha \right] e^{-c_p^2} dc_p d\theta \tag{6.5}
$$

$$
\Pi_{\alpha zz} = \frac{1}{3\pi} \int_0^{2\pi} \int_0^{\infty} c_p \left[R_\alpha \left(1 - c_p^2 \right) + 2S_\alpha \right] e^{-c_p^2} dc_p d\theta \tag{6.6}
$$

$$
\Pi_{\alpha xy} = \frac{1}{\pi} \int_0^{2\pi} \int_0^{\infty} c_p^3 \cos\theta \sin\theta R_\alpha e^{-c_p^2} dc_p d\theta \tag{6.7}
$$

Finally, the quantity δ_α is the rarefaction parameter of each species defined as

$$
\delta_\alpha = \frac{H}{\upsilon_{0\alpha}} \frac{P_\alpha}{\mu_\alpha}, \quad \alpha = 1,2 \tag{7}
$$

where H is the distance between the plates, P_α is the local pressure of each species, $\upsilon_{0\alpha} = \sqrt{2kT_0 / m_\alpha}$ is a characteristic velocity and μ_α the viscosity at temperature T_0. For the purposes of the present work the reference rarefaction parameter is defined as

$$\delta_1 = \frac{H \; P_{01}}{\upsilon_{01} \; \mu_1} \tag{8}$$

Next, the boundary conditions associated with the governing kinetic equations (4) and (5) are described. At the open boundaries of the computational domain defined in figure 1 by the letters A, B, F, G the gas is considered to be in local equilibrium and the distributions representing particles coming inside the computational field are described by their Maxwellian characterized by the local temperature T_0. The distributions representing particles departing from the computational field are unknown and part of the solution. At the wall boundaries of the computation domain defined in figure 1 by the letters C, D, E the Maxwell purely diffuse reflection interaction model is assumed satisfying the no penetration condition. Finally, along the symmetry axis $y = 0$ defined in figure 1 by the letter I the reflection is specular. Based on above the boundary condition are

$$\text{A, B:} \quad R_\alpha^+ = 1, \quad S_\alpha^+ = 0 \tag{9.1}$$

$$\text{F, G:} \quad R_\alpha^+ = 0, \quad S_\alpha^+ = 0 \tag{9.2}$$

$$\text{C, D, E:} \quad R_\alpha^+ = \nu_{\alpha,wall}, \quad S_\alpha^+ = 0 \tag{9.3}$$

$$\text{I:} \quad \begin{aligned} R_\alpha^+ (x,0,c_p,\theta) &= R_\alpha^+ (x,0,c_p,-\theta) \\ S_\alpha^+ (x,0,c_p,\theta) &= S_\alpha^+ (x,0,c_p,-\theta) \end{aligned} \tag{9.4}$$

where $\nu_{\alpha,wall}$, $\alpha = 1,2$ are parameters, which are defined by the impermeability condition. The superscript (+) denotes distributions departing from the boundaries.

The main quantities of practical interest in the present work are the so called "kinetic coefficients" Λ_{PP} and Λ_{CP} [1,2]. The first one is related to the mass flow rate and it corresponds to the classical Poiseuille flow, while the second one is related to the diffusion flux and is known as barodiffusion. Both of them are caused by the pressure difference with Λ_{PP} to be a direct effect, and Λ_{CP} to be a cross effect. The two kinetic coefficients are computed by [7]:

$$\Lambda_{PP} = \left[C_0 \langle u_{1x} \rangle + (1 - C_0) \sqrt{(m_1 / m_2)} \langle u_{2x} \rangle \right] \tag{10.1}$$

$$\Lambda_{CP} = C_0 \left[\langle u_{1x} \rangle - \sqrt{(m_1 / m_2)} \langle u_{2x} \rangle \right] \tag{10.2}$$

where $\langle u_{\alpha x} \rangle$ is obtained by integrating $u_{\alpha x}$ at any cross section of the channel. The objective is to numerically solve equations (4) and (5) subject to boundary conditions (9) and compute the macroscopic quantities of practical interest given by equations (6).

3. Numerical scheme

The implement numerical scheme is based on the discrete velocity method and on a typical second order finite volume scheme. The continuum spectrum of molecular velocities is replaced by a discrete set of velocities which are defined by their magnitude c_p and polar angle θ. The discretized version of equations (4), (5) and (6) is solved in an iterative manner. The convergence is fast at small rarefaction parameters and then is gradually slows down as δ_1 is increased. The set of numerical parameters used in the results section is shown in Table 1.

Table 1. Numerical parameters used in the simulations.

Parameters	Values
Nodes / unit length in dense areas ($N_x = N_y$)	100 ($L/H=0$); 50 ($L/H=1,5$)
Size of inlet and outlet region	16 x 16
Discrete angles N_θ in $(0,2\pi)$	64
Discrete magnitudes M	16
Max. Value of velocity magnitude (max c_p)	5
Convergence criterion	10^{-7}

4. Results and discussion

Results are provided for one mixture, namely He-Ne, with a reference concentration of $C_0 = 0.50$, flowing between two parallel plates in a wide range of the rarefaction parameter $0 \le \delta_1 \le 10$. The ratio of the length L over the height H is taken to be equal to 0, 1 and 5. In all cases the hard sphere intermolecular potential model is applied.

In Table 2 the kinetic coefficients Λ_{PP} and Λ_{CP} are presented for $L/H=0$. The corresponding results from [7] are also tabulated for comparison purposes. It is seen that for all values of δ_1 tested the agreement is very good. The observed small deviations between the present results and the ones in [7] may be contributed the different intermolecular model (in [7] the LJ potential is applied) and in the involved numerical parameters including the size of the computational domain before and after the slit.

In Table 3 the kinetic coefficients Λ_{PP} and Λ_{CP} are shown for the cases of $L/H=1$ and $L/H=5$. Observing the behavior of the two kinetic coefficients with regard to L/H and δ_1 the following remarks can be made. It is clearly seen that as L/H is increased both flow rates (Λ_{PP} and Λ_{CP}) are decreased. It is expected that as L/H is further increased the results will tend to the corresponding fully developed solution [2]. As δ_1 is increased it is seen that the behavior of Λ_{PP} is not monotonic for all lengths. In particular, it is seen that for $L/H=0$ and 1 as δ_1 is increased Λ_{PP} is monotonically increased. However, for $L/H=5$ the kinetic coefficient Λ_{PP} is initially decreased up to δ_1=0.4 and then as δ_1 is further increased Λ_{PP} is also increased. It is clearly seen that a Knudsen minimum is observed for a channel length which is relatively small compared to the corresponding single gas flow [8]. The coefficient Λ_{CP} related to barodiffusion in all cases, as expected, it is reduced as δ_1 is increased. This type of flow is diminishing in the hydrodynamic regime.

Table 2. Kinetic coefficients for various δ_1 with $L/H=0$ and comparison with [7].

	Present work		**[7]**	
δ_1	Λ_{PP}	Λ_{CP}	Λ_{PP}	Λ_{CP}
0	0.203	0.078	0.204	0.078
0.1	0.210	0.072	0.213	0.073
0.4	0.226	0.057	0.229	0.058
1	0.258	0.039	0.262	0.044
10	0.724	0.0040	0.680	0.011

Table 3. Kinetic coefficients for various δ_1 with $L/H = 1$ and 5.

	$L/H=1$		$L/H=5$	
δ_1	Λ_{PP}	Λ_{CP}	Λ_{PP}	Λ_{CP}
0	0.138	0.053	0.072	0.028
0.1	0.142	0.049	0.071	0.025
0.4	0.150	0.039	0.070	0.019
1	0.165	0.027	0.071	0.013
2	0.191	0.018	0.077	0.0081
10	0.396	0.0034	0.139	0.0016

In Figure 2 the x-component of the bulk velocity vector for each species at the center of the channel is shown versus the vertical distance between plates. The velocities of Ne produced by the code have been divided by the factor $\sqrt{m_{Ne}/m_{He}} = 2.24$ in order to be directly comparable with the ones of He. Results are provided for δ_1=0.1, 1, 10 and $L/H = 1$, 5. In all cases the bulk velocity of the light species is higher than the velocity of the heavy one. The differences between the velocities of the two species are increased as δ_1 is decreased, i.e. as the flow becomes more rarefied. This behavior of the two species creates the separation phenomenon which is also more dominant as the length of the channel is increased. Also, as the ratio L/H is increased the magnitude of the velocity in all cases is decreased.

5. Concluding remarks

The flow of a binary gas mixture between two parallel plates of any length due to small pressure differences has been simulated based on the McCormack kinetic model. Results are provided for the binary gas mixture of He-Ne, with a reference concentration of 0.5 and three different channel lengths. A Knudsen minimum is observed at $L/H = 5$. The presence of the separation phenomenon is confirmed by plotting the velocity profiles of each species.

The present work is considered as preliminary and it will be extended in the short future in order to investigate in a more complete and detailed manner the effect of the type of the mixture, its concentration and the length of the channel on the flow characteristics.

Acknowledgements

This work has been supported by the European Communities under the contract of Association EURATOM / Hellenic Republic within the framework of the EFDA GOT program VACU-TEC (Vacuum Technologies and Pumping). The views and opinions expressed herein do not necessarily react those of the European Commission.

References

[1] Sharipov F, Kalemba D, 2002 *J. Vac. Sci. Technol. A* **20** 814
[2] Naris S, Valougeorgis D, Kalemba D, Sharipov F, 2003 *Physica A* **336** 294
[3] Kosuge S, Takata S, 2008 *Eur. J. Mech. B/Fluids* **27** 444
[4] Szalmas L, Pitakarnnop J, Geoffroy S, Colin S, Valougeorgis D, 2010 Microfluid. Nanofluid. **9** 1103
[5] Szalmas L, Valougeorgis D, Colin S, 2011 *9th ICNMM* ASME
[6] McCormack JF, 1973 *Phys. Fluids* **16** 2095
[7] Cercignani C, Sharipov F, 1992 *Phys. Fluids A* **4** 128
[8] Pantazis S, 2011 *Ph.D. thesis* University of Thessaly

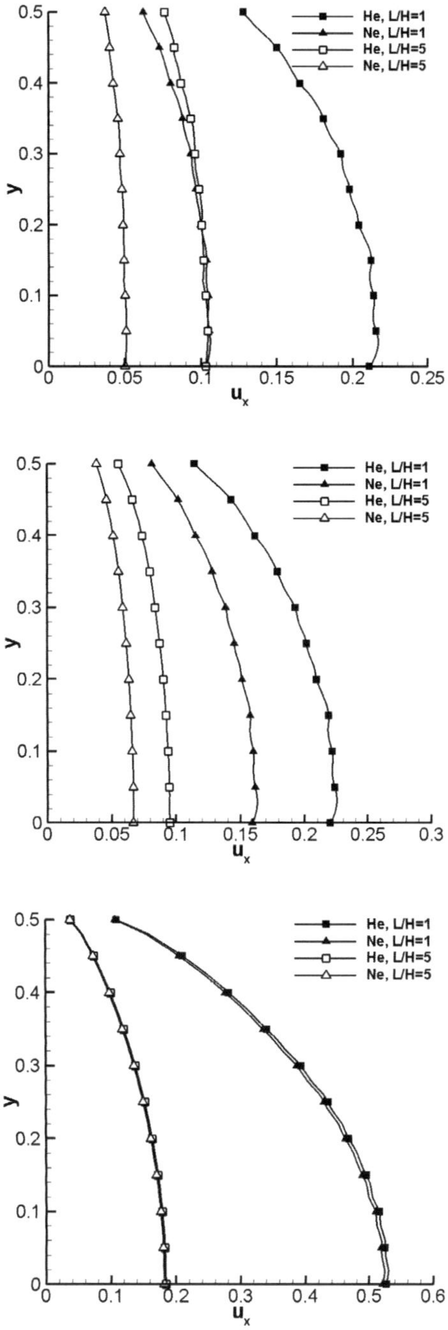

Figure 2. Figure with axial velocity profiles for each species for $L/H=1$ and 5 with $\delta_1=0.1$ (top), $\delta_1=1$ (middle) and $\delta_1=10$ (bottom).

A kinetic switching criterion for hybrid modelling of multiscale gas flows

Jianping Meng, Nishanth Dongari, Jason M Reese, Yonghao Zhang

Department of Mechanical & Aerospace Engineering, University of Strathclyde, Glasgow G1 1XJ, United Kingdom

E-mail: jianping.meng@strath.ac.uk; nishanth.dongari@strath.ac.uk; jason.reese@strath.ac.uk; yonghao.zhang@strath.ac.uk

Abstract. In some important engineering applications, gas flows are often found to be hydrodynamic in one part of device and highly rarefied in the others. To solve this kind of multi-scale flow problems efficiently and accurately, hybrid methods coupling hydrodynamic and kinetic methods are attractive. The successful implementation of hybrid methods relies on the accurate assessment of the level of non-equilibrium (rarefaction) in the local flowfield. Currently available criteria, such as Knudsen and Mach numbers, are based on macroscopic parameters and have been shown to be restrictive. Here, we propose a new kinetic criterion that utilises the fundamental molecular distribution function to assess the local flow field. Through numerical evaluation we show that our criterion provides a reasonable assessment and, in particular, it behaves consistently for both high-speed and low-speed flows, which is not the case for the other criteria based on macroscopic parameters. As our criterion fully utilises the accurate information provided by the molecular distribution function, it is particularly suitable for recently developed multi-scale kinetic methods.

1. Introduction

Physical phenomena may occur over a broad range of spatial and temporal scales in practical engineering problems. In gas flows, for instance, equilibrium and non-equilibrium phenomena may co-exist within the same device [1]. To qualitatively identify the non-equilibrium level of the local flowfield, the Knudsen number Kn can be used, which is the ratio of the mean free path of a gas to the characteristic length scale of the flow process. It is commonly accepted that the conventional hydrodynamic description is only valid for $Kn < 0.001$. When Kn is larger than 0.001, rarefaction effects have to be considered. These rarefied flows can be further classified into the slip ($0.001 \leq Kn < 0.1$), transitional ($0.1 \leq Kn < 10$) and free molecular ($Kn \geq 10$) flow regimes.

Multi-scale methods are designed to model gas flows with a broad range of Knudsen numbers [2]-[13]. Conventional continuum type solvers based on the Navier-Stokes (NS) framework are efficient, but only valid in the hydrodynamic flow regime. Their modelling capabilities may be extended to the slip flow regime by providing appropriate velocity-slip and temperature-jump boundary conditions. By contrast, kinetic solvers, such as the direct simulation Monte Carlo (DSMC) method [14] or direct solvers of the Boltzmann equation [15], can provide accurate solutions. However, they are computationally much more expensive, and may be prohibitively so for realistic 3D simulations of multiscale flows. To take advantages of both methods, the central idea of a multi-scale method is to deploy a kinetic solver in rarefied flow regions only, and apply a hydrodynamic solver in the other regions. Hence it is practically possible to provide not only efficient but also sufficiently accurate solutions for multi-scale problems.

An appropriate model switching criterion is a key factor in the success of a multi-scale method. It should be able to accurately assess the local flowfield in order to deploy appropriate solvers. The global Knudsen number may not be appropriate for this purpose [2], due to its arbitrariness in choosing a characteristic length. Various criteria are proposed ([2, 13, 16, 17, 18] and references therein). A commonly-used one is the so-called local Knudsen number, where the length scale is a formulation based on the local spatial gradients of hydrodynamic variables, i.e., $Kn_l = \frac{\lambda}{\phi}|\frac{d\phi}{dx}|$, where ϕ denotes a significant flow quantity, typically density, temperature or pressure. For hypersonic aerodynamic flows, in particular, it has been suggested that $Ma \cdot Kn_l$, where Ma is the Mach number, is a more appropriate switching parameter [16]. Although these parameters have been proven to be applicable to a range of problems, they may not be generally optimal. For instance, they may become a negligible quantity in low speed flows even for high Knudsen numbers. On the other hand, there are several proposed multi-scale methods that are built purely on the kinetic theory [19, 20]. For these kinds of methods a criterion based on macroscopic quantities does not take advantage of the available kinetic information. So recently, we propose using the molecular distribution function, instead of macroscopic flow properties, to assess the level of thermodynamic non-equilibrium in the local flowfield [21]. Here, we will numerically evaluate this approach as a model switching criterion.

2. Kinetic criterion

The Boltzmann equation describes the dynamical behaviour of dilute gases, and assumes both binary collisions between gas molecules and molecular chaos to enable a single molecular velocity distribution function to describe the gas motion. The conventional macroscopic equations, such as the NS equations, can be considered as approximate solutions of the Boltzmann equation in the hydrodynamic flow regimes, via the Chapman-Enskog approach. In this approach the solution of the Boltzmann equation, is assumed to be of the form

$$f = f^{(0)} + f^{(1)} + f^{(2)} + \cdots + f^{(\alpha)} + \cdots, \tag{1}$$

where the distribution functions $f^{(\alpha)}$ can be asymptotically obtained from the Boltzmann equation. The Maxwell-Boltzmann equilibrium distribution,

$$f^{eq} = \frac{\rho}{(2\pi RT)^{3/2}} \exp\left[-\frac{\varsigma^2}{2RT}\right], \tag{2}$$

is the zeroth-order solution $f^{(0)}$, and leads to the Euler equations. Here, ρ denotes the gas density, T the temperature, R the gas constant, and $\boldsymbol{\varsigma}$ the peculiar velocity of molecules, which is $\boldsymbol{\xi} - \boldsymbol{u}$ where $\boldsymbol{\xi}$ represents the molecular velocity and $\boldsymbol{u}$ is the macroscopic fluid velocity. In Eq.(1), $f^{(1)}$ provides a non-equilibrium correction of order Kn. To recover the Navier-Stokes equations,

$$f^{(1)} = f^{eq}\left[\left(\frac{\sigma_{ij}\varsigma_{<i}\varsigma_{j>}}{2pRT}\right) + \frac{2q_i\varsigma_i}{5pRT}\left(\frac{\varsigma^2}{2RT} - \frac{5}{2}\right)\right], \tag{3}$$

where p is the gas pressure, and the shear stress σ_{ij} and the heat flux q_i are related to the following first-order gradients of velocity and temperature,

$$\sigma_{ij} = -2\mu\frac{\partial u_{<i}}{\partial x_{j>}}, \quad q_i = -\kappa\frac{\partial T}{\partial x_i}, \tag{4}$$

where μ and κ denote the viscosity and thermal conductivity, respectively. We can continue the series to obtain higher-order corrections in terms of the Knudsen number.

As indicated by the Chapman-Enskog procedure, the Euler equations are appropriate for thermodynamically equilibrium flows with $Kn \to 0$, while the Navier-Stokes equations are valid for linear departures from equilibrium, where the Knudsen number is close to zero. In high altitude applications, such as spacecraft re-entry into planetary atmospheres, or vacuum applications, such as low-pressure chemical processes, the flows can be highly non-equilibrium. The Navier-Stokes equations fail to provide an adequate description in such cases. However, the Knudsen number alone may not be sufficient to describe the level of non-equilibrium in the flowfield, and therefore to assess whether the Navier-Stokes equations are applicable or not. Both simulation data [16] and theoretical analysis [22] show that the level of non-equilibrium is also strongly influenced by the Mach number.

In our recent work, we have proposed two kinetic parameters for indicating the level of non-equilibrium in the local flowfield [21]. These parameters can be used as model switching criteria as well. Here, we give a brief summary. Considering Eq.(1), the distribution function may be naturally split into two parts: the equilibrium and the non-equilibrium components, i.e.,

$$f = f^{eq} + f^{neq}. \tag{5}$$

The ratio of the non-equilibrium to the equilibrium parts can be used to measure the departure from equilibrium. However, it is not a practical criterion for low-speed flows, and it may be more appropriate to utilise higher order corrections. If the validity of the Navier-Stokes equations is of interest, the non-equilibrium part f^{neq} may be further divided into two components, i.e.,

$$f^{neq} = f^{(1)} + f^{(H)}, \tag{6}$$

where $f^{(1)}$ is the Navier-Stokes level correction, which is given by Eq.(3), and $f^{(H)}$ is the higher-order non-equilibrium correction beyond the Navier-Stokes level. Obviously, only when $f^{(1)} \gg f^{(H)}$, can the Navier-Stokes equations be a valid approximation of the Boltzmann equation. Moreover, a L^2 norm of

$$C_1 = \sqrt{\frac{\int (f^{(H)})^2 d\boldsymbol{\xi}}{\int (f^{(1)})^2 d\boldsymbol{\xi}}} = \sqrt{\frac{\int (f - f^{eq} - f^{(1)})^2 d\boldsymbol{\xi}}{\int (f^{(1)})^2 d\boldsymbol{\xi}}}, \tag{7}$$

may then be defined to assess the non-equilibrium level, and is a direct indicator of the relative error introduced by using a Navier-Stokes solution on the flowfield. As discussed above, another parameter

$$C_0 = \sqrt{\frac{\int (f^{neq})^2 d\boldsymbol{\xi}}{\int (f^{eq})^2 d\boldsymbol{\xi}}} = \sqrt{\frac{\int (f - f^{eq})^2 d\boldsymbol{\xi}}{\int (f^{eq})^2 d\boldsymbol{\xi}}} = \frac{\sqrt{8}\pi^{3/4}(RT)^{3/4}}{\rho}\sqrt{\int (f - f^{eq})^2 d\boldsymbol{\xi}}, \tag{8}$$

may be defined similarly, as a measure of the relative error of f to the Maxwellian f^{eq}. In the following numerical simulations, we will solve the ellipsoidal statistical BGK (ES-BGK) model [23] to evaluate these two parameters for Couette flows over a broad range of Knudsen and Mach numbers.

3. Numerical evaluation for Couette flow

Due to the simplicity of the Couette flow its characteristics are well understood. Non-equilibrium effect is usually strongest near the wall boundary, and weaker in the bulk, even there is a large global Knudsen number. Therefore it is a suitable benchmark case to numerically examine our proposed model switching criteria.

The ES-BGK kinetic model [23] to obtain f can be written as

$$\frac{\partial f}{\partial t} + \xi_i \frac{\partial f}{\partial x_i} = -\frac{1}{\tau}(f - f^{ES}), \tag{9}$$

where τ is the mean relaxation time, which is assumed to be independent of molecular velocity. The anisotropic Gaussian distribution function can be written as

$$f^{ES} = \rho \frac{1}{\sqrt{\det[2\pi\lambda_{ij}]}} \exp\left[-\frac{1}{2}\lambda_{ij}^{-1}\varsigma_i\varsigma_j\right], \tag{10}$$

where $\lambda_{ij} = RT\delta_{ij} + (b\sigma_{ij})/\rho$, σ_{ij} is the shear stress. The parameters b and τ can be determined using either the Boltzmann integral with a known inter-molecular force law, or through experimental values. The relations between the model parameters and the transport coefficients are $\mu = (p\tau)/(1-b)$ and $\kappa = pR(D+2)\tau/2$ (D is the dimension number), so that $Pr = 1/(1-b)$. Therefore, the Prandtl number can be adjusted via the free parameter b.

To numerically solve Eq.(11), the discrete velocity method is employed. Since this simple Couette flow is actually only one-dimensional, Eq.(11) can be simplified as,

$$\frac{\partial f}{\partial t} + \xi_y \frac{\partial f}{\partial y} = -\frac{1}{\tau}(f - f^{ES}), \tag{11}$$

where the coordinate y is perpendicular to the stream-wise direction. To save computational cost, a two-dimensional ideal gas, where particle motion is limited to a two-dimensional physical space is considered here. Therefore the particle velocity space can be discretized with a polar coordinate system using Simpson's rule. For the physical space, the discretization is accomplished by a second-order total variations diminishing (TVD) scheme, as presented below,

$$f_j^{n+1}|_{c_y\geq 0} = A\left\{f_j^n - \nu\triangle\phi_{j-} - \frac{1}{2}(1-\nu)\nu\left[\mathcal{J}_1\triangle f_{j+} - \mathcal{J}_2\triangle f_{j-}\right] + \frac{1}{\tau_j^n} f_{ES,j}^n dt\right\},$$

$$f_j^{n+1}|_{c_y<0} = A\left\{f_j^n - \nu\triangle f_{j+} + \frac{1}{2}(1+\nu)\nu\left[\mathcal{J}_3\triangle f_{j+} - \mathcal{J}_4\triangle f_{j-}\right] + \frac{1}{\tau_j^n} f_{ES,j}^n dt\right\},$$

with

$$A = \frac{\tau}{\tau + dt},$$

$$\triangle f_{j+} = f_{j+1} - f_j \quad \triangle f_{j-} = f_j - f_{j-1},$$

$$\theta_1 = \frac{f_j - f_{j-1}}{f_{j+1} - f_j} \quad \theta_2 = \frac{f_{j-1} - f_{j-2}}{f_j - f_{j-1}},$$

$$\theta_3 = \frac{f_{j+2} - f_{j+1}}{f_{j+1} - f_j} \quad \theta_4 = \frac{f_{i+1} - f_i}{f_j - \phi f_{j-1}},$$

$$\triangle\phi_{j+} = f_{j+1} - f_j \quad \triangle f_{j-} = f_j - f_{j-1}.$$

The minmod function is

$$\mathcal{J}_i = \max\left[0, \min(1, \theta_i)\right] \quad i = 1 - 4$$

and

$$\nu = \frac{\xi_y dt}{dy}$$

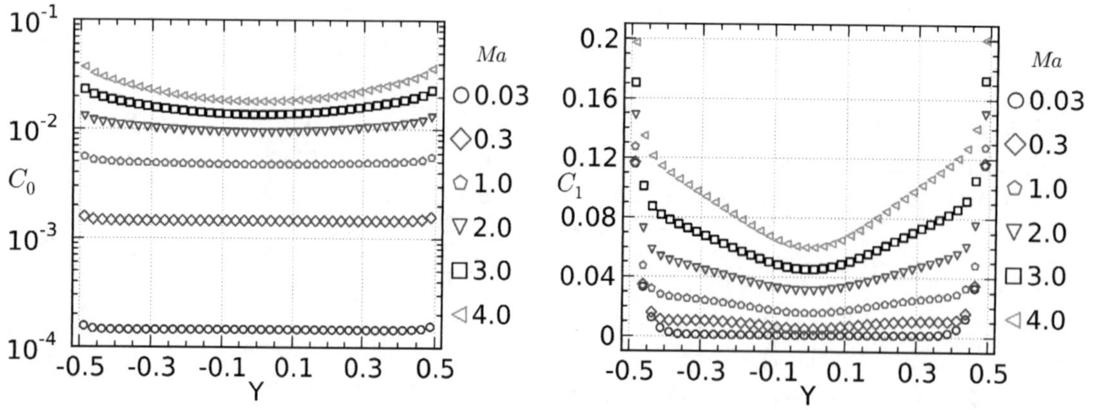

Figure 1. Profiles of C_0 and C_1 across the Couette flow channel at $Kn = 0.01$ and for various Mach numbers.

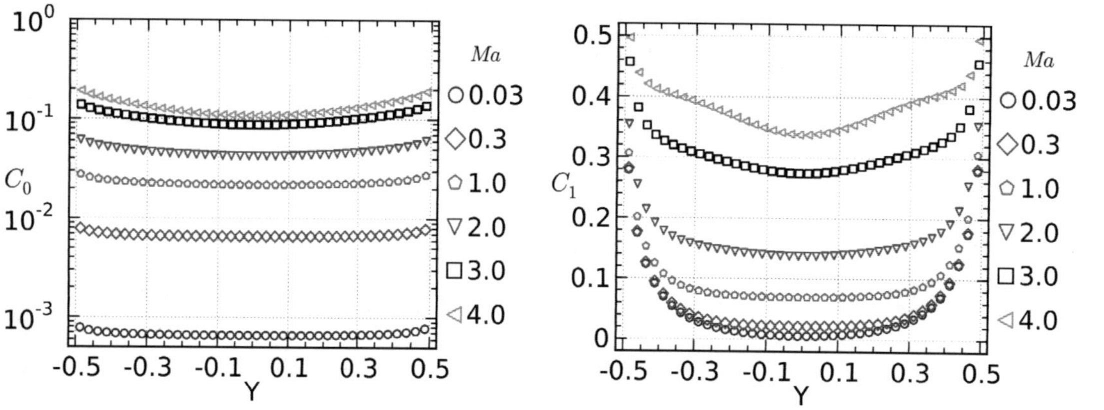

Figure 2. Profiles of C_0 and C_1 across the Couette flow channel at $Kn = 0.05$ and for various Mach numbers.

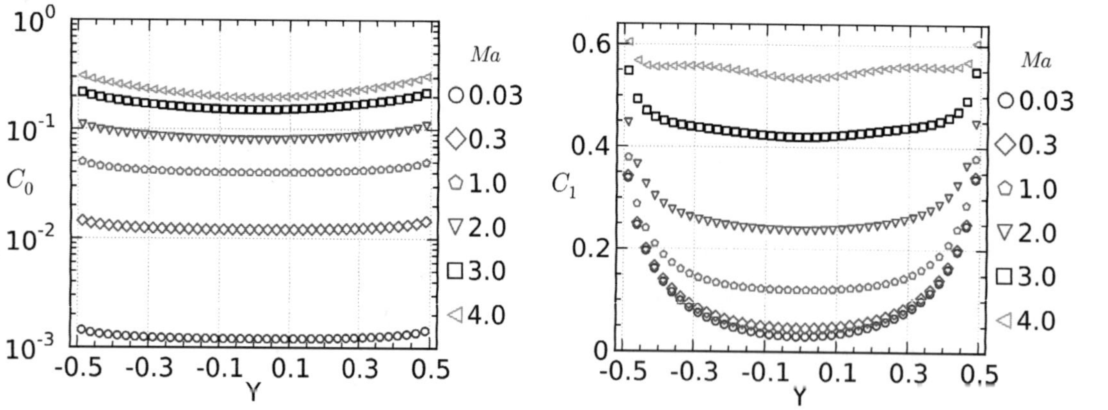

Figure 3. Profiles of C_0 and C_1 across the Couette flow channel at $Kn = 0.1$ and for various Mach numbers.

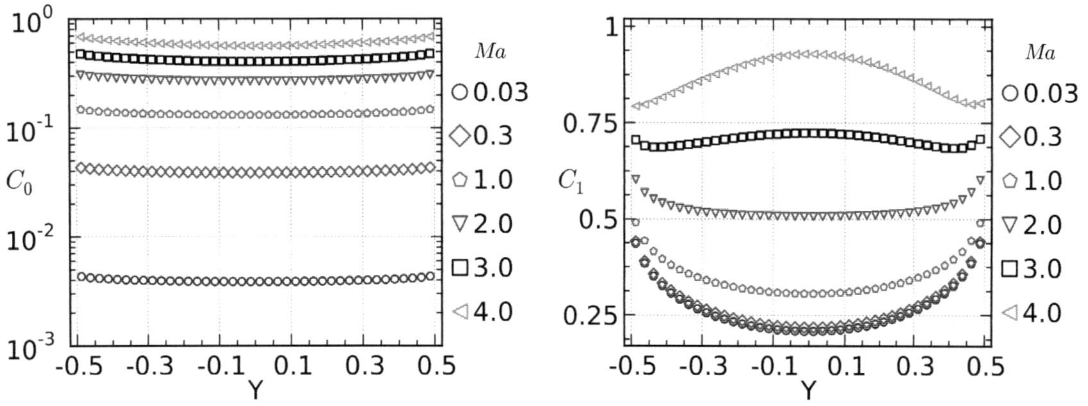

Figure 4. Profiles of C_0 and C_1 across the Couette flow channel at $Kn = 0.5$ and for various Mach numbers.

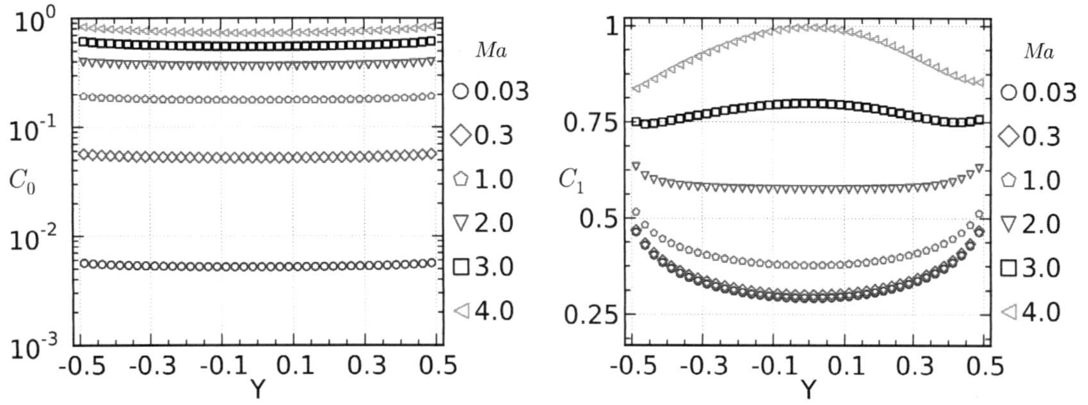

Figure 5. Profiles of C_0 and C_1 across the Couette flow channel at $Kn = 0.9$ and for various Mach numbers.

is the Courant–Friedrichs–Lewy (CFL) number. Alhough the molecular velocity is discretized based on the polar coordinate system, the Cartesian components of the molecular velocity is still used. The transformation between tow systems is accomplished as normal. After the solution is converged, the relevant distribution can be obtained directly via Eqs. (2) and (3).

In the simulations, five global Knudsen numbers ($Kn = [0.01, 0.05, 0.1, 0.5, 0.9]$) and six upper wall speeds ($U_w = [0.03, 0.3, 1.0, 2.0, 3.0, 4.0]\sqrt{RT_0}$) are considered.The variables p_0, T_0 and μ_0 denote the reference pressure, temperature and viscosity coefficient, respectively, while L is the characteristic length. The non-dimensional number Kn and Ma is then defined as $(\mu_0\sqrt{RT_0})/(p_0 L)$ and $U_w/\sqrt{RT_0}$ respectively. While the upper wall is moving with U_w, the bottom one is set to be at rest.

The profiles of C_0 and C_1 are presented in Figs. 1-5. Clearly C_0 is strongly related not only to the global Knudsen but also to the Mach number with an almost linear correlation. This correlation makes C_0 inappropriate for low-speed flows. For instance C_0 in the bulk is less than 0.01 for both two cases ($Kn = 0.9$, $Ma = 0.03$) and ($Kn = 0.05$ $Ma = 0.3$). Although C_0 in the first case is even smaller than in the second one, it is generally accepted that the NS equations can be applicable in the bulk for the second case and but inappropriate for the first one.

However, C_1 gives much more consistent indications for both low-speed and high-speed flows. For instance, C_1 in the bulk is larger than 0.25 at $Kn = 0.9$ for both $Ma = 0.3$ and $Ma = 0.03$, which clearly indicates inappropriateness of the NS model. At lower Knudsen numbers C_1 also reflects the dependency of non-equilibrium on non-linearity, i.e., the Mach number.

The difference in the two parameters lies in the way they assess the departure from equilibrium. According to Eqs. (7) and (8), C_1 is assessing the non-equilibrium level with respect to the NS equation, while C_0 is measuring the departure from Maxwellian equilibrium. Although the Chapman-Enskog expansion, Eq. (1), indicates that both parameters should reflect the order of the non-equilibrium level, simulations data here show that C_0 is greatly influenced by the non-linearity, while C_1 can correctly single out the non-equilibrium effect (which is of particular importance for low-speed flows). This observation supports the argument in [2], where the macroscopic quantities are used to assess the level of non-equilibrium in a 'relative' sense.

An interesting phenomenon is the variation of C_1 for strongly non-linear cases with Knudsen numbers larger than 0.5. Its value in the bulk can be larger than that near the walls. This non-intuitive observation may be explained as follows. As the mean particle speed $\bar{\xi}$ is proportional to $\sqrt{T}$, the mean speed in the bulk may become much larger than the near-wall region due to significant temperature rise caused by viscous heating. On the other hand, when the global Knudsen number is sufficiently large, the wall-molecule rather than molecule-molecule interactions dominate the gas behaviour. Therefore, the molecules in the bulk may actually have a great chance to collide with walls due to their much larger speed. In other words the wall-molecule interactions can be stronger for molecules in the bulk. As the NS solution tends to underestimate this kind of interactions, its error can become more significant in these circumstances.

4. Concluding remarks

We have numerically evaluated the 'kinetic criterion' C_1 for a multi-scale gas flow method, which can assess the level of non-equilibrium with respect to the NS solution by using the molecular distribution function rather than macroscopic quantities. This criterion is found to give consistent indications for flows over a broad range of Knudsen and Mach numbers. In particular, we have found it is an appropriate criterion for both high-speed and low-speed flows.

As the criterion takes advantage of the accurate molecular level information provided by the distribution function, it may be very useful for some recently developed kinetic multi-scale methods (e.g., those proposed in [19] and [20]), where the distribution function can be explicitly obtained. In these kinds of solvers, to satisfy the accuracy requirement of the difference levels of non-equilibrium, one may only to adjust the discrete velocity number (e.g., [20]), while the governing equation can keep the same. For switching from the higher non-equilibrium level to a hydrodynamic (continuum-fluid) level, it is quite easy as all the information is in hand. For switching from a near-continuum level to higher non-equilibrium level, one can run two level of simulations (e.g., nine-velocity and sixteen-velocity simulations for the lattice Boltzmann method) at the same time, then all the required information can be obtained. We note here, though two simulations are running, the costs are still expected to be less in comparison to those for high non-equilibrium level. Nevertheless, the practical performance of the criterion will be the subject of further work.

Acknowledgements

The research leading to these results has received funding from the European Community's Seventh Framework Programme FP7/2007-2013 under grant agreement ITN GASMEMS n° 215504. This work is also financially supported by the Engineering and Physical Sciences Research Council U.K. under Grants No. EP/EP/I036117/1/1.

References

[1] Karniadakis G, Beskok A and Aluru N 2005 *Microflows and Nanoflows: Fundamentals and Simulation (Interdisciplinary Applied Mathematics)* 1st ed (Springer)
[2] Lockerby D A, Reese J M and Struchtrup H 2009 *Proc. R. Soc. A* **465** 1581–1598
[3] Tiwari S, Klar A and Hardt S 2009 *J. Comput. Phys.* **228** 7109–7124
[4] Sun Q, Boyd I D and Candler G V 2004 *J. Comput. Phys.* **194** 256–277
[5] Schwartzentruber T E, Scalabrin L C and Boyd I D 2007 *J. Comput. Phys.* **225** 1159–1174
[6] Schwartzentruber T E and Boyd I D 2006 *J. Comput. Phys.* **215** 402–416
[7] Wijesinghe H S and Hadjiconstantinou N G 2004 *Int. J. Mult. Comp. Eng.* **2** 189–202
[8] Tallec P L and Mallinger F 1997 *J. Comput. Phys.* **136** 51–67
[9] Bourgat J F, Tallec P L and Tidriri M D 1996 *J. Comput. Phys.* **127** 227–245
[10] O'Connell S T and Thompson P A 1995 *Phys. Rev. E* **52** R5792–R5795
[11] Burt J M and Boyd I D 2009 *J. Comput. Phys.* **228** 460–475
[12] Hadjiconstantinou N G 2005 *Bull. Pol. Ac.: Tech* **53** 335–342
[13] Wang W L and Boyd I D 2003 *Phys. Fluids* **15** 91–100
[14] Bird G A 1978 *Annu. Rev. Fluid Mech.* **10** 11–31
[15] Yen S M 1984 *Annu. Rev. Fluid Mech.* **16** 67–97
[16] Macrossan M N 2006 Scaling parameters for hypersonic flow : Correlation of sphere drag data *Twenty-fifth Int. Symp. Rarefied Gas Dynamics,* pp 759–764
[17] Garcia A L and Alder B J 1998 *J. Comput. Phys.* **140** 66–70
[18] Boyd I D, Chen G and Candler G V 1995 *Phys. Fluids* **7** 210–219
[19] Xu K and Huang J C 2010 *J. Comput. Phys.* **229** 7747–7764
[20] Meng J, Zhang Y and Shan X 2011 *Phys. Rev. E* **83**
[21] Meng J, Dongari, N, Reese J M and Zhang Y 2012 *arXiv:1201.2005V1*
[22] Tsien H S 1946 *J. Aerospace Sci.* **13** 342
[23] Holway L H 1966 *Phys. Fluids* **9** 1658–1673

Low speed/low rarefaction flow simulation in micro/nano cavity using DSMC method with small number of particles per cell

Ali Amiri-Jaghargh[1], Ehsan Roohi[1,*], Hamid Niazmand[1] and Stefan Stefanov[2]

[1] Department of Mechanical Engineering, Faculty of Engineering, Ferdowsi University of Mashhad, Mashhad, Iran.
[2] Institute of Mechanics, Bulgarian Academy of Science, Acad. G. Bontchev Str., 1113, Sofia, Bulgaria.

Abstract. The aim of this study is to extend the validity of the simplified Bernoulli-trials (SBT)/dual grid algorithm, newly proposed by Stefanov [1], as a suitable alternative of the standard collision scheme in the direct simulation Monte Carlo (DSMC) method, for solving low speed/low Knudsen number rarefied micro/nano flows. The main advantage of the SBT algorithm is to provide accurate calculations using much smaller number of particles per cell, i.e., $< N > \approx 1$. Compared to the original development of SBT [1], we extend the application of the SBT scheme to the near continuum rarefied flows, i.e., Kn = 0.005, where NTC scheme requires a relatively large sample size. Comparing the results of the SBT/dual grid scheme with NTC, it is shown that the SBT/dual grid scheme could successfully predict the thermal pattern and hydrodynamics field as well as surface parameters such as velocity slip and temperature jump. Nonlinear flux-corrected transport algorithm (FCT) is also employed as a filter to extract the smooth solution from the noisy DSMC calculation for low-speed/low-Knudsen number DSMC calculations. The results indicate that combination of SBT/dual grid and FTC filtering can decrease the total sample size needed to reach smooth solution without losing significant accuracy.

1. Introduction

Heat transfer and fluid flow in Micro/Nano-electro-mechanical systems, MEMS/NEMS, is widely gained importance due to the rapid growth of miniaturization of practical engineering and biomedical devices such as heat exchangers and chemical reactors. It is well established that the fluid behavior in MEMS/NEMS is different from their macroscopic counterpart [2]. However, due to their small dimensions, it is hard to study these behaviors experimentally. Actually, the behavior of gas gradually deviates from the thermodynamic equilibrium as the device length scale approaches the mean free path of the gas. Therefore, the numerical modeling of such devises is also problematic because the traditional Navier-Stocks (NS) equations, consistent with the near-equilibrium state, refuse to follow the realistic flow features. Knudsen number (Kn), defined as the ratio of the gas mean free path (λ) to the characteristic

* Corresponding author, Email: e.roohi@ferdowsi.um.ac.ir

length scale of the flow (L), is a good measure to characterize the departure from the equilibrium. Based on Knudsen number, the flow may appear in four different regimes [3]: continuum, slip flow, transition and free molecular regimes. For the Kn < 0.001, i.e., continuum regime, the NS equation with traditional no-slip boundary condition can be utilized to describe the flow behavior. In slip flow regime, 0.001< Kn < 0.1, the NS equation deviate from experimental results and should be accompanied with velocity slip and temperature jump boundary conditions. In the transition regime, 0.1< Kn <10, flow gradually departs from the equilibrium and the NS equations are no longer valid. Finally, flow is considered as free molecular if it exceeds the limit of Kn > 10. Many investigations have proved the accuracy of the slip boundary conditions in slip flow regime [4]. Some researchers have tried to extend the applicability of NS equations to the early transition regime by means of developing various second order slip boundary conditions [5]. However, there is no indisputable proof to prefer one of these models among others. In addition, the concept of thermodynamic equilibrium fails to work as Kn number increases and the breakdown of the NS equations occurs [6]. Many attempts are reported to extend continuum-based equations to non-equilibrium regimes, for example, considering higher order constitutive relations for stress and heat transfer from the Chapman-Enskog expansion, moment methods are examples of such extension [7-9].

On the other hand, in contrast to continuum-based methods, molecular approach is a powerful tool to study rarefied flows. In this approach, the fluid is modeled as a collection of moving molecules interacting through collisions. DSMC method is known as one of the most successful particle-based methods in analyzing the rarefied gaseous flows. The main feature of the DSMC method, originally proposed by Bird [10], is simplifying the interaction between molecules by decoupling the motion into two successive stages of free molecular movement and binary intermolecular collision within the grid cells in each time step. It is mathematically proofed that the solution of the DSMC method converges to Boltzmann equation solution, if adequate number of particle per cells are used [11]. Bird [12] used the DSMC method to solve homogeneous gas relaxation problem. Celenligil and Moss [13] reported good agreements between the DSMC method results and experimental data of wind-tunnel for hypersonic flow over a delta wing.

Many researchers have tried to modify the DSMC method to reduce its restrictions including statistical fluctuations and computation costs. Fan and Shen [14] proposed IP method to reduce statistical fluctuations inherent in modeling of low speed microflows. Kaplan and Oran [15] used a nonlinear flux-corrected transport (FCT) filtering to exclude high frequency noise from the desired solution. Recently, some correction is suggested on the modification of the collision procedure. Stefanov [1] proposed a two-stage collision process based on the simplified Bernoulli-trials (SBT) algorithm. This method provides reasonable results with much lower number of particles per cell.

In this work, we developed a DSMC code based on Stefanov's SBT/dual grid algorithm to solve the near continuum flows, i.e., Kn = 0.005. Lid driven cavity flow is used as a benchmark test to check the ability of the new collision procedure in prediction of low-Knudsen low-speed flows. Additionally, we employ the nonlinear FCT filtering to reduce the statistical fluctuations of low Kn/low speed flows and reduce the computational time. Our results are compared with NTC solution and performance of the new algorithm is discussed.

2. DSMC method

DSMC is a particle method based on the kinetic theory for simulation of dilute gases. The method is carried out by modeling the gas flow using many independent simulator particles, each one representing a large number of real gas molecules in the flow field. In the DSMC method, the time step (Δt) is chosen so small that the positional changes of particles and their collisions could be decoupled for each time step. During DSMC implementation, flow field must be divided into computational cells which provide geometric boundaries and volumes required to sample macroscopic properties. Thus, the size of each cell should be small enough that any small changes in thermodynamic properties could be captured. Cells also organize a framework to select collision pairs. The selection of collision pairs in the standard DSMC is based on the no time counter (NTC) method [16], in which the computational time is proportional to the number of simulated particles. In the NTC collision procedure, the number of particle pairs (N_c) that should be checked for a collision are

$$N_c = \frac{1}{2} f_{num} N^l < N^l > (\sigma g)^l_{max} \Delta t / \forall^l \quad (1)$$

, where f_{num}, $(\sigma g)^l_{max}$ and $\forall$ are the ratio of number of real molecules to simulated particles, maximum of collision cross-section multiplied by particles' relative velocity and cell volume, respectively. The superscript l refers to cell l. Then, each pair (i,j), $1 \le i < j \le N^l$, chosen randomly from particle subset N^l, is checked for a collision with the probability

$$\mathrm{p_{ij}} = \frac{\sigma_{ij} g_{ij}}{(\sigma g)_{max}} \quad (2)$$

, where σ_{ij} is the effective collision cross-section of pair (i,j) and g_{ij} is the magnitude of relative velocity of pair (i,j). The sequence of NTC collision procedure is shown in Figure 1.

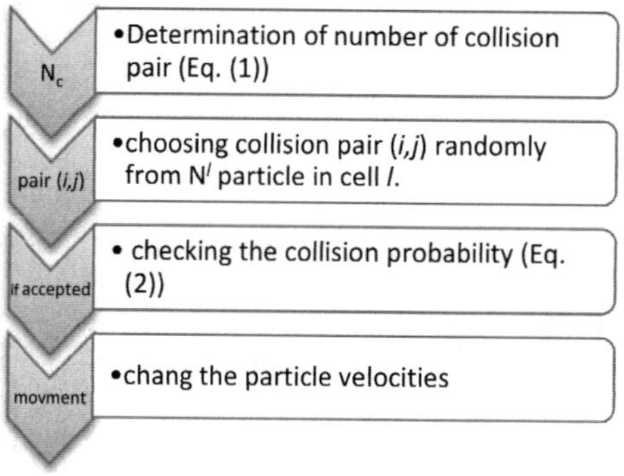

Figure 1. The NTC collision procedure

An insufficient number of particles per cell can be sources of stochastic errors which increases the number of undesirable repeated collisions in cells. However, adjusting enough particles in cells, the NTC scheme is an efficient approach for modeling the intermolecular collisions. But, it is CPU intensive in multidimensional gas flows where a huge mesh size is required for proper flow simulation. At low Knudsen numbers, in particular, using the NTC scheme necessitates the use of large number of simulated particles which greatly increases the computational cost. Recently, Stefanov [1] suggested a simplified Bernoulli trials (SBT) scheme which permits simulation with far less number of particles per cell, $< N > \sim 1$. In the SBT procedure, the particles in the lth cell should be locally indexed in order to form a particle list numbered as 1… N^l. The first particle of the collision pair (i, j), say i, is selected in sequence from the particle list, i.e., i = 1… N^l -1. The second particle, say j, is then selected randomly among $k = N^l - i$ particles taking place in the list after particle i.

$$j = (i+1) + int(k \times rnd) \quad (3)$$

, where rnd refers to a random number between 0 and 1. Then each pair is checked for possible collision with the probability

$$p_{ij} = \frac{1}{2} k f_{num} \Delta t \sigma_{ij} g_{ij} / \forall^l \tag{4}$$

It should be noted that the Δt should be adjusted so that p_{ij} rarely exceeds unity, say

$$prob\{p_{ij} \geq 1\} \rightarrow 0 \tag{5}$$

This procedure avoid the production of at least part of the eventually successively repeated collisions which occurs in Bird's NTC scheme when it is applied with a small number of particles. The theoretical background of this scheme is described in Ref. [1] by more details. The sequence of SBT collision procedure is shown in Figure 2.

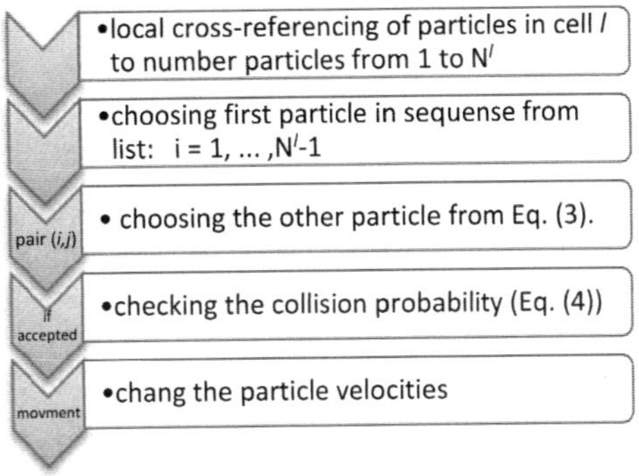

Figure 2. The simplified Bernoulli trials collision procedure

When we use small number of particles per cell, another strategy that improves the collision procedure is the usage of staggered mesh [1]. In fact, the forgoing collision procedure is applied twice in two successive half time steps, $\Delta t/2$, on a dual grid, i.e., in the first step it is applied on a standard grid, in the next step it is applied on a staggered grid which is formed by shifting all cells of the primary grid in each coordinate direction for a distance equal to half-cell size, $\Delta x/2$, $\Delta y/2$. Applying staggered grid provides the particles the chance of collision with other particles which are in a separation distance smaller than a cell size but at the first half time belongs to the adjacent cells. By translocating the cells, they become particles of one cell and will be checked for a possible collision. The steps of implementation of the staggered grid are shown in Fig. 3 where the flowchart of modified DSMC algorithm is sketched.

The results of SBT scheme are also compared with the majorant frequency scheme (MFS) [17]. MFS is an efficient collision model in the DSMC method. In this collision model, the time between two eventual collisions is evaluated from a Poisson distribution while the collisional pair is uniformly chosen from $N(N-1)/2$ available pairs in each cell. A majorant frequency which is defined as

$$\nu_m = \frac{N(N-1)}{2} \frac{(\sigma g)_{max}}{V_c} \tag{6}$$

is used to specify the Poisson distribution. More details of MFS scheme could be found in Ref. [18].

The molecular interaction is modeled by variable hard sphere (VHS) model. Monatomic argon, m=6.63×10^{-26}Kg and d=4.17×10^{-10}m is considered here as the gaseous medium. In order to ensure the satisfaction of the limits on the cell size, the cell dimensions Δx and Δy are considered to be 0.1λ and are

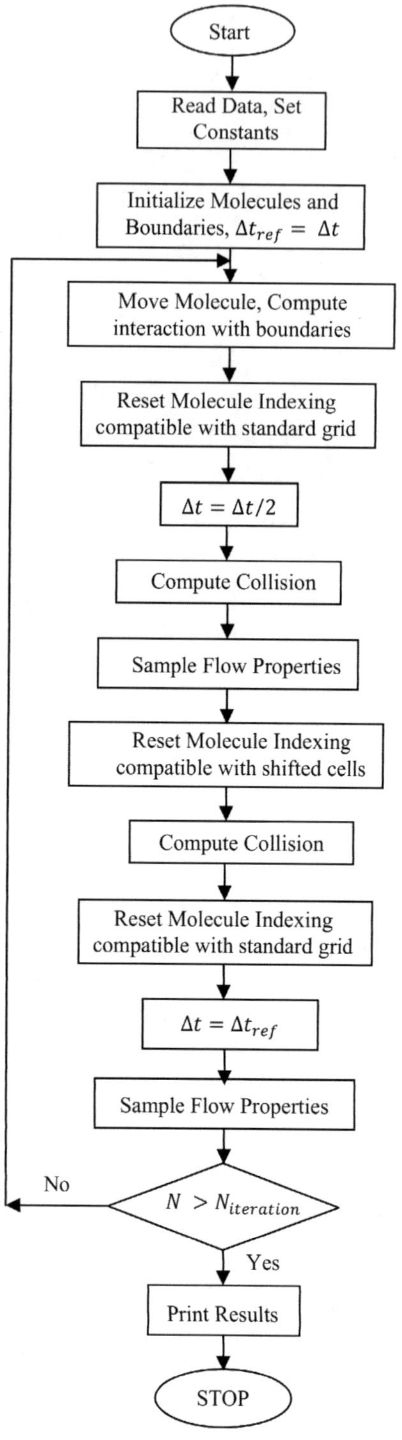

Figure 3.DSMC flowchart for staggered grid

much smaller than that for most cases. Each cell contains two molecules. All walls are treated as diffuse reflectors using the full thermal accommodation coefficient, α =1. Half-range Maxwellian distribution is used to determine the velocity of reflected molecules. After achieving steady flow condition, sampling of flow properties within each cell is fulfilled during sufficient time period to avoid statistical scattering. All thermodynamic parameters such as temperature, density, and pressure are then determined from this time-averaged data. In order to minimize the scattering in the predicted results, particularly temperature, a nonlinear FCT filtering post processor is utilized [15].

3. Flux-Corrected Transport (FCT) Filtering

Modeling of low speed micro/nano flows with DSMC method encounters a serious problem of statistical fluctuations. This occurs when the bulk velocity of the flow is much smaller than the most probable thermal velocity, i.e.,

$$V_{mp} = \sqrt{2kT/m} \tag{7}$$

, where k, T and m are the Boltzmann constant, fluid temperature and molecular mass, respectively. For the present problem $V_{mp} \cong 350\, m/s$, while the lid velocity is $U_{lid} = 100\, m/s$ and the internal field velocity is much smaller below the mid line of the cavity, say of the order of $10\, m/s$. Therefore, the statistical noises are so great that the features of the flow may be easily lost. However, the effects of statistical noises could be reduced by increasing the sample size. In the current work, where just 2 particles in each cell are employed, the run should be continued for longer time to achieve suitable results. Additionally, following Kaplan and Oran [15], the nonlinear FCT filtering is used here as a post processing strategy to reduce statistical fluctuations and CPU run time as a result. The main feature of FCT filter, in comparison with fully diffuse simple filters, is that it makes the noisy flow converge to a solution whose correctness is already proved [15]. The FCT procedure consists of three parts: diffusion, anti-diffusion flux and anti-diffusion flux limiting. Considering the parameter ρ^0 as a noisy field that should be represented in the filtered form ρ^1, we could write

$$\rho^0 \rightarrow \rho^d \rightarrow \rho^{dal} \equiv \rho^1 \tag{8}$$

The steps are as follows:

1) The initial values ρ^0 are diffused, giving ρ^d. This is achieved by the definition of diffusive flux of ρ at the midpoint $i+1/2$ between ρ_i and ρ_i+1, $f^d_{i+1/2}(\rho)$ as

$$f^d_{i+\frac{1}{2}}(\rho^0) = \nu_{i+\frac{1}{2}}(\rho^0_{i+1} - \rho^0_i) \tag{9}$$

, where $\nu_{i+1/2}$ are diffusion coefficients. Then

$$\rho^d_i = \rho^0_i + f^d_{i+\frac{1}{2}}(\rho^0) - f^d_{i-\frac{1}{2}}(\rho^0) \tag{10}$$

2) To remove excess diffusion, anti-diffusion fluxes, $f^{da}_{i+1/2}(\rho)$, are defined which, however, can insert nonphysical overshoots in the solution.

$$f^{da}_{i+\frac{1}{2}}(\rho^d) = \mu_{i+\frac{1}{2}}(\rho^d_{i+1} - \rho^d_i) \tag{11}$$

, where $\mu_{i+1/2}$ are anti-diffusion coefficients.

3) To ensure stability and elimination of nonphysical overshoots, the anti-diffusion fluxes should be limited before applying to the field. This guarantees that no new maxima or minima are added to the solution. The limiter function, following Boris and Book [19] could be defined as

$$f^{dal}_{i+\frac{1}{2}} = S \cdot max\left\{0, min\left[S \cdot (\rho^d_{i+2} - \rho^d_{i+1}), \quad \left|f^{da}_{i+\frac{1}{2}}(\rho^d)\right|, S \cdot (\rho^d_i - \rho^d_{i-1})\right]\right\} \tag{12}$$

, where $S = sgn(\rho^d_{i+1} - \rho^d_i)$. Then

$$\rho^1_i \equiv \rho^{dal}_i = \rho^d_i - f^{dal}_{i+\frac{1}{2}}(\rho^d) + f^{dal}_{i-\frac{1}{2}}(\rho^d) \tag{13}$$

This procedure causes the local peaks to be smoothed while keeps the value of their neighbors nearly unchanged. Actually, the effect of FCT filter can be changed by using different values of diffusion and anti-diffusion coefficients. If $\nu > \mu$ then the filter reveals a diffusive manner and depending on the degree of inequality it can smooth the peaks too fast but may do not converge to a solution. In the other side, if $\nu < \mu$ then it works as a high-frequency filter that removes the sharp peaks and converges to a solution after passing the solution from the filter for multiple times. This is shown in Figure 4 where a noisy square wave (Figure 4-a) with zero velocity is passed through the FCT filter for 1000 times (Figure 4-b) and 10000 times (Figure 4-c) for both of the forgoing cases. It is obvious that in the case of $\nu < \mu$, the filtration converges to a much less-noisier wave that never changes by more passing through the filter.

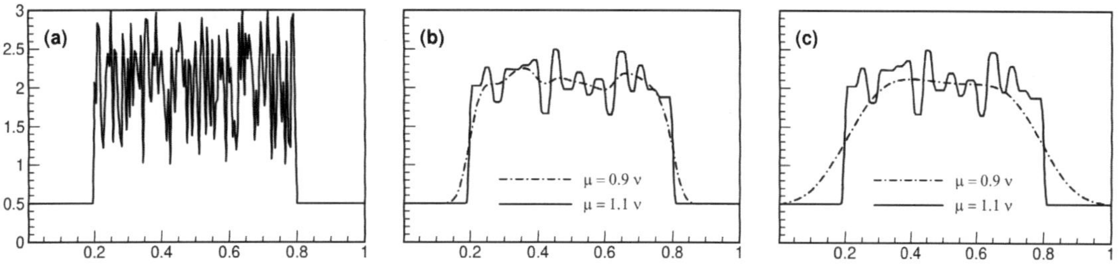

Figure 4. The effect of FCT filter on the zero velocity noisy wave,

(a) before pass **(b)** 1000 passes **(c)** 10000 passes

Also, in the case of $\nu > \mu$, although the filter results in smoother wave, it has a diffusive characteristic and never converges to a solution. Note that when $\nu \approx \mu$, the filter performs in a slightly diffusive manner. In this work, the diffusion and anti-diffusion coefficients are chosen as $\nu = \mu = 1/8$.

4. Grid independency test

The configuration of micro driven cavity considered in this study is shown in Figure 5 which is a square with length $L = 1\ \mu m$ with a moving lid at the velocity of $U_{lid} = 100\ m/s$ and *Kn*=0.005. A grid independency study was carried out using three grids composed of 100×100, 200×200 and 400×400 cells. Figure 6 shows the vertical component of velocity vector along the horizontal axis of the cavity, $Y/L = 0.5$.

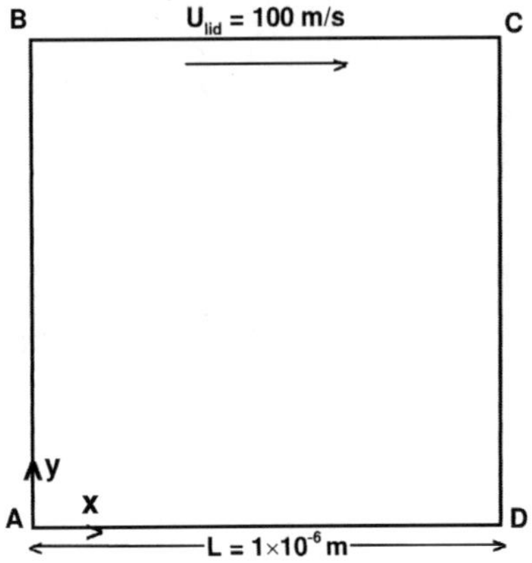

Figure 5. Geometrical configuration of micro cavity.

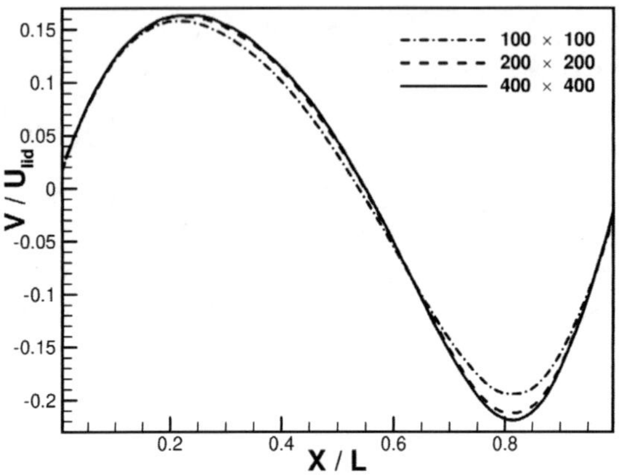

Figure 6. Grid independency test.

The results are almost equivalent for 200×200 and 400×400 grids; however the mesh size of 400×400 is selected to ensure the criterion that limits the cell size as $\Delta x \leq \frac{1}{3}\lambda$ [20].

5. Results and discussions

A micro square cavity of length $L = 1\ \mu m$ with a moving lid at the velocity of $U_{lid} = 100\ m/s$ is considered. The walls temperature is set to the reference temperature, i.e., $T_w = T_0 = 300\ K$. The aim of the article is applying the DSMC method based on SBT/dual grid collision procedure in near continuum flows, i.e., Kn = 0.005 and compare the results with the NTC algorithm solutions. In this regard, NTC scheme is applied with various number of particles per cell to find the minimum value which leads to accurate solution. The comparison of results, which is shown in Figure 7, indicates that $< N > = 20$ could reasonably provide the sufficient accuracy.

The ability of the SBT scheme in prediction of the cavity flow field solution is evaluated in Figures 8 and 9 where SBT results for $< N > = 2$ are compared with those of NTC with $< N > = 20$. In Figure 8, the dimensionless y-component of the velocity vector, V/U_{slip}, is plotted along the horizontal axis of the cavity ($Y/L = 0.5$). Also, the dimensionless x-component of the velocity vector, U/U_{slip}, is plotted along the vertical axis of the cavity at $X/L = 0.5$.The figure shows reasonably good agreement between the SBT and NTC results in prediction of the velocity field. A better assessment of the SBT scheme could be achieved if we plot the temperature field which is a square function of molecular velocity and is more sensitive to the statistical fluctuations. The temperature profile along the vertical cavity axis at $X/L = 0.5$

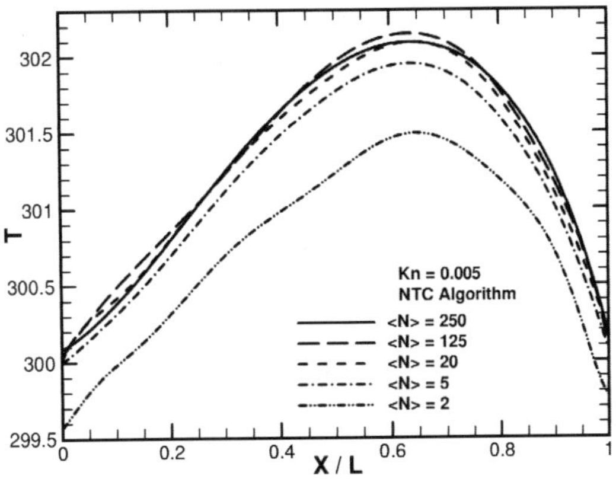

Figure 7. Effect of number of particle per cell in the accuracy of the NTC collision scheme.

is shown in Figure 9. The figure indicates that the SBT scheme with 2 particles per cell could reasonably follow the results of the NTC scheme as well as MFS with more particles. The results of NTC scheme with two particles in each cell is also plotted in this figure to emphasize on the inadequacy of this scheme at low number of particles per cell, which may locally happen in a flow with notable density gradients. In addition, although the MFS also works efficiently with small number of simulators, this figure shows that it does not guaranty the reasonable accuracy in comparison with the SBT scheme. In fact, repeated collisions may occur in the MFS scheme. This could be the source of slight deviations between the results of MFS and SBT for small number of simulators.

It should be noted that although the lid velocity is 100 m/s which correspond to a Mach number of $Ma \approx 0.3$ the flow can be strongly considered as low velocity field. The contours of local Mach number calculated with SBT scheme is shown in Figure 10. The figure indicates that the most parts of the flow field correspond to a Mach number of $Ma \leq 0.1$.

The relative computation times for the runs performed by the NTC and SBT algorithms are shown in Table 1. For the purpose of generality, the reported times are normalized by the required solution time for

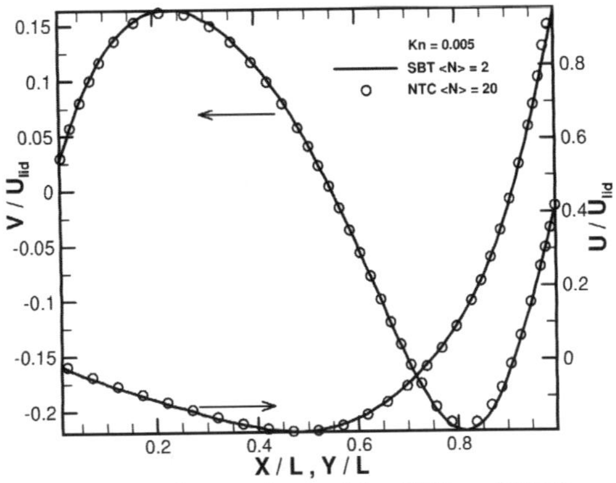

Figure 8. Comparison of the SBT and NTC schemes in prediction of flow field.

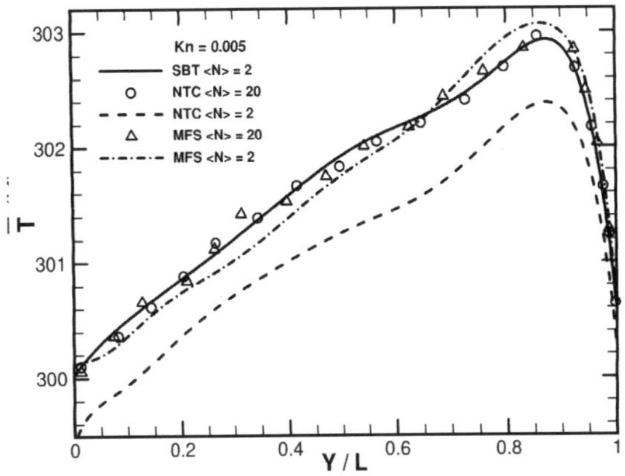

Figure 9. Comparison of the SBT with NTC and MFS schemes in prediction of thermal pattern.

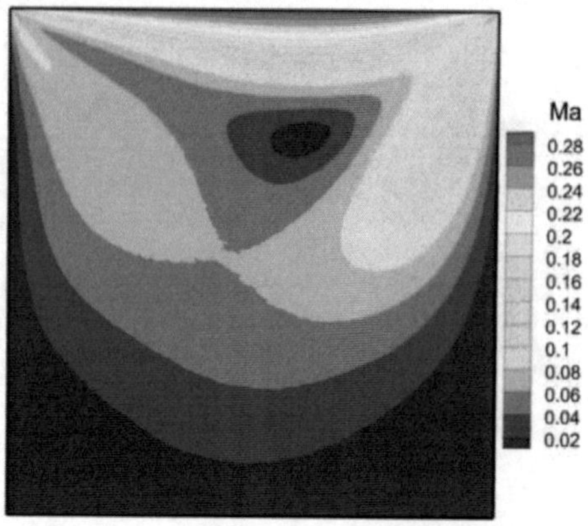

Figure 10. Contours of local Mach number calculated with SBT scheme.

NTC algorithm with $< N > = 20$. The computational time are calculated for equal sample size, i.e., $N_{tot}N_s = cons.$, where N_{tot} and N_s are the total number of particles in the domain and the number of sampling, respectively.

Table 1. Relative computational time

	<N>	*time*
Algorithm SBT	2	0.91
Algorithm NTC	250	1.15
	125	1.07
	20	1
	5	0.98
	2	0.75

The data of Table 1 indicate that the SBT algorithm needs less CPU time compared to the NTC to accumulate the same sample size. In addition, whereas the N_{tot} in SBT algorithm is much smaller than the NTC algorithm, say $N_{tot,SBT} = 0.4N_{tot,NTC}$ for case $< N > = 20$; equality of the sample size in the comparison means that the number of sampling in SBT method is remarkably higher. Our numerical experiences reveal that in case of equal sample size, the more time iterating the less statistical fluctuations remain in the solution. This is consistent with the improvement of the DSMC solution in each time step that means more correct data will be accumulated as iteration advances. For example, for the cases mentioned in Table 1, the NTC algorithm for case $< N > = 20$ was iterated for about 3,400,000 times in a mesh of 200×200 with two subcells in each direction while the SBT algorithm was iterated for about 8,500,000 times in a mesh of 400×400 without any sub-cell (actually SBT scheme needs no sub-cell) to forms the same sample size. However, the SBT does not only take less CPU time, but also results in more smooth solution.

Another feature which should be noted in further developments of the SBT scheme is the percentage of the CPU time that is consumed by different steps of the DCMC algorithm. This is mentioned in Table 2 where the NTC data are presented for comparison purpose. The relative time consumption of each DSMC step with respect to the collision is also reported to introduce a better comparison between each step as well as between SBT and NTC schemes. The most time consuming step in the SBT scheme is indexing, which is due to using the staggered mesh that necessitates the indexing step to be applied three times at each time step, see Figure 3. Therefore, any refinement of indexing process can notably improve the speed of SBT scheme.

Table2. Percentage of CPU time usage in DSMC steps.

	SBT		NTC$\langle N \rangle=20$	
	%	relative	%	relative
Movement	15	0.58	21	0.57
Indexing	32	1.23	18	0.49
Collision	26	1	37	1
Sampling	22	0.85	23	0.62
Other	5	0.19	1	0.03

The contours of x-component of the velocity vector is shown in Figure 11 for both the SBT (Figure 11-a) and NTC (Figure 11-b) approaches. This figure indicates that the results are in good agreement with each other. In addition, the streamlines are plotted in Figure 11. The flow consists of a large primary vortex which is followed by two small secondary vortex at corners A and D. Even though we used small number of particles per cell in SBT, we had detected the secondary vortices as well. Secondary vortices disappear at higher Kn numbers. The position of the main vortex is interested by researchers and usually is used as a validity parameter. Both of the SBT and NTC algorithms predict the primary vortex center at (0.61, 0.75).

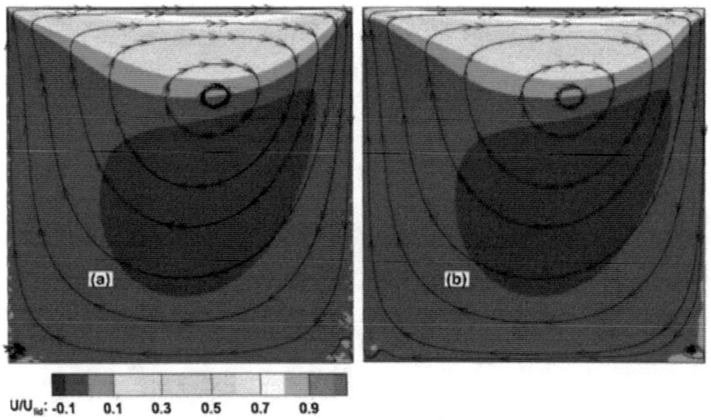

Figure 11. Comparison of x-velocity and stream lines for
(a) SBT scheme (b) NTC scheme $< N > = 20$

Until here, it is proofed that the SBT scheme could predict the thermal and hydrodynamic patterns as accurate as the NTC scheme. But, as a result of improvement of stochastic modeling, the SBT scheme is allowed to use significantly less particle per cell at low Kn/low velocity flows. Another feature of rarefied flows which is usually interested by the researcher is the surface parameters such as velocity slip and temperature jump. In DSMC method, the slip/jump phenomena could be evaluated based on the sampling of the corresponding molecular properties of all the particles that strike the wall surface. Following Ref. [21], the formulations for velocity slip and temperature jump are:

$$u_{slip} = \frac{\sum\left(\frac{m}{|v_p|}u_p\right)}{\sum\left(\frac{m}{|v_p|}\right)} \tag{14}$$

$$T_{gass} - T_{wall} = \frac{1}{3R}\frac{\sum\left(\frac{1}{|v_p|}\|U_p\|^2\right) - \sum\left(\frac{1}{|v_p|}\right)u_{slip}^2}{\sum\left(\frac{1}{|v_p|}\right)} \tag{15}$$

, where p stands for particle and the summation is taken over all particles striking the wall of the regarding cell. $|v_p|$, $\|U_p\|$ and R are the absolute values of the normal velocity, velocity magnitude, i.e., $\|U_p\| = \sqrt{u_p^2 + v_p^2 + w_p^2}$, and gas constant, respectively. Therefore, the slip velocity is accumulated based on the change in particle velocity due to the collision to the surface considering that the particles collide to wall with probability of $1/v_p$. From Equation (14), the temperature jump is proportional to a fraction of total kinetic energy. The velocity slip and temperature jump along the driven lid, computed by SBT and NTC schemes are plotted in Figure 12.

As observed in Figure 11, good conformity of results of SBT scheme with those of NTC indicates that the SBT model is powerful in prediction of the surface parameters as well as inter-domain features.

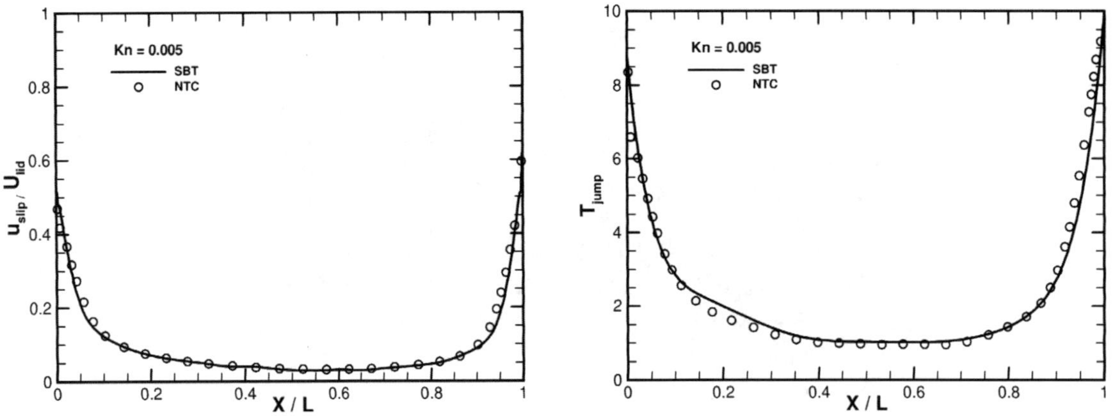

Figure 12. Velocity slip (left) and Temperature jump (right) along the driven lid computed by SBT and NTC schemes.

The performance of FCT filter is evaluated in Figure 13. This figure consists of three contour layers: the flood contour shows an unfiltered intermediate solution, after filtration this solution is shown with bold blackline contours, and the red line contour represent the filtered final solution. The filtered solution is passed through the filter for 5000 times for both cases. From the figure, it is easily realized that the FCT filter has a key role on the removal of statistical scatters from the DSMC solution. The comparison of red and black line contours offers that the solution could be interrupted at early times without dropping much accuracy. Note that it takes us about 240 hours to reach the final solution from the intermediate solution. In the other words, this huge computational cost could be removed if we suitably employ the FTC filtering.

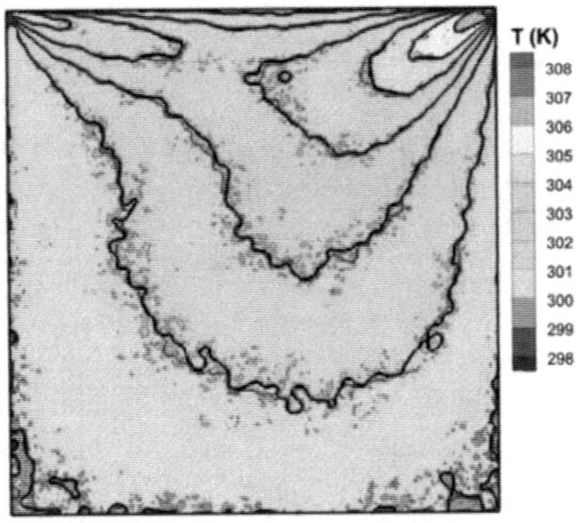

Figure 13. Ability of FCT filtering in removing stochastic noises
Flood contours: unfiltered early solution, Black line: filtered early solution, Red line: final filtered solution

6. Conclusions

DSMC Simulation of low-speed/low-Knudsen rarefied flows at micro/nano scales remains a challenge for researchers due to large computational requirements and noisy final solutions. Even though there are some suggestions to replace the basic DSMC algorithm with alternative schemes, i.e., IP scheme, these alternative schemes are not fully validated for a wide classes of problems and their derivations are case-dependent. In this work, we suggested a combination of simplified Bernoulli trial (SBT) algorithm and dual grid strategy, recently proposed by Stefanov, and nonlinear flux-corrected transport algorithm (FTC) filtering, to simulate low-speed/low-Knudsen rarefied flows at micro/nano scales efficiently. The main advantage of the proposed strategy is that it allows accurate calculations using much smaller number of particles per cell, i.e., $< N > \approx 2$, and less computation time compared to the standard NTC scheme. We tested cavity flow at Kn=0.005 and showed that our SBT solution for velocity and temperature fields agrees well with standard solution with $< N > \approx 20$, while the computational time of the SBT is around 0.91 of the NTC. Additionally, the use of FTC filtering provided correct smooth solution with less computational time compared to the basic SBT algorithm.

REFERENCES

[1] Stefanov, S.K., 2011. "On DSMC calculation of rarefied gas flow s with small number of particles in cells". *SIAM J. Sci. comput.*, Vol. 33 (2), pp. 677-702.

[2] Beskok, A. and Karniadakis, G.E., 1999. "A model for flows in channels, pipes and ducts at micro and nano scale".*J. MicroscaleThermophys. Eng.*,Vol. 3, pp. 43-77.

[3] Schaaf, S.A. and Chambre, P.L., 1961."Flow of Rarefied Gases". Princeton University press, USA.

[4] Karniadakis, G., Beskok, A., and Aluru, N., 2005. "Microflows and Nanoflows: fundamentals and simulation". Springer-Verlag, New York.

[5] Barber, R.W. and Emerson, D.R., 2006."Challenges in modeling gas-phase flow in microchannels: from slip to transition". *Heat Transfer Engineering*, Vol. 27 (4), pp.3-12.

[6] Mizzi, S., Emerson, D.R., and Stefanov, S.K., 2007."Effect of rarefaction on cavity flow in the slip regime". *Journal of computational and theoretical nanoscience*, Vol. 4 (4), pp. 817-822.

[7] Chapman, S. and Cowling, T.G., 1991. "The mathematical theory of non-uniform gases: an account of the kinetic theory of viscosity, thermal conduction and diffusion in gases". 3rd ed., Cambridge university press, Cambridge, UK.

[8] Grad, H., 1949."On the kinetic theory of rarefied gases". *Communication on pure applied mathematics*, Vol. 2 (4), pp. 331-407.

[9] Kogan, M.N., 1969."Rarefied gas dynamics". 1st ed., Plenum press, New York.

[10] Bird, G.A., 1976."Molecular Gas Dynamics". Clarendon Press, Oxford.

[11] Wagner, W., 1992. "A convergence proof for Bird's direct simulation Monte Carlo method for the Boltzmann equation". *J. Statist. Phys.*, Vol. 66, pp. 1011–1044.

[12] Bird, G.A., 1963."Approach to translational equilibrium in a rigid sphere gas". *Physics of fluids*, Vol. 6 (10), pp. 1518.

[13] Celenligil, M.C. and Moss, J.N., 1992."Hypersonic rarefied flow about a delta wing direct simulation and comparison with experiment". *AIAA journal*, Vol. 30 (8), pp. 2017-2023.

[14] Fan, J. and Shen, C., 2002."Micro-scale gas flows". *Adv. Mech.*, Vol. 32, pp. 321-336.

[15] Kaplan, C.R. and Oran, E.S., 2002."Nonlinear filtering for low-velocity gaseous microflows". *AIAA Journal*, Vol. 40, pp. 82-90.

[16] Bird, G.A., 1994."Molecular gas dynamics and the direct simulation of gas flows", 2nd ed., Oxford university press, Oxford.

[17] Ivanov, M.S. and Rogasinsky, S.V., 1988. "Analysis of the numerical techniques of the direct simulation Monte Carlo method in the rarefied gas dynamics". *Soviet J. Numer. Anal. Math. Modelling*, Vol. 3(6), pp. 453-465.

[18] Ivanov, M.S., Markelov, G.N. and Gimelshein, S.F., 1998. "Statistical Simulation of Reactive Rarefied Flows: Numerical Approach and Applications". AIAA Paper 98-2669.

[19] Boris, J.P and Book, D.L, 1997."Flux-corrected transport: I. SHASTA, a fluid transport algorithm that works". *Journal of computational physics*, Vol. 135, pp. 172–186.

[20] Alexander, F.J., Garcia, A.L. and Alder, B.J., 1998. "Cell size dependence of transport coefficients in stochastic particle algorithms". *Physics of Fluids*, Vol. 10 (6), pp. 1540-1542.

[21] Lofthouse, A.J., 2008, PhD Thesis, University of Michigan.

Compressibility and rarefaction effects on entropy and entropy generation in micro/nano Couette flow using DSMC

Omid Ejtehadi, Javad Abolfazli Esfahani, Ehsan Roohi[1]
Department of Mechanical Engineering, Faculty of Engineering, Ferdowsi university of Mashhad, Mashhad, Iran, P.O. Box: 91775-1111

Abstract. In the present work, compressible flow of argon gas in the famous problem of Couette flow in micro/nano-scale is considered and numerically analyzed using the direct simulation Monte Carlo (DSMC) method. The effects of compressibility and rarefaction on entropy and entropy generation in terms of viscous dissipation and thermal diffusion are studied in a wide range of Mach and Knudsen numbers and the observed physics are discussed. In this regard, we computed entropy by using its kinetic theory formulation in a microscopic way while the entropy generation distribution is achieved by applying a semi-microscopic approach and thoroughly free from equilibrium assumptions. The results of our simulations demonstrated that the entropy profiles are in accordance with the temperature profiles. It is also illustrated that the increase of Mach number will result in non-uniform entropy profiles with increase in the vicinity of the central regions of the channel. Moreover, generation of entropy in all regions of the domain stages clear growth. By contrast, increasing the Knudsen number has inverse effects such as: uniform entropy profiles and a falling off in entropy generation amount throughout the channel.

1. Introduction

In recent years, prediction of rarefied gas flows and heat transfer behaviour attracted attention due to the wide application of micro/nano electromechanical system (MEMS/NEMS) devices such as micro/nano- valve, micro/nano- turbine, micro/nano- pump and micro/nano- nozzle. Moreover, many experimental, numerical and theoretical studies have been conducted in the field [1-6]. Rarefied shear-driven flows such as Couette are encountered in micro-motor, comb mechanism, and micro-bearing. To measure the degree of rarefaction the Knudsen number (Kn) is defined as the ratio of mean free path of gas molecules to the characteristic length of the geometry ($Kn=\lambda/L$). When the Knudsen number is sufficiently large, the gas rarefaction is the main parameter to evaluate these systems.

In this study we consider the entropy and entropy generation of the well known problem of planar Couette flow, i.e., a gas confined between two infinite parallel plates which are at the same temperature but moving relative to each other. The advantage of studying Couette flow is that despite it is a relatively simple problem it includes many features found in more complex rarefied gas dynamics problems. The micro/nano-Couette flow has been widely applied in existing literature for different purposes. Some of the sample works can be found in [7-14]. Meanwhile, entropy and entropy generation concepts are wildly used in the literature for different important purposes such as: evaluating a system's irreversibility, optimizing a thermodynamic system or as a means of examining

[1] Corresponding Author: Ehsan Roohi, Assistant Professor, Tel.: +98-511-8805136, Fax: +98-511-8763304, Email: e.roohi@ferdowsi.um.ac.ir

continuum breakdown in rarefied fluids. For example: Reitebuch and Weiss [15] employed a new min-max principle for the entropy production to determine more boundary conditions. They used 26 field equations derived by the moment method to describe the stationary plane Couette flow for dense and rarefied gases. Their new boundary conditions results in a heat flux in the direction of the flow which agrees with molecular dynamics results. Ansumali and Karlin [16] implemented the lattice Boltzmann method based on the *H* theorem (entropic lattice Boltzmann method) for a one-dimensional benchmark shock tube problem. Results of their simulation demonstrate significant improvement of stability, as compared to realizations without explicit entropic estimations. Schrock et al. [17-19] expanded upon the concept of entropy generation as a means of examining the breakdown of the continuum fluid equations in regions of non-equilibrium. They compared the DSMC result with those obtained by numerically integrating the NS equations in normal shock waves. They have shown that significant non-equilibrium may exist upstream of any indication of such phenomena in the NS data in the case of the normal shock. They also stated that any breakdown parameter based on the continuum data will fail to capture the initial onset of the shock front. Carr [20] introduced a new method for calculating entropy generation using the DSMC method to investigate the limits of the continuum constitutive relations. He compared continuum results with the DSMC solution for the hypersonic flow over two axisymmetric geometries and showed that the kinetic method predicts a larger shock region than the continuum method. Erbay et al. [21] investigated the effects of aspect ratio, Reynolds number (Re), Prandtl number (Pr), Brinkman number (Br), and the motion of the lower plate on the entropy generation during the simultaneously developing flow in a parallel-plates channel. They showed that entropy generation has its highest value at channel with the smallest aspect ratio at counter motion of the lower plate while the highest Re, Pr and Br/Ω values considered in the problem. Ozlap [22] numerically investigated entropy generation for laminar fully developed, forced convection in a micro-pipe. The simulations were concentrated on the impact of wall roughness based viscous dissipation on the heat transfer behaviour and the resulting overall and radial entropy generation. In another work, Ozlap [23] investigated entropy generation for wide ranges of pipe diameter ($d = 0.5 - 1mm$), wall heat flux ($q'' = 1000 - 2000$ w/m^2) and Reynolds number ($\mathrm{Re} = 1 - 2000$). He concludes that the cross-sectional total entropy generation is computed to be most influenced by pipe diameter at high wall heat flux and low Reynolds numbers. Ozlap [24] also studied the integrated effects of surface roughness, heat flux, and Reynolds number on the first and second law characteristics of laminar-transitional flow in a micropipe and showed that the frictional entropy is minor and the major portion of the total entropy generation is thermal based. Chigullapalli et al. [25] formulated and applied a discrete version of the Boltzmann's H-theorem for analysis of non-equilibrium onset and accuracy of the numerical modelling of rarefied gas flows. They illustrated the use of entropy considerations in rarefied flow simulations for the normal shock, the Riemann and the two-dimensional shock tube problems. They stated that the entropy generation rate based on the kinetic theory is shown to be a powerful indicator of the onset of non-equilibrium, accuracy of numerical solution as well as the compatibility of boundary conditions for both steady and unsteady problems. Leite and Santos [26] investigated effect of the step back-face height on the thermal non-equilibrium in backward-facing steps in a rarefied hypersonic flow using DSMC. They showed that entropy generation regions are related to the thermal non-equilibrium regions in the flow field. Parlak et al. [27] studied steady-laminar flow of water in adiabatic microtubes experimentally and analytically. Through employing the second law analysis, they showed that the flow characteristics in the smooth microtubes distinguish substantially from the conventional theory for flow in larger tubes with respect to viscous heating/dissipation, temperature rise of flow, total entropy generation rate and lost work.

It is observed that there is a great deal of papers dealing with the entropy generation analysis in high Knudsen number simulations; however, the entropy and entropy generation analysis of the micro/nano-Couette flow is not implemented by DSMC method yet. The present work is following the previous works by authors regarding DSMC simulations [28-30] and second law analysis [31-33]. Here, the compressible micro/nano-Couette flow is studied with a special focus on entropy and entropy generation profiles. Compressibility and rarefaction effects in a wide range of Mach and Knudsen numbers are investigated. The results are then discussed and justified to obtain a better knowledge of the system behaviour regarding the second law analysis.

2. Numerical analysis

2.1. Problem statement

Planar Couette flow involves a gas trapped between two infinite plate at $y = \pm h/2$ moving relative to each other in opposite directions with a constant velocity $\pm U_w$ and maintained at the same constant temperature, T_w. The flow is driven by the shear stress of the moving plates and depends on the y coordinate only. Therefore the flow is considered steady, one dimensional and compressible. The flow velocity generated by shear is assumed to have only an x component varying only in the y direction. Temperature profiles are parabolic and are generated by the viscous dissipation. The amount of heat flux in y direction is zero in the center of the channel and increases towards the plates. The shear stress in the domain is constant. Also the external forces are assumed to be absent. A schematic of the problem is depicted in Fig. 1.

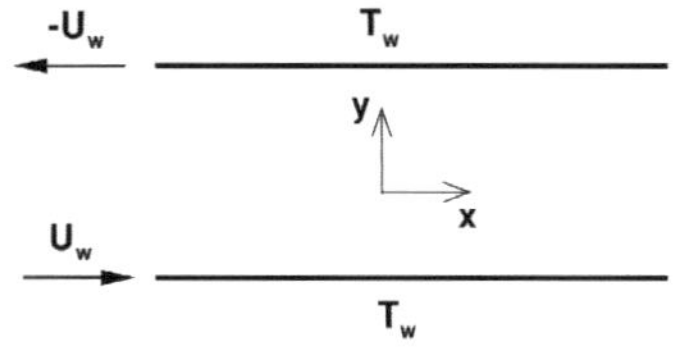

Figure 1. Schematic of the problem

2.2. DSMC method

The Boltzmann equation describes the gas flow behaviour in all the regimes. The DSMC method is proved to be the most accurate numerical tool to solve the Boltzmann equation based on direct statistical simulation of the molecular processes described by the kinetic theory [34]. In fact, DSMC simulates particle behaviour in a manner consistent with what is described by the Boltzmann equation. The algorithm includes four primary steps: moving the particles, indexing them, collision simulation, and sampling the flow-field. The primary principle of DSMC is to decouple the motion and collision of particles during one time step. The implementation of DSMC needs breaking down the computational domain into a collection of grid cells. The cells are divided into subcells in each direction. The subcells are then utilized to facilitate the selection of collision pairs. After fulfilling all molecular movements, the collision between molecules are simulated in each cell separately in a statistical manner. Accurate DSMC solution requires some constraints on the cell size, time step and number of particles. The random selection of the particles from a cell for binary collisions requires that the cell size to be a small fraction of the gas mean free path. The decoupling between the particles movement and collisions is correct if the time step is a small fraction of the mean collision time. Number of particles per cell should to be high enough, around 20, to avoid repeated binary collisions between the same particles. The following procedure is used to solve a stationary problem with DSMC. In the entire computational domain, an arbitrary initial state of gas particles is specified and the desired boundary conditions are imposed at time zero. Particles movement and binary collisions are performed separately. After achieving steady flow condition, sampling of molecular properties within each cell is fulfilled during sufficient time period to avoid statistical scattering. All thermodynamic parameters such as temperature, velocity, density and pressure are then determined from this time-averaged data. More details on DSMC algorithm are given in Refs. [34-35]. In the current study, a modified version of Bird's DSMC2D program is applied and periodic boundary conditions are imposed for left and right boundaries such that one-dimensional Couette flow is simulated. Variable hard sphere (VHS) collision model is used and the collision pair is chosen based on the no time counter (NTC) method. We use diffuse reflection model with the full thermal accommodation coefficient for the walls and set the time step and cell dimensions equal to 10^{-13}(s) and 10^{-8}(m), respectively. These values are set in such a way that they do not exceed one third of collision time and mean free path, respectively.

3. Results and discussion

3.1. Entropy

The second law of thermodynamics states that entropy of an isolated system always increases or remains constant. The second thermodynamic law appears to establish the direction that the heat flux will occur, say, the direction will be that one where $\Delta s > 0$. Because such a process acts to reduce the initial state of the order of the system, entropy is in conjunction with disorder or randomness especially in microscopic interpretation of entropy in statistical mechanics. We can also deal with entropy as of thermal energy dispersal. The relation of entropy as a measure of disorder is Boltzmann relation for the entropy [36]:

$$S = k_B \ln \Omega \tag{1}$$

where k_B is the Boltzmann constant, equal to 1.38065×10^{-23} (J/K) and Ω the number of possible microstates in the system called degeneracy or statistical multiplicity of the gas, respectively. In fact, entropy is a logarithmic measure of the density of microstate which characterizes all molecular details about the system such as molecular position and velocity. It can be stated as a summation over all microstates that a system can be in:

$$S = -k_B f_i \ln(f_i) \tag{2}$$

In the above equation, f_i is the probability that the system is in the ith microstate.

Here, because of the availability of the data in the DSMC solver, we use a velocity distribution function in order to calculate entropy. It is assumed that the velocity distribution could be separated into a product of the component distributions, that is:

$$f(\mathbf{c}) = f_x(c_x) f_y(c_y) f_z(c_z) \tag{3}$$

The velocity spectrum is divided into bins and the number of molecules associated with each bin is counted to form the above distribution function. Entropy is then calculated based on the above procedure and Eq. (2). For detailed information regarding implementing the method see Ref. [17, 18]

3.1.1. Compressibility effects

The wall Mach number, $M_{w,}$ is defined as the ratio of U_w to the parameter $\sqrt{\gamma k_B T_W / m}$ and is specified by varying driving velocity of both plates U_w. Here, γ and m are specific heat ratio of gas and molecular mass, respectively. p

Figure 2 illustrates the variation of temperature in the domain with wall Mach variation. The temperature profiles show that as M_w increases, the temperature difference between the plates and center of channel becomes larger due to higher viscous dissipation. Thus, compressibility effects become significant. The growth of temperature jump at the walls by increasing the Mach number is quite obvious in Fig. 2. The magnitude of the temperature jump at M_w=1.2 is almost 20% of the wall temperature.

Compressibility effects on entropy profiles are illustrated in Fig. 3. Comparison of Figs. 2 and 3 reveals that these two quantities follow a same trend, i.e., a parabolic profile with the maximum value at the center of the channel. This is because the heat is a disordered form of energy and flows with the entropy. Consequently, entropy or molecular chaos increases in the regions of higher temperature. This extremum swells as Mach increases. This is due to the existence of higher molecular chaos as a result of the greater number of microstates in the middle of the channel. In fact, existence of less number of molecules leads to smaller amount of intermolecular collisions. Hence, the molecules are not able to balance their energy level as it is in near-wall regions. It is also apparent that the plates'

Journal of Physics: Conference Series **362** (2012) 012008 doi:10.1088/1742-6596/362/1/012008

movement in the x direction organizes adjacent molecules to the plates in such a way that the implications are less entropy (disorder) in the vicinity of the walls. As this velocity increases, the order of molecules near the plates is more obvious resulting in less entropy compared to the disorder at the channel center.

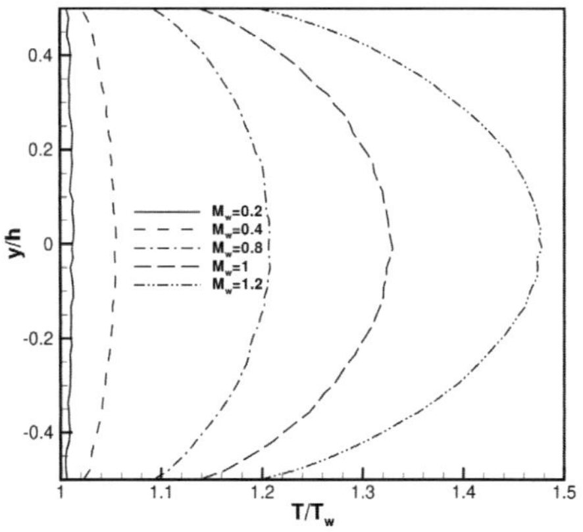

Figure 2. Compressibility effects on non-dimensional temperature profiles (Kn=0.1)

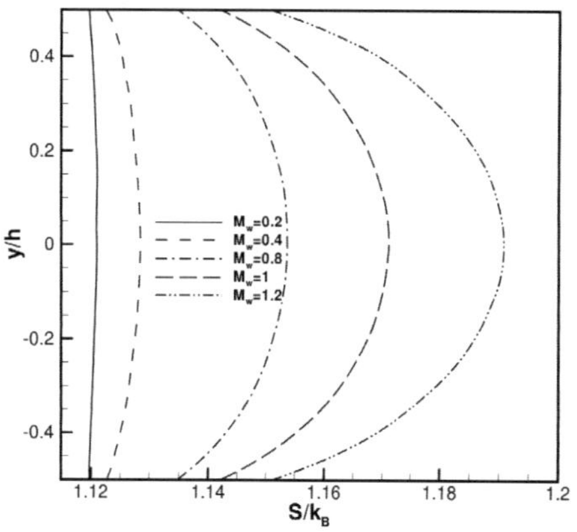

Figure 3. Compressibility effects on non-dimensional entropy profiles (Kn=0.1)

3.1.2. Rarefaction effects

Rarefaction effects gain importance with the reduction in the geometry size, since the sizes of the geometries become comparable to the mean free path of the gas molecules. These effects are specified through the Knudsen number. Now, the plates' velocity is kept constant (U_w=250 m/s) and the corresponding wall Mach numbers is 0.81.

Due to the similarities between entropy and temperature profiles discussed in the previous section, first and foremost, temperature variations with Knudsen number are investigated. Figure 5 illustrates the temperature profiles at different Knudsen numbers. As the Knudsen number increases, temperature in the domain and also temperature jump at the plates becomes more significant, but the curvature of temperature profiles reduces. In a constant wall Mach number flow, the amount of wall kinetic energy is constant. As the flow becomes more rarefied, this constant kinetic energy will be saved in smaller number of the molecules and shows itself in term of temperature rise. However, in a dense gas this amount of constant energy is distributed in larger number of molecules and the temperature rise is smaller than what is observed in rarefied gases. Also for small Knudsen numbers and close to the continuum regime, only near surface molecules collide with the walls and in the middle of channel only the intermolecular collisions happen. For larger Knudsen numbers, as the gas becomes more rarefied, molecules in the middle of channel have the same chance for molecular-surface collision and transferring heat from/to the walls. Consequently, the maximum curvature in Fig. 4 is observed in the Kn=0.01 and by increase of the Knudsen number the temperature in the domain becomes uniform.

Figure 5 shows rarefaction effects on entropy profiles. It can be seen from the graph that the curvature in the entropy diagrams subsides as the Knudsen number increases. This is because of occupation of the domain by the Knudsen layer, a kinetic boundary layer on the order of one mean free path, for more rarefied flows. We can also see that the trend is quite similar to the observed trends in temperature profiles. Therefore, as a general conclusion we could state that the more rarefied flow, more uniform entropy profiles are observed. The justification for this conclusion is the more uniform distribution of the molecules in the domain for rarefied flows rather than dense gases. Therefore, the distribution of energy in molecules is more balanced.

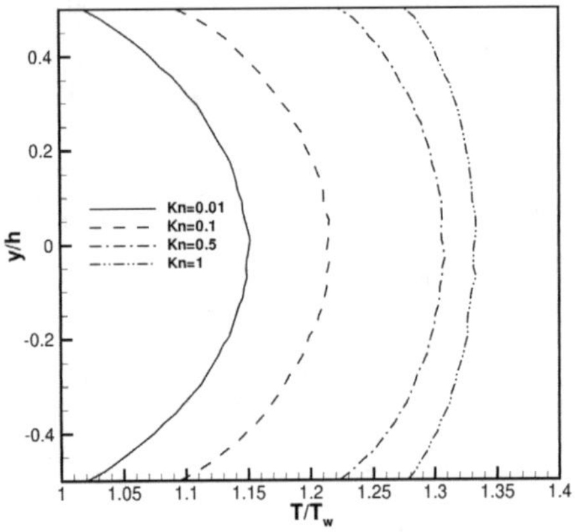

Figure 4. Rarefaction effects on non-dimensional temperature profiles (M_W=0.81)

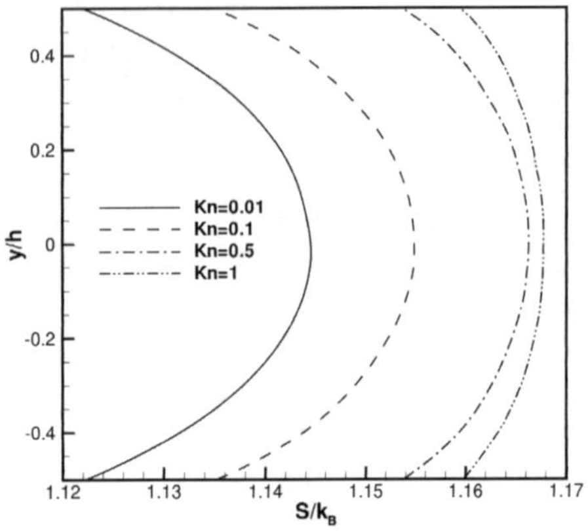

Figure 5. Rarefaction effects on non-dimensional entropy profiles (M_W=0.81)

3.2. Entropy generation

Studies of Schrock et al. [17-19] state that because entropy generation represents non-equilibrium it can be employed as a similar parameter to the Knudsen number. Therefore it will predict the continuum breakdown onset. They also showed that breakdown parameters based upon continuum data may fail to capture the initial non-equilibrium effects. Their investigation showed that it is necessary to develop a formulation for entropy generation based on kinetic theory and without inherent assumptions of equilibrium built in to be able to measure the small perturbations from non-equilibrium observed in the NS equations. Therefore, they proposed expressions for entropy and entropy generation in terms of the distribution functions of velocity, rotational energy, and vibrational energy. The DSMC method was applied in their work to generate these distribution functions. A theoretically accurate but computationally expensive and difficult method due to the necessity of sorting particles to create the distribution functions was employed in their work. For this reason, a different approach for calculating entropy generation proposed by Carr [20] has been used here, eliminating the need to sort particles into distribution functions. In this method a constitutive relation valid in equilibrium is applied. However, by replacing the parameters which are calculated by the DSMC solution, the final relation for entropy generation will be free from assumptions of equilibrium.

$$\dot{S}_{gen} = \frac{\tau_{ij}}{T}\frac{\partial u_i}{\partial x_j} - \frac{q_i}{T^2}\frac{\partial T}{\partial x_i} \tag{4}$$

In the above equation τ_{ij} and q_i are the shear stress and heat flux, respectively. Because of the equilibrium assumptions made in the constitutive relations for the shear stress and heat flux in the NS equation, the corresponding relations for these quantities in the kinetic theory will be used:

$$\boldsymbol{\tau} = \tau_{ij} = -(\overline{\rho c'_i c'_j} - \delta_{ij}\boldsymbol{p}) \tag{5}$$

where c_i' and c_j' are the components of $\boldsymbol{c}'$, the velocity of molecule relative to stream velocity, i.e., thermal velocity and $\boldsymbol{p}$ is the scalar pressure defined as:

$$\boldsymbol{p} = p_{ij} = \overline{\rho c'_i c'_j} \tag{6}$$

The heat flux vector is defined as:

$$\mathbf{q} = \frac{1}{2}\rho\overline{c'^2 \boldsymbol{c}'} + n\overline{\varepsilon_{int}\boldsymbol{c}'} \tag{7}$$

where, ε_{int} is the internal energy of a single molecule.

Finally, the entropy generation derivation for the one-dimensional Couette flow according to [20] becomes:

$$\dot{S}_{gen} = \frac{\tau_{ij}}{T}\frac{\partial u_i}{\partial x_j} - \frac{q_i}{T^2}\frac{\partial T}{\partial x_i}$$
$$= \frac{\tau_{xx}}{T}\frac{\partial u}{\partial x} - \frac{q_x}{T^2}\frac{\partial T}{\partial x} + \frac{\tau_{xy}}{T}\frac{\partial u}{\partial y} - \frac{q_x}{T^2}\frac{\partial T}{\partial x} + \frac{\tau_{yx}}{T}\frac{\partial v}{\partial x} - \frac{q_y}{T^2}\frac{\partial T}{\partial y} + \frac{\tau_{yy}}{T}\frac{\partial v}{\partial y} - \frac{q_y}{T^2}\frac{\partial T}{\partial y} \tag{8}$$

Where, both i and j are representative of x and y, sequentially. Since there is no velocity and temperature gradients in x direction the final expression for entropy generation yields:

$$\dot{S}_{gen} = \frac{\tau_{xy}}{T}\frac{\partial u}{\partial y} - 2\frac{q_y}{T^2}\frac{\partial T}{\partial y} \tag{9}$$

In the above equation the first term is called the generation of entropy due to viscous dissipation shown by S_μ and the second due to thermal diffusion represented by S_D.

3.2.1. Compressibility effects

Owing to the fact that entropy generation can be treated as a means of quantifying non-equilibrium, first density variation are investigated to look for similarities in these quantities. Figure 6 illustrates the normalized density profiles for different Mach numbers. It is clear from the graph that as the Mach increases the density profiles deviate from the uniform distribution. It is also observed that the amount of density is in its minimum at the center of the channel for all Mach numbers.

Figure 7 shows the compressibility effects on the entropy generation profiles. The figure shows that more entropy is generated in near wall regions where the Knudsen layer is formed. Here the flow is less rarefied than the center of the channel. It is also observed that for higher wall Mach numbers production of entropy in whole regions of the domain increases which is the result of viscous heating, caused by the more kinetic energy absorbed from the wall moving in higher speeds.

In Figs. 8 and 9 each term of Eq. (9) is plotted separately. The comparison of these two figures shows that the generated entropy in the domain is mainly influenced by the viscous dissipation term, S_μ rather than thermal diffusion term, S_D. However, the latter is the main cause for non-uniform entropy generation profiles due to large temperature gradients in the vicinity of the walls.

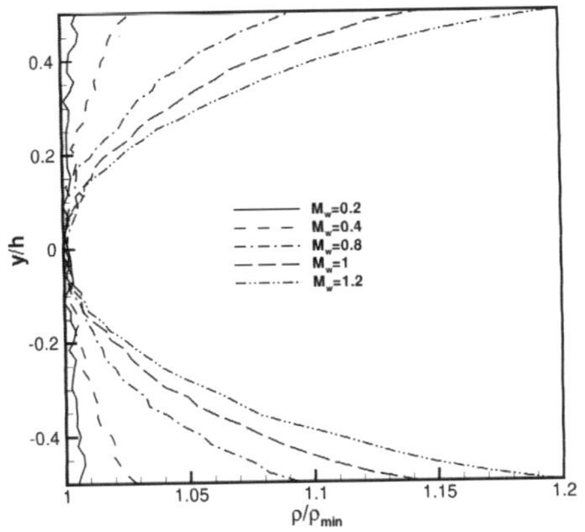

Figure 6. Compressibility effects on non-dimensional density profiles (Kn=0.1)

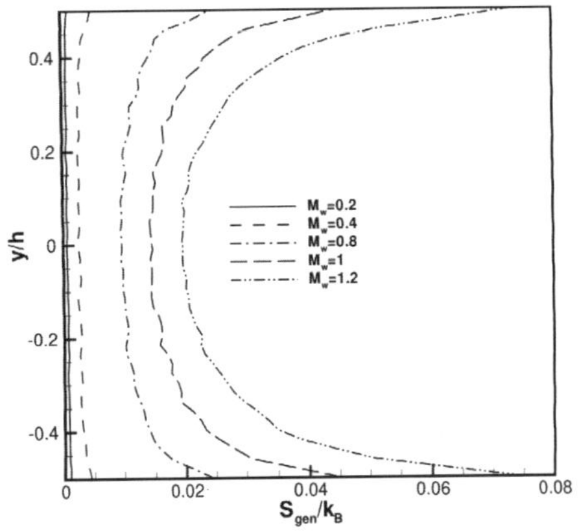

Figure 7. Compressibility effects on non-dimensional entropy generation profiles (Kn=0.1)

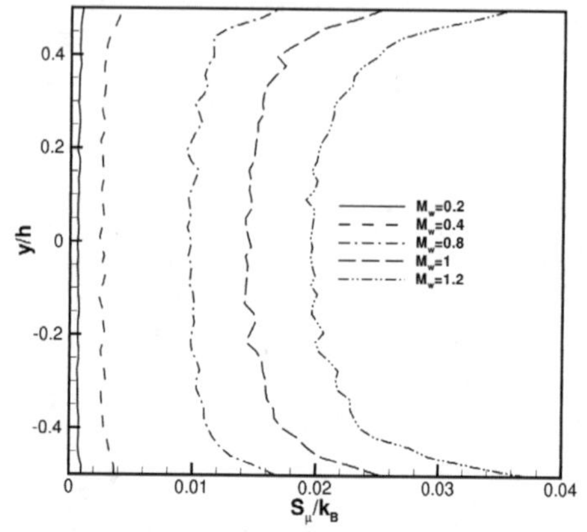

Figure 8. Compressibility effects on non-dimensional viscous dissipation term of the entropy generation profiles (Kn=0.1)

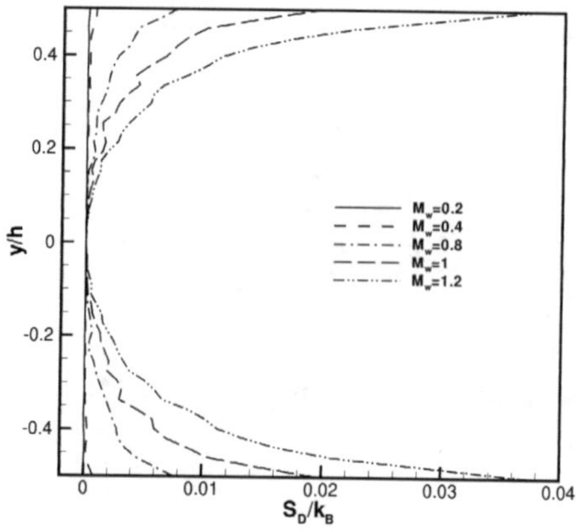

Figure 9. Compressibility effects on non-dimensional thermal diffusion term of the entropy generation profiles (Kn=0.1)

3.2.2. Rarefaction effects

Variation of non-dimensional density profiles for different Knudsen numbers is plotted in Fig. 10. It is obvious that increase in the Knudsen number leads to decrease in density. Also the non-dimensional density profiles tend to uniform profiles in more rarefied simulations as a result of development of the Knudsen layer in the entire domain.

In Fig. 11 the profiles of entropy generation for different Knudsen numbers confirms the expected trends in density profiles. The figure reveals that in a more rarefied flow less entropy is generated. It is also observed that for smaller Knudsen numbers ($Kn < 0.5$) the distribution of entropy generation is non-uniform in the domain. However, as the Knudsen number increases higher percentage of the channel area is occupied by the Knudsen layer resulting in uniform profiles of entropy generation. Figures 12 and 13 show how the generation of entropy is distributed in the viscous dissipation and thermal diffusion terms. It can be concluded that this parameter is mainly a result of the viscous dissipation term than thermal diffusion. This effect is greater in dense gases rather than in rarefied gases. Also it is apparent from Fig. 13 that the thermal diffusion is the cause of non-uniform profiles of entropy generation.

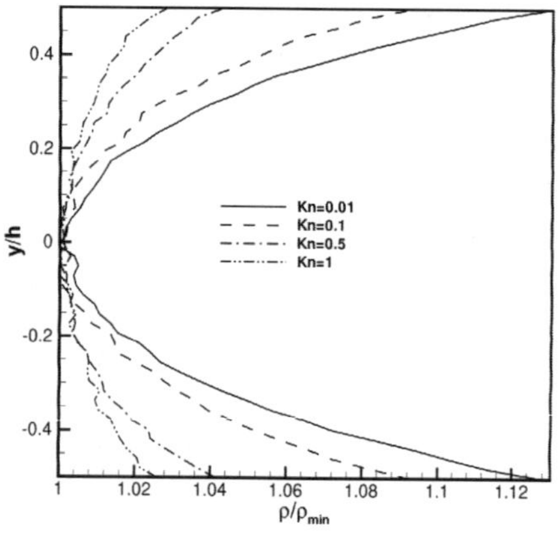

Figure 10. Rarefaction effects on non-dimensional density profiles (M_W=0.81)

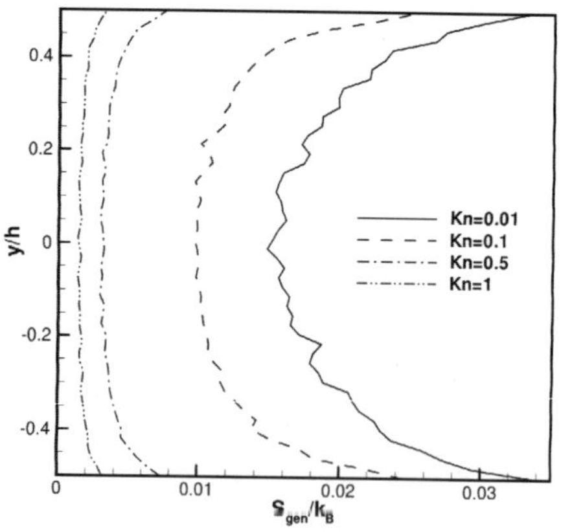

Figure 11. Rarefaction effects on non-dimensional entropy generation profiles (M_W=0.81)

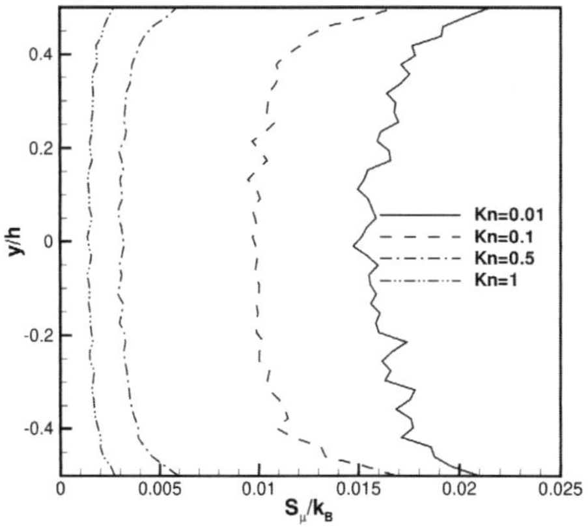

Figure 12. Rarefaction effects on non-dimensional viscous dissipation term of the entropy generation profiles (M_W=0.81)

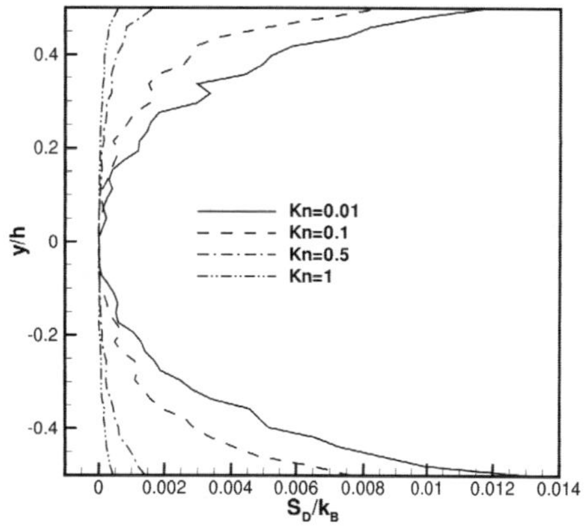

Figure 13. Rarefaction effects on non-dimensional thermal diffusion term of the entropy generation profiles (M_W=0.81)

4. Concluding remarks

Simulating the rarefied flow of argon gas in the famous problem of Couette flow, compressibility and rarefaction effects on entropy and entropy generation profiles are investigated using the DSMC method. The main concluding remarks of the present work include:

- Investigation of compressibility effects shows that increase of wall Mach number results in non-uniform entropy profiles with a peak in the center of the channel and more generation of entropy in whole regions of the channel because of the viscous heating. The results also reveal that entropy and temperature profiles are following an identical trend.
- Investigation of rarefaction effects shows that increase of Knudsen number results in more uniform entropy profiles.
- Considering the entropy generation profiles along with density profiles, it can be perceived that this quantity can be properly applied in quantifying non-equilibrium phenomena not only in hypersonic flows as stated in the existing literature, but also in low Reynolds micro/nano flows.
- It is shown that in the micro-Couette flow the generation of entropy is mainly due to viscous dissipation term observed in the final derived equation. However, the issue that causes more non-uniform profiles is the thermal diffusion term.

Acknowledgement

The authors of the paper would like to acknowledge the financial supports of the Ferdowsi University of Mashhad under grant number 20522.

References

[1] Ho C-M and Tai Y-C 1998 Micro-electro-mechanical-systems (MEMS) and fluid flows *Annual Review of Fluid Mechanics* **30** 579-612

[2] Reese J M, Gallis M A and Lockerby D A 2003 New directions in fluid dynamics: Non-equilibrium aerodynamic and microsystem flows *Philosophical Transactions of the Royal Society of London. Series A: Mathematical, Physical and Engineering Sciences* **361** 2967-2988

[3] Darhuber A A and Troian S M 2005 Principles of microfluidic actuation by modulation of surface stresses *Annual Review of Fluid Mechanics* **37** 425-455

[4] Karniadakis G, Beskok A and Aluru N 2005 *Micro flows and nano flows: Fundamentals and simulation* (New York, Springer-Verlag)

[5] Tabeling P 2005 *Introduction to microfluidics* (New York, Oxford University Press)

[6] Liou W W and Fang Y 2006 *Microfluid mechanics: Principles and modelling* McGraw-Hill)

[7] Willis D 1962 Comparison of kinetic theory analyses of linearized Couette flow *Phys. Fluids* **5** 127

[8] Beskok A, Karniadakis G E and Trimmer W 1996 Rarefaction and compressibility effects in gas microflows *Journal of Fluids Engineering* **118** 448-456

[9] Xue H, Fan Q and Shu C 2000 Prediction of micro-channel flows using direct simulation Monte Carlo *Probabilistic Engineering Mechanics* **15** 213-219

[10] Xue H, Ji H M and Shu C 2001 Analysis of micro-Couette flow using the Burnett equations *International Journal of Heat and Mass Transfer* **44** 4139-4146

[11] Roy S and Chakraborty S 2007 Near-wall effects in micro scale Couette flow and heat transfer in the maxwell-slip regimes *Microfluidics and Nanofluidics* **3** 437-449

[12] Frezzotti A and Gibelli L 2008 A kinetic model for fluid-wall interaction *Proceedings of the Institution of Mechanical Engineers, Part C: Journal of Mechanical Engineering Science* **222** 787-795

[13] Gu X J and Emerson D R 2009 A high-order moment approach for capturing non-equilibrium phenomena in the transition regime *Journal of Fluid Mechanics* **636** 177-216

[14] Kumar, Rakesh, Titov, V. E, Levin and A. D 2010 Reconsideration of planar Couette flows using the statistical bhatnagar-gross-krook approach *Anglais* **24** 9

[15] Reitebuch D and Weiss W 1999 Application of high moment theory to the plane Couette flow *Continuum Mechanics and Thermodynamics* **11** 217-225

[16] Ansumali S and Karlin I V 2000 Stabilization of the lattice Boltzmann method by the h theorem: A numerical test *Physical Review E* **62** 7999

[17] Schrock, C.R. 2005 Entropy Generation as a Means of Examining Continuum Breakdown, Air Force Ins of Tech Wright-Patterson AFB OH School of Engineering and Management

[18] Schrock C R and McMullan R J 2005 *Calculation of entropy generation rates via dsmc with application to continuum/equilibrium onset* 38th AIAA Thermophysics Conference Toronto, Ontario Canada

[19] Schrock C R, McMullan R J and Camberos J A 2005 *Continuum onset parameter based on entropy gradients using Boltzmann's h-theorem* 43rd AIAA Aerospace Sciences Meeting and Exhibit Reno, Nevada

[20] R.W. Carr, 2007 Quantifying non-equilibrium in hypersonic flows using entropy generation, Air Force Institute of Technology.

[21] Erbay L, Yalçin M and Ercan M S 2007 Entropy generation in parallel plate microchannels *Heat and Mass Transfer* **43** 729-739

[22] Ozalp A 2008 Roughness induced forced convective laminar-transitional micropipe flow: Energy and exergy analysis *Heat and Mass Transfer* **45** 31-46

[23] Ozalp A A 2010 Combined effects of pipe diameter, Reynolds number and wall heat flux and on flow, heat transfer and second-law characteristics of laminar-transitional micro-pipe flows *Entropy* **12** 445-472

[24] Ozalp A A 2009 1st and 2nd law characteristics in a micropipe: Integrated effects of surface roughness, heat flux and reynolds number *Heat Transfer Engineering* **30** 973-987

[25] Chigullapalli S, Venkattraman A, Ivanov M S and Alexeenko A A 2010 Entropy considerations in numerical simulations of non-equilibrium rarefied flows *Journal of Computational Physics* **229** 2139-2158

[26] Leite P H M and Santos W F N 2010 Thermal non-equilibrium effects on the flow field structure of hypersonic backward-facing step flow *Mecánica Computacional* **XXIX** 4935-4953

[27] Parlak N, Gür M, ArI V, Küçük H and Engin T 2010 Second law analysis of water flow through smooth microtubes under adiabatic conditions *Experimental Thermal and Fluid Science* **35** 60-67

[28] Ejtehadi O, Roohi E and Esfahani J A 2011 *Dsmc simulation of micro/nano shear driven flows* The 10th Iranian Aerospace Society Conference Tehran Iran
[29] Ejtehadi O, Roohi E and Esfahani J A 2012 Investigation of basic molecular gas structural effects on hydrodynamics and thermal behaviours of rarefied shear driven micro/nano flow using DSMC *International Communications in Heat and Mass Transfer* **39** 439-448
[30] Roohi E and Darbandi M 2012 Recommendations on performance of parallel dsmc algorithm in solving subsonic nanoflows *Applied Mathematical Modelling* **36** 2314-2321
[31] Esfahani J A and Jafarian M M 2005 Entropy generation analysis of a flat plate boundary layer with various solution methods *Scientia Iranica* **12** 233-240
[32] Esfahani J A and Koohi-Fayegh S 2009 Entropy generation analysis in error estimation of an approximate solution: A constant surface temperature semi-infinite conductive problem *THERMAL SCIENCE* **13** 133-140
[33] Esfahani J A and Shahabi P B 2010 Effect of non-uniform heating on entropy generation for the laminar developing pipe flow of a high prandtl number fluid *Energy Conversion and Management* **51** 2087-2097
[34] Bird G A 1994 *Molecular gas dynamics and the direct simulation of gas flows* Clarendon Press, Oxford, United Kingdom)
[35] Cercignani C 2000 *Rarefied gas dynamics. From basic concepts to actual calculations* Cambridge UniversityPress)
[36] Vincenti W G and Kruger C H 1967 *Introduction to physical gas dynamics* (New York, Wiley & Sons, Inc.)

1st European Conference on Gas Micro Flows (GasMems 2012) IOP Publishing
Journal of Physics: Conference Series **362** (2012) 012009 doi:10.1088/1742-6596/362/1/012009

Velocity Inversion In Cylindrical Couette Gas Flows

Nishanth Dongari[1,2], Robert W. Barber[2], David R. Emerson[2], Yonghao Zhang[1] and Jason M. Reese[1]

[1]Department of Mechanical & Aerospace Engineering, University of Strathclyde, Glasgow G1 1XJ, UK

[2]Centre for Microfluidics & Microsystems Modelling, STFC Daresbury Laboratory, Warrington WA4 4AD, UK

E-mail: nishanth.dongari@strath.ac.uk

Abstract. We investigate a power-law probability distribution function to describe the mean free path of rarefied gas molecules in non-planar geometries. A new curvature-dependent model is derived by taking into account the boundary-limiting effects on the molecular mean free path for surfaces with both convex and concave curvatures. In comparison to a planar wall, we find that the mean free path for a convex surface is higher at the wall and exhibits a sharper gradient within the Knudsen layer. In contrast, a concave wall exhibits a lower mean free path near the surface and the gradients in the Knudsen layer are shallower. The Navier-Stokes constitutive relations and velocity-slip boundary conditions are modified based on a power-law scaling to describe the mean free path, in accordance with the kinetic theory of gases, i.e. transport properties can be described in terms of the mean free path.

Velocity profiles for isothermal cylindrical Couette flow are obtained using the power-law model. We demonstrate that our model is more accurate than the classical slip solution, especially in the transition regime, and we are able to capture important non-linear trends associated with the non-equilibrium physics of the Knudsen layer. In addition, we establish a new criterion for the critical accommodation coefficient that leads to the non-intuitive phenomena of velocity-inversion. Our results are compared with conventional hydrodynamic models and direct simulation Monte Carlo data. The power-law model predicts that the critical accommodation coefficient is significantly lower than that calculated using the classical slip solution and is in good agreement with available DSMC data. Our proposed constitutive scaling for non-planar surfaces is based on simple physical arguments and can be readily implemented in conventional fluid dynamics codes for arbitrary geometric configurations.

1. Introduction

In rarefied/micro gas flows, the behavior of the gas near a solid boundary is dominated by surface and near-wall interactions and will lead to the formation of a Knudsen layer. This is a local thermodynamically non-equilibrium region of thickness $O(\lambda)$ from the surface, where λ is the mean free path (MFP) of the gas. In this Knudsen layer, molecular collisions are considerably modified due to the presence of the solid boundary with a corresponding reduction in the mean time between collisions, i.e. the MFP of the gas will effectively be reduced and will also vary non-linearly in this thin layer [1]. Linear constitutive relations for shear stress and heat flux are also no longer valid in the Knudsen layer [2, 3, 4].

The behavior of a rarefied gas can accurately be described by the Boltzmann equation [5, 6]. However, directly solving the Boltzmann equation for practical applications remains computationally challenging due to the complicated structure of the molecular collision term. The direct simulation Monte Carlo (DSMC) method [7] provides an excellent alternative approach for solving high-speed rarefied flows. Unfortunately, the computational cost of the DSMC method for low-speed flows in the slip- and transition-flow regimes is still also formidable. Many researchers developing engineering applications have instead proposed modifications to the velocity-slip boundary condition originally proposed by Maxwell [8]. This has generally led to first- and second-order treatments of the velocity gradient at the wall (see Beskok [9], Barber *et al.* [10] for a review), but with limited success because this approach fails to reproduce the non-linear stress/strain-rate relationship observed in the Knudsen layer.

An alternative strategy is to introduce a slip coefficient that depends on the Knudsen number or to introduce an effective viscosity term. The latter approach has met with some engineering success but has relied on fitting data to solutions from the Boltzmann equation. Cercignani [5] used kinetic theory to propose a "wall-function" that scaled with the mean free path. This introduces a function that complements the constitutive relations, effectively modifying the viscosity, and is able to capture some non-equilibrium behavior of the Knudsen layer. This idea has been adopted and extended [11, 12, 13] with some success.

Recently, Dongari *et al.* [14] carried out molecular dynamics (MD) studies of gases and the results indicate that molecules perform Levy-type flights under rarefied conditions, i.e. the free paths of gas molecules follow a power-law (PL) distribution. Consequently, they hypothesized that the probability distribution function for the molecular free paths of a rarefied gas followed a PL form, and this was validated against the MD data under various rarefied conditions. Using a PL distribution to describe free paths, Dongari *et al.* [15] derived an effective MFP model for flows confined by planar surfaces by taking into account the solid boundary effects and they obtained good agreement with the MD data up to the early transition regime. Further, they modified the Navier-Stokes constitutive relations using this PL-based MFP scaling and were then able to accurately capture the non-equilibrium effects in the Knudsen layer for isothermal pressure-driven gas flows between planar parallel walls.

A fundamental non-planar test case concerns the flow between two concentric cylinders. This is a classical no-slip fluid dynamics problem that is treated in many well-known textbooks [16]. However, under certain conditions, the flow between the cylinders can exhibit highly non-intuitive behavior. For example, if the outer cylinder is stationary and the inner cylinder is rotating, it is possible for the velocity profile to become *inverted* i.e. the velocity will *increase* from the inner to the outer cylinder wall. This unusual phenomenon was first predicted by Einzel *et al.* [17] for the case of liquid helium. Tibbs *et al.* [18] extended the analysis to the case of a rarefied gas and, using DSMC, demonstrated that velocity inversion could occur provided the tangential momentum accommodation coefficient (TMAC) for the surfaces was small. This early work has led to a number of important curvature studies for rotating Couette flow [10, 19, 20]. These confirmed the existence of velocity inversion for small values of the TMAC and also showed that the phenomenon could be related to a critical accommodation coefficient. Indeed, Yuhong et al. [20] derived an analytical criterion for the critical accommodation coefficient and also showed that velocity inversion was solely dependent on the value of the TMAC associated with the outer cylinder.

In this paper, we build on the approach of Dongari *et al.* [15] and derive a PL-based effective MFP model for non-planar surfaces by incorporating the effects of curvature. We develop a curvature-dependent MFP solution for both convex and concave surfaces, and extend this analysis to deduce the effective MFP solution for a gas confined between two concentric cylinders. Further, constitutive relations and velocity-slip boundary conditions are modified in accordance with the kinetic theory of gases, and tested for the classical case of isothermal rarefied Couette flow between two concentric rotating cylinders. Modified governing equations are solved for low-speed gas flows, and semi-analytical solutions are derived for velocity profiles and the critical accommodation coefficient for velocity inversion. Our PL model results are compared with an existing slip model [20], the linearized BGK solution [19], and DSMC simulation data [18, 21]. We show that our PL model captures important non-linear trends associated with the physics of transition regime cylindrical Couette gas flows.

2. Geometry-Dependent Molecular Mean Free Path

We consider a homogeneous gas, where molecules moving at an average speed $\overline{v}$ and experiencing an inter-molecular collision rate of $\dot{\theta}_v$ have a molecular mean free path $\lambda = \overline{v}/\dot{\theta}_v$. Kinetic theory-based classical expressions for the probability distribution of the free paths and the mean free path (λ) are, respectively [22]:

$$\psi(r) = \frac{1}{\lambda}\exp\left(-\frac{|r|}{\lambda}\right), \tag{1}$$

$$\lambda = \frac{m/N_{Av}}{\pi\rho\delta^2\sqrt{2}}, \tag{2}$$

where $|r|$ is the length of free path, given by $\overline{v}t$, that has been traveled by a molecule at time t; $N_{Av} = 6.0221415 \times 10^{23}$ is Avogadro's number; m is the mass of a molecule; ρ is the gas density, and δ is the hard-sphere diameter of the gas molecules.

MD results [14] show that the exponential distribution function for free paths (Eq. 1) is only valid when the gas is under thermodynamic equilibrium conditions. Under non-equilibrium conditions, the MD data exhibit long-tail behavior (i.e. molecules perform Lévy-type flights) whereas the exponential distribution has a faster decay. Montroll and Scher [23] pointed out that a finite moment of the probability distribution function implies an exponential character of the randomness. So results obtained using exponential forms of the distribution functions applicable only to homogeneous media at equilibrium. A distribution function with diverging higher-order moments (such as the standard deviation) is essential to describe non-equilibrium transport [23]. Instead of the classical exponential form of distribution function, Dongari *et al.* [14, 15] have recently hypothesized the following power-law form for the free path distribution function for non-equilibrium gas MFP, which satisfies the requirements for diverging higher-order moments:

$$\psi(r) = C(a + r)^{-n}, \tag{3}$$

where a and C are constants with positive values that are determined through the zero and first moments. The range of values for the exponent n can be obtained by making one of the higher-order moments divergent. Zero and first moments are given as follows:

$$1 = \int_0^\infty C(a + r)^{-n} dr, \tag{4}$$

$$\lambda = \int_0^\infty Br(a + r)^{-n} dr. \tag{5}$$

Equation 4 requires the probability to range only from zero to one. Equation 5 defines the unconfined, conventional MFP value, λ. It then follows that

$$B = (n - 1)a^{n-1}, \tag{6}$$

$$a = \lambda(n - 2). \tag{7}$$

To develop an expression for the effective mean free path, we use the approach of Dongari *et al.* [15], based on an integrated form of the probability distribution function, i.e.

$$p(r) = \int_0^r \psi(r) dr = \left[1 - \left(1 + \frac{r}{a}\right)^{1-n}\right], \tag{8}$$

where $p(r)$ describes the probability a molecule travels a distance r with experiencing a collision.

The probability distribution function has to be real and positive and thus $a > 0$ and $n > 2$. As $n \to \infty$, the distribution function will have finite moments, which is the condition required of an equilibrium distribution function. For a finite n, the distribution function describes a system deviating from equilibrium. So n acts as a decisive parameter to define the extent of the deviation from equilibrium. If one wishes to make the ith moment diverging, then $n_{max} = i + 1$. In the current paper, all our power-law (PL) model based results have been obtained for $n = 3$, unless otherwise explicitly stated. Dongari *et al.* [14] validated the PL free path distribution function (Eq. 3) with MD simulations under various rarefied conditions.

Following a similar approach to that adopted by Dongari *et al.* [15], here, we derive an effective mean free path model for non-planar surfaces based on the PL form of free path distribution. Our model is derived for two generalized non-planar cases: (i) gas placed over a cylinder of radius R_1, i.e. a surface with convex curvature, and (ii) gas inside a cylinder of radius R_2, i.e a surface with concave curvature. Both convex and concave curvature models are then used to deduce an effective MFP model for gas confined between concentric cylinders.

Note that in the following discussion we use the notation R^- if a test molecule is placed over a cylinder/surface with convex curvature, and R^+ if the molecule is placed inside the cylinder/surface with concave curvature. We also use the notations θ^- and θ^+ for the equally probable zenith angle traveling direction of the molecule, towards convex and concave surfaces, respectively.

2.1. Gas placed outside a cylinder

Figure 1a presents the situation of a gas molecule placed outside a solid cylinder at a distance r from the centre of the cylinder. The traveling distance limit for a molecule traveling towards the cylinder is denoted R_u^-, and this is associated with a critical limit for θ^- defined as θ_u^-, above which the molecule passes the cylinder and travels into the bulk. This angle can be calculated by using the geometry of a formed right-angled triangle, with its right angle at the point connecting the radius of the cylinder, and R_u^- which is tangential to the cylinder. The following relation can then be found:

$$r^2 = R_1^2 + \left(R_u^-\right)^2. \tag{9}$$

By using the relation $R_1 = r\,\sin(\theta_u^-)$, the value of θ_u^- is obtained as:

$$\theta_u^- = \arcsin\left(\frac{R_1}{r}\right). \tag{10}$$

The distance of the molecule away from the cylinder surface, R^-, as a function of θ^-, can be determined in terms of r and R_1 by using the cosine law:

$$R_1^2 = r^2 + (R^-)^2 - 2rR^-\cos(\theta^-), \tag{11}$$

which results in:

$$R^-(r,\theta^-) = r\cos(\theta^-) - \sqrt{(r\cos(\theta^-))^2 + R_1^2 - r^2}. \tag{12}$$

The average distance a molecule travels to the cylinder's surface, with respect to the angle θ^-, is achieved by the following integral mean value theorem:

$$\langle R^-\rangle = \frac{1}{\Omega}\int_0^{\Omega} R^-(r,\theta^-)d\theta^-, \tag{13}$$

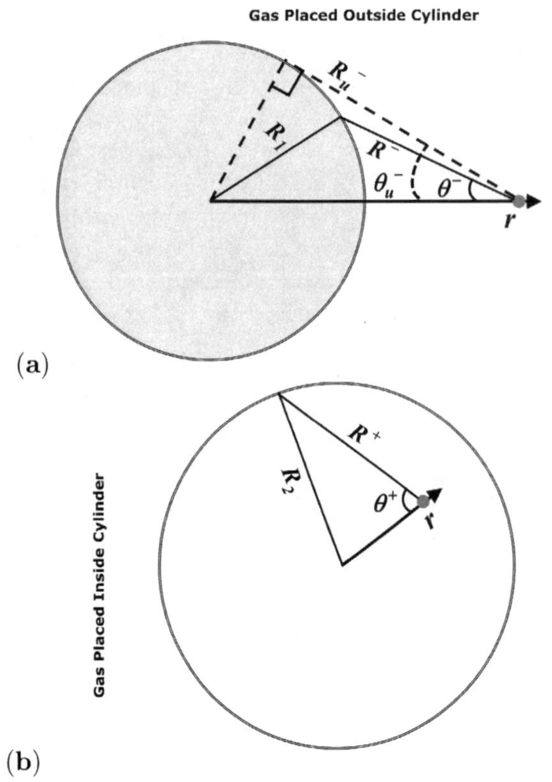

Figure 1. (a) A gas molecule placed outside a solid cylinder and situated at a radial distance r from the centre of the cylinder of radius R_1. R^- is the travelling distance limit for a molecule moving towards the cylinder surface, for a given zenith angle θ^-. The largest travelling distance R_u^- is achieved for the zenith angle direction θ_u^-, above which the molecule by-passes the cylinder surface and travels into the bulk. **(b)** A gas molecule inside a cylindrical cavity of radius R_2, at a wall normal distance of $R_2 - r$, where r is the radial distance of the molecule from the centre of the cylinder. The molecule has a traveling distance of R^+ to the wall for a traveling direction of θ^+, where θ^+ is varied from 0 to π.

where Ω is the solid angle subtended by the cylinder at a point in the base plane, at a distance r ($> R_1$) from the axis of the cylinder of length L, and which can be expressed as [24]:

$$\Omega = \theta_u^- - \frac{1}{2}\left[\left(\left(\frac{r}{R_1}\right)^2 - 1\right)^{\frac{1}{2}} + \theta_u^- - \frac{\pi}{2}\right] L^{-2} + O\left(L^{-4}\right). \quad (14)$$

If we consider the length of the cylinder L tending to infinity, Ω then can be simplified to:

$$\Omega = \theta_u^-. \quad (15)$$

Using Eq. 8, the mean free path of the molecules traveling in the direction of the cylinder (a surface with convex curvature) is then,

$$\lambda_{\text{eff(conv)}} = \lambda\left[1 - \frac{1}{\theta_u^-}\int_0^{\theta_u^-}\left(1 + \frac{R^-(r,\theta^-)}{a}\right)^{(1-n)} d\theta^-\right]. \quad (16)$$

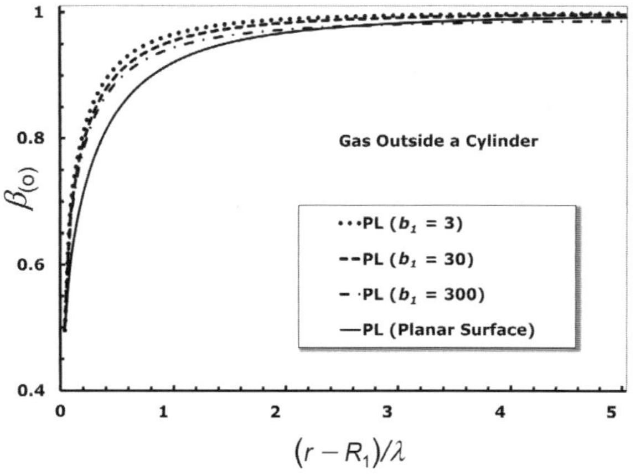

Figure 2. Variation of the effective mean free path of gas molecules outside a solid cylinder (surface with a convex curvature) as a function of normalized wall normal distance $[(r-R_1)/\lambda]$. Here $r \geq R_1$, as the gas is placed outside a cylinder and on the x-axis, r is varied by fixing R_1 and λ, for a given value of b_1. The MFP profiles for various values of b_1 (R_1/λ) are illustrated to show the effect of the convex curvature and their comparison with the planar one-wall solution [15].

The general mean free path expression is obtained by considering all possible molecular traveling directions. This is achieved in a similar manner to the analysis for a gas confined between planar parallel walls presented in Dongari *et al.* [15]. In the case of planar parallel walls, it is equally probable that a molecule travels towards one surface or the other. In the present case, solid angle theory is used to determine the likelihood of a molecule traveling in the direction of the cylinder, as opposed to traveling towards the bulk; the probabilities being expressed as θ_u^-/π and $[1-(\theta_u^-/\pi)]$, respectively. From this weighting, the general expression for the effective mean free path of a gas placed outside a cylinder is:

$$\lambda_{\text{eff(o)}} = \lambda_{\text{eff(conv)}}\left(\frac{\theta_u^-}{\pi}\right) + \lambda\left[1-\left(\frac{\theta_u^-}{\pi}\right)\right], \tag{17}$$

where λ is the unconfined gas mean free path. This equation can also be written as $\lambda_{\text{eff(o)}} = \lambda\,\beta_{(o)}(r, R_1/\lambda)$, where

$$\beta_{(o)} = \left(\frac{\theta_u^-}{\pi}\right)\left[1-\frac{1}{\theta_u^-}\int_0^{\theta_u^-}\left(1+\frac{R^-(r,\theta^-)}{a}\right)^{(1-n)} d\theta^-\right] + \left[1-\left(\frac{\theta_u^-}{\pi}\right)\right], \tag{18}$$

which is the normalized effective MFP based on a power-law distribution function, where the subscript (o) denotes our considerations of gas molecules outside a solid cylindrical surface. Eq. 18 is dependent on r, R_1 and λ, and evaluated as a function of normalized wall normal distance $[(r-R_1)/\lambda]$, using the parameter $b_1 = R_1/\lambda$. It should be noted that a is dependent on the mean free path (see Eq. 4) and $r \geq R_1$, as the gas is placed outside a cylinder. The parameter b_1 denotes the inverse of the normalized curvature and as $b_1 \to \infty$, the solution of Eq. 18 should reduce to the planar case. Eq. 18 can be computed numerically using Simpson's rule, here we use 16 subintervals.

The solution of the effective mean free path for a convex surface must reduce to that for a planar one-wall case when the radius of the cylinder tends to infinity. To verify this limiting case, the normalized effective MFP solutions (i.e.$\beta_{(o)}$) based on Eq. 18, are plotted in Fig. 2 for various values of b_1 and are compared with the planar one-wall solution derived by Dongari *et al.* [15]. For a value of $b_1 = 3$, the non-planar solution predicts a higher β value at the surface and also exhibits sharper gradient compared to the planar one-wall case. Fig. 2, shows that as the radius R_1 of the cylinder increases, our non-planar solution tends to the planar one.

2.2. *Gas inside a hollow cylinder*

Using a similar approach, it is possible to obtain the effective mean free path for gas molecules inside a hallow cylinder of radius R_2, i.e. for a surface with a concave curvature. As illustrated in Fig. (1b), r is a target molecule's radial distance from the centre of the cylinder and R^+ is the molecule's distance to the wall if the molecule has a traveling trajectory of θ^+.
To calculate the effective mean free path, the wall distance of the molecule has first to be calculated using the cosine law:

$$R_2^2 = r^2 + (R^+)^2 - 2rR^+\cos(\theta^+), \tag{19}$$

where the molecule's traveling trajectory θ^+ can vary from 0 to π. The molecular traveling distance to the wall R^+, dependent on r and θ^+, can then be expressed as:

$$R^+(r,\theta^+) = -r\cos(\theta^+) + \sqrt{(r\cos(\theta^+))^2 + R_2^2 - r^2}. \tag{20}$$

The solid angle subtended by the cylinder at a point lying inside the cylinder is 2π. Based on the quarter symmetry, it is sufficient to integrate θ^+ from 0 to $\pi/2$. Using our PL distribution to describe the molecular free paths (Eq. 8), the expression for the effective mean free path of a gas molecule inside a cylinder (i.e. interacting with a surface with concave curvature) is:

$$\lambda_{\text{eff(conc)}} = \lambda\left[1 - \frac{2}{\pi}\int_0^{\pi/2}\left(1 + \frac{R^+(r,\theta^+)}{a}\right)^{(1-n)} d\theta^+\right], \tag{21}$$

or the normalized expression is:

$$\beta_{(\text{i})} = \frac{\lambda_{\text{eff(conc)}}}{\lambda}, \tag{22}$$

where the subscript (i) denotes that the gas molecules are located inside the cylinder. Eq. 22 is dependent on r, R_2 and λ, and evaluated as a function of normalized wall normal distance $[(R_2 - r)/\lambda]$, using the parameter $b_2 = R_2/\lambda$. Here $r \leq R_2$, as the gas is inside the cylinder.

In a similar manner to the convex case, the effective MFP solution for a concave surface must also reduce to the planar one-wall case when the radius of the cylinder tends to infinity. Fig. 3 shows the variation of the normalized effective mean free path $\beta_{(\text{i})}$, for various b_2 values and a comparison with the planar one-wall solution [15]. For a value of $b_2 = 5$, the non-planar solution predicts a lower β value at the surface and exhibits a shallower gradient compared to the planar case. For higher values of b_2, the non-planar concave solution approaches the planar case. However, at the surface, even for $b_2 = 30$, the non-planar solution predicts a lower value of $\beta_{(\text{i})}\sim 0.4691$ compared to the corresponding planar wall value of 0.5. This is due to the concave nature of the surface, where a molecule will encounter more frequent collisions with the wall before it escapes from the concave well.

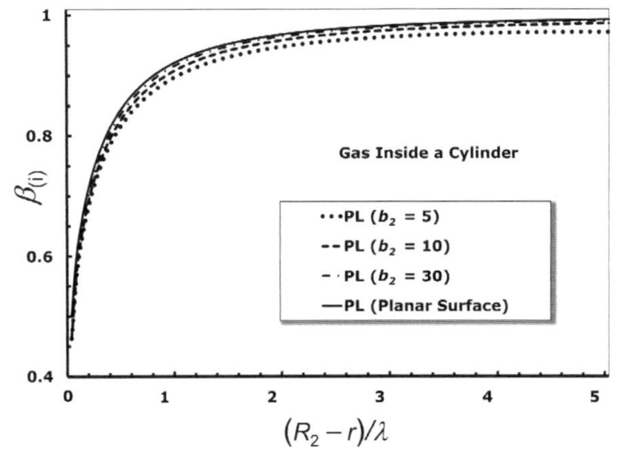

Figure 3. Variation of the effective mean free path of a gas molecule inside a cylinder (i.e.a a bounding surface with concave curvature) as a function of normalized wall normal distance $[(R_2 - r)/\lambda]$. Here $r \leq R_2$, as the gas is placed outside a cylinder and on the x-axis, r is varied by fixing R_2 and λ, for a given value of b_2. The MFP profiles for various values of b_2 (R_2/λ) are illustrated to show the effect of the concave curvature and their comparison with the planar one-wall solution [15].

2.3. Gas confined between concentric cylinders

We now consider a gas between two concentric cylinders, with R_1 and R_2 being the radii of the inner and outer cylinders, respectively. Let us consider a molecule situated at a distance r from the coincident centres of the two cylinders. The inner cylinder presents a surface with convex curvature (Fig. 1a) to the molecule and the outer cylinder presents a surface with a concave curvature (Fig. 1b). The probability of a molecule traveling towards the inner and outer cylinder directions can be evaluated as θ_u^-/π and $[1 - (\theta_u^-/\pi)]$, respectively. Using Eqs. 16 and 21 and accounting for this weighting, the complete expression for the effective mean free path of a gas confined between two concentric cylinders is:

$$\lambda_{\text{eff}} = \lambda_{\text{eff(conv)}} \left(\frac{\theta_u^-}{\pi}\right) + \lambda_{\text{eff(conc)}} \left[1 - \left(\frac{\theta_u^-}{\pi}\right)\right], \tag{23}$$

where $\lambda_{\text{eff}} = \lambda\, \beta$, and so

$$\beta = \left(\frac{\theta_u^-}{\pi}\right)\left[1 - \frac{1}{\theta_u^-}\int_0^{\theta_u^-}\left(1 + \frac{R^-(r,\theta^-)}{a}\right)^{(1-n)} d\theta^-\right] + \left[1 - \left(\frac{\theta_u^-}{\pi}\right)\right]\left[1 - \frac{1}{\theta_u^+}\int_0^{\theta_u^+}\left(1 + \frac{R^+(r,\theta^+)}{a}\right)^{(1-n)} d\theta^+\right], \tag{24}$$

with

$$\theta_u^+ = \pi - \theta_u^- . \tag{25}$$

Here, β is the normalized geometry-dependent MFP based on a power-law distribution function, and is dependent on the Knudsen number Kn through the mean free path, λ and a geometry constraint R_2/R_1, defining Kn as:

$$Kn = \frac{\lambda}{R_2 - R_1}. \tag{26}$$

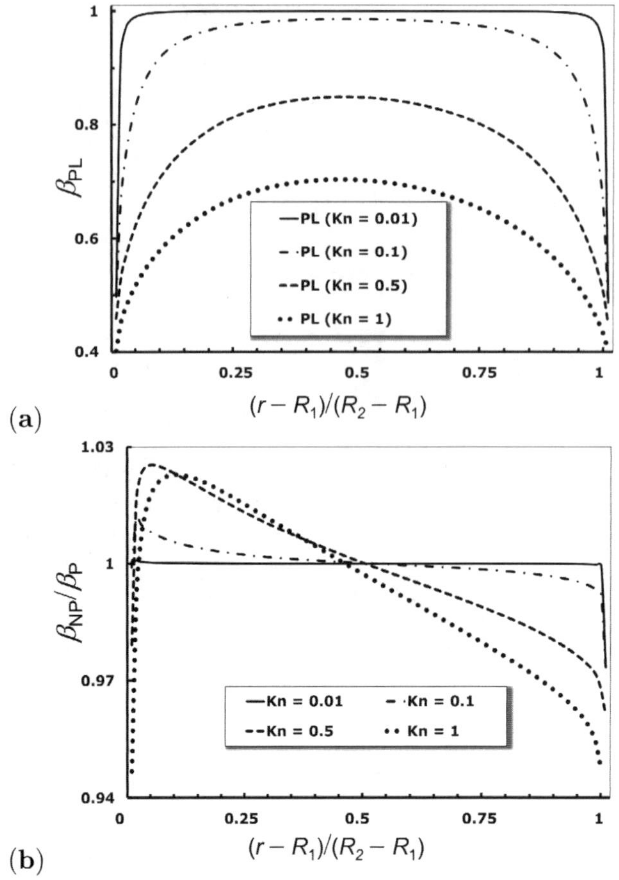

Figure 4. (**a**) Variation of β with normalized radial distance $(r - R_1)/(R_2 - R_1)$ between the inner and outer cylinders for various Knudsen numbers in the slip and transition regimes, for the case of $R_2/R_1 = 5/3$. (**b**) Illustration of the curvature effects through the ratio of the normalized MFP values of gas confined between non-planar (NP) and planar (P) surfaces. The non-planar solution is obtained from Eq. 24 and the planar-surface values are computed using Equation 12 in Dongari *et al.* [15].

Figure (4a) shows the variation of β between inner and outer cylinders (with $R_2/R_1 = 5/3$) for various Knudsen numbers in the slip and early transition regimes. The MFP profiles are sharper at the inner cylinder and shallower at the outer cylinder and they are not symmetric about the midpoint of the gap between the cylinders. This finding is in contrast to the solution for the planar parallel wall case, where the MFP profile is symmetric about the midpoint of the gap between the surfaces [15]. The anti-symmetric effect increases with Knudsen number. The values of β are relatively high at the inner cylinder surface compared to the outer cylinder, due to the convex and concave curvature effects, respectively. The difference in the values of β at the inner and outer cylinders is also dependent on the Knudsen number.

To illustrate further the curvature effects, Fig. (4b) presents the ratio of the normalized effective MFPs non-planar to planar. Non-planar β_{NP} values are obtained from Eq. (24) for $R_2/R_1 = 5/3$ and β_P for the case of planar parallel walls is evaluated using Equation 12 in Dongari *et al.* [15]. For $Kn = 0.01$, the curvature effects are negligible, except in a thin layer close to the outer cylinder surface. As the Knudsen number increases, both the curvature and anti-symmetric effects become more pronounced, whereas the difference in the values of β at the inner and outer cylinders decreased. The ratio β_{NP}/β_P has its minimum at the outer cylinder for all Kn and its maximum at a position close to the surface of the inner-cylinder. However, this point moves away from the inner cylinder with increase in Kn, as the thickness of the Knudsen-layer increases.

3. Cylindrical Couette Flow

3.1. Problem description, governing equations and results

To test the merits of our PL-based effective MFP scaling for non-planar cases, we consider a rarefied gas confined between two concentric rotating cylinders. The flow is assumed to be fully developed, two-dimensional, isothermal, laminar and steady, with a low Reynolds number (Re) so that inertial effects may be neglected.

With these assumptions, the governing flow equation in polar coordinates is

$$\frac{1}{r^2}\frac{d}{dr}\left(r^2\tau_{r\phi}\right) = 0, \tag{27}$$

where r is the radial coordinate, ϕ the tangential coordinate and $\tau_{r\phi}$ the stress, which is defined as:

$$\tau_{r\phi} = \mu\left(\frac{du_\phi}{dr} - \frac{u_\phi}{r}\right), \tag{28}$$

where μ is the fluid dynamic viscosity and u_ϕ is the velocity of the fluid in the tangential direction.

From the kinetic theory of gases, the fluid viscosity can be explained in terms of the collisions between gas molecules, and of the free paths which the molecules describe between collisions. The unconfined MFP is then related to the shear viscosity [5]:

$$\mu = \rho\frac{\lambda}{\sqrt{\pi/2RT}}, \tag{29}$$

where ρ is the gas density, R the specific gas constant and T the gas temperature.

Equation (29) is assumed to be valid only for flows that are quasi-equilibrium. Within the Knudsen layer the flight paths of gas molecules are affected by the presence of a solid wall. If we wish to use Eq. (29), we need to take into account the MFP affected by gas molecular collisions with surfaces. If the unconfined expression for the MFP, λ, is replaced by our effective and geometry-dependent mean free path, λ_{eff} (Eqs. 23 and 24), we obtain a non-constant, geometry-dependent, effective viscosity, $\mu_{\text{eff(PL)}}$, that can then be used to deduce a non-linear stress/strain-rate relation:

$$\tau_{r\phi} = \underbrace{\mu\beta}_{\mu_{\text{eff}}}\left(\frac{du_\phi}{dr} - \frac{u_\phi}{r}\right). \tag{30}$$

Substituting Eq. (30) into Eq. (27) results in the modified governing equation:

$$\frac{\mu}{r^2}\frac{d}{dr}\left[r^2\beta\left(\frac{du_\phi}{dr} - \frac{u_\phi}{r}\right)\right] = 0. \tag{31}$$

This needs to be solved in conjunction with the following slip boundary conditions at the inner and outer cylinder surfaces, respectively:

$$u_\phi\,|_{r=R_1} = \omega_1 R_1 + \underbrace{\frac{2-\sigma_1}{\sigma_1}}_{\alpha_1}\lambda\left[\beta\left(\frac{du_\phi}{dr}-\frac{u_\phi}{r}\right)\right]|_{r=R_1}, \tag{32}$$

$$u_\phi\,|_{r=R_2} = \omega_2 R_2 - \underbrace{\frac{2-\sigma_2}{\sigma_2}}_{\alpha_2}\lambda\left[\beta\left(\frac{du_\phi}{dr}-\frac{u_\phi}{r}\right)\right]|_{r=R_2}, \tag{33}$$

where ω_1 and ω_2 are the angular velocities and σ_1 and σ_2 are the tangential moment accommodation coefficients of inner and outer cylinders, respectively.

The solution for the velocity profile can then be obtained as:

$$u_\phi(r) = r\left[\omega_1 + \frac{\alpha_1\lambda C}{R_1^3} + G(r)C\right], \tag{34}$$

where

$$G(r) = \int_{R_1}^{r}\frac{dr}{r^3\beta_{\mathrm{PL}}}, \tag{35}$$

$$C = \frac{\omega_2-\omega_1}{\left[H + \frac{\alpha_1\lambda}{R_1^3} + \frac{\alpha_2\lambda}{R_2^3}\right]}, \tag{36}$$

$$H = \int_{R_1}^{R_2}\frac{dr}{r^3\beta_{\mathrm{PL}}}. \tag{37}$$

When the effects of the Knudsen layer are neglected, i.e. the mean free path has no geometry dependence, then Eq. (34) simply reduces to the velocity profile based on the classical slip solution, as presented in Equation (6) of Yuhong et al. [20]. Normalized velocity profiles $[U^* = u_\phi/(\omega_1 R_1)]$ are presented in Fig. 5) for the case of $Kn = 0.5$, a stationary outer cylinder ($\omega_2 = 0$), $R_2/R_1 = 5/3$, and various combinations of accommodation coefficients.

To illustrate the phenomenon of velocity profile inversion, the outer cylinder is stationary ($\omega_2 = 0$) and the inner cylinder is rotating. The radii of the inner and outer cylinders are chosen to be 3λ and 5λ, respectively, and the accommodation coefficients are assumed to be equal at the inner and outer cylinder surfaces ($\sigma_1 = \sigma_2 = \sigma$). Figure. (5a) presents a comparison between our solution based on the PL effective MFP (Eq. 34) and the DSMC data reported by Tibbs *et al.* [18]. The PL model is in very good quantitative agreement with the DSMC results for the $\sigma = 1.0$ and 0.7 cases. The results show that the DSMC data and PL formulation follow the same basic trends and predict an inverted velocity profile when the accommodation coefficient of 0.1. However, slight deviations are discernible at the surface of the outer cylinder for the $\sigma = 0.4$ and 0.1 cases.

It is interesting to note that, for the specific case of the accommodation coefficients of the inner and outer cylinders having the same value, the family of velocity profiles all pass through a common point that is independent of the value of the accommodation coefficient. This intersection point in the PL profiles is fairly close to the point predicted by DSMC data, whereas the classical slip solution predicts this point closer to the outer cylinder, see Fig. 5b. This figure presents the results based on the classical governing equation together with the conventional slip boundary conditions, i.e. without considering any Knudsen layer effects (see Equation 6 in Yuhong *et al.* [20]), alongside the PL model results. It is evident that the classical slip solution fails to capture non-linear effects associated within the Knudsen layers at the inner and outer cylinders. The discrepancies are greatest at $\sigma = 1.0$, although at lower values of σ the discrepancies are smaller. At smaller values of σ, the PL and classical slip solution are identical and yield the same solid body rotation solution when $\sigma = 0$.

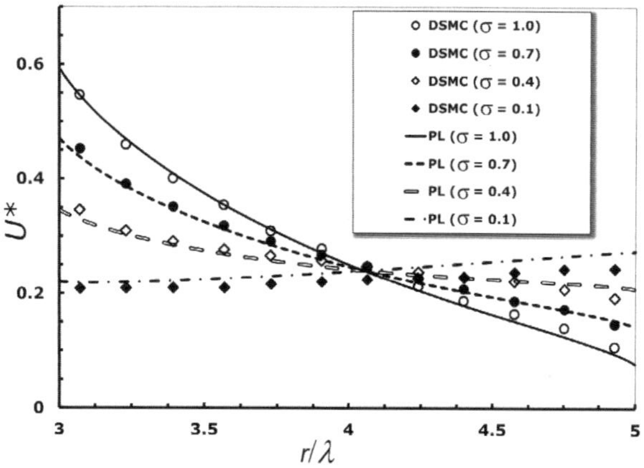

(a)

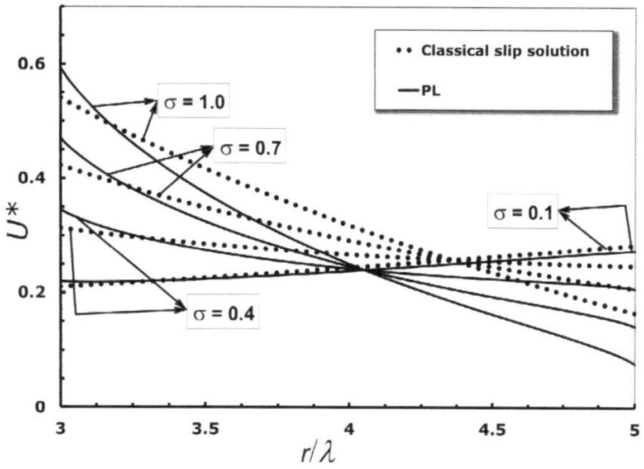

(b)

Figure 5. Cylindrical Couette flow, non-dimensional velocity $[U^* = u_\phi/(\omega_1 R_1)]$ versus normalized radial distance for $\sigma_1 = \sigma_2 = \sigma$. Comparison of PL model results against (**a**) DSMC data [18] and (**b**) the classical slip solution [20]. The results are presented for $Kn = 0.5$ and $R_2/R_1 = 5/3$.

3.2. Critical accommodation coefficients

In this section, we identify limiting analytical criterion for critical accommodation coefficients that produce (i) normal, (ii) fully-inverted, and (iii) partially-inverted velocity profiles. A normal (non-inverted) velocity profile occurs when the tangential velocity decreases monotonically from the inner to the outer cylinder. In contrast, a fully-inverted case occurs when the velocity monotonically increases. Finally, a partially-inverted profile occurs when there is a decrease and increase at multiple locations between the inner and outer cylinders.

The velocity profile is given in Eq. (34); the gradient of this can be used as a criterion to judge whether the function is monotonically decreasing or increasing. After some simplifications, the velocity gradient for the case of a rotating inner cylinder and stationary outer cylinder can be expressed as:

$$\frac{du_\phi}{dr} = \omega_1 \left[1 - \frac{1}{H + \frac{\alpha_1 \lambda}{R_1^3} + \frac{\alpha_2 \lambda}{R_2^3}} \left(\frac{\alpha_1 \lambda}{R_1^3} + G(r) + \frac{1}{r^2 \beta_{PL}} \right) \right]. \quad (38)$$

If $du_\phi/dr > 0$ a fully inverted velocity profile exists, and this is only satisfied when:

$$\frac{\left(\frac{\alpha_1\lambda}{R_1^3}+G(r)+\frac{1}{r^2\beta_{PL}}\right)}{H+\frac{\alpha_1\lambda}{R_1^3}+\frac{\alpha_2\lambda}{R_2^3}}<1, \tag{39}$$

which can only hold throughout the annular clearance ($R_1 \leq r \leq R_2$), when $[G(r)+1/(r^2\beta_{PL})]$ is at its maximum value, which occurs at $r = R_1$. Using this condition, the criterion for the critical accommodation coefficient for a fully inverted velocity profile (i.e. positive velocity gradient) can be expressed as:

$$(\sigma_2)_{\mathrm{P}} = 2\left[1+\frac{R_2^3}{R_1^2\beta_{PL}\lambda}-\frac{R_2^3}{\lambda}H\right]^{-1}. \tag{40}$$

Equation 40 provides an upper bound, so for $\sigma_2 < (\sigma_2)_{\mathrm{P}}$ a fully inverted velocity will always occur, and this phenomenon is independent of the value of the accommodation coefficient at the inner cylinder. In the absence of any Knudsen layer in the flow, i.e. a constant mean free path between the inner and outer cylinders, Eq. 40 simply reduces to the classical slip solution presented in Equation 17 of Yuhong *et al.* [20].

From Eq. 38, if $du_\phi/dr < 0$ the velocity will be a decreasing function of r and no inverted velocity profile will be observed. This situation can only exist when:

$$\frac{\left(\frac{\alpha_1\lambda}{R_1^3}+G(r)+\frac{1}{r^2\beta_{PL}}\right)}{H+\frac{\alpha_1\lambda}{R_1^3}+\frac{\alpha_2\lambda}{R_2^3}}>1, \tag{41}$$

which can only hold throughout the annular clearance ($R_1 \leq r \leq R_2$), when $G(r)+1/(r^2\beta_{PL})$ is at its minimum value, which occurs for $r = r_c$ ($R_1 < r_c < R_2$). The value of r_c can be obtained numerically for a given Kn and R_2/R_1 and then the critical accommodation coefficient for no velocity inversion (i.e. negative velocity gradient) follows as:

$$(\sigma_2)_{\mathrm{N}} = 2\left[1+\frac{R_2^3}{R_c^2\beta_{PL}(r_c)\lambda}-\frac{R_2^3}{\lambda}\left(G(r_c)-H\right)\right]^{-1} \tag{42}$$

Equation (42) provides a lower bound, so for $\sigma_2 > (\sigma_2)_{\mathrm{N}}$ no inverted velocity profile will occur and this phenomenon is also independent of the value of the accommodation coefficient at the inner cylinder, as in the case above. When there is no mean free path variation between the inner and the outer cylinders, Eq. 42 also reduces to the corresponding classical slip solution, (see Equation 19 of Yuhong *et al.* [20]).

Considering these two cases of Eqs. 40 and 42, where the velocity is monotonically decreasing or increasing, it is evident that a *partially-inverted* velocity profile will occur when the accommodation coefficient at the outer cylinder lies within the range:

$$(\sigma_2)_{\mathrm{P}} < \sigma_2 < (\sigma_2)_{\mathrm{N}}. \tag{43}$$

This also does not depend on the accommodation coefficient of the inner cylinder.

Figure 6a presents the variation of the critical accommodation coefficient $(\sigma_2)_{\mathrm{N}}$ as a function of Knudsen number for the case of no velocity inversion. Our PL model results are compared with DSMC data [21], the classical slip solution (Equation 19 of Yuhong *et al.* [20]) and the linearized BGK model [19], in the flow regime of $0 < Kn < 1$. In the continuum regime ($Kn < 0.01$), all three theoretical models converge to similar values. In the slip flow regime ($0.01 < Kn < 0.1$), the PL model lies below the classical slip solution and the BGK model. The PL model results show very good quantitative agreement with the DSMC data in the early transition regime ($0.1 < Kn < 0.5$), whereas the other two models show significant deviations. Although DSMC

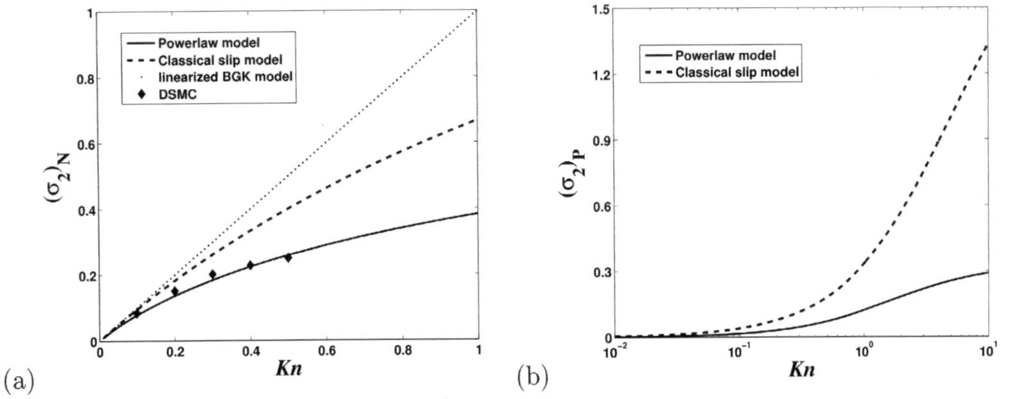

Figure 6. Variation of critical accommodation coefficient with Knudsen number for the geometric configuration of $R_2/R_1 = 5/3$; (**a**) $(\sigma_2)_\mathrm{N}$ the lower bound for no velocity inversion, i.e. negative velocity gradient throughout the annular clearance ($R_1 < r < R_2$); (**b**) $(\sigma_2)_\mathrm{P}$ the upper bound for full velocity inversion, i.e. positive velocity gradient throughout the annular clearance.

data is not available beyond $Kn > 0.5$, by extrapolating the trend up to $Kn = 0.5$, the PL model may slightly overpredict $(\sigma_2)_\mathrm{N}$ in the later part of the transition regime. In the free-molecular regime, the PL model asymptotically reaches a constant value (~ 0.7) and both the classical slip solution and the BGK model diverge (not shown in Fig. 6a).

Figure 6b shows the variation of the critical accommodation coefficient of fully inverted velocity $(\sigma_2)_\mathrm{P}$, as a function of Knudsen number. Our PL model results are compared with the classical slip solution (Equation 17 of Yuhong *et al.* [20]) from the slip to the free-molecular regime ($0.01 < Kn < 10$), although the continuum fluid assumptions are highly questionable for $Kn > 1$. The classical slip solution significantly overpredicts the PL model in the transition regime and beyond. For $Kn >> 1$, the slip solution indeed shows a diverging and unphysical behavior with $(\sigma_2)_\mathrm{P}$ values greater than unity. Conversely our PL model reaches a constant value for accommodation coefficient of ~ 0.3.

4. Conclusions

The non-equilibrium flow physics of rarefied gases interacting with non-planar surfaces has been described using a power-law (PL) probability distribution function for the free-paths of the gas molecules. We have developed new geometry-dependent mean free path solutions for both convex and concave surfaces that take into account the termination of the free paths of gas molecules at the curved surfaces. Subsequently an hypothetical constitutive scaling approach to model the Knudsen layer within a conventional continuum fluid dynamics framework has been proposed, in accordance with the kinetic theory of gases. This has been tested for the case of isothermal rarefied Couette flow between two concentric rotating cylinders.

The gas velocity profiles show good agreement with DSMC simulation data in the transition regime for various values of the accommodation coefficients of the inner and outer cylinder surfaces. When the accommodation coefficients of the inner and outer cylinders have the same value, the family of velocity profiles all pass through a common point. This intersection point for our PL model is very close to the point predicted by the DSMC data, whereas the classical slip solution predicts this point much closer to the outer cylinder. New analytical criteria have been developed for the critical accommodation coefficients for non-inverted, fully inverted and partially inverted velocity profiles. The PL model shows excellent agreement with the DSMC data for these, too.

Our new PL model is more accurate than conventional continuum fluid solutions in capturing the non-linear effects associated with Knudsen layers in the slip and transition flow regimes. With the current approach we can also model the velocity profile at a fixed Knudsen number by modifying the power-law exponent, to provide accurate results in both the Knudsen layer and bulk flow. Whereas using the conventional slip modeling technique we can only modify the slip velocity at the surface by tuning the value of the slip coefficient. This PL mean free path scaling can be readily extended to more complex geometries and straightforwardly incorporated into existing CFD codes to solve for low-speed rarefied gas flows in arbitrary geometries.

Acknowledgments

The research leading to these results has received funding from the European Community's Seventh Framework Programme FP7/2007-2013 under grant agreement ITN GASMEMS n° 215504. Additional support was provided by the UK Engineering and Physical Sciences Research Council (EPSRC) under the auspices of Collaborative Computational Project 12 (CCP12).

References

[1] Stops D W 1970 *J. Phys. D Appl. Phys.* **3** 685
[2] Burnett D 1935 *Proc. Lond. Math. Soc.* **40** 382
[3] Grad H 1949 *Comm. on Pure and App. Math.* **2** 325
[4] Chapman S and Cowling T G 1970 *Mathematical theory of non-uniform gases* (Cambridge University press)
[5] Cercignani C 1988 *The Boltzmann Equation and its Applications* (Springer-Verlag, New York)
[6] Sone Y 2002 *Kinetic theory and fluid dynamics* (Birkhauser, Boston)
[7] Bird G A 1994 *Molecular gas dynamics and the direct simulation of gas flows* (Oxford University Press, New York)
[8] Maxwell J C 1879 *Philos. Trans. R. Soc. 1* **170** 231
[9] Beskok A 2001 *Numerical Heat Trans. Part B: Fund.* **40**(6) 451
[10] Barber R W *et al.* 2004 *Vacuum* **76** 73
[11] Lockerby D A and Reese J M 2008 *J. Fluid Mech.* **604** 235
[12] Lilley C R and Sader J E 2007 *Phys. Rev. E.* **76** 026315
[13] Guo Z L *et al.* 2007 *Euro. Phys. Lett.* **80** 24001
[14] Dongari N *et al.* 2011 *J. Phys. D. App. Phys.* **44** 125502
[15] Dongari N *et al.* 2011 *J. Fluid. Engg.* **133**(7) 071101
[16] Schlichting H 1979 *Boundary-layer theory* (7th ed. McGraw-Hill, New York)
[17] Einzel D *et al.* 1990 *Phys. Rev. Lett.* **64** 2269
[18] Tibbs K W *et al.* 1997 Phys. Rev. E **56** 2282
[19] Aoki K *et al.* 2003 *Phys. Rev. E* **68** 016302
[20] Yuhong S *et al.* 2005 *Phys. Fluids* **17** 047102
[21] Stefanov S K *et al.* 2007 *Proc. 25th Int. Symp. on Rarefied Gas Dynamics* 1146
[22] Kennard, E. H., 1938, *Kinetic Theory of Gases with an Introduction to Statistical Mechanics*, McGraw-Hill, New York.
[23] Montroll E W and Scher H 1973 *J. Stat. Phys.* **9** (2) 101
[24] Guest P G 1961 *Rev. Sci. Instrum.* **32** 164

Heat transfer enhancement of a horizontal microchannel using ferro-nanoparticles injection

M Shams[1] and S Tolou[2]
[1] Faculty of Mech. Engng., K.N. Toosi University of Tech., Tehran, Iran
[2] Faculty of Mech. Engng., Islamic Azad University-South Tehran Branch, Tehran, Iran

E-mail: shams@kntu.ac.ir

Abstract. A numerical study of Ferro-nanoparticles dispersion has been performed inside a micro-channel by using two phase method. Euler-Lagrange approach was conducted to simulate the presence of Ferro-nanoparticles in the horizontal micro-channel under laminar flow condition. Considering slip velocity and temperature jump due to the effect of decreasing hydrodynamic diameter of channel, temperature and velocity profiles were evaluated and effect of some different parameters on heat transfer has been discussed. Drag, Gravity and Brownian forces were considered using a Lagrangian approach. Furthermore, in order to consider slip effects in micro-channels, Cunningham slip correction factor has been pondered in correcting drag force influence.

1. Introduction

Solid-Gas two-phase flows are encountered frequently in many environmental, industrial and pharmaceutical uses. Furthermore, over the last decade there has been an enormous interest in micro and nano-technology. As systems approach microscopic scales, it has been reported that some phenomena may differ from those in macroscopic counterparts. Therefore, a number of these factors that are dominant in micro scale have been identified through various experimental, analytical and numerical works. Among them, noncontinuum, compressibility, and different surface effects have been under vigorous investigation [1].

Noncontinuum effect: According to different investigation, it would be possible classifying flows into four regimes in order to consider nonequilibrium effects by using a dimensionless number named Kn. Also, Kn number can be defined as the ratio of molecular mean free path to characteristic length of channel. The regime $0<Kn<0.1$ is referred to as slip flow; no-slip is captured by Kn=0. For $Kn \geq 0.1$ the continuum description is expected to fail, and the regime $0.1<Kn \leq 10$ is referred to as the transition regime because the molecular motion undergoes a transition from diffusive (continuum) for $Kn \leq 0.1$, to ballistic for $Kn \geq 10$ (free molecular flow) [2].

Compressibility effect: In a microchannel, the high Mach number and large pressure drop can be reached even at low Reynolds numbers [3]. Therefore in order to not consider compressibility effect, as we have done in this paper, paying attention to flow Reynolds number would be completely vital.

Roughness effect: At the microscale level, it is impossible to obtain a completely smooth wall surface. According to the traditional knowledge for macrosystems, when the relative roughness is less than 5%, its effect on the friction factor is negligible. But for microscale channels, previously reported experimental and computational results have drawn a conclusion that surface roughness has a Significant influence on flow and heat transfer [3, 4].

To analyze microscales gas flows, a number of different analytical and numerical studies have been conducted. Some of these investigations have been done using Direct-Simulation Monte Carlo (DSMC) and the others have been considered Navier-Stokes methods with slip boundary condition. However, the DSMC is still believed to be inefficient tool for low-speed microscale gas flows, due to its huge computational load and large inevitable statistical noise [5]. Therefore, the Navier-Stokes approach including suitable slip boundary condition can analyze the velocity and temperature slip effect in an economical way without losing any accuracy in flow modeling [6]. Inman was the first investigator who studied the constant-wall-temperature Nusselt number in the slip flow regime using continuum theory subject to slip boundary conditions (both velocity and temperature) [7]. Inman's work shows that for fully accommodating walls the Nusselt number reduces due to slip boundary condition. Kavehpour et al. conducted a series of calculations in the slip flow regime examining the relative effects of rarefaction and compressibility [8]. Furthermore, as it has been mentioned before, for microscale channels, previously reported experimental and computational results have shown a considerable influence of roughness on flow and heat transfer. As an example, Kandlikar et al. have conducted an experiment indicating that for a 0.62 mm tube with relative roughness of 0.355%, the effect of roughness on friction factor and heat transfer is significant [4]. Mala and Li studied the effect of roughness in a number of channels with diameters ranging from 50 to 254 mm and relative roughness height 0.7-3.5%. They have stated that in their model, the pressure gradient was higher than that expected through classical theory. In addition, the friction factor increased by increasing Reynolds number and an early transition from laminar to turbulent flow happened at Reynolds number less than 2300. Usami et al. considered gas flow through a two-dimensional channel using a DSMC approach by changing the surface roughness distribution and Kn number. They have indicated the reduction of flow conductivity as a result of surface roughness in transition flow [9]. Karniadakis et al used a more accurate gas flow model and discovered that surface roughness can show a more significant effect with increasing the Kn number [10].
In addition, to analyze the second part of this paper's investigation, injection of Ferro-nanoparticles, it would be necessary to have a look at some invaluable experiments which have done by a number of investigators in the field of two-phase flow in micro-channels. About a century ago, Cunningham [11] derived a correction factor for Stokes drag coefficient to consider slip effects on the drag force acting on a spherical particle. He presented his conclusion in the form of (1 + AKn). Another experiment was done by Millikan who identified the linear dependency of Cunningham correction factor on the gas mean free path [12]. Knudsen and Weber investigated more rarefied flow regimes. They expressed the A parameter at Cunningham correction factor as a function of Knudsen number.

$$
\begin{aligned}
&C(Kn) = 1 + AKn \\
&A = \alpha + \beta \exp\left(\frac{-\gamma}{Kn}\right) \qquad (1) \\
&Kn = \frac{\lambda}{L}
\end{aligned}
$$

In the above equation, λ is gas mean free path and L is flow characteristic length. The constants α , β and γ are determined experimentally. Until recently, researchers have tried to propose slip correction factors based upon above equation. Moshfegh et al brought a total look at different existed conclusions in their paper [13].

All in all, Rapid Progress in the development of micro-fluidic devices has led to a big interest in developing of investigation in micro-channels and totally micro-fluidics. The various relevant applications include devices for flow control, biomedical systems, and chemical reaction systems. These devices, consisting of many micrometer-sized components such as channels, nozzles, and, valves, still need to be improved to enhance performance [6]. Also, the fluid which past these devices can be in two-phase flow in many industrial or environmental applications. Therefore, any kind of progress in the field of micro-channels besides two-phase flow can result in huge advantages in environmental, industrial and pharmaceutical issues.

2. Problem definition and CFD model

2.1. Model development

In this paper we are trying to have a combination of two aforementioned concepts, Gas-Solid two-phase flow and micro-channel. Therefore, previously reported investigation and governing equations which can include the whole geometry would be applied. Here we consider the constant-wall-temperature heat-transfer characteristics of two-dimensional channels that are sufficiently long for flow to be fully developed [2]. Actually, a two-dimensional plane duct geometry has been analyzed in order to simulate Gas-Solid two-phase flow in a microscale channel. To remove compressibility effect, inflow Reynold number respecting the characteristic length of channel has been considered less than 1. In addition, to study roughness effect, Walls have been studied in both smooth and rough situation. A schematic of two described geometries (smooth and rough) has been shown in figure 1. L is the duct length and H defines its height. Also, roughness on channel surfaces is considered as triangular peaks with height r and width w. These elements are uniformly and symmetrically distributed on the top and bottom surfaces. Although this geometry is not exactly the same as actual rough surfaces, it is deliberated as a close approximation to study the roughness effect on flow heat transfer. A parameter called "relative roughness height" is defined as $e = r/H$. In addition, another parameter called "peak density (Pd)" has been defined for the number of roughness elements in each 0.1 mm of microchannels length. In most micro-fluidic systems, the relative roughness heights are evaluated at 0.1-6% [1]. Therefore, in this study, we consider the mentioned range for analyzing the roughness effect.

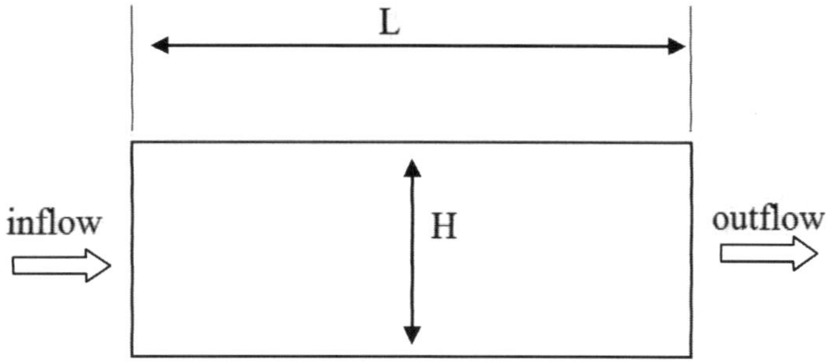

Figure 1. Microchannel with smooth walls

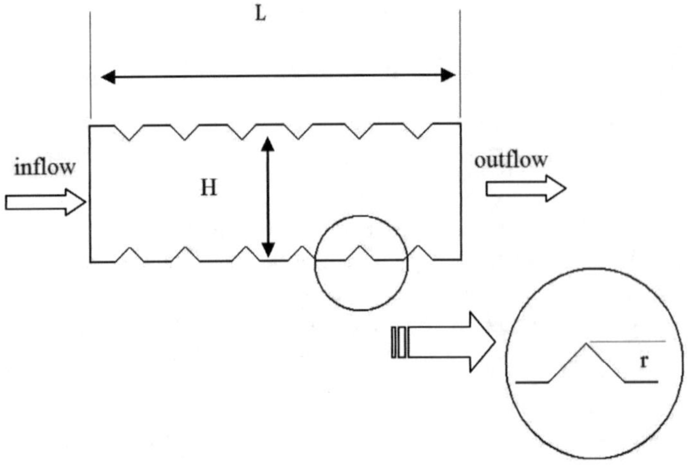

Figure 2. Microchannel with triangular roughness elements

2.2. *Governing equations*

As mentioned before, this work is a combination of two different concepts, Gas-Solid two-phase flow and microchannels. Therefore to have a better look at the whole problem, the governing equations describing the problem geometry will be presented in two separated section as below.

2.2.1. Governing equations describing single-phase micro-channel. As we know, the Navier-Stokes equation is the first order approximation to the Chapman-Enskog solution and can be extended to the slip flow region with the Maxell's slip boundary condition and Smoluchowskyi's temperature jump. In this paper flow has been considered as a steady laminar incompressible one with constant thermophysical properties. Therefore, the needed governing equations are continuity, momentum and energy that must be solved with suitable boundary conditions which will be stated in the next section. Equations (2)-(5):

$$\frac{\partial(\rho u)}{\partial x}+\frac{\partial(\rho v)}{\partial y}=0 \tag{2}$$

$$\rho u\frac{\partial u}{\partial x}+\rho v\frac{\partial u}{\partial y}=-\frac{\partial P}{\partial x}+\mu[\frac{\partial^2 u}{\partial x^2}+\frac{\partial^2 u}{\partial y^2}+\frac{1}{3}(\frac{\partial^2 u}{\partial x^2}+\frac{\partial^2 v}{\partial x\partial y})] \tag{3}$$

$$\rho u\frac{\partial v}{\partial x}+\rho v\frac{\partial v}{\partial y}=-\frac{\partial P}{\partial y}+\mu[\frac{\partial^2 v}{\partial x^2}+\frac{\partial^2 v}{\partial y^2}+\frac{1}{3}(\frac{\partial^2 v}{\partial y^2}+\frac{\partial^2 u}{\partial x\partial y})] \tag{4}$$

$$\begin{aligned}\rho C_p u\frac{\partial T}{\partial x}+\rho C_p v\frac{\partial T}{\partial y}\\ =\frac{\partial}{\partial x}(k\frac{\partial T}{\partial x})+\frac{\partial}{\partial y}(k\frac{\partial T}{\partial y})+u\frac{\partial P}{\partial x}+v\frac{\partial P}{\partial y}\\ +\mu[2(\frac{\partial u}{\partial x})^2+2(\frac{\partial v}{\partial y})^2+2(\frac{\partial u}{\partial y}+\frac{\partial v}{\partial x})^2]-\frac{2}{3}\mu(\frac{\partial u}{\partial x}+\frac{\partial v}{\partial y})^2\end{aligned} \tag{5}$$

Where u, v, T and P are gas streamwise velocity, normal velocity, temperature and pressure. Respectively, C_p, μ and k are the specific heat, viscosity, thermal conductivity of the gas that are considered as constant. The above equations can be non-dimensionalized with respect to the channel height H, inlet mean velocity u_i, inlet temperature T_i, inlet mean density ρ_i and inlet pressure P_i.

$$\frac{\partial(\rho^* U)}{\partial X} + \frac{\partial(\rho^* V)}{\partial Y} = 0 \tag{6}$$

$$\mathrm{Re}_i\,\rho^*(\xi U\frac{\partial U}{\partial X} + V\frac{\partial U}{\partial Y}) = -\frac{\xi\,\mathrm{Re}_i}{\gamma Ma_i^2}\frac{\partial P^*}{\partial X} + [\xi^2\frac{\partial^2 U}{\partial X^2} + \frac{\partial^2 U}{\partial Y^2} + \frac{1}{3}(\xi^2\frac{\partial^2 U}{\partial X^2} + \xi\frac{\partial^2 V}{\partial X\partial Y})] \tag{7}$$

$$\mathrm{Re}_i\,\rho^*(\xi U\frac{\partial V}{\partial X} + V\frac{\partial V}{\partial Y}) = -\frac{\mathrm{Re}_i}{\gamma Ma_i^2}\frac{\partial P^*}{\partial Y} + [\xi^2\frac{\partial^2 V}{\partial X^2} + \frac{\partial^2 V}{\partial Y^2} + \frac{1}{3}(\frac{\partial^2 V}{\partial Y^2} + \xi\frac{\partial^2 U}{\partial X\partial Y})] \tag{8}$$

$$\xi U\frac{\partial\theta}{\partial X} + V\frac{\partial\theta}{\partial Y} = \frac{1}{Pe_i\rho^*}[\xi^2\frac{\partial}{\partial X}(\frac{\partial\theta}{\partial X}) + \frac{\partial}{\partial Y}(\frac{\partial\theta}{\partial Y})] = Ec_i\frac{P_i}{\rho_i u_i^2\rho^*}[\xi U\frac{\partial P}{\partial X} + V\frac{\partial P}{\partial Y}] + \frac{Ec_i}{\mathrm{Re}_i\,\rho^*}[2(\frac{\partial U}{\partial X}\xi)^2 + 2(\frac{\partial V}{\partial Y})^2 + (\frac{\partial V}{\partial X}\xi + \frac{\partial U}{\partial Y})^2 - \frac{2}{3}(\xi\frac{\partial U}{\partial X} + \frac{\partial V}{\partial Y})^2] \tag{9}$$

Where the parameters for non-dimensionalizing are as belows:

$$\theta = \frac{T - T_w}{T_i - T_w},\quad \xi = \frac{H}{L},\quad X = \frac{x}{L},\quad Y = \frac{y}{H},\quad \rho^* = \frac{\rho}{\rho_i}$$
$$P^* = \frac{P}{P_i},\quad U = \frac{u}{u_i},\quad V = \frac{v}{u_i},\quad C_p^* = \frac{C_p}{C_{p_i}},\quad \alpha_i = \frac{K_i}{\rho_i C_{p_i}},\quad K^* = \frac{K}{K_i} \tag{10}$$

Also during non-dimensionalizing equations some important dimensionless number appear. Respecting above process, the dimensionless numbers will be as belows:

$$\mathrm{Re}_i = \frac{\rho_i u_i H}{\mu},\quad Ma_i = \frac{u_i}{\sqrt{\gamma R T_i}},\quad Pe_i = \frac{u_i H}{\alpha_i},\quad Ec_i = \frac{u_i^2}{C_p(T_i - T_w)},\quad Kn_i = \sqrt{\frac{\pi\gamma}{2}}\frac{Ma_i}{\mathrm{Re}_i} \tag{11}$$

2.2.2. Governing equations describing two-phase Euler-Lagrange approach. In Euler-Lagrange approach, the fluid is considered as a continuous phase while particles are dispersed in it. Therefore, the effect of particles on fluid is encountered as a source term in momentum equation and other conservative equations will be kept the same. As mentioned before, the two following equations can be considered as momentum equations.

$$\rho u\frac{\partial u}{\partial x}+\rho v\frac{\partial u}{\partial y}=-\frac{\partial P}{\partial x}+\mu[\frac{\partial^2 u}{\partial x^2}+\frac{\partial^2 u}{\partial y^2}+\frac{1}{3}(\frac{\partial^2 u}{\partial x^2}+\frac{\partial^2 v}{\partial x\partial y})] \quad (12)$$

$$\rho u\frac{\partial v}{\partial x}+\rho v\frac{\partial v}{\partial y}=-\frac{\partial P}{\partial y}+\mu[\frac{\partial^2 v}{\partial x^2}+\frac{\partial^2 v}{\partial y^2}+\frac{1}{3}(\frac{\partial^2 v}{\partial y^2}+\frac{\partial^2 u}{\partial x\partial y})] \quad (13)$$

We can write the above momentum equations in the total form of below:

$$\nabla.(\rho_f V_f V_f)=-\nabla P+\nabla.(\mu_f \nabla V_f) \quad (14)$$

Consequently, the momentum equation considering the source-term describing the effect of suspended particles on continues phase is written as below:

$$\nabla.(\rho_f V_f V_f)=-\nabla P+\nabla.(\mu_f \nabla V_f)+S_p \quad (15)$$

Where S_p in equation (15) is the source term that represents the momentum transfer between the fluid and particles and is found by computing the momentum variation of particles as they pass through the fluid phase control volume, is calculated by:

$$S_p=\sum m_p F\Delta t \quad (16)$$

Where m_p is the particle mass and F is the total force per unit particle mass acting on the dispersed phase elements and will be discussed later. Also, F (total force per unit particle mass) is defined as the rate of momentum change within the control volume.

$$F=\frac{dV_p}{dt} \quad (17)$$

Where V_p is the particle velocity.

In this paper we have considered gravity, drag and Brownian forces as acting forces on particles. So, we can describe F as below:

$$F=F_G+F_D+F_B \quad (18)$$

Where F_G is the gravity force and can be calculated by below formula:

$$F_G=\frac{(\rho_p-\rho)g}{\rho_p} \quad (19)$$

Furthermore, F_D is the drag force acting on particles and as mentioned before there are several investigations to find a suitable value for Cunningham correction factor that can be used to correct

Stokes drag coefficient for sub-micron particles. Here we will use Davis slip correction factor that can be obtained from below formula:

$$C_c = 1 + Kn\left[1.257 + 0.400\exp\left(-1.100/Kn\right)\right] \tag{20}$$

Where Kn number is one of the most important dimensionless number in microscale channels. It can be defined as below:

$$Kn = \frac{\lambda}{L} \tag{21}$$

$$\lambda = \frac{kT}{\sqrt{2}\pi\sigma^2 P} \tag{22}$$

In the equation (21) λ is molecular mean free path and L can be considered as characteristic length of channel. For two-dimensional plane duct as we have deliberated here, the hydraulic diameter of channel can be considered as channel characteristic length and it is defined as 2H. Also in the definition of molecular mean free path, k, T, σ and P are Boltzmann Constant, gas temperature, collision diameter of the molecules and gas pressure.
Respecting all aforementioned definitions, we will use the following formula in order to calculate drag force acting on particles.

$$F_D = \frac{18\mu}{d_p^2 \rho_p C_c}(V - V_p) \tag{23}$$

Where d_p, ρ_p, C_c, μ, V, V_p are particle diameter, particle density, Cunningham correction factor, fluid dynamic viscosity, fluid velocity and particle velocity.

Another force that can be considered in our analysis is Brownian force. If the size of suspended particles is almost submicron or less than that, particles motion will be affected by the discrete nature of molecular motion, exhibiting a random motion due to collisions of fluid molecules with the particles. This is called Brownian motion. For sub-micron particles, the effects of Brownian motion can be optionally included in the additional force term. The components of the Brownian force F_B are modeled as a Gaussian white noise process with spectral intensity $S_{n,ij}$ given by (24):

$$S_{n,ij} = S_0 \delta_{ij} \tag{24}$$

Where δ_{ij} is the Kronecker delta function, and

$$S_0 = \frac{216 \nu k T}{\pi^2 \rho d_p^2 (\frac{\rho_p}{\rho})^2 C_c} \tag{25}$$

T is the absolute temperature of the fluid, ν is the kinematic viscosity, and k is the Boltzmann constant. Also amplitudes of the Brownian force components are of the form:

$$F_{bi} = \zeta_i \sqrt{\frac{\pi S_0}{\Delta t}} \tag{26}$$

Where ζ_i are zero-mean, unit-variance-independent Gaussian random numbers. The amplitudes of the Brownian force components are evaluated at each time step.

2.3. Boundary condition

In this paper slip flow regime ($0.001 \le Kn \le 0.1$) between inlet and outlet of a two-dimensional microscale channel is studied. To analyze this problem slip boundary conditions have to be used besides Navier-Stokes equations. Until now there have been several models in order to examine slip regime. As an example we can state second order slip model [3], Langmuir[6], and a number of other first and second order boundary conditions. Here we will use Maxwell's model for slip velocity and Von Smoluchowski's one for temperature jump. Cartesian forms of these boundary conditions have been shown below:

$$u_{fluid} - u_{wall} = \frac{2-\sigma_v}{\sigma_v}\lambda\frac{\partial u}{\partial y} + \frac{3}{4}\frac{\mu}{\rho T}\frac{\partial T}{\partial x} \tag{27}$$

$$T_{fluid} - T_{wall} = \frac{2-\sigma_T}{\sigma_T}\frac{2\gamma}{\gamma+1}\frac{\lambda}{\Pr}\frac{\partial T}{\partial y} \tag{28}$$

Where σ_v is momentum accommodation coefficient and σ_T represents energy accommodation coefficient. Considering $\sigma_v = 1$ can make a good agreement with a number of experimental works. But all these works have done on smooth surfaces. Therefore here we will choose $\sigma_v = 0.9$ respecting Ji et al's work [3] in their study on rectangular roughness elements. Also it is necessary to consider some other boundary conditions at inlet and outlet of the channel. We can state these boundary conditions as bellow:

$$\begin{aligned} &at\ X=0 \quad U=1, \quad V=0, \quad \theta=1, \\ &at\ X=1 \quad \frac{\partial U}{\partial X}=0, \quad \frac{\partial V}{\partial X}=0, \quad \frac{\partial \theta}{\partial X}=0, \quad P^* = \frac{P_{out}}{P_i} \end{aligned} \tag{29}$$

In addition, to reduce computational work, only one half of the channel has been considered as a result of symmetrical conditions that exists at Y=H/2.

Equations (27) & (28) for slip and temperature jump apply to flat surfaces in non-rotating domains. As it has been mentioned by other investigators, neglecting the effect of surface curvature or rotating motion on slip behavior leads to false evaluation. Therefore introducing modifications for curved surfaces, following formulas have been stated:

$$u_{fluid} - u_{wall} = \frac{2-\sigma_v}{\sigma_v}\lambda\left(\frac{\partial u}{\partial n} + \frac{\partial u}{\partial t}\right) + \frac{3}{4}\frac{\mu}{\rho T}\frac{\partial T}{\partial t} \tag{30}$$

$$T_{fluid} - T_{wall} = \frac{2-\sigma_T}{\sigma_T}\frac{2\gamma}{\gamma+1}\frac{\lambda}{\Pr}\frac{\partial T}{\partial n} \tag{31}$$

Where n and t stand for normal and tangential to surface at each point of the wall.

2.4. Numerical Simulation Procedure and validation

Our simulation is based on five cases as shown in table 1. As stated before, rarefaction, compressibility and roughness play important roles in gaseous flows and will result in varied flow characteristics and heat transfer behavior. Here we have removed compressibility effect by choosing inflow Reynolds number for all cases less than 1. We have considered five sub-groups for each of our five cases. On the other hand, we have divided our five cases into five different parts respecting the inflow Reynolds number. Also for each of these cases (except of case 1 and 4) we have examined three various wall roughnesses.

Furthermore, Air has been chosen as the working gas. For all cases the inlet flow temperature is 250 K and the constant-wall-temperature has been considered as 350 K. The injection volume fraction in our different cases varies from 0 to 10% according to studies situation. Where the volume fraction has been considered as the ratio of injected nanoparticles volume to the total volume of duct. Also the influence of these values of Ferro-nanoparticles injection has been studied on heat transfer enhancement. Constant properties of Ferro-nanoparticles have been encounter as $447\, {}^{J}\!/_{KgK}$, $80.4\, {}^{W}\!/_{mK}$ and $7870\, {}^{Kg}\!/_{m^3}$ for specific heat, thermal conductivity and density. Nanoparticles diameter have been chosen as $50\,nm$ for all different cases and they are released from inflow in continuously-distributed way.

Aforementioned governing equations with suitable boundary conditions have been solved by employing the SIMPLE algorithm, a finite volume method. The entire microchannel length has been taken as the computational domain to study the effect of roughness and Ferro-nanoparticles injection on heat transfer enhancement. The coupling between fluid and nanoparticles is two-way and discrete phase elements are updated at every flow iteration. Because of triangular shapes of the roughness elements, triangular meshes have been used all over the computational domain. Also the grids are refined near the wall region to obtain highly accurate numerical solutions around the roughness elements. To evaluate the grid size effect on the accuracy of numerical solutions, grid-independence tests have been performed in the way that grid size has been made refined until acceptable differences between the last two grid sizes were found. Four different grid sizes ($480\times 64, 280\times 48, 160\times 32$ *and* 128×24) have been examined. For a model with Kn=0.0033 and relative roughness=1.25% the maximum differences in h (heat transfer coefficient) were found to be 0.75%, 1.8% and 4%. By balancing between the computation time and cost besides accuracy, the grid size 280×48 has been selected.

Table 1. Details of five cases using in this study

Cases	L/H	Re	Kn	Relative roughness (%)	Peak density	Injection volume fraction (%)
1	20	0.1, 0.3, 0.5, 0.7, 1	0.0033	Smooth	-	Without Injection, 0.5
2	20	0.1, 0.3, 0.5, 0.7, 1	0.0033	Smooth, 1.25, 2.5	10, 50	Without Injection
3	20	0.1, 0.3, 0.5, 0.7, 1	0.0033	Smooth, 1.25, 2.5	10, 50	Without Injection, 0.5
4	20	0.1, 0.3, 0.5, 0.7, 1	0.0033	1.25	10, 50	Without Injection, 0.5, 3, 10
5	20	0.1, 0.3, 0.5, 0.7, 1	0.0033, 0.01, 0.033, 0.067	Smooth, 1.25	10	Without Injection

To demonstrate the validity and precision of the model and numerical procedure, comparison with the reported numerical and experimental data has been done. It is necessary to mention it that respecting to our studies there is not any previously-reported data which can cover the effect of roughness and nanoparticles injection on heat transfer. Therefore we have used the available data about examining the influence of slip boundary condition on heat transfer of single-phase micro-channel by Hadjiconstantin and Simek [2] to almost validate our numerical results. Figure 3 shows graphical conclusions that are calculated by us in comparison with Hadjiconstantin and Simek's results. The range of numbers and process for both of them are almost the same.

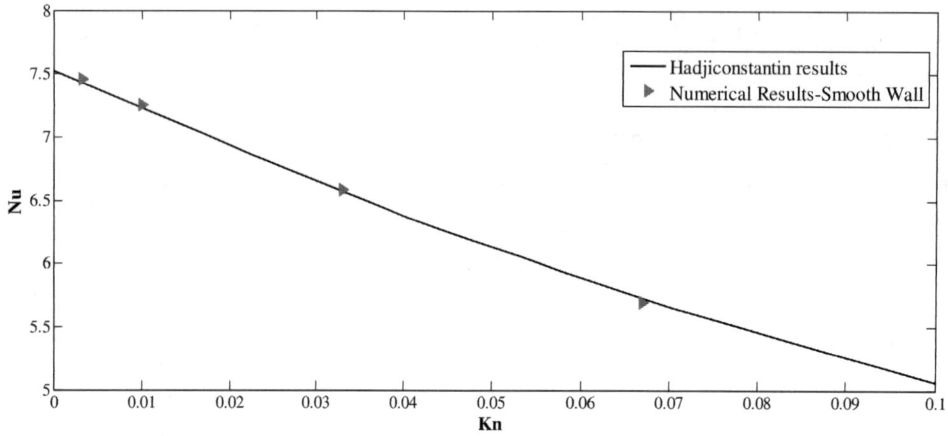

Figure 3. Variation of fully-developed Nusselt number with Knudsen number

3. Results and discussion

In this section the effect of wall roughness and Ferro-nanoparticles injection on heat transfer of rarefied, incompressible flows are investigated. In order to have a better look at influences of these parameters, the related results have been presented in four distinct parts.

3.1. Effect of Ferro-nanoparticles injection on smooth rarefied flow

To study heat transfer the most important factor is dimensionless Nusselt number. Here the most conventional definition of Nusselt number is followed as equation below:

$$Nu = \frac{q'' D_h}{(T_w - T_b)k} \tag{32}$$

Where q'', D_h, T_w, T_b, k are heat flux, hydraulic diameter, wall temperature, gas bulk temperature and thermal conductivity.

Also we can define hydraulic diameter for two-dimensional plane duct as below:

$$D_h = \frac{4(cross\ section\ area)}{(cross\ section\ perimeter)} = 2H \tag{33}$$

In addition, we calculate bulk temperature by using below equation:

$$T_b = \frac{\int_0^{H/2} \rho u T dy}{\int_0^{H/2} \rho u dy} \tag{34}$$

Where ρ is the gas density.

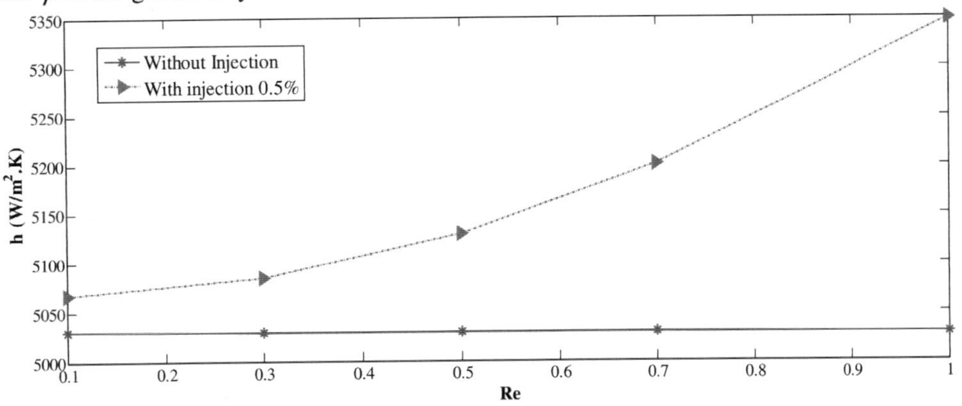

Figure 4. Variation of heat transfer coefficient with Re number for smooth-wall microchannel with and without injection at Kn=0.0033

Figure 4 Shows the heat transfer enhancement of rarefied micro-channel (Kn=0.0033) by injecting of Ferro-nanoparticles (volume fraction= 0.5%). As we are considering gas-solid two-phase flow, we have to know the equivalent thermal conductivity of this two-phase flow to calculate Nusselt number. According to our studies there is no information about equivalent thermal conductivity of micro-channels. Existed results are related to macro-channels that in this area most of the current investigations have been considering the equivalent thermal conductivity as a combination of two single phase thermal conductivity or finding more accurate results by experimental methods [14, 15]. As we could not find suitable equivalent thermal conductivity we have presented the influence of nanoparticles injection on heat transfer enhancement in the form of h (convective heat transfer coefficient). It is clear in the figure 4 that we can have improvement in heat transfer by using constant volume fraction as a result of increasing Re number. In addition we should pay attention to this fact that increasing Re number do not affect heat transfer in single-phase significantly.

Figures 5&6 show the internal temperature distribution of previously-described microchannel at Re=1. As it has shown in the figures, We can have a smoother temperature distribution by using nano-particles injection. Also as a result of heat transfer enhancement concept the gas bulk temperature has been increased at output from 289.69 to 315.064.

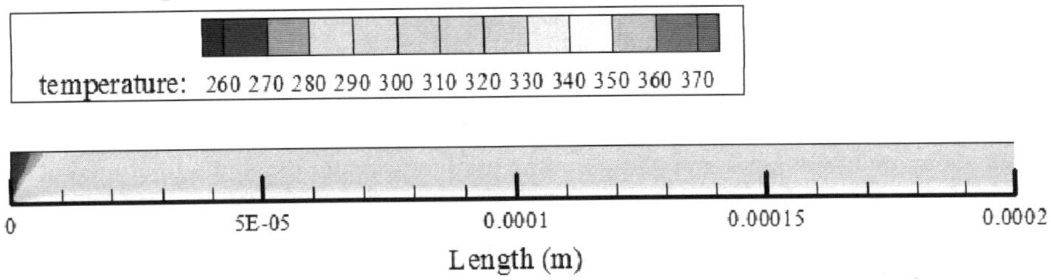

Figure 5. Distribution of microchannel interior tempreture (K) without injection at Re=1

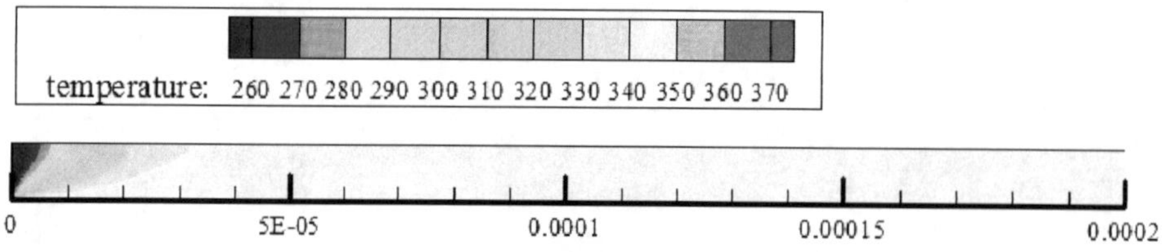

Figure 6. Distribution of microchannel interior tempreture (K) with injection (0.5%) at Re=1

Figs.7&8 show the internal relative pressure distribution of previously-described microchannel at Re=1. According to the figs, it can be clear that with small amount of injection (as an example volume fraction=0.5%) we will not have a big change in pressure drop. But still we have a bit earlier start in decreasing of the stream-wise pressure due to nano-particles injection.

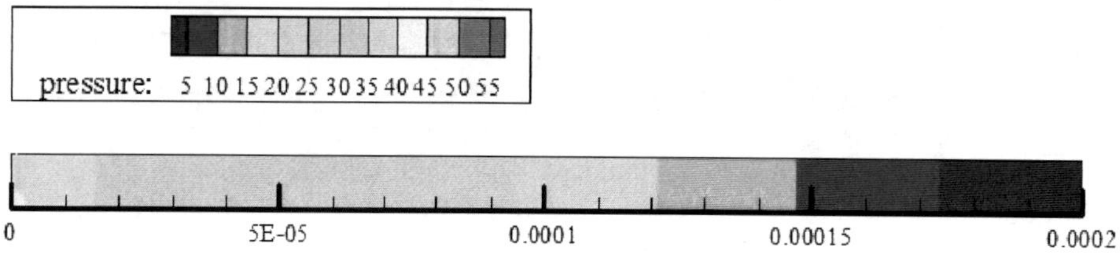

Figure 7. Distribution of microchannel interior relative pressure (Pa) without injection at Re=1

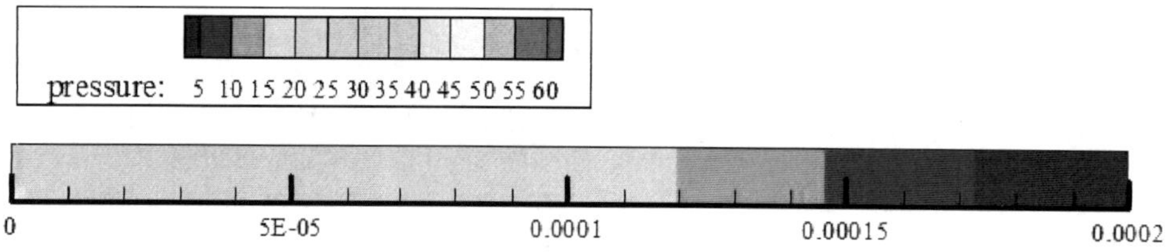

Figure 8. Distribution of microchannel interior relative pressure (Pa) with injection (0.5%) at Re=1

3.2. Effect of Ferro-nanoparticles injection on rough rarefied flow

Here we want to study wall roughness on heat transfer and then investigating combination of both roughness and Ferro-nanoparticles injection on Heat transfer. First it is better to have a conclusion about wall roughness on heat transfer. As figure 9 Shows, increasing wall roughness can improve heat transfer in an effective way. The only exception is peak density=10 and relative roughness of 1.25%. In this case we can see a reduction in heat transfer due to wall roughness. It seems logical considering the geometry of problem. Since the number of elements are not a lot. Increasing of relative roughness can make a hole-like shape at wall. Therefore this can make a number of preventive circulations inside the hole- shape. As a conclusion these probable circulations can prevent from wall heat flux penetration through flow and clear decrease in heat transfer. Figure 10 represents one of these preventive circulations. But totally we can have a conclusion which increasing wall roughness in an optimized way can improve heat transfer and an increase in numbers of wall roughness (peak density) will result in a better heat transfer enhancement in comparison with increasing of relative roughness.

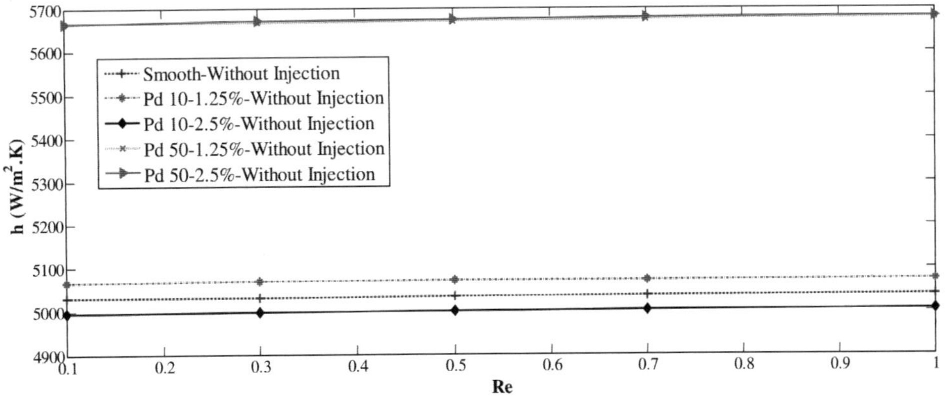

Figure 9. Variation of heat transfer coefficient with Re number as a result of roughness increase

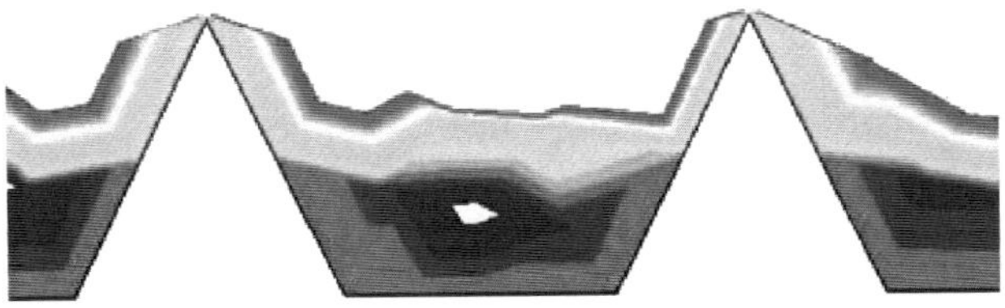

Figure 10. A sample of hole-shape roughness elements and preventive circulations at peak density=10 & relative roughness=2.5%

Now we want to examine what we have pondered in the previous section. It means that studying the influence of Ferro-nanoparticles injection on heat transfer. But the difference is studying injection effect on rough-wall cases instead of smooth-wall ones.

Figure 11 presents some information on the discussed issue. As it has been shown, like smooth-wall cases, injection of Ferro-nanoparticles can improve heat transfer. But there is an obvious difference with smooth-wall cases. Rough-wall cases show more resistance to the effect of nanaoparticles injection. It means we cannot have heat transfer enhancement related to nanoparticles injection as before. Also another issue which can absorb our attention is comparing two various cases, relative roughness of 1.25% and 2.5% besides peak density of 50. As it has been shown in the figure, for both cases without injection heat transfer coefficient is the same. But since increasing wall roughness can cause more resistance to heat transfer improvement with nanoparticles injection, case with relative roughness of 1.25% and peak density of 50 shows a better improvement as a result of injection in comparison with relative roughness of 2.5%. Therefore, finally we have a better heat transfer coefficient in the mentioned case (relative roughness=1.25% and peak density=50).

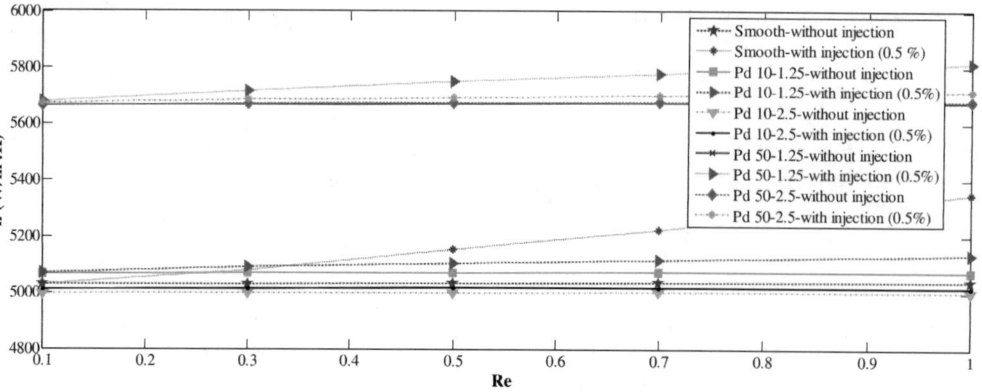

Figure 11. Effect of Ferro-nanoparticles injection on smooth and rough-wall microchannels density=10 & relative roughness=2.5%

The last studied factor in this paper is considering rarefaction effect on heat transfer. As we have stated in the section 2.4., Hadjiconstantin and Simek's results show that with increasing rarefaction effect we will have a reduction in Nusselt number and therefore in h (heat transfer coefficient). Here we want to consider roughness effect and rarefaction together. According to our studies, roughness in an optimized way can increase heat transfer coefficient although the increasing of rarefaction shows a reverse effect. It means we have an increase as a result of roughness and a decrease due to rarefaction progress. It is clear that the final result must be as a result of superposition of these two effects. According to our investigations, roughness plays a more important role by increasing rarefaction. On the other hand, increasing heat transfer because of a constant roughness will increase by rarefaction progress. Consequently, decreasing heat transfer due to increasing rarefaction will be hidden behind a huge increase in heat transfer coefficient because of roughness effect. Figure 12 shows the discussed issue clearly.

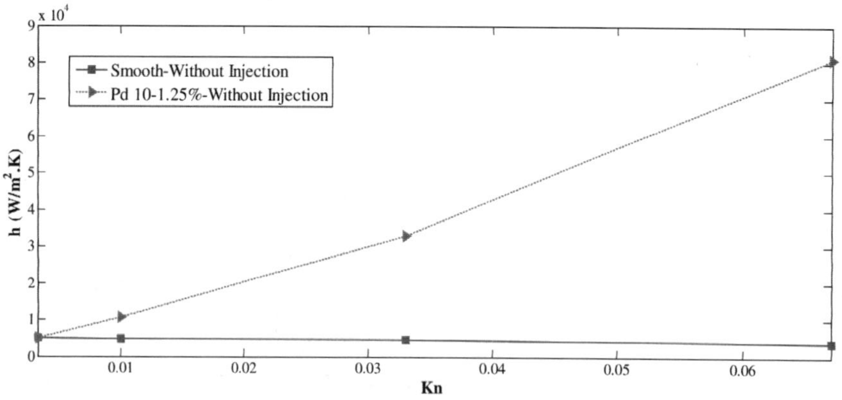

Figure 12. Effect of wall roughness and rarefaction together

4. Conclusions

The effect of wall roughness and Ferro-nanoparticles injection on heat transfer of rarefied gas flows were investigated and the following conclusions have been reached:

- Nanoparticles injection can cause an absolute increase in the heat transfer coefficient of smooth-wall cases.

- Increasing roughness in an optimized way can result in the increase of heat transfer coefficient. Although in some cases as a result of recirculation, roughness will cause a heat transfer reduction, totally it can help heat transfer.
- Rough-wall cases in comparison with smooth-wall ones show more resistance to increasing heat transfer coefficient as a result of nanoparticles injection. Therefore, we can have a better enhancement in the less roughness.
- Increasing rarefaction will cause a decrease in the heat transfer. Also, roughness will play a more important role with increasing rarefaction in the heat transfer enhancement. Therefore, we will have a huge increase in the heat transfer as a result of increasing rarefaction in a constant roughness.

References

[1] Khadem K, Shams M and Hossainpour S 2009 *Int. Communication in Heat and Mass Transfer* Numerical simulation of roughness effects on flow and heat transfer in microchannels at slip flow regime **36** 69-77

[2] Hadjiconstantinou N G and Simek O 2002 *ASME J. Heat Transfer* Constant-wall-temperature Nusselt number in micro and nano-channels **124** 356–364.

[3] Ji Y, Yuan K and Chung J N 2006 *Int. J. Heat Mass Transfer* Numerical simulation of wall roughness on gaseous flow and heat transfer in a microchannel **49** 1329–1339.

[4] Kandlikar S G, Joshi S and Tian S 2001 *Proceedings of NHTC_01 35th National Heat Transfer Conference, Anaheim, CA* Effect of channel roughness on heat transfer and fluid flow characteristics at low Reynolds numbers in small diameter tubes 1–10.

[5] Gad-el-Hak M 1999 *The Freeman Scholar Lecture, J. Fluids Eng.* The Fluid Mechanics of Microdevices **121** 5–33

[6] Hyung-il Ch, Dong-ho L and Dohyung L 2005 *Numer. Heat Transf, Part A: Appl Complex microscale flow simulations using Langmuir slip condition* ***48*** 407–425.

[7] Inman R M 1964 *NASA TN D-2213* Heat Transfer for Laminar Slip Flow of a Rarefied Gas in a Parallel Plate Channel or a Circular Tube With Uniform Wall Temperature

[8] Kavehpour H P, Faghri M and Asako Y 1997 *Numer. Heat Transfer, Part A* Effects of Compressibility and Rarefaction on Gaseous Flows in Microchannels **32** 677–696.

[9] Usami M, Fujimoto T and Kato S 1988 *Trans. Jpn. Soc. Mech. Eng.* Mass-flow reduction of rarefied flow roughness of a slit surface **54** 1042–1050.

[10] Karniadakis G E and Beskok A 2002 *Micro Flows, Fundamental and emulation (Springer, Berlin)*

[11] Cunningham E 1910 *Proc. R. Soc. Lond.* On the velocity of steady fall of spherical particles through fluid medium **83** 357–365.

[12] Millikan R A 1910 *Science* The isolation of an ion, a precision measurement of its charge, and the correction of Stokes's law **32** 436–448.

[13] Moshfegh A, Shams M, Ahmadi G and Ebrahimi R 2009 *Colloids and Surfaces A: Physicochemical and Engineering Aspects* A novel surface-slip correction for microparticles motion **345** 112–120

[14] Chon C H, Kihm K D, Lee S P and Choi S U-S 2005 *Applied Physics Letters* Empirical correlation finding the role of temperature and particle size for nanofluid (CuO) thermal conductivity enhancement **87** 1–3.

[15] Angue Minsta H, Roy G, Nguyen C T and Doucet D 2009 *Int. J. Therm. Sci.* New temperature and conductivity data for water-based nanofluids **48** 363–371.

Uniform Polynomial Equations Providing Higher-order Multi-dimensional Models in Lattice Boltzmann Theory

Jae Wan Shim[1]

KIST, 39-1 Hawolgok Seongbuk 136-791 Seoul, Korea

jae-wan.shim@polytechnique.org

Abstract. We present a set of polynomial equations that provides models of the lattice Boltzmann theory for any required level of accuracy and for any dimensional space in a general form. We explicitly derive two- and three-dimensional models applicable to describe thermal compressible flows of the level of the Navier-Stokes equations.

1. Introduction

Since Ulam and von Neumann's concept of the cellular automata, where the collective behavior of the entire system constituted of each automaton governed by a simple rule that the finite number of states of each automaton at time t_{n+1} is determined by that of its neighbors at time t_n could be sufficiently complex under certain conditions, the applications of the idea were developed in many areas [1]. In the field of fluid flows, Frisch, Hasslacher, and Pomeau [2] introduced the lattice-gas automata as a model of the Navier-Stokes equations, which led to the development of the lattice Boltzmann theory [3]. Recently, various attempts have been performed to derive higher-order accuracy models including the works of [4-6]. However, their frameworks are insufficient to derive higher-order accurate and efficient models in systematical ways.

Here, we present a set of polynomial equations in a single form. This enables us to obtain multi-dimensional models having accuracy up to any required level even beyond the Navier-Stokes equations by a systematical way for the purpose of increasing efficiency by reducing the number of discrete velocities with keeping accuracy. We explicitly derive efficient multi-dimensional models having smaller number of discrete velocities than the previously known.

2. Uniform polynomial equations

Fluid flows described by the lattice Boltzmann equation use the notion of fictitious particles moving their positions and changing their velocity distribution according to a simple rule $f_i(\mathbf{x}+\mathbf{v}_i, t+\Delta t) = (1-\omega) f_i(\mathbf{x},t) + \omega f_i^{eq}(\mathbf{x},t)$ where $f_i(\mathbf{x},t)$ is the density of particles having discrete velocities $\mathbf{v}_i$ at position $\mathbf{x}$ at time t. The density distribution $f_i^{eq}(\mathbf{x},t)$ is the equilibrium state of $f_i(\mathbf{x},t)$ and ω adjusts viscosity. For the continuous velocity space which is not in this case, the equilibrium state is the Maxwell-Boltzmann distribution $f^{eq} = \rho(\pi\theta)^{-D/2} \exp\left(-\|\mathbf{v}-\mathbf{u}\|^2\right)$ where ρ is the density, D the dimension of space , and the dimensionless variables defined by $\theta \equiv 2k_B T / m_g$,

[1] To whom any correspondence should be addressed.

$\mathbf{v} \equiv \theta^{-1/2}\mathbf{V}$, and $\mathbf{u} \equiv \theta^{-1/2}\mathbf{U}$. The symbol T is the temperature, k_B the Boltzmann constant, m_g the molecular mass, $\mathbf{V}$ the microscopic (or molecular) velocity, and $\mathbf{U}$ the macroscopic velocity that is a local average of the microscopic velocity. For the discrete velocity space, we use the following constraints to obtain $f_i^{eq}(\mathbf{x},t)$;

$$\mathbf{M}(n) \equiv \int \mathbf{v}^n f^{eq}(\mathbf{v})d\mathbf{V} = \sum_i \mathbf{v}_i{}^n f_i^{eq}(\mathbf{v}_i) \text{ for } n = 0,1,...,m \tag{1}$$

where $\mathbf{v}^n \equiv v_{x_1}^{a_1} v_{x_2}^{a_2} ... v_{x_D}^{a_D}$ and $\mathbf{v}_i{}^n \equiv v_{i,x_1}^{a_1} v_{i,x_2}^{a_2} ... v_{i,x_D}^{a_D}$ such that $a_1 + a_2 + ... + a_D = n$ and $a_\alpha \in \mathbb{N}_0$ for $\alpha = 1,2,...,D$. The subscripts x_i for $i = 1,2,...,D$ signify the coordinates in D-dimensional Cartesian coordinate system. In Formula (1), the maximum order m of the momentum $\mathbf{M}$ is one of the factors determining the accuracy of the discrete models to be obtained. The constraints satisfy the conservation of physical properties such as mass, momentum, pressure tensor, energy flux, and the change rate of the energy flux *etc.* as the order of $\mathbf{M}$ increases. If we express f^{eq} by series expansions such as the Taylor and the Hermite having the form of $f^{eq}(\mathbf{v}) \approx \exp(-v^2)P(\mathbf{v})$ where $P(\mathbf{v})$ is a polynomial in $\mathbf{v}$ and $v^2 \equiv \mathbf{v}\cdot\mathbf{v}$, we can obtain f_i^{eq} in the form of

$$f_i^{eq}(\mathbf{v}_i) = w_i P(\mathbf{v}_i) \tag{2}$$

where w_i are constant weight coefficients. With the use of Formula (2), Formula (1) is satisfied up to the order of the velocity momentum m for any $P(\mathbf{v}_i)$ of degree k, if and only if

$$\int \mathbf{v}^n \exp(-v^2)d\mathbf{v} = \sum_i w_i \mathbf{v}_i{}^n \text{ for } n = 0,1,...,m+k \tag{3}$$

The left side of Formula (3) can be evaluated by

$$\int \mathbf{v}^n \exp(-v^2)d\mathbf{v} = \begin{cases} \prod_{\alpha=1}^{D} \Gamma\left((a_\alpha+1)/2\right) & \text{for even } a_\alpha, \\ 0 & \text{otherwise} \end{cases} \tag{4}$$

where Γ is the Gaussian Gamma function which can be expressed by the double factorial as $\Gamma\left((n+1)/2\right) = \sqrt{\pi}(n-1)!!/2^{n/2}$. If we assume that the discrete equilibrium distribution f_i^{eq} is isotropic when $\mathbf{u} = \mathbf{0}$ like the Maxwell-Boltzmann distribution, Formula (3) is satisfied when n is odd. Therefore, Formula (3) can be written by

$$\sum_i w_i \mathbf{v}_i{}^n = \prod_{\alpha=1}^{D} \Gamma\left((a_\alpha+1)/2\right) \text{ for } n = 0,2,4,...,n_{\max} \le m+k \text{ and even } a_\alpha. \tag{5}$$

Note that m and k are two factors determining the accuracy of the discrete models. If we assume that we expand the Maxwell-Boltzmann distribution by the Hermite expansion for Formula (2), the degree k of $P(\mathbf{v})$ should be m at least to satisfy up to the m-th order momentum $\mathbf{M}(m)$ in Formula (1). Therefore, the weight coefficients w_i and the discrete velocities $\mathbf{v}_i$ should satisfy Formula (5) up to $n_{\max} = 2m$. Let us define $\pi(q,D)$ by the number of possible ways of representing q as a sum of natural numbers such that the number of terms are equal or less than D and we additionally define $\pi(0,D) = 1$. Then, the number of equations provided from Formula (5) up to $n_{\max} = 2m$ can be calculated by $\sum_{q=0}^{m} \pi(q,D)$.

3. Examples of the lattice Boltzmann models

For a thermal compressible flow of the level of the Navier-Stokes equations, Formula (1) should be satisfied up to $m=4$. Then, the numbers of equations for two- and three-dimensional spaces are $\sum_{\bar{n}=0}^{4}\pi(\bar{n},2)=9$ and $\sum_{\bar{n}=0}^{4}\pi(\bar{n},3)=11$, respectively. In the case of two-dimensional space, we have 9 equations to satisfy, therefore, we make models having 9 unknown variables as in Table 1. To satisfy isotropic condition, we generate discrete velocities from the representative ones by the symmetries about y-axis, x-axis, and $y=\pm x$ lines in the Cartesian coordinate system. For example, from $(1,1)$, we generate $(-1,1)$ by the symmetry about y-axis, $(1,-1)$ by the symmetry about x-axis and $(-1,-1)$ by the symmetry about $y=-x$ line. Then, we obtain the two-dimensional 33-velocities model. For three-dimensional models, we present an example, the 95-velocities model in Table 2. Note that the models derived here have smaller sets of discrete velocities than the previously known models such as the two-dimensional 37-velocities model and the three-dimensional 107-velocities model [4] without losing accuracy. Of course, we can obtain the low-order **M**(2) accuracy models such as the three-dimensional 15- and 19-velocities ones as well [7].

Table 1. Detail of the two-dimensional 33-velocities model.

group	number of discrete velocities	representative discrete velocity ($c=.819381$)	weight coefficient
1	1	$(0,0)$	.161987
2	4	$c(1,0)$	.143204
3	4	$c(1,1)$	.0338840
4	4	$c(2,0)$	.00556112
5	4	$c(2,2)$	8.44799×10^{-5}
6	4	$c(3,0)$	.00113254
7	8	$c(2,1)$	.0128169
8	4	$c(4,4)$	3.45552×10^{-6}

Table 2. Detail of the three-dimensional 95-velocities model.

group	number of discrete velocities	representative discrete velocity ($c=.421803$)	weight coefficient
1	1	$(0,0,0)$	.206847
2	6	$c(2,0,0)$	.00442257
3	12	$c(2,2,0)$	.0333341
4	8	$c(2,2,2)$	.0128902
5	6	$c(3,0,0)$	.0287920
6	12	$c(3,3,0)$	.00264319
7	8	$c(3,3,3)$	.000927908
8	24	$c(2,2,5)$	.00106078
9	12	$c(4,4,0)$	.000804376
10	6	$c(5,0,0)$	.00274697

4. Simulations

We demonstrate the accuracy and the stability of the three-dimensional 95-velocities model by the shock tube problem. The dimension of the nodes of the shock tube is $X \times Y \times Z = 11 \times 11 \times 800$. The two homogeneous initial conditions $C_L = \{\rho = p = 4, \theta = 1, \mathbf{u} = \mathbf{0}\}$ and $C_R = \{\rho = p = 1, \theta = 1, \mathbf{u} = \mathbf{0}\}$ are applied to the left ($1 \le Z \le 400$) and the right ($401 \le Z \le 800$) half spaces. The boundary conditions on the XZ-and YZ-planes are periodic and those on the left and the right XY-planes are C_L and C_R, respectively. Figure 1 shows the density and the temperature profiles along the central longitudinal axis of the shock tube. The excellent agreements of the values between the analytical solutions and the simulation results are observed on the plateaus. Note that the result of the shock tube simulation with the model having the third-order momentum accuracy shows slight mismatches with respect to the analytical solution on the plateaus [8]. The difference of the steepness on the shock front comes from the difference of the viscosity. The analytical solution of the Riemann problem, which deals with the zero viscosity flow, is steeper than the result of the simulation dealing with the viscous flow with $\omega = 1.5$.

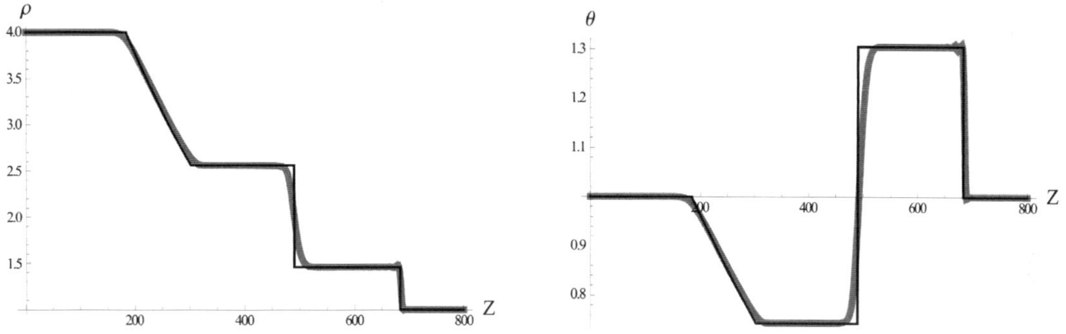

FIGURE 1. The density and the temperature profiles of the shock tube simulation are shown on the left and the right, respectively. The analytical solution (thin black line) of the Riemann problem and the simulation results of $\omega = 1.5$ (thick red line) are drawn.

5. Conclusion

We have obtained a set of polynomial equations that provides models of the lattice Boltzmann theory for any required level of accuracy and for any dimensional space in a general form. We have explicitly derived two- and three-dimensional models applicable to describe thermal compressible flows of the level of the Navier-Stokes equations. We conclude this paper with a remark that we can obtain the models having higher-order momentum accuracy for the purpose of the level of accuracy of the Burnett or the super Burnett equations from the set of polynomial equations Formula (5) expressed in a general form.

Acknowledgments

This work was partially supported by the KIST Institutional Program.

References

[1] Rothman D H and Zaleski S 1997 *Lattice-Gas Cellular Automata: Simple Models of Complex Hydrodynamics* (Cambridge: Cambridge University Press)
[2] Frisch U, Hasslacher B, and Pomeau Y 1986 *Phys. Rev. Lett.* **56**, 1505
[3] Benzi R, Succi S, and Vergassola M 1992 *Phys. Rep.* **222**, 145
[4] Philippi P C et al. 2006 *Phys. Rev. E* **73**, 056702
[5] Rubinstein R and Luo L-S 2008 *Phys. Rev. E* **77**, 036709

[6] Shim J W 2011 arXiv:1103.5184v1

[7] Qian Y H, Succi S, and Orszag S A 1995 "Recent Advances in Lattice Boltzmann Computing" in Annual Reviews of Computational Physics III, edited by D. Stauffer, Singapore: World Scientific, pp. 195-242

[8] Shim J W and Gatignol R 2011 *Phys. Rev. E* **83**, 046710

Hydraulic resistance and convective heat transfer within independent power generation micro sources (IPM) channels

Sudarev A.V.[1], Sudarev B.V., Suryaninov A.A.

«Research Center «Ceramic Engines» named after A.M. Boyko»
Polyustrovsky Av., 15, block 2, 195221, St-Petersburg, Russia

e-mail: soudarev@boykocenter.spb.ru

Abstract. The introduction of new structural materials and technologies contributes to the efficiency increase for the compact **IPMs** used in various branches of engineering. Use of a driving high-temperature (**TIT**>1600K), regenerative (the regeneration ratio is E>85%) micro gas turbine engine **μGTE**, major components which are made of structural ceramics, allows not only to maintain the effective efficiency at η_e=26-30%, but, also, sharply reduce the material consumption rate for the micro source as a whole.
Application of the laser prototyping technique to manufacture the air heater, which is a part of μGTE, increases the IPM compactness. Miniaturization of the air heater, manufactured by the structural ceramics laser fusion, can significantly reduce the hydraulic diameter ($d_h \leq 1.0$ mm) of the channels, designed to transport the working media inside it. Reducing d_h leads to a significant increase in the hydraulic resistance of the micro channels. The associated increase in the energy consumption for **μGTE**'s own needs is compensated by increasing the **TIT**, **E**, and heat transfer coefficients in micro channels, and by eliminating the need in cooling for high temperature **IPM** components.

Introduction

The continued growth in demand for compact mobile and efficient decentralized energy power systems, robotics, transportation, gas pipelines, aircraft micro hardware, energy and electronics cooling provision, etc., requires creation of independent power micro sources **(IPM).** As an electric drive for IPM, various types of heat engines are used, though only the micro gas turbine engine **(μGTE)** can dramatically increase the specific energy density, increase independency, power, and improve the environmental performance of IPM **(Fig. 1).**

However, small sizes of the electric generator **(EG)** gas turbine engine will inevitably reduce the efficiency of its basic components: compressor, turbine, combustion chamber **(CC)** and, consequently, the efficiency of IPM as a whole. It is known that the efficiency of **μGTE** working on a simple thermodynamic cycle does not exceed 3-6%, which makes the problem of improving its efficiency extremely urgent, especially for micro-engines of power N_e less than 10 kW.

[1] Corresponding author

Published under licence by IOP Publishing Ltd

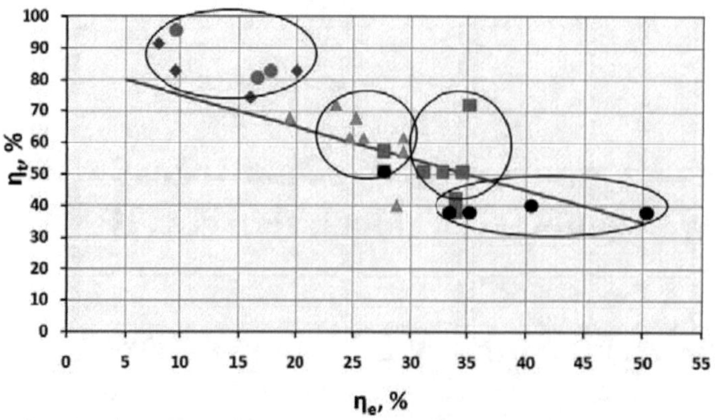

Figure 1. Effect of technologies applied on the heat and electric efficiency of the existing IPMs **[1]**.
Designations: ◆ – Stirling engines; ▲ – internal combustion technologies (ICE, GTE);
● – high temperature solid oxide fuel cell (SOFC),
■ – low temperature FC; ● – Rankine cycle engines;
—— – Line showing the total efficiency, 85%.

The most promising way to increase the efficiency of **μGTE**, as shown in **[2] (Fig. 2)**, is a combination of increasing the initial temperature (**TIT**) of the working media and using the heat recovery E in the thermodynamic cycle of the engine.

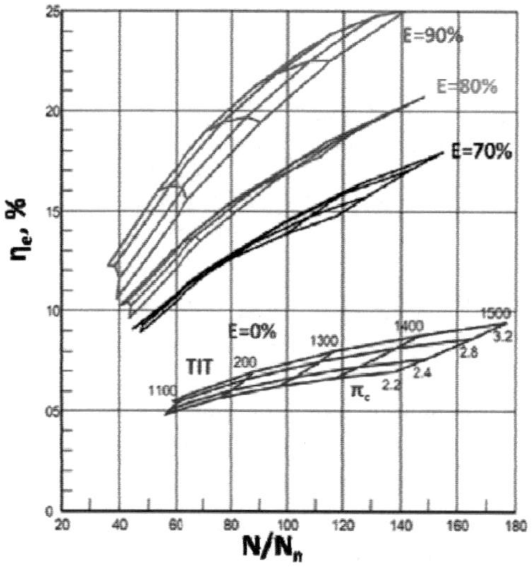

Figure 2. Effect of the initial cycle temperature, TIT, pressure ratio π_k, regeneration ratio E and loading for μGTE N/N_n on the thermal efficiency η_e of μGTE **[2]**.

The growth of the initial cycle temperature **TIT** limits the thermal strength of materials used to manufacture the basic **IPM** components.

The thermal strength increase is achieved through:

- the use of new high-temperature structural materials, including ceramics;
- the development of laser fusion technologies, suitable to manufacture IPMs of structural ceramics.

The presence of the air heater in the **IPM** not only reduces the fuel consumption, but also imposes a positive effect on its environmental performance.

The ceramics density is several times less that of steel alloys, which dramatically reduces the material consumption of the micro source. Its high thermal stability allows abandoning cooling of the high-temperature IPM parts, reducing the energy consumption for its own needs, and improving the reliability of μGTE by:

- the reduction of the thermal stress in the furnace and the firing sections in the combustion chamber,
- the improvement of the flame stability due to a "hotter" air feeding to the combustion zone,
- the reduction of the temperature gradient in all the stator parts and components of the engine, etc.

In addition, the regenerative **μGTE** allows:

- to reduce significantly the pressure ratio π_c in its cycle [**3**],
- to simplify the design of the engine compressor through increasing its compactness,
- to improve the rigidity of the turbocharger rotor through shifting its critical speeds towards the zone of speeds which are significantly higher the operating values.

Recuperative heat exchangers are now widely used as μGTE air heaters. Given their high thermal efficiency, they are characterized by a low compactness and a relatively high material consumption rate. Significant improvements in terms of the weight and size parameters of the recuperative air heaters can be achieved by a **miniaturization**, which is realized through:

- having micro channels made in their matrices with d_h<1.0 mm,
- using such a manufacturing technology as the laser fusion.

1. Hydraulic resistance at gas motion in microchannels.

Development of the air heaters with micro channels makes it necessary to establish the applicability of the "classical" equations of the thermal and hydrodynamic similarity to its heat hydraulic calculation. The careful analysis of experimental and theoretical works carried out by a number of researchers and dedicated to this problem is presented in **[4].** The authors note the existence of significant variations regarding an impact of some factors on the hydrodynamics and heat transfer during the coolant motion in the micro channels; these factors are as follows:

- the absolute diameter of the channel,
- the relative roughness ($\boldsymbol{\Delta=\delta/d_h}$) of its walls, where $\boldsymbol{\delta}$ is the absolute roughness,
- the cross-sectional shape and its relative dimensions,
- the turbulence ratio at inlet, etc.

An ambiguous and contradictory nature of the experimental results is largely due to the difference in performance of the design execution of operation areas for pilot plants, experiments conducting methodologies, and accuracy of the measurements performed.

The results of many experiments show that:

- correlations developed for the calculation of the convective heat transfer at the coolant motion in "conventional" tubes [**5**] are not fully suitable for calculation of the heat exchange in the micro channels;
- there is a need in more systematic researches to determine the mechanisms responsible for the flow structure and heat exchange variations in such channels;
- under the laminar condition in the micro channels, the friction coefficients are in good agreement with the dependencies of Hagen-Poiseuille only within the zone, where the Reynolds number does not exceed Re <600;
- at Re>600, the experimental data considered in the review [**4**] deviate from the Hagen-Poiseuille's law in the direction of higher values of the friction coefficient ξ_f;
- as in the case for the coolant flow in rough industrial tubes, the transition from the laminar to the turbulent flow in an isothermal flow in the micro channels occurs at 1,900<Re<2,500.

Therefore, some experimental data of studies over the hydraulic resistance and heat transfer at the coolant motion in the micro channels as applied to the matrix type heat exchangers, manufactured using innovative nonconventional technologies **[6]** are reported in the given paper.

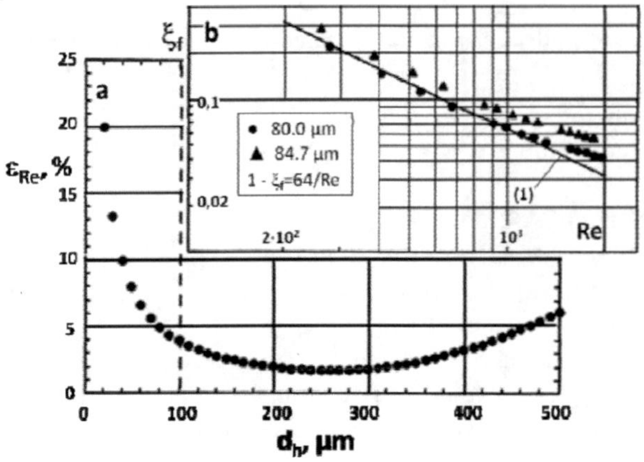

Figure 3. Effect of the absolute size of the hydraulic diameter d_h of micro channels on the relative deviation ε_{Re} of the Re number from the critical $\mathbf{Re_{c1}}$ (a) and the friction coefficient ξ_f (b) from the Hagen-Poiseuille's dependency ξ_f=64/Re (1) **[4]**.

Researchers of the transition processes taking place in micro channels indicate that with decreasing the channel diameter $\mathbf{d_h}$ it is required to change the heat exchanger design as a whole, since, to maintain constant the total pressure drop $\mathbf{\Delta P_\Sigma}$ in it at the heat output **Q**, temperature differences **ΔT** of the coolant mass flows $\mathbf{G_i}$, it would be necessary to change the number **n**, the channel length **L**, and the heat exchange surface area **F.** For to the most probable laminar flow condition, it is possible to assess potent changes in the heat exchange design using graphics from the paper by Palm and Peng **[6]** (**Fig. 4**).

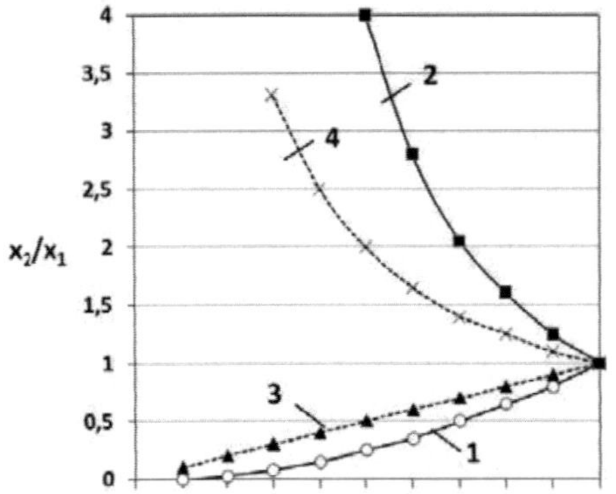

Figure 4. Effect of changes of hydraulic diameter d_h in micro channels on the structural and geometrical parameters of the heat exchanger: 1 - channel length, 2 - number of channels in the matrix; 3 – heat exchange surface, 4 – convective heat transfer coefficient in the heat exchanger **[6]**.

Along with the conclusions reported in [4], studies conducted by Ruckes [7] have shown that there is no deviation from the "classical" theory [5] in channels with the diameter d_h=0.123-2.0 mm for the isothermal air flow, and that there is no impact by the critical Reynolds numbers Re_{c1}, Re_{c2} on the absolute size of d_h.

With the compressed air flowing in the capillary channels (d_h=0.28-0.622 mm), M. Valdhagen revealed a shift of Re_{cp1} toward the lower values (Re_{c1}=1,900) **[8]**. It was established in the same paper that the air flow turbulization in the channels of a smaller diameter begins at the lower Reynolds numbers. This is consistent with a deviation of the resistance coefficient from the Hagen-Poiseuille's law established above; this law being at the same time valid for "conventional" engineering tubes with a non-uniform roughness. Results of an experimental study of isothermal flow of nitrogen in glass capillaries (d_h=0.08-1.4 mm) **[6]** serve as a confirmation of this phenomena. It is stressed upon in the study that even at Re=1,500 the coefficient of hydraulic resistance ceases to obey the Hagen-Poiseuille's law. These experiments were performed under isothermal conditions without the flow turbulization by the convective forces.

A systematic study of hydrodynamics and heat exchange was carried out at the Institute for Thermal Power Engineering, Academy of Sciences of Ukraine **[9,10]** with the compressed air flow in capillaries of stainless steel (d_h=0.39-2.0 mm) and in complicated cross-section slotted channels (slot lock joints of blades with the gas turbine disc, **Fig. 5**). The hydraulic diameters of the slotted channels are close to those of the capillary tubes studied earlier [**9,10**].

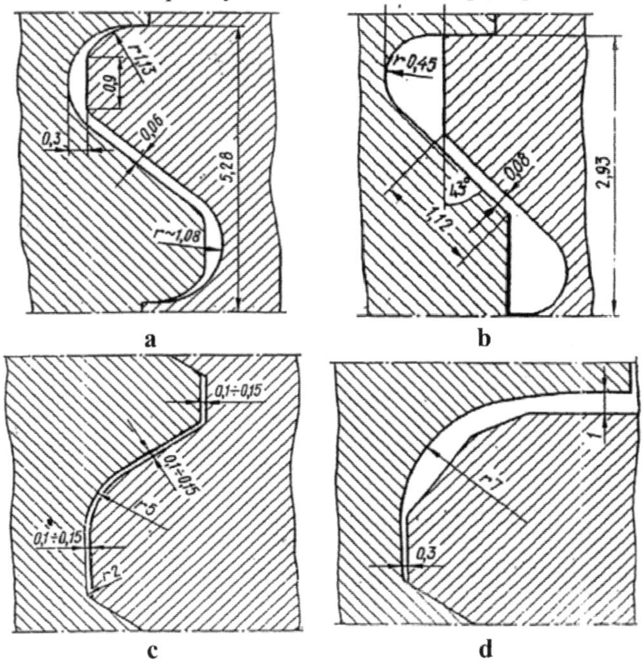

Figure 5. Micro channels of various configuration studied in **[9,10]**.
Designations: a, b - mounting clearances of herringbone shanks, c- the same for shanks of GT-6-750 TMZ (Turbomotor Works) unit type; d - cooling channels for shanks of GT-6-750 TMZ.

The heating of the capillary tubes walls in the experimental study **[9]** is electric. Experiments were carried out at relatively low temperature gradients (temperature ratio T_f/T_w=0.9), so the impact of the free convection and the direction of heat flow is small. The air flow rate did not exceed 0.15 M, where M is the Mach number, so there was no need in accounting for the air compressibility.

In all channels studied: flat gap thickness δ_s=0.18-0.05 mm, δ_s=1.0-4.0 mm at Re<2,300, the channel length averaged resistance coefficient ξ_f is described by the known Hagen-Poiseuille's relation, while at Re>2,300 it is described by the Blasius dependence (2).

$$\xi_f = 0.3164/Re^{0.25}. \qquad (2)$$

Change of ξ_f=f(Re) for the air flow in the channels with a different cross-sectional configuration and different size of d_h is shown in **Fig. 6**

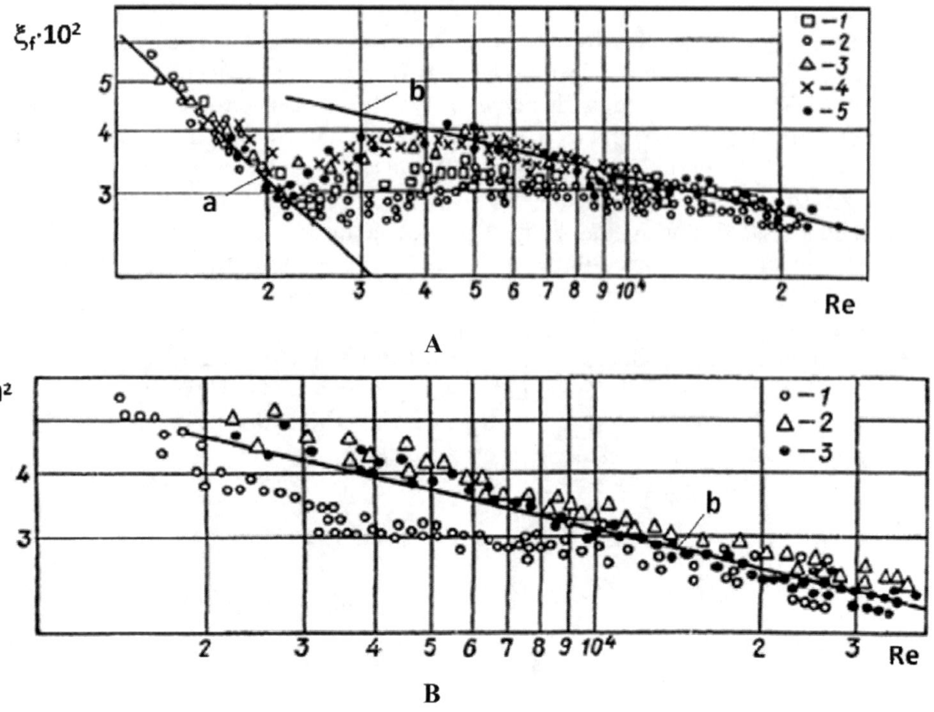

Figure 6. The friction resistance coefficient ξ_f=f(Re) with the air moving in the shank mounting clearances of rotor blades (1, 2) and in capillary tubes with diameter: d_h=0.39 mm (3); 0.51 mm (4) and 2 mm (5) for laminar, transitional and turbulent (A) conditions and under the turbulent condition for mounting clearance of rotor blades (1); capillary tubes with diameter 0.51 mm (2) and 2 mm (3) - (B) **[10].**
Designations : a – Hagen-Poiseuille's dependence (1); b – Blasius dependence (2).

It is seen that at Re<2,000, the experimental data are placed near the line "a", while with increasing the Reynolds number they are displaced in the direction of the line "b", as it is peculiar to tubes with non-uniform wall roughness. At Re=6,000, the resistance coefficient obeys the Blasius law [**11,12**] (line b).

Results of experiments with micro channels d_h=0.39 and 2.0 mm are in good agreement with the "cold" test tube where d_h=0.51 mm.

The laminar gas flow in micro channels is maintained if Re<2,000, with the channel roughness imposing no effect on the friction coefficient ξ_f, the latter determined by the Hagen-Poiseuille's law

$$\xi_f = A/Re \tag{3}$$

where A is a value dependent on the configuration and cross-sectional dimensions of the channel.

In the transition zone:

- for the first section (Re=2,000...4,000), ξ_f increases sharply with increasing the Reynolds number, regardless of the relative roughness walls:

$$\mathbf{\Delta = \delta / d_h;} \tag{4}$$

- for the second section, it obeys the Blasius law for smooth channels, which is valid in accordance with the following regularity: the greater the Δ, the smaller is the range of validity;

- for the third section, ξ_f deviates from the Blasius law (2) and with Δ increasing ξ_f increases as well.

With a heat supply to the air flow and the increased turbulence at inlet, there is a smooth "crisis-free" transition from the Hagen-Poiseuille's dependence (a) to the Blasius dependence (b) (**Fig. 7**).

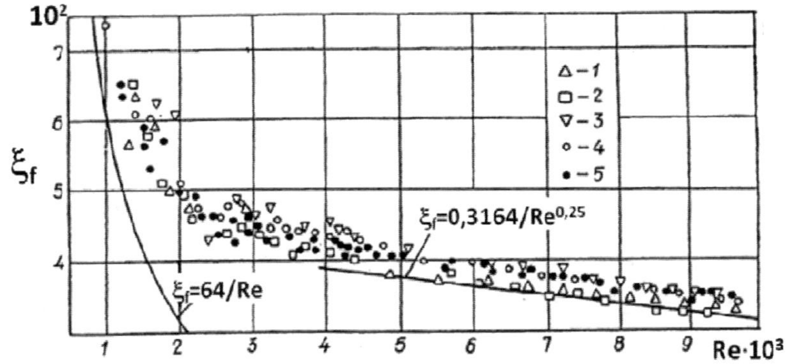

Figure 7. Effect of initial gas flow turbulence (Tu) on the ξ_f=f(Re) dependence [**10**].
Designations: 1,2 – Tu=1.0-1.5%, 3-5 – Tu=5.0-18.0%.

The surface roughness of channels, posing a significant effect on the pressure drops and heat transfer in the micro channels differs in structure and its absolute size from the roughness of "conventional" tubes, for which $d_h >> \delta_b$ (thickness of the boundary layer).

However, for the micro channels with the non-uniform roughness as well as for "conventional" tubes, the calculation of the boundary values Re_{c1} and Re_{c2} can be done by the formulas reported in the Reference Book by I.E. Idelchik [**11**], according to which:

$$Re_{c1}=1160(\delta_s/d_h)^{0.11} \text{ at } (\delta_s/d_h)>0.007; \quad (5)$$

$$Re_{c2}=2090(\delta_s/d_h)^{0.0636} \text{ (for tubes with any roughness).} \quad (6)$$

In the Reynolds number transition zones reported by Celata [**4**], the following dependences are recommended to calculate ξ_f in the micro channels:

$$Re_0<Re<Re_{c1} \; \xi_h=4.4\, Re^{-0.595}\exp(-0.00275/(\delta_s/d_h)), \text{ where } Re_0=1{,}900; \quad (7)$$

$$Re_{c1}<Re<Re_{c2} \; \xi_f=(\xi_{f2}-\xi_{f1})\exp(-(Re_{f2}-Re))^2+\xi_{f1}. \quad (8)$$

$$\text{Here: at } (\delta_s/d_h)\le 0.007 \; \xi_{f1}=0.032; \; \xi_{f2}=7.244\, Re^{-0.643}; \quad (8a)$$

$$\text{at } (\delta_s/d_h)>0.007 \; \xi_{f1}=0.0758-0.0109/(\delta_s/d_h)^{0.286}; \; \xi_{f2}=0.145//(\delta_s/d_h)^{-0.244}. \quad (8b)$$

With the self-similar turbulent condition, ξ_f in the micro channels, as in "conventional" tubes, does not depend on the Reynolds number and is determined solely by the RR value for the channel walls.

For the micro channel in the transition zone, ξ_f depends on the Re number and RR. At the transition to the turbulent condition, ξ_f gradually decreases with increasing the Re number, reaching the lowest values at the quadratic (self-similar) mode. For the laminar flow in the channel with an equivalent RR over 0.7%, a deviation from the Hagen-Poiseuille' law may occur in the direction of ξ_f increasing. The greater is RR, with a smaller Re number the deviation occurs.

On the basis of the analysis of results of the experiments carried out [**9,10,13,14**], we can conclude that there is no influence of the absolute sizes of the microchannel on the transition from the laminar to the turbulent flows, which allows to apply the criteria relations given in the reference literature **[5,11,12,15]** for the calculation of hydrodynamics in the capillary channels.

2. Convective heat transfer in micro channels

The use of lengthy ($(l/d_h)>100$) channels of small diameter leads to generation of the laminar flow condition. Here, we see a relatively low flow rates and hence small values of the complex «Pe d_h/l» (Pe $d_h/l<10$, where Pe=RePr is the Peclet number; l is the length of the channel). As a result, the Nusselt number Nu ceases to depend on this complex and tends to a constant value Nu_{min}. That is, the convective heat transfer coefficient is independent of the inertial forces (working media flow rate) **[16]**.

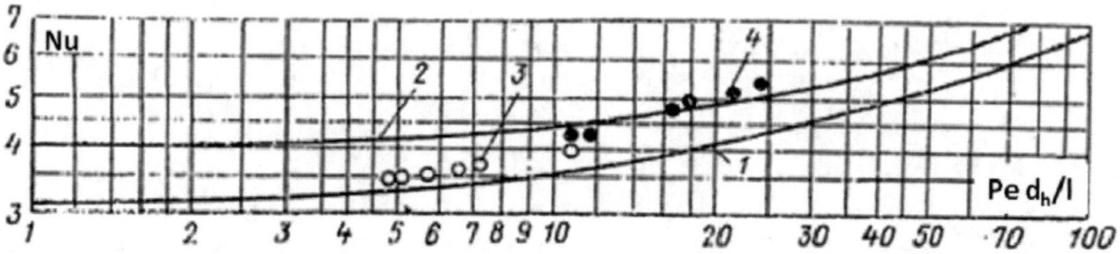

Figure 8. Heat transfer in micro channels with Pe $d_h/l<10$ for channels of rectangular cross-section with a different aspect ratio ψ=h/b (where h, b are thickness and width of channel **[16]**).
Designations: 1.3 - ψ=1; 3 is the calculation result; 2.4 - ψ=1; 2.62 is the test result [17].

Here, the number of Nu_{min} and the hydraulic resistance coefficient of friction ξ_f is significantly affected by the cross-sectional shape of the channel. For the capillary tubes of circular cross-section $Nu_{min,t}$=3.66, and the rectangular $Nu_{min,t}$=7.5-8.24, as a function of the ratio ψ=h/b. Here, $Nu_{min,t}$ is the minimum Nusselt number at the constant wall temperature t_w. With reduction of ψ, $Nu_{min,t}$ increases. The reduction of the complex "Pe d_h/l» is promoted by the Prandtl number Pr, which lowers in a gas coolant (e.g. air) and becomes less 1.0. **[16,18].** With respect to the slotted micro channel, to calculate ξ_f and $Nu_{min.t}$, you can use the following approximate formulas (9) and (10):

$$\xi_f Re=57.736\,\psi^2-95.072\,\psi+94.792; \tag{9}$$

$$Nu_{min,t}=8.7699\,\psi^2-13.524\,\psi+7.8092. \tag{10}$$

At the laminar gas flow in the micro channel (opposite to conventional channels **[19]**), the convective heat transfer is strongly affected by the Mach bumber M (**Fig. 9**) **[4].**

It is seen that the effect of the Mach number M on the heat transfer in the micro channel is already apparent at M>0.2, and a situation is likely to take place when the heat flow is directed away from the wall towards the "hot" gas, i.e. $T_{aw}> T_W$.

Here:

$$T_{aw}=T_w(1+(r(k-1)/2)M^2), \tag{11}$$

T_{aw} is the adiabatic wall temperature,
T_g is the gas stream temperature;
r is the temperature recovery coefficient at the gas motion in the channel, equal to:

$$r=Pr^{0.33}-\Delta r_x. \tag{11a}$$

$$\Delta r_x= 7.16*10^5 Re_x{}^{0.4}\, f\,(x/d_h); \tag{11b}$$

x is the characteristic size;

$$f(x/d_h)=1 \text{ at } x/d_h=0\text{-}15 \text{ (initial area);} \tag{11c}$$

$$f(x/d_h)=1+0.0413(x/d_h-15) \text{ at } x/d_h=15\text{-}27 \text{ [20].} \tag{11d}$$

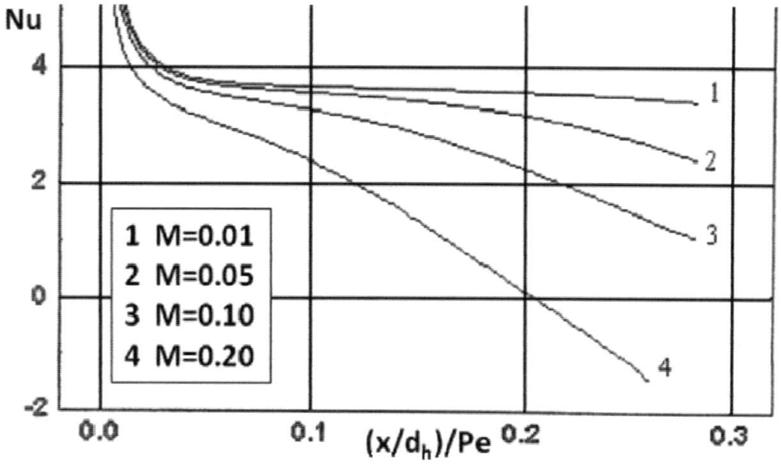

Figure 9. Mach number effect on the convective heat transfer at gas motion in micro channels [**4**].

The cross-sectional shape of micro channels is very diverse (**Fig. 10**). It has a significant effect on the rate of heat transfer in channels, especially under the laminar condition.

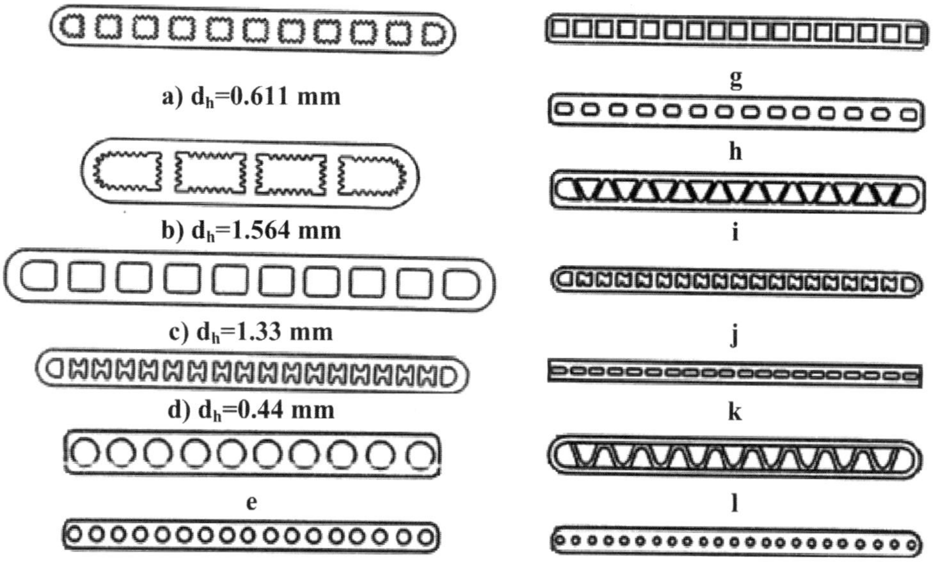

Figure 10. Applied shape and sizes of cross section of micro channels in flat multi-channel tubes used to manufacture heat exchangers.

The difference in heat transfer is due to the secondary flows generated in such channels. They lead to significant differences in the coefficients of heat transfer in comparison with a round channel of the same hydraulic diameter. In publications devoted to the study of heat transfer in micro channels (d_h<1.0 mm), it is shown that the technology of manufacturing a trial channel with a noncircular cross section causes a number of effects leading to significant differences in heat transfer compared with the round section micro tubes. Available information on the heat transfer in circular and noncircular micro channels is not enough for reliable recommendations for carrying out thermo-hydraulic calculations of compact heat exchangers with micro channels. More systematic researches are needed to consider each of the parameters that affect the transfer processes at the gas motion in micro channels. At the same time, you should not neglect the experimental information already available and summarized in the works by Palm and Peng [6] for heat transfer in noncircular micro channels with d_h=133-367 μm. The paper's authors found a significant difference in the intensity of heat transfer in the micro channels and "normal" channels.

For the laminar flow in micro channels, the experimental data for a series of parallel rectangular channels are summarized by the equations

$$\mathbf{Nu=C_1\,Re^{0.62}Pr^{0.3}} \quad \textbf{(12)}$$

$$\textbf{или } \mathbf{Nu=0.1165\,Re^{0.62}Pr^{0.33}\,(d_h/b)^{0.81}\,(h/b)^{-0.79}}, \quad \textbf{(13)}$$

which are applicable at $900\geq Re\geq 80$ and $367\geq d_h\geq 133$μm. Here, C_1 is the empirical coefficient dependent both on the hydraulic diameter d_h and the ratio of channel thickness h to its width b (**Fig. 11**).

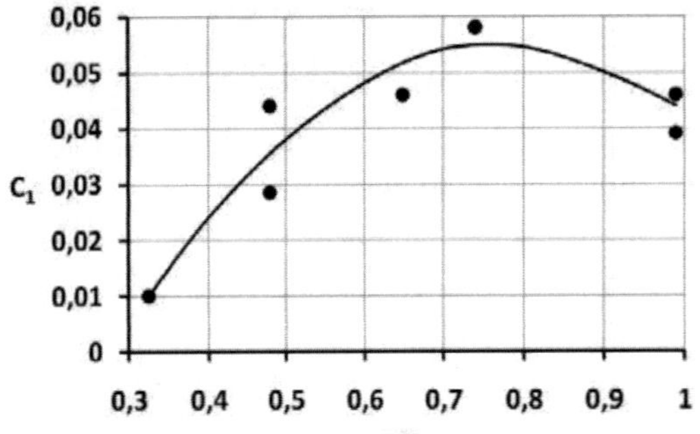

Figure 11. Effect of relative height (h/b) and d_h on convective heat transfer in micro channels [**6**].

It is evident that there is an optimal height (h/b), which minimizes the thermal resistance in the rectangular channel. In Peng's experiments [**6**], this height is h/b≈0.75.

Table 1. The geometrical parameters of investigated micro channels and empirical coefficients in the Peng's criteria equations

№	Width b, mm	Height h , mm	Length L, mm	Hydraulic diameter d_g, mm	Ratio h/b	C_1
1	0.4	0.3	50	0.343	0.75	0.058
2	0.3	0.3	50	0.3	1.0	0.0384
3	0.4	0.2	50	0.267	0.50	0.0426
4	0.3	0.2	50	0.24	0.667	0.0472
5	0.2	0.2	50	0.20	0.1,0	0.0468
6	0.3	0.1	50	0.15	0.333	0.0104
7	0.2	0.1	50	0.133	0.50	0.0285

For conditions in which Re≥1,000, Palm and Peng [**6**] suggested a correlation (13)

$$Nu=C_1Re^{0.8}Pr^{0.33} \quad (13)$$

for the turbulent flow. Unfortunately, this correlation features a considerable scatter of experimental data.

On the basis of the analysis of experimental data on heat transfer in the micro channels under the turbulent flow condition, it is established that the hydraulic diameter for the micro channels is $d_h \approx 1.2$ mm. This is the limiting value of d_h at which you can use the "classical" equations of thermal similarity [**12,15,20**] to calculate the convective heat transfer.

In N.M. Zotov's study **[21]** over the air flow in micro slotted channels (h=0.048...0.18 mm), a stratification of the experimental data was revealed, depending on the relative roughness of the channel walls regardless of a relatively small height of irregularities (δ_s not more than 1.2 μm). In some experiments, the air flow rate at the channel outlet reached the sound speed. Results of this work are consistent with the materials by Watts **[22]**, where the hydraulic resistance of friction was studied for ξ_f with nitrogen, argon and hydrogen moving in planar channels of glass and silicon of height h= 0.028...0.065 mm, with the average height of roughness bulges on the surface being less δ=0.004 mm and RR ratio Δ=0.143...0.0615. With such RR, greatly exceeding the limit value Δ_{lim}=15/Re=0.03...0.015 for any commercial-grade tubes, the stratification should take place as noted in the handbook by I.E. Idelchik [**11**].

Influence of the isotropic surface roughness on the intensity of heat transfer was examined in the experimental study by E.P. Dyban et al [**13**] at the compressed air motion (P=0.01...0.7 MPa) in micro slotted channels. Slotted channels of stainless steel and glass were employed. The absolute roughness of the channel walls was: δ_s=0.1, 1, 2, 3, 4, 7, 10 mm, while the thickness of pads that form the channel is h=0.05; 0.06; 0.09 mm. The thickness of the pads was measured within the 1 μm accuracy. Channel width was b=3.1 mm, length L=13.5; 22.6 mm. To calculate ξ_f over the average length of the slotted channel, a relationship was employed applicable to an object that has no heat exchange with the environment, with the resistance coefficient being constant along the length of the channel.

On the basis of the analysis of results of the experiment, we can conclude that there is no influence of the absolute size of the micro channels on the transition from the laminar to turbulent flows, which allows for calculation of the heat transfer in the capillary channels of the conventional criteria relations **[11,12,15,20]**.

All the capillary tubes, where the air flow was investigated, are hydraulically smooth [*23*], so no experimental data stratification under the turbulent condition was revealed. Increasing the level of the initial turbulence reduces Re_{cl} to 1,700...1,800, and practically does not change the friction resistance coefficient in the turbulent air flow. At the higher turbulence (Tu>5...7%), a crisis-free transition from the laminar to turbulent flow occurs (**Fig.** 7), and at Re=1,000...1,900 ξ_f is 20...40% higher that which corresponds to the Hagen-Poiseuille's law ξ_f=64/Re for circular cross-section tubes.

The experimental materials **[21-27]** confirm the applicability of the classical equations of similarity to calculate the convective heat transfer at the motion of gases in the micro channels, and the applicability of the formulas of Hagen-Poiseuille and Blasius for determination of ξ_f.

3. Heat exchange enhancement

To intensify the heat exchange within the micro channels, some known methods **[28]** can be used, namely the coolant flow twisting **[29,30]** and variation of the longitudinal channel configuration, the latter contributing to generation of the pressure fluctuations in the moving stream **[31-33]**.

At the gas/air motion along curvilinear channels, an uneven distribution of velocities and pressures is established in its sections causing a secondary vortex which is superimposed on the main stream. The stream lines are closed in the channel cross section.

The secondary flows consist of two streams, which are near the walls and are directed to the convex surface, while in the center of the channel they are directed to the concave surface and have a symmetrical – helical nature.

The structure of the secondary flow and the ensuing loss of energy is the function of the geometrical shape of the channel and the flow condition, defined by the Reynolds and Mach numbers. The secondary flows enhance the heat transfer in any mode of flow **[29].**

With the Dean number $De=Re\sqrt{d/D_{ben}}$=26-7,000 and (d/D_{ben})=6.2-62.5, where D_{ben} is the bending diameter, the laminar flow occurs with the macro vortexes, at which:

$$Nu=0.0575\ Re^{0.8}Pr^{0.43}\ (d/D_b)^{0.21}(Pr_f/Pr_w)^{0.25}. \tag{14}$$

The scatter of the experimental data relative to this common equation is not above ±(12-15%).

The ξ_f calculation could be carried out using the Tschukin's formula **[29]**:

$$\xi_f=0.0385(D_{ben}/d_h)^{0.5}+0.312Re^{0.25}. \tag{15}$$

For the heat transfer enhancement, longitudinally profiled tubes with the configuration of the "convergent-divergent" type are applied. According to the Mighay's data [**32**], their thermal efficiency at $Re>10^4$ is significantly higher that achieved by other methods of heat transfer improvement in tubes, which is confirmed by more recent systematic experimental studies of models of convergent-divergent and divergent-convergent channels, made by E.F. Kuznetsov and B.P. Ivanov [**33**] in a wide range of the Reynolds numbers $5.0 < Re < 2\cdot10^5$.

4. Conclusions

1. Experimental data on the mass- and heat exchange at the coolant motion in micro channels (d_h<1.0) are ambiguous and often contradictory. These data do not provide specific guidance on the thermal- hydraulic calculation of miniature heat exchangers to which matrix the similar micro channels are applied.
2. For the thermal-hydraulic calculation of miniature heat exchangers, it is much to the point to use "classical" empirical equations of thermal and hydrodynamic similarity obtained for "conventional" type tubes and channels for which $d_h >> \delta_b$.
3. When using "classical" equations of similarity with respect to the channels d_h<1.0 mm, the accuracy of the calculated data will be significantly lower those for "conventional" tubes.
4. When you develop a micro heat exchanger for μGTE, there should be borne in mind that results of the calculation of "classical" equations of similarity are to be compared with the recommendations of individual researchers; their empirical equations are placed in this paper.
5. To improve the heat transfer coefficients in both paths of the miniature heat exchanger matrices, micro channels with slotted or flat-oval cross-sectional configuration should be used.
6. Reduction of the hydraulic resistance of micro channels of compact heat exchangers for IPMs is promoted by the use of convergent-divergent type channels.
7. Twisting of the coolant flow and a rational management of variations in the longitudinal configuration of micro channels for miniature heat exchangers lead to a significant enhancement of heat transfer in them.

References

[1] Sudarev A.V., Konakov V.G., Boziev P.S., Ceramic gas turbine micro plant as a green source of the alternative c-generation power supply, paper at the XXth annual international congress «Advanced Technologies for Gas, Oil Industries, Power and Communication Systems», August 23-26, 2011, Irkutsk, 12 p.
[2] Visser V.J., Shakaryants S.A., Ostwin M., Development of 3KW micro turbine for the CHP applications, ASME Turbo Expo 2010: Power for Land, Sea and Air, GT2010 June14-18, 2010, UK, GT2010-22007, 10 p.
[3] Maslov L.A., Ship's gas turbine propulsion systems, L.:Sudostroyenie, 1973, 400 p.

[4] Celata G.P., Heat Transfer and Fluid Flow in Microchannels (Series in Thermal and Fluid Physics and Engineering) Begell House,Inc., 145 Madison Avenue, New York, NY 10016, 2004, 251 p.
[5] Demenyuk S.L. Heat exchange and hydraulic resistance in tubes and channels. St-Petersburg.: N-Ind Bureau -2012.-304 p.
[6] Palm B., Peng X.F., Single-Phase Convective Heat Transfer, Celata G.P. Heat Transfer and Fluid Flow in Microchannels (Series in Thermal and Fluid Physics and Engineering) Begell House Inc., 145 Madison Avenue, New York, NY 10016, 2004, pp. 31-52.
[7] Ruckes W., Untersuchungen uber den Ausfluss kompimierter Luft aus Kapillaren und die dabai auftretenden Turbulenzerscheinungen, Ann. d. Phys., XXV, №4, 1908, s. 983.
[8] Waldhagen M., Uber den Stromungswiederstand hochverdichteter Luft in Rohrleitung. ZUMM, III, 1923, s.181.
[9] Shvetz I.T., Dyban E.P., Study over energy exchange in mounting clearances of rotor blade shanks, Pr. ITE AN UkSSR, issue 13, 1956.
[10] Shvetz I.T., Dyban E.P., Air cooling for gas turbine rotors, Kiev, Pub. KDU, 1959, 352 p.
[11] Idelchik I.E., Reference Book for hydraulic resistances, M. Mashinostroeinie. 1975, 559 p.
[12] Bazhan P.И., Konevetz G.E., Seliverstov V.M., Reference Book for heat exchangers, M., Mashinostroenie, 1989, 366 p.
[13] Dyban E.P., Zotov N.M., Khoroshavin A.A., Resistance in micro slotted channels with rough walls, Promyshlennaya teplotekhnika, 1990, v. 12, №2, pp. 2-9.
[14] Shvetz I.T., Dyban E.P., Air cooling for gas turbine parts, Kiev, Naukova Dumka, 1974, 488 p.
[15] PTM-24.031.05-72 Methodologies and relationships for theoretical calculation of heat exchange and hydraulic resistance in heat exchange hardware of nuclear power stations. M., MTE and TM, 1972, 125 p.
[16] Mighay V.K., Increasing the efficiency of current heat exchangers, L.:LO, Energhia, 1980, 144 p.
[17] Klark S.H., Kays E.M., Transactions of the ASME, Bd.75, 1953, №5.
[18] Kutateladze S.S., Basics of heat exchange theory. 5th edition, M.: Atomizdat, 1979, 416 p.
[19] Deytch M.E., Technical gas dynamics, M., Energhia, 1974, 592 p.
[20] Isachenko V.P., Osipova V.A., Sukomel A.C., Heat transfer, M.: Energoizdat, 1981, 416 p.
[21] Zotov N.M., Experimental studies over heat exchange at the air flow in micro slotted channels, IVUS,Mashinostroenie, 1969, №8, pp. 74-76.
[22] Watts J.H., Microturbines: a new Class of Gas Turbine Engine. Global Gas Turbine News, vol. 39; 1999, №1, pp. 4-8.
[23] Wallick G.C., The Steady Rate From of Gas through Glass Capillary Tubes, Journal Petrol. Technol, Vol.11, sec.I, pp. 20-23; sec. II, p.4; 1953.
[24] Celata G.P., Gumo M., Guglielmi M., Zummo G., Experimental Investigation of Hydraulic and Single Phase Heat Transfer in 0,130 mm.
[25] Celata G.P., Guo Z.Y.,Lu Z.X., Zummo G., Heat Transfer and Fluid Flow in Microchannels (Series in Thermal and Fluid Physics and Engineering), Begell House Inc., 2004, New York, Wallingford (U.K.), pp.3-29.
[26] Bucci A.,Celata G.P., Serra E., Zummo G., Analisi Microscopica Tridimensionale Della Superficie Interna e Carratterizzazione Termoidraulica di Capillari in acciaio da 172, 290 e 520 micron, ENEA Internal Report (in Italian), 2001.
[27] Morini G.L., Lorenzini M.,Colin S., Geoffroy S., Experimental Investigation of the Compressibility Effects on the Friction Factor of Gas Flows in Microtubes, Proceedings of ICNMM2006 Fourth International Conference on Nanochannels, Microchannels and Minichannels, June 19-21, 2006, Limerick, Ireland, ICNMM2006-96112, pp. 1-8.
[28] Dreytzer G.A., Current challenges of analysis of effectiveness, design, production and operating of compact tube heat exchangers, Proceedings of the XIIIth School-Seminar

«Physical basics for experimental and mathematical simulation of gas dynamics and heat mass exchange processes in power equipment », v. 2, M.: Pub. MEI, 2001, pp.299-306.

[29] Tschukin V.K., Heat exchange and hydrodynamics for internal streams in mass forces fields, M.: Masinostroenie, 1970, 332 pp.

[30] Dementieva K.V., Teleghina I.I., Experimental study over hydrodynamics and heat exchange in curvilinear slotted section channels, Teploenerghetika 1979, №1, pp51-54.

[31] Kirpikova V.A., Orlov V.K., Prikhod'ko V.F., Development of compact heat exchange surface on the basis of introduction of the pressure non-uniformity into the flow, Teploenerghetika, 1977, №4, pp. 29-33.

[32] Mighay V.K., Simulation of heat exchange power generation equipment, L. Energoatomizdat, LO, 1987, 264 p.

[33] Ivanov B.P., Kuznetsov E.F., Convective heat transfer and hydraulic resistance in circular section channels, Turbines and Compressors, 1998, №5, pp.15-21.

Monte Carlo analysis of thermal transpiration effects in capacitance diaphragm gauges with helicoidal baffle system

M Vargas[1], M Wüest[2] and S Stefanov[1]

[1] Institute of Mechanics, Bulgarian Academy of Sciences, Acad. G. Bonchev str., bl. 4, 1113, Sofia, Bulgaria
[2] INFICON Ltd, Alte Landstrasse 6, 9496, Balzers, Principality of Liechtenstein

Email: manuel.vargas@imbm.bas.bg

Abstract. The Capacitance Diaphragm Gauge (CDG) is one of the most widely used vacuum gauges in low and middle vacuum ranges. This device consists basically of a very thin ceramic or metal diaphragm which forms one of the electrodes of a cap acitor. The pressure is determined by measuring the variation in the capacitance due to the deflection of the diaphragm caused by the pressure difference established across the membrane. In order to minimize zero drift, some CDGs are operated keeping the sensor at a higher temperature. This difference in the temperature between the sensor and the vacuum chamber makes the behaviour of the gauge non-linear due to thermal transpiration effects. This effect becomes more significant when we move from the transitional flow to the free molecular regime. Besides, CDGs may incorporate different baffle systems to avoid the condensation on the membrane or its contamination. In this work, the thermal transpiration effect on the behaviour of a rarefied gas and on the measurements in a CDG with a helicoidal baffle system is investigated by using the Direct Simulation Monte Carlo method (DSMC). The study covers the behaviour of the system under the whole range of rarefaction, from the continuum up to the free molecular limit and the results are compared with empirical results. Moreover, the influence of the boundary conditions on the thermal transpiration effects is investigated by using Maxwell boundary conditions.

1. Introduction

Capacitance Diaphragm Gauges (CDG) consist basically of a very thin ceramic or metal diaphragm which forms one of the electrodes of a cap acitor. The pressure is determined by measuring the variation in the capacitance due to the deflection of the diaphragm caused by the pressure difference established across the membrane. In order to minimize zero drift, some CDGs are operated keeping the sensor at a higher temperature [1]. This difference in the temperature between the sensor and the vacuum chamber makes the behavior of the gauge non-linear due to thermal transpiration effects. Besides, CDGs may incorporate different baffle systems to avoid the condensation on the membrane or its contamination. In figure 1 a schematic representation of a CDG with the helicoidal baffle system is shown.

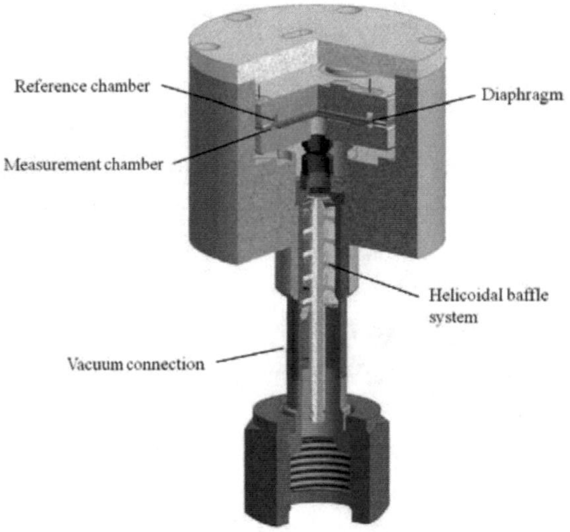

Figure 1. Capacitance diaphragm gauge as built by INFICON.

The thermal transpiration or thermal creep is a rarefaction effect which appears when there is a temperature gradient along the boundaries of a channel, which creates a flow next to the walls from the colder to the hotter side. The tube of the gauge connecting vacuum chamber and membrane is such a channel filled with gas under rarefied gas conditions. The rarefaction of the gas is defined by the Knudsen number (Kn) as the ratio of the mean free path and the characteristic size of the channel $Kn = \lambda/L$, where the mean free path for a hard sphere gas is defined as $\lambda = \left(\sqrt{2}\pi d^2 n\right)^{-1}$, d is the diameter of the molecules and n is the number density of the gas. The mean free path can be related to the viscosity of the gas μ by:

$$\lambda = \frac{\sqrt{\pi}\mu v_0}{2P} \tag{1}$$

where $v_0 = \sqrt{2k_B T/m}$ is the most probable velocity and k_B is the Boltzmann constant. In the free molecular limit the pressure ratio created by the temperature gradient along the channel is:

$$\frac{P_H}{P_C} = \left(\frac{T_H}{T_C}\right)^{0.5} \tag{2}$$

where T_H and T_C are the temperatures of both ends of the channel ($T_H > T_C$).

An excellent explanation of the physical mechanism of the thermal creep flow can be found in Sone's book [2]. A critical review of the numerical data and analytical results available was made by Sharipov and Seleznev [3]. The thermal creep flow and the thermomolecular pressure difference are described and experimental and analytical solutions are given for different configurations, rarefaction conditions and pressure and temperature drops. Several authors have performed experiments in order to measure the thermal transpiration either in capacitance manometers or in microchannels. Hobson [4] carried out the experiments under very low pressures matching the ideal limiting value in the free molecular regime when using a leached pyrex tube joining the cold and hot regions. However, the results did not match the theoretical value when the tube was smooth due to the fact that the reflection of the molecules in this case in not purely diffusive. Storvick et al. [5] measured the absolute thermal transpiration differential pressure in a capillary tube for different gases and compared their measurements with their numerical results obtained by solving the BGK kinetic model. Recently experimental measurements of the mass flow rate caused by thermal transpiration due to a temperature gradient along a microchannel have been reported in [6]. The calibration of CDGs with heated heads has been analyzed by Poulter et al. [7], studying the thermal transpiration effect in such devices and comparing their results with some empirical equations proposed by Liang [8] for different pressures and gases. Additional empirical equations relating the pressure measured at the sensor to the actual pressure for various gases can be found in [9]. These results have been widely used for calibrating CDGs for different gases to those employed in the calibration procedure. Complete descriptions on the operation of CDGs and analysis of their deviations and calibration errors have been done by Jousten [10]. Šetina suggested [11] later a universal expression for the thermal transpiration correction in CDGs as a function of normalized pressure using viscosity and molecular mass of the gas to simplify the expression. More recently Nishizawa and Hirata [12] applied the DSMC method to compute numerically the pressure distribution along the connecting tube of the gauge in the whole range of rarefaction and Iwata and Kusumoto [13] analysed by using DSMC the limits of the Takaishi-Sensui formula and suggested a rescaling factor for short pipes.

In the present work, the thermal transpiration effect on the behavior of a rarefied gas and on the measurements in a CDG with a helicoidal baffle system is investigated by using the Direct Simulation Monte Carlo method (DSMC). The study covers the behavior of the system under the whole range of rarefaction, from the continuum up to the free molecular limit for the channel confined in the double helix which acts as a baffle system to protect the diaphragm of the instrument. The results obtained are compared with the empirical Takaishi-Sensui formula and the effect of different accommodation coefficients is studied.

2. Problem formulation and numerical simulation

The traditional Direct Simulation Monte Carlo (DSMC) method proposed by Bird [14] is used for simulation of the gas flow in gauge system. This method combines deterministic aspects for modeling particle motions and statistical aspects for computing collisions between particles. The collision-sample technique that we use is the "No Time Counter" scheme suggested by Bird, except for a slight modification in the calculation of maximum collision number in a cell, as described by Stefanov et al. [15]. The simulation is carried out in three-dimensional computational domain defined by the geometry of the baffle system.

The geometry of the studied baffle system is a double helix as it is shown in figure 2. The inner radius of the helix is R_{in} = 1.7 mm, its outer radius R_{out} = 4.8 mm, the vertical distance between the ramps is Δz = 4.0 mm and the thickness ε = 1.0 mm. The helix contains two channels with rectangular cross section (width 3.1 mm and height 4.0 mm) and, as each channel corresponds to 1.5 turns, the total length of the channel is 34.1 mm.

Only one of the two channels confined within the helix is taken into consideration in the numerical calculations under the assumption that the behavior of the gas in both channels is the same because of the system symmetry. The calculations are realized in cylindrical coordinates, by using the procedure for axially symmetric flows, and the 3D-space domain is then divided into uniform cells in the radial direction r, in the axial direction z and in the polar angular direction θ. The cell size is always kept smaller than the mean free path, the time step smaller than the cell traversal time and the average number of particles in cells was never smaller than 15. The most demanding calculation which has been carried out employed 50x300x50 basic cells with approximately $2.1 \cdot 10^7$ simulators.

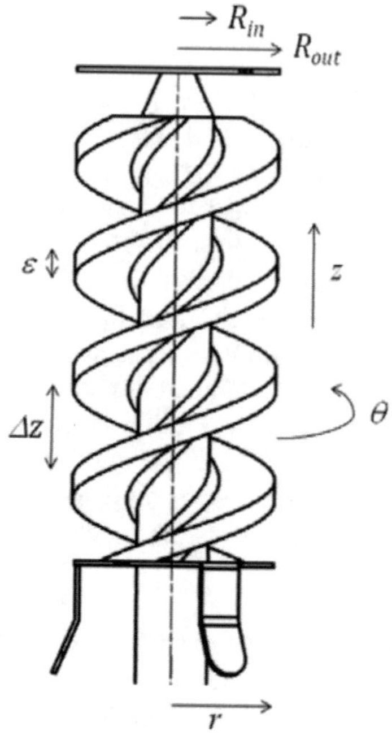

Figure 2. Scheme of the investigated helicoidal baffle system.

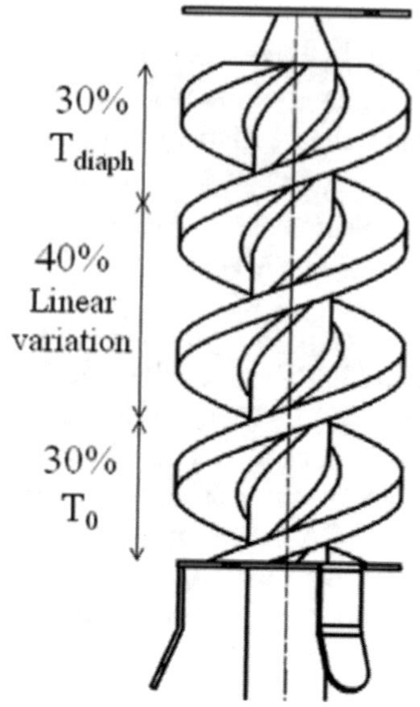

Figure 3. Temperature distribution within the helix.

The two ramps which form the lower and upper boundaries of the channel are situated at $z_1 = \frac{\Delta z + \varepsilon}{\pi}\theta$ and $z_2 = \Delta z + \frac{\Delta z + \varepsilon}{\pi}\theta$. In general, the boundary conditions are chosen to be purely diffusive at the temperature of the wall and the hard sphere model is used to compute the interaction between molecules. The simulations are carried out in a dimensionless manner and the results expressed in non-dimensional form can be applied for any gas which is simulated as a hard sphere gas. The reference quantities are taken to be n_0, T_0 and $P_0 = n_0 k_B T_0$, which define the number density, temperature and pressure at the inlet and are kept constant during the calculations. Moreover, we define the reference Knudsen number as:

$$Kn = \frac{\lambda}{R_{out} - R_{in}} \tag{3}$$

For the dimensional results, the gas used for the calculations was nitrogen, with molecular diameter and molecular mass $d = 3.75 \cdot 10^{-10}$ m and $m = 4.65 \cdot 10^{-26}$ kg respectively. The slope of the upper and lower boundaries must be taken into account when computing the reflection of the particles, due to the fact that the slope of the ramp varies depending on the radial coordinate, being larger next to the inner cylinder and smaller near the outer cylinder. This slope (m) can be determined by $m = \arctan\left(\frac{\Delta z + \varepsilon}{\pi r}\right)$ as a function of the radial coordinate for each molecules hitting any of those two boundaries. Besides, the inlet is defined as an open boundary where particles crossing it are automatically removed, while, at the same time, particles are produced at a given rate and with a certain molecular velocity depending on the pressure and temperature defined at that boundary. For this reason, all mentions to inlet properties in this work are equivalent to the properties at the vacuum chamber in the reality.

3. Steady calculations of thermal transpiration in CDGs with helicoidal baffle system

In this section the effect of the thermal transpiration on the pressure distribution along the channels at equilibrium is investigated in the whole range of rarefaction. The vacuum chamber is maintained at $T_0 = 23°\text{C}$, while the diaphragm is kept at a higher temperature of $T_{diaph} = 82°\text{C}$. This temperature difference between the inlet and the diaphragm is large enough to reduce the statistical fluctuations and hence the sample size of the calculations. The temperature is considered to vary linearly along the *z*-axis in a certain region of the helix, as defined in figure 3.

The calculations have been performed for different values of the pressure at the inlet of the helix. In figure 4 the pressure distribution inside the helix is shown for a pressure in the vacuum chamber of $P_0 = 0.01$ Pa. Under these conditions the Knudsen number takes a value of $Kn \approx 200$, which corresponds practically to free molecular regime. As it can be seen, a pressure variation along the helix appears in the steady state as a consequence of the temperature difference between the diaphragm and the vacuum chamber.

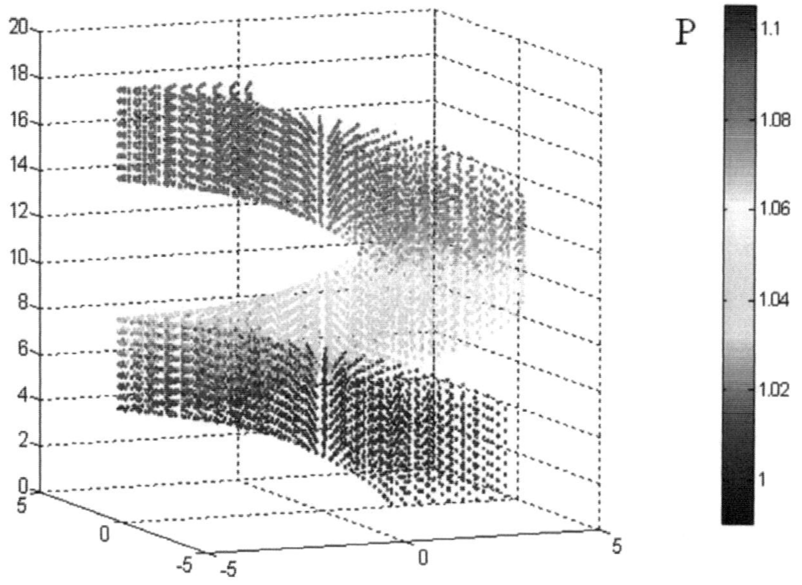

Figure 4. Pressure distribution in the helicoidal baffle system for T_{diaph} = 82°C, T_0 = 23°C and $P_0 = 10^{-2}$ Pa.

In order to obtain a better quantitative representation of the results, the distributions of the macroscopic properties along the central line of the channel, at $r = (R_{in} + R_{out})/2$ and $z = \frac{\Delta z}{2} + \frac{\Delta z + \varepsilon}{\pi}\theta$, are plotted in figure 5 for different values of the reference pressure at the inlet.

The effect of the thermal transpiration can be clearly seen. The ratio between the pressure measured at the diaphragm at equilibrium and the pressure at the vacuum chamber is increased as the reference pressure in the vacuum chamber is decreased (the Kn is increased). It can be seen that for $Kn = 208$, which defines practically free molecular conditions, the numerical results match the theoretical value of this pressure ratio due to thermal transpiration effect in the free molecular limit.

$$\frac{P_{diaph}}{P_0} = \sqrt{\frac{273+82}{273+23}} = 1.095 \tag{4}$$

This value tends to unity when the conditions are close to continuum regime. It can be seen that the temperature distributions remain unchanged for the different values of the pressure at the inlet. As for the density distributions, its value at the hotter region is always smaller than at the inlet region. This difference becomes smaller as the flow regime approaches the free molecular regime and it becomes larger in the continuum limit.

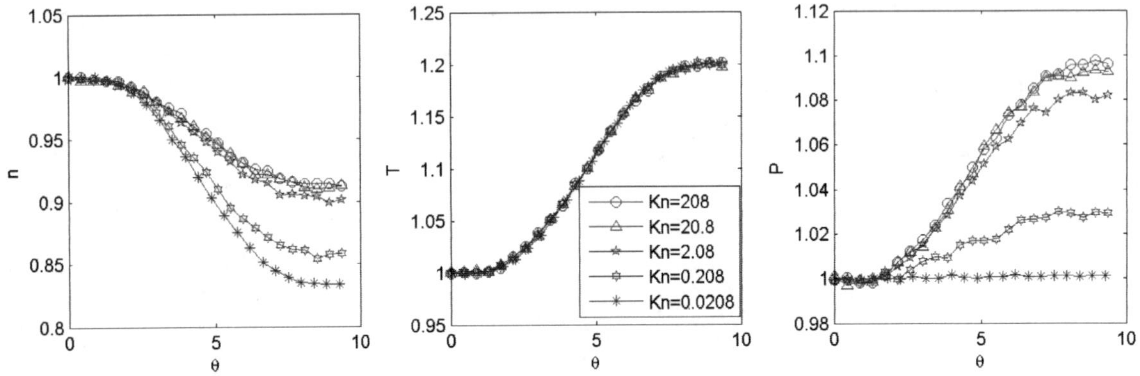

Figure 5. Density, temperature and pressure distribution in the central line of the channel defined by the helicoidal baffle system for T_{diaph} = 82°C, T_0 = 23°C and different values of the pressure at the inlet.

In figure 6 the ratio between the pressure measured at the diaphragm at equilibrium and the pressure at the vacuum chamber is represented as a function of the reference pressure. The results obtained by using DSMC are compared with those obtained by using the Takaishi-Sensui (TS) formula [9]:

$$\frac{\left(P_{diaph}/P_0\right)-1}{\sqrt{T_{diaph}/T_0}-1} = \frac{1}{A^* X^2 + B^* X + C^* \sqrt{X} + 1} \tag{5}$$

where $X = P_0 D/\overline{T}$, D is the hydraulic diameter of the channel (3.49 mm) and $\overline{T}$ is the mean temperature along the channel. The empirical constants (A^*, B^* and C^*) for nitrogen are taken from table 1 of [9]. The two curves are shown in figure 6. Although near the free molecular regime the DSMC results deviate slightly from the TS curves, in general the agreement between the results is good.

In the following part of this section, the effect of the accommodation coefficient (α) on the pressure difference created by thermal transpiration effects is investigated. The specular-diffuse reflection model is used for this purpose, where α defines the fraction of particles reflected diffusively and $(1-\alpha)$ is the fraction of particles which reflect specularly. The results for various values of the accommodation coefficient and different rarefaction conditions are shown in table 1. It can be deduced from the results that, in the free molecular regime, the value of the accommodation coefficient in the range analyzed here has no effect on the induced pressure gradient. However, in the transition and slip flow regimes, the induced pressure gradient depends slightly, but clearly, on the values of the accommodation coefficient included in this work. As shown in table 1, the pressure gradient for the same rarefaction conditions becomes larger as the accommodation coefficient is increased. It is expected that in the continuum limit the accommodation coefficient has again no effect on the pressure, because the thermal transpiration effect is negligible under those conditions.

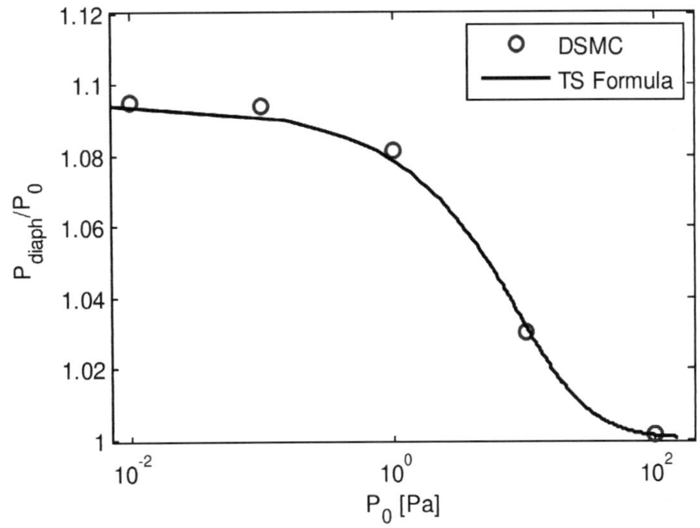

Figure 6. Variation of the ratio between the pressure at the diaphragm and at the vacuum chamber as a function of the reference pressure obtained by using DSMC (blue circles) and TS formula (black solid line) for $T_{diaph} = 82°C$, $T_0 = 23°C$.

Table 1. Pressure measured at the diaphragm for various accommodation coefficients and rarefaction conditions.

α	P_0 = 0.01 Pa (Kn =208)	P_0 = 0.1 Pa (Kn = 20.8)	P_0 = 1.0 Pa (Kn = 2.08)	P_0 = 10 Pa (Kn = 0.208)
0.7	1.0958	1.0905	1.0754	1.0217
0.8	1.0963	1.0937	1.0772	1.0277
0.9	1.0959	1.0955	1.0806	1.0294
1	1.0952	1.0941	1.0812	1.0302

Finally, it is interesting to have an estimation of the computational cost needed for solving the problem under various conditions. As mentioned above, the minimum number of particles per cell which was used was 15 in order to obtain accurate results. This minimum number of particles occurs in the region near the inner cylinder, while the number of particles per cell becomes larger in cells close to the outer cylinder. The required CPU time needed to obtain a sample size of at least 150000 particles is given in table 2 for two different cases: the first near free molecular conditions and the second close to the continuum regime. The calculations were run in a single core with processor Intel(R) Core(TM) i7 CPU 860 at 2.80 GHz. The time which is shown in table 2 is the time needed to obtain the sample size described above, without taking into account the transient time until the steady state solution is achieved, which is of the same order of magnitude.

Table 2. CPU time for the indicated conditions.

P_0 [Pa]	Cells	Number of particles	Time steps	CPU time [h]
0.01	25x200x25	$3.45{\cdot}10^6$	10000	4.52
100	50x300x50	$21.1{\cdot}10^6$	10000	47.8

4. Conclusions

In order to improve the stability of the CDGs, these devices operate usually keeping the sensor at higher temperature. In this work, the thermal transpiration effect caused by this temperature difference between the sensor and the vacuum chamber on the behavior of a r arefied gas and on the measurements in a CDG with a helicoidal baffle system has been investigated by using the Direct Simulation Monte Carlo method (DSMC). The ratio between the pressure measured at the diaphragm and the pressure at the vacuum chamber has been obtained under the whole range of rarefaction, from the continuum up to the free molecular limit. The results have been compared with those obtained from the empirical Takaishi-Sensui formula showing good agreement. The effect of the accommodation coefficient on the induced pressure gradient was investigated by using the Maxwell boundary conditions model.

Acknowledgements

The research leading to these results has received funding from the European Community's Seventh Framework Programme (ITN - FP7/2007-2013) under grant agreement n° 215504.

References

[1] Jousten K 2008 *Handbook of Vacuum Technology* (Wiley-VCH)
[2] Sone Y 2007 *Molecular Gas Dynamics: Theory, Techniques and Applications* (Birkhäuser, Boston)
[3] Sharipov F, Seleznev V 1998 *J. Phys. Chem. Ref. Data* **27**(3) 657-706
[4] Hobson J P 1969 *J. Vac. Sci. Technol.* **6** 257-259
[5] Storvick T S and Park H S 1978 *J. Vac. Sci. Technol.* **15**(6) 1844-52
[6] Rojas Cardenas M, Graur I, Perrier P and Meolans J G 2011 *Phys. Fluids* **23**(3) 031702
[7] Poulter K F, Rodgers M J, Nash P J, Thompson T J and Perkin M P 1983 *Vacuum* **33** 311-16
[8] Liang S C 1951 *J. Appl. Phys.* **22** 148-153
[9] Takaishi T and Sensui Y 1963 *Trans. Farad. Soc.* **59** 2503-2514
[10] Jousten K 1998 *Vacuum* **49** 81-87
[11] Setina J 1999 *Metrologia* **36** 623-626
[12] Nishizawa S I and Hirata M 2002 *Vacuum* **67** 301-306
[13] Iwata K and Kusumoto Y 2009 *Vacuum* **84** 729-733
[14] Bird G A 1994 *Molecular Gas Dynamics and the Direct Simulation of Gas Flows* (Clarendon, Oxford)
[15] Stefanov S, Gospodinov P and Cercignani C 1998 *Phys. Fluids* **10** 289-300
[16] Sharipov F 1996 *J. Vac. Sci. Technol. A* **14**(4) 2627-35

DSMC simulation of the gas flow through a bend and a short microchannel

N K Kulakarni[1] and S K Stefanov
Institute of Mechanics-BAS, Acad. G. Bonchev str., bl.4, Sofia 1113, Bulgaria

E-mail: naveen.kulakarni@imbm.bas.bg, stefanov@imbm.bas.bg

Abstract. The present work is related to the study of the gas flows through micro-channels with 90^0 bend, and short straight channels. The direct simulation Monte Carlo (DSMC) method has been used to study the flow. The DSMC simulations of the gas flow through micro-channels of real dimensions, which are used in the laboratory or in industry, are computationally very expensive, requiring large amounts of the computer resources and time. This fact places a limit on the applicability of the DSMC method while considering the real channel dimensions. It is a good practice to approximate the three dimensional micro-channel into two dimensions, and to study the flow characteristics, when the channel width is essentially large in comparison with their heights. However, the simulation of gas flows in more complicated channels with rectangular cross-section having comparable height and width unavoidably requires a 3D consideration. In the present paper, the 3D flow through a bend and short straight channels are presented, and the flow characteristics studied using the DSMC method. The complete 3D simulations were carried out for the case the height H at 19.83e-6m considering to the same as the one used in the laboratory (without taking similarity of the flow into consideration for the reduction of the actual dimensions of the channel). However, the other dimensions (length and width) were chosen different from the real channel due to the computational requirements. Two types of pressure boundary treatment have been implemented in three dimensions using the DSMC method

1. Introduction

Micro-channel is the one of the basic elements of many micromechanical systems; in recent years, the gas flows through straight and uniform micro-channels have been extensively studied. However, the most applications involve more complex geometries including elements such as channel junctions, bends, sudden expansions and contractions, etc. [15, 16]. The numerical modelling plays an important role in understanding the gas flow behaviour inside the micro-channel systems. The dimensions of the micro-channel systems are such that it is necessary to consider rarefaction, which often requires an application of both continuum and kinetic numerical approaches. Particularly, this is important for micro scale gas flows in micro devices with a complex geometry, which are, in general, non-equilibrium and cannot be well described by using only equations and models of the classical fluid dynamic theory [17]. Typical examples for such flows are the gas flows through different micro-junction combinations [4, 5, 13, 14].

[1] To whom any correspondence should be addressed.

Several researchers have studied [3-6, 19] the gas flow through the 90^0 bend micro-channel, at relatively high pressures, of the order of atmospheric pressure, and a very little work has been done for a lower pressure. It is a usual practice to approximate a three dimensional micro-channel into a 2D one assuming the flow to be uniform in the third direction. However, this approximation will be valid when the sidewalls are far apart, and do not influence the flow in the middle of the channel. In the present study a three dimensional flow is simulated.

One of the main objectives of this paper is a numerical study of the behaviour of the gas flows [3-6, 19] through 90^0 bend micro-channel using the DSMC method [1, 2, 9-12] applying a pressure difference between the inlet and outlet of the channel. To simulate such flows a given pressure is applied at the open ends of the micro-channel.

Two types of approaches have been considered for maintaining a given pressure at the pressure boundaries [8], and the flow evolution through the micro-channel has been computed for various Knudsen numbers in a range from 0.043 to 0.087. The simulations are carried out in three dimensions considering short micro-channels and 90^0 bend micro-channels (for the sake of simplicity hereafter referred to as bend micro-channel). The height of micro-channel is taken to be the same as the one that were used in the laboratory (UP/IUSTI) [20]. However, the computational time and memory requirements have placed a limit on choosing the other dimensions (length and width) to be the same as the real micro-channel. The results of the DSMC based code are first compared for validation with the finite volume method SIMPLE-TS [7] results provided by Shterev K and Stefanov S, allowing for a successful assessment of the accuracy of the results. The details of the flow structure near the corner are analyzed for some selected cases, and a comparison of the pressure boundary treatment considering the straight channel is presented. The investigation has a preliminary character.

2. Problem formulation

We consider a pressure driven isothermal monatomic rarefied gas flow through the micro-channel with 90^0 bend. The Knudsen number (*Kn*), which is the ratio of the mean free path to the characteristic length [1, 2], determines the rarefaction of the gas: $Kn = \frac{\lambda}{H}$, λ is the gas mean free path and H is characteristic length (the minimum of the distances between the opposite side walls). The DSMC simulations were carried out for a steady state of nearly isothermal rarefied gas flow through the micro-channels with 90^0 bend, and straight channels with different aspect ratios 1/36, and 1/60. Macroscopic field variables have been studied. All the simulations were carried out in three dimensions. In the present study is related to the flows of pressures about 1000Pa to 5000Pa. The particles were treated using the hard sphere model.

In defining Knudsen number, the mean free path is defined as $\lambda = \frac{1}{\sqrt{2}\pi d^2 n}$, where d is the diameter of the molecule, and n is the reference number density. In the paper all macroscopic variables are given in non-dimensional form by using the scales: mean free path of gas molecules for length, thermal velocity $V_{th} = (2RT_w)^{0.5}$ for velocity, and T_w for temperature and the time values are scaled by the ratio of mean free path and molecular thermal velocity λ / V_{th}.

Two types of boundary conditions were considered for maintaining the pressure at the inlet and outlets. The first is by injecting the model particles at a constant rate with a semi-Maxwellian distribution of velocity with zero bulk velocity [18, 19], and the second approach as described below under the subsection approach 2. Several methods of the pressure boundary treatments for the DSMC simulation of microchannel pressure driven flow are discussed in the literature [1, 6, 8, 18]. In the present work, two types of approaches have been implemented corresponding to a given pressure at the open boundaries.

2.1. Approach 1

At the open boundaries, the particles are injected according to a 'semi-Maxwellian velocity distribution with zero bulk velocity for particle flux at a constant' as [18, 1]:

$$\frac{\dot{N}}{A} = n\,V_{mp}\,\frac{\{\exp(-q^2)+\sqrt{\pi}\,q[1+\mathrm{erf}(q)]\}}{2\sqrt{\pi}} \tag{1}$$

Where, $q = \frac{V\cos\theta}{V_{mp}}$, V_{mp} is the most probable velocity, A is the area of an element. A particle would be removed from the list of particles whenever it is supposed to leave the computational domain via open boundaries.

2.2. Approach 2

In this approach the treatment of the upstream boundary is different from the downstream boundary. The first Monte Carlo (MC) sweep in approach 2 is the same as approach 1, and for all subsequent MC sweeps the rate of the incoming particles is calculated using equaion1 using an updated speed of the flow.

2.2.1. Upstream boundary. It is based on the first-order extrapolation to determine the inlet mean velocity from the flow inside the computational domain [21, 8]. At the upstream boundary, the pressure p_{in} and temperature T_{in} are the given parameters to the flow and the number density n_{in} could be obtained from the equation of state as following:

$$n_{in} = \frac{p_{in}}{k\,T_{in}} \tag{2}$$

The velocity components are updated after every MC sweep as:

$$(u_{in})_{\mathrm{m}} = u_m \tag{3}$$

$$(v_{in})_{\mathrm{m}} = v_m \tag{4}$$

$$(w_{in})_{\mathrm{m}} = w_m \tag{5}$$

Where the subscript m denotes average cell values and the subscript '*in*' signifies the inlet boundary.

2.2.2. Downstream boundary. At the downstream boundary, the approach is similar to the one developed by Wang and Li [8], based on the given flow parameter, pressure p_e at the exit. The other mean properties of the flow are to be determined from the preceding calculations, using the following equations

$$(\rho_e)_{\mathrm{m}} = \rho_m + \frac{p_e - p_m}{(a_m)^2} \tag{6}$$

$$(u_e)_{\mathrm{m}} = u_m - \frac{p_e - p_m}{\rho_m\,a_m} \tag{7}$$

$$(v_e)_{\mathrm{m}} = v_m \tag{8}$$

$$(w_e)_{\mathrm{m}} = w_m \tag{9}$$

$$T_e = \frac{p_e}{(\rho_e)_m \, \mathrm{R}} \tag{10}$$

a_m is the local sound speed, and ρ is the density. The equations 6,7, 8, and 10 are similar to the one developed by Wang and Li [8].

The first approach of treatment of boundaries with a 'semi-Maxwellian velocity distribution with zero bulk velocity for incoming particle flux at a constant rate' has been one of the simplest for implementation. However, the flow properties predicted by it are less accurate at the boundaries when used with the short channels, however, for an aspect ratio of 1/60 and of lower value they give an acceptable result as shown in figures 1, 2 and 3.

3. Validation: comparison with the SIMPLE-TS

First the DSMC simulations were performed using the approach 1, maintaining the pressure by injecting the model particles at a constant rate with a semi-Maxwellian distribution of velocity with zero bulk velocity at both the inlet and outlet. The DSMC simulations were carried out for a case of straight channel for an outlet pressure of 5025.547 Pa, and for inlet pressure of 15784.7 Pa (chosen from an experimental measurement) giving rise to the ratio of pressures as 3.1409 [20]. The corresponding outlet mean free path was used (by K Shterev et al) in the finite volume based SIMPLE-TS calculations, for an inlet to an outlet pressure ratio of 3. The DSMC simulations were carried out with the temperature of the walls and I/O boundaries at 297.15 K.

Figure 1.a shows the comparison of the pressure profile along the channel and along the normal directions in the middle section, for an aspect ratio of 1/60, the slightly higher value of the pressure predicted by the DSMC could be due to the higher pressure applied at the inlet. Figure 1.b shows a zoom of the comparison of the pressures obtained from both the methods near the inlet.

Figure 2 shows the comparison of the *u*, the x- component of velocity along the centreline of the channel along the flow direction. The value predicted by approach 1 for the pressure boundary treatment slightly differs from the SIMPLE-TS calculations near the outlet. It falls short to predict the same level of gain in the velocity component *u* as that of the SIMPLE-TS by some observable an amount. However, a more careful investigation is required to quantize this deviation.

Figure 3 shows the temperature profiles obtained using the DSMC method for an aspect ratio 1/60, applying semi-Maxwellian boundary condition with zero bulk velocity. The temperature was largely the same along the channel.

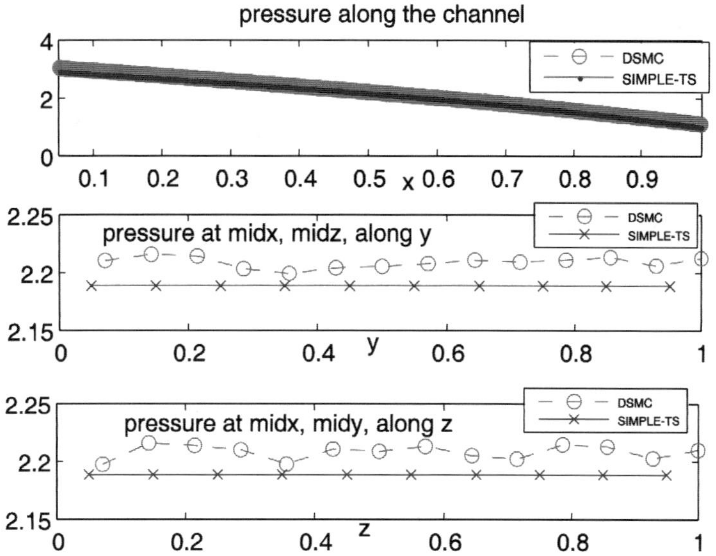

Figure1.a: A three dimensional DSMC based code validation with finite volume SIMPLE-TS method with pressure boundary with semi Maxwellian velocity distribution with zero bulk velocity, for straight channel of aspect ratio 1/60, comparison of pressure profiles.

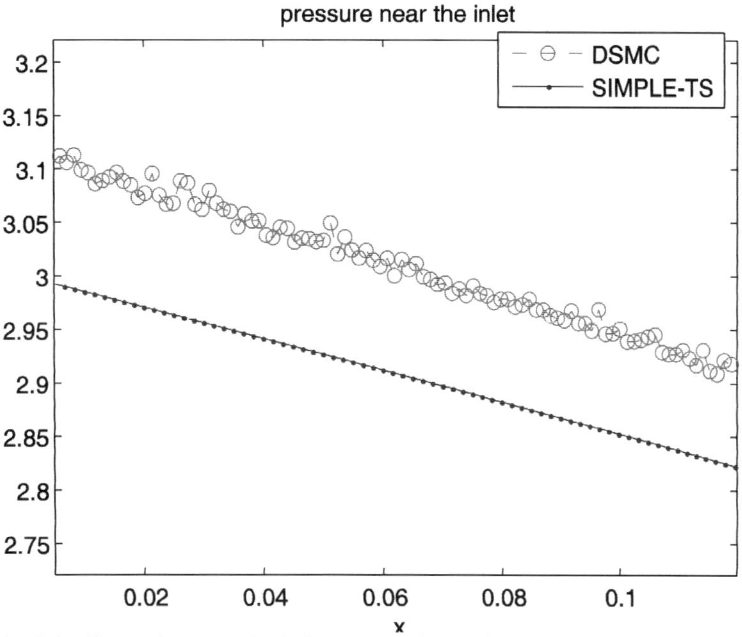

Figure1.b: A zoom of the figure 1.a near the inlet, comparison of the non-dimensional pressures along the channel, obtained from three dimensional DSMC simulations with finite volume SIMPLE-TS method; DSMC results using boundaries with semi Maxwellian velocity distribution with zero bulk velocity, for straight channel of aspect ratio 1/60

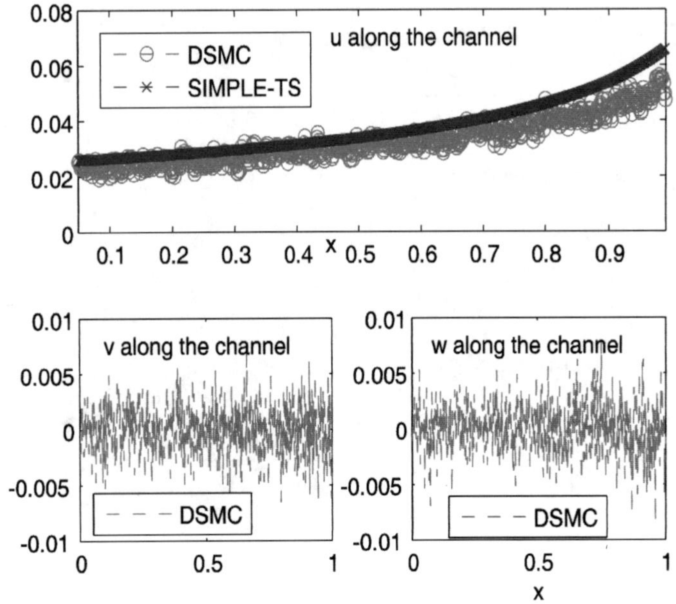

Figure2: comparison of centerline u, the x-component of velocity obtained using DSMC and SIMPLE-TS methods, and v, w (y and z components of velocity) for straight channel of aspect ratio 1/60, using semi-Maxwellian boundary condition with zero bulk velocity.

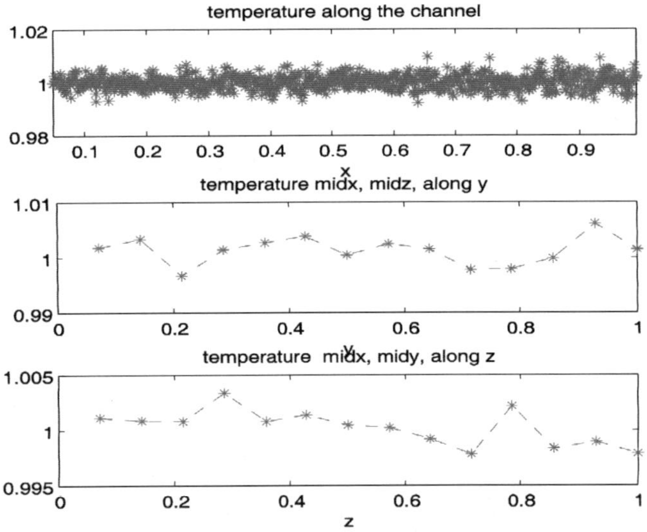

Figure3: Temperature profiles obtained using the DSMC method for aspect ratio 1/60, semi-Maxwellian boundary condition with zero bulk velocity

4. Application of approach 1 to the bend micro-channel

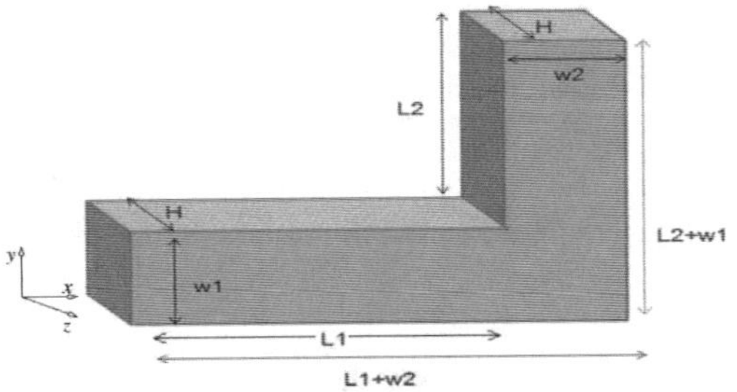

Figure 4: A three dimensional bend micro-channel

We apply the approach 1 (a semi-Maxwellian distribution of velocity with zero bulk velocity for injection of particles at both the inlet and outlet, and injecting the particles at a constant rate) to a 90 degree bend micro-channel with the length of each leg of the channel equal to the height 19.83e-6m and also the width equal to the height. In this case the centreline distance between the inlet and outlet is 3 times the height of the micro-channel. The pressure applied at the outlet p_{out} is equal to 1e4 Pa and the pressure applied at the inlet p_{in} is 3e4 Pa.

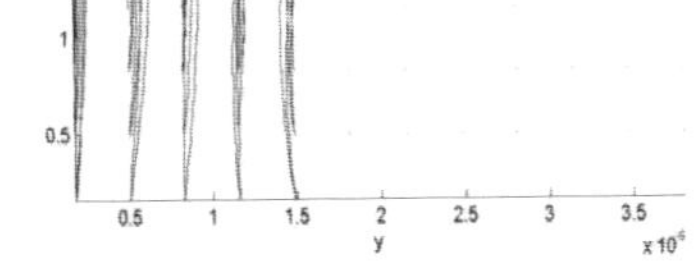

Figure 5.a: Streamlines of the gas flow through a sharp 90 degrees bend , for p_{out} = 1e4Pa, p=3, H=19.83e-6m

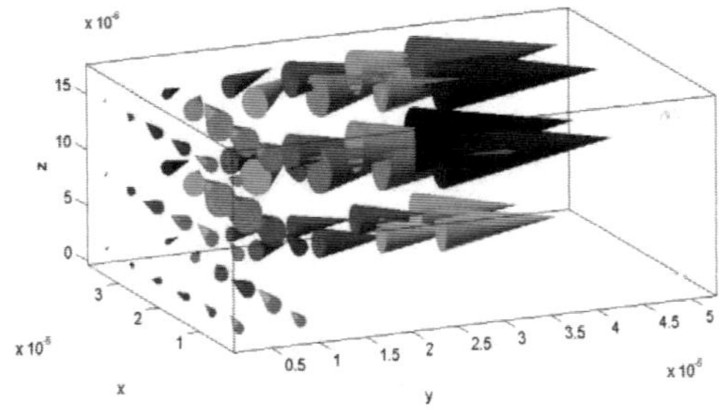

Figure 5.b: Cone plot of the gas flow through sharp 90 degrees bend , for p_{out} =1e4Pa, p=3, H=19.83e-6m

The figure 5.a shows the streamlines obtained from the gas flow through the bend micro-channel, applying outlet pressure 1e4 Pa, and inlet pressure 3e4 Pa, using the height of the micro-channel at 19.83e-6 m and with the centreline length of 3 times of the height, and width equal to the height. The stream is branching out means the flow is expanding and is compressible. Figure 5.b shows the corresponding cone plot. It also shows that the flow has the highest value of speed at the outlet.

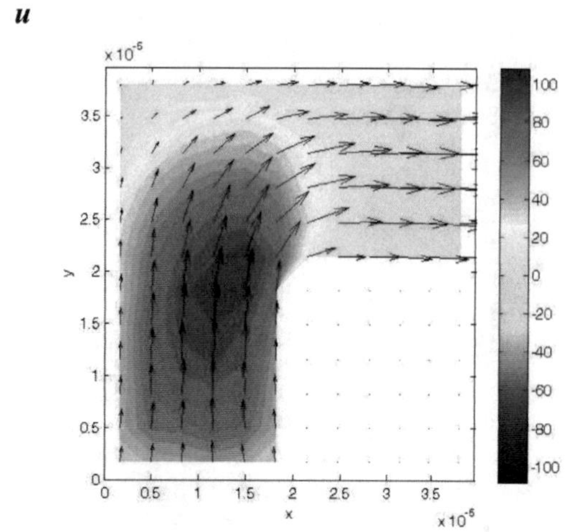

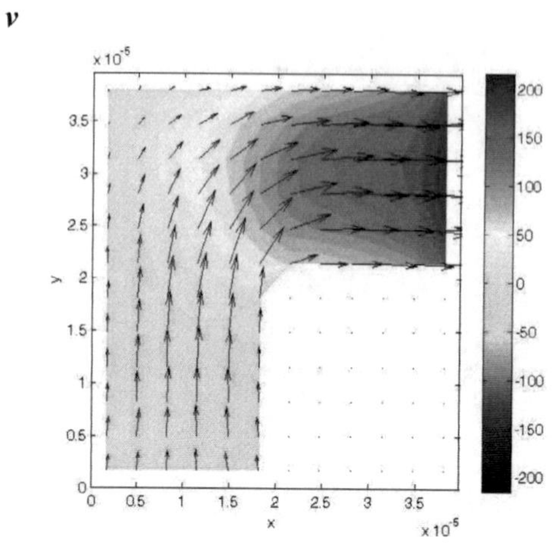

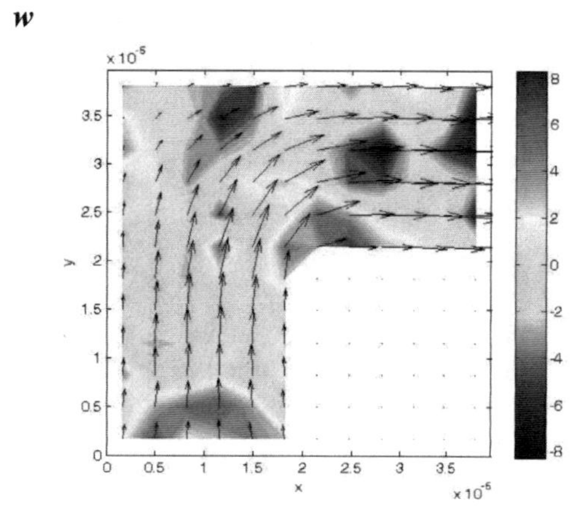

Figure 6: *u, v, w* (x, y and z components of velocity) for the gas flow through sharp 90 degrees bend micro-channel, at mid section along the z direction;

For p_{out} =1e4Pa, $p_{in} : p_{out}$ =3,

H=19.83e-6m, the colour bar shows values in m/sec.

Figure 6 shows the surface plots of *u, v, w* (x, y and z components of velocity) for the gas flow through bend micro-channel, on the middle section along the z direction. It can be seen that the flow velocity is the highest at the outlet, which in the present case is about 0.6 mach. The surface plots for *u* and *v* show that the flow speeds up along the channel.

5. DSMC results using the approach 2 for the boundary conditions

The following are the results from the DSMC simulation using the approach 2 for a pressure boundary condition at the inlet and outlet.

Figure 7.a shows *u, v, w* (x, y and z components of velocity), for the gas flow through straight channel of aspect ratio 1/36 for a ratio of inlet to outlet pressures 3.1409, there is an improvement of in the prediction of the velocity component *u* at the outlet. The other components of velocity *v*, and *w* are (fluctuating) around zero, along the channel as expected. The temperature (figure 7.b) was largely the same along the channel, but for some cooling near the outlet.

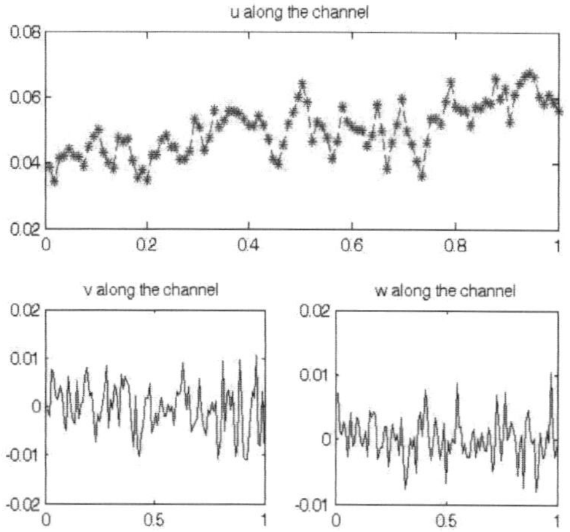

Figure 7.a: *u, v, w* (x, y and z components of velocity) along the channel, for the gas flow through straight channel of aspect ratio 1/36 for a ratio of inlet to outlet pressures 3.1409

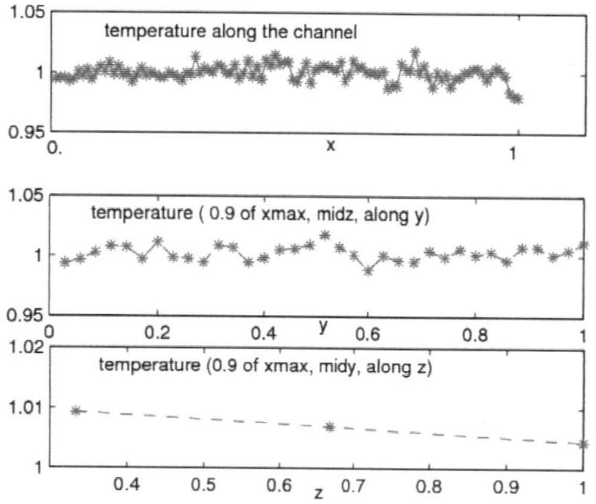

Figure 7.b: temperature along the channel, and normal to the flow direction at 90 percent of the channel length along both the normal directions, for the gas flow through straight channel of aspect ratio 1/36 for a ratio of inlet to outlet pressures 3.1409

Figure 8.a shows the *u* (x components of velocity) along the channel, for the gas flow through straight channel of aspect ratio 1/36 for a fixed ratio of inlet to outlet pressures 2.4, for various conditions of outlet *Kn*, the various *Kn* were obtained by varying the pressure at the outlet. The figure 8.a shows that the flow velocity changes with the (outlet/inlet) rarefaction (or *Kn*) and it also shows the percentage gain in the velocity is along the channel is almost the same for the range of *Kn* considered 0.043÷0.087. The figure 8.a shows that the flow velocity has a maximum around the *Kn* 0.0553 in the range of *Kn* considered.

Figure 8.b shows that the non-dimensional pressure along channel obtained, for the gas flow through straight channel of aspect ratio 1/36 for a fixed ratio of inlet to outlet pressures 2.4, for various conditions of outlet *Kn*, the pressure profile is similar for the range of *Kn* considered.

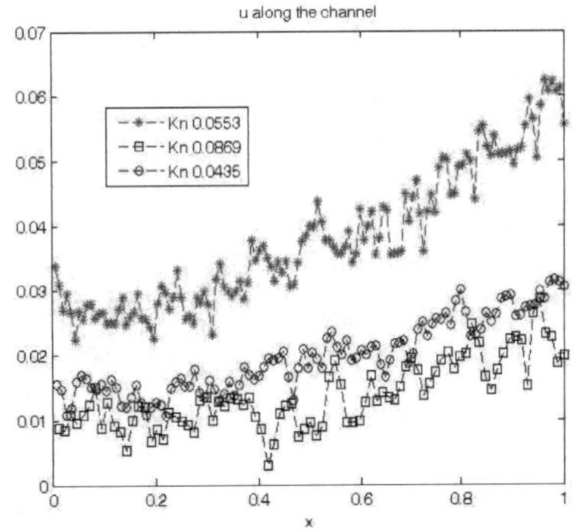

Figure 8.a: *u* along the channel, for the gas flow through straight channel of aspect ratio 1/36 for a fixed ratio of inlet to outlet pressures 2.4, for various conditions of outlet.

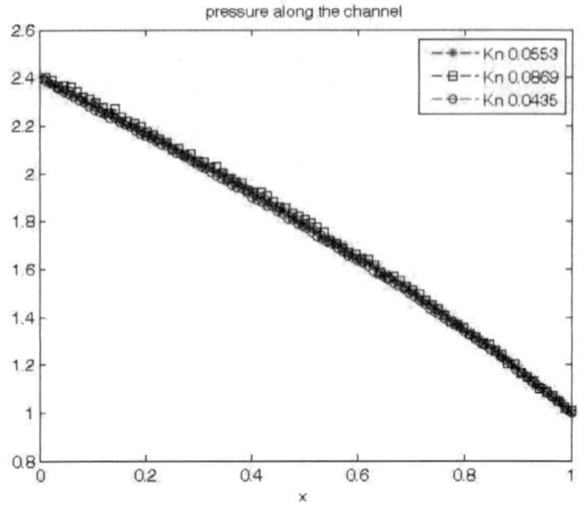

Figure 8.b: centerline pressure along the channel, for the gas flow through straight channel of aspect ratio 1/36 for a fixed ratio of inlet to outlet pressures 2.4, for various conditions of outlet.

6. Conclusions

DSMC simulations were carried out in three dimensions considering two basic elements: straight channel and 90 degrees bend micro-channel. The simulations were carried out for straight channel for various aspect ratios and for different inlet and outlet conditions.

A comparison of the results from the DSMC simulation with the results from the finite volume method SIMPLE-TS was performed, considering a short micro-channel of aspect ratio 1/60.

Two types of pressure boundary treatments were implemented using DSMC method in three dimensions. The first approach was using a 'semi-Maxwellian velocity distribution with zero bulk velocity for the incoming particle flux' with particles injected at a constant rate and the second approach was based on the mean properties of the incoming flow being determined from the preceding calculations inside the computational domain.

The flow structure around the bend was shown for a short bend micro-channel, the flow expansion and compressible behaviour was shown using the streamlines and cone plot.

It was shown that the flow velocity changes with the (outlet/inlet) rarefaction (or *Kn*) and it also shows the percentage gain in the velocity along the channel is almost the same for the range of *Kn* considered 0.043÷0.087.

The flow velocity has a maximum around the *Kn* 0.0553 in the range of *Kn* considered 0.043÷0.087, for short micro-channels of aspect ratio 1/36, with the height of the micro-channel equal to width, for height equal to 19.83e-6m. The pressure profile along the channel was similar in the range of *Kn* considered.

Acknowledgements

The research leading to these results has received funding from the European Community's Seventh Framework Programme (ITN - FP7/2007-2013) under grant agreement n° 215504.

The author N K Kulakarni would like to thank Dr. K Shterev for the SIMPLE-TS data provided to the validation of the DSMC based code.

References

[1] Bird G A 1994 *Molecular Gas Dynamics and the Direct Simulation of Gas flows* (Oxford Science Publications)
[2] Cercignani C 2000 *Rarefied Gas dynamics* (Macmillan)
[3] Rovenskaya O 2010 Numerical analysis of rarefaction and compressibility effects in bent Microchannel *proc. of ASME 2010*, FEDSM-ICNMM2010
[4] Kulakarni N K, Stefanov S and Markelov G N July 9-10, 2010 DSMC simulation of rarefied gas flow through Microchannel junctions *proc. of 2nd GASMEMS workshop* (Les Embiez, France)
[5] Agrawal A, Djenidi L and Agrawal A 2009 Simulation of gasflow in microchannels with a single 90 degree bend, *Computers & Fluids* **38** 1629-1637
[6] Ye J, Yang J, Zhang J, Xu P, Lam C, wong I and Ma Y January 5-8 2009 Effects of wall temperature on the heat and mass transfer in microchannels using the DSMC method *proc. of the 4th IEEE Int. Conf. on Nano/Micro Engineered and Molecular systems* (Shenzhen, China)
[7] Shterev K and S. Stefanov 2010 Pressure based finite volume method for calculation of compressible viscous gas flows *J. Computational Physics* 229 461-480
[8] Wang M and Li Z 2004 Simulations for gas flows in microgeometris using the direct simulation Monte Carlo method *Int. J. of heat and Fluid flow* 25 975-985
[9] Serikov V V and Nanbu K 1994 Methodological aspects of MCDS approach as applied to 3D flow in the sputtering chamber, *Rep. Inst. Fluid Sci.* **6** (Tohoku Univ.)
[10] Koura K 1990 A sensitivity test for accuracy in evaluation of molecular collision number in the direct-simulation Monte Carlo method, *Phys. of Fluid* **2** No. 7 p 1287-1289
[11] Ivanov M S and Rogazinsky SV 1991 Theoretical analysis of traditional and modern schemes of the DSMC method, *Proc. 17th Int. Symp. On Rarefied Gas Dynamics* p 629-642 (Aachen, ed. AE Beylich, VCH)
[12] Stefanov S, Gospodinov P and Cercignani C 1998 Monte Carlo simulation and Navier-stokes finite difference calculation of unsteady-state rarefied gas flows *Phys. Fluids* **10** (1) 289-300
[13] Gat A D and Frankel I September 9-11 2009 A higher-order Hele-Shaw approximation with application to gas flows through shallow micro-channels *J. Fluid Mech.* **638** 141-160
[14] Gad-El-Hak M 1999 The fluid mechanics of microdevices – the Freeman scholar lecture. *J. Fluids Engng* **121**, 5–33
[15] Gat A D, Frankel I and Weihs D Gas flows through shallow micro-configurations II. Microchannel junctions and networks September 9-11 2009 *1st Gasmems Workshop* (Eindhoven, NL)
[16] Gad-el-Hak M 2006 MEMS: Design and Fabrication, CRC Press Taylor & Francis Group
[17] Colin S, Aubert C and Caen R 1998 Unsteady gaseous flows in rectangular microchannels: frequency response of one or two pneumatic lines connected in series, *Eur. I. Mech. B/Fluids* **17** No. 1, p 79-104
[18] Wu J, Lee F, and Wong S 2001 Pressure boundary treatment in micro-channel devices using the Direct Simulation Monte Carlo method *JSME Int. Journal*, series **B** 44 (3)

[19] Markelov G N, Kulakarni N K and Stefanov S K November 1-6, 2010 DSMC simulation of unsteady rarefied gas flows through microchannel combinations *Proc. 15th Int. Conf. on the Methods of Aerophysical Research (ICMAR 2010)* p 1-8 (Novosibirsk, Russia)

[20] Kulakarni N K, Rovenskaya O, Perrier P and Graour I, Measurement of gas flows through 90 degrees bend micro-channel (work in progress)

[21] Fang Y and Liou W W 2002 Computations of the flow and heat transfer in micro devices using DSMC with implicit boundary *J. Heat Transfer Transactions of ASME 124,* 338-345

Oscillatory heating in a microchannel at arbitrary oscillation frequency in the whole range of the Knudsen number

O Buchina, D Valougeorgis

Department of Mechanical Engineering, University of Thessaly, Volos 38334, Greece

diva@mie.uth.gr, buchina@mie.uth.gr

Abstract. The oscillatory heating in a microchannel is investigated on the basis of linear kinetic theory. In particular, the linearized unsteady Shakhov kinetic model, subject to Maxwell boundary conditions is numerically solved in a fully deterministic manner based on finite differencing scheme in the physical space and in the discrete velocity method in the molecular velocity space. The solution of the problem is determined by two parameters: the Knudsen number and the ratio of the intermolecular collision frequency over the oscillation frequency. The numerical calculations are carried out for a wide range of both parameters and results are presented for the amplitude and the phase of all macroscopic quantities of physical interest.

1. Introduction

Recent interest in investigating unsteady heating processes in rarefied gases is motivated by their applications in several fields including micro-electromechanical systems, microelectronics and laser industry. Time dependent heat transfer configurations are common in gaseous micro devices and may be produced by time dependent boundary cooling or/and heating.

A transient flow of a gas caused by a sudden change in the boundary temperature, which is the counterpart of the periodic time-dependent problem, is a type of basic problem of a rarefied gas flow [1]. Recently in [2] the transient heat transfer in a gas confined in a small-scale slab due to the instantaneous change of a wall temperature has been investigated. In this work semi-analytical approaches have been applied in the free molecular and hydrodynamic limits, while the DSMC method has been used in the transition regime.

The correct description of heat flow through a rarefied gas for arbitrary rarefaction (the ratio of a characteristic length of the flow to the molecular mean free path, i.e. it is inversely proportional to the Knudsen number) and oscillation speed (the ratio of the collision frequency to the frequency of the temperature on the plate) is very important in many field including micro-electromechanical systems, microelectronics and laser industry. Current interest in analyzing unsteady heating processes is motivated by the common occurrence of time-varying boundary temperatures in a wide scope of microelectromechanical and nanoelectromechanical applications, ranging from microprocessor chip heating [3] to ultrafast temperature variations encounted in the laser industry [4-5]. The oscillatory heating in a microchannel has been recently investigated by the low-variance deviational simulation Monte Carlo method motion for an arbitrary time variation of the boundary temperature [6-7]. Periodic time-dependent behaviour of a rarefied gas between two parallel planes caused by an oscillatory heating of one plane has been numerically studied based on the linearized Boltzmann equation for a hard-sphere molecular gas [8].

In the present study, an analysis of the oscillatory heating of a rarefied gas confined between two parallel plates is based on linear kinetic theory. The implementation of a kinetic solution provides reliable results in the whole range of the Knudsen number with modest computational effort. In particular, the time dependent heat transfer is modelled by the linearized nonstationary kinetic equation, subject to Maxwell purely diffuse boundary conditions. The Shakhov model of the Boltzmann equation is chosen as the most appropriate one because it provides the correct Prandtl number so that both viscosity and thermal conductivity are taken into account in the solution of the kinetic equation. The solution of the problem is determined by two parameters: the Knudsen number and the ratio of the intermolecular collision frequency over the oscillation frequency. It is assumed that the oscillation is fully established and the dependence of the solution on the time is harmonic, while its dependence on the spatial coordinate will be obtained numerically.

The numerical solution of the problem is fully deterministic and based on finite differencing scheme in the physical space and in the discrete velocity method in the molecular velocity space. The aim of the present work is to calculate the heat flow induced by oscillatory heating of one of the parallel plates over a wide range of both parameters. Results will be presented for the amplitude and the phase of all macroscopic quantities of physical interest.

2. Statement of the problem

A monoatomic gas is considered to be confined between two infinite parallel plates located at $y'=0$ and $y'=L'$. The plate placed at $y'=0$ is heated with the wall temperature oscillating harmonically, while the temperature of the second plate is kept constant and equal to the ambient temperature T_0. It is assumed the oscillation to be fully established through the heat flow and the solution is harmonic with respect to time. The oscillating temperature can be presented as

$$T_w(t') = \Delta T_w \operatorname{Re}\left[\exp(-i\omega t')\right] + T_0 \tag{2.1}$$

where ω is the frequency, t' is the time, Re denotes the real part of a complex expression and ΔT_w is the amplitude of the oscillating temperature. It is assumed that $\Delta T_w \ll T_0$. The temperature oscillation on the plate causes a gas flow in the y direction characterized by a number density $n(t',y')$, bulk velocity $U(t',y')$, temperature $T(t',y')$ and heat flux $q(t',y')$ which depend on the time harmonically as

$$n(t',y') = \operatorname{Re}\left[\tilde{n}(y')\exp(-i\omega t')\right] + n_0, \tag{2.1}$$

$$U(t',y') = \operatorname{Re}\left[\tilde{U}(y')\exp(-i\omega t')\right], \tag{2.3}$$

$$T(t',y') = \operatorname{Re}\left[\tilde{T}(y')\exp(-i\omega t')\right] + T_0, \tag{2.4}$$

$$Q(t',y') = \operatorname{Re}\left[\tilde{Q}(y')\exp(-i\omega t')\right]. \tag{2.5}$$

Here, $\tilde{n}(y')$, $\tilde{U}(y')$, $\tilde{T}(y')$ and $\tilde{Q}(y')$ are complex functions and n_0 is the equilibrium number density.

The solution of the problem is determined by two parameters. The first one is the gas rarefaction defined as the ratio of a characteristic length of the flow to the molecular mean free path, i.e. it is inversely proportional to the Knudsen number:

$$\delta = \frac{P_0 L'}{\mu \upsilon_0} \sim \frac{1}{Kn}, \tag{2.6}$$

where $P_0 = n_0 k_B T_0$ is the gas pressure, μ is the viscosity of the gas at T_0 and $\upsilon_0 = \sqrt{2RT_0}$ is the most probable molecular velocity, with R denoting the gas constant. The second parameter characterizes the oscillation speed and is equal to the ratio of the intermolecular collision frequency over the oscillation frequency

$$\theta = \frac{P_0}{\mu\omega}. \tag{2.7}$$

Regarding the values of the parameters δ and θ, two limit regimes can be distinguished. Large values of both parameters, i.e., $\delta \to \infty$ and $\theta \to \infty$, correspond to the hydrodynamic regime. Then, the problem can be solved on the basis of the continuum mechanics equations. When both parameters are very small, i.e., $\delta \ll 1$ and $\theta \ll 1$, the intermolecular collisions can be neglected and the kinetic equation is significantly simplified. This regime is called free molecular. For arbitrary values of the both parameters the problem must be solved on the basis of the nonstationary Boltzmann equation.

Now it is convenient to introduce the dimensionless coordinate, distance and time

$$y = \frac{\omega}{\upsilon_0} y', \quad L = \frac{\omega}{\upsilon_0} L' = \frac{\delta}{\theta}, \quad t = \omega t'. \tag{2.8}$$

The macroscopic quantities (2.2)-(2.5) are also written in the dimensionless form as

$$\rho(y) = \frac{\tilde{n} - n_0}{n_0} \frac{T_0}{\Delta T_w}, \quad u(y) = \frac{\tilde{U}}{\upsilon_0} \frac{T_0}{\Delta T_w}, \tag{2.9}$$

$$\tau(y) = \frac{\tilde{T} - T_0}{\Delta T_w}, \quad q(y) = \frac{\tilde{Q}}{P_0 \upsilon_0} \frac{T_0}{\Delta T_w}. \tag{2.10}$$

Since these quantities are complex, they can be written as

$$\rho(y) = \alpha_\rho(y) \exp\left[i\varphi_\rho(y)\right], \quad u(y) = \alpha_u(y) \exp\left[i\varphi_u(y)\right], \tag{2.11}$$

$$\tau(y) = \alpha_\tau(y) \exp\left[i\varphi_\tau(y)\right], \quad q(y) = \alpha_q(y) \exp\left[i\varphi_q(y)\right], \tag{2.12}$$

where α_ρ, α_u, α_τ and α_q are amplitudes of the macroscopic quantities, while φ_ρ, φ_u, φ_τ and φ_q are their phases. We are going to calculate the functions $\alpha_i(y)$ and $\varphi_i(y)$, $i = \rho, u, \tau, q$ for various values of the parameters δ and θ.

3. Kinetic equation

In order to consider arbitrary values of the both parameters the problem must be solved on the basis of the nonstationary Boltzmann equation. By considering a limited temperature difference between the plates according to $\Delta T_w \ll T_0$, the heat flow is modelled by the unsteady linearized Shakhov model equation [9-10]. By assuming an oscillatory behaviour of the distribution function and following a typical procedure [11] two coupled integrodifferential equations describing the problem are:

$$(\theta - i)\phi + c_y \frac{\partial \phi}{\partial y} = \theta \left[\rho + 2c_y u + \tau \left(c_y^2 - \frac{1}{2} \right) + \frac{4}{15} q c_y \left(c_y^2 - \frac{3}{2} \right) \right], \tag{3.1}$$

$$(\theta - i)\psi + c_y \frac{\partial \psi}{\partial y} = \theta \left[\tau + \frac{4}{15} q c_y \right]. \tag{3.2}$$

Here, $\phi = \phi(y, \upsilon_y)$ and $\psi = \psi(y, \upsilon_y)$ are the reduced complex distributions functions, with $c_y \in (-\infty, \infty)$ denoting the y-component of the molecular velocity, while the macroscopic moments at the right side of equations (3.1)-(3.2) are given by

$$\rho(y) = \frac{1}{\sqrt{\pi}} \int \phi e^{-c_y^2} dc_y, \tag{3.3}$$

$$u(y) = \frac{1}{\sqrt{\pi}} \int c_y \phi e^{-c_y^2} dc_y, \tag{3.4}$$

$$\tau(y)=\frac{1}{3\sqrt{\pi}}\int\left[2\psi+\phi\left(2c_y^2-1\right)\right]e^{-c_y^2}dc_y\,, \tag{3.5}$$

$$q(y)=\frac{1}{\sqrt{\pi}}\int\left[\psi+\phi\left(c_y^2-\frac{3}{2}\right)\right]c_y e^{-c_y^2}dc_y\,. \tag{3.6}$$

The interaction between the particles and the walls is modelled according to Maxwell purely diffuse boundary conditions, which are written for the outgoing distributions at the walls as

$$\phi(y,c_y)=\begin{cases}c^2-1+v_0 & \text{at } y=0 \text{ and } c_y>0,\\ v_L & \text{at } y=L \text{ and } c_y<0,\end{cases} \tag{3.7}$$

$$\psi(y,c_y)=\begin{cases}1 \text{ at } y=0 \text{ and } c_y>0,\\ 0 \text{ at } y=L \text{ and } c_y<0,\end{cases} \tag{3.8}$$

where the constants v_0 and v_L are

$$v_0=2\int_0^\infty c_y\phi(0,-c_y)e^{-c_y^2}dc_y\,, \tag{3.9}$$

$$v_L=2\int_0^\infty c_y\phi(L,c_y)e^{-c_y^2}dc_y\,. \tag{3.10}$$

4. Numerical scheme

To reduce the number of the velocity nodes the complex perturbation functions $\phi(y,c_y)$ and $\psi(y,c_y)$ are split into two parts as follows

$$\phi(y,c_y)=\tilde{\phi}(y,c_y)+\phi_0(y,c_y), \tag{4.1}$$

$$\psi(y,c_y)=\tilde{\psi}(y,c_y)+\psi_0(y,c_y), \tag{4.2}$$

where the functions $\phi_0(y,c_y)$ and $\psi_0(y,c_y)$ satisfies the following equations

$$(\theta-i)\phi_0+c_y\frac{\partial\phi_0}{\partial y}=0, \tag{4.3}$$

$$(\theta-i)\psi_0+c_y\frac{\partial\psi_0}{\partial y}=0, \tag{4.4}$$

with the following boundary conditions

$$\phi_0(y,c_y)=\begin{cases}c_y^2-1+v_{00} & \text{at } y=0 \text{ and } c_y>0,\\ v_{0L} & \text{at } y=L \text{ and } c_y<0,\end{cases} \tag{4.5}$$

$$\psi_0(y,c_y)=\begin{cases}1 \text{ at } y=0 \text{ and } c_y>0,\\ 0 \text{ at } y=L \text{ and } c_y<0.\end{cases} \tag{4.6}$$

Euations (4.3)-(4.4) have the following analytical solutions

$$\phi_0(y,c_y)=\begin{cases}\left(c_y^2-1+v_{00}\right)\exp\left[-(\theta-i)y/c_y\right], & c_y>0,\\ v_{0L}\exp\left[-(\theta-i)(y-L)/c_y\right], & c_y<0,\end{cases} \tag{4.7}$$

$$\psi_0(y,c_y)=\begin{cases}\exp\left[-(\theta-i)y/c_y\right], & c_y>0,\\ 0, & c_y<0.\end{cases} \tag{4.8}$$

The constants v_{00} and v_{0L} are obtained from equations (3.9)-(3.10) and (4.5)-(4.6) as

$$v_{00} = 4I_1\left[(\theta-i)L\right]\frac{I_3\left[(\theta-i)L\right]-I_1\left[(\theta-i)L\right]}{1-4I_1^2\left[(\theta-i)L\right]}, \tag{4.9}$$

$$v_{0L} = 2\frac{I_3\left[(\theta-i)L\right]-I_1\left[(\theta-i)L\right]}{1-4I_1^2\left[(\theta-i)L\right]}, \tag{4.10}$$

where the special functions $I_n(x)$ [12] are defined as

$$I_n(x)=\int_0^{\infty} c^n \exp\left(-c^2-\frac{x}{c}\right)dc. \tag{4.11}$$

The kinetic equations for the functions $\tilde{\phi}(y,c_y)$ and $\tilde{\psi}(y,c_y)$ are

$$(\theta-i)\tilde{\phi}+c_y\frac{\partial\tilde{\phi}}{\partial y}=\theta\left[\rho+2c_y u+\tau\left(c_y^2-\frac{1}{2}\right)+\frac{4}{15}qc_y\left(c_y^2-\frac{3}{2}\right)\right], \tag{4.12}$$

$$(\theta-i)\tilde{\psi}+c_y\frac{\partial\tilde{\psi}}{\partial y}=\theta\left[\tau+\frac{4}{15}qc_y\right], \tag{4.13}$$

with the boundary conditions

$$\tilde{\phi}(y,c_y)=\begin{cases}2\int_{-\infty}^{0} c_y\tilde{\phi}(0,c_y)e^{-c_y^2}dc_y & \text{at } y=0 \text{ and } c_y>0,\\ 2\int_0^{\infty} c_y\tilde{\phi}(L,c_y)e^{-c_y^2}dc_y & \text{at } y=L \text{ and } c_y<0,\end{cases} \tag{4.14}$$

$$\tilde{\psi}(y,c_y)=\begin{cases}0 \text{ at } y=0 \text{ and } c_y>0,\\ 0 \text{ at } y=L \text{ and } c_y<0.\end{cases} \tag{4.15}$$

The macroscopic moments (3.3)-(3.6) are also decomposed into two parts as

$$\rho(y)=\tilde{\rho}(y)+\rho_0(y), \tag{4.16}$$

$$u(y)=\tilde{u}(y)+u_0(y), \tag{4.17}$$

$$\tau(y)=\tilde{\tau}(y)+\tau_0(y), \tag{4.18}$$

$$q(y)=\tilde{q}(y)+q_0(y), \tag{4.19}$$

where the terms $\rho_0(y)$, $u_0(y)$, $\tau_0(y)$ and $q_0(y)$ are obtained by substituting the solutions (4.7)-(4.8) into (3.3)-(3.6) to yield

$$\rho_0(y)=\frac{1}{\sqrt{\pi}}\left\{v_{0L}I_0\left[-(\theta-i)(L-y)\right]+I_2\left[(\theta-i)y\right]+(v_{00}-1)I_0\left[(\theta-i)y\right]\right\}, \tag{4.20}$$

$$u_0(y)=\frac{1}{\sqrt{\pi}}\left\{v_{0L}I_1\left[-(\theta-i)(L-y)\right]+I_3\left[(\theta-i)y\right]+(v_{00}-1)I_1\left[(\theta-i)y\right]\right\}, \tag{4.21}$$

$$\tau_0(y)=\frac{1}{3\sqrt{\pi}}\left\{\begin{aligned}&v_{0L}\left(2I_2\left[(\theta-i)(L-y)\right]-I_0\left[(\theta-i)(L-y)\right]\right)+\\&+(v_{00}+1)\left(2I_2\left[(\theta-i)y\right]-I_0\left[(\theta-i)y\right]\right)+\\&+2\left(I_4\left[(\theta-i)y\right]+I_0\left[(\theta-i)y\right]\right)-I_2\left[(\theta-i)y\right]\end{aligned}\right\}, \tag{4.22}$$

$$q_0(y)=\frac{1}{\sqrt{\pi}}\left\{\begin{array}{l} v_{0L}\left(I_2\left[(\theta-i)(L-y)\right]-\frac{3}{2}I_0\left[(\theta-i)(L-y)\right]\right)+ \\ +(v_{00}+1)\left(I_2\left[(\theta-i)y\right]-\frac{3}{2}I_0\left[(\theta-i)y\right]\right)+ \\ +I_4\left[(\theta-i)y\right]+I_0\left[(\theta-i)y\right]-\frac{3}{2}I_2\left[(\theta-i)y\right] \end{array}\right\}. \tag{4.23}$$

Finally, the quantities $\tilde{\rho}(y)$, $\tilde{u}(y)$, $\tilde{\tau}(y)$ and $\tilde{q}(y)$ in equations (4.16)-(4.19) are calculated by equations (3.3)-(3.6) using $\tilde{\phi}$ and $\tilde{\psi}$ instead of ϕ and ψ, respectively.

The governing equations (4.12) and (4.13) subject to the boundary conditions (4.14-4.15) are solved numerically in the whole range of the rarefaction δ and oscillation speed θ parameters. The computational scheme is fully deterministic and all spaces are accordingly discretized. In particular, the discretization in the molecular velocity space is performed by using the discrete velocity method. The continuum spectrum is substituted by a discrete set of velocities υ_m, $m=1,2,...M$, which are taken to be the roots of the Legendre polynomial, with M denoting the degree of the polynomial, accordingly mapped into the interval of interest. The physical space is divided into I segments and it is consisting of $i=1,2,...I+1$ nodes, while the discretization is performed by a second order central difference scheme.

The discretized set of equations (4.12-4.13) is solved for each molecular velocity marching through the physical space, while the macroscopic quantities defined by equations (4.16-4.19) are estimated by a Gauss-Legendre quadrature. Based on the above the results presented in the next section have been obtained with $M=2000$ and $I=10^5$.

5. Results

The amplitudes $\alpha_q(y)$ and phases $\varphi_q(y)$ of the normal heat flux $q(y)$ at the surfaces, i.e., at y = 0 and $L = \delta/\theta$, are given in Tables 1, 2. From these data it is seen that at $\delta = 0$ the value of $\alpha_q(y)$ remains as expected constant. For $\delta > 0$ at any fixed value of the rarefaction parameter the heat flux amplitude $\alpha_q(0)$ at the oscillatory heated plate decreases by increasing the oscillation speed parameter θ, while $\alpha_q(L)$ increases. Also, this behaviour can be observed in Figure 1, where the graph of the heat flux amplitude vs. y is shown.

Table 1. Amplitudes of the heat flux vs. δ and θ at $y = 0$ and $y = L$.

δ	α_q						
	$\theta=0.1$		$\theta=1$		$\theta=10$		$\theta=\infty$
	$y=0$	$y=L$	$y=0$	$y=L$	$y=0$	$y=L$	kin.eq.
0.	0.564	0.564	0.564	0.564	0.564	0.564	0.564
0.01	0.560	0.554	0.560	0.560	0.560	0.560	0.560
0.1	0.537	0.395	0.537	0.530	0.535	0.535	0.528
0.5	0.541	0.165	0.507	0.413	0.466	0.458	0.433
1	0.561	0.082	0.530	0.292	0.406	0.398	0.362
10	0.564	0.0000	0.556	0.017	0.402	0.053	0.099
20	0.564	0.0000	0.555	0.001	0.436	0.017	0.055

For a fixed value of oscillation parameter θ the amplitude $\alpha_q(0)$ has a nonmonotonical dependence on the rarefaction parameter δ and it is seems that there is a minimum at the transition regime in the interval $0.1\theta < \delta < \theta$. Then, in every case of $\delta > \theta$ the amplitude $\alpha_q(0)$ tends to and finally reaches a constant value. Next, it is interesting to note that in the low speed oscillation regime, i.e. at large values of θ, and in highly rarefied atmosphere ($\theta = 10$, $\delta < 0.1$ in Table 1) due to the small number of collisions there is a very small variation of the heat flux between the plates. For comparison purposes, in the last column of Table 1 the values of the heat flux are presented for the case of stationary heat flow between two plates ($\theta = \infty$).

Table 2. Phase of the heat flux vs. δ and θ at $y = 0$ and $y = L$.

δ	φ_q					
	$\theta = 0.1$		$\theta = 1$		$\theta = 10$	
	$y = 0$	$y = L$	$y = 0$	$y = L$	$y = 0$	$y = L$
0.	0.000	0.000	0.000	0.000	0.000	0.000
0.01	-0.0067	0.134	-0.0005	0.0171	0.000	0.0017
0.1	-0.0336	1.007	-0.0072	0.128	-0.0025	0.0148
0.5	-0.0255	2.95	-0.104	0.554	-0.0220	0.0637
1	-0.0094	5.42	-0.170	0.984	-0.0634	0.122
10	-0.010	31.5	-0.102	7.223	-0.507	1.04
20	-0.005	51.5	-0.103	15.6	-0.362	3.87

In Table 2 at any value of δ the phase $\varphi_q(0)$ on the oscillatory heated plate is always negative and tends to zero in the free-molecular regime ($\delta \to 0$). Also, the phase of the heat flux $\varphi_q(0)$ similar to its amplitude has a nonmonotonical behaviour with respect to δ and there is a maximum approximately in the interval $0.1\theta < \delta < \theta$. The phase $\varphi_q(L)$ increases by increasing the rarefaction parameter δ when the oscillation speed parameter θ is fixed.

Some of the results presented in Table1 have been compared with the tabulated results of Doi [8] and very good agreement has been found for the amplitude of the heat flux for the specific cases of $(\delta,\theta) = (0.1,0.1)$, $(1,1)$ and $(10,10)$.

As explained in Section 2 in Figures 1–4 the amplitudes $\alpha_i(y)$ and phases $\varphi_i(y)$, $i = q, \tau, u, \rho$ of the heat flux, temperature, bulk velocity and number density are shown. Three values of the rarefaction parameters $\delta = 0.1$, 1, and 10 and three values of the oscillation parameters $\theta = 0.1$, 1, and 10 are considered. One can see that for $\delta = 0.1$ the dependence of the amplitudes $\alpha_i(y)$ and phases $\varphi_i(y)$ on the dimensionless coordinate y/L is qualitatively the same for all values of θ. They vary just quantitatively by increasing the oscillation parameter θ. However, for $\delta = 1$ and 10 the behaviours of $\alpha_i(y)$ and $\varphi_i(y)$ change both qualitatively and quantitatively by increasing the oscillation parameter θ. Also, as it was shown above at $\delta = 0.1$ and $\theta = 10$ the heat flow reaches nearly the nonstationary heat flow with a zero bulk velocity and constant heat flux.

Figure 1 shows the heat flux amplitudes $\alpha_q(y)$ and phases $\varphi_q(y)$. Due to the oscillatory heating of the plate at $y = 0$, an oscillating heat flow propagates towards the second plate. For every δ the number of collisions results in the decrease of the amplitudes and increase of phases with respect to the distance towards the nonheated plate located at $y = L$. Since $L = \delta/\theta$ then, for instance, at $\delta = 0.1$ (Figure 1a) and $\theta = 10$ the distance $L = 0.01$ is the shortest, therefore at the relatively high rarefaction the oscillations of the heat flux have almost no delay in propagating, hence the constant amplitude and phase are observed. In general the low speed oscillation regime, i.e. $\theta \to \infty$, provides less increase of the phase and decrease of the amplitude at smaller rarefaction parameter δ, and the heat flux as well as the other macroscopic parameters tends to ones corresponding to the stationary heat flow. So, at $\delta = 1$ (Figure 1b) and $\theta = 10$ the heat flux is also constant.

The phases of the heat flux vary significantly for the small value of the parameter θ, i.e. when the temperature oscillation frequency ω is large. This variation of the phases becomes larger as the rarefaction δ is increased (Figure 1c: $\theta = 10$), providing very fast decay of the amplitude. It is noted that due to the dependency of the macroscopic parameters on the distance $L = \delta/\theta$ the profiles of the amplitudes and phases of all macroscopic parameters are quantitatively the same, when values of δ and θ give the same value of L (e.g. Figure 1b: $\theta = 0.1$ and Figure 1c: $\theta = 1$).

The temperature amplitudes $\alpha_\tau(y)$ and phases $\varphi_\tau(y)$ are given in Figure 2 as functions of the dimensionless coordinate y/L. The behaviour of the temperature is similar to that of the heat flux, but as opposed to the negative values of the heat flux phases at the plate y = 0, the $\varphi_\tau(0)$ is positive. The negative values of $\varphi_q(y)$ indicate that $q(y)$ is slower relatively to the phase of the temperature of the plate.

In Figure 3 the bulk velocity amplitudes $\alpha_u(y)$ and phases $\varphi_u(y)$ are shown. As it was expected for the large value of θ, where the heat flow tends to the stationary behaviour, the phases of the bulk velocity vary weakly. At $\theta = 10$ they are nearly constant, while the amplitude is equal to zero, what is practically corresponds to the stationary heat flow between plates. Compared to the phases of the heat fluxes and temperature at $\theta = 10$, the phase of the bulk velocity is much more far from zero.

The behaviour of the density distribution (Figure 4) is quite different from those for the bulk velocity, temperature and the heat flux. The density waves propagate and reflect off the other plate at $y = L$, and forward and backward waves are superimposed. It also causes and defines the bulk velocity distribution profiles. It is noted that the distribution of the number density at any moment obeys the mass conservation law, which implies that the integral of the perturbed density between the plates is zero.

6. Conclusion

The oscillatory heat flow of a rarefied gas contained between two parallel plates has been investigated on the basis of the linearized unsteady Shakhov kinetic model, subject to Maxwell diffuse boundary conditions. The kinetic equation has been solved numerically in a fully deterministic manner based on finite differencing scheme in the physical space and on the discrete velocity method in the molecular velocity space. Numerical calculations have been carried out for a wide range of the rarefaction parameter and for typical small, moderate and large values of the oscillation speed. Results are presented for the amplitude and the phase of all macroscopic quantities of physical interest in graphical and tabulated forms. Both the amplitude and the phase of all macroscopic quantities strongly depend on the two main parameters (δ, θ) and their dependency is further increased as the gas rarefaction is decreased and as the oscillation speed is increased. When the oscillation speed parameter tends to infinity the flow field correspond to that for the steady flow. Tabulated results for the amplitude of the heat flux has been compared with the results available in the literature and very good agreement has been found.

Acknowledgments

The research leading to these results has received funding from the European Community's Seventh Framework Programme (ITN - FP7/2007-2013) under grant agreement n° 215504.

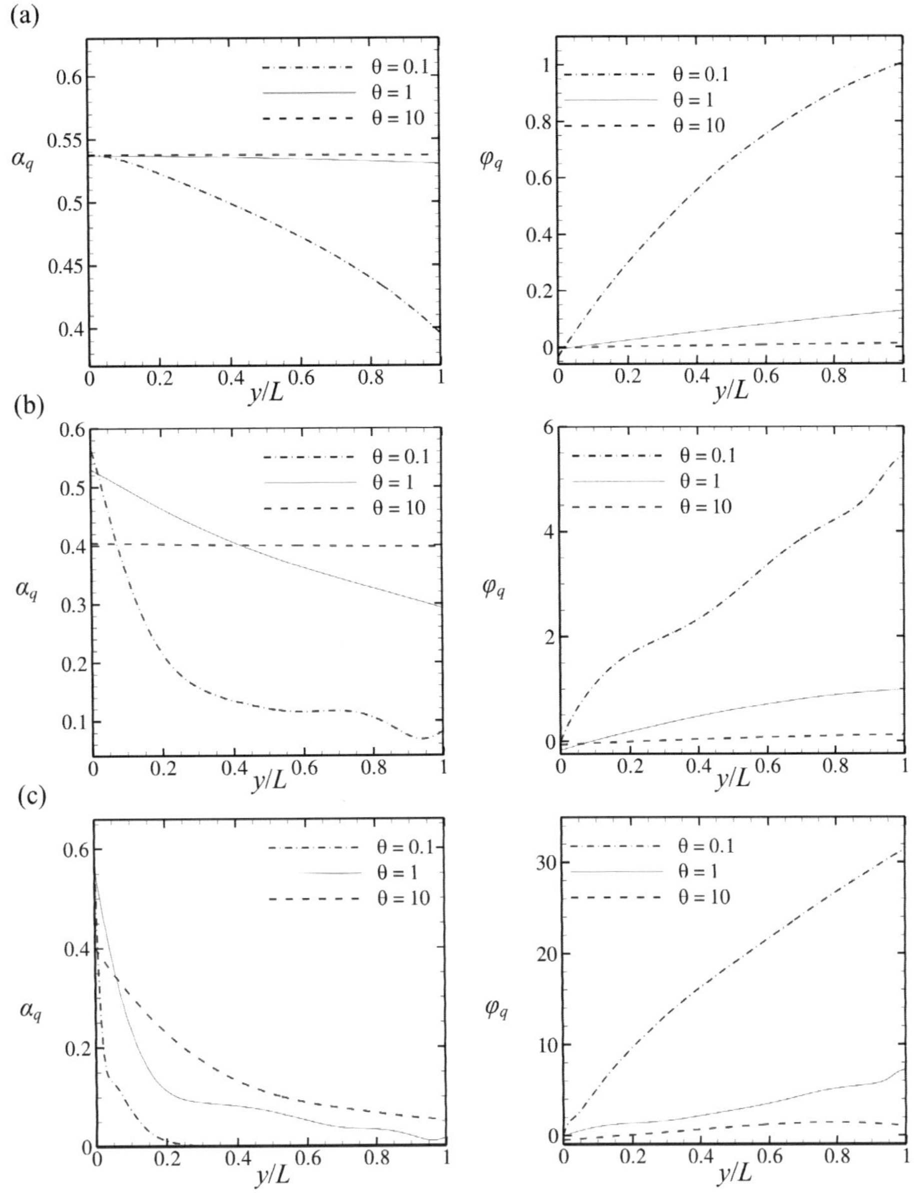

Figure 1. Heat flux amplitudes and phases at (a) $\delta = 0.1$, (b) 1, (c) 10 and θ = 0.1, 1, 10.

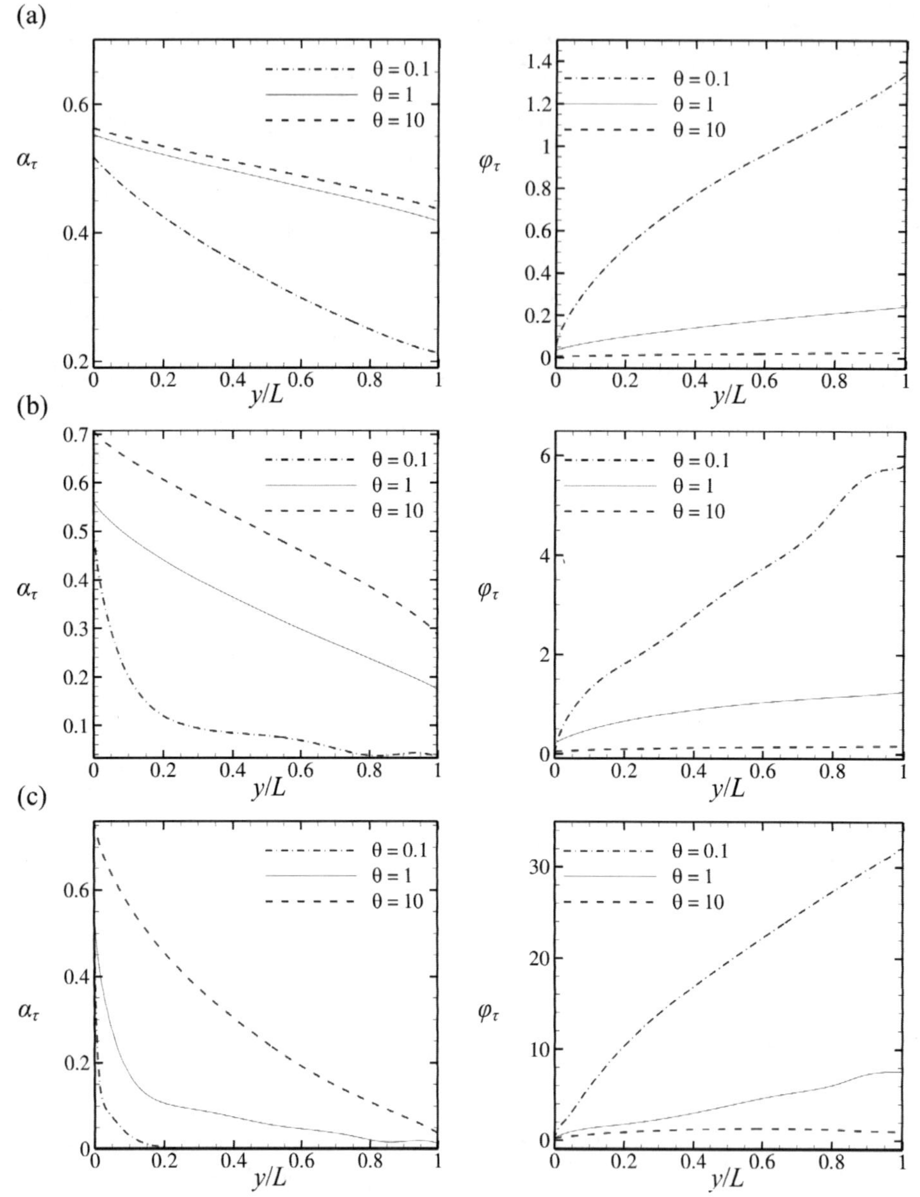

Figure 2. Temperature amplitudes and phases at (a) $\delta = 0.1$, (b) 1, (c) 10 and θ = 0.1, 1, 10.

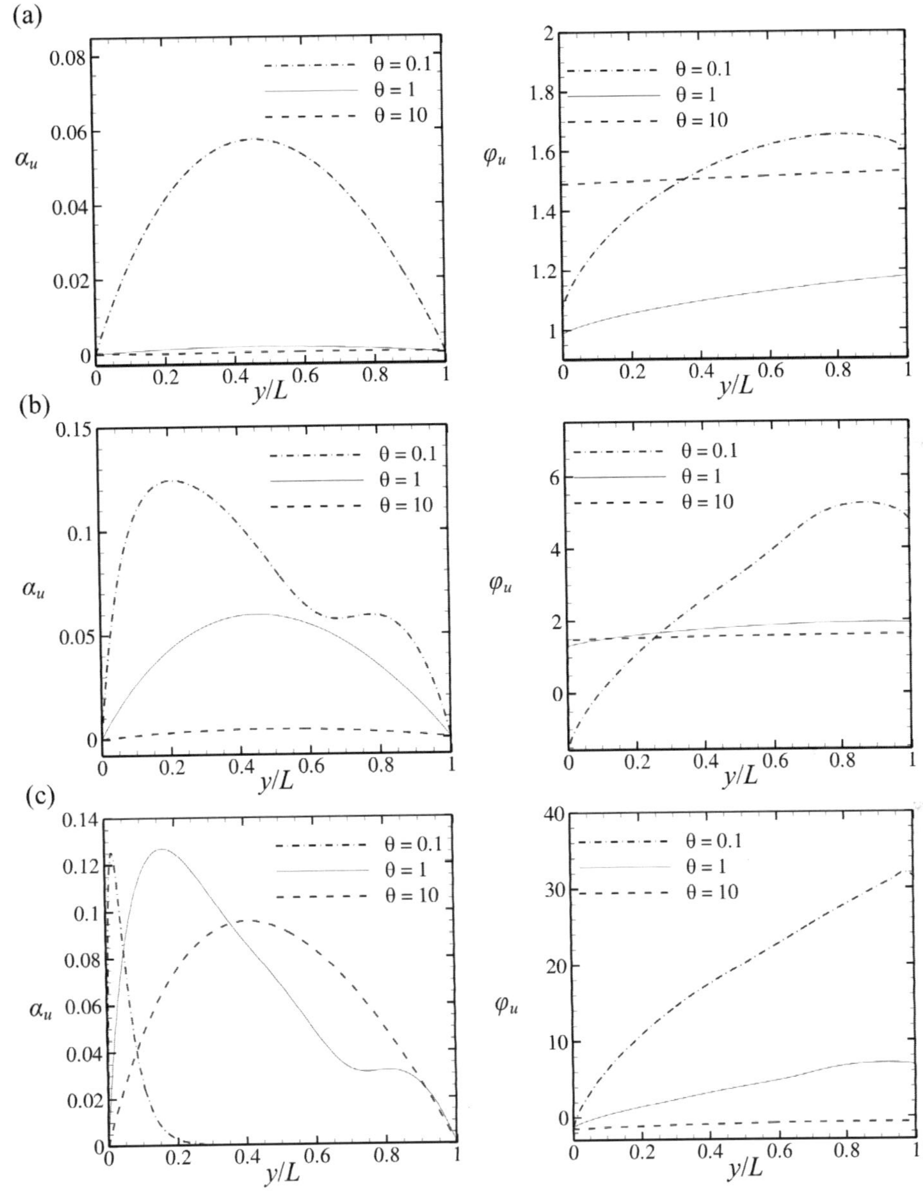

Figure 3. Velocity amplitudes and phases at (a) $\delta = 0.1$, (b) 1, (c) 10 and $\theta = 0.1$, 1, 10.

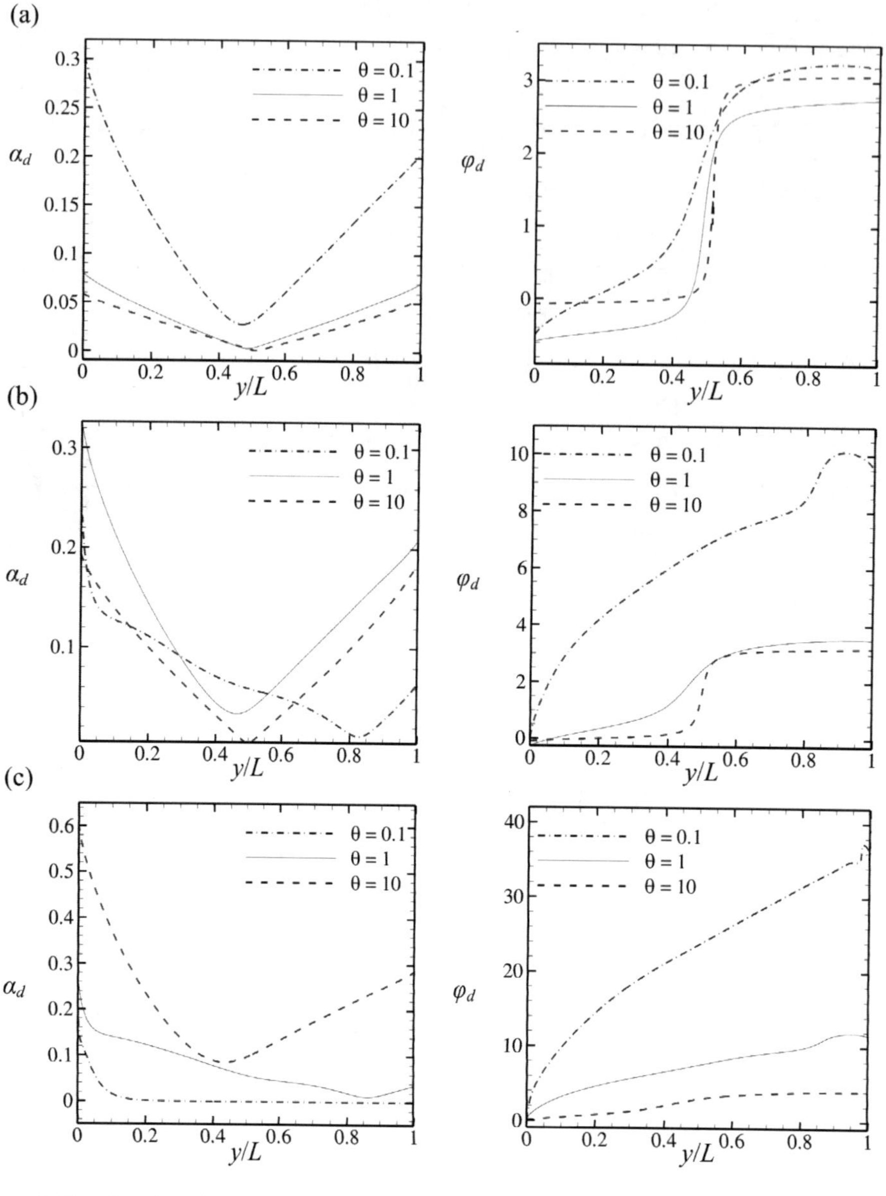

Figure 4. Number density amplitudes and phases at (a) $\delta = 0.1$, (b) 1, (c) 10 and $\theta = 0.1$, 1, 10.

References

[1] Sone Y 1965 Effect of sudden change of wall temperature in a rarefied gas *J. Phys. Soc. Jpn.* 20
[2] Manela A Hadjiconstantinou N G 2007 On the motion induced in a small-scale gap due to instantaneous boundary heating *J. Fluid Mech.* 593 453-462
[3] Ho C M Tai Y C 1998 Micro-electro-mechanical systems and fluid flows *Annu. Rev. Fluid Mech.* 30 579
[4] Tzou D Y 1997 *Macro-to-Microscale Heat Transfer* (Taylor and Francis Washington DC)
[5] Tzou D Y Beraun J E Chen J K 2002 Ultrafast deformation in femtosecond laser heating *ASME J. Heat Transfer* 124 284
[6] Manela A Hadjiconstantinou N G 2008 Gas motion induced by unsteady boundary heating in a small-scale slab, Phys. Fluids 20 117104
[7] Manela A Hadjiconstantinou N G 2010 Gas flow animation by unsteady heating in a microchannel, *Phys. Fluids* 22 062001
[8] Doi T 2011 Numerical analysis of the time-dependent energy and momentum transfers in a rarefied gas between two parallel planes based on the linearized Boltzmann equation *J. Heat Transfer* 133 022404
[9] Shakhov E M 1968 Approximate kinetic equations in rarefied gas theory *Fluid Dyn* 3 156-161
[10] Shakhov E M 1968 Generalization of the Krook kinetic relaxation equation *Fluid Dyn* 3 142-145
[11] Sharipov F Kalempa D 2009 Sound propagation through a gas confined between source and receptor at arbitrary Knudsen number and sound frequency *Phys. Fluids* 21
[12] Abramowitz M Stegun I 1965 *Handbook of Mathematical Functions* (Dover New York) p 1001
[13] Buchina O Vargas M Stefanov S Valougeorgis D 2011 Transient micro heat transfer in a gas confined between parallel plates due to a sudden increase of the wall temperature *3rd Micro and Nano Flows Conference* Thessaloniki Greece August 22-24

Accounting for rotational non-equilibrium effects in subsonic DSMC boundary conditions

Craig White, Matthew K. Borg, Thomas J. Scanlon and Jason M. Reese

Department of Mechanical and Aerospace Engineering,
University of Strathclyde, Glasgow G1 1XJ, UK

E-mail: craig.white@strath.ac.uk

Abstract. From study of the kinetic theory of dilute gases, it becomes clear that the subsonic, implicit boundary conditions which are commonly used in direct simulation Monte Carlo simulations of micro-scale channel flows do not take any form of rotational non-equilibrium into account. The consequences of not considering non-equilibration of the rotational mode are discussed and a new form of the boundary conditions which attempt to take this effect into account are proposed. It is shown that for a highly compressible test case involving these implicit boundaries in the literature that a degree of rotational non-equilibrium is present. The changes in the recovered macroscopic properties when the new boundary is applied in the same case include an improved rotational temperature distribution and a slightly lower velocity at the channel exit. A further test case exhibiting no rotational non-equilibrium was performed and it is shown that the macroscopic properties which are recovered are essentially the same for both boundaries. The results indicate that the newly proposed outlet boundary condition increases the range of applicability of these types of boundary conditions, without adding any significant computational expense.

1. Introduction

Gaseous flows in micro-scale geometries are an increasingly important area of research in the design of micro-electro-mechanical system (MEMS) technologies such as micro-pumps, micro-turbines and micro-fuel cells. The operation of MEMS devices can differ greatly to that of their macro-scale companions because the fluid flow characteristics can be vastly different. The micro-metre length scales involved can invalidate the continuum assumption altogether and rarefaction effects such as velocity slip, temperature jump and thermal transpiration may have to be accounted for [1]. The degree of gas rarefaction is normally defined through the Knudsen number

$$Kn = \frac{\lambda}{L}, \tag{1}$$

where λ is the mean free path of the gas molecules or atoms and L is a characteristic length scale.

It is the transition Knudsen number regime ($10^{-1} \leq Kn \leq 10$) which presents the greatest difficulty to model numerically. In this regime, the mean free path is becoming significant compared to the characteristic length scale, but molecular collisions are still important. MEMS devices involving gas flows tend to operate in the slip ($10^{-3} \leq Kn \leq 10^{-1}$) and transition regimes, even at relatively high pressures (i.e. around atmospheric). The direct simulation Monte Carlo (DSMC) method [2], based on the kinetic theory of dilute gases, has emerged as the dominant numerical technique for solving gaseous flows in the transition regime.

A variety of non-intuitive physics in high Knudsen number flows have been observed experimentally and been predicted both analytically and numerically. For example, non-linear pressure gradients have been observed for micro-channel flows [3], along with large velocity slip and temperature jump values. The movement of radiometer vanes when exposed to sunlight is caused by rarefied gas flow effects, i.e. uneven gas heating over an area bounded by one mean free path length from the edges of the vanes and gas creeping from a colder region towards a hotter region [4].

The set of pressure driven, implicit boundary conditions for subsonic flows are well known for use in conjunction with DSMC and have been used to successfully simulate a variety of rarefied gas flows in micro-scale geometries. Several different forms of these boundary conditions have been proposed; particle flux conservation was initially popular [5], but suffered from a high degree of statistical noise. Theory of characteristics based forms of the boundary conditions, using time averaged macroscopic properties obtained from the boundary cells, have emerged as the most efficient and widely used versions. However, these are essentially continuum computational fluid dynamics (CFD) boundary conditions that have been adapted for use with the DSMC algorithm. As such, they do not take into account some of the possible phenomena which may be present in rarefied gas flows, but are not observed in the continuum regime. In the current work the effect of non-equilibration of the rotational energy mode with the translational energy mode, in the context of these subsonic boundary conditions, is investigated. As such, the current work deals with diatomic or polyatomic gases in which the rotational mode is active, but the vibrational mode is not considered as the working temperatures are low enough that the vibrational mode is not likely to be active. It is always considered that the translational mode is in equilibrium, in order to keep the analysis focused on only one form of non-equilibrium for now.

2. Boundary conditions

Gaseous flows in MEMS flows are often characterised as being of low speed and being driven by a pressure gradient. Unlike more traditional hypersonic DSMC boundaries, it is likely that particles will have thermal velocities that will be orders of magnitude larger than the stream velocity. This makes it likely that particles will be able to enter from the outlet or leave the system at the inlet. The velocities at the inlet and outlet are not known due to current experimental limitations at the micro-scale. Pressure and temperature are usually the only known flow conditions at the boundaries.

When using DSMC, it is vital to know the macroscopic velocity, temperature and density in order to obtain the correct number flux at a boundary, since the particle flux $\dot{N}$, at an angle θ to the surface normal, across a boundary with area A, is calculated from the equilibrium Maxwell-Boltzmann distribution

$$\frac{\dot{N}}{A} = \frac{nV_{mp}\left\{\exp\left(-q^2\right) + \sqrt{\pi}q\left[1 + \text{erf}\left(q\right)\right]\right\}}{2\sqrt{\pi}}, \tag{2}$$

where

$$q = \frac{V}{V_{mp}}\cos\theta, \tag{3}$$

$$V_{mp} = \sqrt{\frac{2kT}{m}}, \tag{4}$$

2.1. Upstream boundaries

A theory of characteristic based inlet boundary condition was first proposed by Wang & Li [6]. In this implementation the user defines values of pressure P_{in} and temperature T_{in} for the gas at an inflow boundary and the perfect gas law is used to calculate the inlet number density n_{in}

$$n_{in} = \frac{P_{in}}{kT_{in}}. \tag{5}$$

Based on the theory of characteristics, the stream-wise u_{in} and tangential v_{in} velocities at an inlet boundary face f are calculated as

$$(u_{in})_f = u_j + \frac{P_{in} - P_j}{\rho_j a_j} \tag{6}$$

and

$$(v_{in})_f = v_j, \tag{7}$$

where u_j and v_j are first order extrapolations from the cells attached to the relevant boundary face, ρ is mass density and a is the local speed of sound. The subscript j denotes values in the boundary cells.

2.2. Downstream boundaries

At the exit boundaries, only the pressure P_{out} is defined and the boundary conditions as proposed by Nance *et al.* [7] are:

$$(\rho_{out})_f = \rho_j + \frac{P_{out} - P_j}{(a_j)^2}, \tag{8}$$

$$(u_{out})_f = u_j + \frac{P_j - P_{out}}{\rho_j a_j}, \tag{9}$$

$$(v_{out})_f = v_j, \tag{10}$$

$$(T_{out})_f = P_{out} / \left[R \left(\rho_{out} \right)_f \right]. \tag{11}$$

Using the implementations of these boundary conditions found in the literature, the values of P_j are calculated using the overall temperature, i.e.

$$P_j = \rho_j R \left[\left(\frac{\xi_{trans} T_{trans} + \xi_{rot} T_{rot}}{\xi_{trans} + \xi_{rot}} \right) \right]_j, \tag{12}$$

where R is the gas constant and ξ_{trans}, ξ_{rot}, T_{trans} and T_{rot} are the number of translational degrees of freedom, the number of rotational degrees of freedom, the translational temperature and the rotational temperature respectively. When particles are introduced at either the inlet or outlet boundaries, their translational and kinetic energies are also based on the overall temperature.

According the kinetic theory of dilute gases [2, 8, 9], the pressure should be measured using the translational temperature, as it depends only on the translational kinetic motions of the particles. If the gas is in perfect thermodynamic equilibrium, the overall temperature will be equal to both the translational and rotational temperatures and so this is not an issue. However, it is not certain that a gas will be remain in equilibrium along the length of a micro-channel, particularly if the inlet to outlet pressure ratio is high or the aspect ratio of the micro-channel is small.

As a gas travels downstream under the influence of a pressure gradient, the density decreases and the gas expands and cools. This decrease in thermal energy per unit mass initially appears in the translational mode and is redistributed to the rotational mode through inter-molecular collisions, only after a corresponding relaxation time [10]. Although the rotational relaxation time is not large, if the gas is sufficiently rarefied, there may not be enough molecular collisions occurring for the translational and rotational energies to equilibrate with one another. If the energy is not being redistributed to the rotational mode effectively enough, there will be a local deficit of energy in the translational mode compared to an equilibrium gas of the same density and total energy. This in turn means that the pressure will be lower than in the corresponding equilibrium gas.

A new form of the characteristics based boundary conditions is proposed here. At the inlet, the boundary conditions remain unchanged from those of Wang and Li [6], which means that it is assumed that the gas is always in rotational equilibrium at the inlet. At the outlet, a state of rotational equilibrium is not assumed and the boundary conditions have been modified as follows to account for a degree of non-equilibrium between the translational and rotational modes. All of the following macroscopic properties are measured after the movement and collision stages and are time averaged for the entire simulation duration. The pressure P_j in outlet boundary cells is measured as

$$P_j = \rho_j R \left(T_{trans}\right)_j . \tag{13}$$

The mass density and velocities at the exit boundaries are then calculated as in the Nance *et al.* [7] implementation, but with the new definition of pressure included in the correction terms

$$\left(\rho_{out}\right)_f = \rho_j + \frac{P_{out} - P_j}{\left(a_j\right)^2}, \tag{14}$$

$$\left(u_{out}\right)_f = u_j + \frac{P_j - P_{out}}{\rho_j a_j}, \tag{15}$$

$$\left(v_{out}\right)_f = v_j. \tag{16}$$

The translational temperature at the exit boundary is calculated using the perfect gas law

$$\left[\left(T_{trans}\right)_{out}\right]_j = P_{out} / \left[R \left(\rho_{out}\right)_f\right], \tag{17}$$

and the rotational temperature at the exit boundary is simply taken as a first order extrapolation of the measured rotational temperature from the boundary cell, i.e.

$$\left[\left(T_{rot}\right)_{out}\right]_f = \left(T_{rot}\right)_j . \tag{18}$$

When particles are introduced at the boundaries, they are assigned translational energies based only on the value of $\left(T_{trans}\right)_{out}$ by substituting this term in equation (4) for T. Use of equation (2) to calculate the entering particle flux means that translational equilibrium is being assumed at both the inlet and outlet. Their rotational energies are assigned only from the value of $\left(T_{rot}\right)_{out}$ and the usual rotational energy selection routine. In this way, it is believed that a more realistic representation of gas flows with a degree of rotational non-equilibrium can be obtained.

The boundary conditions of Wang & Li [6] and the new boundaries proposed here have been implemented within the free and open-source DSMC solver, *dsmcFoam*. This solver is released as part of the C++ fluid dynamics toolbox, OpenFOAM, which can be downloaded from [11] and has previously been rigorously validated for a variety of benchmark cases [12].

3. Rotational non-equilibrium

In order to demonstrate that a gas may be out of rotational equilibrium inside a micro-channel geometry, a test case from the literature is chosen and velocity distribution functions at various locations in the channel are measured and compared to the equivalent equilibrium distribution function.

The chosen test test case is a micro-Poiseuille flow from Liou & Fang [13]. The channel has a length of 2 μm and a height of 0.4 μm. The inlet gas temperature and the wall temperatures are all constant at 300 K. The inlet pressure is 72640 Pa, with an inlet to outlet pressure ratio of 4.54. The working gas is nitrogen and the Knudsen number based on the channel height and the variable hard sphere (VHS) mean free path ranges from 0.19 at the inlet to 0.72 at the outlet; therefore the problem lies within the transition Knudsen number regime. The boundary conditions used for the current problem are those described in [6].

To remain consistent with the test case in the literature, 100 x 60 regular rectangular computational cells are used to mesh the computational domain and the system has around 180,000 DSMC simulator particles at steady state. Groups of cells were joined together and used as zones in which a variety of macroscopic or microscopic properties can be measured. Figure 1 shows the mesh and measurement zones. The overall temperature and the velocity distribution functions were measured in each of the control zones in order to determine whether the system is in local rotational equilibrium at these points.

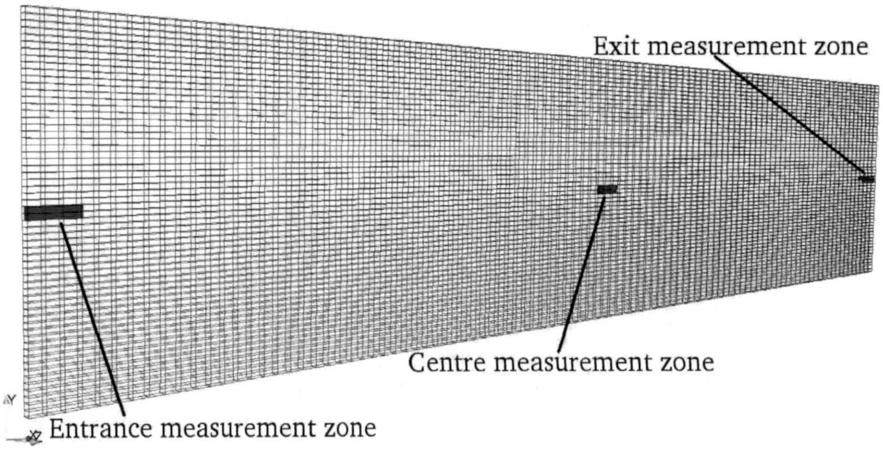

Figure 1: Snapshot of the mesh including measurement zones.

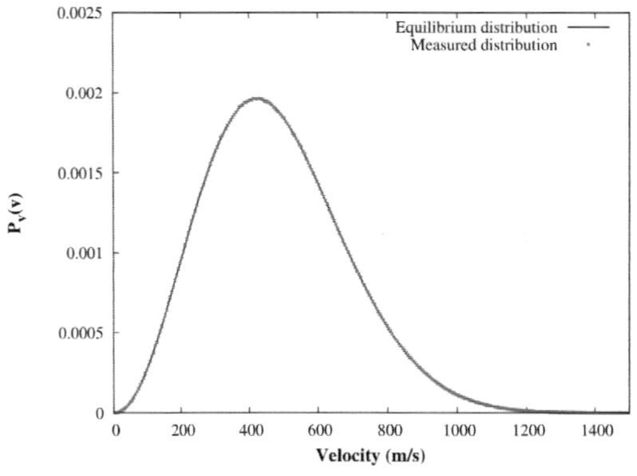

(a)

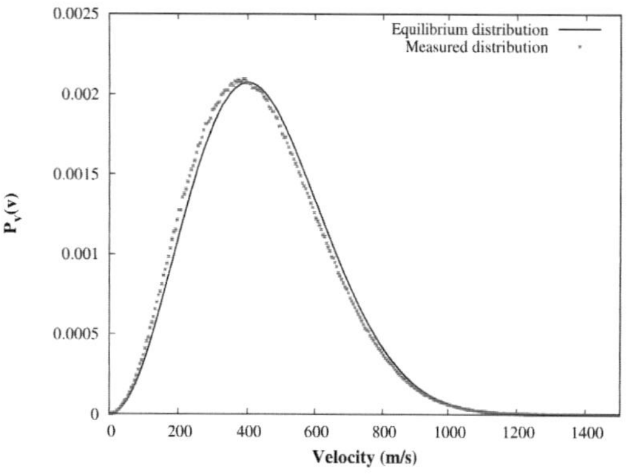

(b)

Figure 2: Comparison of measured and equilibrium distribution functions: (a) Entrance measurement zone (b) Exit measurement zone

Around 1 million statistical samples were employed to obtain the following results and the problem was solved in parallel on 2 processors. The total run time was 47 hours. Figure 2 shows the measured velocity distribution functions in the entrance and exit measurement zones and compares them with the equilibrium velocity distributions based on the measured value of overall temperature in the respective measurement zones. If the gas is in rotational equilibrium, the measured distribution should match the theoretical one exactly. In the entrance measurement zone this is the case, as it is in the centre measurement zone (not shown here); however, in the exit measurement zone the measured distribution is shifted slightly to the left of the theoretical one and has a slightly higher peak value. This is consistent with a cooler distribution function and comes about because a degree of rotational non-equilibrium is present in this measurement zone.

As discussed previously, the gas expands and the thermal energy per unit mass decreases as it travels downstream. The translational mode cools first and the rotational mode comes into equilibrium with this only after a certain relaxation time. This relaxation process occurs through inter-molecular collisions, but in this moderately rarefied flow there are not enough collisions occurring for the rotational mode to come into equilibrium with the translational mode. This means that the translational temperature in the exit measurement zone is significantly lower than the rotational temperature. The overall temperature is a weighted average of the translational and rotational temperatures, so this will also be higher than the translational temperature; hence the theoretical distribution function is that of a slightly warmer gas compared to the measured one.

Table 1: Simulation details

Case	Length (μm)	Height (μm)	P_{in} (Pa)	P_{out} (Pa)	Boundaries
1	2	0.4	72,640	16,000	Wang & Li [6]
2	2	0.4	72,640	16,000	Current work
3	20	0.4	100,000	50,000	Wang & Li [6]
4	20	0.4	100,000	50,000	Current work

The results in figure 2 show that it can be necessary to account for rotational non-equilibrium at the exit of a micro-channel and a series of cases are performed in order to compare results obtained using the boundary conditions from [6] and the new boundary conditions described above. The micro-channel dimensions and inlet/outlet pressures used in each case are detailed in table 1. Cases 1 and 2 were meshed with 100 x 60 computational cells, while cases 3 and 4 were meshed using 1000 x 60 computational cells. The working gas is nitrogen in all cases, where the inlet gas temperature is 300 K and all walls are fully diffuse at 300 K. All simulations use the VHS collision model with the phenomenological Larsen-Borgnakke internal energy redistribution technique [14].

4. Results

Cases 1 and 2 were run in parallel on 2 cores each and sampled for 1 million time steps. At steady state, case 1 had roughly 182,000 DSMC simulator particles and case 2 had slightly more at 183,000. The total run time was around 47 hours for each case, with case 2 being around 3% slower than case 1. This indicates that the new outlet boundary condition proposed here does not introduce any significant computational expense with respect to that found in [6].

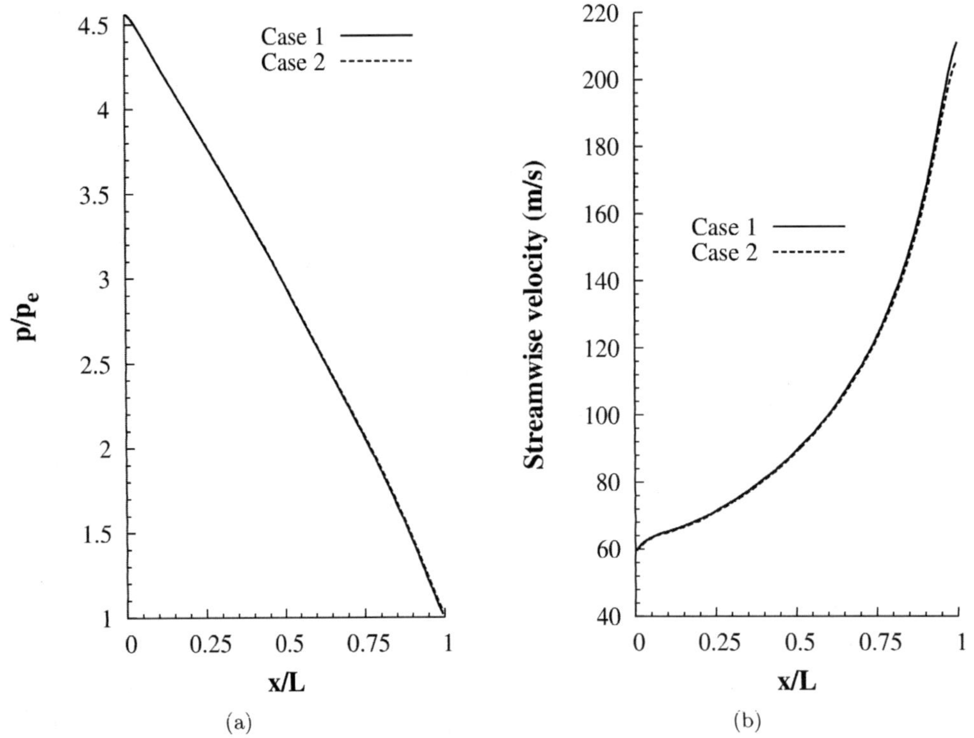

Figure 3: Centre-line distributions in cases 1 and 2 for (a) Normalised pressure (b) Velocity.

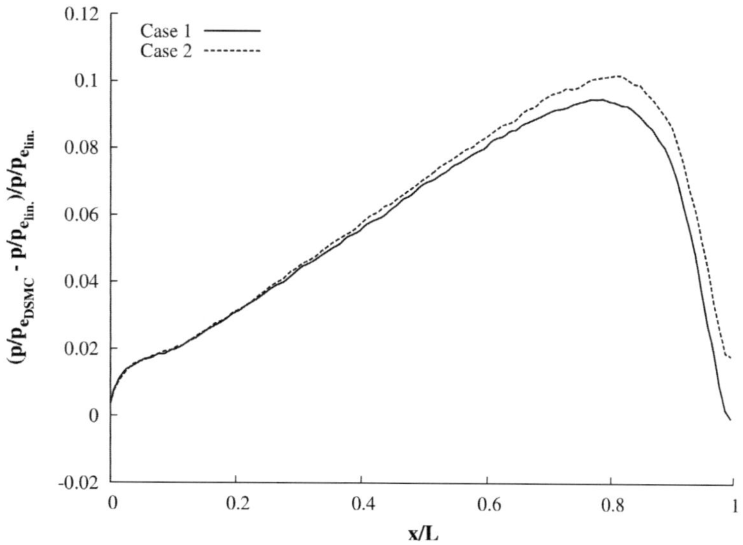

Figure 4: Deviation from a linear pressure profile for cases 1 and 2.

Figure 3a shows the centre-line pressure distributions obtained from cases 1 and 2. In general, the pressure profiles match very well, with the new implementation boundaries resulting in slightly higher pressures towards the channel exit. Figure 3b shows the centre-line velocity profiles from cases 1 and 2. The velocity profiles match well until near the channel exit, where the new boundaries achieve a slightly lower value (around 6 m/s slower). Figure 4 shows how the pressure profile varies from that of a linear pressure drop along the length of the centre-line for both the old and new boundary conditions. There is a difference in the pressure profile towards the exit when using the new boundary conditions, because the measured pressure at the outlet is based on the translational temperature rather than the overall temperature.

The translational, rotational and overall temperature profiles at the channel centre-line in cases 1 and 2 are shown in figure 5. The translational temperature profiles match well for both boundary conditions, however the rotational temperature near the exit is increased slightly when using the newly developed boundary conditions. As a result, the overall temperature profiles match well until near the exit, and because the overall temperature is a weighted average of the three translational modes and the two rotational modes, the magnitude of the difference in overall temperature is smaller than that observed for the rotational temperatures. The temperature profiles from a similar case ran with the pressure difference being generated in a high and low pressure reservoir at either end of the micro-channel are included for reference here.

The temperature profiles highlight the fact that the overall temperature can be significantly higher than the translational temperature at the channel exit (around 15 K in these cases) and demonstrates why it is important to measure pressure using only the translational temperature and introduce new particles based on the respective translational and rotational temperatures at the face they are introduced at, as opposed to a single value of overall temperature.

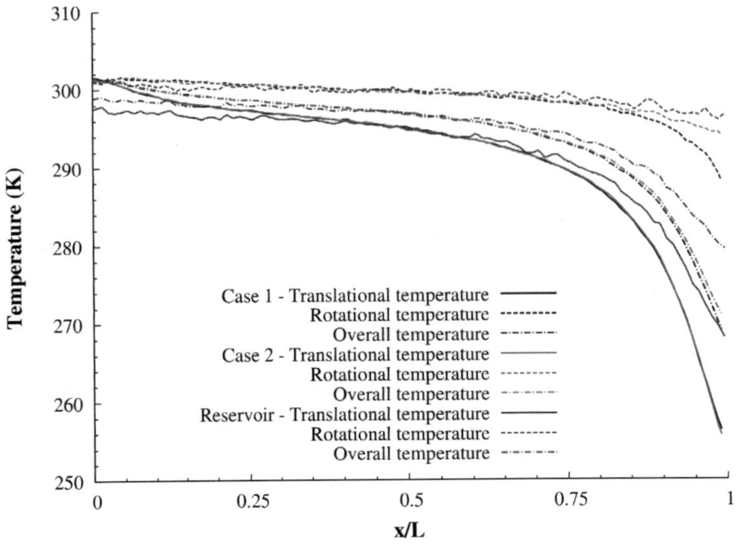

Figure 5: Centre-line temperature distributions for cases 1 and 2.

The translational temperatures achieved when using both types of boundary conditions are lower than that found when using the reservoirs, but the rotational temperature achieved using the new boundary conditions is much closer to that in the reservoir simulation compared to the Wang & Li implementation of the boundaries. It is to be expected that the implicit boundary conditions will be unable to capture the same translational temperature as the reservoir case here because of the rapid expansion which occurs as the gas exits the channel and enters the reservoir. The implicit boundaries are unable to capture this effect as effectively as the reservoir, which is a more realistic representation of the problem, but they do have the advantage of being much more computationally efficient. For example, the reservoir case displayed here required over a week of run time on 64 cores.

Cases 3 and 4 were run in parallel on 8 cores for a total of 2 million time steps. The domain was meshed using 1000 x 60 regular rectangular cells. The expected signal-to-noise ratio was small in these two cases because the channel is much longer and the pressure difference is also reduced with respect to cases 1 and 2. Two million statistical samples were employed to try and reduce the error to an acceptable level. Using equations from [15], the estimated error in the macroscopic velocity at the channel exit centre-line is less than 1%. Both cases contained around 920,000 DSMC simulator particles at steady state.

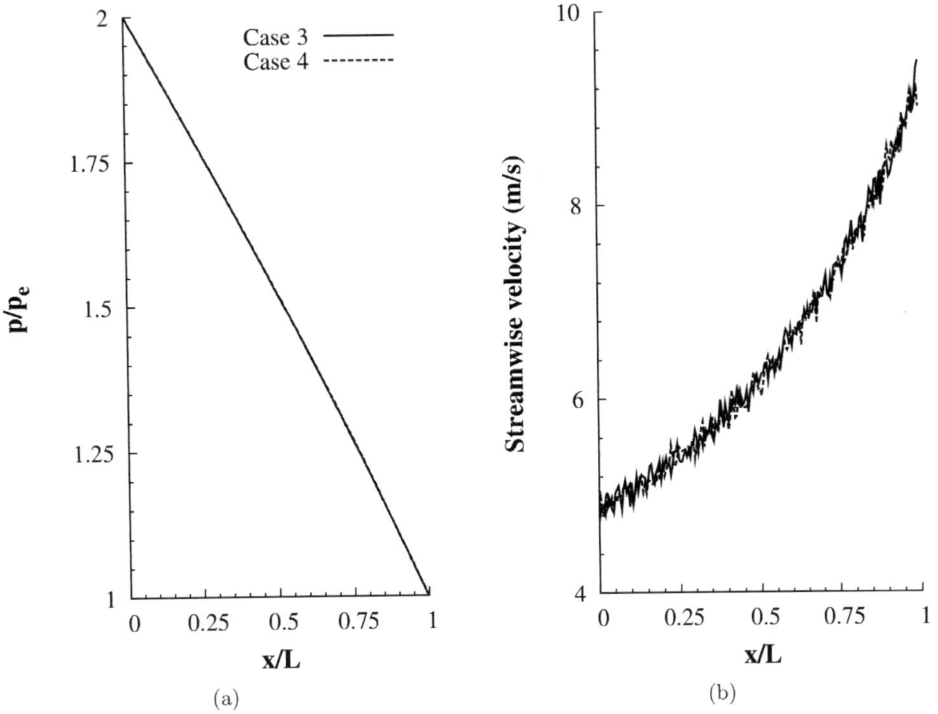

Figure 6: Centre-line distributions in cases 3 and 4 for (a) Normalised pressure (b) Velocity.

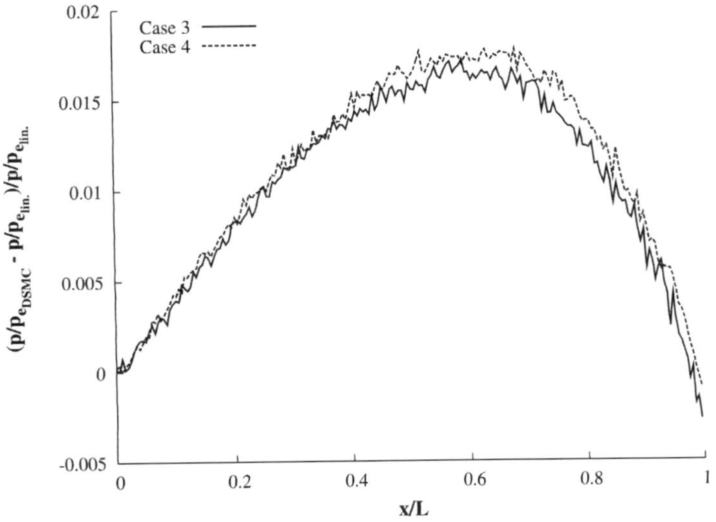

Figure 7: Deviation from a linear pressure profile for cases 3 and 4.

Figure 6a shows the centre-line pressure distributions obtained from cases 3 and 4. The pressure profiles are in excellent agreement along the whole length of the micro-channel. The centre-line streamwise velocity profiles for cases 3 and 4 are shown in figure 6b and good agreement is found between between the two cases, although the relatively high amount of statistical scatter makes it difficult to see exactly how well these profiles match. Figure 7 shows how the pressure profile varies from that of a linear pressure drop along the length of the centre-line for both the old and new boundary conditions. The magnitude of the deviation from a linear profile is much reduced with respect to the results shown in figure 4 and the results when using the new boundary conditions (case 4) are in much closer agreement with those achieved when using the Wang and Li implementation of the boundary conditions (case 3).

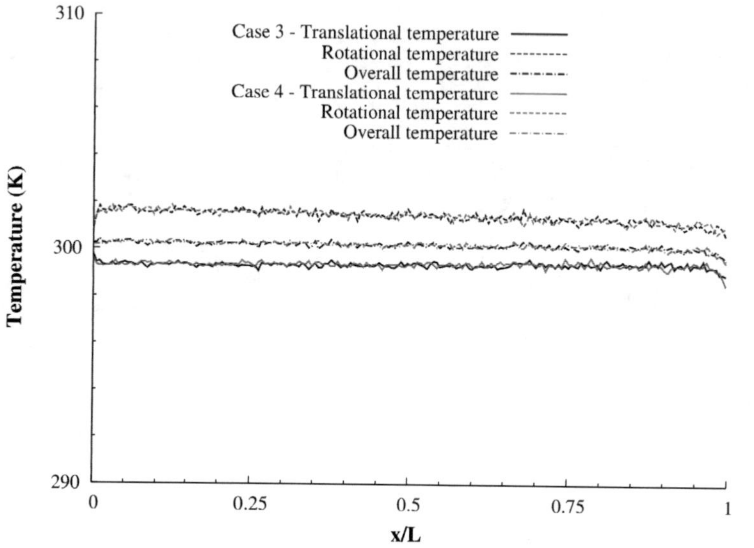

Figure 8: Centre-line temperature distributions for cases 3 and 4.

Figure 8 shows the translational, rotational and overall temperature profiles at the channel centre-line in cases 3 and 4. All of the temperature profiles are in excellent agreement between cases 3 and 4.

The results for cases 3 and 4 are consistent with a gas which is in (or at least very close to) rotational equilibrium. This would be expected because the micro-channel length is longer than in cases 1 and 2 and the pressure difference is also smaller. There will be less compressibility effects and so the gas does not expand and cool, as illustrated by the translational temperature profiles in figure 8. The deviation from a linear pressure profile is also expected to be smaller when compressibility effects are reduced, as was shown in figure 7. The results from cases 3 and 4 show that the new boundary conditions give the same results as the those proposed by Wang & Li [6] when there is no, or very little, rotational non-equilibrium present in a gaseous micro-channel flow.

5. Conclusions

It has been shown, through measurement of velocity distribution functions, that a rarefied gas flow in a micro-channel may not remain in rotational equilibrium along the channel length, especially if compressibility effects are significant. An implicit boundary condition to be applied at the exit of a micro-channel geometry has been proposed which attempts to account for non-equilibration of the rotational energy mode with the translational mode. A comparison, for a highly compressible case, of the new boundary conditions with a previous implementation that does not account for rotational non-equilibrium has shown that the new boundaries lead to a slightly more non-linear pressure profile than the old boundaries. The velocity near the micro-channel exit is slightly reduced when using the new boundaries.

It is in the rotational temperature profiles where the most significant change can be observed between the two boundaries. The new boundaries result in an improved rotational temperature profile along the micro-channel centre-line, with respect to the old boundaries, compared to the temperature profile from an equivalent micro-channel simulation performed with high and low pressure reservoirs rather than implicit boundary conditions.

When the boundary conditions are compared for a case with significantly less compressibility, and therefore no rotational non-equilibrium effects, the two different boundaries achieve an almost identical solution. The new outlet boundary proposed here provides additional functionality by considering a more complete form of kinetic theory and does so without incurring any significant computational expense, as has been shown in the run times for the simulations performed in the current work.

References

[1] Gad-el-Hak M 1999 *Journal of Fluids Engineering* **121** 5–33
[2] Bird G 1994 *Molecular Gas Dynamics and the Direct Simulation of Gas Flows* (Oxford University Press Inc, New York: Oxford Science Publications)
[3] Arkilic E, Breuer K and Schmidt M 2001 *Journal of Fluid Mechanics* **437** 29–43
[4] Gimelshein S F, Gimelshein N E, Ketsdever A D and Selden N P 2011 *AIP Conference Proceedings* **1333** 693–700
[5] Wu J S and Tseng K C 2001 *Computers & Fluids* **30** 711–735
[6] Wang M and Li Z 2004 *International Journal of Heat and Fluid Flow* **25** 975–985 ISSN 0142-727X
[7] Nance R, Hash D and Hassan H 1997 *Journal of Thermophysics and Heat Transfer* **12** 119–128
[8] Kennard E 1938 *Kinetic Theory of Gases* (New York and London: McGraw-Hill Book Company, Inc.)
[9] Patterson G N 1956 *Molecular Flow of Gases* (New York: John Wiley & Sons, Inc.)
[10] Vincenti W G and Kruger C H 1965 *Introduction to Physical Gas Dynamics* (Malabar, Florida: Krieger Publishing Company)
[11] OpenCFD http://www.openfoam.com
[12] Scanlon T, Roohi E, White C, Darbandi M and Reese J 2010 *Computers & Fluids* **39** 2078–2089
[13] Liou W and Fang Y 2000 *Computer Modeling in Engineering & Sciences* **1** 119–128
[14] Borgnakke C and Larsen P 1975 *Journal of Computational Physics* **18** 405–420
[15] Hadjiconstantinou N, Garcia A, Bazant M and He G 2003 *Journal of Computational Physics* **187** 274–297

1st European Conference on Gas Micro Flows (GasMems 2012) IOP Publishing
Journal of Physics: Conference Series **362** (2012) 012017 doi:10.1088/1742-6596/362/1/012017

On the role of surface shape in a micro-scale heat conduction problem

A Dinler[1*], I A Graur[2], R W Barber[1], D R Emerson[1] and P Perrier[2]

[1] Centre for Microfluidics and Microsystems Modelling, STFC Daresbury Laboratory, Warrington WA4 4AD, United Kingdom
[2] Universite de Provence, Ecole Polytechnique Universitaire de Marseille, UMR CNRS 6595, 5 rue Enrico Fermi, 13453 Marseille, France

Abstract. The present study investigates the importance of the surface shape in a micro-scale heat conduction problem. A heated infinitely-thin cylindrical shell is positioned in the middle of two concentric cylinders, and the heat transfer through a rarefied gas between the shell and the confining inner (or outer) cylinder is investigated. The study initially considers the solution of the first- and second-order temperature-jump models (*i.e.* the conventional heat equation with temperature-jump boundary conditions). The study then examines the numerical solution of the nonlinear Shakhov model kinetic equation subject to the Maxwell boundary condition using the discrete velocity method (DVM). The variable-hard-sphere molecular interaction model is taken into account in the temperature-jump models allowing the presence of significant temperature differences between surfaces to be considered. Anomalous temperature profiles near the convex (or concave) side of the shell are attributed to the effects of surface shape.

1. Introduction

Understanding the role of heat flow is an essential part of designing various MEMS and NEMS devices. Several microelectronic devices are exposed to a significant quantity of heat in operation, while many microdevices are based on heat transfer phenomena such as thermoelastic actuators and thermal anemometers [1,2]. Some other MEMS devices are designed to enhance the heating or cooling performance by incorporating more complex manifold geometries. However, the role of surface shape and curvature in the heat transfer between a microsystem and its environment has not been fully understood.

The present study investigates the importance of the surface shape in micro-scale heat conduction around an infinitely-thin cylindrical shell. A heated cylindrical shell is positioned in the middle of two concentric cylinders, and the heat transfer through a rarefied gas between the shell and the confining inner (or outer) cylinder is investigated. Since the shell has both concave and convex sides with the same surface area and same degree of curvature, any anomalous behaviour in the heat flow over the concave or convex side can be attributed directly to the effects of surface shape. The study initially considers the solution of the first- and second-order temperature-jump models (*i.e.* the conventional heat equation with temperature-jump boundary conditions) by taking into account the presence of significant temperature gradients. Then, the study examines the numerical solution of the nonlinear

* ali.dinler@stfc.ac.uk Tel: +44 (0)1925 603663

Shakhov model kinetic equation subject to the Maxwell boundary condition using the discrete velocity method (DVM).

The present study reveals the remarkable effects of the surface shape and curvature. The results show that the degree of temperature jump over a convex surface is markedly larger than that over a concave surface. The results clearly demonstrate the difference in the magnitudes of the heat fluxes over the concave and convex surfaces of the shell. The present study also investigates the accuracy of temperature-jump models over nonplanar surfaces in comparison with both the solution of the nonlinear Shakhov model kinetic equation and direct simulation Monte Carlo (DSMC) results. The DSMC data and the nonlinear Shakhov kinetic solution are found to be in excellent agreement for the cases considered in this study.

2. Continuum description of the problem

The present study considers a heated infinitely-thin cylindrical shell positioned in the middle of a gap between two concentric cylinders. The energy equation for the heat transfer through a rarefied gas between the shell and the confining inner (or outer) cylinder can be written in a cylindrical-polar coordinate reference frame as

$$\frac{\partial}{\partial r}\left[k(T)\frac{\partial T}{\partial r}r\right]=0 \tag{1}$$

where T is the temperature and r is the radial distance. The thermal conductivity $k(T)$ depends on temperature T with a variable hard sphere (VHS) power law $k(T)=k_0(T/T_0)^{\omega}$, where T_0 is the reference temperature and k_0 is the thermal conductivity at T_0. In the presence of a significant temperature difference between the surface and its environment (*i.e.* in the presence of significant temperature gradients), the variable hard sphere power law must be taken into account in the modelling. The exponent, ω, typically lies in the range between 0.5 (for hard sphere molecules) and 1 (for Maxwell molecules).

The general solution of equation (1) is given by

$$T(r)=(\omega+1)\left[C_1+C_2T_0^{\omega}\ln r\right]^{1/(\omega+1)} \tag{2}$$

where C_1 and C_2 are the constants determined by the temperature-jump boundary conditions. The heat flux can then be written as

$$q(r)=-k(T)\frac{dT}{dr}=-\frac{k_0}{r}\left[C_2\Sigma^{-1+1/(\omega+1)}(1+\omega)^{\omega}\Sigma^{\omega/(\omega+1)}\right] \tag{3}$$

where $\Sigma=C_1+C_2T_0^{\omega}\ln r$.

Several first- and second-order temperature-jump boundary conditions are proposed in the literature for modelling the heat transfer over planar surfaces, as reviewed by Colin [3]. The temperature-jump solutions for the planar geometry are reported to be in good agreement with DSMC data up to a Knudsen number of 0.4 [4]. However, relatively little research has been conducted on their accuracy at nonplanar surfaces. The Knudsen number can be defined in terms of the annular gap between the shell and the confining inner (or outer) cylinder, *i.e.* $Kn=2\lambda/(R_2-R_1)$, where R_1 and R_2 are the radii of the inner and outer cylinders, and λ is the mean free path, defined as $\lambda=(\mu/p)(\pi\bar{R}T_0/2)^{1/2}$ where p is the pressure, μ is the dynamic viscosity and $\bar{R}$ is the specific gas constant. When the mean free path is considered to be globally constant throughout the annular gap, the Knudsen number can be interpreted as a scale parameter which increases monotonically down to the micro/nanoscale.

The second-order temperature-jump boundary condition can be written as

$$T_{jump} = T_{gas} - T_{wall} = \frac{2-\sigma_T}{\sigma_T}\frac{2\gamma}{(\gamma+1)\mathrm{Pr}}\left(\pm\lambda\left.\frac{\partial T}{\partial n}\right|_{wall} - \frac{\lambda^2}{2}\left.\frac{\partial^2 T}{\partial n^2}\right|_{wall}\right) \tag{4}$$

where σ_T is the thermal accommodation coefficient, γ is the specific heat ratio, Pr is the Prandtl number and $\partial/\partial n$ is the derivative in the direction normal to the surface. The specific heat ratio, γ, is assumed to be 5/3 and the Prandtl number, Pr, is taken as 2/3 whilst the thermal accommodation coefficient is specified as $\sigma_T = 1.0$ [5]. We employ a second-order temperature-jump boundary condition with a minus sign in front of the second term of order λ^2 in equation (4), instead of the positive sign originally proposed by Karniadakis *et al.* [6]. The first-order temperature jump condition can be obtained by omitting the second term in equation (4).

The solution of equation (2) can be found in the continuum regime for $T(R_1)=T_1$ and $T(R_2)=T_2$ as

$$T_{cont}(r) = (1+\omega)\left[\frac{T_1(T_1/(1+\omega))^{\omega}\ln(r/R_2) - T_2(T_2/(1+\omega))^{\omega}\ln(r/R_1)}{(1+\omega)\ln(R_1/R_2)}\right]^{1/(1+\omega)} \tag{5}$$

and the heat flux is given as

$$q_{cont}(r) = -k_0\frac{(1+\omega)^{\omega-1}\left[T_2(T_2/(1+\omega))^{\omega} - T_1(T_1/(1+\omega))^{\omega}\right]}{rT_0^{\omega}\ln(R_2/R_1)}. \tag{6}$$

When equation (2) is subjected to temperature-jump conditions, the constants C_1 and C_2 cannot be obtained straightforwardly. These constants are obtained numerically using a technique for the solution of the system of nonlinear equations. The exponent, ω, is set to 0.5 (*i.e.* hard sphere molecules) unless otherwise stated.

3. Formulation of the problem by the nonlinear Shakhov kinetic model

For the heat conduction problem through a rarefied gas between the heated shell and confining (cold) cylinder, the nonlinear Shakhov model kinetic equation is solved numerically and compared with first- and second-order temperature-jump models. The nonlinear Shakhov model kinetic equation is obtained by replacing the problematic collision term of the Boltzmann equation by a simpler collision model [7,8]. The Shakhov model is generally accepted to be more accurate for flows associated with heat transfer compared to the BGK (Bhatnagar, Gross and Krook) model, since the BGK model gives an incorrect Prandtl number equal to unity (*i.e.* Pr=1) instead of Pr=2/3 for monatomic gases [9,10]. In the present study, the Maxwell (specular/diffusive) boundary condition has been employed with the nonlinear Shakhov kinetic equation. It is important to note that the Boltzmann equation describes gas behaviour for all flow regimes and it is valid at all Knudsen numbers. The relation between the microscopic and macroscopic descriptions of gas is postulated as follows: all macroscopic quantities such as density, velocity and temperature are described in terms of the microscopic states, which are the solution of the Boltzmann equation. The state of the gas is given by the distribution function $f(\mathbf{x},\mathbf{v},t)$, where $\mathbf{x}$ is the position vector and $\mathbf{v}$ is the velocity vector of a molecule at time *t*. Once the distribution function is known, all the moments, for example gas density, pressure or heat flux can be computed.

By taking into account the axial symmetry of the shell problem problem in cylindrical coordinates (r, φ, z) in physical space, and considering the cylinders to be long enough to neglect the effects in the z-direction, the Shakhov model kinetic equation (*i.e.* the Boltzmann equation with the Shakhov collision model) can be written at steady-state as [10-12]

$$v_p \cos\theta \frac{\partial f}{\partial r} - \frac{v_p \sin\theta}{r}\frac{\partial f}{\partial \theta} = \frac{(f^S - f)}{\bar{\tau}} \tag{7}$$

using $v_r = v_p \cos\theta$, $v_\theta = v_p \sin\theta$ with $v_p = \sqrt{v_r^2 + v_\varphi^2}$ and $\theta = \arctan(v_\varphi / v_r)$, where the distribution function does not depend on φ (*i.e.* $\partial f / \partial \varphi = 0$) due to the axial symmetry. The term on the right-hand side of equation (7) is the Shakhov collision model

$$Q_S(f, f^S) = \frac{(f^S - f)}{\bar{\tau}} \tag{8}$$

where $\bar{\tau} = \mu / p$ is the relaxation time, μ is the dynamic viscosity and p is the gas pressure, and

$$f^S(r, \mathbf{v}_c) = f^M(r, \mathbf{v}_c)\left[1 + \frac{4m}{15n(r)(k_B T(r))^2} q(r) v_p \cos\theta \left(\frac{m\mathbf{v}_c^{\ 2}}{2k_B T(r)} - \frac{5}{2}\right)\right] \tag{9}$$

with

$$f^M(r, \mathbf{v}_c) = n(r)\left(\frac{m}{2\pi k_B T(r)}\right)^{3/2} \exp\left(-\frac{m\mathbf{v}_c^{\ 2}}{2k_B T(r)}\right) \tag{10}$$

where $\mathbf{v}_c$ is the molecular velocity vector, r is the radial distance, f^M is the local Maxwellian distribution, m is the molecular mass, k_B is the Boltzmann constant, $q(r)$ is the heat flux profile in the radial direction, and $n(r)$ and $T(r)$ are the macroscopic number density and temperature profiles, respectively.

The unknown distribution function, f, in equation (7) depends on four variables (three velocity components and the radial distance). Collisions occur in this four dimensional phase space. However, the dependence of the distribution function on the velocity in the z-direction can be eliminated using a projection procedure. The reduced distribution functions ϕ and ψ are introduced [13] as

$$\phi(r, v_p, \theta) = \int f(r, \mathbf{v}_c) dv_z \text{ and } \psi(r, v_p, \theta) = \int f(r, \mathbf{v}_c) v_z^2 dv_z . \tag{11}$$

The dimensionless variables in the present study are specified as

$$\hat{r} = \frac{r}{R_2}, \quad \hat{T} = \frac{T}{T_2}, \quad \hat{\mathbf{v}} = \frac{\mathbf{v}}{v_0}, \quad \hat{n} = \frac{n}{n_0}, \quad \hat{p} = \frac{p}{p_2}, \quad \hat{\mu} = \frac{\mu}{\mu_2} = \hat{T}^\omega, \quad \hat{f} = \frac{f v_0^2}{n_0}, \quad \hat{q} = \frac{2}{n_0 m v_0^2} q \tag{12}$$

where $v_0 = \sqrt{2k_B T_2 / m}$ is the most probable molecular velocity at T_2 and n_0 is the average numerical density, which is

$$n_0 = \frac{2}{R_2^2 - R_1^2} \int_{R_1}^{R_2} n(r) r dr . \tag{13}$$

The rarefaction parameter is defined as $\delta = R_2 p_2 / (\mu_2 v_0)$ where μ_2 is the dynamic viscosity at T_2. According to equations (12) and (13), the rarefaction parameter can be related to the Knudsen number as $\delta = \sqrt{\pi} R_2 / (2Kn(R_2 - R_1))$, where the rarefaction parameter is inversely proportional to the Knudsen number.

Multiplying equation (7) by 1 and v_z^2, and integrating over v_z according to the relations in equation (11), two reduced kinetic equations are obtained in terms of the reduced distribution functions as

$$\hat{v}_p \cos\theta \frac{\partial \phi}{\partial \hat{r}} - \frac{\hat{v}_p \sin\theta}{\hat{r}} \frac{\partial \phi}{\partial \theta} = \delta \hat{n} \hat{T}^{1-\omega} (\phi^S - \phi) \tag{14}$$

$$\hat{v}_p \cos\theta \frac{\partial \psi}{\partial \hat{r}} - \frac{\hat{v}_p \sin\theta}{\hat{r}} \frac{\partial \psi}{\partial \theta} = \delta \hat{n} \hat{T}^{1-\omega} (\psi^S - \psi) \quad (15)$$

where

$$\phi^S = \phi^M \left[1 + \frac{4}{15} \frac{1}{\hat{n}\hat{T}^2} q(\hat{r}) \hat{v}_p \cos\theta \left(\frac{\hat{v}_p^2}{\hat{T}} - 2 \right) \right] \quad (16)$$

$$\psi^S = \psi^M \left[1 + \frac{4}{15} \frac{1}{\hat{n}\hat{T}^2} q(\hat{r}) \hat{v}_p \cos\theta \left(\frac{\hat{v}_p^2}{\hat{T}} - 1 \right) \right] \quad (17)$$

with

$$\phi^M = \frac{\hat{n}}{\pi \hat{T}} \exp\left(-\frac{\hat{v}_p^2}{\hat{T}} \right), \quad \psi^M = \frac{\hat{T}}{2} \phi^M, \quad \text{and} \quad \hat{v}_p^2 = \hat{v}_r^2 + \hat{v}_\theta^2. \quad (18)$$

The macroscopic quantities can be found in terms of the reduced distribution functions, ϕ and ψ, as [10-12]

$$\hat{n}(\hat{r}) = \int_0^{2\pi} \int_0^{\infty} \phi \hat{v}_p d\hat{v}_p d\theta \quad (19)$$

$$\hat{T}(\hat{r}) = \frac{2}{3\hat{n}(\hat{r})} \int_0^{2\pi} \int_0^{\infty} (\hat{v}_p^2 \phi + \psi) \hat{v}_p d\hat{v}_p d\theta \quad (20)$$

$$\hat{q}(\hat{r}) = \frac{2}{3\hat{n}(\hat{r})} \int_0^{2\pi} \int_0^{\infty} (\hat{v}_p^2 \phi + \psi) \hat{v}_r \hat{v}_p d\hat{v}_p d\theta \quad (21)$$

with the equation of state $p = nk_B T$.

The present study implements the Maxwell (specular/diffusive) gas-wall interaction model. In the Maxwell model, a proportion of the molecules, α, are reflected diffusively from the surface while the remaining proportion, $1-\alpha$, are reflected specularly from the surface. The specular reflection model is straightforward. When a molecule strikes a wall, the normal velocity component of the molecule is simply reversed and directed back into the flow domain; the other velocity components remain unchanged. In other words, the molecules bounce off the wall with the same velocity distribution except their reversed normal velocity components, while their other velocity components remain unchanged. For the diffuse reflection, the reflected tangential velocity of the molecule is considered to be uncorrelated with its impinging velocity and the molecules bounces off the surface according to the local Maxwell distribution. Therefore, for equation (14), the Maxwell boundary conditions at the inner and outer cylinders can be written as

$$\phi(r_1, \hat{v}_p, \theta) = (1-\alpha)\phi(r_1, \hat{v}_p, \pi - \theta) + \alpha \frac{n_1}{\pi \hat{T}_1} \exp\left(-\frac{\hat{v}_p^2}{\hat{T}_1} \right), \quad -\pi/2 \le \theta \le \pi/2 \quad (22)$$

$$\phi(1, \hat{v}_p, \theta) = (1-\alpha)\phi(1, \hat{v}_p, \pi - \theta) + \alpha \frac{n_2}{\pi} \exp(-\hat{v}_p^2), \quad \pi/2 \le \theta \le 3\pi/2 \quad (23)$$

associated with impermeability assumption [10]:

$$n_1 = 2\sqrt{\frac{\pi}{\hat{T}_1}} \int_0^{\infty} \hat{v}_p^2 d\hat{v}_p \int_{-\pi/2}^{\pi/2} \phi \cos\theta d\theta \quad \text{and} \quad n_2 = -2\sqrt{\pi} \int_0^{\infty} \hat{v}_p^2 d\hat{v}_p \int_{\pi/2}^{3\pi/2} \phi \cos\theta d\theta \quad (24)$$

where $0 \le \alpha \le 1$ is the accommodation coefficient, $r_1 = R_1/R_2$ is the ratio between cylinder radii and $\hat{T}_1 = T_1/T_2$ is the ratio between the inner and outer cylinder temperatures. The boundary conditions for equation (15) can be written in a very similar manner. In the present study, the accommodation coefficient, α, is specified as unity (*i.e.* $\alpha = 1$), and therefore only fully diffusive reflections are considered.

The heat transfer problem under consideration is solved using a computational scheme that discretizes equation (14) (and equation (15) in a similar manner) into

$$v_{p_k} \cos\theta_m \frac{\phi^j_{i,k,m} - \phi^j_{i-\beta,k,m}}{\beta \Delta \hat{r}} - v_{p_k} \frac{\sin\theta_m}{\hat{r}_i} \frac{\phi^j_{i,k,m} - \phi^j_{i,k,m-1}}{\Delta\theta} = \delta \hat{n}_i^{j-1} (\hat{T}_i^{1-\omega})^{j-1} ((\phi^S_{i,k,m})^{j-1} - \phi^j_{i,k,m}) \quad (25)$$

where i and m are the indices in the r and θ directions, respectively, j is the iteration index, $\beta = sign(\cos\theta_m)$ is the sign of $\cos\theta_m$; and k is the index for the Gauss quadrature points. In the present study, the annular clearance between the cylinders $[R_1/R_2, 1]$ and the angular period $[0, \pi]$ are divided into 100 equal intervals. For the velocity space, the magnitudes of the molecular velocity, c_{p_k}, are mapped onto 20 Gauss quadrature points with their associated weights. Integrals (19)-(21) for calculating the macroscopic quantities are obtained using the Gauss quadrature rule at each iteration. When the difference between macroscopic temperatures of successive iterations is smaller than a certain tolerance value, the convergence criterion is assumed to be fulfilled.

4. Results and discussions

The second-order temperature profiles of the shell problem are shown in figure 1. The temperature-jump models predict the profound effects of the surface shape (*i.e.* convex/concave) on the degree of temperature jump. The bold vertical lines in the middle of the figures represent the location of the shell. The profiles on the left (blue) and right (red) sides of the shell are created by the concave and convex sides of the heated shell, and they have obviously identical curvatures and surface areas. The remarkable differences in the degrees of temperature jump are attributed to the effects of surface shape. Figure 1 clearly reveals the role of the surface shape, since the degrees of temperature jump would be equal on both sides of the shell in an equivalent geometry between parallel plates.

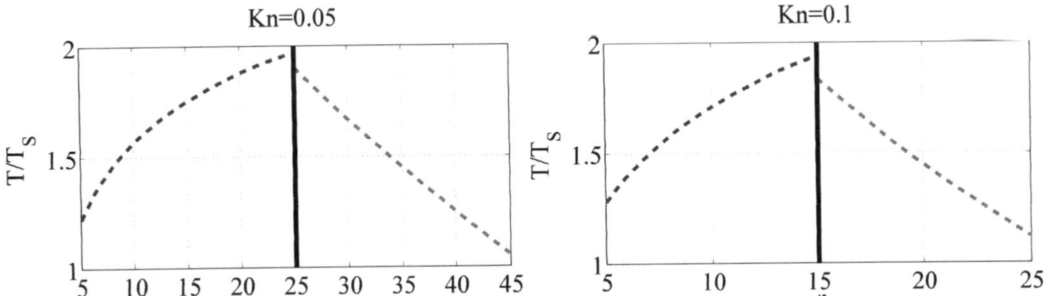

Figure 1. Second-order temperature profiles over the concave and convex surfaces of a heated cylindrical shell. The profiles on the left (blue) and right (red) sides of the shell are created by the concave and convex shell surfaces, respectively. The shell is positioned midway between two cold cylinders and it has a radius of (a) 25λ and (b) 15λ, where λ is the mean free path. The ratio of the temperature of the shell to the temperature of the inner (or outer) cylinder is 2. (a) *Kn*=0.05 and (b) *Kn*=0.1.

The effect of the Knudsen number on the temperature jumps over the concave and convex surfaces of the heated cylindrical shell are demonstrated in figure 2. The temperature jumps are obtained from first- and second-order jump models and the nonlinear Shakhov model solution. Figure 2 presents the results in terms of the Knudsen number, where the corresponding rarefaction parameter can be identified using the formula $\delta = \sqrt{\pi}R_2/(2Kn(R_2 - R_1))$. According to this formulation, the rarefaction parameter is $\delta_{convex} = \sqrt{\pi}R_2/(2Kn(R_2 - R_S))$ on the convex side, whilst it is $\delta_{concave} = \sqrt{\pi}R_S/(2Kn(R_S - R_1))$ on the concave side of the shell, where R_S the radius of the shell. Therefore, the rarefaction parameter is slightly higher (*i.e.* *Kn* is slightly lower) on the convex side of the shell; in fact, this supports our important conclusions about the significant effects of curvature.

In figure 2, the concave and convex sides of the shell have the same surface area and same degree of curvature. The results remarkably show that the degree of temperature jump over a convex surface is markedly larger than that over a concave surface. Moreover, the jump models predict the amount of temperature jump relatively well up to *Kn*=0.3. However, beyond *Kn*=0.3, the predictions of the temperature-jump models start to deviate from the nonlinear Shakhov model solution. Figure 2 also shows that the second order jump solution is slightly in better agreement with the solution of nonlinear Shakhov model equation compared to the first-order solution in predicting the amount of temperature jump over the shell.

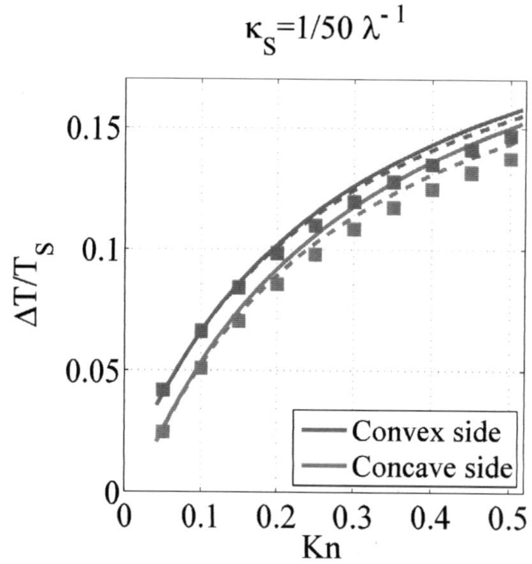

Figure 2. Temperature jumps over the concave and convex surfaces of a heated cylindrical shell against the Knudsen number. The red and blue curves illustrate temperature jumps over the convex and concave sides of the shell, respectively. The solid and dashed curves are the first- and second-order temperature-jump solutions, respectively. The square symbols illustrate the temperature jumps obtained from the solution of nonlinear Shakhov model kinetic equation. The shell curvature is $\kappa_S = 1/50\lambda^{-1}$ (*i.e.* the shell radius is $R_S = 50\lambda$).

Figure 3 shows that the DSMC data and the nonlinear Shakhov model solution are in excellent agreement. On the other hand, the accuracy of temperature-jump models deteriorates when the cylinder surface is cooler than the surrounding gas. This has been reported before for the planar heat conduction problem by Pan *et al.* [14] when trying to determine the temperature-jump coefficient of a first-order temperature-jump boundary condition. Using the DSMC method, Pan *et al.* [14] deduced different coefficients for the heat flow from a hot gas to a colder wall compared to the heat flow from a hot wall to a colder gas.

The magnitudes of the heat fluxes over the concave and convex surfaces of the shell are demonstrated in figure 4. The profiles are obtained from the solution of the nonlinear Shakhov model kinetic equation. The shell curvature is $\kappa_S = 1/6\lambda^{-1}$ (*i.e.* the shell radius is $R_S = 6\lambda$). The results remarkably show the dramatic effect of the surface shape on the heat flux; the heat flux is significantly higher over the convex surface compared to the concave surface of the shell.

The variation of the normalized heat fluxes over the concave and convex surfaces are shown in figure 5 for increasing values of curvature of the shell. The heat fluxes are obtained from the first-order temperature-jump model and normalized by the heat flux of the continuum regime as obtained in equation (6). For Knudsen numbers of 0.05 and 0.1, the first and second-order order jump solutions are expected to be very close to each other. Figure 5 shows that, as the curvature of the shell increases, the heat fluxes over the concave and convex surfaces of the shell start to show significant differences. It can be seen that the heat flux over the concave side becomes considerably lower than the heat flux over the convex side of the shell, beyond a certain curvature value depending on the Knudsen number.

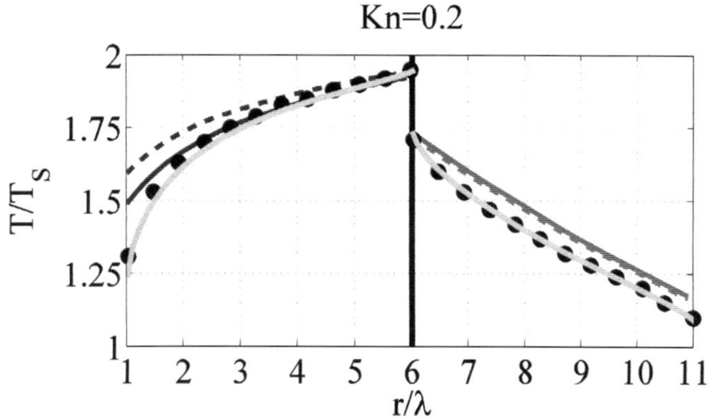

Figure 3. Temperature profiles over the concave and convex surfaces of the heated cylindrical shell. The profiles on the left- and right-hand sides of the shell are created by the heated concave and convex shell surfaces, respectively. The red and blue curves illustrate the temperature profiles obtained from the jump models over the convex and concave sides of the shell, respectively. The solid and dashed curves are the first- and second-order temperature-jump solutions, respectively. The green curves illustrate the temperature profiles obtained from the solution of nonlinear Shakhov model kinetic equation. The black dots are the DSMC data.

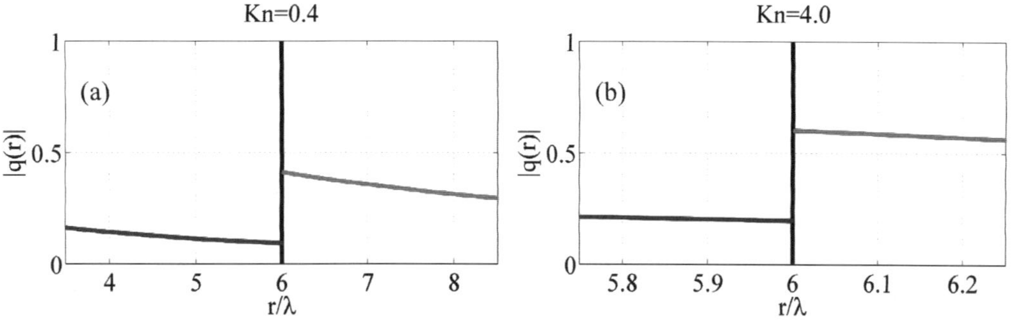

Figure 4. Magnitudes of the heat fluxes in the radial direction over the concave and convex surfaces of the shell. The profiles are obtained from the solution of the nonlinear Shakhov model kinetic equation. (a) *Kn*=0.4, (b) *Kn*=4.0.

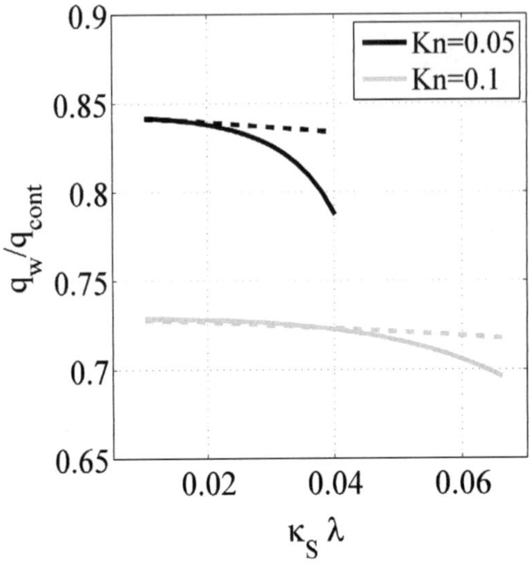

Figure 5. Normalized heat fluxes at the shell surface predicted by the first-order temperature-jump model against the curvature of the shell. The dashed and solid curves are the heat fluxes over the convex and concave surfaces of the shell, respectively, and q_{cont} is the heat flux in the continuum regime as $\mathrm{Kn} \to 0$.

Acknowledgments
The authors would like to thank Manual Vargas and Stefan Stefanov from the Institute of Mechanics, Bulgarian Academy of Sciences, for allowing the authors to use their DSMC code. The first author would also like to thank Irina Graur and Pierre Perrier for their warm welcome during his stay in Marseille. The research leading to these results has received funding from the European Community's Seventh Framework Programme (ITN - FP7/2007-2013) under the GASMEMS project with a grant agreement no: 215504.

References

[1] Pelesko J A and Bernstein D H 2003 *Modeling MEMS and NEMS* (Boca Raton, FL: Chapman & Hall/CRC)
[2] Nguyen N -T and Wereley S T 2006 *Fundamentals and Applications of Microfluidics* (Boston: Artech House)
[3] Colin S 2012 *Journal of Heat Transfer* **134** 020908
[4] Wadsworth D C 1993 *Phys. Fluids A* **5** 1831
[5] Sharipov F 2011 *J. Phys. Chem. Ref. Data* **40** 023101
[6] Karniadakis G E, Beskok A and Aluru N 2005 *Microflows and Nanoflows: Fundamentals and Simulation* (Berlin: Springer)
[7] Chu C K 1965 *Phys. Fluids* **8** 12
[8] Shakhov E M 1968 *Fluid Dynamics* **3** 156
[9] Shakhov E M 1968 *Fluid Dynamics* **3** 142
[10] Graur I A and Polikarpov A P 2009 *Heat Mass Transfer* **46** 237
[11] Sharipov F and Bertoldo G 2006 *J. Vac. Sci. Technol. A* **24** 2087
[12] Pantazis S and Valougeorgis D 2010 *Eur. J. Mech. B/Fluids* **29** 494
[13] Zhuk V I, Rykov V A and Shakhov E M 1973 *Fluid Dynamics* **8** 620
[14] Pan L S, Ng T Y, Xu D, Liu G R and Lam K Y 2002 *J. Micromech. Microeng.* **12** 41

1st European Conference on Gas Micro Flows (GasMems 2012) IOP Publishing
Journal of Physics: Conference Series **362** (2012) 012018 doi:10.1088/1742-6596/362/1/012018

Carbon Dioxide Transport in Carbon Nanopores

Dimitrios Mantzalis, Nikolaos Asproulis, Dimitris Drikakis
Fluid Mechanics and Computational Science Department, Cranfield University, UK
E-mail: n.asproulis@cranfield.ac.uk

Abstract. In the present study atomistic simulations are employed for investigating the CO_2 transport properties through pores that are of slit, cylindrical and scroll configuration. We aim to investigate any possible differences in the adsorption process among the three different geometries focusing on the adsorption type that occurs along with pore filling and pore emptying mechanisms. Carbon-slit pores are of widths between 0.8 and 2.0 nm. The simulated Single-Walled Carbon Nanotubes (SWNTs) are of 1.08 and 2.17 nm ((8,8) and (16,16) respectively) while for the Carbon Nanoscrolls (CNSs) the inner diameter corresponds to a (6,6) SWNT with an intralayer distance of 0.4 to 1.0nm.

1. Introduction

Climate change due to global warming is one of the main challenges that modern world faces. The development of novel techniques and methods able to capture and separate green house emissions and primarily carbon dioxide from fossil fuels, atmosphere and natural gas has attracted the industrial and scientific interests[1, 2]. Nanotechnology and more specifically carbon nanostructures as a relative new scientific field can provide solutions based on materials designed to offer superior properties, compared to traditional filtering approaches, in terms of selectivity, capacity and flux rates[3, 4].

Over the last two decades a number of studies, both experimental and computational, have been carried out aiming to investigate the transport properties of gases through carbon nanostructures such as carbon nanotubes (CNTs), carbon slit pore (CSP) and carbon nanoscrolls (CNS) [5, 6, 7, 8]. These different forms of nanostructures have been utilised for developing novel carbon capture devices able to operate under a wide range of conditions. Carbon Nanoscrolls (CNS), firstly fabricated in 2003 [9], are novel carbon based materials initially reported to be promising for H_2 storage [10]. CNS display a spiral form and are schematically obtained by wrapping a graphite sheet in a helical formation.

Computational and experimental evidence show that both gases and liquids can be transported through carbon nanopores, such as CNTs, orders of magnitude faster compared to ordinary microporous materials such as zeolites [11]. Their nearly frictionless walls combined with their very small pore size offer increased transport efficiency making them ideal adsorbents [3, 4].

Although great amount of studies that have been devoted in confined liquids within CNTs and CSPs, pure or functionalized, little attention has been attracted to CNSs and to comparative studies between the three carbon allotropes[12, 13, 8]. In the present work, we draw our attention in the CO_2 transport and adsorption properties of three carbon allotropes and specifically on carbon nanotubes, carbon slit pores and carbon nanoscrolls. Specifically, we perform a direct comparison among the three carbon allotropes in terms of CO_2 capture and we aim to investigate any possible differences in the adsorption process among the three different geometries focusing in the adsorption type that occurs along with pore filling and pore emptying mechanisms. The present article is structured as follows: Section 2 outlines the simulation methodology and the models employed, Section 3 provides the results of the study along with the corresponding discussion and Section 4 the main conclusions together with future work.

2. Simulation Method

The nanotubes considered in the present work are of type armchair while their carbon atoms, C^*, are modeled as Lennard-Jones spheres, holding them in their equilibrium crystallographic positions during the simulations. In practice, carbon nanotubes, mainly those with large pores, are flexible materials; however the choice of frozen curvature reduces the computational cost. Several studies have shown that the deviations observed into these two approaches can be initially ignored; self-diffusion computed for simple fluids in rigid tubes has been found to predict higher values compared with cases that flexibility is involved in the computations [14, 8]. This effect are negligible in narrower pores in cases where the confined liquid reaches high densities. The choice of frozen curvature in combination with the fixed relevant motion between C^* atoms, reduces the computational cost compared with a flexible model. However several studies have shown that the disagreements between the two techniques can be ignored in first place. In the present work, the nanotubes have been assumed to be rigid as in case of the narrow (8,8) SWNT, the C^* vibrations are not expected to affect the diffusion mechanism of carbon dioxide molecules bearing in mind the small overestimation of self-diffusion as stated by Skoulidas et al. [15]. The two indices used for describing a SWNT, such as (8,8), characterize the way that the graphene sheet is wrapped and the CNTs diameter is given by:

$$d = \frac{a}{\pi}\sqrt{(n^2 + nm + m^2)} \tag{1}$$

where $a = 0.246nm$.

The microcanonical ensemble (NVE) [16] has been used for the integration of the equations of motion, where the number of molecules in the system (N), the volume of the simulations box (V) and the system's total energy (E) are kept constant. The simulations have been carried out with LAMMPS [17] with the integration time step is $\delta t = 0.5\ fs$ for avoid large fluctuations of the total energy and the temperature of the system. The system has been equilibrated for $200 - 600\ ps$, period that has been disregarded for the further calculations according to the charge of the CNTs. To ensure that the temperature fluctuations will not considerably deviate from the set value of 300 K, a temperature rescaling has been employed during the equilibration period every 1000 time steps.

The carbon dioxide molecules are simulated through the 3-site model suggested by Murthy et al. [18]. Previous computational studies regarding the interaction potential between carbon nanostructures and CO_2 have shown that 3-site models offer more realistic representations with respect to 1-site models and the differences between the available linear models in the literature is negligible[19]. A Lennard Jones (LJ) potential is employed to mimic the Van der Waals interactions with the ε and σ parameters being $\varepsilon_{CC}/k_B = 29.0\ K$, $\varepsilon_{OO}/k_B = 83.1\ K$, $\sigma_{CC} = 2.785\ \mathring{A}$ and $\sigma_{OO} = 3.014\ \mathring{A}$ and with CO_2 mass being equal with 44.010 g/mol[18]. At carbon and oxygen sites partial charges have been introduced with $q_C = 0.5957$ and $q_O = q_C/2$, respectively, and mixed LJ parameters are computed based on the Lorentz-Berthelot rule. For the bonds and angles of carbon dioxide molecules the SHAKE algorithm [17] is utilised with the equilibrium length and angular values being reset at every time step to their equilibrium figures $l_{CO_2} = 1.16\ \mathring{A}$ and $C - O - C = 180°$ respectively. The interaction LJ parameters used between the nanopores and the carbon dioxide are $\varepsilon_{C^*}/k_B = 28.0068\ K$ and $\sigma_{C^*} = 3.4\ \mathring{A}$ [20], where C^* corresponds to carbon atom of the nanopores.

3. Results and Discussion

Figure 1 shows for pressure $P = 5\ bar$the layering formation close to the nanopores surface for a (8,8) CNT, a CNS with interlayer distance $7.0\mathring{A}$ and a CSP with the graphene layers being $8.1\mathring{A}$ away from each other. The volume close to the nanopore in each case is divided to bins parallel to the nanostructure graphitic walls with characteristic CO_2 structures being formed in the proximity of the nanopores surface independently to the shape of the pore. The number and the relative strength of the layers formed near depend upon a number of parameters such as diameter in case of CNTs, interlayer distances, partial charges and gas-structure interactions.

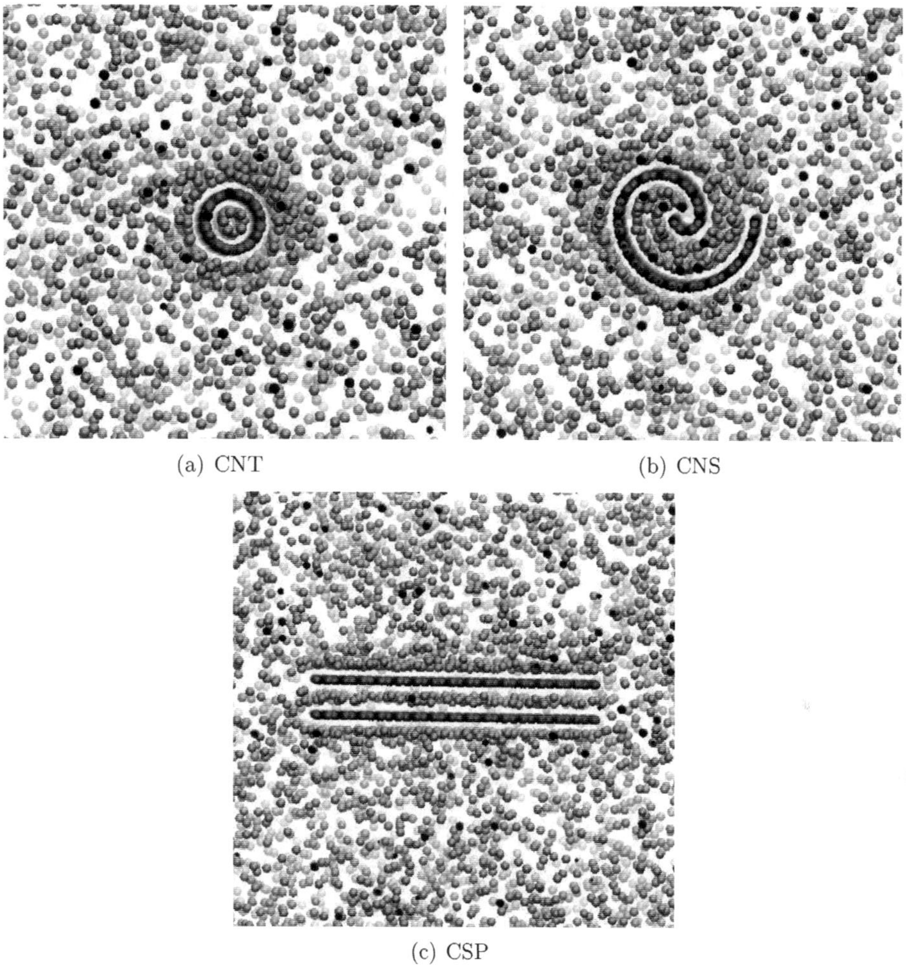

(a) CNT (b) CNS

(c) CSP

Figure 1. Projections of CO_2 distributions (a) (8,8) CNT (b) $CNS - 7.0\mathring{A}$ (c) $CSP - 8.1\mathring{A}$

Figure 2 shows the number of CO_2 molecules adsorbed per carbon atom C^* as a function of time for all the three CNTs ((8,8), (12,12) and (16,16) respectively) that have been simulated. It is noticed that for the three cases as time evolves the number of carbon dioxide molecules adsorbed both in the internal and external surface of the CNT approaches a mean value with an additional fluctuating component that varies with the radius of the CNT and gets higher values for the (16, 16) case. The smaller adsorbing capacity is equal to 0.045 CO_2 molecules per carbon atom C^* and is noticed for $5.37\mathring{A}$ radius. For the $(8,8)$ nanotube its radius compared to the van der Waals one is 1.58 time higher and therefore does not allow the formation of a secondary layer within its internal volume explaining the reduced adsorbing performance. As the CNT diameter increases to $8.10Å$, for the $(12, 12)$ CNT, the non-dimensional adsorption obtains its maximum mean value 0.068; however the standard deviation of the fluctuating component is slightly increased compared to the previous case mainly due to the smaller binding energy[21]. As the larger nanotubes such as (16,16) (see Fig. 2) the surface-to-volume ratio of the CNT increases leading to smaller number of carbon dioxide molecules adsorbed per C^*[14].

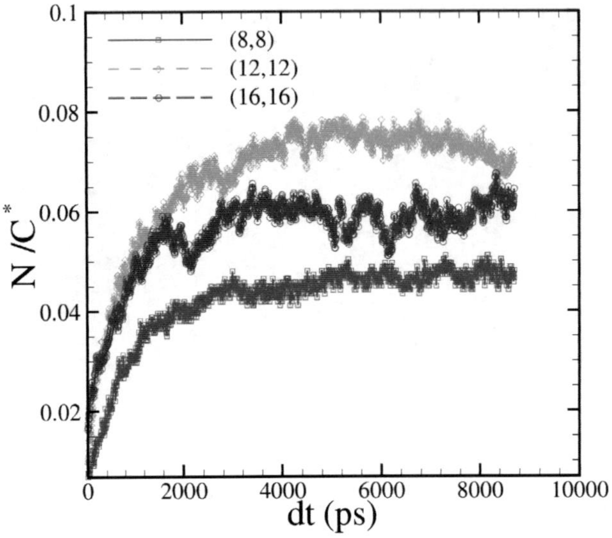

Figure 2. CO_2 molecules adsorbed per carbon atom C^* as a function of time for (8,8), (12,12) and (16,16) CNTs

Figure 3 shows the number of CO_2 molecules adsorbed per carbon atom C^* as a function of time for CNSs with $5.6Å$, $7.0Å$ and $8.3Å$ layering distances respectively. Compared the CNT case larger adsorbing values are reported mainly due to the higher surface-to-volume ratio that the nanoscrolls present compared to the CNTs and the lower energy needed for a molecule to enter the scrolls internal volume. As the interlayer distance of the nanoscroll increases from $5.6Å$ to $7.0Å$ the average carbon dioxide adsorption decreases and then as it further increases to $8.3Å$ it settles back to its initial values. It can be noticed that as the interlayer increases the surface-to-volume ratio and the standard deviation of the fluctuating component decreases showing a preference for smaller binding energies.

Figure 4, similarly to Figs. 2 and 3, demonstrates the number of carbon dioxide molecules adsorbed as function of time for the three CSPs employed in the present study. The main points regarding the adsorbing capacity of the slit pores with respect to time and interlayer distances are similar to previous cases. A main difference though that can be identified is the almost zero adsorption of the slit pore with $5.3Å$ distance between the graphene sheets. For this interlayer distance the carbon dioxide molecules cannot enter in the interlayer region and the binding energy of those that lay in the proximity of the outer surface is small explaining therefore the poor adsorption performance.

Figures 5 and 6 demonstrate the gravimetric and volumetric carbon dioxide uptake of the different nanopores simulated as a function of pressure. As pressure increases the number of

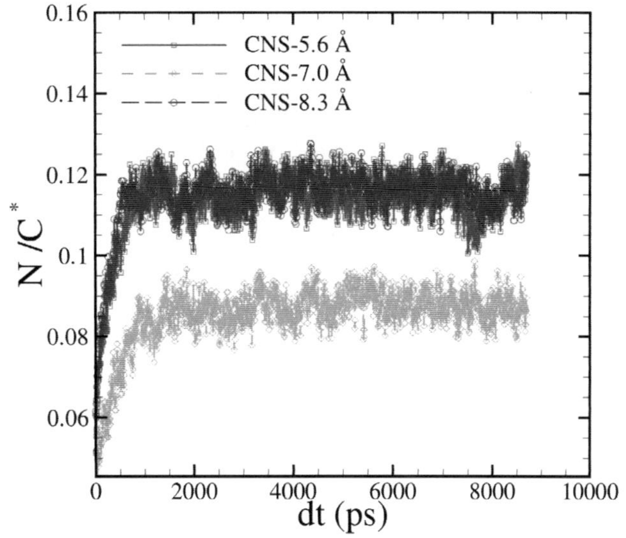

Figure 3. CO_2 molecules adsorbed per carbon atom C^* as a function of time for CNSs with $5.6\mathring{A}$, $7.0\mathring{A}$ and $8.3\mathring{A}$ layering distances respectively

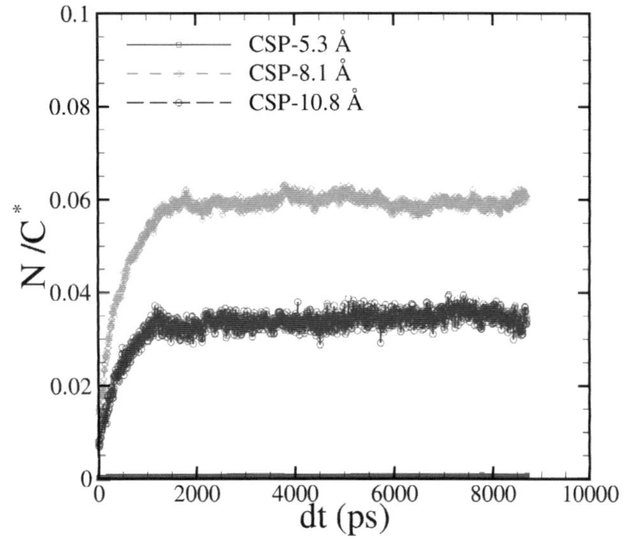

Figure 4. CO_2 molecules adsorbed per carbon atom C^* as a function of time for CSPs with graphene sheets being $5.3\mathring{A}$, $8.1\mathring{A}$ and $10.8\mathring{A}$ apart respectively

CO_2 increases leading to a more closely packed arrangements of carbon dioxide molecules and consequently to increased numbers near or within the nanopores. CSP arrangements in both cases (Fig. 5(c) and 6(c)) present lower adsorption capacity among the structures examined and as pressure increases it looks to approach their saturation point much faster as the slope of the graph continuously decreases. Furthermore it has to be noticed that the binding energies values reported for slit pores are lower compared to nanotubes and nanoscrolls making CSPs less attractive structures for gas storage and filtering.

The volumetric and gravimetric uptake CNTs and CNSs looks to scale almost linearly with pressure and, apart from $CNS - 7\mathring{A}$ case, and we do not notice any saturation indications. For the carbon nanotubes the $(8,8)$ is the one that presents higher uptake values for both measures since its radius allows only one very strong internal carbon dioxide layer leading at the same time to higher adsorption energies as has been reported in previous studies in the literature [14]. In the volumetric uptake as the diameter of nanotube increases the surface-to-volume ratio decreases as well contributing to lower adsorbing levels. The same applies to CNSs with larger interlayer distance as shown in Fig. 6(b) as the interlayer distance increases from $5.6\mathring{A}$ to $8.0\mathring{A}$ the uptake decreases in a non-linear fashion.

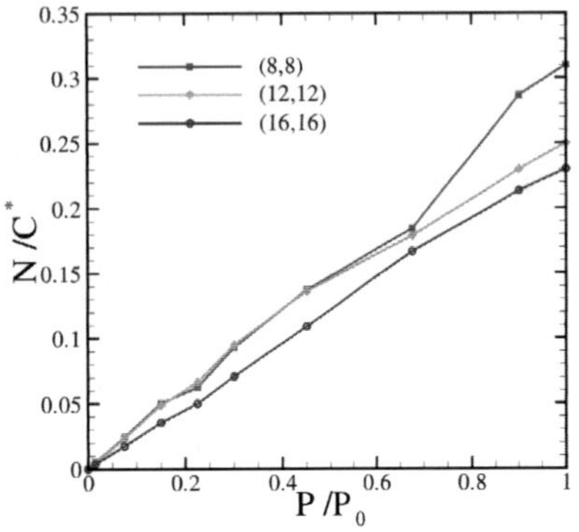

(a) CNT

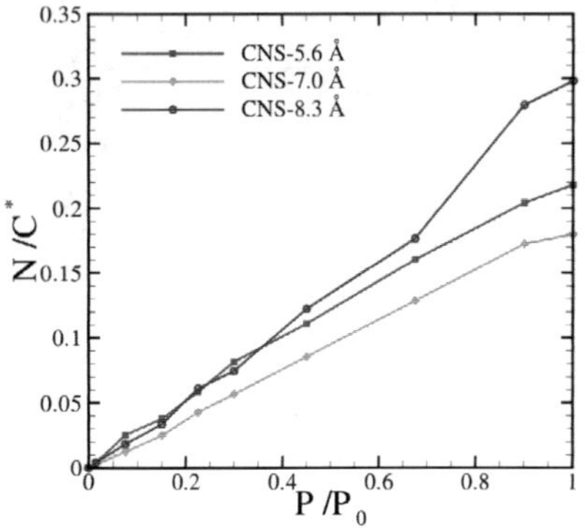

(b) CNS

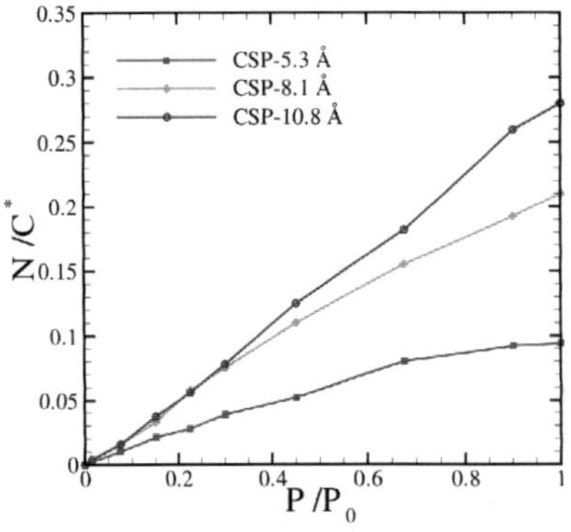

(c) CSP

Figure 5. Gravimetric CO_2 uptake for (a) a carbon nanotube (b) carbon nanoscroll (c) carbon slit pore at 300 K

4. Conclusions

In conclusion, in the present paper the effect of the nanopore structure in terms of carbon dioxide uptake is examined for a number of pressures. Three types of nanopore are considered:

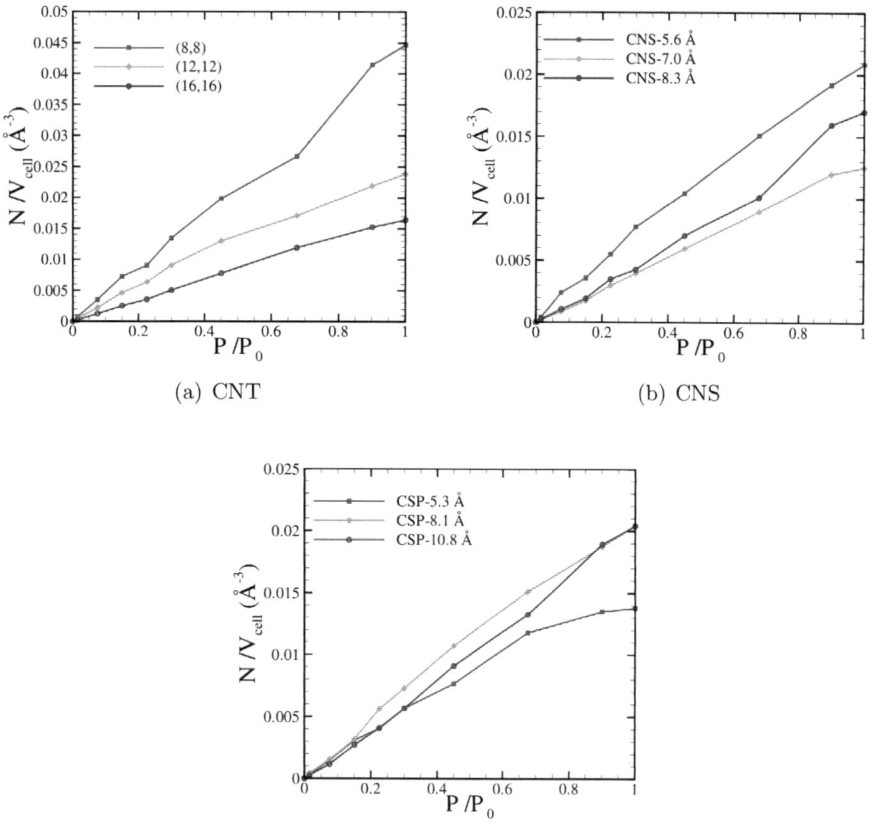

(a) CNT (b) CNS

(c) CSP

Figure 6. Volumetric CO_2 uptake for (a) a carbon nanotube (b) carbon nanoscroll (c) carbon slit pore at 300 K

nanotubes, nanoscrolls and slit pores with equivalent geometrical characteristics and in all cases the formation of carbon dioxide layers in the vicinity of the pores graphitic walls is noticed. The layers formed close to the nanopores wall vary based on the structure and its properties, such as radius or interlayer distance, showing that the geometry of the structure has be cautiously selected for optimum performance. In terms of uptake the lower values have been reported for slip pores whereas for nanotubes and nanoscrolls higher figures have been obtained for both volumetric and gravimetric uptake. A clear selection between CNTs and CNSs cannot be easily performed and additional studies have to be carried out for exploring the performance of these nanopores for pressure and temperature conditions. Towards this direction maybe Grand Canonical Monde Carbo simulations have to be employed for faster scanning of the entire phase-space.

5. References

[1] Aaron D and Tsouris C 2005 *Separation Science and Technology* **40** 321–348
[2] Karagiannis I C and Soldatos P G 2010 *Energy Policy* **38** 3891–3897
[3] Hinds B J, Chopra N, Rantell T, Andrews R, Gavalas V and Bachas L G 2004 *Science* **303** 62–65
[4] Majumder M, Chopra N, Andrews R and Hinds B J 2005 *Nature* **438** 44

[5] Huang L, Zhang L, Shao Q, Lu L, Lu X, Jiang S and Shen W 2007 *Journal of Physical Chemistry C* **111** 11912–11920

[6] Ellison M D, Crotty M J, Koh D, Spray R L and Tate K E 2004 *Journal of Physical Chemistry B* **108** 7938–7943

[7] Ackerman D M, Skoulidas A I, Sholl D S and Johnson J K 2003 *Molecular Simulation* **29** 677–684

[8] Mantzalis D, Asproulis N and Drikakis D 2011 *Physical Review E - Statistical, Nonlinear, and Soft Matter Physics* **84** 066304

[9] Viculis L H, Mack J J and Kaner R B 2003 *Science* **299** 1361

[10] Mpourmpakis G, Tylianakis E and Froudakis G 2006 *Journal of Nanoscience and Nanotechnology* **6** 87–90

[11] Skoulidas A and Sholl D 2003 *Journal of Physical Chemistry A* **107** 10132–10141

[12] Palmer J C, Moore J D, Roussel T J, Brennan J K and Gubbins K E 2011 *Physical Chemistry Chemical Physics* **13** 3985–3996

[13] Mpourmpakis G, Froudakis G E, Lithoxoos G P and Samios J 2006 *Nano Letters* **6** 1581–1583

[14] Mantzalis D, Asproulis N and Drikakis D 2011 *Chemical Physics Letters* **506** 81–85

[15] Skoulidas A I, Ackerman D M, Johnson J K and Sholl D S 2002 *Physical Review Letters* **89** 185901

[16] Allen M and Tildesley D 1987 *Computer Simulation of Liquids*

[17] Plimpton S 1995 *Journal of Computational Physics* **117** 1–19

[18] Murthy C, Singer K and McDonald I 1981 *Molecular Physics* **44** 135–143

[19] Mantzalis D, Asproulis N and Drikakis D 2010 *2nd European Conference on Microfluidics*

[20] Alexiadis A and Kassinos S 2008 *Chemical Physics Letters* **460** 512–516

[21] Chkhartishvili L and Murusidze I 2010 *Material Sciences and Applications* **1** 223–46

Characterization of CO_2 Flow Through Charged Carbon Nanotubes

Dimitrios Mantzalis, Nikolaos Asproulis, Dimitris Drikakis
Fluid Mechanics and Computational Science Department, Cranfield University, UK
E-mail: n.asproulis@cranfield.ac.uk

Abstract. The equilibrium transport of CO_2 through infinitely long charged and uncharged single-walled carbon nanotubes has been examined by employing molecular dynamics simulations. It has been shown that the molecular transport concludes into a Fickian diffusion for all the examined cases. Ballistic and single-file diffusion mechanisms have been met especially in cases where the degree of loading is low since at higher pressures the electrostatic interactions derived by the charge of the nanotubes is canceled by the stronger SWNT-CO_2 interactions.

1. Introduction

In 1991, Iijima [1] enriched the family of carbon based materials with a forth member known as Carbon Nanotubes (CNT). Numerous studies have been carried out since then, both numerical and experimental, aiming to shed lights in nanofluidics topics. The investigation of confined fluids to nanopores encompasses various aspects with the diffusive and adsorptive behavior of gas molecules in nanosized porous materials is considered of vital importance for a series of applications spanning from materials science and gas separation [2, 3, 4, 5] to nanoelectronics and biology [6, 7]. Liquids in confined nanoflows exhibit considerable differences in the diffusion mechanisms compared with flows in bulk. The differences can become more eloquent when the molecular diameter of the confined fluid obtains a size comparable to the characteristic length of the confining environment [8, 9, 10]. Over the past years, a great number of studies have been focused in studying the potential capabilities of CNTs in the fields of gas filtering, separation and storage. Both computational and experimental works have attempted to show that fluids can experience a transport inside single walled carbon nanotubes (SWNTs) characterized by orders of magnitude faster compared to traditional microporous materials [11] such as the zeolites, which are considered among the conventional means of filtering and storage. Their small pore size along with the nearly frictionless curved graphitic surfaces provide a rare combination of high selectivity and transport efficiency [12] making them ideal gas adsorbents [13].

However, the amount of studies relating with gasses' nanofluidics is much smaller compared with water which is the greatest "rival" [14, 15, 9, 8]; hydrogen keeps the greatest portion within the gas studies for energy reasons [16, 3] with carbon dioxide [5, 17, 18] to possess an even lower position in the "preferences". This tendency altered partially its direction since the green house effect exerted extra pressure in the already large volume of environmental problems that modern world faces. Carbon dioxide is known as the main gas responsible for the increase of temperature with the green house effect to be the most obvious result. The atmospheric concentration is continuously rising [19] with billions of tones of CO_2 to be produced each year basically originated from the fossil fuel plants and the vehicles. The increasing quantities of CO_2 dispersed in the atmosphere have emerged the scientific community in developing new CNT-based technologies aiming to reduce its concentration through sequestration and storage techniques [11, 20, 5, 18]. Recent numerical studies, have investigated how CO_2 is formulated in regions close to the CNTs [17, 5]. Both groups showed that carbon dioxide accumulated in the proximity of the carbon walls forming concentric layers with the tubes. This behavior has been considered as a CO_2-philic feature accompanying the hydrophobic properties of the curved graphitic walls comprising into repelling for example water molecules from the carbon surface while they tend to stay together because of the stronger hydrogen bonding network that has been developed. The former property is explainable by the stronger attachment in case of CO_2-CNT compared to $CO_2 - CO_2$ molecules. In one of our previous studies [5] we have shown that well defined patterns of CO_2 layers formations have been predicted, with the parameters that affect the adsorbate's behavior to be concluded in the pore diameter and the gas-structure interactions. Molecular simulations have been carried out indicating that besides the primary layers formed around the carbon surface, secondary and some times tertiary layers have been observed in the internal regions of the tubes depending on the pore's width. Further focusing in the permeability and the selectivity of several porous membranes dedicated in gas separation both the adsorption and the diffusion processes have to be explored for the captured gas species. In this context, Skoulidas et al. [21] investigated the adsorption and the diffusion mechanisms of CO_2 and N_2 within SWNTs at room temperature for different pore widths. The employed techniques allowed them to study how two different gas species behave confined into the interior of rigid tubes. In case of carbon dioxide, both spherical and linear models have been studied predicting qualitatively equivalent performances for the adsorption isotherms, especially in cases where the loadings correspond to low pressures indicating that the most favorable adsorption area for the CO_2 molecules, at room temperature locates near the tube walls. These differences in the diffusivities tend to dissolve both for the transport and the self-diffusion showing that the computed diffusion mechanism is not of Knudsen-type but configurational. In this framework, Arora et al. [2] place in the adsorbate's role nitrogen to investigate the adsorption and the diffusion within SWNTs using molecular simulations means for a wide range of tube diameters at 100 K and 298 K . They proposed a steric resistance that may lead in a core structure of nitrogen molecules characterized by a higher mobility compared with the molecules located in the annular layer enhancing molecular diffusion. Detailed visualizations revealed a worm-like topology when pressure approaches the saturation point. The narrowest tubes demonstrated the highest diffusivities in case of loadings corresponding to saturation pressure. Confined flows of H_2 and CH_4 within carbon nanotubes and zeolites with comparable pore sizes have been reported using molecular dynamics simulations [11]. Skoulidas et al. focused into the adsorption by the interior of the tubes and not by the interstices nor the external surfaces. Equilibrium adsorption isotherms have been calculated for pure hydrogen and methane in (10,10) SWNTs, namely, silicalite and ZSM-12 through molecular simulations with the three materials to present similar adsorption performances in a volumetric basis. Skoulidas et al. [11] have calculated self (D_s) and transport diffusion (D_t) by performing a series of simulations both canonical Molecular Dynamics (MD) and Monte Carlo (MC). Alike isotherms, self diffusion presented a difference of 3-4 orders of magnitude higher than in the zeolites, with the aforementioned deviations to presumably attributed in the innate frictionless carbon surfaces while transport diffusion to comply with the same strikingly difference among the three adsorbents. Higher pressures lead in a decrease of self diffusivities, with this drop to be more pinpointed in case of the SWNTs, as the fluid-fluid collisions include a dominated mechanism of momentum decorrelation over the fluid-tube collisions. Finally, unlike the calculated adsorption performance, the gas transport of the adsorbed gases is not significantly affected by the fluid-adsorbate interaction potentials since the rapid transport is induced by the no-corrugation in the potential energy surface (PES).

The same group [22], performed a work for D_s and D_t of argon and neon in (8,8),(10,10) and (12,12) SWNTs for a wide range of loadings related to bulk phase pressures between 0 and 100 bar. To perform comparable results with the diffusivities of Ar and Ne in silicalite they used the same pressure range. Once more, Ackerman et al. [22] spotted the better performance of SWNTs over any known microporous adsorbent considering diffusion within them.

Although the great amount of studied that have been devoted in confined liquids within CNTs, pure or functionalized, little attention has been attracted for the charged CNTs. An electric field may generate important effects in nanofluid properties such as the transport behavior, the filling rate, the topology of the developed layers, while some claim that it may affect the fluid velocity [23]. Huang et al. [24] studied the developing water structures inside a charged (10,10) SWNT both negatively and positively. In case negative charges have been applied, they showed that water molecules favor the regular structures compared with uncharged SWNTs [25], while for positively charged tubes different orientations have been predicted. Benerjee et al. [26] have also studied how the charge influence apart of the structure of the water molecules the dynamics, proposing that the hydrophobic properties that characterize uncharged CNTs alter to hydrophilic. A change like this, leads to higher rates of filling but at the same time hobbles the flow through them.

In the present article, we focus in the diffusion mechanism of carbon dioxide in charged and uncharged (8,8) SWNTs, close to the room temperature. We investigate how the different degrees of filling within infinitely long SWNTs affects the diffusion mechanism in charged and uncharged narrow cylindrical pores. Furthermore, an extensive investigation has been made how the presence of different charges, both in sign and in absolute values, may affect the transport properties of the confined liquid. The results are presented with respect to the MSD as a function of time validating the existence of transition periods from ballistic to Fickian diffusion through a single file mechanism. In section 2 we present the simulation methodology along with the models that have been used while in section 3 a thorough discussion is offered. In section 4 we outline the main conclusions of the study.

2. Simulation methods and model

The governing system of equations for molecular dynamics is a system of Newton's equation of motion

$$m_i\ddot{r}_i = -\frac{\partial V_i}{\partial r_i} \tag{1}$$

for each atom i. These are modelled as mass points with position r_i and mass m_i. Each atoms potential energy V_i is the sum of semi-empirical analytical functions that model the real interatomic forces. The atomic trajectories are calculated by time integration of Equation 1 for all atoms. The time integration is performed by a finite difference method such as the predictor-corrector method or the Verlet algorithm [27, 28]. Despite the apparent simplicity, the simulations are extremely demanding computationally due to huge number of atoms involved, even in the smallest systems. The nanotubes considered in the present work are of type armchair while their carbon atoms,C^*, are modeled as Lennard-Jones spheres, holding them in their equilibrium crystallographic positions during the simulations. In practice, carbon nanotubes are flexible materials mainly those with large pores; the choice of frozen curvature reduces the computational cost compared with a flexible model. Several studies have shown that the deviations observed into these two approaches can be initially ignored; self-diffusion computed for simple fluids in rigid tubes has been found to predict higher values compared with cases that flexibility is involved in the computations. This effect is negligible in narrower pores in cases where the confined liquid reaches high densities. The choice of frozen curvature in combination with the fixed relevant motion between C^* atoms, reduces the computational cost compared with a flexible model. However several studies have shown that the disagreements between the two techniques can be ignored in first place. In the present work, the nanotubes have been assumed to be rigid as in case of the narrow (8,8) SWNT, the C^* vibrations are not expected to affect the diffusion mechanism of carbon dioxide molecules bearing in mind the small overestimation of self-diffusion as stated by Skoulidas et al. [11].

The present study is primary focusing in exploring the transport properties of carbon dioxide confined in uncharged and charged (8,8) SWNTs studying how the presence of a charge may affect the type of diffusion. Basing our work in previous studies that had investigated the above in confined water molecules, here we will adopt uniformly charged CNTs namely as $\pm0.01e, \pm0.1eand \pm 0.5e$. The microcanonical ensemble (NVE) [28] has been used for the integration of the equations of motion, where the number of molecules in the system (N), the volume of the simulations box (V) and the system's total energy (E) are kept constant. The calculations have been performed with LAMMPS [29] while the integration timestep is $\delta t = 0.5\ fs$ for the uncharged SWNT while for the charged the time step chose to be $\delta t = 0.25\ fs$ to avoid large fluctuations of the total energy and the temperature of the system arising from the presence of more electrostatic interactions. The system has been equilibrated for $200 - 600\ ps$, period that has been disregarded for the further calculations according to the charge of the CNTs.

To ensure that the temperature fluctuations will not will not considerably deviate from the set value of $300\ K$, a temperature rescaling has been employed during the equilibration period every 1000 timesteps achieving the temperature fluctuations to never exceed $34\ K$ in case of the charged SWNTs and $36\ K$ for the charged cases. After equilibrating the system, the temperature is allowed to fluctuate while CO_2 positions were stored every $5.0\ ps$ a value that has been found to minimize the correlation effects of the mean square displacements (MSD). For studying the diffusion mechanism only the MSD in the axial direction has been taken into consideration. The modeling of carbon dioxide has been implemented through the 3-site model proposed by Murthy et al. [30]. A previous computational study [31] focusing in the transport properties of CO_2 in bulk has investigated that deviations among the linear models [30, 32, 33, 34] used in conjunction with MD simulations lead to small discrepancies in these properties; for example, the diffusion coefficient differs less than 5%. The SHAKE algorithm [35] has been applied so both bonds and angles in carbon dioxide are being reset to their equilibrium length and angular values, retaining the geometry of each adsorbate molecule rigid in practice, allowing longer time steps to be employed. $CO_2 - CO_2$ interactions have been described through a $12-6$ LJ potential, to calculate the dispersive terms, augmented by the electrostatic interactions. The employed parameters are $\varepsilon_{CC}/k_B = 29.0\ K$, $\varepsilon_{OO}/k_B = 83.1\ K$, $\sigma_{CC} = 2.785\ \mathring{A}$ and $\sigma_{OO} = 3.014\ \mathring{A}$ with the CO_2 mass being ($44.010\ g/mol$). Partial charges have been assumed at carbon and oxygen sites with ($q_C = 0.5957e$) and ($q_O = -q_C/2$), respectively.

To represent the interactions between the CNTs and carbon dioxide, the AMBER force field [36] has been employed as described in our earlier publications [5] with the parameters to be ($\varepsilon_{C^*}/k_B = 28.0068\ K$) and ($\sigma_{C^*} = 3.4\ \mathring{A}$). Additionally, cross interactions have been calculated through the Lorentz-Berthelot mixing rules. The long range electrostatic interactions are treated according to the the particle-particle particle-mesh solver (PPPM) [37]. The length of both nanotubes is $150\ \mathring{A}$ however to ensure that the finite size effects this choice do not affect our calculations, as proposed by Striolo [9], the MD configurations have been replicated four times towards the axial direction. The diameter of the (8,8) SWNTs have been defined as the distance between the centers of two carbon atoms laying opposite in the tube. To implement the PPPM solver, the nanotubes have been placed along the Y axis in a rectangular parallelepiped computational boxes with the X- and Z- directions with size of $50\ \mathring{A}$. The size in the Y-direction equals with the length of the nanotubes, while periodic boundary conditions have been assigned to all three directions.

3. Results and Discussion

The transport mechanism within a CNT is mainly affected by the interactions between the surface and the confined molecules along with the pore width of the membrane [38] while the ratio of the fluid's molecular size to the pore's diameter [39] is a factor with equivalent influence. To explore the diffusion mechanism of fluid molecules through a CNT the scaling behavior between the MSD and the time can be employed. The diffusion transport in narrow pores is broadly classified in the following mechanisms; The normal-mode or Fickian diffusion where the molecules are able to surpass each other in the flow direction with the MSD to scale linearly with time noting that this type of diffusion is an innate fluid property in bulk systems. The single-file diffusion (SFD) where the MSD scales with the square root of time being met in cases where the degree of confinement is higher and the molecules are not allowed in passing each other. The last type of diffusion is the ballistic mechanism where it scales proportional to the square of time and the flow can be characterized by a higher degree of coordination. The three previously described diffusion mechanisms can be expressed through the following equations as [9, 8]

$$\lim_{t\to\infty} \langle [y(t) - y(0)]^2 \rangle = 2Dt \tag{2}$$

$$\lim_{t\to\infty} \langle [y(t) - y(0)]^2 \rangle = 2S\sqrt{t} \tag{3}$$

$$\lim_{t\to\infty} \langle [y(t) - y(0)]^2 \rangle = 2Bt^2 \tag{4}$$

where $\langle [y(t) - y(0)]^2 \rangle$ is the axial MSD of the adsorbed molecules [40], $\langle\rangle$ refers to the average over the molecules into the interior of the SWNT and $y(t)$ is the position of the a molecule within the nanotube in its axial direction as a function of time t. In Eqs. 2-4 the coefficients D, S and B denote the Fickian, single-file and ballistic diffusion mobilities respectively [41, 42].It has been shown that the constrained molecular diffusion is the main transport mechanism of gasses in micro- and nanopores, channels with pore widths up to 2 nm [43], expecting to obey the Fickian dependence in general. However several studies have shown that the diffusion mechanisms presents a more than one stage mechanisms with the final observation to be in most of the cases the Fickian one. The type of diffusion is more complicated in narrow nanotubes whereas in wider tubes things are more lucid. Aiming to enhance our understanding in diffusion mechanism within the uncharged and charged (8,8) SWNTs, the double logarithmic plots of the axial MSD against time along embraced with the typical plots for the three diffusion mechanisms are illustrated in Figs. 1-4 for different CO_2 loadings.

Figure 1 illustrates the MSD obtained for 10 CO_2 confined molecules in differently charged (8,8) SWNTs corresponding to +0.01e,-0.01e, +0.1e,-0.1e and 0.0e; however more simulations have been carried out for additional charges. The conclusions generated here can be easily extended in all cases that have been simulated. The mechanism of diffusion is explained through a double logarithmic plot of the MSD as a function of time. For all the cases the results illustrate a complex diffusion mechanism when 10 molecules have been confined. For the uncharged SWNT, as it has been described earlier, a ballistic-type mechanism is illustrated 1 persisting for $\approx$ 100 ps followed by a short period of SFD while it concludes to the expected Fickian diffusion. Interestingly, compared to previous studies where the single-file mechanism is not observed, in the present work we find evidence of it as it has been proposed by others [44, 45] to explain the anomalous transport of nanoconfined carbon dioxide molecules mainly in short nanotubes. However it still persists for short periods indications that can be explained though the same reasons that the no evidence of SFD has been. The length of nanotubes can be recognized as a factor that may influence the computations. Charging the nanotube with $-0.01e$, we observe almost the same behavior with the uncharged case. The diffusion initiates

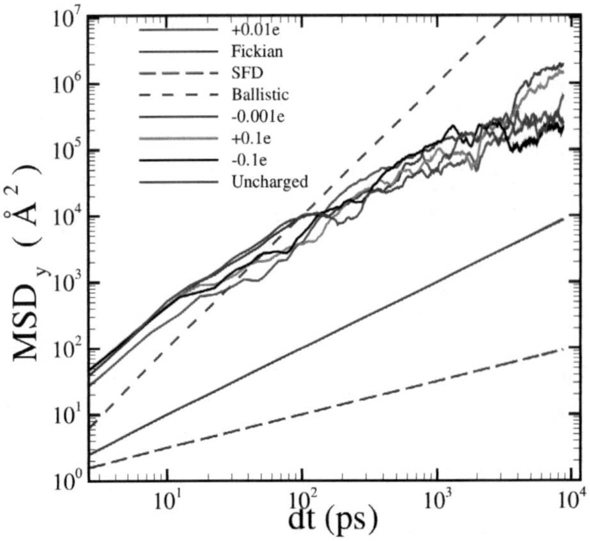

Figure 1. Mean square displacement along the pore axis for 10 carbon dioxide molecules confined within a (8,8) SWNT. The grey line represent the typical diffusion mechanisms of the Fickian type. The red, blue, green, black, purple lines are for +0.01e,-0.01e, +0.1e,-0.1e, Uncharged SWNTs respectively.

with a ballistic mechanism followed by a SFD and a Fickian type. The only difference that can be marked is located in the time interval where the Fickian diffusion is observed obtaining higher values of displacements. Increasing the charge to $-0.1e$ the electrostatic interactions affect the diffusion within the SWNT; the ballistic mechanism is restricted to ≈ 10 ps while the SFD is visible for up to 90 ps of simulation time with the Fickian mechanism at longer periods. Positively charging the SWNT, we observe once more some worth noting changes. In case a $+0.01e$ charged is obtained, the ballistic mechanism persists for ≈ 400 ps accompanied by a SFD for an additional period of ≈ 700 ps suggesting that the combination of the positively charged SWNT with the charged sites located in CO_2 produce a considerable force component in the already acting forces between the adsorbate and the adsorbent. Further increase in the charge does not produce a significant change in how the 10 molecules diffuse within the tube apart of adding a short period inside the ballistic mechanism where the diffusion acquires a SFD-like type which in practice is a deceleration which requires further studying.

Increasing CO_2 loading to 30 molecules (see Fig. 2) the type of diffusion alters with the impact of the charged CNTs to start reducing. In the uncharged case, the collective motion persists for a considerable shorter period (≈ 20 ps) while it is followed by a SFD and Fickian diffusion. Charging a SWNT with $-0.01e$, as in case of 10 molecules, does not have a practical impact in the diffusion mechanism compared with the uncharged SWNT. However, increasing the charge to $-0.1e$ significantly alters the diffusion of the 30 CO_2 molecules within the SWNT. The ballistic period is extended to ≈ 100 ps with additional prolonged period with a single-file type. On the contrary, when the 30 molecules have been confined in a positively charged environment, we observe the greatest changes; the ballistic period is further suppressed compared not only to the rest cases in the 30 molecules framework but also with the 10 molecules case. The 400 ps of ballistic diffusion in case of 10 molecules, with a $+0.01e$ applied, becomes ≈ 10 ps

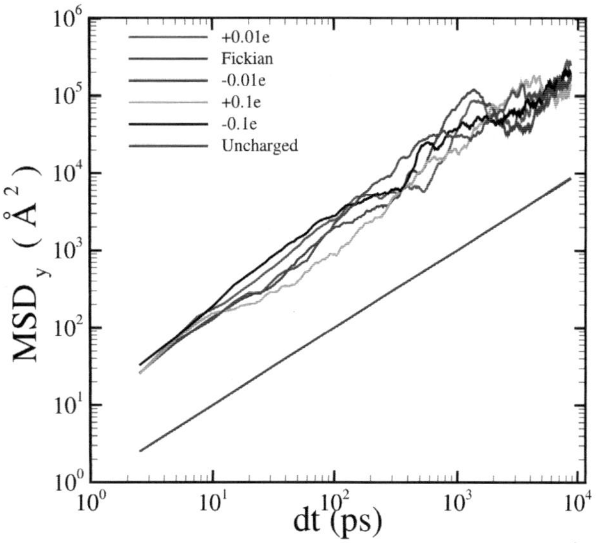

Figure 2. Mean square displacement along the pore axis for 30 carbon dioxide molecules confined within a (8,8) SWNT.The grey line represent the typical diffusion mechanisms of the Fickian type. The red,blue, green, black, purple lines are for +0.01e,-0.01e, +0.1e,-0.1e, Uncharged SWNTs respectively.

followed by a Fickian diffusion. Increasing to $+0.1e$ give birth to SFD even for a short period lasting $\approx 10 - 20$ ps. When larger loadings (60 molecules see 3) are studied corresponding in higher pressure, the effect of the charge tends to diminish. The latter is mainly explicated by directly comparing the uncharged scenario with the charged cases. When $0.0e$ is employed there is a short ballistic period while in the charged cases this period is either vanished or remains unaffected ($-0.1e$). The aforementioned suggests that as the amount of the adsorbate increases the effect of the electrostatic interactions between the pore and the CO_2 molecules decreases; with the electrostatic effect to be canceled. The aforementioned is more lucid when the loading of the adsorbate approaches the number of adsorbed molecules near the saturation pressure where in practice there are no differences (see 4).

4. Conclusions

To summarize, we have investigated the axial diffusion of carbon dioxide within infinitely long narrow charged and uncharged SWNTs. It has been shown that in case of the uncharged (8,8) SWNT independently on the degree of filling a considerable long-time diffusion mechanism occurring for up to 400 ps. In first place the MSD follows a ballistic mechanism as a function of time except in the case where the adsorbate corresponds to close to saturation pressures. As the variable of charge is inserted into the simulation we observe that there is an influence in the diffusion mechanism within the tube. However the degree of influence is not of the same importance as the amount of loading especially in cases of higher amount of adsorbate. It seems that at higher pressures the CNT-CO_2 electrostatic interactions are canceled by the stronger van der Waals interactions between them. The results exhibited in the present study can be proved valuable in engineering CNT-based devices including carbon dioxide that require an in depth understanding of its flow in molecular level.

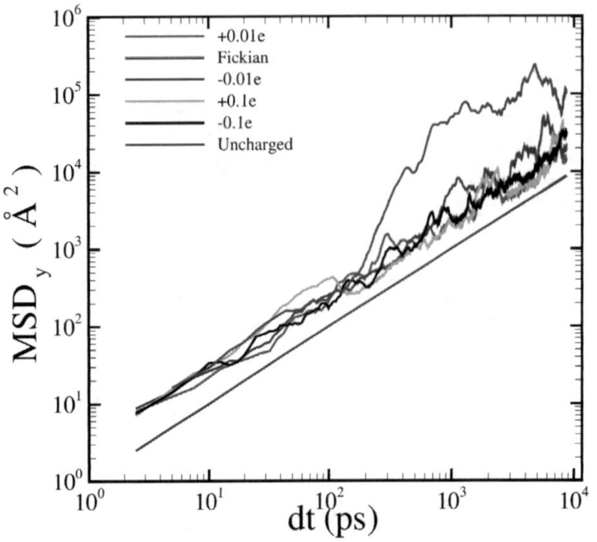

Figure 3. Mean square displacement along the pore axis for 60 carbon dioxide molecules confined within a (8,8) SWNT.The grey line represent the typical diffusion mechanisms of the Fickian type. The red,blue, green, black, purple lines are for +0.01e,-0.01e, +0.1e,-0.1e, Uncharged SWNTs respectively.

5. References

[1] Iijima S 1991 *Nature* **354** 56–58
[2] Arora G, Wagner N J and Sandler S I 2004 *Langmuir* **20** 6268–6277
[3] Adisa O O, Cox B J and Hill J M 2011 *Physica B* **406** 88–93
[4] Drikakis D and Asproulis 2010 *International Journal for Numerical Methods for Heat and Fluid Flow* **5**
[5] Mantzalis D, Asproulis N and Drikakis D 2011 *Chemical Physics Letters* **506** 81–85
[6] Kato K and Kitajima M 2010 *Journal of the Vacuum Society of Japan* **53** 317–326
[7] Zheng Mand Jagota A, Semke E, Diner B, McLean RSand Lustig S, Richardson R and Tassi N 2003 *Nature Materials* **2** 338–342
[8] Ye H, Zhang H, Zheng Y and Zhang Z 2011 *Microfluidics and Nanofluidics* **10** 1359–1364
[9] Striolo A 2006 *Nano Letters* **6** 633–639
[10] Liu Y, Wang Q, Wu T and Zhang L 2005 *Journal of Chemical Physics* **123** 1–7
[11] Skoulidas A I, Ackerman D M, Johnson J K and Sholl D S 2002 *Physical Review Letters* **89** 185901
[12] Noy A, Park H G, Fornasiero F, Holt J K, Grigoropoulos C P and Bakajin O 2007 *Nano Today* **2** 22 – 29
[13] Shi W and Johnson J K 2003 *Physical Review Letters* **91**
[14] Striolo A, Gubbins K E, Chialvo A A and Cummings P T 2004 *Molecular Physics* **102** 243–251
[15] Striolo A, Chialvo A and Gubbins KEand Cummings P 2005 *Journal of Chemical Physics* **122** 1–14
[16] Mpourmpakis G, Tylianakis E and Froudakis G 2006 *Journal of Nanoscience and Nanotechnology* **6** 87–90
[17] Alexiadis A and Kassinos S 2008 *Chemical Physics Letters* **460** 512–516
[18] Mantzalis D, Asproulis N and Drikakis D 2011 *Physical Review E - Statistical, Nonlinear, and Soft Matter Physics* **84** 066304
[19] Aaron D and Tsouris C 2005 *Separation Science and Technology* **40** 321–348
[20] Matranga C, Chen L, Smith M, Bittner E, Johnson J and Bockrath B 2003 *Journal of Physical Chemistry B* **107** 12930–12941
[21] Skoulidas A, Sholl D and Johnson J 2006 *Journal of Chemical Physics* **124** 1–7
[22] Ackerman D M, Skoulidas A I, Sholl D S and Johnson J K 2003 *Molecular Simulation* **29** 677–684
[23] Alexiadis A and Kassinos S 2008 *Chemical reviews* **108** 5014–5034
[24] Huang B, Xia Y, Zhao M, Li F, Liu X, Ji Y and Song C 2005 *Journal of Chemical Physics* **122** 1–8

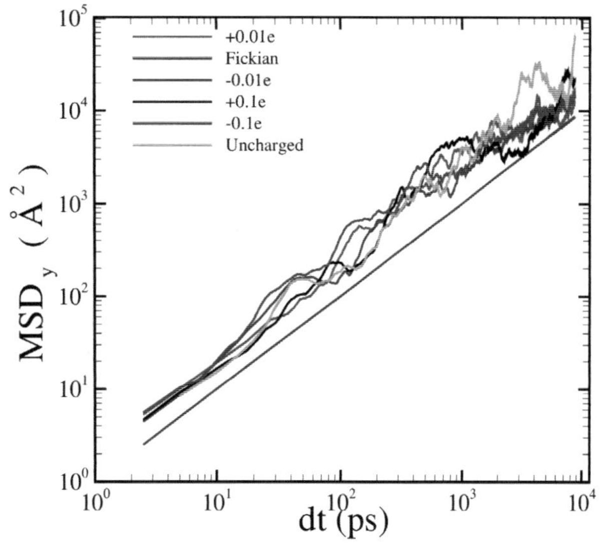

Figure 4. Mean square displacement along the pore axis for 120 carbon dioxide molecules confined within a (8,8) SWNT.The grey line represent the typical diffusion mechanisms of the Fickian type. The red,blue, green, black, purple lines are for +0.01e,-0.01e, +0.1e,-0.1e, Uncharged SWNTs respectively.

[25] Koga K, Gao G T, Tanaka H and Zeng X C 2001 *Nature* **412** 802–805
[26] Banerjee S, Murad S and Puri I K 2007 *Chemical Physics Letters* **434** 292–296
[27] Verlet L 1967 *Physical Review* **159** 98–103
[28] Allen M and Tildesley D 1987 *Computer Simulation of Liquids*
[29] Plimpton S 1995 *Journal of Computational Physics* **117** 1–19
[30] Murthy C, Singer K and McDonald I 1981 *Mol. Phys.* **44** 135–143
[31] Mantzalis D, Asproulis N and Drikakis D 2010 *2nd European Conference on Microfluidics*
[32] Harris JGand Yung K 1995 *Journal of Physical Chemistry* **99** 12021–12024
[33] Zhang Z and Duan Z 2005 *Journal of Chemical Physics* **122** 1–15
[34] Potoff JJ Siepmann J 2001 *American Institute of Chemical Engineers* **47** 1676–1682
[35] Ryckaert J, Ciccotti G and Berendsen H J C 1977 *Journal of Computational Physics* **23** 327–341
[36] Pearlman D, Case D, Caldwell J, Ross W, Cheatham III T, DeBolt S, Ferguson D, Seibel G and Kollman P 1995 *Computer Physics Communications* **91** 1–41
[37] Hockney R and Eastwood J 1989 *Computer Simulation Using Particles*
[38] Verweij H, Schillo M and Li J 2007 *Small* **3** 1996–2004
[39] Mao Z and Sinnott S B 2000 *Journal of Physical Chemistry B* **104** 4618–4624
[40] Devi R, Sood J, Srivastava S and Tankeshwar K 2010 *Microfluidics and Nanofluidics* **9** 737–742
[41] Bouchaud J , Georges A, Koplik J, Provata A and Redner S 1990 *Physical Review Letters* **64** 2503–2506
[42] Lee K H and Sinnott S B 2005 *Nano Letters* **5** 793–798
[43] Koros W, Ma Y and Shimidzu T 1996 *Pure Appl. Chem.* **86** 14791489
[44] Hummer G, Rasaiah J C and Noworyta J P 2001 *Nature* **414** 188–190
[45] Beckstein O, Biggin P C and Sansom M S P 2001 *Journal of Physical Chemistry B* **105** 12902–12905

Pressure Calculations in Nanochannel Gas Flows

JH Kim[1] A J H Frijns, S V Nedea, A A van Steenhoven
Department of Mechanical Engineering, Eindhoven University of Technology
Eindhoven, The Netherlands

E-mail: j.h.kim@tue.nl

Abstract. In this research, pressure driven flow within a nanochannel is studied for argon in rarefied gas states. A Molecular Dynamics simulation is used to resolve the density and stress variations. Normal stress calculations are based on Irving-Kirkwood method, which divides the stress tensor into its kinetic and virial parts. The kinetic component recovers pressure based on the ideal-gas law. The particle-particle virial increases with increased density. The kinetic part and the virial part of the pressure is studied from Kn = 0.1 to Kn = 0.5 at 300 K. This pressure calculation was also compared with a simulation for a static fluid. Pressure driven gas flow gives a different result compared to static situation, which has a uniform particle distribution across the system. For a flow situation, the virial part is no longer the same. This can also be predicted from the virial equation and by the fact that the compressible fluid flow causes a temperature and density difference along the flow direction. The relation between the virial and distribution of particles has been studied further to understand the nature of virial component along the flow direction in compressible flows.

1. Introduction

The development of new technologies in micro- and nano engineering has a major impact on many disciplines including fluid dynamics and transport phenomena. When dealing with flows in the scale of microns or smaller, many phenomena such as velocity slip and temperature jumps have been observed [1]. The flow characteristics at the macro and micro levels are not the same and the features of micromechanics require additional basic and applied research work.

The Knudsen number is used to characterize the gas flow regimes. The Knudsen number is defined as $Kn = \frac{\lambda}{L}$, where L is the characteristic length scale of the physical system, and λ is the molecular mean free-path. Mean free path λ is defined as $\lambda = \frac{1}{\pi\sqrt{2}\sigma^2 n}$, where $n = \frac{N}{Vol}$. N is number of particles and Vol is the volume of the system. It is commonly assumed that failure of continuum approach occurs at Knudsen number around 0.01 [2]. In most cases, gas flows on the nano level have been studied by direct simulation Monte Carlo (DSMC) or Molecular Dynamics (MD).

Local pressures are important in analysing micromechanical behaviour. The local pressure in the DSMC method may be evaluated from the ideal gas law or by explicitly evaluating the pressure tensor for calculating the pressure near the wall.

However, the ideal-gas law is known to fail as the pressure increases. Unlike the ideal-gas law, MD simulations can predict a pressure using the Irving-Kirkwood method [3]. Less than 10 % difference from experimental results are even shown for the liquid state, while ideal-gas law gives an error larger than 100% near the liquid state [4].

According to the Irving Kirkwood method the contribution of each particle to the pressure tensor consists of two parts, a configuration part and a kinetic part:

$$P_{kl} = \frac{1}{Vol}\langle \sum_i^N m^i \left(V_k^i - \overline{V}_k^i\right)\left(V_l^i - \overline{V}_l^i\right) - W_{kl}\rangle \quad (1)$$

$$W_{kl} = \frac{1}{2}\sum_{i,j}^N \left(r_k^j - r_k^i\right) f_l^{i,j} \quad (2)$$

[1] To whom any correspondence should be addressed.

Published under licence by IOP Publishing Ltd

where the first term on the right-hand side of equation 1 represents the kinetic component and the W_{kl} term represents the virial component. In the kinetic part, m^i is the atomic mass of particle *i*, whereas subscripts *k* and *l* are the axes of the Cartesian coordinate system, V_k^i and V_l^i are the peculiar velocity component of particle *i* in the *k* and *l* directions, and $\overline{V}_k^i$ and $\overline{V}_l^i$ are the local average streaming velocities at the position of particle *i*, in the *k* and *l* directions, respectively. The term $(r_k^j - r_k^i)$ in Equation 2 is k^{th} component of the distance vector between particles *i* and *j*, and $f_l^{i,j}$ is the lth component of the intermolecular force exerted on particle i by particle j.

The first term, $\sum_i^N m^i \left(V_k^i - \overline{V}_k^i\right)\left(V_l^i - \overline{V}_l^i\right)$, in the Irving-Kirkwood expression is related to the ideal-gas law. It calculates ideal gas pressure exactly by considering the momentum resulting from the particle velocities, whereas the second term, W_{kl}, the virial component are the corrections to the ideal-gas law because of the interaction of particles having non-zero volumes and force field.

For flow situations, the virial part is no longer isotropic. It will have a different virial component in the flow direction compared to other axial components. This can also be predicted from equation 2 and by the fact that the compressible fluid flow causes a density difference along the flow direction.Therefore, investigation of the pressure of a gaseous flow requires careful consideration of the density gradient along the channel.

In the present work, we mainly focus on two aspects; in the first part we investigate characteristics of the pressure from $Kn \cong 0.1$ to $Kn \cong 0.5$ of the gas flow inside the nanochannel since there have been no detailed study on the pressure in nanochannel gas flow. The Knudsen numbers were chosen where the transition from a CFD method to the DSMC method lies. In the second part, we investigate the effect of temperature and density gradient on virial component, so to suggest a better estimation of pressure without using a computationally expensive Irving-Kirkwood method.

In order to achieve these goals we will perform detailed 3D molecular dynamics simulations .

2 Three-dimensional MD simulation details

A 3D Molecular Dynamics simulation method is used to simulate the flow of argon (Ar) in a nanochannel (figure 1). Molecular Dynamics is a discrete, deterministic method that simulates the temporal and spatial dynamics of matter (solid, liquid, gases). Matter is explicitly modeled by atoms or molecules, which interact together through potentials. In this research a truncated and shifted Lennard-Jones potential, V_{tr} is used to model the interaction:

$$V_{tr}(r_{ij}) = 4\varepsilon\left(\left(\left(\frac{\sigma}{r_{ij}}\right)^{12} - \left(\frac{\sigma}{r_{ij}}\right)^{6}\right) - \left(\left(\frac{\sigma}{r_c}\right)^{12} - \left(\frac{\sigma}{r_c}\right)^{6}\right)\right) \quad (3)$$

for $r_{ij} \le r_c$, and $V_{tr}(r_{ij}) = 0$ for $r_{ij} > r_c$.

The walls are modeled by a thermal wall condition. The interaction between two argon molecules is described by the Lennard Jones potential of equation (3) with parameters as given in Table 1.

Table 1. Molecular Dynamics properties for argon (Ar) [5].

argon particle		
σ_{Ar}	0.340	[nm]
m_{Ar}	$6.63 \cdot 10^{-26}$	[kg]
ε_{Ar}/k_B	121	[K]
Cut-off Radius	2.5	[σ]

The Molecular Dynamics code *PumMa*[6] has been used here. This code has also been used by previous researchers in the area of nano flow and heat transfer [7, 8].

In the present research, two different sets of simulations will be carried out in order to achieve two different goals of the work mentioned in the preceding section. The first simulation set (section 2.1) will be focused on the nanochannel gas flow while the second simulation set (section 2.2) will be focused on the relationship between the virial component and density gradient.

2.1 Pressure in nanochannel gas flow

A developing flow inside a slit channel with a height H has been simulated (Figure 1). The characteristic length of this system is the separation distance H between the two parallel plates. In the flow direction, x direction, the channel length is kept constant, L = 100 nm, and while H is changed from 10nm to 100nm (H = 10 , 25, 50, and 100 nm).

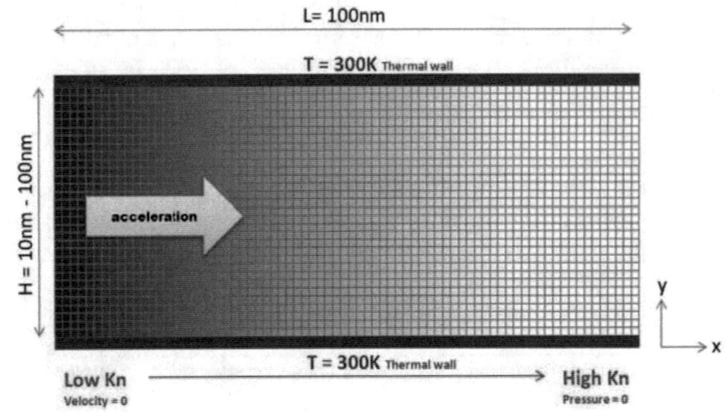

Figure 1. Developing nanochannel flow by using gravity fed method.

The simulation is started from the Maxwell-Boltzmann velocity distribution at 300 K. The walls were kept at a constant temperature T_w = 300 K. A thermal wall condition was applied as a boundary condition at the top and bottom plates.

The simulation box domain in the XY plane is divided by a 100x100 mesh. 10000 bins are extended across the Z direction. The local pressure properties are calculated in each bin.

Flow was induced by the gravity fed method with a gravity of gx = m/ in the whole domain.

Time steps were chosen = 6.43 fs.

A typical computation in our simulations requires 5,000,000 time steps for flows to reach a steady flow state. We spent about 7,000,000 additional time steps for computing the averaged macroscopic properties: flow velocity, pressure, temperature, etc.

2.2 Virial vs. density & temperature gradient

In the second simulation the effect of a density and temperature gradient is investigated. Therefore a temperature gradient between two parallel walls was presented.

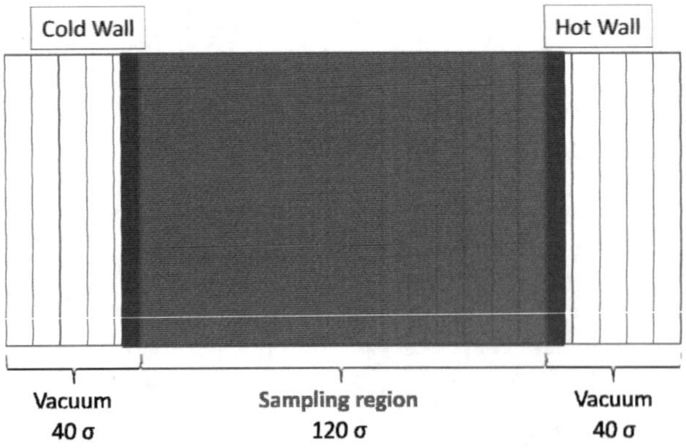

Figure 2. Density gradient caused by a temperature gradient

The Knudsen number of the sampling region was chosen to be 0.1, 0.2, 0.3, 0.4, and 0.5. Different numbers of particles were considered depending on the *Kn*. Number of particles varies from 1910($Kn = 0.5$) to 121593 ($Kn = 0.1$).

In the simulation the two walls are kept at a different constant temperature. The cold wall was kept at constant temperatures $T_{cold} = 60.5$ K, 121 K, 181.5 K and 242 K, and the hot wall at constant temperatures $T_{hot} = 242$ K, 301.5 K, 363 K, 424.5 K. A thermal wall condition was applied as a boundary condition.

Table 2. Combinations of two temperature walls to produce density gradients.

T_{cold} [K]	T_{hot} [K]
242.0	*242.0*
181.5	*302.5*
121.0	*363.0*
60.5	*423.5*

Density gradient is produced by using a temperature gradient caused by the different temperature of the walls. Four different density gradients have been simulated with the same total density in the sampling region.

A typical computation in our simulations requires 5,000,000 time steps to reach a steady state. We spent about 7,000,000 additional time steps for computing the averaged macroscopic properties.

3 Results and Discussion

3.1 Pressure in nanochannel gas flow

We first focus on argon gas flow at a temperature of 300 K that is developing inside the slit channel. Simulations have been carried out for 4 different channel heights H = 10 nm, 25 nm, 50 nm, and 100 nm. The results for the total pressure and the virial component are shown in figure 3. Figure 3a clearly shows the pressure increase when the channel height decreases for the same given Knudsen numbers. It is the same for the virial components (figure 3b). The absolute value for the virial component increases as the channel height decreases. The ideal-gas law is known to fail as the pressure increases, and understanding of the virial component becomes more important in nanochannel flows compared to microchannel flows for the same Knudsen number.

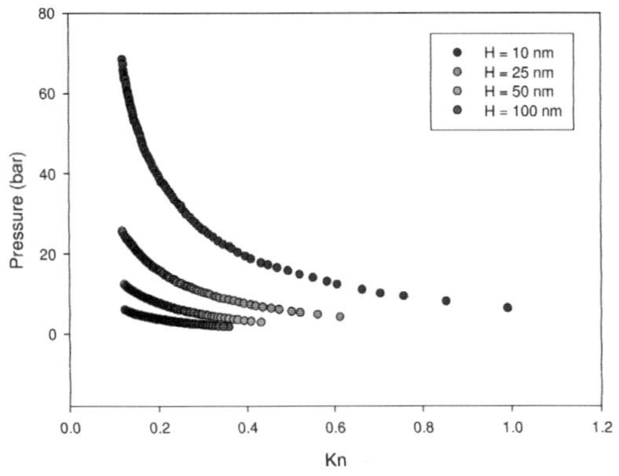

Figure 3a. Pressure of an argon gas flow inside a nanochannel

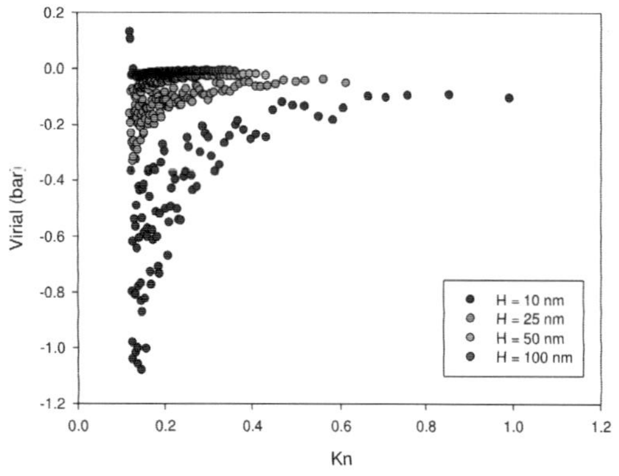

Figure 3b. Virial component of an argon gas flow inside a nanochannel

Figure 4 shows the distribution of the kinetic component and the virial component along the channel. While the kinetic term is equally distributed along the y-axis (height), the virial component shows a gradient due to the density gradient caused by the temperature gradient due to the expansion flow near the outlet.

Since the virial component is not proportional to kinetic component, it is important to understand the nature of virial component since this component can have a dominant role in the pressure calculation.

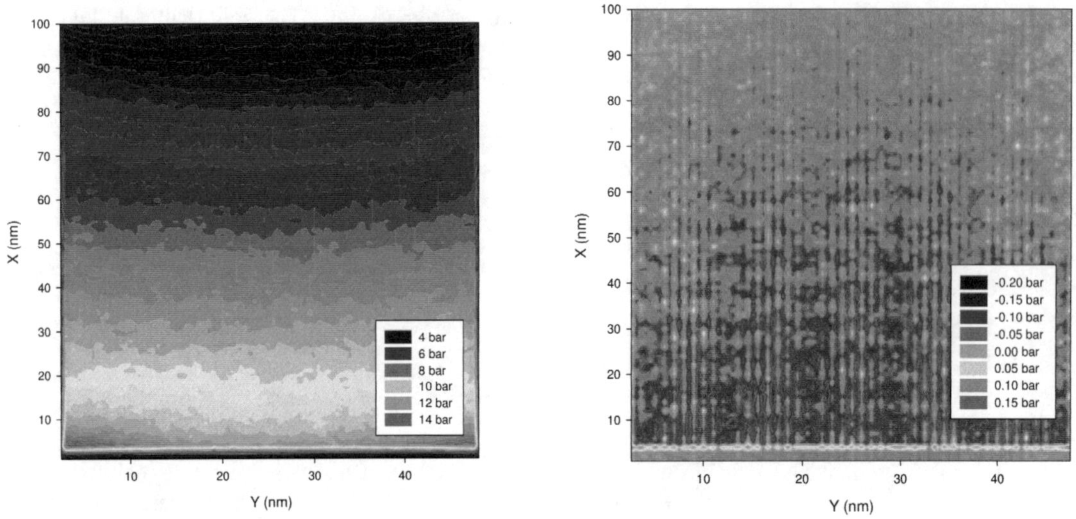

Figure 4. Contour graph of the kinetic component (left) and virial component (right) of the pressure inside the 50nm height nanochannel

One of the situations where a virial component becomes important is a case with a low temperature. Gas channel flows often deal with an expansion flow caused by an outlet effect. Temperature drop is one of the widely known features of such expansion flow [9]. There are two reasons why the virial component becomes more important for low temperature. Equation 1 shows that the virial component takes a larger role in overall pressure at low temperature since the kinetic component is proportional to the temperature according to the Maxwell distribution. The other reason is the increase in the absolute value of the virial component at low temperature.

Density is a very important factor of understanding the nature of the virial component. In Figure 5, virial components of the simulations with different channel heights have been plotted against the number density. It shows an exponential decay in virial component of all channel heights. The result near the density 1.6 has a large error due to small number of ensembles caused by the small channel height.

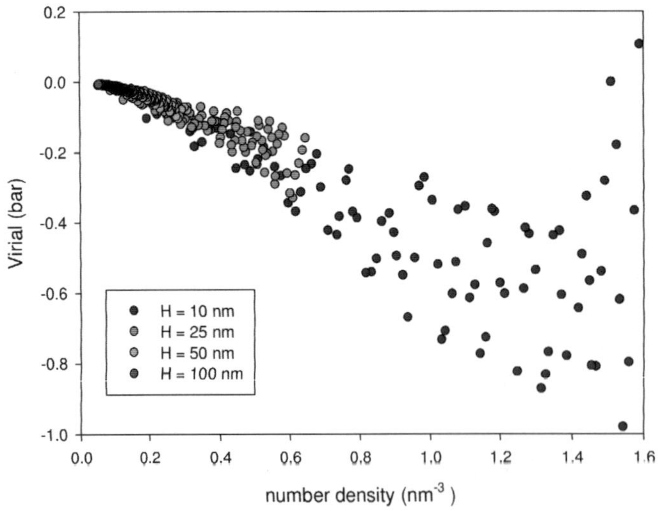

Figure 5. Virial pressure component as a function of the number density.

The cut-off radius should be carefully chosen for the larger density simulations. As for example, in a 10 nm channel, different cut-off radii are used, a large difference in the virial calculation can be produced. Figure 6 shows a comparison of two different cut-off radii. For the smaller cut-off radius (2.5 σ) , the simulation produces a smaller absolute virial component value than it should have.

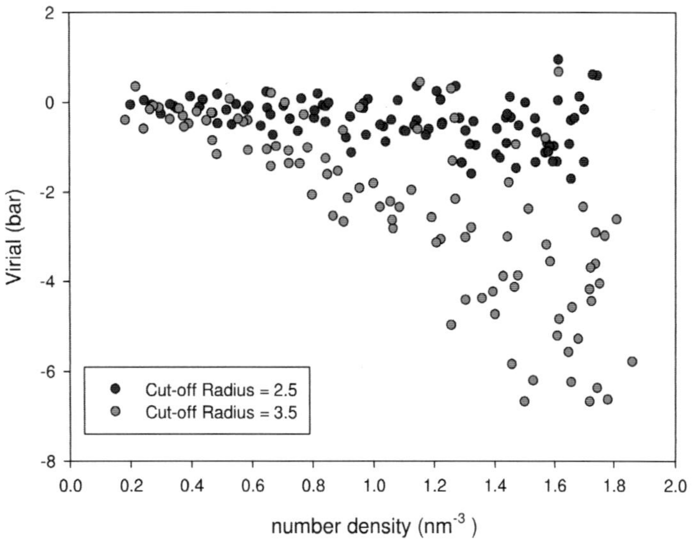

Figure 6. H = 10 nm channel simulations with a different cut-off radius. 2.5 σ (red) and 3.5 σ (orange).

3.2 Static situation

To understand the nature of virial component in flow cases, it is necessary to understand non flow situation in advance. Figure 7 shows the simulation results of static situations.

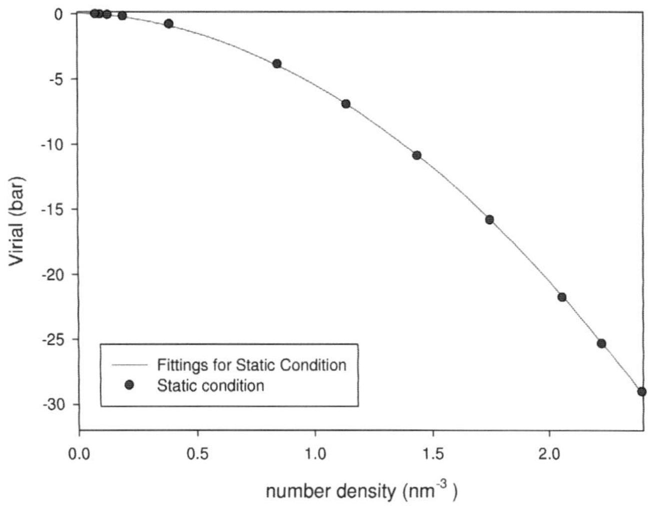

Figure 7. Virial component as function of number density in the case without any density gradient at 121 K.

A 2nd order polynomial function was fitted through the data using *Origin 8.5* Software [10].

$$W_{\text{static}} = 0.080 - 17.499\,\rho - 3108.418\,\rho^2 \quad \text{[bar]} \qquad (5)$$

where W_{static} is a virial component in the static condition, and ρ is a number density (σ^{-3}) of the system. This corresponds well with the second virial coefficient. For a certain temperature, the virial expansion can be shown in the form of

$$P = C_1\,\rho - C_2\,\rho^2 \quad \text{[bar]} \qquad (6)$$

where $C_1 = k_B T$ and $C_2 = B_2(T) k_B T$. B_2(T) is the second virial coefficient. In table 2, the C_2 values are compared with other literature values [11, 12].

Table 3. Comparison between Irving-Kirkwood method and the second virial coefficient at 121 K. Coefficient C_2 in equation 6 is compared.

This work	-3108.418 $[\sigma^6/\text{bar}]$
Experiments [11]	-3111.896 $[\sigma^6/\text{bar}]$
Leonhard [11]	-2972.592 $[\sigma^6/\text{bar}]$
Hloucha [12]	-3222.866 $[\sigma^6/\text{bar}]$

Our simulation results in static situation perfectly match with the literature values. This result validates our MD code while using the Irving-Kirkwood method.

3.3 Static situation vs. flow situation

A 10nm channel was chosen to compare a static situation with a flow situation. Figure 8 shows the virial component of two flow situations (red, orange) and two static situations (green, light green). The results of two static situations are calculated based on the literature values [11].

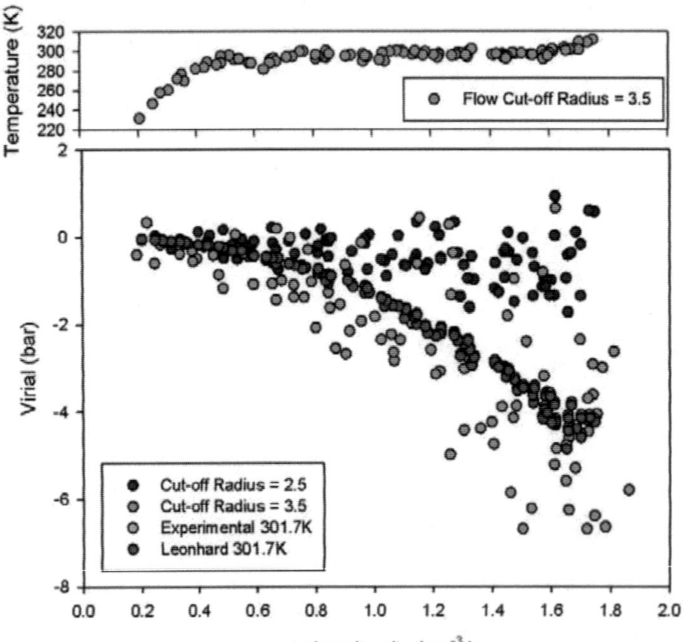

Figure 8. Comparison of flow vs. static virial component in the H = 10 nm channel.

The 10nm channel gas flow simulation results with cut-off radius 3.5σ were compared to the static situation at 301.7 K. It indicates that the virial with a flow situation has a larger absolute value compared to the no-flow situation (static situation). This difference can be caused by the following effects:

1. Temperature
2. Effect of cut-off radius
3. Boundary effect (confinement effect)
4. Velocity of flow
5. Density gradient.

Since the temperature is nearly constant at 300 K for number densities ranging from 0.4 nm^{-3} to 1.8 nm^{-3} for this 10nm channel, this difference can hardly be caused by temperature effects.
It is not due to the cut-off neither. If it was caused by the choice of the cut-off radius, it should have a smaller absolute value than literature value. This is not observed in our simulations.
For smaller nanochannels, boundary effect cannot be neglected. For our simulation, a vacuum region is modelled next to the nanochannel. This vacuum region near the channel boundaries is the one of the main cause for a smaller absolute virial component value. According to Barisik's results [4], these boundary effects last up to 0.5 nm. For 10 nm slit channel the effect will be 10% of whole channel. Table 3 indicated the importance of this boundary effect. This table shows a change in virial when the height of the channel changes from 60 σ to 9 σ. It shows the decrease in absolute virial component value with the decrease in channel height. But this result does not also give an explanation of larger absolute virial component value compared to static situation.

Table 3. Virial component of a static fluid inside the slit of different heights.

Density $= 0.3896\ nm^{-3}$	
$H = 60\ \sigma$	-0.4040 [bar]
$H = 30\ \sigma$	-0.3964 [bar]
$H = 15\ \sigma$	-0.3826 [bar]
$H = 9\ \sigma$	-0.3634 [bar]

The velocity might also be the cause of the difference. However, the simulation has a near zero velocity near the inlet (number density = 1.8 nm^{-3}). Therefore, we suspect that the density gradient plays a major role on the virial component.

3.4 Virial component vs. density & temperature gradient
To study this nature of the virial component and density gradient in more detail, simulations were performed for different temperature gradients as mentioned in section 2.2. Figure 9 shows the density gradient produced by this method, and figure 10 shows the temperature gradient.
We have chosen a position X = -5σ(red transparent bar in figures 9 and 10) for a comparison. This position was chosen to ensure that the temperature effect can be neglected since it is approximately 240 K for all cases. However we can see that the number density for combination T_{cold} = 60.5 K, T_{hot}= 423.5 K (indictaed by the green symbols in figure 9 and 10), has a slightly lower density compared to other 3 combinations.
According to the equation of state, combination T_{cold} = 60.5 K, T_{hot}= 423 .5 K should have approximately 0.1 bar difference with the given density difference with the other combinations. However at this position, it all had a virial near -0.3 bar.
This results indicates that steeper density gradient gives a larger absolute virial component. This founding agrees well with the definition of virial component proposed by Zhou [13]: he proposed that 'the virial component is a stress-like measure for momentum change in space'. Density gradient in our simulation can be simply converted into the form of momentum gradient in space.

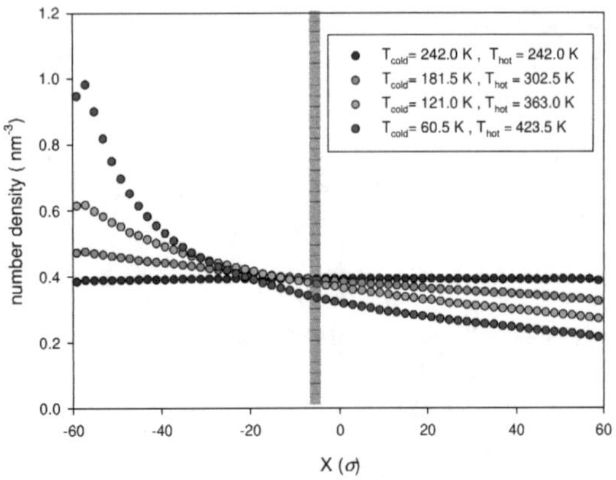

Figure 9. Density gradients caused by a temperature difference between the two channel walls

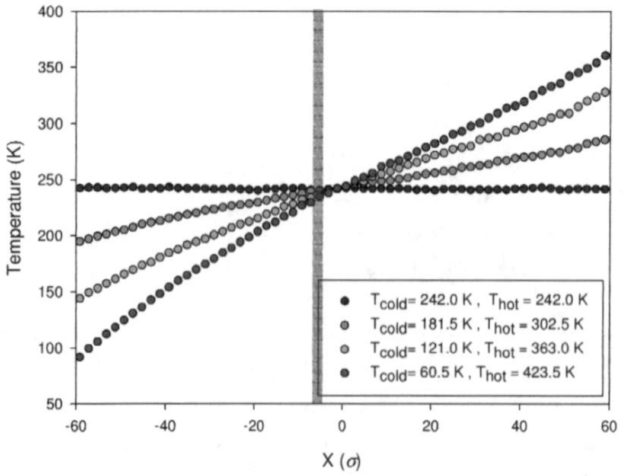

Figure 10. Temperature gradient caused by a temperature difference between the two channel walls.

This finding can also be an opening step for the study why virial component increases with a decrease in temperature. The Irving-Kirkwood equation shows that the virial component is independent on the velocities of particles but only dependent on the positions of the particles. One of the hypothetical explanations to this contradiction is that density distribution of a given system becomes more non-uniform with lower temperature. If this hypothesis is true, density gradient of a given system will also work in the same manner as a temperature change in the system.

4. Conclusion

In this article the pressure in a nanaochannel gas flow has been studied by using MD simulation and Irving-Kirkwood method. The numerical data mainly covers the transition regime where the boundary between DSMC method and Navier Stokes Equation calculation lies. An analysis of the numerical data shows (i) the importance of virial component in the pressure calculation of a nanochannel gas flow and (ii) a relationship between the virial component and density gradient. The study suggests that steeper density gradient gives a larger absolute virial component.

5 Acknowledgements

The research leading to these results has received funding from the European Community's Seventh Framework Program FP7/2007-2013 under grant agreement n° 215504.

6 References

[1] Hamdan M A, Al-Nimr M A, Hammoudeh V A, 2010 *J. Fluids Eng.* **132** 0734503
[2] Alexander F, Garcia A, 1997 *Computers in Physics* **11** 6, 588-593
[3] Irving J H, Kirkwood J G, 1950 *J. Chem. Phys.* **18** 817
[4] Barisik M, Beskok A, 2012 *Microfluid Nanofluid* **11** 269-282
[5] Maruyama S, 2004 Molecular dynamics methods in microscale heat transfer, Ch 2.13.7. Begell House
[6] http://www.pumma.nl , BioModelling and BioInfomatics TU/e
[7] Nedea S V, Markvoort A J, van Steenhoven A A, Hilbers P A J, 2009 *J. Heat Transfer* **131** 103-111
[8] Markvoort A, Hilbers P, Nedea S, 2005 *Phys. Rev. E.* **71** 066702
[9] Arlemark E, Makelov G, Nedea S V, 2012 *1st European Conference on Gas Micro Flow,* Skiathos Island, Greece
[10] http://www.originlab.com, Origin Lab Corporation
[11] Leonhard K, Dieters U K, 2000 *Molecular Physics* **98** 20, 1603-1616
[12] Hloucha M, 1998 *PhD thesis*, University of Koln
[13] Zhou M, 2003, *Proc. R. Soc. Lond.* **459** 2347-2392

Integrated temperature microsensors for the characterization of gas heat transfer

Alice Vittoriosi, Juergen J Brandner and Roland Dittmeyer
Karlsruhe Institute of Technology, Institute for Micro Process Engineering.
Hermann-von-Helmholtz-Platz 1, 76344 Eggenstein-Leopoldshafen, Germany

E-mail: alice.vittoriosi@kit.edu

Abstract. A new method for the integration of silicon-based miniaturized sensors in microchannel systems is proposed. Unlike existing designs realized with IC-technologies, the new design allows flexible integration of temperature sensors in a variety of microchannel materials. Since no bonding technology is required, it is possible to employ the same sensor array with different test sections, broadening the possible applications of the experimental device. The designed experimental device has been employed for the characterization of gaseous heat transfer in microchannels. Conventional temperature measuring technique and CFD simulations have been employed to compare the obtained results and validate the new experimental approach. The local measurement technique enables investigating the influence of the channel material and local surface characteristics (i.e., roughness) on the heat transfer performance at varying flow conditions.

1. Introduction

Thanks to the recent technological advances of micro- and nano-machining, the miniaturization of mechanical and electrical devices has become a common procedure to achieve process intensification and efficiency enhancements. To fully understand the governing physical mechanisms at small scales, new integrated tools for direct in-situ observation and control are required [1]. With the development of IC fabrication and silicon planar technologies it has been possible to create miniaturized transducers for sensing and actuation [2]. Small scale devices integrating microsensing elements with transduction units are the core components of micro electro mechanical systems (MEMS) [3]. The employment of MEMS-based sensors with microfluidic devices is particularly appealing since they present high frequency responses and small sizes not achievable with conventional measuring techniques [1]. In particular, the possibility of a local flow characterization allows investigating those phenomena which are driven by surface effects, such as the interaction of the fluid with the system walls.

Along with "scaling effects" such as viscous dissipation, conjugate heat transfer effect and compressibility [4], slip effects may also appear for gas flows in microchannels [5]. When scales are reduced, some assumptions at the basis of the conventional model equations (namely the Navier-Stokes and energy balance equations for compressible flows) do not hold anymore. Rarefaction, and in particular slip flow conditions, refer to a non continuous behaviour of the gas flow in proximity of the gas-wall interface [5]. In these cases the boundary conditions of the Navier-Stokes equations must be modified [6]. As the discontinuities first appear near the wall, it is clear that the gas-surface interactions play a major role in determining the flow behaviour in this region. Several experimental studies have been done to characterize gas flows in microchannels under slip flow (some examples can be found in the works proposed by Arkilic [7], Turner [8] and Pitakarnnop [9]). Yet, most of the produced results are based on the indirect evaluation of the rarefaction parameters with measurements performed outside the channels and do not offer a complete overview of the flow characteristics. Even if these studies are useful for the validation of theoretical models, they do not help in identifying the actual role of the surface interactions on the flow. This is a typical local scale phenomenon which requires direct measurements inside the channels and could therefore benefit from the measuring technique proposed in this work.

In literature, many designs of integrated systems have been proposed. A common example of integrated sensors are miniaturized flow sensors, proposed, e.g., by Lammerink et al. for both liquids [10] and gases [11]. A suspended microchannel with embedded temperature sensors has been developed for high pressure application studies by Wu et al. [12]. Internal temperature measurements have been presented also by Park et al. [13], who developed a silicon channel bonded to a glass cover with integrated platinum resistance temperature detectors (RTD). Micro sensor fabrication technologies have been developed also on glass substrates. Xue and Qiu [14] presented an integrated array of temperature sensors for a glass microchannel generated by planar technology. Low temperature manufacturing processes for integrated sensors employing particular materials for heat transfer studies have also been developed and described, i.e., by Liu et al. [15] and Ko and Gau [16].

Despite the progresses in miniaturized sensor fabrication techniques and the high number of configurations proposed over the years, a full implementation of these systems in terms of integrating the sensors into a multi-objective micro-scale device has not been achieved yet.

The highly process-tailored designs available from literature limit the use of the sensors to the single application they have been designed for. Moreover, planar fabrication technologies, which offer great manufacturing potentials, limit the employable materials to a few (basically silicon and its compounds and glass). In most applications, the microchannel is fabricated along with the sensors, or bonded to the sensor substrate. This means that, whenever the microchannel configuration needs to be changed (i.e. channel geometry, material, surface finish, etc.), a new complete sensor-microchannel assembly becomes necessary.

To overcome some of these limitations, a sensor integration system has been developed and tested in a new microchannel device. This work presents the validation of the integrated sensor prototypes. The thermal characterization of a gas flowing inside a microchannel, in terms of internal temperature distribution, has been considered as relevant application to test the sensors functioning. Possible implementations of the newly developed system are however not limited to gas flows only. The main advantage of the present configuration is the absence of material limitations for the microchannel. The sensor integration can be broadened to those fields which typically do not employ silicon or glass and for which the available integrated sensors designs are not suitable. A representative example is the area of micro process engineering where usually metallic microstructured devices are used. In these cases the monitoring of the local temperature could avoid deviations from the process optimal conditions, allowing a better exploitation of the resources. The analysis has been performed using two experimental approaches. The first one is based on integrated microsensors, the second employs conventional measuring systems. Finally, comparisons with a CFD numerical model have been performed, too.

2. Experimental setup

The experimental device design is based on a previously developed system employed for microchannel heat transfer studies [15]. It has a multilayer configuration where each part is designed for a specific function and works independently from the others.

The device is showed in Fig. 1 and includes (from the bottom of the figure), gas inlet and outlet fittings, a base frame of stainless steel, three separated heating/cooling units in copper, a channel layer, a sensor layer and a stainless steel cover.

The heating/cooling copper blocks present two receptacle holes each, for the installation of electrically powered heating cartridges or tubes connected with a cooling loop. The blocks are separated by an air gap, which ensures an insulation layer between them. The power source for the electrical heating elements is provided by three separate channels (one per block), using three independent temperature controllers. On top of each block two grooves for the installation of thermocouples are included. These can be employed to record the block temperature and can be connected to the temperature controller as feedback sensors. This configuration allows the imposition of controlled temperature boundary conditions below the channel sections and the possibility of achieving steep temperature profiles along the microchannel bottom wall [17].

The microchannel section installed above the heating/cooling unit can be easily removed and replaced after opening the device. This allows investigating different materials, and different microchannel configurations with the same measuring system and in the same housing. The microchannels are realized on planar foils of the chosen material and by using the desired manufacturing technique. All the microchannels are open to the top, to allow the installation of the sensors as a cover.

The sensor layer is constituted by a polytetrafluoroethylene (PTFE) support frame with a recess for the positioning of a silicon chip. On top of it an array of temperature sensors have been realized with planar fabrication technologies. The chip, once installed, covers the microchannels entire length and constitutes its upper wall. The PTFE support is used also as channel sealing, thanks to the elastic properties of the polymeric material. Once the device is closed, the channel foil presses directly on the PTFE frame, ensuring leak tightness of the assembly, even when the sensor chip is installed.

Finally, the structure is closed by a cover and fixed by screws, which allow for an easy disassembling of the device and the exchange of the test sections. The different layers constituting the experimental device are shown in Fig. 2.

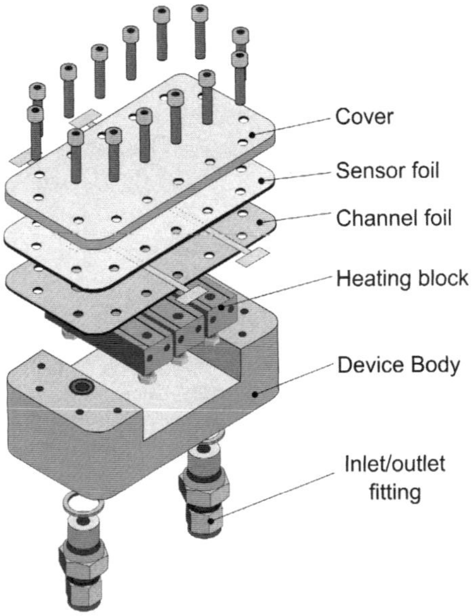

Figure 1. Scheme of the multilayer experimental device with integrated measuring system (4) and exchangeable test sections (3).

Figure 2. Photographs of the different parts constituting the experimental device, including the vacuum vessel for the thermal insulation of the device.

The outer metallic frame gives the device a large thermal inertia compared to the small gas volume flowing inside the microchannel. This results in relatively long transitories (depending on the set temperature, the warm-up time ranges between some minutes and one hour) but also in very stable boundary conditions for the microchannel (once the system reaches the steady state).

The device is installed into an experimental rig (see Fig. 3) including a vacuum pump for the regulation of the pressure levels at the channel inlet and outlet and a complete measuring system to monitor the pressure and the temperature of the gas at the inlet and outlet of the device. To eliminate the influence of thermal losses to the environment, the device is installed into a vessel which can be evacuated during the experiments.

The leak tightness of the device and of the whole experimental circuit has been verified with vacuum tests. Overall leakage rates of about $6.63 \cdot 10^{-3}$ mbar·l·sec^{-1} have been measured. Even though this values is higher than for typical gas-tight systems it represents less than 1 % of the mass flow rates established during the experiments and was therefore considered acceptable. Leakage occurs mainly where the cables exit the device. However to maintain an easy exchangeability of structures the design chosen was considered a good compromise.

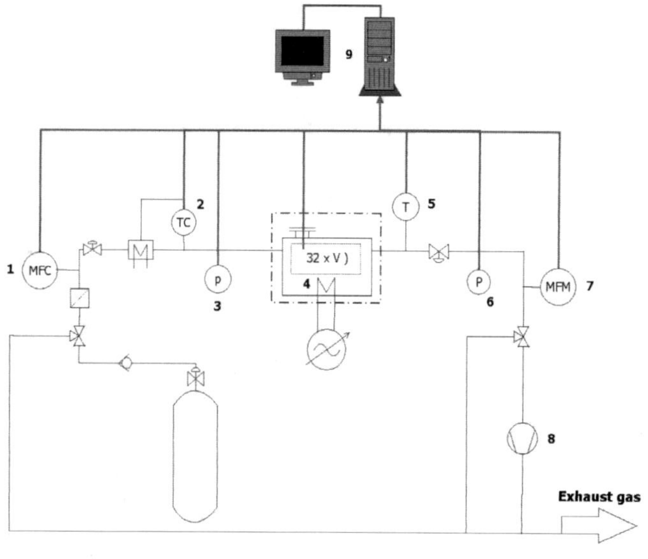

Figure 3. Scheme of the experimental rig. Mass flow controller (1); inlet temperature controller (2); inlet pressure measurement (3); flange vacuum system with the device and the integrated sensors (4); outlet temperature measurement (5); outlet pressure measurement (6); mass flow meter (7); vacuum pump (8).

Four different kinds of microchannel sections have been realized; their geometrical and surface characteristics are listed in Table 1. The average roughness has been measured with a chromatic white light sensor along the entire microchannel length. This non-contact technique is very precise and convenient for surface characterization at small scale.

2.1. Integrated measuring system

The manufactured microsensors present a thermopile configuration which exploits the Seebeck thermoelectric effect and presents several junctions in parallel to achieve an acceptable sensitivity [18]. In the present case the thermopiles are constituted by ten junctions between aluminum and n-doped polysilicon. To avoid thermal shortcuts (spoiling the sensor sensitivity), the sensors are fabricated on top of thin film membranes (10-15 μm thick) realized by wet etching of the chips back side and arranged along the chips longitudinal axis. Since thermopiles are temperature difference transducers, a reference temperature measurement is needed. This is provided by placing, in addition, a series of resistance temperature detectors (RTD) on the chip (outside the membranes), corresponding to the thermopiles "cold" junctions. The "hot" junctions

Table 1. Microchannel test sections main characteristics.

Material	Manufacturing technique	Section Shape	Average dimensions		Average roughness
			Width [μm]	Heigth [μm]	[μm]
Stainless steel	Micro-milling	Rectangular	400 ± 4	110 ± 4	0.08
Copper	Micro-milling	Rectangular	405 ± 4	80 ± 4	0.300
PEEK	Micro-milling	Rectangular	401 ± 4	108 ± 4	0.140
Stainless steel	Wet-chemical etching	Semi-elliptical	422 ± 4	140 ± 15	0.460

are positioned on the membranes and, once the chip is installed, face the inner part of the microchannel. The thermopile signal must be interpreted then as the temperature difference between the chip and the gas inside the channel. The thermistors measure the absolute chip temperature (directly proportional to the contact resistance) and allow the calculation of the absolute gas temperature, as described by Eq. 1. The layout of the integrated sensing unit is shown in Fig. 4.

$$T_{gas} = T_{chip} + \Delta T \tag{1}$$

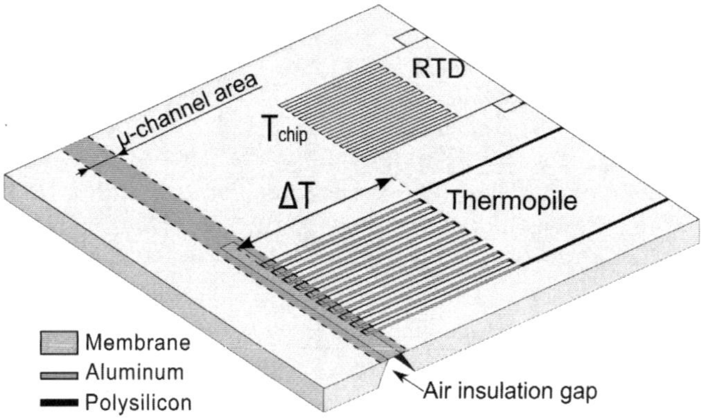

Figure 4. Schematic layout of a temperature sensing unit. The thermopile sensing junctions are included in the microchannel with a membrane configuration, while the reference temperature is measured by the thermistors (RTD) on the chip.

The sensor chip comprises 8 membranes 160 μm wide, each containing two thermopiles (one per side). This configuration allows a symmetric distribution of the heat conduction along the sensor leads and a redundant measure of the temperature at each position. Finally 16 RTDs have been manufactured in between the membranes. To collect the data registered by the sensors, flexible cables using polyimide as hose material are employed. These are bonded at four positions to the chip sides, where the sensor contacts end. The cables exit the device laterally, where printed circuit board (PCB) adapters and connectors are installed for the data recording. Despite the presence of the PTFE chip-holder, the major leaks occur at the four locations where the cables exit the device. However, the position of the cables prevents the use of conventional O-rings around the microchannels. To preserve the possibility of disassembling the different parts, no additional bonding has been foreseen, given also the small weight of the measured leakages on the overall mass flow rates. The cable bonding to the chip is realized by precise positioning of the two parts and by application of a two-step pressure/temperature cycle. This procedure is very delicate as proper contact between chip and cable is required for good signal propagation. Too high contact pressures or improper temperature cycles might result in the spurt of the contact material out of place and in shortcuts between the sensors. To verify the good functioning of the sensors, the resistances of all contacts have been tested before implementation of the chip. Typical values of some kΩ are characteristic for the sensor contacts, while an infinite resistance should be measured between non communicating contacts.

Figure 5 shows a sensor chip with thin cables bonded to the sides, installed in the PTFE holder.

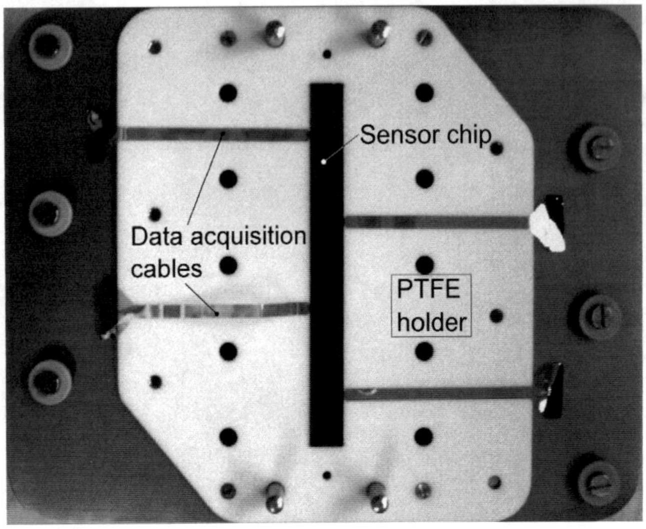

Figure 5. PTFE support foil with installed sensor chip and thin cables for data acquisition.

2.2. Conventional measuring system

Normally, the integration of conventional temperature sensors (e.g., commercial thermocouples) is not feasible for microchannel systems. Indeed, since even the smallest available sizes for the sensing tips are comparable to the microchannel dimensions, major flow disturbances may occur, which may negatively affect the measurements.

For the present case, a special design has been created, allowing the precise positioning of thermocouples in a microchannel, with minimal interaction with the flow. The microchannel sections are covered with a Polyether-ether-ketone (PEEK) cover along the axis of which 6 type-K thermocouples with 125 μm sensing tips have been installed with vertical orientation. The sensors are inserted in pre-fabricated holes on the cover and fixed with precision gluing. The sensor tips face the inner part of the microchannel, where they are in contact with the gas and measure the temperature distribution. The actual capability of sensing the gas temperature with this layout must be assessed before its implementation.

A photograph of the PEEK cover with embedded thermocouples is shown in Fig 6.

3. Numerical model

CFD simulation was used to model the behavior of the microchannel-sensors assembly.

Based on symmetry considerations, only one half of the microchannel geometry is studied. To reproduce the actual thermal boundary conditions encountered during the experiments, the channel has been split into five parts, corresponding to the unheated entrance and exit lengths and the three heated lengths, respectively.

Both geometry and mesh have been created with GAMBIT®, while the flow simulations have been realized with FLUENT®.

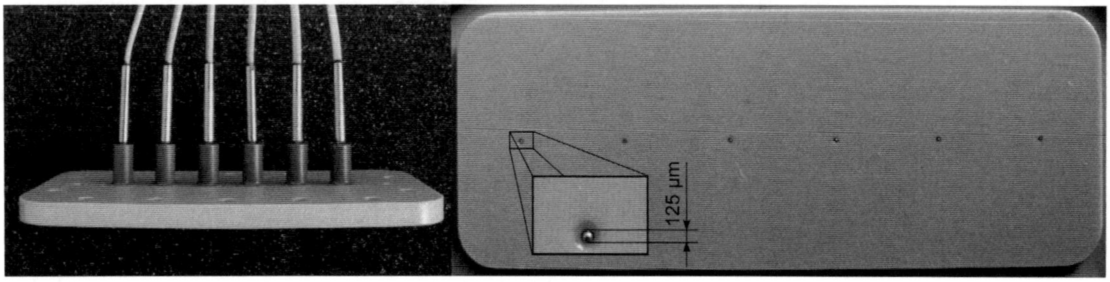

Figure 6. PEEK cover with embedded thermocouples. Cover overview (left) and details of the embedded sensors (right).

The boundary conditions for the problem (i.e., inlet, and outlet pressures, inlet temperature and wall temperatures) have been directly taken from the experimental results, and the calculated temperature distribution for the gas has been compared to the corresponding data.

The model includes compressibility effects for the flow, conjugate heat transfer, and slip flow at the microchannel walls.

4. Results

Before use of the integrated sensors the measuring principle and the calibration of the transducers was validated. This has been performed with a series of different tests described in the following.

4.1. Calibration of the RTDs

The thermistors are absolute temperature sensors, and the calibration can be performed by registering the signal for given temperature set points on the heaters and no imposed gas flow.

As reference temperatures for the RTDs the signal recorded with the embedded thermocouples on the PEEK cover have been used. In particular, as the wall temperature is to be measured, a blind foil (without the channel) has been installed during the calibration tests. For every temperature set point on the power supply, the steady state signals of the RTDs and the embedded thermocouples have been recorded separately. The results have been then reported in a resistance-temperature plot, where, with a least square fitting method, the calibration coefficients for each sensor have been calculated. The ranges for the temperature set points have been chosen within the limits encountered for the mechanical stability of the chip. The chip has been heated up to about 350 K without major fractures occurring. However, a confidence interval from ambient temperature up to 330 K has been chosen for the calibration. The relative fragility of the chip is not connected to the materials themselves, which could stand higher temperatures, but is rather related to the mechanical and thermal stresses occurring during the tests. The whole chip structure is indeed relatively large and presents several points for crack growth, especially concerning the membranes. For these reasons, particular attention had to be paid during the mounting of the chip in the PTFE holder and during device assembly. Sudden thermal stresses together with severe pressure gradients along the chip also had to be avoided.

Figure 7 presents an example of a calibration fitting for one RTD. A good linear behavior in the tested temperature range has been found for all thermistors. This trend is consistent with commercial RTDs showing a linear response over relatively small temperature ranges, as in the present case.

Considering the followed calibration procedure for the RTDs, the uncertainty associated with their signals has been calculated based on three different contributes as shown in Eq. 2:

$$\Delta T_{tot} = \pm\sqrt{\Delta T_{ref}^2 + \Delta T_{fit}^2 + \Delta T_{calib}^2} \quad (2)$$

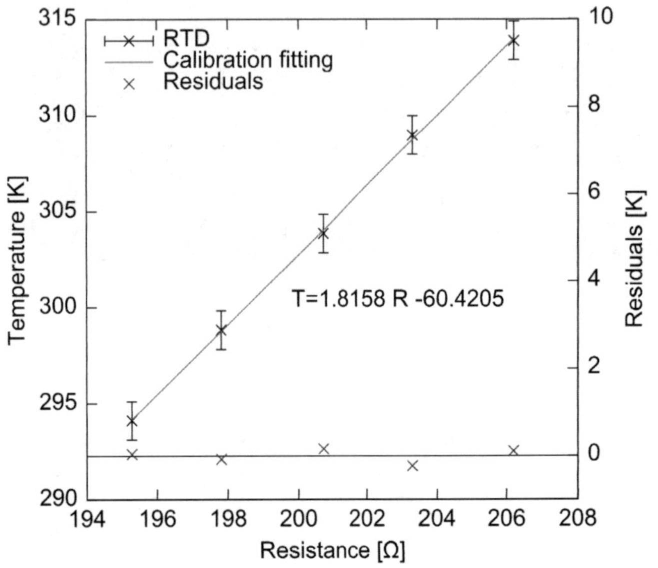

Figure 7. Calibration experimental point for a RTD and linear fitting.

The different error contribution are classified as follows:

- ΔT_{ref} : uncertainty associated with the reference sensors (contribution of the reference temperature measurements);
- ΔT_{fit} : uncertainty associated with the fitting procedure (contribution considering the deviations between the experimental points and the fitted line);
- ΔT_{calib} : calibration uncertainty associated with error propagation of the resistance measurement (contribution that takes into account the uncertainty of the resistance measurements and how this is converted into an uncertainty for temperature evaluation).

Typical uncertainties in the range of ± 0.7 K - ± 0.8 K have been found for the RTDs, where the main error contribution is given by the temperature reference sensors ($\Delta T_{ref} \approx 0.5$ K).

4.2. Thermopile tests

Whenever a temperature difference is established along the membranes, the thermopiles produce a proportional voltage signal. The sensors calibration requires the imposition of controlled temperature differences across each sensor, the recording of the corresponding voltage signal and the fitting of the experimental data. As a first step, to assess the functioning of the thermopiles and understand the answer to different boundary conditions, a series of tests have been realized.

In Fig. 8 the results of one of the performed sequences are shown for two different thermopiles. In the first case (Fig. 8-a) the thermopile is positioned directly above the heated area, while in the second case (Fig. 8-b), the sensor lay on the unheated length close to the microchannel exit. The test has been performed as follows:

- (i) The thermopile signals have been recorded at rest (i.e., neither the heating nor the gas flow were activated).
- (ii) A gas flow is established without heating the microchannel.
- (iii) The heating of the copper blocks is turned on, and the desired temperature is set on the power supply controllers.
- (iv) The heating is turned off.

(v) The gas flow is closed.

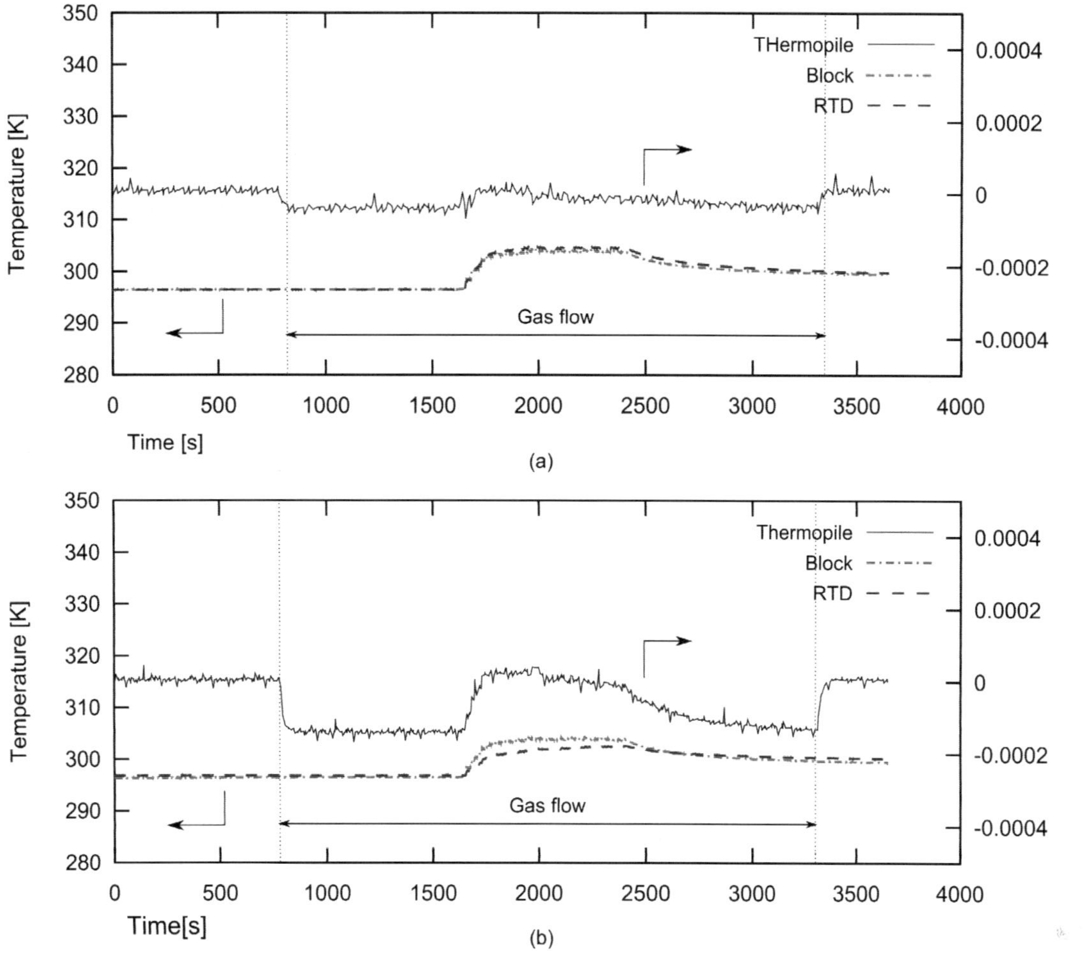

Figure 8. Comparison of the results of two thermopiles for a test sequence. (a) thermopile positioned over the heated area. (b) thermopile positioned outside the heated area and close to the microchannel exit.

For both thermopiles, the absence of any gas flow (at the beginning and at the end of the test) corresponds to a zero signal, independent of the microchannels absolute temperature. In this case, indeed, the membrane and the chip have the same temperature and $\Delta T = T_{gas} - T_{chip} = 0$.

When the gas flow is activated, the thermopiles record a negative step. This means that the gas flowing into the microchannel decreases the temperature of the membrane due to convection effects (which has a much lower thermal inertia with respect to the chip), while the chip temperature remains unvaried. The negative step is more evident for the thermopile close to the exit which suggests that the gas is cooler close to the exit. This can be an influence of the inlet and outlet microchannel openings, which act as sort of plena for the gas and are at constant ambient temperature.

When the heating is activated, the two thermopiles show a quite different behavior. For the sensor positioned over the heated area (Fig. 8-a), the signal has a small positive variation only at the beginning of the heating. The average value remains, however, almost unchanged and around zero. This means that the temperature difference between the gas and the chip is almost constant. The gas at the sensor position has increased its temperature from the upstream heated area, but also the chip is heated according to the temperature set at the heating blocks.

On the contrary, the thermopile closer to the exit (Fig. 8-b) shows a big positive step of the voltage signal in correspondence to the beginning of the heating. As the temperature of the block increases, the thermopile signal rapidly falls back to zero (the gas reaches the temperature of the chip), then progressively decreases as the chip is heated by conduction from the microchannel central area and a negative signal is established again.

Besides the different behavior of the thermopiles according to their relative position on the microchannel, some common characteristics for the integrated sensors-microchannel assembly can be highlighted. As expected, the miniaturized sensors show very fast responses in transient regime. The thermopiles have a high sensitivity (due to the membrane configuration) and thus very small temperature variations can be detected. Axial heat conduction takes place along both the microchannel foil and the chip layer from the directly heated to the entrance and exit regions.

Since the gas temperature along the channel rapidly changes with the position, direct calibration of the sensors by assuming a uniform temperature distribution can not be performed (even if constant temperature boundary conditions are imposed). A possible solution is an indirect calibration, achieved by comparing the results of two different measuring methods. For this purpose, the experimental data obtained with the PEEK cover with embedded thermocouples have been used.

4.3. Comparison with commercial sensors

Two separate tests have been performed with the PEEK cover, repeating the same test procedure and imposing the same boundary conditions, one with the blind foil (without the channel) and one with the microchannel foil and an imposed gas flow.

From the two tests the temperature difference between the top wall (blind foil results) and the gas (microchannel foil results) along the channel is calculated. This is then compared to the thermopile signals recorded for the same test sequence and boundary conditions with the sensors chip installed in the device.

The main drawback of this method is the low sensitivity of the PEEK-thermocouple assembly, and its higher thermal inertia compared to the chip micro sensors. In particular, at low mass flow rates (around 100 ml/min), the thermocouple signals show no detectable differences between the tests with the blind foil and those with the microchannel and the gas flow. For these reasons a comparison between the two methods can only be done at high mass flow rates and relatively high temperatures.

In Fig. 9 the comparison between the PEEK cover and the silicon chip results is shown for two sensors. The plots refer to a test where, after an initial waiting period (phase 1), the gas flow has been activated (phase 2), and only the copper block close to the channel exit has been heated to 40 °C (phase 3). A short cut-off of the gas flow has been imposed (phase 4), to compare the signals in a transient regime, before turning off the heating (phase 5).

From the plots it is evident that the two techniques produces comparable data, as the trend of the thermopile signals can be reproduced by the calculated temperature differences recorded with the PEEK cover.

These results suggest that an indirect calibration technique for the thermopile could work with the present configuration, provided that it is performed within the range of conditions at which the reference system (PEEK cover) is reliable. Once calibrated, the thermopiles measuring range can be extended, as they demonstrated good sensitivity also for smaller temperature differences or flow rates.

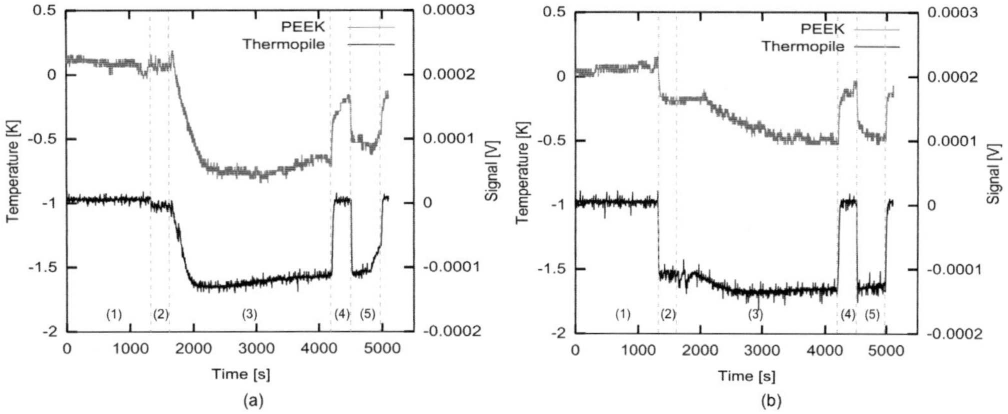

Figure 9. Comparison between conventional thermocouples and thermopile signals for two positions. (a) data recorded on the heated area. (b) data recorded outside the heated area and close to the microchannel exit. The data refers to the temperature difference between the wall and the gas in the microchannel for the thermocouples (red lines) and between the membranes and the chip for the thermopiles (black lines).

4.4. Numerical results

As pointed out, the validity of the temperature measurements with the PEEK cover embedded thermocouples must be verified. In particular, the ability of capturing the actual gas temperature must be assessed. For this purpose some of the measurements taken with the PEEK cover have been compared to numerical simulation results obtained from the CFD model. The boundary conditions for the channel bottom wall have been set at the temperatures registered by the thermocouples installed on the heating blocks. The upper wall temperatures have been derived from measurements with a blind foil. Finally, the inlet and outlet parameters have also been measured experimentally. The simulations have been run for steady flow, with the ideal-gas model for the density calculation. The actual thickness of the walls has also been taken into account, to include conjugate heat transfer effects in the calculations.

Figure 10 shows the results for a temperature set point at the heating blocks of 50 °C. The plot presents the axial temperature distribution for the top wall and for the gas flowing below it (namely 10 and 20 μm away from the wall). These profiles are compared to the measured values. The agreement is quite good, taken into account that the model dismisses the influence of the device surrounding the microchannel and the heat losses to the ambient (even though these have been minimized by the evacuated chamber).

5. Conclusions

A new sensor integration technique for studying microchannels fabricated from different materials has been presented. The developed prototype device allows overcoming the material limitations of previously developed integrated systems. Moreover, it allows the exchangeability of the microchannel test sections, broadening the flexibility and the range of possible applications for the integrated measuring system.

As a study case, the heat transfer of gas flowing in a microchannel has been investigated. This represents a typical case where the characterization of the local temperature distribution could be beneficial (especially when it comes to rarefied conditions where gas-surface interactions become important).

A special layout for the integration of commercial thermocouples in a microchannel has been developed, too. This was employed for validation and calibration of the integrated sensor

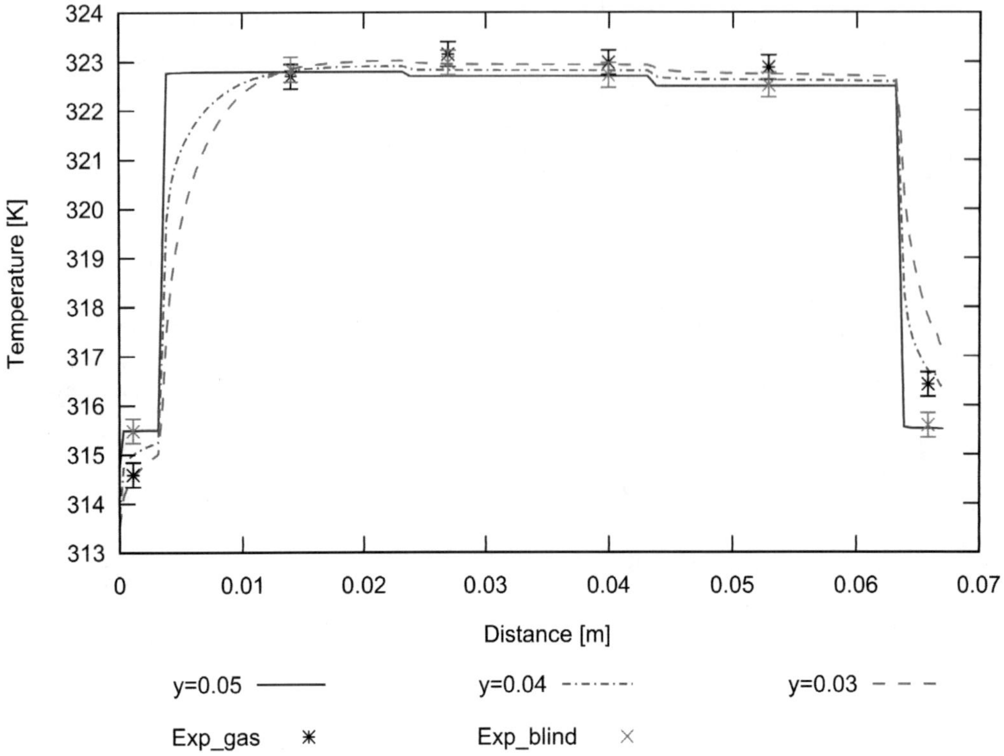

Figure 10. Comparison between experimental data and CFD simulations for a wall temperature set at 50 °C, an inlet pressure of 150000 Pa and an outlet pressure of 83000 Pa (same as in the experimental conditions). Measures taken with the PEEK cover and embedded thermocouples. CFD simulation realized with FLUENT®.

prototypes.

The results showed the performance of the integrated sensors and highlighted the main issues encountered during the tests. These can be summarized as follows:

- The integrated measuring system allows the monitoring of the internal temperature distribution of a gas flow in a microchannel.
- The sensors show a good time response (compared to commercial sensors), thanks to the thin membrane configuration.
- The sensor calibration is not trivial and requires the implementation of indirect reference systems for temperature differences (thermopile calibration).
- The chip on top of which the membranes and the sensors have been manufactured presents major fragility issues. Defects of the chip created during the fabrication processes and the membrane itself can constitute crack starting points when the chip is mounted and exposed to thermal and mechanical stresses.
- The integration system for the commercial thermocouples allows a broader range of working conditions (thanks to a higher intrinsic mechanical stability). However, the sensitivity is lost at low mass flow rates and lower temperature.
- The validity of the results obtained with the commercial thermocouple array has been validated with CFD simulations.

As a final remark it can be said that, whenever local phenomena are investigated, the employment of integrated sensors allows a closer insight into microchannel flows. Commercial sensors can be employed in some cases, but presents some major limitations (e.g. can not be employed for rarefied gas flows due to the small quantities to be detected). The presented microsensor integration system showed the high potential of silicon-based sensors, and the possibility of broadening their use to applications not addressed yet. Further optimizations for the sensor layout are still needed (especially concerning the fragility during handling).

Acknowledgments

The research leading to these results has received funding from the European Community's Seventh Framework Program (FP7/2007-20013) under grant agreement n 215504.

The authors wish to acknowledge Dr. Patrick Ruther from IMTEK, University of Freiburg, for his work and cooperation in the design and fabrication of the integrated microsensors.

References

[1] Ho C M and Tai Y C 1998 *Annu. Rev. Fluid. Mech.* **30** 579–612
[2] van Oudheusden 1988 *IEE Proceedings* vol 135-D
[3] Hsu T 2008 *MEMS and Microsystems. Design Manufacture and Nanoscale Engineering* (John Wiley & Sons, Inc.)
[4] Morini G 2006 *Heat Transfer Engineering* **27** 64–73
[5] Gad-el-Hak M 1999 *J. Fluids Eng.* **121** 5–33
[6] Colin S 2006 *Heat Transfer and fluid flow in minichannels and microchannels* ed Kandlikar S (Elsevier)
[7] Arkilic E B, Breuer K S and Schmidt M A 2001 *J. Fluid Mech.* **437** 29–43
[8] Turner S E, Lam L C, Faghri M and Gregory O 2004 *J. Heat Transfer* **126** 753–763
[9] Pitakarnnop J, Varoutis S, Valougeorgis D, Geoffroy S, Baldas L and Colin S 2010 *Microfluid Nanofluid* **8** 57–72
[10] Lammerink T S J, Tas N R, Elwenspoek M and Flultman J H J 1993 *Sens. Actuators, A* **38** 45–50
[11] Lammerink T S J, Dijkstra F, Houkes Z and Kuijk J V 1995 *Sens. Actuators A* **46-47** 380–84
[12] Wu S, Mai J, Zohar Y, Tai Y and Ho C 1998 *Proceddings of the 11th International Workshop on Micro Electro Mechanical Systems* (Piscataway, NJ) pp 87–92
[13] Park H, Pak J J, Son S Y, Lim G and Song I 2003 *Sens. Actuators, A* **103** 317
[14] Xue Z and Qiu H 2005 *Sens. Actuators, A* **122** 189–195
[15] Liu C W, Gau C, Ko H S, Yang C S and Dai B T 2006 *Sens. Actuators, A* **131** 575–582
[16] Ko H, Liu C W and Yang C G C S 2009 *J. Micromech. Microeng.* **17** 983–993
[17] Maikowske S 2010 *Appl. Therm. Eng.* **30** 1872–76
[18] van Herwaarden A, van Duyn D, van Oudheusden B and Sarro P 1989 *Sens. Actuators, A* **23** 621–630

1st European Conference on Gas Micro Flows (GasMems 2012) IOP Publishing
Journal of Physics: Conference Series **362** (2012) 012022 doi:10.1088/1742-6596/362/1/012022

Flow and Heat Transfer Characteristics of Turbulent Gas Flow in Microtube with Constant Heat Flux

Chungpyo Hong[1], Yutaka Asako[2], Shinichi Matsushita[1] and Ichiro Ueno[1]
[1]Dept. of Mechanical Engineering, Tokyo University of Science, Chiba, Japan
[2]Dept. of Mechanical Engineering, Tokyo Metropolitan University, Tokyo, Japan

E-mail: hong@rs.noda.tus.ac.jp

Abstract. Local friction factors for turbulent gas flows in circular microtubes with constant wall heat flux were obtained numerically. The numerical methodology is based on arbitrary-Lagrangian-Eulerian method to solve two-dimensional compressible momentum and energy equations. The Lam-Bremhorst's Low-Reynolds number turbulence model was employed to calculate eddy viscosity coefficient and turbulence energy. The simulations were performed for a wide flow range of Reynolds numbers and Mach numbers with different constant wall heat fluxes. The stagnation pressure was chosen in such a way that the outlet Mach number ranged from 0.07 to 1.0. Both Darcy friction factor and Fanning friction factor were locally obtained. The result shows that the obtained both friction factors were evaluated as a function of Reynolds number on the Moody chart. The values of Darcy friction factor differ from Blasius correlation due to the compressibility effects but the values of Fanning friction factor almost coincide with Blasius correlation. The wall heat flux varied from 100 to 10000 W/m^2. The wall and bulk temperatures with positive heat flux are compared with those of incompressible flow. The result shows that the Nusselt number of turbulent gas flow is different from that of incompressible flow.

1. Introduction

The number of applications of micro-electro mechanical systems (MEMS) has increased the need for understanding heat transfer in micro-geometries. In the case of micro-tube gas flows, it is found that the flow accelerates due to the gas expansion and thermal energy converts to kinetic energy. This results in a static temperature decrease of the gas [1].

Since the experimental work by Wu and Little [2], who measured the friction coefficient and Nusselt number for nitrogen, argon and helium flows in silicon or glass micro-channels, many experimental and numerical investigations have been undertaken.

Chen and Kuo [3] solved numerically the compressible turbulent boundary layer equations with using Baldwin-Lomax model. Their calculated friction factor is higher than that of Blasius correlation and calculated temperature dropped well below the wall temperature at the outlet due to energy conversion into the kinetic energy.

Hara et al. [4] experimentally investigated heat transfer for high pressure ratio of air flow in square mini-channels whose size was 0.3 to 2.0 mm and length was 10 mm to 100 mm. They reported that

Address correspondence to Dr. Chungpyo Hong, Dept. of Mechanical Engineering, Tokyo University of Science, 2641 Yamazaki, Noda, Chiba 278-8510, Japan. E-mail:hong@rs.noda.tus.ac.jp the heat transfer coefficient was about 7.3 times greater than Dittus and Boelter correlation for fully developed turbulent pipe flow.

Turner et al. [5] performed experimental investigations to measure the inlet and outlet temperatures and the micro-channel wall temperature in thermal boundary conditions of constant temperature gradient along the micro-channel length. The experimental measurements of the inlet and outlet gas temperature and the micro-channel wall temperature were used to validate a two dimensional numerical model for the laminar gaseous flow in micro-channel. Asako [6], Asako and Toriyama [7] and Hong and Asako [8] performed numerical investigations to obtain the heat transfer characteristics of laminar gas flows in a micro-channel and in a micro-tube with constant wall temperature, whose wall temperature is lower or higher than the inlet temperature. In the case of slow flow (Ma_{out}<0.3), identical heat transfer coefficients are obtained for both heated and cooled cases. However, in the case of fast flow, different heat transfer coefficients are obtained for each cooled and heated case. The way to predict of the heat transfer rate of gaseous flow in a micro-channel and in a micro-tube has been proposed.

Recently, Hong et al. [9] measured total temperatures at the outlet of a micro-tube of D=163-242 μm with constant wall temperature whose temperature is higher than the inlet temperature in the laminar flow regime in order to validate numerical results by Hong and Asako [8]. The bulk temperatures obtained experimentally are in excellent agreement with those of numerical results. They found that measured total temperatures at the outlet are higher than the bulk temperatures of incompressible due to the additional heat transfer from the wall to the gas.

Chen et al. [10] investigated experimentally forced convective heat transfer characteristics of air and CO_2 in micro stainless steel tube with inside diameter of 920 μm. The tube wall was heated by electric current. They used a non-contacted liquid crystal thermography (LCT) temperature measurement method and thermocouples to measure the surface temperature of micro-tube. They concluded that the conventional heat transfer correlation of laminar and turbulent flow can be well applied for prediction of heat transfer performance in a micro-tube. However, Yang et al. [11] investigated experimentally and numerically the characteristics of nitrogen gas convective heat transfer in commercial stainless steel micro tubes with inner diameter of 175μm and 750 μm.

Hong et al. [12] and Isobe et al. [13] investigated experimentally heat transfer characteristics of turbulent gas flow in micro tube with constant wall temperature. The experiments were performed for nitrogen gas flow through a micro-tube with 243 μm and 342 μm in diameter and 50 mm in length. The wall temperature was maintained at 310-350 K. They reported that heat transfer characteristics of turbulent gas flow in micro-tube with constant wall temperature are different from those of the incompressible flow. Additional heat transfer for turbulent flow is more dominant than that for laminar flow.

As can be seen from the literature survey, there seems to be few numerical investigations on the heat transfer of turbulent gas flow though a micro-tube. Chen and Kuo [3] numerically investigated the heat transfer of turbulent gas flow through a micro-tube. However, in their paper, the effect of compressibility on turbulent heat transfer is not clear. Therefore, in turbulent flow regime the effect of compressibility on heat transfer has not been clarified totally yet. This has motivated the present numerical study to obtain heat transfer characteristics of turbulent gas flow in microtubes with constant heat flux.

2. Formulation

2.1. Description of the problem and conservation equations

The problem modelled in this study is depicted schematically in Figure. 1. The characteristics of turbulent gas flow in a microtube will be determined. A reservoir at the stagnation temperature, T_{stg}, and the stagnation pressure, p_{stg}, is attached to the upstream section of the microtube. The gas flows into the microtube and flows out to atmosphere. The wall temperature of the tube is assumed to be constant. The

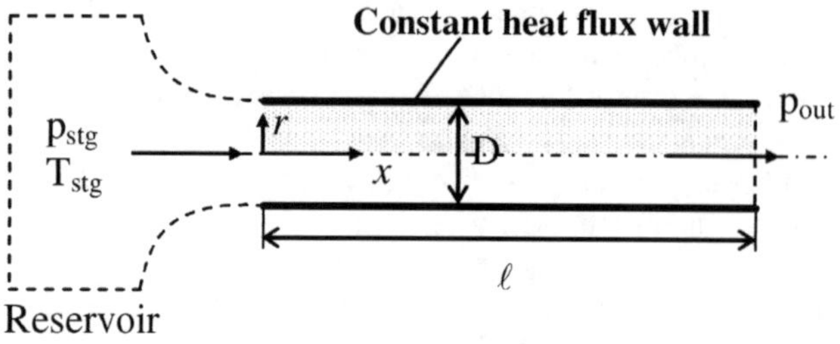

Fig. 1 Schematic diagram of problem

flow assumed to be steady, axisymmetric and turbulent. Two-dimensional coordinates are used for computations. The fluid is assumed to be an ideal gas. The governing equations to be considered are the time-averaged compressible continuity, momentum, and energy equations [14]. An eddy viscosity model is used to account for the effect of turbulence. The model chosen is the Lam-Bremhorst low-Reynolds number form of the k-ε (turbulence kinetic energy-turbulence dissipation rate) model [15]. Constant thermophysical properties are assumed. The Reynolds stress of compressible flow is modeled by

$$-\overline{\rho u_i' u_j'} = \mu_t \left\{ \left(\frac{\partial \tilde{u}_i}{\partial x_j} + \frac{\partial \tilde{u}_j}{\partial x_i} \right) - \frac{2}{3}\delta_{ij}\frac{\partial \tilde{u}_k}{\partial x_k} \right\} - \frac{2}{3}\delta_{ij}\bar{\rho}\tilde{k} \tag{1}$$

The governing equations can be expressed as

$$\frac{\partial \bar{\rho}\tilde{u}}{\partial x} + \frac{1}{r}\frac{\partial \bar{\rho} r \tilde{v}}{\partial r} = 0 \tag{2}$$

$$\frac{\partial \bar{\rho}\tilde{u}\tilde{u}}{\partial x} + \frac{1}{r}\frac{\partial \bar{\rho} r \tilde{u}\tilde{v}}{\partial r} = -\frac{\partial}{\partial x}\left(\bar{p} + \frac{2}{3}\bar{\rho}k \right) + \frac{\partial \tau_{xx}}{\partial x} + \frac{1}{r}\frac{\partial}{\partial r}(r\tau_{rx}) \tag{3}$$

$$\frac{\partial \bar{\rho}\tilde{u}\tilde{v}}{\partial x} + \frac{1}{r}\frac{\partial \bar{\rho} r \tilde{v}\tilde{v}}{\partial r} = -\frac{\partial}{\partial r}\left(\bar{p} + \frac{2}{3}\bar{\rho}k \right) + \frac{\partial \tau_{rx}}{\partial x} + \frac{1}{r}\frac{\partial}{\partial r}(r\tau_{rr}) - \frac{\tau_{\theta\theta}}{r} \tag{4}$$

$$\frac{\partial \bar{\rho}\tilde{u}\tilde{i}}{\partial x} + \frac{1}{r}\frac{\partial \bar{\rho} r \tilde{v}\tilde{i}}{\partial r} = -\left(\bar{p} + \frac{2}{3}\bar{\rho}K \right)\left(\frac{\partial \tilde{u}}{\partial x} + \frac{1}{r}\frac{\partial r\tilde{v}}{\partial r} \right) + \frac{\partial}{\partial x}\left\{ (\lambda + \lambda_t)\frac{\partial \tilde{T}}{\partial x} \right\} + \frac{1}{r}\frac{\partial}{\partial r}\left\{ (\lambda + \lambda_t) r \frac{\partial \tilde{T}}{\partial r} \right\} + (\mu + \mu_t)\phi \tag{5}$$

where the components of viscous stress tensor are

$$\tau_{xx} = (\mu + \mu_t)\left\{ 2\frac{\partial \tilde{u}}{\partial x} - \frac{2}{3}\left(\frac{\partial \tilde{u}}{\partial x} + \frac{1}{r}\frac{\partial r\tilde{v}}{\partial r} \right) \right\},\quad \tau_{xr} = \tau_{rx} = (\mu + \mu_t)\left(\frac{\partial \tilde{v}}{\partial x} + \frac{\partial \tilde{u}}{\partial r} \right)$$
$$\tau_{rr} = (\mu + \mu_t)\left\{ 2\frac{\partial \tilde{v}}{\partial r} - \frac{2}{3}\left(\frac{\partial \tilde{u}}{\partial x} + \frac{1}{r}\frac{\partial r\tilde{v}}{\partial r} \right) \right\},\quad \tau_{\theta\theta} = (\mu + \mu_t)\left\{ 2\frac{\tilde{v}}{r} - \frac{2}{3}\left(\frac{\partial \tilde{u}}{\partial x} + \frac{1}{r}\frac{\partial r\tilde{v}}{\partial r} \right) \right\} \tag{6}$$

ϕ is dissipation function described as follows

$$\phi = 2\left[\left(\frac{\partial \tilde{u}}{\partial x} \right)^2 + \left(\frac{\tilde{v}}{r} \right)^2 + \left(\frac{\partial \tilde{v}}{\partial r} \right)^2 \right] - \frac{2}{3}\left(\frac{\partial \tilde{u}}{\partial x} + \frac{1}{r}\frac{\partial r\tilde{v}}{\partial r} \right)^2 + \left(\frac{\partial \tilde{v}}{\partial x} + \frac{\partial \tilde{u}}{\partial r} \right)^2 \tag{7}$$

Furthermore, the working gas is assumed to be an ideal gas so that the equation of state is valid

$$\tilde{i} = \frac{R}{\gamma - 1}\bar{T} = \frac{1}{\gamma - 1}\frac{\bar{p}}{\bar{\rho}} \tag{8}$$

2.2. Turbulence model

In *k*-ε turbulence model, turbulence kinetic energy *k* and viscosity coefficient μ_t are determined solving the equations of transportation below

$$\frac{\partial \bar{\rho}\tilde{u}k}{\partial x}+\frac{1}{r}\frac{\partial \bar{\rho}r\tilde{v}k}{\partial r}=\frac{\partial}{\partial x}\left\{\left(\mu+\frac{\mu_t}{\sigma_k}\right)\frac{\partial k}{\partial x}\right\}+\frac{1}{r}\frac{\partial}{\partial r}\left\{\left(\mu+\frac{\mu_t}{\sigma_k}\right)r\frac{\partial k}{\partial r}\right\}+(P)_k-\bar{\rho}\varepsilon \tag{9}$$

$$\frac{\partial \bar{\rho}\tilde{u}\varepsilon}{\partial x}+\frac{1}{r}\frac{\partial \bar{\rho}r\tilde{v}\varepsilon}{\partial r}=\frac{\partial}{\partial x}\left\{\left(\mu+\frac{\mu_t}{\sigma_\varepsilon}\right)\frac{\partial \varepsilon}{\partial x}\right\}+\frac{1}{r}\frac{\partial}{\partial r}\left\{\left(\mu+\frac{\mu_t}{\sigma_\varepsilon}\right)r\frac{\partial \varepsilon}{\partial r}\right\}+c_1f_1\frac{\varepsilon}{k}P_k-c_2f_2\bar{\rho}\frac{\varepsilon^2}{k} \tag{10}$$

$$P_k=\mu_t\varphi-\frac{2}{3}\bar{\rho}k\left(\frac{\partial\tilde{u}}{\partial x}+\frac{1}{r}\frac{\partial r\tilde{v}}{\partial r}\right) \tag{11}$$

$$\mu_t=C_\mu f_\mu\bar{\rho}\frac{k^2}{\varepsilon} \tag{12}$$

The various ways have been presented to model the coefficients appearing these equations [16]. In this paper, the Lam- Bremhorst low Reynolds number model was adopted and the following values are used for coefficients

$$C_\mu=0.09,\quad \sigma_k=1.0,\quad \sigma_\varepsilon=1.3,\quad C_1=1.44,\quad C_2=1.92 \tag{13}$$

$$f_1=1+\left(\frac{0.05}{f_\mu}\right)^3,\quad f_2=1-e^{-Rt^2}\quad f_\mu=\left(1-e^{-0.0165R_y}\right)^2\left(1+\frac{20.5}{R_T}\right) \tag{14}$$

$$R_T=\frac{\bar{\rho}k^2}{\varepsilon\bar{\mu}},\qquad R_y=\frac{\bar{\rho}\sqrt{k}y_W}{\bar{\mu}} \tag{15}$$

where y_W is defined as the minimum distance from the tube wall. Turbulent thermal conductivity in Eq. (5) is

$$\lambda_t=\frac{c_p\mu_t}{Pr_t} \tag{16}$$

where Pr_t is the turbulent Prandtl number and its value is 0.9.

2.3. Boundary conditions

In the range of the micro tube sizes considered in this paper, Knudsen number is less than 0.001 for the atmospheric pressure and the room temperature so that the effect of rarefaction can be neglected. Therefore no-slip condition at the wall is used for velocity. Also, it is assumed that the tube wall is constant heat flux and at the tube inlet the distributions of velocity, pressure, temperature and density are uniform. From the assumptions above, the boundary conditions are described as follows

on the wall: $u=v=0,\quad \bar{\lambda}\frac{\partial T}{\partial r}=\dot{q}$

on the symmetric axis: $\frac{\partial u}{\partial r}=v=0,\quad \frac{\partial T}{\partial r}=0$

at the inlet: $u=u_{in},\quad v=0,\quad p=p_{in},\quad \rho=\rho_{in},\quad T=T_{in}$

at the outlet: $p=p_{out}$

The values of velocity, pressure, temperature and density at the inlet are evaluated by the method presented by Kariki [17] as described below. For the cells adjacent to the symmetric axis, pressure of the n-th cell from the tube inlet is expressed as p_n. At the time t, pressure of the cell adjacent to the inlet, $(p_{in})_{guess}$, is extrapolated from p_2 and p_3. Then using this $(p_{in})_{guess}$ and $(p_1)_{old}$ at the time t, the inlet pressure p_{in} at the time $t+\Delta t$ is calculated as

$$p_{in}=\omega(p_1)_{old}+(1-\omega)(p_{in})_{guess} \tag{17}$$

where ω is relaxation factor and the value of 0.9 is chosen in this paper. Assuming isentropic change from stagnation point to the tube inlet, ρ_{in}, T_{in} and u_{in} at the time $t+\Delta t$ are calculated using p_{in} of Eq. (17) as

$$\frac{p_{in}}{p_{stg}} = \left(\frac{\rho_{in}}{\rho_{stg}}\right)^{\gamma} = \left(\frac{T_{in}}{T_{stg}}\right)^{\gamma/\gamma-1} \tag{18}$$

$$u_{in} = \sqrt{\frac{2\gamma RT_{stg}}{\gamma-1}\left\{1-\left(\frac{p_{in}}{p_{stg}}\right)^{(\gamma-1)/\gamma}\right\}} \tag{19}$$

k and ε at the inlet is determined as follows

$$k_{in} = \frac{3}{2}\left(t_u\, u_{in}\right)^2 \tag{20}$$

$$\varepsilon_{in} = 0.1k_{in}^2 \tag{21}$$

where t_u is ranges from 1% to 6%.

2.4. Dimensionless variables

Attention will now be focused on the calculation of the Reynolds and Mach numbers that will be defined as

$$Re = \frac{2\dot{m}}{\mu} = \frac{\rho_{ave}u_{ave}D}{\mu}, \quad Ma = \frac{u_{ave}}{\sqrt{\gamma(\gamma-1)i_{ave}}} \tag{22}$$

where $\dot{m}$ is the mass flow rate per unit depth and D is the tube diameter. u_{ave}, ρ_{ave} and i_{ave} are the average velocity, density and specific internal energy at a cross-section.

$$u_{ave} = \frac{1}{A}\int udA, \quad \rho_{ave} = \frac{1}{A}\int \rho udA \Big/ \int udA, \quad p_{ave} = \frac{1}{A}\int pdA, \quad i_{ave} = \frac{1}{\gamma-1}\frac{p_{ave}}{\rho_{ave}} \tag{23}$$

The product of friction factor and Reynolds number is called Poiseuille number. The friction factors based on the Darcy's and Fanning definitions will be introduced. The Darcy friction factor is defined as

$$f_d = \frac{-2D}{\rho_{ave}u_{ave}^2}\left(\frac{dp_{ave}}{dx}\right) \tag{24}$$

The modified Fanning friction factors (four times of Fanning friction factor) is based on the wall shear stress and is defined as

$$f_f = \frac{4\tau_w}{(1/2)\rho_{ave}u_{ave}^2} = 2D\left(\frac{dp_{ave}}{dx}\right)\left\{\frac{1}{p_{ave}} - \frac{1}{\rho_{ave}u_{ave}^2}\right\} \tag{25}$$

2.5. Numerical solution

The numerical methodology is based on the Arbitrary-Lagrangian-Eulerian (ALE) method proposed by Amsden et al. [18]. The details of the ALE method are documented in the literature and will not be given here. The computational domain is divided into quadrilateral cells. The velocity components are defined at the vertices of the cell and other variables such as pressures, specific internal energy and density are assigned at the cell centers. The number of cells in the *x*-direction was 200. The cell size gradually increased in the x-direction to the mid of the tube and it gradually decreases to the outlet of the tube. The number of cells in *r*-direction was fixed at 40 for all the components. The ALE method

Table 1 Tube diameter, length, p_{stg} Re and Ma

#	D (μm)	L (m)	p_{stg} (kPa)	p_{out} (kPa)	Re	Ma_{in}	Ma_{out}	flow
1	200	0.04	1450	1400	4497	0.070	0.075	slow
2			300	100	2947	0.228	0.660	fast
3			350		3575	0.238	0.778	
4			400		4208	0.246	0.889	
5	400	0.08	200	100	3770	0.218	0.446	fast
6			250		5180	0.242	0.591	
7			300		6541	0.255	0.725	
8			350		7862	0.264	0.846	
9			400		9143	0.269	0.951	

is a time marching method. The value of 10^{-4} was used for convergence criterion of Newton-Raphson iteration in the interior loop of the ALE method.

3. Results and discussion

The computations were performed for the tube with the constant heat flux of $\dot{q}=10^2$, 10^3 and 10^4 W/m^2. The working fluid was nitrogen and it was assumed to be ideal gas. The thermo-physical properties of $R = 296.7$ J·(kg·K)$^{-1}$, $\gamma = 1.399$, $\mu = 1.787\times10^{-5}$ Pa·s and $\lambda = 0.0260$ W·(m·K)$^{-1}$ at 300K were used for the computations. The diameter of the tube was 200 and 400 μm and the aspect ratio of the tube height and length is 200. The stagnation temperature was kept at T_{stg} = 300 K. The stagnation pressure, p_{stg} was chosen in such away that the Mach number at the exit ranges from 0.1 to 1.0. The outlet pressure maintained at atmospheric condition, p_{out} =100 kPa. Additionally, in order to obtain turbulent slow flow, the higher outlet pressure maintained with the 1400 kPa for D=200 μm was computed. The tube diameter, the tube length, the stagnation pressure and the corresponding Reynolds & Mach numbers at the inlet and outlet for $\dot{q}=10^4$ W/m^2 are listed in table 1. The Reynolds number obtained ranges from 2947 to 9143, and the Mach number at the outlet ranges from 0.070 to 0.951 widely.

3.1. Velocity vectors and contour plot of temperature

Velocity vectors for the case #1 and #4 in table 1 are presented in Figure 2. Contour plots of temperature are also presented in Figure 3. These are typical velocity vectors and temperature contour plots for the combination of slow and fast flows. Figure 2 shows the velocity vectors of gas flow with uniform velocity at the inlet. The flows are turbulent since the shape of the velocity profile except near the inlet is almost flat. Also, obtained Reynolds number is well lager than the critical Reynolds number. The gaseous flow is accelerated in a microtube. Therefore, Mach number at the outlet is greater than that at the inlet, $Ma_{in}<Ma_{out}$. The flow whose Ma at the outlet is less than 0.3 is called as "slow flow" and the flow whose Ma at the outlet is greater than 0.3 is called as "fast flow". As can be seen in Figure 3 (a), in the case of the slow flow (Ma_{out} < 0.3), the temperature rises gradually along the tubes through the influence of the constant heat flux. This is the similar temperature contour to that of the incompressible flow since the temperature contour begins to develop along the tube length with parabolic curves depending on the constant heat flux. On the other hand, in the case of the fast

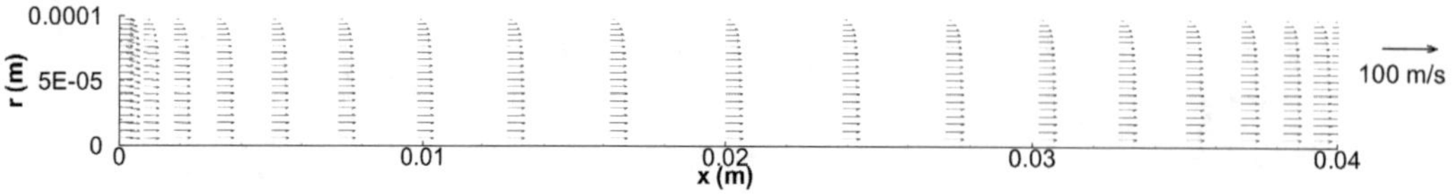

(a) #1

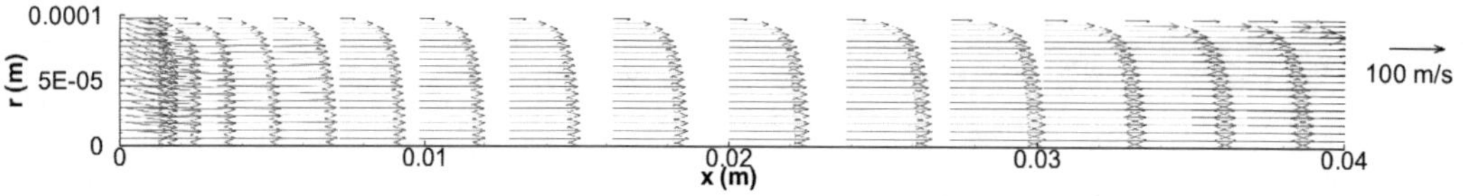

(b) #4

Figure 2. Velocity vector for slow (a) and fast flow (b)

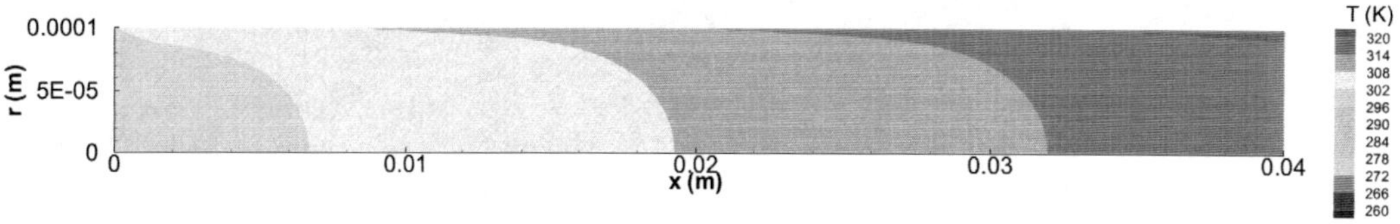

(a) #1

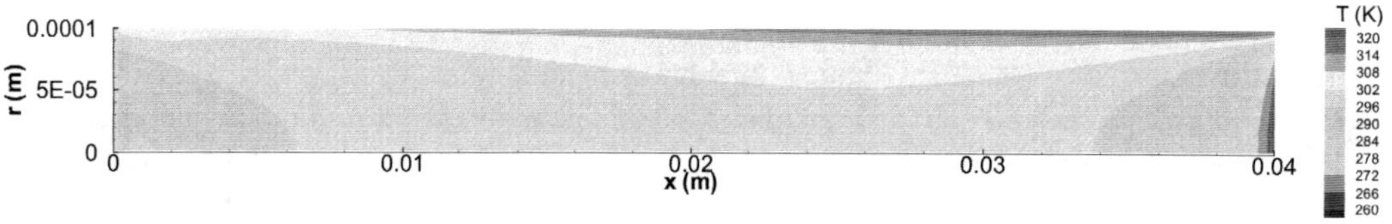

(b) #4

Figure 3. Contour plots of temperature for slow (a) and fast flow (b)

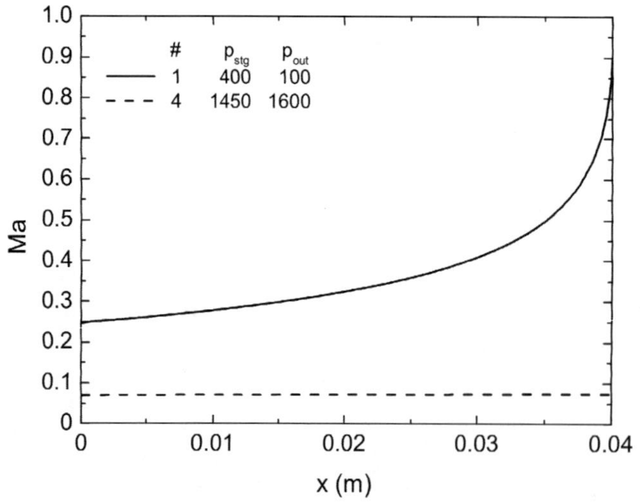

Figure 4 Ma as a function x

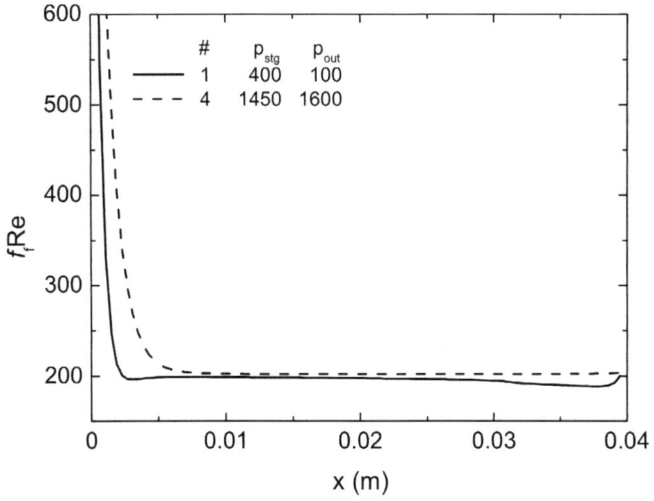

(a) $f_f \cdot$Re

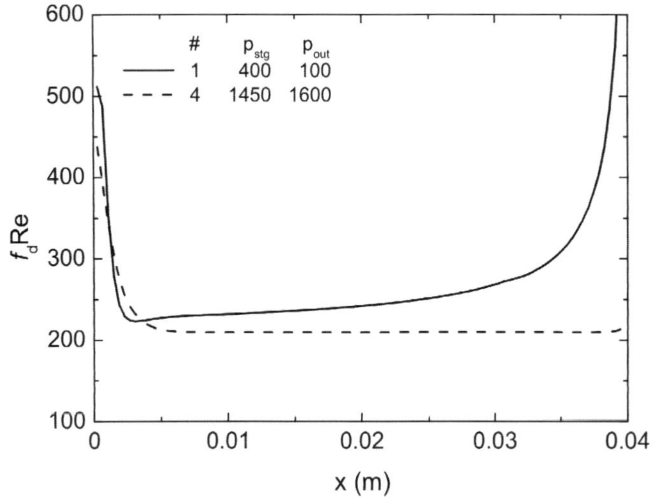

(b) $f_d \cdot$Re

Figure 5 $f \cdot$Re as a function of x

flow ($Ma_{out} > 0.3$), the temperature fall can be seen near the outlet due to the energy conversion into the kinetic energy caused by the flow acceleration in the core region of the tube (Figure 3 (b)). The wall temperature decreases approaching to the outlet due to the temperature fall in the core region. This is the typical temperature contour of the compressible flow.

3.2. Mach number and friction factor

The product of the friction factor and Reynolds number, $f_f \cdot$Re and $f_d \cdot$Re of two cases for Figures 2 and 3 are plotted as a function of x in Figures 4 (a) and (b), respectively. Corresponding Mach numbers are also plotted in Figure 5. As can be seen in Figures 4 (a) and (b), the tube can be divided into two

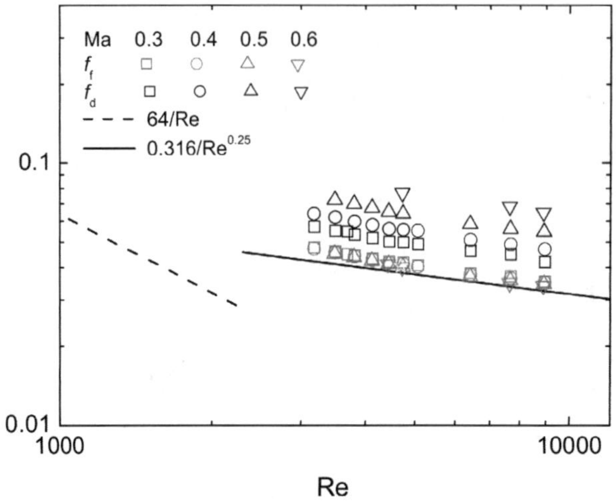

Figure 6 f on Moody chart

regions. The region where the friction factor decrease sharply for both cases can be considered the entrance region. In the case slow flow, the region after the entrance region can be considered as fully developed region. However, in the case of fast flow, the region after the entrance region can be considered as the quasi-fully developed region. Then the Mach number increases along the tube due to acceleration of the flow. The value of $f_f \cdot$Re of fast flow almost coincide with that of slow flow since the viscous wall friction loss for turbulent flow depends on Reynolds number. The value of f_d $\cdot$Re of fast flow is higher than that of slow flow since $f_d \cdot$Re includes the acceleration loss. And the value of $f_f \cdot$Re of fast flow agrees with that of slow flow since the $f_f \cdot$Re do not include flow acceleration loss. The f_f and f_d obtained for all case of the present numerical calculations plotted as a function on Moody chart in Figure 6. Then, the f_f and f_d at the locations where Mach number indicated just 0.3, 0.4, 0.5, 0.6 and 0.7 are extracted from numerical results. The solid line and dashed line represent conventional correlation of friction factor for incompressible flow (f = 64/Re for laminar flow and $f = 0.316\mathrm{Re}^{-0.25}$ (Blasius correlation) for turbulent flow). f_f is almost coincide with conventional value given by Blasius correlation. As can be seen in Figure 2, gradient of the normalized velocity at the wall is almost constant even if Mach number changes, so f_f depends only on Reynolds number. However, the plotted values of f_d at constant Mach number are aligned on a straight lines. These lines are parallel to the line of Blasius correlation ($f = 0.316\mathrm{Re}^{-0.25}$) and shift upwards as Mach number increases. The ratio of f_d to f_f or Blasius correlation is dependent on only Mach number and independent on Reynolds number.

3.3. Wall temperatue and Nusselt number of incompressible flow

In the case of the incompressible flow, the local Nusselt number, based on the temperature difference between the wall and the bulk temperatures, of a simultaneously developing flow in a duct with the constant heat flux is defined as

$$\mathrm{Nu}_x = \frac{\dot{q}D}{(T_w - T_b)k} \tag{26}$$

If the flow is incompressible, the bulk temperature is expressed by the function of x as

$$T_b = T_{in} + \frac{2\dot{q}x}{\int \rho C_p u dA} \tag{27}$$

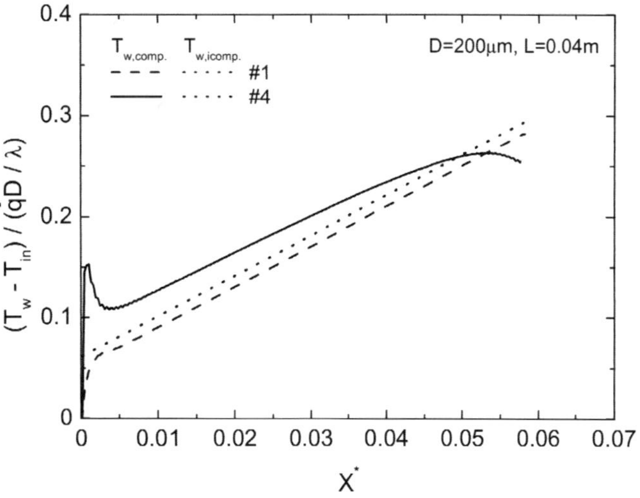

Figure 7 Dimensionless wall temperature as a function of X^*

The wall temperature of an incompressible flow in a duct is expressed as a function of a local Nusselt number and the location as

$$T_w = \left(4X^* + \frac{1}{Nu_x}\right)\left(\frac{\dot{q}D}{k}\right) + T_{in} \tag{28}$$

where X^* is the inverse of Graetz number defined by

$$X^* = \frac{x}{D\ Re\ Pr} \tag{29}$$

Equation (28) can be rewritten as

$$\frac{T_w - T_{in}}{\dot{q}\ D\ /k} = 4X^* + \frac{1}{Nu_x} \tag{30}$$

The dimensionless wall temperature of an incompressible flow in a duct is a function of X^* and Nu_x. The turbulent heat transfer characteristics for incompressible duct flows has been investigated by many researchers and Nusselt numbers were reported in literatures (e.g., Kays and Crawford [19]). In the present study, the obtained Nu for the turbulent fully developed flow by Kays and Crawford [19] defined by

$$Nu = 0.022\,Re^{0.8}Pr^{0.5} \tag{31}$$

was used for the calculation of the wall temperature for the incompressible flow.

3.4. Estimation of wall temperature

The dimensionless wall temperature, $(T_w - T_{in})/(\dot{q}\ D/k)$, for two cases of the tubes of D=200 μm for $\dot{q}=10^4$ W/m^2 is plotted as a function of X^* in Figure. 7. The dimensionless wall temperature for the incompressible flow obtained by Equation (30) is also plotted in the figure. As can be seen in Figure 2 in the case of slow flow (#1), the dimensionless wall temperature increases along the dimensionless length and it coincides with that of incompressible flow. On the other hand, in the case of fast flow (#4), the dimensionless wall temperature increases gradually along the dimensionless length and level off and decreases quickly approaching to the outlet due to the conversion of the thermal energy into the kinetic energy. The qualitative same tendency can be seen for the tubes with D=400μm. Therefore, in the case of fast flow (Ma_{out}>0.3), the wall temperature of the gas flow in the microtube

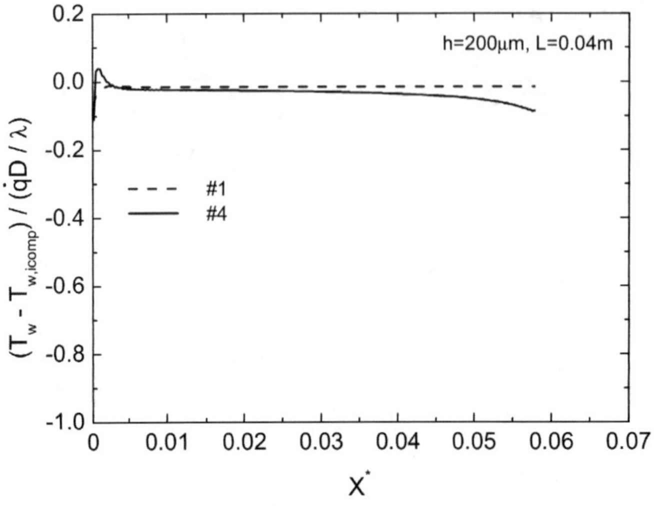

(a) Dimensionless wall temperature difference

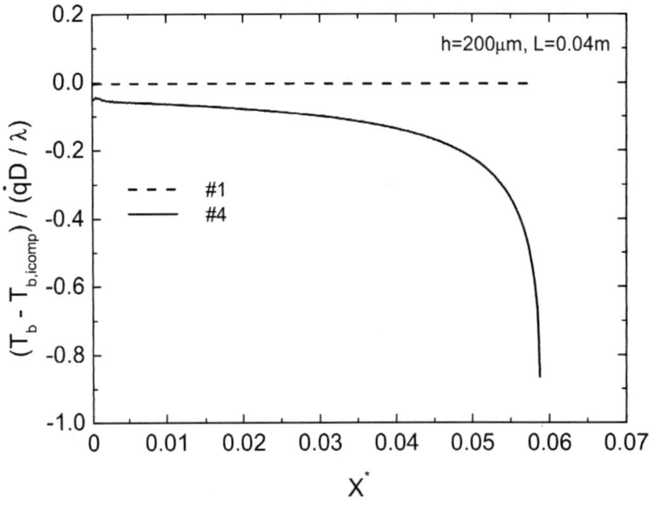

(b) Dimensionless bulk temperature difference

Figure 8 Dimensionless wall and bulk temperature differences as a function of X^*

can not be directly estimated from the correlation for the incompressible flow. In addition, the computed dimensionless temperatures of several cases are slightly higher than that of the incompressible flow near the inlet since the effect of viscous dissipation is relatively prevalent before the flow acceleration occurs.

The dimensionless wall temperature differences, $(T_w - T_{w,icomp})/(\dot{q}D/k)$, between the gas flow and incompressible flow for two cases of the tubes of D=200 μm for $\dot{q}=10^4$ W/m^2 are plotted as a function of X^* in Figure 8 (a). Also, the dimensionless bulk temperature differences, $(T_b - T_{b,icomp})/(\dot{q}D/k)$, between the gas flow and incompressible flow are plotted as a function of X^* in Figure 8 (b). In the case of slow flow, the dimensionless wall and bulk temperatures coincide with those of incompressible flow. Therefore, there is no difference between them as shown in Figure 8 (a) and (b). On the other hand, in the case of fast flow, the dimensionless wall and bulk temperature difference decreases approaching to the outlet.

Note that the bulk temperature difference between the gas flow and incompressible flow represents the dynamic temperature of the fluid, T_k, defined as

$$T_k = \frac{\int \rho u \frac{u^2}{2} dA}{\int \rho C_p u dA} \tag{32}$$

The gas temperature decreases near the outlet due to the energy conversion into the kinetic energy as shown in Figure 3. If the thermal conductivity of the gas is extremely high, the wall temperature falls corresponding to the fall of the bulk temperature. However, in actual situation, the thermal conductivity of the gas is low, therefore the fall of the wall temperature is smaller than that of the bulk temperature as shown in Figure 8 (a) and (b). Then, the ratio of wall temperature difference between the gaseous flow and incompressible flow and the dynamic temperature, ζ is defined as

$$\zeta = \frac{T_{w,icomp} - T_w}{T_k} \tag{33}$$

The ratio, ζ calculated from Equation (33) for fast flow cases of Table 1 is plotted in Figure 9 as a function of the Mach number at the outlet. The ratio calculated for $\dot{q}=10^2$ W/m^2 and $\dot{q}=10^3$ W/m^2 are plotted in the figure. For the case of slow flow, the flow can be assumed to be an incompressible flow, and the temperature differences between the wall temperature and those of incompressible flow and the dynamic temperature are quite small, therefore, it is not plotted in Figure 9. The dashed line represent the ratio for laminar flow obtained by the previous study (Hong et al. [20]). The ratio, ζ is independent of hydraulic diameter and decreases with increasing the Ma at the outlet. Eq. (33) can be rewritten as

$$T_w = T_{w,icomp} - \zeta T_k \tag{34}$$

Then, the wall temperature of tubes with constant heat flux can be predicted from the wall temperature of the incompressible flow, the ratio, ζ and the dynamic temperature. However, the ratios for $\dot{q}=10^2$ W/m^2 and $\dot{q}=10^3$ W/m^2 scattered since their heat flux are relatively not large for the turbulent fast flow ($Ma_{out} > 0.3$) of $D \geq 200\mu$m.

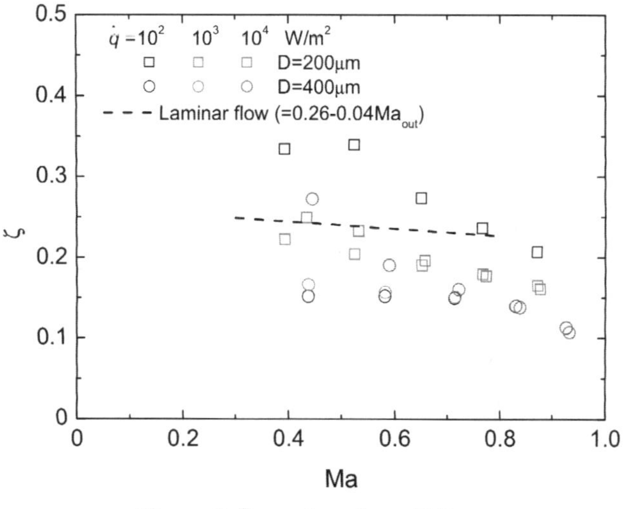

Figure 9 ζ as a function of Ma_{out}

4. Concluding remarks

Two-dimensional compressible momentum and energy equations are solved to obtain the heat transfer characteristics of turbulent gas flow in microtubes with constant hest flux. The Lam-bremhorst Low Reynolds number turbulence model was adopted to simulate turbulent flow cases. The following conclusions are obtained.

(1) For turbulent flow, Fanning friction factor, f_f is almost equal to the value given by Blasius correlation. However, Darcy friction factor, f_d with acceleration loss is higher than f_f or Blasius correlation.

(2) In the case of slow flow, the identical temperature profiles normalized by heat flux are obtained with those of incompressible flow. However, in the case of fast flow in a microtube, different temperature profiles are obtained.

(3) The wall temperature of a microtube can be predicted from the wall temperature of the incompressible flow, the ratio, ζ and the dynamic temperature.

$$T_w = T_{w,icomp} - \zeta\, T_k$$

5. References

[1] Asako, Y., Nakayama, K. and Shinozuka, T., 2005, Effect of Compressibility on Gaseous Flows in Micro-tube, *International Journal of Heat and Mass Transfer*, vol. 48, pp. 4985-4994.

[2] Wu, P. and Little, W.A., 1984, Measurement of the Heat Transfer Characteristics of Gas Flow in Fine Channel Heat Exchanger used for Microminiature Refrigerators, *Cryogenics* vol. 24, pp. 415-420.

[3] Chen, C.S. and Kuo, W.J., 2004, Numerical Study of Compressible Turbulent Flow in Microtubes, *Numerical Heat Transfer*, *Part A*. Vol. 45, pp. 85-99.

[4] Hara, K, Inoue, M and Furukawa, M., 2007, Heat Transfer in Mini-Channel Gaseous Cooling, *J. Environ. Eng.*, Vol. 2(3), pp. 525-534.

[5] Turner, S.E., Asako, Y. and Faghri, M., 2007, Convection Heat Transfer in Microchannels with High Speed Gas Flow, *Journal of Heat Transfer*, 129, pp.319-328.

[6] Asako, Y., 2004, Heat transfer characteristics of gaseous flow in a Micro-tube, *Thermal science & engineering*, 12(5), pp. 31-37.

[7] Asako, Y. and Toriyama, H., 2005, Heat Transfer Characteristics of Gaseous Flows in Micro-channels, *Microscale Thermophysical Engineering*, 9, pp. 15-31.

[8] Hong, C. and Asako, A., 2007, Heat Transfer Characteristics of Gaseous Flows In a Microchannel and a Micortube With Constant Wall Temperature, *Numerical Heat Transfer, Part A* , 52, pp. 219-238.

[9] Hong, C., Uchida, Y., Yamamoto, T., Asako, Y. and Suzuki, K., 2010, Heat Transfer Characteristics of Turbulent Gas Flow through Micro-Tubes, 2010 ASME International Heat Transfer Conference, Washington, DC, USA, IHTC14-22355.

[10] Chen, C., Lin, T., Yang, C. and Kandlikar, S.G., 2011, An experimental investigation on heat transfer characteristics of air and CO_2 in microtubes, *Ninth International Conference on Nanochannels, Microchannels, and Minichannels*, Paper No. ICNMM2011-58102.

[11] Yang, Y., Morini, G.L., Chalabi, H. and Lorenzini, M., 2011, Experimental Nusselt number determination for gas flows through commercial microtubes, *Third GASMEMS Workshop*, Paper No. GASMEMS11-21.

[12] Hong, C., Yamamoto, T., Asako Y. and Suzuki, K., 2012, Heat Transfer Characteristics of Compressible Laminar Flow Through Microtubes, *Journal of Heat Transfer*, Vol. 134, p. 011602 (8 page)

[13] Isobe, K., Hong, C., Asako, Y. and Ueno, I., 2011, Heat transfer characteristics of gas flow in micro-tubes with constant wall temperature, 4th International Conference on Heat Transfer and Fluid Flow in microscale, HTFFM-IV-068.

[14] Patel, C.V., Rodi, W. and Scheuerer, G., 1985, Turbulent Models for Near-Wall and Low Reynolds Number Flows: A Review, *AIAA Journal*, Vol.23 , pp. 1308-1319.

[15] Lam, C.K.G. and Bremhorst, K., 1981, A Modified Form of the k-ε Model for Predicting Wall Turbulence, *Transactions of American Society of Mechanical Engineers*, Vol. 103, pp. 456-460.

[16] Murakami, S. and Asako, Y., 2011, Local Friction Factor of Compressible Laminar or Turbulent Flow in Micro-Tubes, ASME 2011 9th International Conference on Nanochannels, Microchannels and Minichannels, ICNMM2011-58036.

[17] Kariki, C.K., 1986, A Calculation Procedure for Viscous Flows at All Speeds in Complex Geometries, PhD Thesis, *University of Minnesota*.

[18] Amsden, A.A., Ruppel, H.M. and Hire, C.W., 1980, SALE a Simplified ALE computer Program for fluid Flow at all Speeds, Los Alamos Scientific Laboratory Report No. LA-8095.

[19] Kays, W.M. and Crawford, M.E., 1980, Convective heat and Mass Transfer, 2nd ed. New York McGraw-Hill.

[20] Hong, C., Asako, A. and Lee, J-H., 2007, Heat transfer characteristics of gaseous flows in micro-channel with constant heat flux, *International Journal of Thermal Sciences*, Vol. 46, pp. 1153-1162.

Hydraulic and thermal design of a gas microchannel heat exchanger

Yahui Yang[1, 2], Juergen J. Brandner[1], Gian Luca Morini[2]

[1] IMVT, Karlsruhe Institute of Technology, Eggenstein-Leopoldshafen, Germany
[2] DIENCA, Alma Mater Studiorum Università di Bologna, Bologna, Italy
E-mail: yahui.yang@kit.edu; juergen.brandner@kit.edu; gianluca.morini3@unibo.it

Abstract. In this paper investigations on the design of a gas flow microchannel heat exchanger are described in terms of hydrodynamic and thermal aspects. The optimal choice for thermal conductivity of the solid material is discussed by analysis of its influences on the thermal performance of a micro heat exchanger. Two numerical models are built by means of a commercial CFD code (Fluent). The simulation results provide the distribution of mass flow rate, inlet pressure and pressure loss, outlet pressure and pressure loss, subjected to various feeding pressure values. Based on the thermal and hydrodynamic analysis, a micro heat exchanger made of polymer (PEEK) is designed and manufactured for flow and heat transfer measurements in air flows. Sensors are integrated into the micro heat exchanger in order to measure the local pressure and temperature in an accurate way. Finally, combined with numerical simulation, an operating range is suggested for the present micro heat exchanger in order to guarantee uniform flow distribution and best thermal and hydraulic performances.

1. Introduction

With the fast development of micro fabrication technologies in the recent years, many conventional devices in which heat transfer and chemical reactions take place, can be greatly miniaturized with precise control of their dimension [1]. Among these micro-structured devices, the micro heat exchangers are of increasing interest, for their improved heat exchange capacities due to very large surface-to-volume ratio [2].

The applied research on micro channel heat exchangers using liquid (mostly deionized water) as working media can be considered extensive, even though some of the results are not in agreement. Based on experiments with cross-flow micro heat exchangers, Harris et al. [3] found that the volumetric heat transfer in these micro-sized devices is improved at least five times compared with their conventional-sized counterparts. Improved heat exchanger devices of different microstructure designs and manufactured in metal have been described by Brandner et al. [4, 5]. Alm et al. [6] reported larger heat transfer coefficients and larger pressure losses in ceramic microchannel heat exchangers available for both counter flows and cross flows. Cao et al. [7] proposed experimental correlation for the prediction of the average value of Nusselt number and pressure drop of microchannel heat exchangers. On the contrary, the investigation of Garcia-Hernando [8] showed no heat transfer enhancement or higher pressure drop through micro heat exchangers due to the small scale of the microchannels. More recently, Mathew and Hegab [9] provided a universal correlation to estimate the heat transfer performance of microchannel heat exchangers subjected to external heat flux, regardless of the cross-sectional geometry of the microchannels.

Compared with research on liquid flow micro heat exchangers, experimental investigations on these devices objected to gas flows is scarce. The previous works available in literature for gas micro heat exchangers are merely reported by Bier at al. [10], Meschke et al. [11] and Koyama and Asako [12]. The detailed parameters of the gas micro heat exchangers analyzed in their works and their test conditions are listed in Table 1.

Table 1. Previous work on gas-to-gas microchannel heat exchangers

Authors	Channel width [μm]	Channel height [μm]	Materials	Conductivity [$WK^{-1}m^{-1}$]	Partition wall thickness [μm]	Gas species	Flow arrangement
Bier et al. [10]	100	78	copper	401	22	He, N_2, Ar	cross flow
Bier et al. [10]	90	75	stainless steel	14.7	25	He, N_2, Ar	cross flow
Meschke et al. [11]	1500	1500	ceramic SiC	125	2000	N_2	cross flow
Koyama and Asako [12]	300	200	stainless steel	15	300	air	counter-current

The first work on gas micro heat exchangers made of metal was reported by Bier et al. [10] for cross flows of helium, nitrogen and argon. The researchers found that in laminar regime the overall heat transfer coefficients to be greatly reduced for low mass flow rates. In addition they proposed a homogeneous model to account for a dominant influence of longitudinal heat conduction in the solid wall. As the thermal conduction in metal (copper and stainless steel) overwhelms that of the fluid, actually in such a case the micro heat exchanger works as a temperature mixer for cold and hot flows. This is probably the reason why under this model the thermal performance of heat exchangers does not differ between cross flow, co-current flow and counter-current flow, which is quite different from the typical features of conventional-sized heat exchangers. The heat transfer rate in this work was reported highly dependent on the arrangement of flow. Accordingly it was suggested by the authors that, in order to obtain larger heat transfer rates, material of lower thermal conductivity should be used both for microchannel plates and for the shell of the micro heat exchanger.

The investigation by Meschke et al. [11] on a ceramic micro heat exchangers for gases also stressed lower values of heat transfer coefficients than the theoretical predictions based on the conventional theory. According to the authors, the reduced heat transfer coefficients were subjected to the strong axial heat conduction in the solid wall. In addition, the heat loss due to conduction from the ceramic core to the steel housing was found to become significant on the thermal performance of the micro heat exchanger. The influence of thermal conduction in the solid on the overall thermal performance was that large that the impact of flow arrangement (cross, co-current, counter-current) becomes comparably insignificant.

Similar conclusion was drawn by Koyama and Asako [12] from experiments with a stainless steel micro heat exchanger. In this work no significant difference in the total thermal performance was found between co-current and counter-current flows. The investigators finally used an isothermal model with T boundary condition to explain their experimental data.

Summarizing the above experimental research of gas flow micro heat exchangers, it is worthy to note that all the authors did not carefully consider the possible effect of material choice on the thermal performance of the micro heat exchangers in the design phase of these devices.

Another issue raised on the research of micro heat exchangers is the flow distribution among the channels, which is linked to the hydraulic performance of these devices. Most of the micro heat exchangers described in literature are based on parallel microchannels built on a plate; different arrangements of the plates in the stacking process enable different geometries for the flow configuration of the micro heat exchangers. One of the common features of such plates is that the flows through all the parallel microchannels share the same feeding source; the source flow requires a distribution chamber in order to spontaneously feed all the microchannels as uniformly as possible. In the work of Bier et al. [10] and Meschke et al. [11] this issue of flow distribution was not addressed. Koyama and Asako [12] found the experimentally measured pressure drop to be several times higher than prediction, and they regarded non-uniform distribution of mass flow as one reason to this disagreement. However, the exact impact of flow maldistribution on additional pressure drop was not estimated in the work of Koyama and Asako [12].

The flow distribution is very important for micro heat exchangers, because it does not only influence the hydraulic behaviour of the micro heat exchanger, but also changes the thermal behaviour of the device. The optimal design for uniform distribution of flow was studied by Commenge et al. [13] who proposed an approximate pressure drop model. Chein and Chen [14] compared the velocity and temperature distribution along 11 parallel microchannels by changing the position of distribution/collecting ports. Tonomura et al. [15] developed an automatic optimization algorithm which is able to tailor iteratively the shape of contributing/collecting manifold until satisfying mass flow distribution. Balaji and Lakshminarayanan [16] proposed a new design of the distribution chamber in order to obtain uniform flow distribution among the parallel microchannels. In their design the flow enters the distribution chamber from one port while the exhausted flow is collected by two ports. An analytical model for optimization of flow distribution in a manifold has been reported by Renault et al. [17]. The model is able to calculate the flow distribution faster than CFD simulation without losing too much accuracy.

The methods described in the works available in the open literature provide convenient instruction for the design of micro heat exchangers. However, up to now it is difficult to find a systematic way to optimize the flow distribution for multi-plate microchannel heat exchangers, mainly due to the limited possibilities to balance all the parameters in the design. These parameters include dimension of micro channels (width, height and length), shape of distributing/collecting manifolds, curvature of the corners, length of connections between the heat exchanger core and manifolds, position of the feeding/exhaust ports, direction of the inflow/outflow, etc. Furthermore, in practical tests and applications of micro heat exchangers the parameters mentioned above may be subjected to different types of restriction coupled with higher priorities (assembling, working condition, vulnerability to possible damages, pricing etc.). Therefore, experimental investigators should carefully look for compromises in the design phase, and it is recommended that the influence, such as solid wall conduction and flow distribution, are accurately estimated or even simulated with CFD methods in order to interpret experimental data under specific conditions. This applies more truly to micro heat exchangers using gas as working fluid, because up to now the experimental research in this field is weak if compared with results of numerical works and/or with the experimental works on liquid micro heat exchangers.

In this study the choice of the best material for a gas micro heat exchanger was quantitatively evaluated in order to optimize its dynamic and thermal performances. With an appropriate design some scaling effect, like the axial thermal conduction along the solid walls of the micro heat exchanger, can be greatly suppressed, while the overall heat transfer efficiency can be improved. The uniformity of mass flow and pressure distribution was obtained for different flow conditions by means of a commercial CFD code. The local pressure losses at both distribution and collecting chambers are numerically evaluated for different configurations of the manifolds.

2. Experiment

Figure 1 shows schematics of the stacked core for a double-layered microchannel heat exchanger and a multi-layered one. The arrangement of the plates enables co-current and counter-current gas flows within large ranges of pressure and mass flow rate. The cores are stacked in such a way that the layer of hot flow and the layer of cold flow are alternated. The core is housed in a shell made of polymer. The low heat conductivity of polymer (compared with metal) ensures low heat losses. In addition, for the double-layered micro heat exchanger the thin plate between the hot and cold flow can be exchanged. The material and thickness of the plate can be chosen more or less freely based on various purposes.

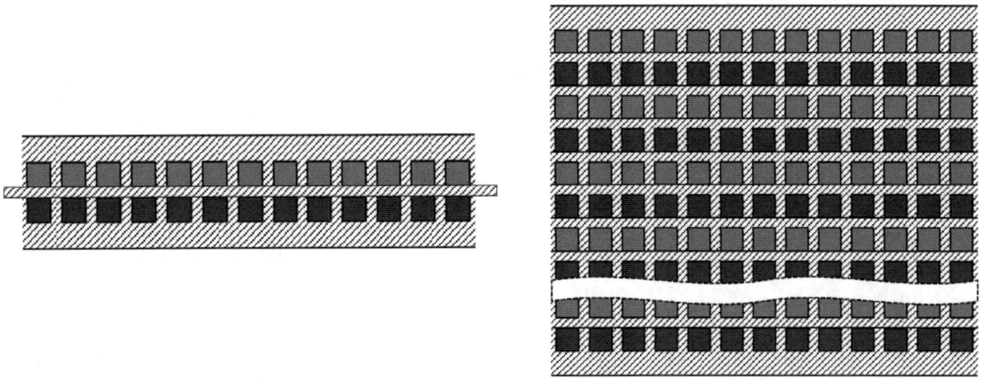

(a) (b)

Figure 1. Schematic of the micro heat exchanger core. (a) double-layered; (b) multi-layered.

The detailed parameters of the micro heat exchanger are listed in Table 2. The microchannels have square cross section with a length of 39.8 mm. The bottom thickness of the thin layer is 100 μm. On each layer 133 parallel microchannels are manufactured. These parallel microchannels are seperated by side walls with a thickness of 100 μm. The micro heat exchanger has a surface-to-volume ratio of 20000, larger than the typical values of the conventional compact heat exchangers (<1000). The entire core is made of PEEK with a thermal conductivity of 0.25 $WK^{-1}m^{-1}$. The reason for such a choice will be detailed in the discussion section.

Table 2. Parameters of the micro heat exchanger.

Micro heat exchanger	**Value of parameters**
Channel width W: [μm]	200
Channel height H: [μm]	200
Hydraulic diameter d_h: [mm]	200
Channel length L: [mm]	39.8
Surface roughness of channel ε: [μm]	0.15
Thickness of channel side wall s_1: [μm]	100
Thickness of partition wall s_2: [μm]	100
Material of channel wall	PEEK
Thermal conductivity of channel wall λ_w: [$WK^{-1}m^{-1}$]	0.25
Material of shell	PEEK
Thermal conductivity of shell: [$WK^{-1}m^{-1}$]	0.25
Number of channels per layer	133
Number of layers for hot/cold flow	1 and 10
Ratio of (heat transfer) surface to (flow) volume: [m^2/m^3]	20000

A typical layer with 133 parallel microchannels, distribution/collecting chambers and feeding/exhaust ports is shown in Figure 2(a). The layer has a width of 39.8 mm, which is identical to the length of the microchannels. For cross flow the alternative layers can be rotated by 90 degree, while the hot layer and cold layer of the core are still in perfect thermal contact. As the micro heat exchangers are designed to work in large pressure range (up to 40 bar), 22 pillars with a diameter of 0.8 mm are added to the distribution/collecting chamber in order to maintain mechanical stability and avoid possible deformation under high pressure. With the presence of these supporting pillars, different methods on flow distribution in literature are not able to predict the flow distribution. Actually such supporting pillars are quite common in the application of microdevices (such as micro evaporators [18]), if the dimension and thickness of microstructure are greatly reduced while the mechanical stability has to be maintained.

The micro heat exchanger was manufactured using mechanical precision machining [19, 20]. CNC-controlled tooling systems and miniaturized tools made from natural diamond have been used to create the microchannels into polished PEEK platelets. The precise control of the design allows manufacturing of microchannels with an absolute dimension uncertainty of about +/- 1μm in all three dimensions (namely: width, depth and length of the microchannels). The average surface roughness ε inside the microchannels was not measured but estimated to be in the range of 150 nm. This was based on experiences with the manufacturing of microstructure devices made of PEEK as well as on measurements undertaken by Hecht et al. [21] and Vittoriosi et al. [22]. The ratio between the average roughness and the hydraulic diameter is about $7*10^{-4}$. Due to this low value the estimated relative roughness is most likely small enough to be neglected in the evaluation of frictional losses and convective heat transfer coefficients along the microchannels.

Figure 3(b) shows a magnified area of the distribution chamber. Two K-type temperature sensors and one pressure sensor are delicately planted in the top surface of the chamber. In the previous experimental work on gas micro heat exchangers, fluid temperature and pressure were measured before the feeding ports, or after the exhaust ports [12]. This underestimates the local pressure loss and temperature loss by far. The large disagreements between experimental data and theory were attributed to the pressure or temperature losses [12]. To solve those measurement problems, in the present study pressure and temperature are measured at a position very close to the ends of microchannels.

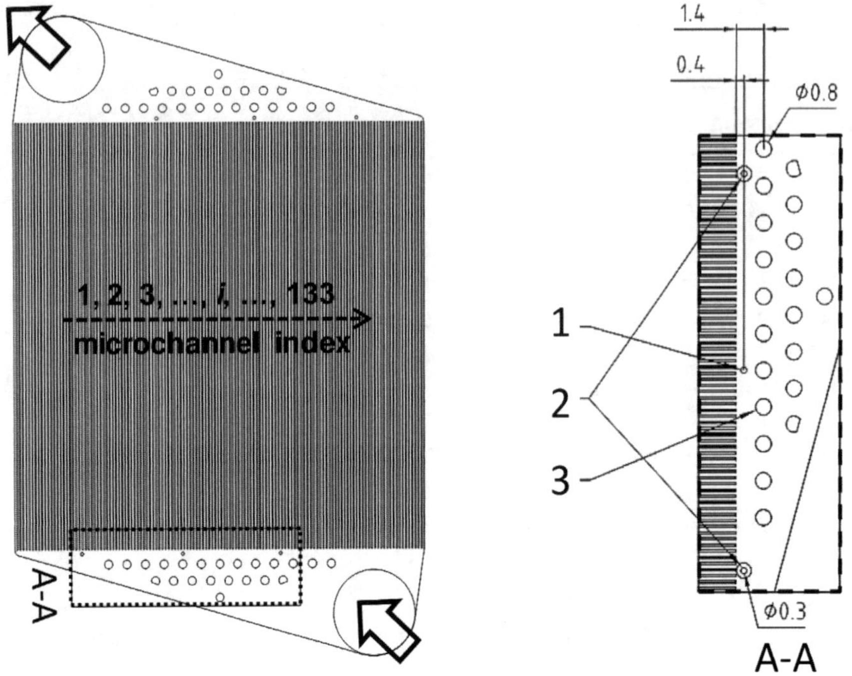

(a) (b)

Figure 2. Typical layer (with 133 microchannels) of the micro heat-exchanger (a); in-situ measurement in distribution/collecting chambers (b) (1- pressure sensor; 2- temperature sensor; 3- pillars. unit: mm).

3. Numerical model

The existing models in literature on uniform flow distribution cannot be applied to the present pillared chambers, not only due to the existence of pillars, but due to the different location and direction of inflow and outflow voids. Therefore, a CFD model was created using the finite volume solver of ANSYS (FLUENT 6.3). The model is able to calculate the exact mass flow and pressure distribution of the present design. The dependence of mass flow and pressure distribution on absolute feeding pressure was examined using this model. In addition, the local pressure drop at both distribution and collecting chambers was computed.

The CFD model adopted the following assumptions:

(1) Two-dimensional model;
(2) The flow is incompressible;
(3) The walls have smooth surfaces;
(4) No heat transfer is involved;

The influence of the grid dependence on the result was carefully examined. Three sets of meshes with different coarseness were built, as listed in Table 3. To reduce the computation time, only the distribution chamber with a very short length of the microchannels was meshed, as suggested by Renault et al. [17]. The grid dependence was checked for four different pressure drops under which the mass flow rate was monitored. The finest mesh was taken as a reference. From Table 3 it can be noted that with the fine mesh the relative difference in the mass flow rate is less than 1% for all the pressure drops in computation. Therefore, the model was built with a fine mesh.

Two different CFD models were built. The first one is only for the distribution chamber with a short length of microchannels; the second one is a full simulation of the entire layer, including the full length of the 133 parallel microchannels and the collecting chamber. The numerical results from these two models have been compared.

Table 3. Validation of grid dependence.

p_{in} [kPa]	p_{out} [kPa]	Mass flow rate [kg/s] and relative diff. [%]		
		Coarse mesh (95 970 cells)	Fine mesh (213 335 cells)	Very fine mesh (280 734 cells)
110	100	1.63(1.37%)	1.60(-0.41%)	1.60(ref.)
120	100	2.32(2.42%)	2.26(-0.29%)	2.27(ref.)
130	100	2.87(2.82%)	2.77(-0.86%)	2.79(ref.)
150	100	3.70(3.17%)	3.58(-0.08%)	3.59(ref.)

4. Results and discussion

4.1. Heat conduction in the solid wall

As stated in the introduction section, all the available experimental works on gas flow micro heat exchangers concluded that conventional theory on heat exchangers failed to explain the experimental data due to heat conduction in the solid wall. Instead, an isothermal assumption was used in order to better explain the experimental trends. Even the flow arrangements (cross flow, co-current flow and counter-current flow) make no difference on the convective heat transfer coefficients by using highly conductive material (copper, stainless steel and ceramic), as evidenced by the work of Bier et al. [10], Meschke et al. [11] and Koyama and Asako [12]. However, in the previous works no statement was made on how the thermal conductivity of wall affects the performance of the gas micro heat exchangers.

To estimate the axial heat conduction in the wall, Maranzana et al. [23] proposed a wall axial conduction parameter, the ratio of conductive heat flux in the solid wall to the convective heat flux in the microchannel. For a circular tube with inner diameter *d* and external diameter *D*, the wall axial conduction parameter can be expressed as:

$$M = \frac{\phi_w}{\phi_f} = \frac{\lambda_w}{\lambda_f} \frac{D^2 - d^2}{dL} \frac{1}{RePr} \tag{1}$$

where ϕ_w is the heat flux due to thermal conduction in the wall, ϕ_f the convective heat flux, λ_w the thermal conductivity of the solid wall, λ_f the fluid thermal conductivity, *L* the length of microchannel, *Re* the Reynolds number and *Pr* the Prandtl Number of the fluid, respectively.

Generally if $M<0.01$, which means the conductive heat flux is less than 1% of the convective heat flux, the influence of axial heat conduction in the wall can be neglected. When this parameter is larger than 0.01, the conductive heat flux becomes significant in the overall heat transfer. The influence of axial wall heat conduction on heat transfer was investigated in the work of Yang et al. [24] for a single microtube with thick walls.

For multi-layered microchannel heat exchangers, the ratio of wall conductive heat flux to convective heat flux can be calculated by

$$M = \frac{\phi_w}{\phi_f} = \frac{\lambda_w}{\lambda_f} \frac{(W + s_1/2)(H + s_2/2)}{WH} \frac{2WH}{(W + H)L} \frac{1}{RePr} \tag{2}$$

where *W* is the width of the microchannel, *H* the microchannel height, s_1 the thickness of sidewall between parallel microchannels and s_2 the thickness of the partition wall separating the cold flow and the adjacent hot flow.

In order to have an insight of the wall thermal conduction influence, the data points measured by Bier at al. [10], Meschke et al. [11] and Koyama and Asako [12] for gas flow micro heat exchangers were depicted in Figure 3, showing the axial conduction parameter as a function of Reynolds number. The value of the axial conduction parameter was calculated based on the exact dimension of the microchannels, thicknesses of the walls, thermal conductivity of the materials, thermal properties of the fluids as well as specific flow conditions from each author. The dashed line in Figure 3 indicates the threshold indicated by Maranzana et al. [23] for the axial conduction parameter. From Figure 3 it is surprising to note that almost all the data points by these investigators fall in the region where axial heat conduction in the wall is of significance. In some cases the axial conductive heat flux in the wall is comparable to or even larger than the convective heat flux. Thus, the temperature profile tends to flatten out throughout the micro heat exchangers and the heat transfer was deteriorated. This is probably the reason why these researchers found an isothermal model or a T boundary condition (uniform wall temperature) more appropriate to explain the experimental trends. If the tests carried out by Bier et al. for micro heat exchangers made of different materials are compared, it can be found that for the same gas species (e.g. helium) the axial heat conduction in copper micro heat exchangers is approximately one order of magnitude higher than that in stainless steel micro heat exchangers. Actually this is easily reflected in Eq. (2). For the same material (stainless steel), the axial conduction parameter is larger in the test of Koyama and Asako [12] than that in the investigation by Bier at al. [10]. This is because of the much thicker wall used in the micro heat exchanger of the former. By observing these data points, it can be found that the axial conduction parameter has a decreasing trend with the increase of Reynolds number. Therefore, for micro heat exchangers made of more conductive material, larger Reynolds numbers or higher mass flow rates are suggested in order to avoid large wall axial heat conduction.

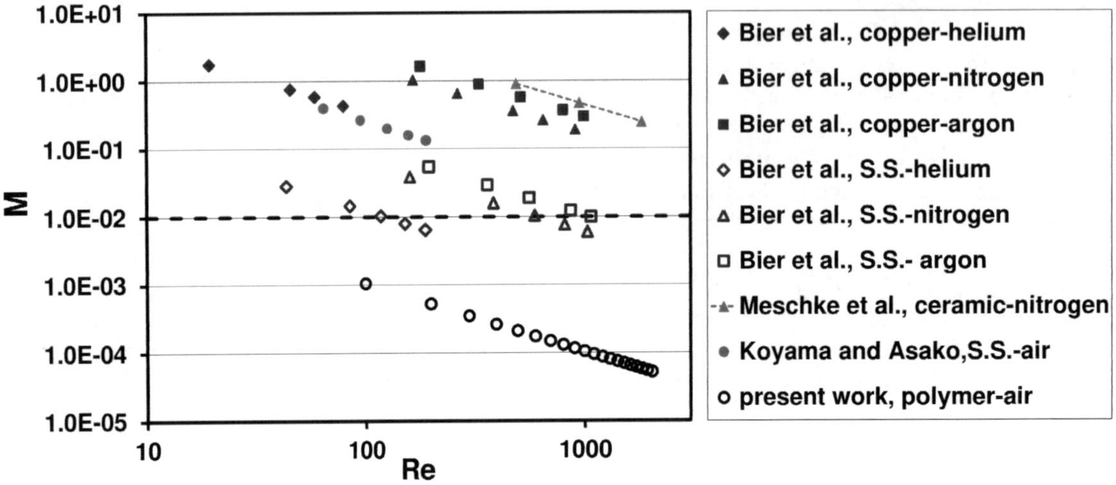

Figure 3. Wall conduction parameter in the previous and present work (S.S.-stainless steel).

Polymer (PEEK) was selected for the micro heat exchanger in the present work. With the information of microchannel size and thermal properties listed in Table 2, the magnitude of the axial conduction parameter is shown in Figure 4 for the Reynolds number range proposed for the experiments of the current study. In the design of the present work, the significance of axial heat conduction in the wall becomes very slight compared with convective heat transfer. Therefore the present design should be able to use the thermal analysis model proposed for conventional heat exchangers. Also it is expected that with the present design the dependence of the thermal performance on flow arrangements (co-current, counter-current and cross flow) can be distinguished. Experimental data from these heat exchangers will be reported as soon as possible in another paper.

Although all the experimental research on gas flow micro heat exchangers reported strong axial conduction in the wall, similar problem seems not encountered in the study ([3, 9]) on liquid flow through these devices. In order to know the possible influence of fluid types on the wall axial conduction, the data points by Bier at al. [10], Meschke et al. [11] and Koyama and Asako [12] are again depicted in Figure 4. The difference to Figure 4 is that the working gases are changed into water. In Figure 4 the data series (a) corresponds to the Reynolds number in the work of Bier at al. [10]; while the data series (b) refers to the Reynolds number from Koyama and Asako [12]. It can be seen from Figure 4 that most of the data points fall below the threshold when the working fluid is changed from gas to liquid, except for the copper micro heat exchanger working under very low Reynolds numbers. Therefore, in the research of liquid micro heat exchangers the influence of axial conduction in the wall is widely reduced. This can directly be noted from the ratio of axial conduction parameter for gas (i.e. air) to that for liquid (i.e. water), as shown in Eq. (3).

$$\frac{M_{air}}{M_{water}} = \frac{\lambda_{water}\,\mathrm{Pr}_{water}}{\lambda_{air}\,\mathrm{Pr}_{air}} \tag{3}$$

Supposing the same micro heat exchanger working at the same Reynolds number, the ratio in Eq. (3) generates a value of about 220. This draws great attention to axial conduction in the wall when the working fluid is changed from liquid to gas.

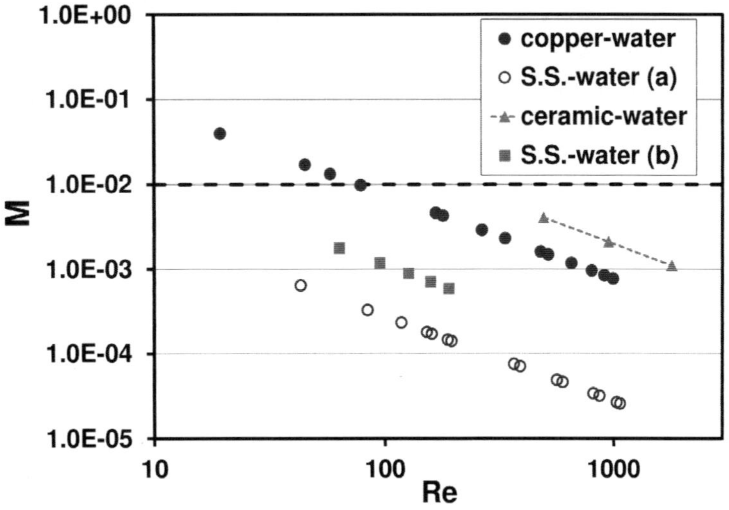

Figure 4. Wall conduction parameter by changing the working gases in previous work into water.

The overall efficiency of heat exchangers, which can be calculated by Eq. (4) ([25]), may also be influenced by the wall thermal conductivity.

$$\varepsilon = \frac{q_c c_{p,c}\left(T_{c,out} - T_{c,in}\right)}{(qc_p)_{\min}\left(T_{h,in} - T_{c,in}\right)} = \frac{q_h c_{p,h}\left(T_{h,in} - T_{h,out}\right)}{(qc_p)_{\min}\left(T_{h,in} - T_{c,in}\right)} \tag{4}$$

where q is the gas mass flow rate on the hot (h) and cold (c) side of the heat exchanger, c_p is the fluid specific heat and T is the temperature at the inlet (in) and at the outlet (out) of the device.

The optimal thermal conductivity of material for micro heat exchangers was numerically studied by Stief et al. [26] in 1999. Figure 5 shows their results for the overall efficiency of a counter-current heat exchanger as a function of the solid thermal conductivity. If the thermal conductivity approaches 0, the solid wall functions as thermal isolation material between hot and cold flows. Therefore the heat transfer is blocked by the wall, and the heat exchanger efficiency is close to 0. If the wall is made from a very good heat conductive material, the strong axial conduction in the wall reduces the heat exchanger efficiency as well. Due to this reason thermal conductivity in the order of 1 W/mK offers best efficiency for a gas-gas heat exchanger. For PEEK with a thermal conductivity of 0.25 W/mK, the micro heat exchanger presented in this work is expected to have relatively high overall efficiency (see Figure 5).

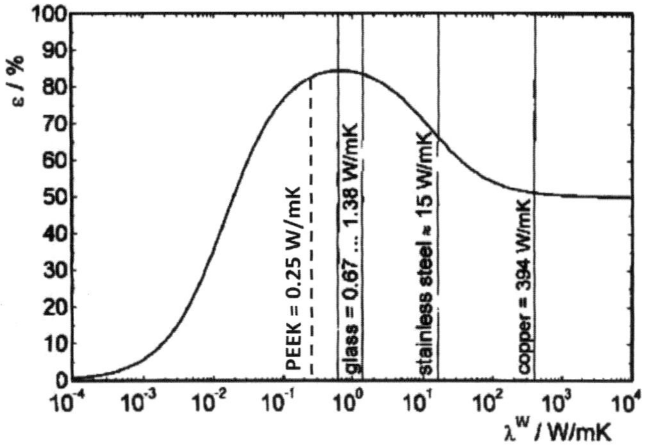

Figure 5. Heat exchanger efficiency in dependence on the wall thermal conductivity (from Stief et al. [26]).

4.2. Flow distribution in the distribution chamber

Numerical simulation may provide useful information on local distribution of flow in the design phase of micro heat exchangers. The main concern on numerical means is that it requires a lot of efforts on the geometry modeling and computation time to simulate a full layer, which is composed of several hundreds of microchannels, distribution chamber as well as collecting chamber. A common treatment for simplification is to simulate how the flow distributes from the feeding port to the inlets of microchannels based on some assumed pressure difference (local pressure loss). This provides a quick solution to know the general flow distribution, and sometimes used by engineers and technicians to modify the designs of micro devices. However, this method is not generally verified in order to know whether it is able or not to predict the real flow distribution, if the flow conditions far downstream the distribution chamber are not considered. In this work, two numerical models are built: the first, in which a single distribution chamber is modeled and the second in which a full layer is taken into account. The numerical results of these two models will be compared, in order to check the real need of an extensive computation of the full-layer model.

For the numerical model of a single distribution chamber, the feeding pressure was fixed at four different absolute values, namely 110 kPa, 160 kPa, 250 kPa and 500 kPa. The absolute value of the exhaust pressure is set at 100 kPa, which is atmospheric pressure. The mass flow rates through every microchannel are the unknown variables to be calculated. The outlet pressure of the single chamber is assumed to be 99% of the feeding pressure, which results in an assumption of 1% pressure drop.

Figure 6 shows the air flow velocity distribution in the distribution chamber under feeding pressure of 160 kPa. The disturbance of the pillar array is distinct in the velocity field. However, the wakes of the disturbance downstream the pillar fade out in a short flow path, and the velocity redistributes immediately before the flow enters the parallel microchannels. In addition, there is a regiion with relatively high flow velocity in the distributingdistribution chamber where there is absence of pillars in the flow path.

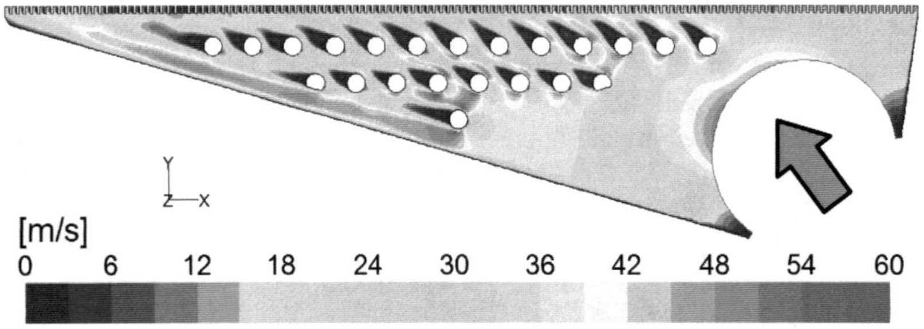

Figure 6. Flow velocity field in the distributingdistribution chamber at feeding pressure of 160 kPa.

To have an insight into the mass flow distribution through each microchannel, the mass flow rate through every microchannel under different feeding pressure values was calculated. The result is shown in Figure 7 for a dimensionless mass flow rate. The average mass flow rate q_{ave} is calcuted by

$$q_{ave} = \frac{1}{n}\sum_{i=1}^{n} q_i, \quad n = 133 \tag{5}$$

where q_i is the mass flow rate of microchannel i.

Figure 7 indicates quite ununiform mass flow distritution for all the parallel microchannels. For micochannels indexed from 115 to 132, the mass flow rate is more than twice as high as the average one. In addition, it is worthy to note that the distribution of the mass flow rate has very slight dependence on the feeding pressure. It is yet unkown whether the trend displayed here is representitative for a full layer.Therefore, another CFD model with a full layer was built in order to have a more detailed insight of the real flow situation.

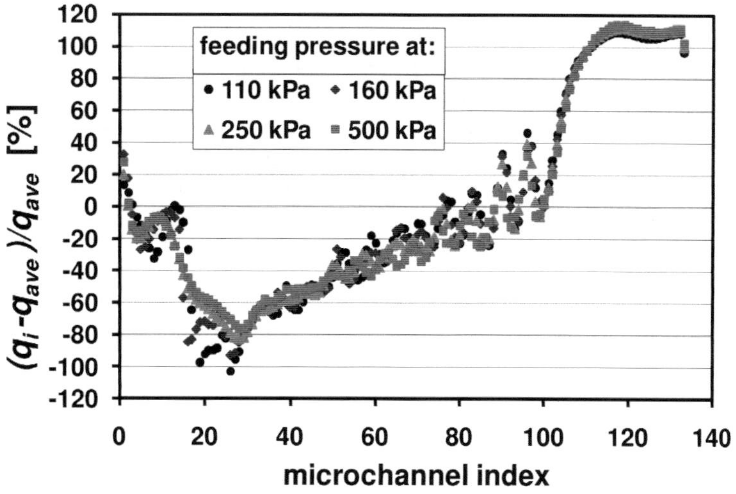

Figure 7 Mass flow rate of each microchannel under different feeding pressure.

4.3. Flow distribution and pressure drop of a full layer

The numerical results from the simulation of a full layer (shown in Figure 2) are discussed in this section for feeding pressure set at 110 kPa, 160 kPa, 250 kPa and 500 kPa. The flow is vented from the exhaust port at atmospheric pressure, namely 100 kPa. With this model the mass flow rate and pressure drop through all the microchannels will be self-balanced.

The velocity field in the distribution chamber and the microchannels is given in Figure 8 for a feeding pressure of 160 kPa. As the velocity profile in the microchannels does not change after the hydraulic entry length, only a part of the microchannel length is shown in Figure 8. The same will be applied thereafter for the downstream part of the layer. From Figure 8 it can be noted that the velocity in the distribution chamber is generally higher than that in the microchannels. This is because the total flow area normal to the streamline in the distribution chamber is smaller than the total flow area normal to the streamline of all the parallel microchannels. This is a common feature of most microchannel devices using distribution connections between macro piping lines and microchannels. The velocity contour of the downstream part of the layer is shown in Figure 9. By observing the velocity distribution, it can be observed that the flow velocity through the microchannels closest to the feeding port (or with short flow path) is unexpectedly the lowest. This is related to the pressure distribution throughout the whole layer.

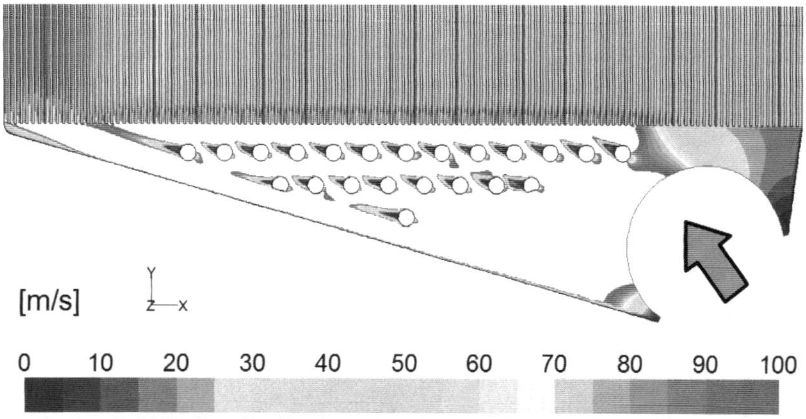

Figure 8. Flow velocity field in the distribution chamber and microchannels at feeding pressure 160 kPa (length of microchannel partly displayed, velocity>100 m/s in white region).

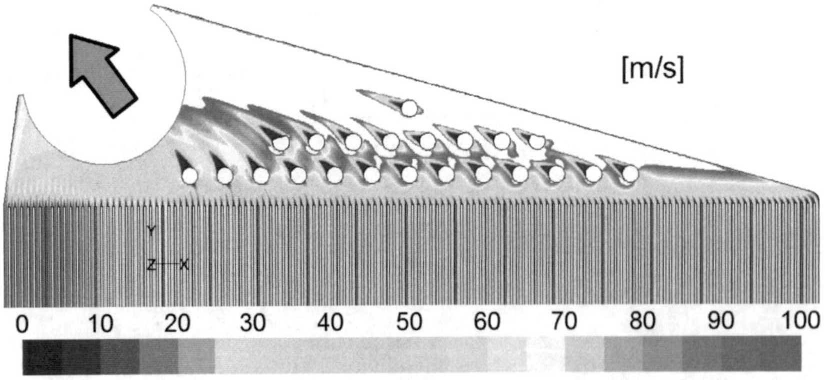

Figure 9. Flow velocity field in the collecting chamber and microchannels at feeding pressure of 160 kPa (length of microchannel partly displayed, velocity>100 m/s in white region).

The pressure contour for the upstream part of the layer is shown in Figure 10 , for the downstream part in Figure 11. The pressure at the inlets of the microchannels close to the feeding port is very high. This is caused by the high pressure at the outlets of these microchannels, as it is shown in Figure 11. The pressure drop from the microchannel outlets to the exhaust port is larger for microchannels further away from the exhaust port. This affects the pressure development inside the microchannels all the way from outlet to inlet. Acordingly the pressure distribution at the inlet region of the microchannels will be influenced. This is quite different from the simulation of a single distributing chamber.

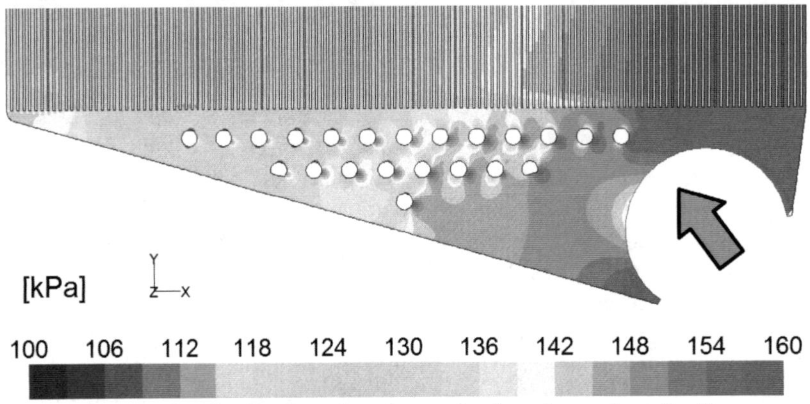

Figure 10. Static pressure in the distribution chamber and microchannels at feeding pressure of 160 kPa (length of microchannel partly displayed).

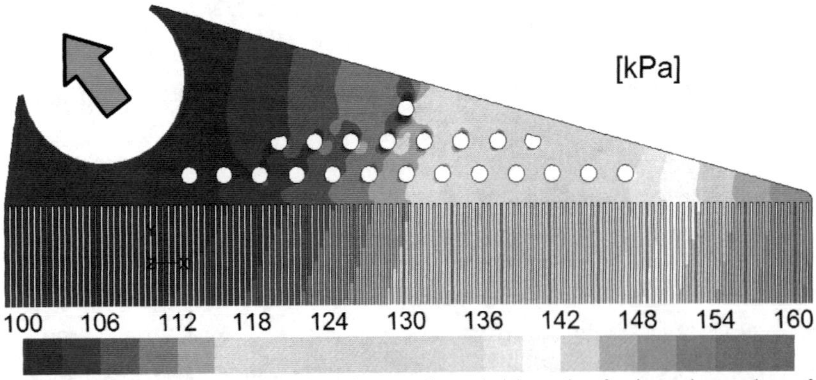

Figure 11. Static pressure in the collecting chamber and partial length of microchannels at feeding pressure of 160 kPa (length of microchannel partly displayed).

Figure 12 shows the distribution of the mass flow rate through all the parallel microchannels for different feeding pressure values. The microchannels furthest from the feeding port show the highest mass flow rate.

The mass flow rate decreases in the microchannels approaching the feeding port, with small fluctuations in between due to disturbances created by the pillars. Acutually from Figure 12 it can be seen that the existence of the pillars seems to even the mass flow distribution by creating an average flow velocity. In the region far from the pillars, the mass flow rate gradient is steeper. In additon, by increasing the feeding pressure, the uniformity of the mass flow rate distribution becomes worse.

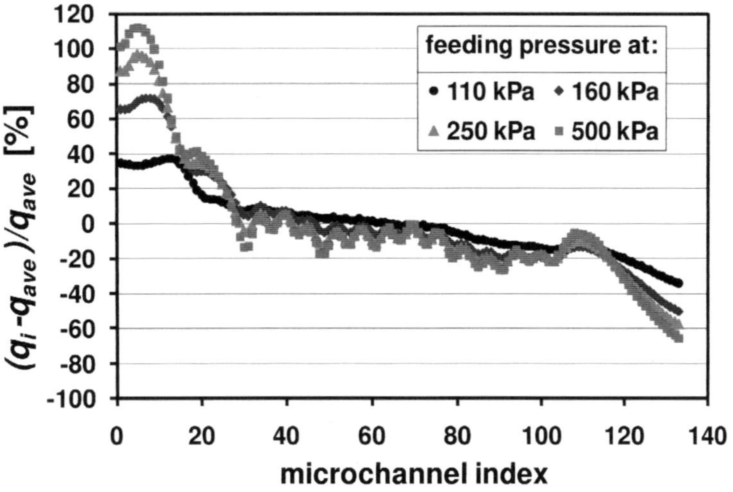

Figure 12. Mass flow rate of each microchannel under different feeding pressure.

By comparison between Figure 12 and Figure 7, the simulation of the full layer generates exhibits a mass flow distribution completely different from the simulation of a single distribution chamber. Therefore, without full simulation the numerical results might be misleading.

The pressure at the inlet and outlet of each microchannel is compared for all the parallel microchannels in a dimensionless way. Similarly to the average mass flow rate, the mean pressure averaged over all the 133 microchannels is calculated by

$$p_{ave} = \frac{1}{n}\sum_{i=1}^{n} p_i, \quad n = 133 \tag{6}$$

where p_i is the pressure in microchannel i.

In Figure 14 the pressure distribution at the inlet of the parallel microchannels is depicted. Higher feeding pressure generates a more uniform distribution of the inlet pressure. For the microchannels in the middle area of the layer, the inlet pressure is quite close to the average one. This provides some hints for the measurement of the average pressure at inlet. According to these simulation result, it seems to be better to place the pressure sensor in the middle part of the total width of all the microchannels.

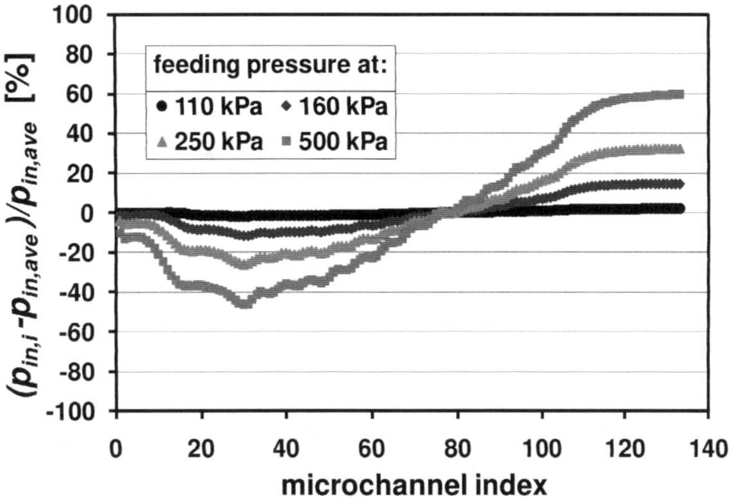

Figure 13. Pressure distribution at inlet of each microchannel.

Besides pressure distribution, the pressure loss at the distribution chamber is calculated for all the microchannels as a subject of different feeding pressure values, as shown in Figure 14. Generally, with higher pressure at the feeding port a higher fraction of pressure is lost due to the flow distribution and the supporting pillars. Therefore, if the pressure measured at the feeding port or piping line (which is the easist way to measure) is considered as the inlet pressure of the microchannels, a large uncertainty may be faced especially at high pressure values or larger mass flow rates.

For a fixed feeding pressure, a negligible pressure loss was found for microchannels very close to the feeding port, indexed from around 110 to 133 for the present micro heat exchanger. For these microchannels, the flow path from the feeding port is very short and free of obstructing pillars. For several microchannels furthest away from the feeding port, the pressure loss is also relatively small due to the fact that there are no obstacles in form of pillars within the flow path.

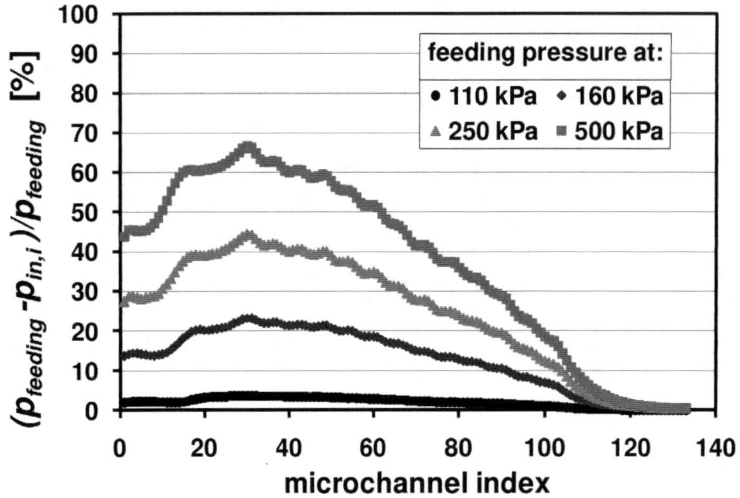

Figure 14. Pressure drop in the distribution chamber (from feeding port to inlet of microchannel).

The pressure distribution at the outlets of the parallel microchannels is shown in Figure 16, calculated for different feeding pressure values. Compared with that at the microchannel inlets, the distribution at the outlets is relatively linear. The further the microchannel is away from the exhaust port, the larger the outlet pressure becomes. Similarly, higher feeding pressure values increases the non-uniformity of outlet pressure distribution. The outlet pressure of the microchannel located at the middle of the whole panel is most representative of the average pressure.

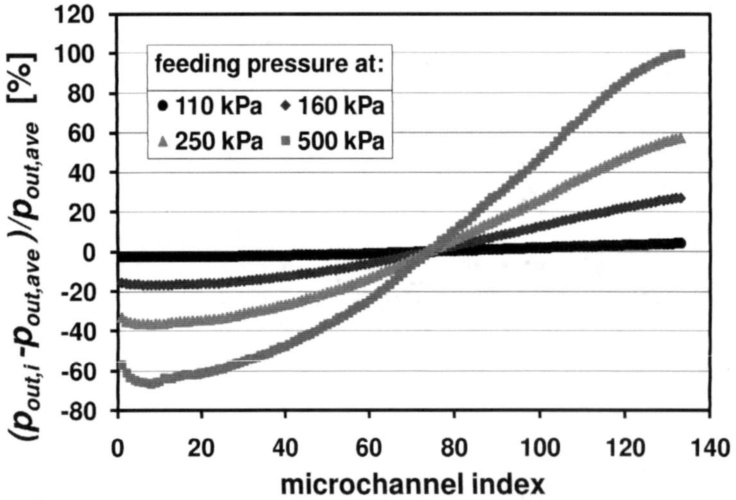

Figure 15. Pressure distribution at outlet of each microchannel.

Figure 16 shows the pressure loss from microchannel outlet to the exhaust port. In the pillar-free region with the shortest flow path, the pressure loss is very small. An exception is the feeding pressure of 500 kPa. At feeding pressure higher than 250 kPa, the microchannel outlet pressure can go above 1.5 times of atmospheric pressure. Thus, simply regarding the outlet pressure to be ambient might lead to large uncertainties.

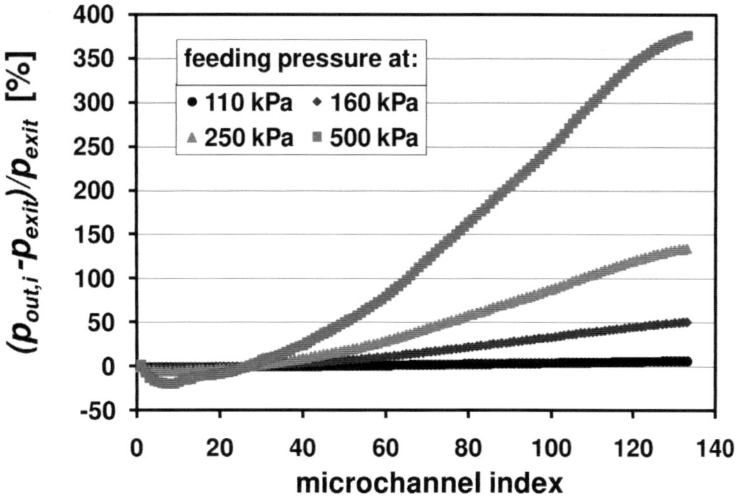

Figure 16. Percentage of pressure drop in the collecting chamber (from outlet to exhaust port).

The statistical data of the numerical simulation on the whole layer are listed in Table 4, in terms of mass flow distribution, microchannel inlet pressure distribution and outlet pressure distribution. The number of microchannels showing a mass flow rate within a band of ±20% of the average value is counted. The same has been performed for the pressure at inlet and outlet. It can be noted that at low feeding pressure the distribution of these parameters is very uniform. Uniformity is decreased with increasing inlet pressure. For a feeding pressure of 500 kPa and above, more than half of all the microchannels show magnitudes for these parameters falling outside the range. Therefore, for the present micro heat exchanger, it is suggest that the feeding pressure should be kept lower than 250 kPa in order to maintain uniform flow distribution.

Table 4. Uniformity of mass flow rate and pressure distribution in terms of microchannel number.

p_{in} [kPa]	p_{out} [kPa]	**Mass flow rate** Num. of channels with q_i <(1±20%)q_{ave}	**Pressure at inlet** Num. of channels with $p_{in,i}$ <(1±20%)$p_{in,ave}$	**Pressure at outlet** Num. of channels with $p_{out,i}$ <(1±20%)$p_{out,ave}$
110	100	102/133	133/133	133/133
160	100	87/133	133/133	115/133
250	100	83/133	81/133	43/133
500	100	79/133	40/133	23/133

5. Conclusions

In this paper the design of a gas flow microchannel heat exchanger was studied from both hydrodynamic and thermal aspects. The main conclusions are drawn as follows:

- For microchannel heat exchangers used with gas flows, the thermal conductivity of the solid material is extremely important. With highly conductive material, the overall heat transfer efficiency is very low and the dependency of thermal performance on flow arrangements is lost.
- A gas flow micro heat exchanger made of polymer is designed and manufactured. Local measurement of pressure and temperature is achieved by integrating sensors very close to the microchannels.
- Simulation of a full layer shows quite different distribution of mass flow and pressure compared with that of a single chamber. This means a mere simulation of the distribution chamber is not sufficient

to gain real insights into the flow distribution, and thus may be misleading in the design of micro heat exchangers.

- The full numerical simulation uncovers the distribution of mass flow rate, inlet and outlet pressure in dependence of the feeding pressure. Based on the results, the present micro heat exchanger is found to have relatively uniform flow distribution if working at a feeding pressure of 250 kPa or lower.
- The pressure drop (losses) in the pillared distribution/collection chamber is obtained for different feeding pressure. The distribution of pressure loss in terms of microchannel index is investigated for both distribution chamber and collecting chamber; this analysis gives useful hints for the optimal hydraulic design of a gas micro heat exchanger.

6. Acknowledgements

The research leading to these results has received funding from the European Community's Seventh Framework Programme (ITN-FP7/2007-2013) under grant agreement n° 215504. In addition, the authors would like to thank the microfabrication group at IMVT, Karlsruhe Institute of Technology, for manufacturing and assembling of the micro heat exchangers.

7. Nomenclature

c_p	Gas specific heat	(J/kgK)
d	Inner diameter	(m)
D	External diameter	(m)
d_h	Hydraulic diameter	(m)
H	Height of microchannel	(μm)
L	Length of microchannel	(μm)
M	Axial heat conduction parameter	(-)
p	Pressure	(Pa)
Pr	Prandtl number	(-)
q	Mass flow rate	(kg/s)
Re	Reynolds number	(-)
s_1	Thickness of channel sidewall	(μm)
s_2	Thickness of partition wall	(μm)
T	Temperature	(K)
W	Width of microchannel	(μm)
Greek symbols		
ε	Absolute roughness	(m)
ϕ	Heat flux	(W/m^2)
λ	Thermal conductivity	(W/mK)
μ	Dynamic viscosity	(kg/ms)
Subscript		
c	cold flow	
f	fluid	
h	hot flow	
in	inlet value	
out	outlet value	
w	wall	

8. References

[1] P. Tabeling, Introduction to microfluidics, Oxford University Press, 2005.

[2] K. Schubert, J. Brandner, M. Fichtner, G. Linder, U. Schygulla, A. Wenka, Microstructure devices for applications in thermal and chemical process engineering, Microscale Thermophysical Engineering 5 (1) (2001) 17–39.

[3] C. Harris, K. Kelly, T. Wang, A. McCandless, S. Motakef, Fabrication, modeling, and testing of micro-cross-flow heat exchangers, Journal of Microelectromechanical Systems 11 (6) (2002) 726–735.

[4] J. Brandner, E. Anurjew, L. Bohn, E. Hansjosten, T. Henning, U. Schygulla, A. Wenka, K. Schubert, Concepts and realization of microstructure heat exchangers for enhanced heat transfer, Experimental Thermal and Fluid Science 30 (8) (2006) 801 – 809.

[5] J. J. Brandner, L. Bohn, T. Henning, U. Schygulla, K. Schubert, Microstructure heat exchanger applications in laboratory and industry, Heat Transfer Engineering 28 (8-9) (2007) 761–771.

[6] B. Alm, U. Imke, R. Knitter, U. Schygulla, S. Zimmermann, Testing and simulation of ceramic micro heat exchangers, Chemical Engineering Journal 135 (1) (2008) S179 – S184.

[7] H. Cao, G. Chen, Q. Yuan, Testing and design of a microchannel heat exchanger with multiple plates, Industrial & Engineering Chemistry Research 48 (9) (2009) 4535–4541.

[8] N. Garcia-Hernando, A. Acosta-Iborra, U. Ruiz-Rivas, M. Izquierdo, Experimental investigation of fluid flow and heat transfer in a single-phase liquid flow micro-heat exchanger, International Journal of Heat and Mass Transfer 52 (2009) 5433–5446.

[9] B. Mathew, H. Hegab, Experimental investigation of thermal model of parallel flow microchannel heat exchangers subjected to external heat flux, International Journal of Heat and Mass Transfer 55 (2012) 2193 – 2199.

[10] W. Bier, W. Keller, G. Linder, D. Seidel, K. Schubert, H. Martin, Gas to gas heat transfer in micro

heat exchangers, Chemical Engineering and Processing: Process Intensification 32 (1) (1993) 33–43.

[11] F. Meschke, G. Riebler, V. Hessel, J. Schuerer, T. Baier, Hermetic gas-tight ceramic microreactors, Chemical Engineering & Technology 28 (4) (2005) 465–473.

[12] K. Koyama, Y. Asako, Experimental investigation of heat transfer characteristics on a gas-to-gas parallel flow microchannel heat exchanger, Experimental Heat Transfer 23 (2010) 130–143.

[13] J. M. Commenge, L. Falk, J. P. Corriou, M. Matlosz, Optimal design for flow uniformity in microchannel reactors, AIChE Journal 48 (2) (2002) 345–358.

[14] R. Chein, J. Chen, Numerical study of the inlet/outlet arrangement effect on microchannel heat sink performance, International Journal of Thermal Sciences 48 (8) (2009) 1627–1638.

[15] O. Tonomura, S. Tanaka, M. Noda, M. Kano, S. Hasebe, I. Hashimoto, CFD-based optimal design of manifold in plate-fin microdevices, Chemical Engineering Journal 101 (2004) 397 – 402.

[16] S. Balaji, S. Lakshminarayanan, Improved design of microchannel plate geometry for uniform flow distribution, The Canadian Journal of Chemical Engineering 84 (6) (2006) 715–721.

[17] C. Renault, S. Colin, S. Orieux, P. Cognet, T. Tzedakis, Optimal design of multi-channel microreactor for uniform residence time distribution, Microsystem Technologies 18 (2012) 209–223.

[18] W. Wibel, U. Schygulla, J. J. Brandner, Micro device for liquid cooling by evaporation of r134a, Chemical Engineering Journal 167 (2-3) (2011) 705 – 712.

[19] M. Madou, Fundamentals of microfabrication: the science of miniaturization, CRC Press, 2002.

[20] J. J. Brandner, T. Gietzelt, T. Henning, M. Kraut, H. Moritz, W. Pfleging, Microfabrication in Metals and Polymers, Wiley-VCH Verlag GmbH, 2008, pp. 267–319

[21] K. J. Hecht, S. Kohl, P. Pfeifer, B. Kraushaar-Czarnetzki, Boundary conditions in multiphase microchannel flows: The role of surface roughness, in: Jahrestreffen Reaktionstechnik, Wuerzburg, 10-12 May 2010.

[22] A. Vittoriosi, J. Brandner, R. Dittmeyer, Experimental design with integrated temperature sensor in mems: an example of application for rarefied gases, in: 3rd Micro and Nano Flows Conf., Thessaloniki, GR, August 22-24, 2011.

[23] G. Maranzana, I. Perry, D. Maillet, Mini- and micro-channels: influence of axial conduction in the walls, International Journal of Heat and Mass Transfer 47 (17-18) (2004) 3993 – 4004.

[24] Y. Yang, G. L. Morini, H. Chalabi, M. Lorenzini, Experimental Nusselt number determination for gas flows through commercial microtubes, in: Proceedings of the 3rd GASMEMS Workshop, June 9-11, Bertinoro, Italy, 2011.

[25] W. Kays, A. London, Compact Heat Exchangers, McGraw-Hill, New York, third ed., 1984.

[26] T. Stief, O.-U. Langer, K. Schubert, Numerical investigations of optimal heat conductivity in micro heat exchangers, Chem. Eng. Technol. 22 (4) (1999) 297–303.

Design and optimization of a Holweck pump via linear kinetic theory

Steryios Naris, Eirini Koutandou, Dimitris Valougeorgis

University of Thesaly, Department of Mechanical Engineering, Pedion Areos, 38334, Volos, Greece

E-mail: `snaris@mie.uth.gr`, `irkoutad@hotmail.com`, `diva@mie.uth.gr`

Abstract. The Holweck pump is widely used in the vacuum pumping industry. It can be a self standing apparatus or it can be part of a more advanced pumping system. It is composed by an inner rotating cylinder (rotor) and an outer stationary cylinder (stator). One of them, has spiral guided grooves resulting to a gas motion from the high towards the low vacuum port. Vacuum pumps may be simulated by the DSMC method but due to the involved high computational cost in many cases manufactures commonly resort to empirical formulas and experimental data. Recently a computationally efficient simulation of the Holweck pump via linear kinetic theory has been proposed by Sharipov et al [1]. Neglecting curvature and end effects the gas flow configuration through the helicoidal channels is decomposed into four basic flows. They correspond to pressure and boundary driven flows through a grooved channel and through a long channel with a T shape cross section. Although the formulation and the methodology are explained in detail, results are very limited and more important they are presented in a normalized way which does not provide the needed information about the pump performance in terms of the involved geometrical and flow parameters. In the present work the four basic flows are solved numerically based on the linearized BGK model equation subjected to diffuse boundary conditions. The results obtained are combined in order to create a database of the flow characteristics for a large spectrum of the rarefaction parameter and various geometrical configurations. Based on this database the performance characteristics which are critical in the design of the Holweck pump are computed and the design parameters such as the angle of the pump and the rotational speed, are optimized. This modeling may be extended to other vacuum pumps.

1. Introduction

The choice of the equipment that is used for the creation and maintenance of vacuum conditions depend on various parameters such as the required pressure, the throughput and the available time for the process. In many cases, the use of a single vacuum pump is not enough and a combination of pumps is needed. It is common to have a first stage where rough vacuum conditions are created and a second one for the achievement of the desired pressure.

The optimization of the design and the operational parameters of the pumps has lead to the development of numerical tools for the simulation of the flow in the pump. In many approaches the Navier-Stokes equations have been used with the corresponding slip boundary conditions. This method is well tested but its range of applicability is limited to rough vacuum conditions. This is due to the fact that in lower pressures the assumption of the continuum medium collapses and the recovered results are not reliable [2].

Published under licence by IOP Publishing Ltd

Another method that can be used for the simulation of flow systems in high vacuum conditions is the mesoscopic approach with either stochastic or deterministic tools. The first one is the Direct Simulation Monte Carlo method (DSMC) [3] where computational molecules move, reflect from solid boundaries and collide to each other so as to statistically mimic the behavior of real molecules. Each model particle in the simulation represents a large number of real molecules in the physical system. The methodology is stochastic in nature since several modules in the algorithm, such as intermolecular collisions, are modeled in a probabilistic manner using random numbers. The state of the system is defined by the position and velocity vectors of the model particles. The most important drawback of DSMC is that it is appropriate for relatively high Mach numbers since the statistical noise can be significant.

The deterministic approach is based on the solution of the Boltzmann equation or the corresponding kinetic equations [4, 5]. The main unknown is the distribution function, while the macroscopic quantities can be recovered as its moments. The most common and computationally efficient method is by discretizing the kinetic equation in the molecular velocity space by the discrete velocity method (DVM) and by a finite differencing scheme in the physical space. This approach is superior to the DSMC method when linearized flows are tackled.

In the present paper the simulation of the Holweck pump based on linear kinetic theory is presented. This is a vacuum pump that is used either as a single apparatus or as a first stage for a pumping system. Simulations of the Howeck pump have been carried out especially in the last years. Most of them are using either the DSMC [6, 7] or the Navier-Stokes [8, 9] equation in order to recover the results and refer to various geometries. Recently, the deterministic approach has been implemented [1] but the details of the pump dimensions and the function characteristics are not provided. The methodology proposed in this paper is followed here and more specifically the discrete velocity method (DVM) is applied to solve the BGK kinetic equation, which is valid for the case under investigation since there are no significant temperature variations. In addition, the equations are linearized, something that is justified by the fact that the length of the pump channels is much bigger than the characteristic length of the cross section and the local forcing term is relatively small. Results are presented for the mass flow rates and the characteristic curves, while a preliminary parametric study is performed for the optimization of the design parameters.

2. Statement of the problem

The Holweck pump consists of two coaxial cylinders. One of them is stationary and the other one is rotating, while the inner cylinder has helicoidal grooves printed on it. The rotation causes a flow that can result to a pressure difference between the two ends of the cylinder. The pressure at the high vacuum end is P_h while the corresponding pressure at the fore vacuum end is symbolized as P_f.

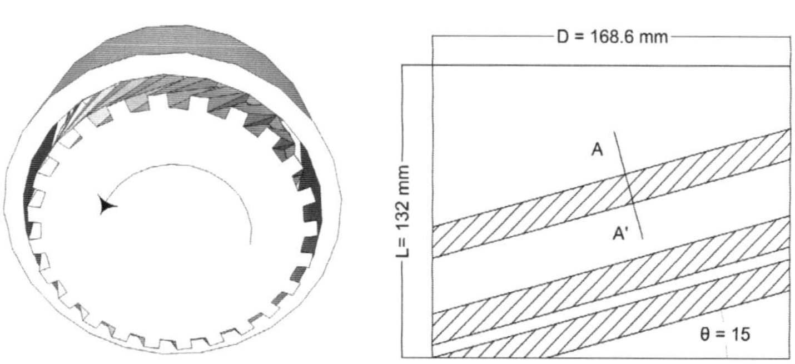

Figure 1. Typical design of the Holweck pump and basic dimensions of the inner cylinder.

The flow is fully three-dimensional, but a simplification can be achieved if the effect of the curvature of the cylinder is neglected as well as the end effects at the inlet and outlet of the pump. This approach can be justified by the fact that the ratio of the characteristic length of the grooves and the radius of the inner cylinder is less than 5% while the length of the channels is big compared to the characteristic length of its cross section.

This approach gives the opportunity to have the solution of the whole flow field by integrating the partial solutions of every cross section of the channel. On the other hand, if the equations are non-dimensionalized by the local forcing term, then only one cross section has to be solved, since the flow can be assumed as a fully developed 3d flow in a duct. This procedure simplifies the solution and decreased drastically the computational time.

The exact geometry of the pump is shown in Figs. 1 and 2 and Table 1. The geometry of the pump is identical to the one used in [7] and it corresponds to a typical design of the Holweck pump. It is obvious that the same approach can be easily applied to any other configuration.

In the present work, this configuration is kept constant except of the angle θ of the channels. This is one of the parameters (the other one is the rotational speed) which are optimized by using the methodology presented in the next paragraphs.

Table 1. Basic parameters of the Holweck pump under investigation.

Parameter	Value
Inner Diameter (D_{in})	168.6 mm
Outer Diameter (D_{out})	170.4 mm
Pump Length	132.0 mm
Number of Grooves	7
Rotational speed (n)	1200-4800 rpm
Angle (θ)	4^0 - 24^0
Gas molecular mass	4.002602 (He)
Boltzmann constant	8.314462 $\frac{J}{mol \times K}$

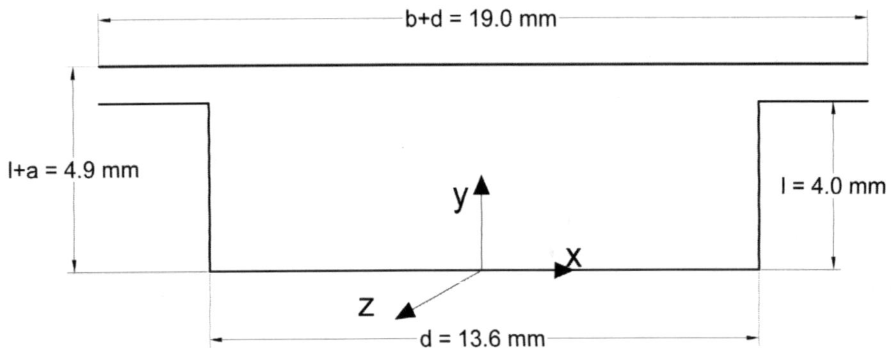

Figure 2. Cross section A-A' of the grooves with dimensions and the coordinate system with its origin

3. Formulation

As it has been already stated, the method followed in the present work is the simulation of the flow by the BGK kinetic equation, which can be deduced from the Boltzmann equation if the collision term is replaced by the BGK model. Then the equation takes the form

$$\boldsymbol{\xi} \cdot \frac{\partial f}{\partial \mathbf{r}} = \frac{P}{\mu}(f - f^{eq}) \tag{1}$$

where $\boldsymbol{\xi}$ is the microscopic velocity, P the pressure, $\mathbf{r}$ the position vector and $f = f(\boldsymbol{r}, \boldsymbol{\xi})$ the distribution function. Finally, $f^{eq} = f^{eq}(\boldsymbol{r}, \boldsymbol{\xi})$ represents the local Maxwellian which is

$$f^{eq} = n(\mathbf{r})\left[\frac{m}{2\pi kT(\mathbf{r})}\right]^{\frac{3}{2}}\left\{-\frac{m(\boldsymbol{\xi} - \hat{\mathbf{u}}(\mathbf{r}))^2}{2kT(\mathbf{r})}\right\} \tag{2}$$

Here m is the molecular mass, k is the Boltzmann constant, $n(\mathbf{r})$ is the number density, $\mathbf{u}(\mathbf{r})$ is the macroscopic velocity and $T(\mathbf{r})$ is the temperature. These quantities can be calculated as moments of the distribution function f

$$n(\mathbf{r}) = \int f(\mathbf{r}, \boldsymbol{\xi})d\boldsymbol{\xi} \qquad \hat{\mathbf{u}}(\mathbf{r}) = \int \boldsymbol{\xi} f(\mathbf{r}, \boldsymbol{\xi})d\boldsymbol{\xi} \qquad T(\mathbf{r}) = \frac{m}{3n(\mathbf{r})k}\int (\boldsymbol{\xi} - \hat{\mathbf{u}}(\mathbf{r}))^2 f(\mathbf{r}, \boldsymbol{\xi})d\boldsymbol{\xi} \tag{3}$$

while the shear stress tensor can be calculated as

$$P_{i,j}(\mathbf{r}) = m\int (\xi_i - u_i)(\xi_j - u_j)f(\mathbf{r}, \boldsymbol{\xi})d\boldsymbol{\xi} \quad i,j = x,y,z \tag{4}$$

The basic parameter of the flow is the Knudsen number which determines the rarefaction of the flow. In this work, the rarefaction parameter δ is used, which is proportional to the inverse Knudsen number and is defined as

$$\delta = \frac{P_0 D_h}{\mu_0 u_m} \sim \frac{1}{Kn}, \qquad u_m = \sqrt{\frac{2k_B T_0}{m}} \tag{5}$$

where P is the local pressure, D_h the hydraulic diameter, μ_0 the viscosity at the reference pressure T_0 and u_m is the most propable molecular velocity. Since the length of the channels is big enough compared to D_h and the speed of the outer cylinder is much smaller than u_m, the kinetic equation can be linearized, non-dimensionalized and solved numerically. On the other hand, taking into account the fact that the flow is linear, a decomposition can be applied and solve four subproblems: Longitudinal Poiseuille and Couette flow and Transversal Poiseuille and Couette flow. Then the results can be combined properly in order to have the full solution. This approach which has been proposed by Sharipov [1] , has been proved efficient and is followed in the present work. Applying this procedure simplifies more the numerical solution of the flow since instead of a 3-d problem one has to solve four 2-d problems. In addition, it gives flexibility to the solution because parameters such as the angle of the channel and the velocity of the cylinder are taken in to account only when the results of the subproblems are combined. So, for a given cross section, a data base of the subproblems results can be created and an optimization with respect to the angle or the velocity can be achieved.

For all the four problems, the non-dimensional parameter in the physical space is the hydraulic diameter of the channel cross section

$$D_h = 4\frac{A}{P'} = 2\frac{d \times l + a \times (b + d)}{(b + d + l + a)} \tag{6}$$

while the velocity vectors are non-dimensionalized as $c = \xi/u_m$. In the following paragraphs, the formulation for each of the four subproblems and the procedure for the results combination are presented. It is noted that the following approach is included here for completeness, eventhough someone can find it in [1]. In addition, here the parameter for the non-dimensionalization of length is the hydraulic diameter instead of the height of the groove and for the two transverse flows the no penetration boundary condition is used while in [1] it is not clear which is the boundary condition.

3.1. Longitudinal Poiseuille flow

The present flow is caused by a pressure difference along the z-axis of the channel. The distribution function is linearized as

$$f(\mathbf{r}, \boldsymbol{\xi}) = f^{eq}(\mathbf{c})\Big[1 + (h(x, y, \mathbf{c}) - z)X_z\Big] \qquad X_z = \frac{D_h}{P}\frac{dP}{dz} << 1 \tag{7}$$

and by taking into account that the flow is considered fully developed and the assumption that the density and the temperature over the whole cross section remain constant, the kinetic equation takes the form

$$c_x\frac{\partial h}{\partial x} + c_y\frac{\partial h}{\partial y} = \delta(2c_z u_z - h) + c_z \tag{8}$$

In addition, the fact that the distribution function h does not change along the z-axis, allows us to eliminate one of the microscopic velocity's components. If the projected distribution function is defined as

$$\Phi(x, y, c_x, c_y) = \frac{1}{\sqrt{\pi}}\int h(x, y, \mathbf{c})c_z e^{-c_z^2}dc_z, \tag{9}$$

the kinetic equation then takes the form

$$c_x\frac{\partial \Phi}{\partial x} + c_y\frac{\partial \Phi}{\partial y} = \delta(u_z - \Phi) + \frac{1}{2} \tag{10}$$

which is solved in order to recover the macroscopic quantities. Diffuse boundary conditions are used for all the solid boundaries and periodic boundary conditions at $-\frac{b+d}{2D_h}$ and $\frac{b+d}{2D_h}$. It is noted that this is also the treatment for the boundary conditions at the longitudinal Couette flow.

The dimensionless velocity u_z and stress tensor Π_{yz} are given as

$$u_z(x, y) = \int\int \Phi e^{-c_x^2 - c_y^2}dc_x dc_y \qquad P_{yz}(x, y) = \int\int c_y\Phi e^{-c_x^2 - c_y^2}dc_x dc_y \tag{11}$$

The dimensional velocity and stress tensor are

$$\hat{u_z} = -u_z u_m X_z \qquad \hat{P}_{yz} = -2P_{yz}PX_z \tag{12}$$

Finally the dimensionless mass flow rate and the reduced drag coefficient on the outer cylinder surface is given as

$$G_z^P = \int_A u_z dxdy \qquad \Pi_z^P = \int_{-\frac{b+d}{2l}}^{\frac{b+d}{2l}} P_{yz}dx \tag{13}$$

and the dimensional flow rate is

$$\dot{M}_z^P = \rho u_m D_h^2 X_z G_z^P \tag{14}$$

3.2. Longitudinal Couette flow

When the flow due to the motion of the upper plate in the z-direction is considered, the distribution function is linearized as follows

$$f(\mathbf{r},\boldsymbol{\xi}) = f^{eq}(\mathbf{c})\left[1 + h(x,y,\mathbf{c})\frac{U_z}{u_m}\right] \tag{15}$$

in this case the kinetic equation becomes

$$c_x\frac{\partial h}{\partial x} + c_y\frac{\partial h}{\partial y} = \delta(2c_z u_z - h) \tag{16}$$

again, c_z is eliminated by projecting the kinetic equation on the velocity space

$$c_x\frac{\partial \Phi}{\partial x} + c_y\frac{\partial \Phi}{\partial y} = \delta(u_z - \Phi) \tag{17}$$

The dimensionless macroscopic quantities are given by Eq.(11). The dimensional velocity and stress tensor are

$$\hat{u_z} = u_z U_z \qquad \hat{P}_{yz} = -2P_{yz}P\frac{U_z}{u_m} \tag{18}$$

The dimensionless mass flow rate and the reduced drag coefficient on the outer cylinder surface are

$$G_z^C = \int_A u_z dxdy \qquad \Pi_z^C = \int_{-\frac{b+d}{2l}}^{\frac{b+d}{2l}} P_{yz}dx \tag{19}$$

while the dimensional mass flow is

$$\dot{M}_z^C = \rho u_m D_h^2 \frac{U_z}{u_m} G_z^C \tag{20}$$

It has to be noted that by using the Onsager-Casimir theory the following relation is recovered

$$G_z^C = 2\,\Pi_z^P \tag{21}$$

3.3. Transversal Poiseuille flow

When the flow is caused by a pressure difference along the x direction, the first difference is that there are two components of the velocity ie. u_x and u_y. In addition, the density variations can not be neglected. On the contrary, the temperature perturbations are small and the flow can be assumed isothermal [10]. Finally the distribution function is linearized as

$$f(\mathbf{r},\boldsymbol{\xi}) = f^{eq}(\mathbf{c})\left[1 + (h(x,y,\mathbf{c}) - x)X_x\right] \qquad X_x = \frac{D_h}{P}\frac{dP}{dx} << 1 \tag{22}$$

the new form of the kinetic equation after the projection is

$$c_x\frac{\partial h}{\partial x} + c_y\frac{\partial h}{\partial y} = \delta(\rho + 2(c_x u_x + c_y u_y) - h) + c_x \tag{23}$$

Again, using the fact that the distribution function h does not change along the z-axis, allows us to eliminate one of the microscopic velocity's components. The projected distribution function is defined as

$$\Phi(x,y,c_x,c_y) = \frac{1}{\sqrt{\pi}}\int h(x,y,\mathbf{c})e^{-c_z^2}dc_z \tag{24}$$

and the kinetic equation then takes the form

$$c_x \frac{\partial \Phi}{\partial x} + c_y \frac{\partial \Phi}{\partial y} = \delta(\rho + 2c_x u_x + 2c_y u_y - \Phi) + c_x \tag{25}$$

The existence of density variations makes the use of the typical Maxwell boundary conditions inappropriate for the flow under investigation and the no-penetration boundary condition is used. According to it, on the walls a new parameter ρ_w is calculated in order to satisfy the equilibrium of the momentum on the wall in the vertical direction. This is a parameter without a physical meaning, but it allows us to ensure that no momentum is crossing the solid boundaries. The exact expressions for the estimation of ρ_w can be found in [11].

The dimensionless velocity u_x and stress tensor P_{xy} are given as

$$u_x(x,y) = \int\int c_x \Phi e^{-c_x^2 - c_y^2} dc_x dc_y \qquad u_y(x,y) = \int\int c_y \Phi e^{-c_x^2 - c_y^2} dc_x dc_y$$
$$P_{xy}(x,y) = \int\int c_x c_y \Phi e^{-c_x^2 - c_y^2} dc_x dc_y \tag{26}$$

The dimensional velocity and stress tensor are

$$\hat{u_x} = -u_x u_m X_x \qquad \hat{u_y} = -u_y u_m X_x \qquad \hat{P}_{xy} = -2P_{xy} P X_x \tag{27}$$

Finally the dimensionless mass flow rate and the reduced drag coefficient on the outer cylinder surface is given as

$$G_x^P = \int_{\frac{l}{D_h}}^{\frac{l+a}{D_h}} u_x dy \qquad \Pi_x^P = \int_{-\frac{b+d}{2l}}^{\frac{b+d}{2l}} P_{xy} dx \tag{28}$$

the dimensional flow rate is

$$\dot{M}_x^P = \rho u_m D_h^2 X_x G_x^P \tag{29}$$

3.4. Transversal Couette flow

The last flow that has to be considered is the one due to a motion of the upper plate in the x-direction. As in the corresponding flow due to pressure gradient, three macroscopic quantities are involved in the kinetic equation ie. ρ, u_x and u_y. The linearization of the distribution function is

$$f(\mathbf{r}, \boldsymbol{\xi}) = f^{eq}(\mathbf{c}) \left[1 + h(x, y, \mathbf{c}) \frac{U_x}{u_m} \right] \tag{30}$$

and the kinetic equation becomes

$$c_x \frac{\partial h}{\partial x} + c_y \frac{\partial h}{\partial y} = \delta(\rho + 2c_x u_x + 2c_y u_y - \Phi) \tag{31}$$

and y using the projection procedure

$$c_x \frac{\partial \Phi}{\partial x} + c_y \frac{\partial \Phi}{\partial y} = \delta(\rho + 2c_x u_x + 2c_y u_y - \Phi) \tag{32}$$

Here, the dimensionless macroscopic quantities are given by Eq.(26) and the corresponding dimensional by Eq.(27). The boundary conditions used, are identical with the corresponding of the Transversal Poiseuille flow. The dimensionless mass flow rate and the reduced drag coefficient on the outer cylinder surface are

$$G_x^C = \int_{\frac{l}{D_h}}^{\frac{l+a}{D_h}} u_x dy \qquad \Pi_x^C = \int_{-\frac{b+d}{2l}}^{\frac{b+d}{2l}} P_{xy} dx \tag{33}$$

while the dimensional mass flow is

$$\dot{M}_x^C = \rho u_m D_h^2 \frac{U_x}{u_m} G_x^C \tag{34}$$

It has to be noted that by using the Onsager-Casimir theory again for the two transversal flows the following relation is recovered

$$G_x^C = 2\ \Pi_x^P \tag{35}$$

4. Numerical scheme

For the four subproblems the kinetic equation has to be solved in order to recover the macroscopic quantities and the dimensionless flow rate. In order to do so in the present work the DVM method is used.

The main idea of the method is that the kinetic equation is solved for a set of discrete microscopic velocity vectors. Then numerical integration is applied in order to recover the moments of the distribution function, which are in fact the macroscopic quantities. The discrete velocities are chosen carefully and most often they are the roots of an orthogonal polynomial, at least as far as the magnitude of the velocities is considered, in order to recover the integrals with the best accuracy for a given number of velocities. Depending on the rarefaction of the flow, different number of velocities is required. In general, the more dense a flow is, the less velocities are required.

In the present work, 16 values have been used for the magnitude of the velocities and 400 angles, since a polar coordinates system is used for the microscopic velocities space. On the other hand, when the rarefaction parameter is larger than 15, the discrete angles are reduced to 80.

In the physical space, the grid used was uniform with $\Delta x = \Delta y = 0.1mm$. For more dense flows ($\delta > 1$) the grid lattice was tripled ($\Delta x = \Delta y = 1/30$) since it is known that higher values of δ require more dense grids. The numerical scheme used for all the four syb-probloems, is a typical central-difference scheme but it is applied on the characteristic of the microscopic velocity since it gives more accurate results due to the Lagransian nature of the Boltzmann equation and is described in detail in [11].

5. Overall quantities

For the characterization of a pump, one of the quantities required is the pressure difference created and the corresponding throughput. The most important problem is that the dimensional quantities G recovered by the numerical solution of the four subproblems depend on the local pressure and on the local rarefaction parameter δ. This is also the reason for the variance of their values while the mass flow rate has to be constant for every cross section of the pump.

In order to solve this problem the quantity G_η is defined which is related to the mass flow rate as

$$\dot{M}_\eta = \frac{2P_h}{u_m} D_h^2 G_\eta \tag{36}$$

with P_h being the pressure on the high vacuum chamber. Since all the other quantities are constant, G_n should not be dependent on the position of the cross section.

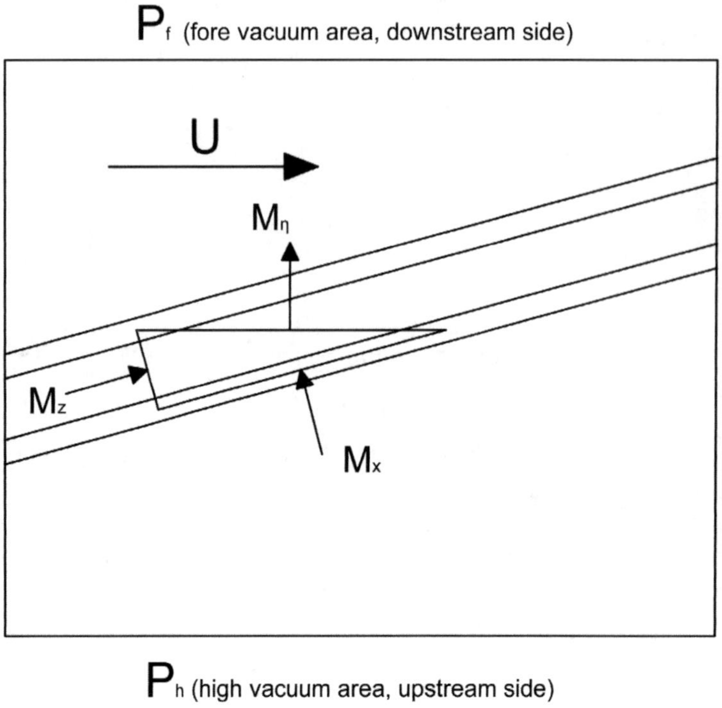

Figure 3. Control volume for mass equilibrium.

Application of the mass conservation law in the triangle of Fig.3 gives

$$\dot{M}_\eta = \dot{M}_x + \dot{M}_z. \tag{37}$$

In addition, from Eqs (14,20)it is deduced that

$$\dot{M}_z = \dot{M}_z^P + \dot{M}_z^C = \rho u_m D_h^2 \left\{ -G_z^P X_z + G_z^C \frac{U_z}{u_m} \right\} \tag{38}$$

and since

$$X_z = -\frac{D_h}{P}\frac{dP}{d\eta} sin(\theta), \qquad U_z = -U cos(\theta) \tag{39}$$

it is concluded that

$$\dot{M}_z = \rho u_m D_h^2 \left\{ \frac{D_h}{P}\frac{dP}{d\eta} sin(\theta) G_z^P - \frac{U cos(\theta)}{u_m} G_z^C \right\} \tag{40}$$

Accordingly for the x-direction

$$\dot{M}_x = \dot{M}_x^P + \dot{M}_x^C = \rho u_m D_h^2 \frac{l_x}{D_h} \left\{ -G_x^P X_x + G_x^C \frac{U_x}{u_m} \right\} \tag{41}$$

with

$$X_x = \frac{D_h}{P}\frac{dP}{d\eta} cos(\theta), \qquad U_x = U sin(\theta) \tag{42}$$

and finally

$$\dot{M}_x = \rho u_m D_h^2 \frac{l_x}{D_h} \left\{ \frac{D_h}{P} \frac{dP}{d\eta} cos(\theta) G_x^P + \frac{U sin(\theta)}{u_m} G_x^C \right\} \tag{43}$$

l_x is the dimensional length of side x. By substituting Eq.(40) and (43) to Eq. (37) a differential equation for the local pressure is recovered

$$D_h \frac{dP}{d\eta} = \frac{sin(\theta) \left\{ P \frac{U}{u_m} cos(\theta) \left[G_z^C - \frac{l_x}{D_h} G_x^C \right] - G_\eta P_h \right\}}{G_z^P sin^2(\theta) + \frac{l_x}{D_h} cos^2(\theta) G_x^P} \tag{44}$$

For a known P_h, Eq. (44) can be solved and the pressure distribution for the whole length of the pump recovered. It has to be noted that on every cross section, the values of the dimensionless flow rates for the local δ have to be used. Finally the pumping speed S and the throughput Q can be found as

$$S = u_m D_h^2 G_\eta N_{gr} \qquad Q = SP_h = P_h u_m D_h^2 G_\eta N_{gr} \tag{45}$$

where N_{gr} is the number of grooves.

6. Results and Discussion

6.1. Partial solutions

The first step for the recover of the overall quantities is the creation of a full database including the dimensionless flow rate for the four subproblems described above. The results are presented in Figs. 4 and 5. They cover the range $0 \leq \delta \leq 200$. All the quantities have been obtained by using the developed kinetic numerical codes. As it can be seen, the dimensionless flow rate for the

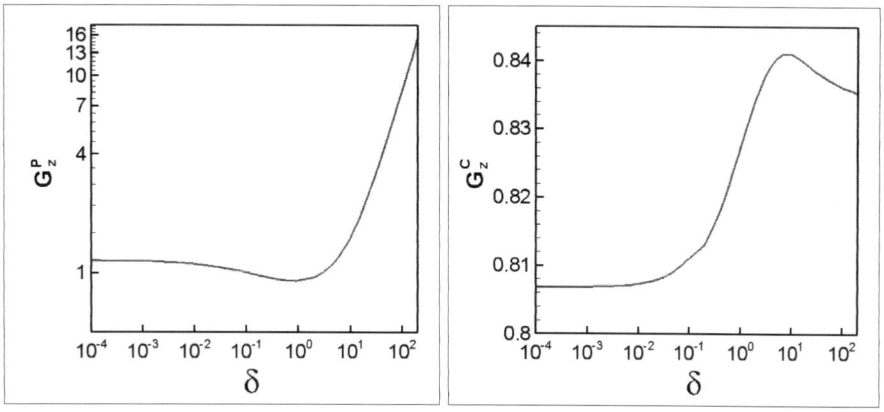

Figure 4. Flow rate for longitudinal Poiseuille (left) and Couette(right) flow

longitudinal flows is an order of magnitude larger than the corresponding transversal flows. This is due to the chosen geometry and more specifically to the small gap between the two cylinders. The relative small effect of the transversal compared to the longitudinal phenomena is desirable because these flows in fact decrease the efficiency of the pump since these are secondary flows where the gas returns back towards the high vacuum chamber. On the other hand it is obvious that for $\delta \leq 10$ the Poiseuille and the Couette flow rate are of the same magnitude but for larger values the Poiseuille flow becomes dominant. Of course, the exact contribution of each flow can be seen only when the operational parameters, such as rotational speed and angle of the channels, are defined.

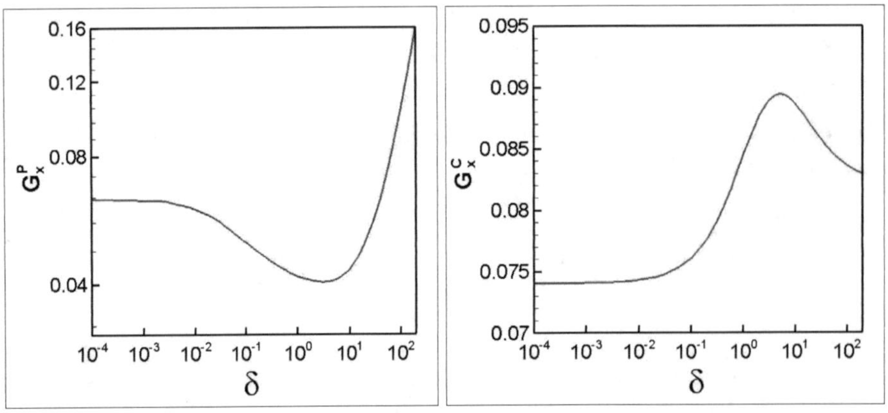

Figure 5. Flow rate for transversal Poiseuille (left) and Couette(right) flow

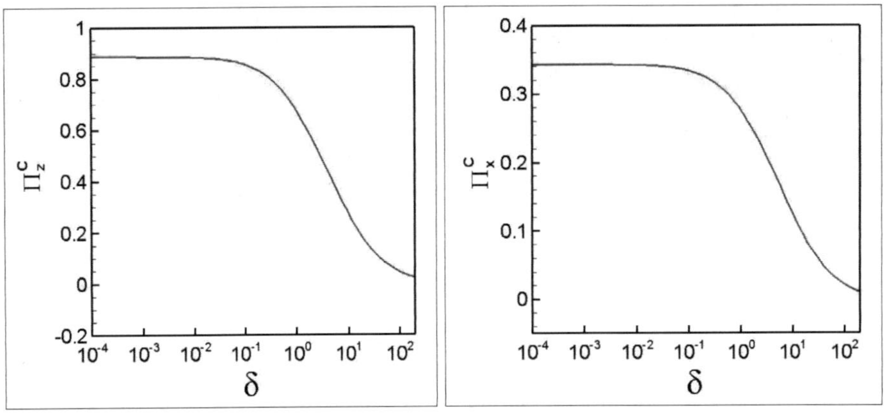

Figure 6. Drag coefficient for longitudinal (left) and transversal (right) Couette flow

In Fig. 6 the drag coefficient is plotted for the two Couette flows. Again the phenomena of the transversal flow are mach smaller than the corresponding of the longitudinal. It has to be mentioned here that the drag coefficient for the Poiseuille flows can be easily deduced if Eqs. (21) and (35) are used.

6.2. Parametric analysis

One of the parameters that are expected to affect the performance of the pump is the angle of the grooved channels with respect to the front surface of the pump (see Fig. 1). The angle in the present work is assumed to be constant for the whole length of the pump. In Fig.7 the dependence of the exit pressure on the angle of the pump is examined. As it can be seen, the influence of angle is strong and the optimum angle depends on the rarefaction at the inlet of the pump. For $\delta_h = 1$ it seems that the optimum angle is close to $\phi = 12^0$, while for $\delta_h = 0.01$ it is about $\phi = 15^0$. This correlation is expected eventhough the exact value of the optimal angle can not be anticipated without numerical or experimental results. It can be deduced that a gradual change of the angle along the pump or a multi-stage pump of equal length but different angles for each stage can be much more efficient.

Finally, on Fig. 8 the influence of the rotating cylinder speed is examined. The increase of the rotational speed is increasing the pressure difference which is the expected behavior.

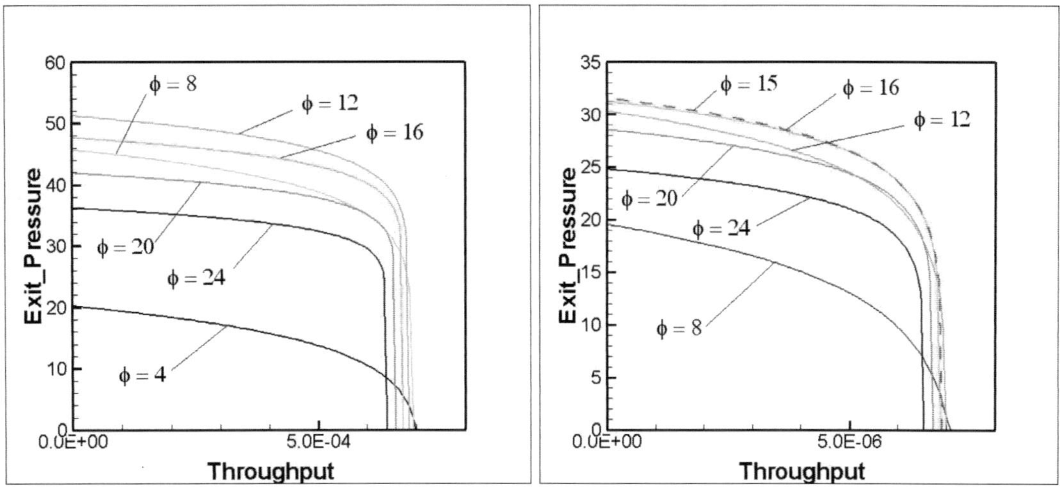

Figure 7. Characteristic curves for various values for angle, $n = 2400$, and $\delta_h = 1$ (left) and $\delta_h = 0.01$ (right)

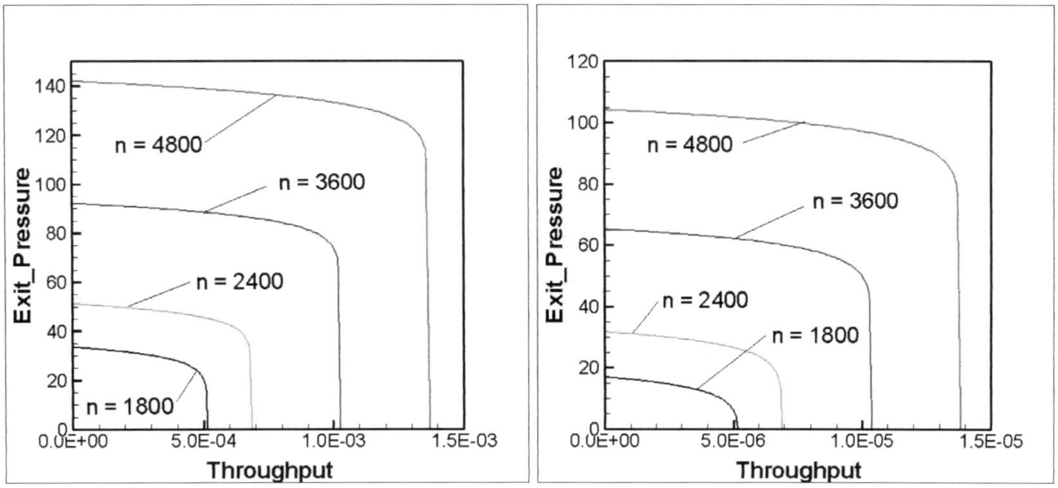

Figure 8. Characteristic curves for $\theta = 12^0$, $\delta_h = 1$ (left) and $\theta = 15^0$, $\delta_h = 10^{-2}$ (right) and various values for n

7. Concluding remarks

The flow in a Holweck pump has been simulated by solving the linearized Boltzmann-BGK equation using the discrete velocity method in the velocity space and an equi-distributed Cartesian grid in the physical space. A decomposition method has been applied in order to reduce the computational cost. Results containing the dependence of the flow on the rarefaction parameter have been provided while the influence of the grooves' angle and the rotational speed on the pressure difference produced by the pump has been examined. It has been shown that for the sets of parameters tested, there is an optimum angle where the pump can produce the highest pressure difference for almost the whole range of the pump throughput, while the increase of the rotational speed seems to increase the pressure difference in any case.

Future studies can include a detailed investigation of the dependence of the pump performance on the design parameters such as the grooves' dimensions, the pump length, the gas type.

In addition the angle of the grooves has to be optimized for various sets of the other design parameters and the overall efficiency of the pump has to be taken into account. Finally the same approach can be modified and applied in other kind of pumps such as the Gaede pump.

References

[1] Sharipov F, Fahrenbach P, Zipp A 2005 *J. Vac. Sci. Technol. A* **23(5)** 1331.
[2] Kandlikar S, Garimella S, Li D, Colin S and King MR 2006 *Heat Transfer and Fluid Flow in microchannels and microchannels* (Oxford: Elsevier)
[3] Bird G A 1994 *Gas Dynamics and the Direct Simulation Monte Carlo* (Oxford University Press)
[4] Ferziger J H and Kaper H G 1972 *Mathematical Theory of Transport Processes in Gases* (North-Holland Publishing Company)
[5] Cercignani C 1988 *The Bolzmann Equation and its Applications* (Springer)
[6] Heo J S and Hwang Y K 2000 *J. Vac. Sci. Tech. A* **18(3)** 662
[7] Huang Y K and Heo J S 2001 *Rarefied Gas Dynamics:* 22nd *Int. Symp.* (Sydney) vol 1 ed T J Bartel and M A Gallis (American Institute of Physics) p 893
[8] Boulon O and Mathes R 2001 *Vacuum* **60** 73
[9] Cheng H P, Jou R Y, Chen F Z, Chang Y W, Iwane M and Hanaoka T 2000 *J. Vac. Sci. Tech A* **18(2)** 543
[10] Naris S and Valougeorgis D 2007 *Phys Fluids* **19** 067103
[11] Naris S and Valougeorgis D 2005 *Phys Fluids* **17** 097106

EXPERIMENTAL STUDY OF THE GAS FLOWS THROUGH CHANNELS WITH CIRCULAR CROSS SECTIONS

M. Hadj-Nacer[1], P. Perrier[1], J. G. Méolans[1], I. Graur[1] and M. Wüest[2]

[1]Aix-Marseille Université, Ecole Polytechnique Universitaire de Marseille, Département de Mécanique Energétique, UMR CNRS 7343, 5 rue Enrico Fermi, 13453 Marseille cedex 13, France.
[2] INFICON Ltd, Alte Landstrasse 6, LI-9496 Balzers, Liechtenstein.

E-mail: mustafa.hadjnacer@polytech.univ-mrs.fr

Abstract. The experimental setup based on the constant volume technique is developed to measure the mass flow rate through the microtubes under the isothermal flow conditions. Four different gases: Helium, Nitrogen, Argon and Carbon dioxide, and two surface materials (Stainless stainless Steel steel and Sulfinert) are considered. In this study the Knudsen number varies from $\sim 10^{-4}$ to 0.3. In this range the approach based on the analytical solution of the Stokes equation subjected to the first and second order velocity slip boundary conditions is used. The tangential momentum accommodation coefficient (TMAC) is extracted from the experimental data on the mass flow rate using its analytical expression. The results are summarized in the tables representing the accommodation coefficients for the couples corresponding gas-surface material combinations. The influence of the molecular mass on the tangential momentum accommodation is discussed.

1. Introduction

The effect of an interaction between the gas flow and the solid surface increases by the decreasing of the channel size, due to the increasing the surface-to-volume ratio when channel size decreases. For the isothermal gas flow the interaction gas-surface is essentially characterized by the Tangential Momentum Accommodation Coefficient (TMAC). Its value varies in the range $[0, 1]$, $\alpha = 0$ corresponds to a specular reflexion, $\alpha = 1$ characterizes a diffuse reflexion. The tangential momentum accommodation coefficient is used in the expressions of the boundary conditions when one uses the continuum or kinetic approaches for the simulation of the flow characterizing by a relatively large Knudsen number. A better knowledge of the mechanism of the gas-solid surface interaction allows us to improve the understanding of the gas flow behavior in the slip and transitional flow regimes and to a better design of the industrial devices which work under rarefaction conditions.

Many researchers carried out different experiments to determine the value of the TMAC for various couples gas-solid surface. The mains experimental techniques used to measure the TMAC were reviewed in [1]. This review shows that generally the accommodation coefficient depends on large number of parameters such as the type of the gas, the surface material, its cleanliness and its roughness.

The main experimental techniques proposed in the literature to measure the mass flow rate through the microchannels are [1]:

- The liquid drop method (LD) [2], [3], [4], [5], [6], [7]. This method permits a direct determination of the volume flow rate by tracking in time the interface of oil drop moving in a calibrated tube by a speed camera.
- The accumulate dual tank method was developed by Arkilic and al in [8] and it consists in measuring relative pressure variation between a reference and an accumulation tank. The mass flow rate is deduced from the law of perfect gas.
- The constant volume technique (CV) is used by following authors: [2], [7], [9], [10], [11]. This method is similar to the accumulate dual tank method, but in this case the absolute pressure variation in the tanks is measured. It needs a very high thermal stability.
- An other method is the constant pressure technique [12], [13], [14]. Contrarily to the constant volume technique this method was not used a lot because of the difficulties in measuring the volume variation in time while the pressure is maintained constant.

All these methods have their own advantages and disadvantages. From the authors point of view, the simplest and the more accurate technique is the constant volume technique (CV) which was adopted in this study.

Various gases: Helium (He), Nitrogen (N_2), Argon (Ar) and Carbon-dioxide (CO_2) and two materials of the surface (stainless steel and Sulfinert) are used in the present study. The mass flow rate through the microtubes is measured using the constant volume technique under the isothermal conditions. The Knudsen number varies from 0.0002 (hydrodynamic regime) to 0.3 (slip regime). The velocity slip and accommodation coefficients are deduced by using the experimental data of the mass flow rate and the Navier-Stokes model with the first and second order slip boundary conditions.

2. Experiment

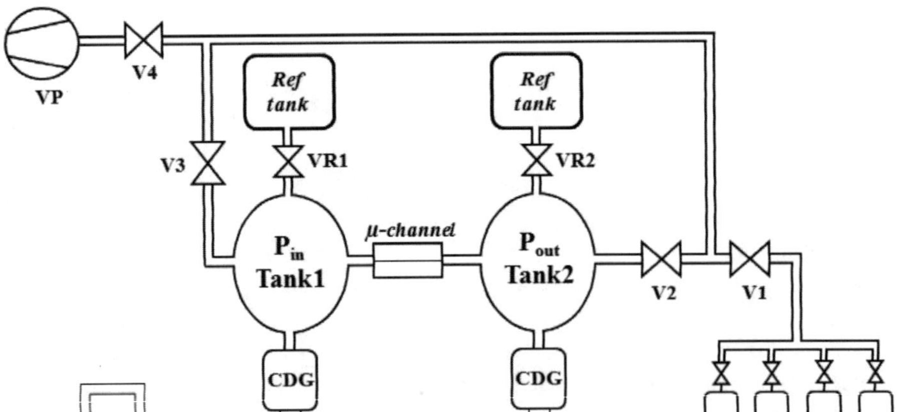

Figure 1. Schematic of the experimental setup

Channel	Material	D $[10^{-6}m]$	L $[10^{-3}m]$
T1	Sulfinert	275.	2000.*
T2	Stainless steel	239.	2013.*

Table 1. Dimensions of the microchannels. Indicative values of the diameter was given by provider equal to $250\mu m$, these values were corrected using the measurements in the hydrodynamic flow regime, the third column in Table. The length of the microtubes $T1$ and $T2$ were reduced to around $10cm$ for the experiments yield with CO_2 gas*.

2.1. Methodology
The experimental setup is represented schematically in Figure 1. To measure the mass flow rate through the microtubes the constant volume technique was implemented. This technique involves the use of two constant volume tanks connected by a microchannel. The flow through this microchannel is generated by setting a pressure difference between the inlet and outlet tanks. The volume of the tanks has to be much greater than the microchannel volume to guarantee that the flow parameters are independent of time, but remain detectable. The pressure and temperature variation during the experiments are measured and the mass flow rate is deduced from the equation of state. The mass variations occurring in the tanks during the experiments do not call into question the stationary assumption because the pressure variations are maintained less than 2% during the experiments.

The experimental methodology used for the mass flow rate measurements is the following:

- First, the experimental loop (Figure 1) is vacuumed using the vacuum pump (VP) by opening the valves $V2$, $V3$, and $V4$ during a time period (from one to two hours).
- Then, the valve $V4$ is closed and the system is filled with the working gas from the high pressure tank by opening the valve $V1$ until having the desired inlet pressure value (p_{in}).
- After that, the valves $V1$ and $V3$ are closed and the valve $V4$ is opened again to decrease the pressure in the outlet tank until obtaining the desired value (p_{out}).
- Finally, the valves $V2$ and $V4$ are closed and the acquisition of the pressure and temperature in both tanks can be started.

The registration of the pressure variations in the inlet and outlet tanks is carried out using two INFICON Capacitance Diaphragm Gauges (CDG), see Figure 1, chosen according to their pressure range and connected to the upstream and downstream tanks.

The temperature of the gas inside the system is assumed to be equal to the room temperature. This assumption was checked by the mean of three temperature sensors PT100. Two of them measured the temperature variations in the inlet and outlet tanks and one measured the room temperature variations. The difference between the measurements of the three temperature sensors was less than their uncertainty $0.3K$. The relative variation of the temperature during the experimental time length was less than $0.1K$. The thermal stabilization is checked before and during each experiment. The purity of the gases was higher than 99.999%.

The volumes of the inlet and outlet tanks are measured using a simple and accurate method. Using a reference tank with known volume the ratio between the measured volume and the reference volume was calculated from the mass conservation. This method is repeated five times and the maximum deviation from the mean value was 0.16%.

2.2. Microchannels characteristics
The two microtubes $T1$ and $T2$, see Table 1, used in this study are made from stainless steel and the internal surface of the microtube $T1$ is coated with Sulfinert. The Sulfinert coating is the

leading passivation technique for the storage and transfer of low-level organo-sulfurcontaining samples. The Sulfinert layer is applied to the stainless steel surface and is completely inert to organo-sulfur compounds. Typically stainless steel will adsorb or react with sulfur compounds such as hydrogen sulfide. The Sulfinert layer will prevent sulfur compounds from contacting the reactive stainless surface. The roughness of the microtubes is not known. The both tubes were provided from RESTEK company. The details of the fabrication process can be found on the company website. *www.restek.com*

The indicative diameter of the microtubes given by the provider is $250\mu m$ and its value was corrected (third column of the Table 1) after comparing the simulated and measured flow rate in the hydrodynamic regime when $Kn < 0.001$, see Section 4 for the details.

2.3. Mass flow rate calculation

Using the constant volume technique, the mass flow rate through the microchannels can be calculated from the equation of state for ideal gas by assuming steady conditions

$$pV = mRT, \tag{1}$$

where V represents the volume of the inlet or outlet tanks and R is the specific gas constant, p, T and m are, respectively, the pressure, temperature and mass of the gas. If we assume small variation of the gas properties (p, m and T) in the tanks, equation (1) can be written as follows

$$\frac{dm}{dt} = \frac{d}{dt}\left(\frac{pV}{RT}\right). \tag{2}$$

As it was shown in [7], [11] the previous relation may be transformed into

$$\frac{dm}{dt} = \frac{V}{RT}\frac{dp}{dt}(1-\varepsilon), \qquad \varepsilon = \frac{dT/T}{dp/p}. \tag{3}$$

If ε is small enough compared to 1, then the ratio dm/dt can be considered as the mass flow rate Q_{exp} through the microchannel.

As it is clear from equations (3) the constant volume technique requires very high-thermal stability, that means that temperature relative variations during the experiments must be much smaller than the pressure relative variations. In our experimental conditions the deviation of the temperature from the initial value is smaller than $0.1K$ during the experimental time length. The relative variation of the temperature dT/T is in order of 2×10^{-4} against 10^{-2} for the relative variation of the pressure dp/p, therefore ε is clearly less than 2×10^{-2} and may be neglected. Thus, the mass flow rate Q_{exp} can be written under the following form

$$Q_{exp} = \frac{V}{RT}\frac{dp}{dt}. \tag{4}$$

The measurement of Q_{exp} is affected by a specific relative error of 2×10^{-2} due to the temperature variation.

In expression of the mass flow rate (4) the parameters V and R are known and remain constant during the experiments. To calculate the term (dp/dt) in equation (4) we use the registered recorded data of the pressure variation p_i at different instants t_i. This variation is of order of 1% and the data of the pressure can be fitted by the first order polynomial form using the least square method as follows

$$p(t_i) = at_i + b, \tag{5}$$

where the slope a of the function $p(t_i)$ is equal to the derivative dp/dt and it is used in the equation (4) to calculate the experimental mass flow rate. The number of the measured pressure values varies between 40 and 2000 depending on the tanks pressure.

The uncertainty on the measurement of the mass flow rate is evaluated using the following expression

$$\frac{\Delta Q_{exp}}{Q_{exp}} = \frac{\Delta V}{V} + \frac{\Delta T}{T} + \frac{\Delta a}{a}, \tag{6}$$

where the uncertainty on the coefficient $a = dp/dt$ is estimated from standard deviation following the method used in [7] and its value is less than 0.5%.

The uncertainty on the temperature measurement is very small but we have to take into account the systematic error made on the stationary assumption previously evaluated at 2% and we affect it to the error on the temperature. The total uncertainty on the experimental mass flow rate is less than $\pm 4.1\%$ ($\Delta V/V = \pm 1.6\%$, $\Delta T/T = \pm 2\%$, $\Delta a/a = 0.5\%$).

3. Analytical solution

The Knudsen number considered in this paper is calculated from the mean pressure between two tanks (upstream and downstream the microchannel) and ranges from ~ 0.0002 to 0.3. This range covers the hydrodynamic and slip flow regimes. Brief presentation of the analytical expressions used for the mass flow rate, the velocity slip and tangential momentum accommodation coefficients is given below.

The continuum approach based on the Navier-Stokes equation has received a particular attention. Its application was extended to the slip regime implementing the first and second order slip boundary conditions for the velocity at the wall [15], [16], [17].

Using the second order slip boundary condition the mass flow rate through microchannel with circular cross section can be expressed in the following form

$$\dot{M} = \frac{\pi D^4 \Delta p p_m}{128 \mu R T L}\left(1 + 8A_1 Kn_m + 16A_2 \frac{\mathcal{P}-1}{\mathcal{P}+1} \ln \mathcal{P} Kn_m^2\right), \tag{7}$$

where D is the tube diameter, $\Delta p = p_{in} - p_{out}$, $\mathcal{P} = p_{in}/p_{out}$, Kn_m is the mean Knudsen number, based on the mean pressure $p_m = 0.5(p_{in} + p_{out})$ and on the tube diameter D. In this theoretical frame, the coefficient A_1 and A_2 may be expressed in the form

$$A_1 = \frac{\sigma_p}{k_\lambda}, \qquad A_2 = \frac{\sigma_{2p}}{k_\lambda^2}, \tag{8}$$

where σ_p and σ_{2p} are the first and second velocity slip coefficients, respectively, k_λ is the coefficient which depends on the molecular interaction model. The value of $\sqrt{\pi}/2$ for hard sphere (HS) model [18] is retained in this paper.

Furthermore, the non-dimensional mass flow rate may be deduced from (7)

$$S = \dot{M}/\dot{M}_P = 1 + 8A_1 Kn_m + 16A_2 \frac{\mathcal{P}-1}{\mathcal{P}+1} \ln \mathcal{P} Kn_m^2, \qquad \text{where} \qquad \dot{M}_P = \frac{\pi D^4 \Delta p p_m}{128 \mu R T L}, \tag{9}$$

here $\dot{M}_P$ is the classical Poiseuille mass flow rate. Equation (9) may be rewritten in more compact form:

$$S^{theor} = 1 + A_1^{theor} Kn_m + A_2^{theor} Kn_m^2. \tag{10}$$

The mass flow rate calculated using the analytical expression (9) will be compared with the appropriate experimental values.

Microchannel	T1 (Sulfinert)		T2 (SS)	
Quantity	Min	Max	Min	Max
Mass flow rate ($10^{-12} kg/s$)	2.7	279 000	2.2	403 000
Inlet pressure (Pa)	147	123 600	173	127 450
Outlet pressure (Pa)	29	78 700	35.5	80 200
Average Knudsen number Kn_m	2.2 10^{-4}	0.29	2.6 10^{-4}	0.30

Table 2. Experimental conditions

Gas	Order	A_0^{exp}	A_1^{exp}	A_2^{exp}	s_r	r^2	$E_s(\%)$
He	First	0.995± 0.002	10.14± 0.07	0	0.010	0.998	0.8
	Second	1.005± 0.004	9.16± 0.17	6.35± 0.78	0.023	0.998	1.7
N_2	First	0.997± 0.001	11.01± 0.05	0	0.008	0.999	0.6
	Second	1.008± 0.004	9.42± 0.14	16.74± 0.562	0.021	0.999	1.4
Ar	First	1.008± 0.002	10.99± 0.08	0	0.011	0.998	0.9
	Second	1.019± 0.004	9.30± 0.16	18.57± 0.65	0.023	0.999	1.5
CO_2	First	1.000± 0.002	10.21± 0.08	0	0.010	0.998	0.9
	Second	0.999± 0.003	10.15± 0.11	3.01± 0.45	0.018	0.999	1.1

Table 3. Experimental coefficients A_0^{exp}, A_1^{exp} and A_2^{exp} obtained from the first and second order polynomial fitting for microtube T1 (Sulfinert)

4. Results and discussion

The flows of four gases (He, N_2, Ar and CO_2) through the microtubes with different surface materials are studied for Knudsen number range [0.0002, 0.3]. The experimental conditions for both microtubes $T1$ and $T2$ are summarized in Table 2. The pressure ratio $\mathcal{P} = p_{in}/p_{out}$ between upstream and downstream tanks was fixed at 1.5 in the hydrodynamic regime then it was increased to reach the value of ~ 5 in the slip flow regime. It was shown in Ref. [9] that the non-dimensional mass flow rate does not depend practically on the pressure ratio and therefore this change in the pressure ratio does not have any influence on the extracted velocity slip coefficient.

The measured mass flow rate, normalized according to (9), was fitted using a second order in Kn_m polynomial form, similar to the analytical expression (10), implementing the least square method detailed in [9]:

$$S_f^{exp} = A_0^{exp} + A_1^{exp} Kn_m + A_2^{exp} Kn_m^2, \tag{11}$$

where the coefficients A_0^{exp}, A_1^{exp} and A_2^{exp} are the least-square method fitting coefficients. From the comparison of the analytical expression (10) and the polynomial fit of the experimental mass flow rate (11), the coefficients A_i^{exp} (i=0, 1 or 2) may be expressed under the following form

$$A_0^{exp} = 1, \qquad A_1^{exp} = 8A_1, \qquad A_2^{exp} = 16A_2 \frac{\mathcal{P}-1}{\mathcal{P}+1} \ln \mathcal{P}. \tag{12}$$

The fitting of the experimental data is done with the first order polynomial form in the range of Knudsen number [0, 0.1] and with the second order polynomial form in the Knudsen number range [0, 0.3]. The fitting of the experimental data are done without fixing the coefficient A_0^{exp} to the value one, even if its "theoretical value" is equal to one, see equation (9). The values of the coefficient A_0^{exp}, obtained from experimental fitting, were used to verify the data of the tube diameter given by the provider. In the hydrodynamic flow regime, when the Knudsen number is very small, the mass flow rate must be equal to the Poiseuille mass flow rate $\dot{M}_P$, see (9).

Gas	Order	A_0^{exp}	A_1^{exp}	A_2^{exp}	s_r	r^2	$E_s(\%)$
He	First	1.000± 0.002	11.09± 0.05	0	0.007	0.999	0.6
	Second	1.002± 0.003	10.82± 0.08	4.12± 0.35	0.012	0.999	0.7
N_2	First	0.996± 0.002	11.46± 0.08	0	0.013	0.998	1.0
	Second	0.999± 0.005	10.53± 0.17	15.69± 0.70	0.026	0.999	1.6
Ar	First	1.008± 0.003	11.62± 0.09	0	0.014	0.997	1.1
	Second	1.019± 0.002	10.15± 0.09	17.41± 0.37	0.013	0.999	0.8
CO_2	First	1.008± 0.002	11.04± 0.07	0	0.010	0.998	0.9
	Second	1.012± 0.002	10.46± 0.08	5.80± 0.35	0.012	0.999	0.8

Table 4. Experimental coefficients A_0^{exp}, A_1^{exp} and A_2^{exp} obtained from the first and second order polynomial fitting for microtube T2 (SS)

If we suppose that all other measured quantities in this Poiseuille expression do not have any errors, then the deviation of the coefficient A_0^{exp} from one may be attributed to an error in the diameter measurements. This technique was used in the present study to adjust the tube diameter measurements given by provider and equal to $250\mu m$. The values of the tube diameter adjusted from the Poiseuille mass flow rate are given in Table 1.

After this first step of the diameter adjustment, the measured mass flow rate was fitted with the first order polynomial form in the Knudsen number range $[0., 0.1]$ and by the second order polynomial form in the range $[0., 0.3]$. The obtained fitting coefficients and the statistical characteristics of the fitting procedure such as the determination coefficient r^2, the standard error E_s and the squared residual sum s_r are given in Tables 3 and 4. The squared residual sum s_r is defined as following $s_r = \sqrt{\frac{1}{n-k}\sum e_i^2}$, where n is the number of the measured points, $e_i = S_i^{exp} - S_{f_i}^{exp}$ is the local difference between measured S_i^{exp}, normalized according to (9), and fitted values $S_{f_i}^{exp}$ (11), k is the number of the unknown coefficients of the fitting model, $k = 3$ for the second order fit. The standard error is $E_s = \sqrt{\frac{1}{n}\sum e_i^2}/S_m^{exp}$, here S_m^{exp} is the average value of the measured quantities.

As it is clear from the results presented in Tables 3 and 4, that the procedure of the adjustment of the tube's diameter works well: the values of the A_0^{exp} coefficients are now very close to their theoretical values equal to one. From Tables 3 and 4 one can see that the values of the determination coefficients r^2 are very close to 1. The worst value is obtained for the microtube $T2$ (SS) and Argon and it is equal to 0.997 which confirms the quality of fitting for both microtubes and for both fits. Whereas, the residual variance s_r is smaller than 0.026 for both microtubes $T1$ and $T2$. The maximum value of the standard error made on the fittings is less than 2%. Therefore, analyzing the statistical characteristics of both fittings we can conclude that the both expressions (of the first and of the second order according to the Knudsen number) represent well the experimental data.

The errors on the coefficients σ_p and σ_{2p} in Tables 5 and 6 derive from the fitting error on the coefficients A_i^{exp}, see relations (8), (12), and do not take into account the systematic uncertainty coming from the errors on the mass flow rate measurements, see (6), and the errors related to the Poiseuille term (9).

When the slip coefficient is known the accommodation coefficient may be calculated. The authors of Ref. [19] calculated the values of the slip coefficient using the BGK kinetic model and the Maxwell diffuse-specular scattering kernel for different values of the accommodation coefficient α. After that the authors of [19] proposed a simple expression associating the slip coefficient and the accommodation coefficients. Later, this expression was improved in Ref. [20].

Gas	Molar mass (g/mol)	Order	σ_p	α	σ_{2p}
He	4.00	First	1.123± 0.007	0.947± 0.003	0
		Second	1.015± 0.019	1.001± 0.010	0.148± 0.018
N_2	28.02	First	1.219± 0.006	0.903± 0.002	0
		Second	1.044± 0.015	0.986± 0.008	0.378± 0.013
Ar	39.95	First	1.218± 0.009	0.904± 0.004	0
		Second	1.031± 0.018	0.992± 0.009	0.418± 0.015
CO_2	44.01	First	1.131± 0.009	0.943± 0.004	0
		Second	1.125± 0.012	0.946± 0.006	0.068± 0.010

Table 5. Experimental accommodation and slip coefficients for microtube T1 (Sulfinert)

Gas	Molar mass (g/mol)	Order	σ_p	α	σ_{2p}
He	4.00	First	1.228± 0.005	0.899± 0.002	0
		Second	1.198± 0.009	0.912± 0.004	0.092± 0.008
N_2	28.02	First	1.270± 0.009	0.882± 0.004	0
		Second	1.166± 0.019	0.927± 0.009	0.348± 0.016
Ar	39.95	First	1.287± 0.010	0.875± 0.004	0
		Second	1.124± 0.010	0.946± 0.005	0.387± 0.008
CO_2	44.01	First	1.223± 0.008	0.902± 0.003	0
		Second	1.159± 0.009	0.930± 0.004	0.132± 0.008

Table 6. Experimental accommodation and slip coefficients for microtube T2 (SS)

In the present work we use this improved expression from [20]:

$$\sigma_p(\alpha) = \frac{2-\alpha}{\alpha}(\sigma_p(1) - 0.1211(1-\alpha)), \tag{13}$$

where $\sigma_p(1) = 1.016$ is the slip coefficient for $\alpha = 1$ calculated first by Cercignani in Ref. [15] using the BGK kinetic model with assumption of full accommodation of the molecule at the wall. The values of the experimental slip coefficients (see Tables 5 and 6) are different from those obtained theoretically in [15] ($\sigma_p = 1.016$), therefore the accommodation is not complete in our experimental conditions.

Analyzing the values of TMAC obtained for both microtubes $T1$ and $T2$ in the range of Knudsen number $[0, 0.1]$ we notice that the highest and similar values for two gases, Helium and Carbon-dioxide, and the lowest values for the two other Nitrogen and Argon are obtained. The tendency of the TMAC to increase with the decrease of the molar mass reported in Ref. [21] can be observed in our results if we do not consider the Carbon-dioxide gas. We have to remind that the length of the microtubes $T1$ and $T2$ was not the same when experiments were carried out for the Carbon-dioxide gas. Its length was reduced by factor 20 and therefore the experimental conditions for this gas were different. The two degree of freedom of the Carbon-dioxide molecule can have also effect on the gas-surface interaction. In general, the values of TMAC obtained for the microtube $T1$ (with sulfinert surface) are significantly higher than those obtained for the microtube $T2$ (with SS surface), see Tables 5 and 6, which means that the Sulfinert surface tends to be more diffusive than the stainless steel surface.

In the $[0, 0.3]$ Knudsen number range, the experimental data are fitted with second order polynomial form. The results show that the TMAC increases comparing to the first order analysis for both microtubes, slightly less for the microtube $T2$ than for the microtube $T1$ (see Fig. 2). The dependence of the TMAC on the molecular mass of the gas for the microtube $T1$

is less observed in this range. The values obtained for the three gases Helium, Nitrogen and Argon (see Table 5) lie in the range [0.985, 1]. In this case the diffuse reflection assumption can be justified for the three gases when the Knudsen number increases. For the microtube $T2$ the TMAC obtained is in the range [0.912, 0.946]. Considering the error made on the TMAC plus some other eventual errors (on the mass flow rate) we can conclude that the value of TMAC tends to a constant value and does not depend any more on the molecular mass, when the Knudsen number increase. Nevertheless, the dependence of the TMAC on the material surface is also observed in this range. The Sulfinert surface can be considered as diffuse surface for the Helium, Nitrogen and Argon molecules.

In Table 7 a comparison of the accommodation coefficients calculated by other authors [22], [21], [23] with our results is given. The authors of these papers did experimental investigation on TMAC in microtube with silica surface. The following differences between our experimental conditions and those of Refs. [22], [21], [23] are:

- The ratio L/D is more than 4 times larger for our microtubes.
- The diameter of the microtubes are different: Ref.[22] $D = 100\mu m$, Ref. [21] $D = 25\ \mu m$ and Ref. [23] $D = 50\ \mu m$.
- The material of the microtube's surfaces are different: glass in Ref.[22] and silica in Refs. [21], [23].

The treatment of the experimental data was also different in Refs. [22], [21] and [23]:

- Only [0, 0.1] Knudsen number range was considered in Ref. [22] and the accommodation coefficient was calculated without taking into account the influence of the Knudsen layer.
- The Variables Hard Sphere model was used in Ref. [23] in the definition of the molecular mean free path.

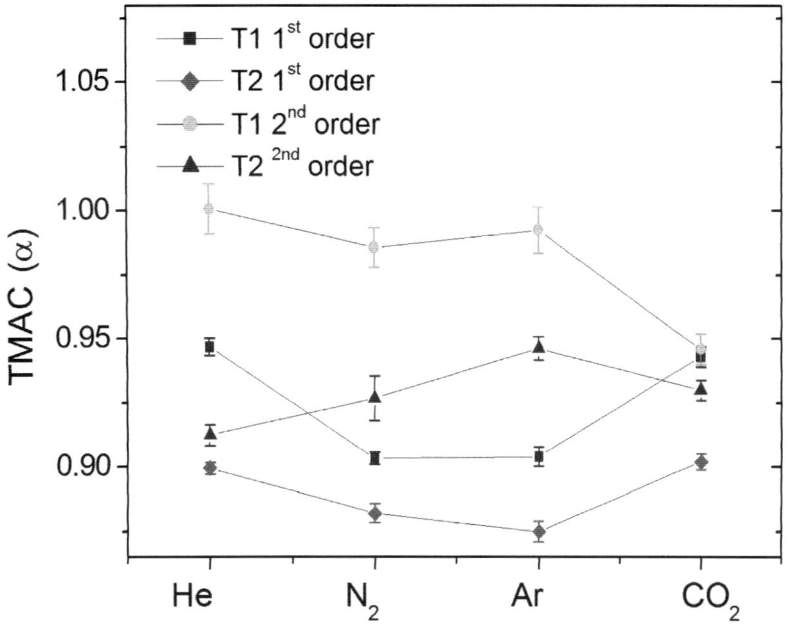

Figure 2. Comparing of TMAC for the microchannels T1 and T2

	Helium	Nitrogen	Argon
Porodnov and al [22]	0.895± 0.004	0.925± 0.014	0.927± 0.028
Ewart and al [21]	0.986± 0.009	0.981±0.041	0.942±0.017
Perrier and al [23]	1.00± 0.019	0.961±0.005	0.954± 0.010
present results ($T1$)	1.00±0.010	0.986±0.008	0.992±0.009
present results ($T2$)	0.912±0.004	0.927±0.009	0.946±0.005

Table 7. The tangential momentum accommodation coefficient (TMAC) obtained from the present experiments (in the $[0, 0.3]$ Knudsen number range using the second order fitting) and in Refs. [22], [21] and [23].

In Table 7 the accommodation coefficient obtained from the second order polynomial fit in the Knudsen number range $[0, 0.3]$ is presented. A good agreement of the TMAC values found for the microtube $T2$ with the results of Ref. [22] may be observed from Table 7. It is well shown that the TMAC depends on the material of the surface. The roughness of the surface may play important role in determining the TMAC. But unfortunately the value of the roughness for the tubes used in the experiments is not known, therefore no conclusion on its role can be made. The decrease of the TMAC with the increase of the molar mass of the gas is observed in the experiments of Refs. [21] and [23], while in our results and the results of Ref. [22] the opposite behavior is observed. Nevertheless, considering the error on the TMAC values (Table 7) one can see that the influence of the molecular mass is not very significant.

5. Conclusion

The experimental investigations on the flow through circular microchannels are presented. The constant volume technique was used to measure the mass flow rates of the gases (Helium, Nitrogen, Argon and Carbon-dioxide).

The continuum approach (Navier-Stokes equations) with first and second order velocity slip boundary conditions was used in the slip flow regime to obtain the experimental velocity slip and tangential momentum accommodation coefficients associated to the Maxwell diffuse-specular boundary condition.

The effect of the materials surface on the interaction gas-solid surface was highlighted by comparing the values of the accommodation coefficients of two surface materials (Sulfinert and stainless steel). It appears that Sulfinert gives more diffuse reflection and tends to make insignificant the influence of the nature of gas interacting with the wall.

The results of the slip and accommodation coefficients were compared to the results obtained by other authors.

Acknowledgments

The research leading to these results has received funding from the European Community's Seventh Framework Program (ITN - FP7/2007-2013) under grant agreement n°215504.

Bibliography

[1] Agrawal A and Prabhu S V 2008 *Journal of Vac. Sci. Technol.* **A26** 634–645
[2] Colin S, Lalonde P and Caen R 2004 *Heat Transf Eng* **25** 23–30
[3] Harley J, Huang Y, Bau H and Zemel J 1995 *Fluid Mechanic* **284** 257–274
[4] Lalonde P 2001 *Etude expérimentale d'écoulements gazeux dans les microsystèmes à fluides* Ph.D. thesis INSA, Toulouse, France
[5] Maurer J, Tabeling P, Joseph P and Willaime H 2003 *Phys. Fluids* **15** 2613–2621
[6] Zohar Y, Lee S Y K, Lee W Y, Jiang L and Tong P 2011 *Fluid Mechanics* **472**
[7] Ewart T, Perrier P, Graur I A and Méolans J G 2006 *Experiments in Fluids* **41** 487–498

[8] Arkilic E B, Breuer K S and Schmidt M A 2001 *Journal of fluid mechanics* **437** 29–43
[9] Ewart T, Perrier P, Graur I A and Méolans J G 2007 *Journal of fluid mechanics* **584** 337–356
[10] Pitakarnnop J, Varoutis S, Valougeorgis D, Geoffroy S, Baldas L and Colin S 2010 *Microfluid and Nanofluid* **8** 57–72
[11] Hadj-Nacer M, Graur I and Perrier P 2011 *La Houille Blanche* 49–54
[12] Jousten K, Messer G and Wandrey D 1993 *Vacuum* **44** 135–141
[13] Peksa L, Gronych T, Vicar M, Jerab M, Repa P, Tesar J, Prazak D and Krajicek Z 2011 *Measurement* **44** 1143–1152
[14] Wangkui L, Dixin Z, Qiang L, Shiliang L, Detian L and Yang M 1996 *Vacuum* **47** 519–522
[15] Cercignani C and Daneri A 1963 *Phys. Fliuds* **6** 993–996
[16] Karniadakis G and Beskok A 2002 *Microflows: Fundamentals and Simulation* (Springer)
[17] Graur I A, Méolans J G and Zeitoun D E 2006 *Microfluidics and Nanofluidics* **2** 64–77
[18] Chapman S and Cowling T G 1952 *The Mathematical Theory of Non-Uniform Gases* (Cambridge: University Press)
[19] Loyalka S K, Petrellis N and Stvorick S T 1975 *Phys. Fluids* **18** 1094
[20] Sharipov F and Seleznev V 1998 *J. Phys. Chem. Ref. Data* **27** 657–706
[21] Ewart T, Perrier P, Graur I A and Méolans J G 2007 *Microfluidics and Nanofluidics* **3** 689–695
[22] Porodnov B T, Suetin P E, Borisov S F and Akinshin V D 1974 *J. Fluid Mech.* **64** 417–437
[23] Perrier P, Graur I A, Ewart T and Méolans J G 2011 *Physics of Fluids* **23**

Velocity field measurements in gas phase internal flows by molecular tagging velocimetry

F Samouda[1], J J Brandner[2], C Barrot[3] and S Colin[4]

[1,3,4] Université de Toulouse; INSA, UPS, Mines Albi, ISAE; ICA (Institut Clément Ader) ; 135 avenue de Rangueil, F-31077 Toulouse, France
[2] Karlsruhe Institute of Technology, Institute for Micro Process Engineering, Campus North, Hermann-von-Helmholtz-Platz 1, D-76344 Eggenstein-Leopoldshafen, Germany

[1] fsamouda@insa-toulouse.fr; [2] juergen.brandner@kit.edu; [3] christine.barrot@insa-toulouse.fr ; [4] stephane.colin@insa-toulouse.fr

Abstract. The goal of the present work is to implement Molecular Tagging Velocimetry (MTV) for the analysis of internal gas flows in mini-channels. A MTV experimental setup has been designed. Tagging and detecting steps are respectively insured by a UV laser and a CCD camera coupled to intensified relay optics. A specific channel with 1 × 5 mm^2 rectangular cross section has been designed and equipped with integrated temperature sensors along its 20 cm length. It has been manufactured in PEEK (PolyEtherEther-Ketone) and Suprasil® optical windows have been integrated for the tagging access. Image processing allows extraction of velocity profiles for a pressure driven steady flow of argon through this channel. These profiles are compared to the theoretical profiles of laminar flows and the accuracy of the method is discussed. The MTV potential for the analysis of internal gaseous flows is commented on, with a discussion on perspectives for velocity measurement in rarefied flows and direct access to slip velocity at the walls.

1. Introduction

Fluidic microsystems working with gases experienced an important development over the last years. An interesting effort of research groups concerned the development of models for the slip flow regime, frequently encountered in gaseous microflows. These models based on different boundary conditions need to be compared to experimental data in order to discuss their accuracy and limits of applicability. The experimental data on gas flows available in the literature, however, mainly concern mass flowrate measurements [1-4]. There is a crucial need for local data on pressure, velocity and concentration.

Particle Image Velocimetry (PIV) is widely used at macro-scales for flow characterization. At micro-scales (µPIV), three main aspects have to be taken into account [5]. First, as the spatial resolution is in the order of micrometers, particles of 100-300 nm are therefore needed. Second, due to their very small size, particles are submitted to the influence of the Brownian motion, which makes difficult the detection of a particle pattern by cross- or auto-correlation in an interrogation window. Third, in µPIV the whole flow volume is illuminated and a microscope lens (with high numerical aperture and narrow depth of field) is necessary to analyse a thin slice of the domain in which the particles image is sharp. This volume illumination increases the background noise due to out-off focus particles, especially if the particle density is high, which is generally the case in µPIV.

The use of Molecular Tagging Velocimetry (MTV) for local gas microflows analysis is a relatively recent step. MTV is an interesting alternative to Particle Image Velocimetry (PIV) to get local velocity information. However, MTV also presents some limitations at micro scales which will be highlighted in this paper.

MTV is a non-intrusive technique for measuring fluid velocities. It is based on the properties of gaseous molecules, such as acetone or biacetyl, used as tracers. These molecules can experience relatively long lifetime luminescence once excited by the appropriate laser wavelength. This luminescence exhibits two steps. First, fluorescence appears very rapidly after the laser excitation: its lifetime duration is in the order of a few nanoseconds. It is followed by phosphorescence, with lifetime duration of the order of hundred microseconds. A pulsed laser is used to tag these molecules (along a grid pattern or a single line in the simplest configuration) and a luminescent signal is then emitted. In the example illustrated in figure 1, the single line is detected at two successive times: a straight line is acquired a few nanoseconds after the laser beam ($t = t_0$) and a distorted one after a delay ($t = t_0 + \Delta t$). The velocity profile is determined from the Lagrangian displacement between the two images during the time Δt, assuming a flow parallel to the channel axis.

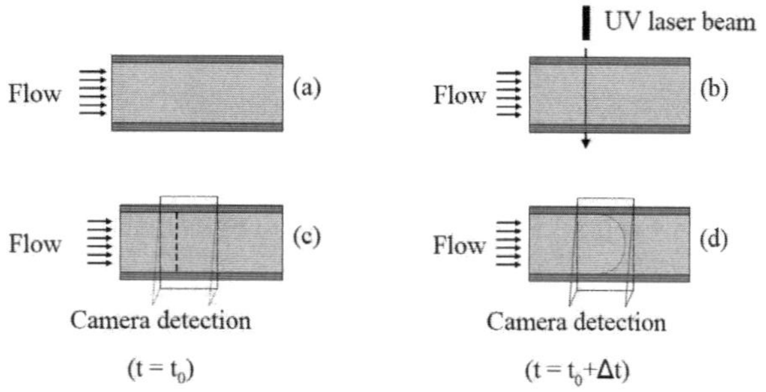

Figure 1: MTV principle: (a) luminescent molecules have been added to the gas flow, (b) molecules are excited with the UV laser beam, (c) fluorescence or early phosphorescence signal is detected (initial tagged line), (d) phosphorescence signal is detected after a Δt delay (distorted line).

MTV does not require any seeding particle and its resolution is defined by the laser beam diameter (the lowest value cited in the literature is in the order of a few tens of μm [6]). This technique is efficient for liquid mini [7-9]- and micro-flows [10-12]. For gas flows, MTV has been successfully used at millimetric scale for external gas flow configurations [13-17]. It is considered as an excellent candidate for the analysis of rarefied gas flows in internal configurations.

MTV by direct phosphorescence is implemented in this paper. This specific MTV technique was already applied in external gas flows at millimetric scales [6]. The goal of the present work is to demonstrate the potentiality of MTV for the analysis of internal gas flows in mini-channels. It is a first step toward the visualisation of velocity profiles in rarefied flow conditions, and the direct measurement of velocity slip at the channel walls.

The experimental setup as well as the tracer properties are presented in Section 2, followed by a description of the flow cell. Then, details about data acquisition are provided. Finally, first experimental data are presented and compared to theoretical velocity profiles.

2. Experimental setup

2.1. Overall setup

A global view of the experimental setup is presented in figure 2. This setup consists of a MTV acquisition system (figure 3) and a flow/ pressure control system (figure 4).

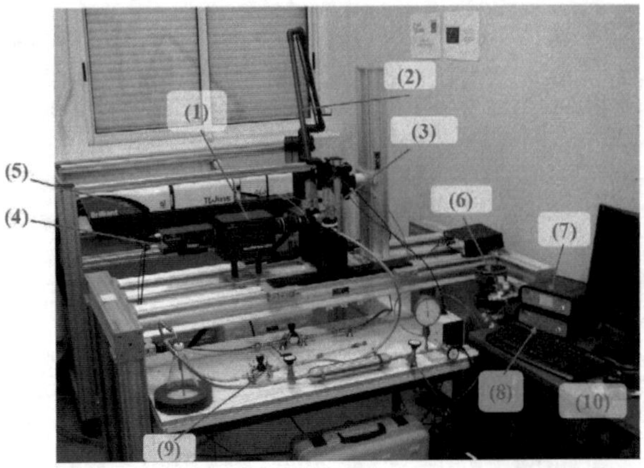

Figure 2: Global view of the experimental setup: (1) Intensified Relay Optics (IRO), (2) Laser guiding arm, (3) Channel, (4) Progressive scan camera, (5) Quadrupled Nd: YAG laser, (6) Argon bottle, (7) IRO controller unit, (8) Programmable Timing Unit (PTU), (9) Gas circuit and (10) Computer.

The MTV setup includes tagging and detecting elements (figure 2), as well as a data processing unit. Tagging is made by a Quantel Twins Brillant Nd: YAG laser (5) with a wavelength of 266 nm. The laser generates 4 ns pulses, and provides a beam, 6 mm in diameter, which is focused with an optical system in order to generate a microbeam of 35 μm diameter, which will tag acetone molecules. The available energy of the laser is reduced from 30 mJ to 1 mJ, in order to avoid damaging the channel walls due to the high energy density of the focused beam. A La Vision laser guiding arm (2) is used to safely guide the laser beam perpendicularly to the channel (3). Detection is insured by an acquisition system (figure 3) which consists in a La Vision 12-bit progressive scan camera with CCD detector (4) coupled to a 25 mm intensified relay optics -IRO- (1) . The CCD has 1376 × 1040 pixels. It is operating at a frequency of 10 frames per second. The IRO is equipped with a S20 type photocathode with a P46 phosphor. An IRO controller unit (7) allows the control of the IRO timing setup, in terms of delay, gate and gain. The detected signal is sent to a computer (10) equipped with an acquisition and processing software. The whole system is externally synchronized by a programmable timing unit -PTU- (8). A protective tent is added around the channel during the acquisition to prevent possible signal perturbation by external light sources; it allows a significant improvement of the image quality by increasing the signal to noise ratio.

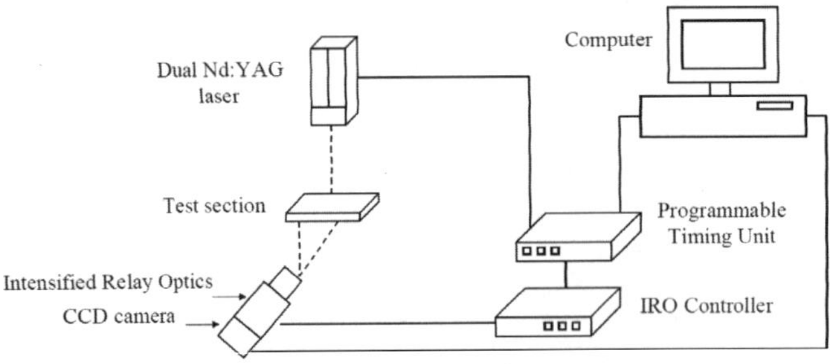

Figure 3: MTV acquisition system.

A gas circuit (9) supplies the channel with argon (6). It is presented in figure 4. A small intermediate reservoir is used to create a buffer volume for smoothing possible pressure fluctuations. Acetone molecules are added to the flow by bubbling argon gas through a liquid acetone bath. Argon seeded with acetone molecules enters the channel, equipped with capacitive pressure gauges (CGA and CGB) at its inlet and outlet. The downstream pressure can be reduced by the vacuum pump 2 connected to the channel outlet and controlled by regulating valves.

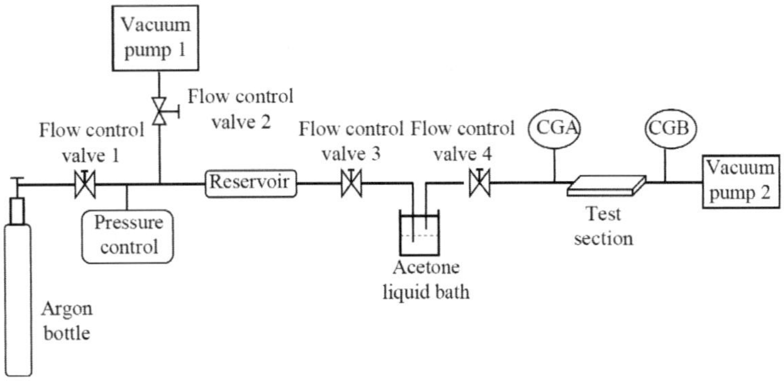

Figure 4: Gas circuit and pressure control.

2.2. Acetone properties

Also called 2-propanone, acetone (CH_3-CO-CH_3), which molar weight is 58 g/mol, absorbs UV light between 200 and 320 nm (figure 5a), with an absorption peak at 275 nm [18]. With a wavelength of 266 nm, a quadrupled Nd-YAG laser is appropriate for acetone molecules excitation.

Figure 5(b) represents the normalized intensity of acetone luminescence once excited with a 266 nm wavelength. Both fluorescent and phosphorescent signals are in the blue spectrum, between 300 and 550 nm with a peak around 400 nm [18]. Phosphorescence, however, exhibits a much lower intensity; about 10 times lower than for fluorescence. As fluorescence and phosphorescence are observed in the same wavelength range, both signals can not be separated by simple spectral filtering. They appear, however, at different times after excitation. In addition, fluorescence and phosphorescence lifetime durations are typically 4 ns and 200 μs [19], respectively. These values are sensitive to pressure and temperature conditions. Additional properties of acetone can be found in [18-20].

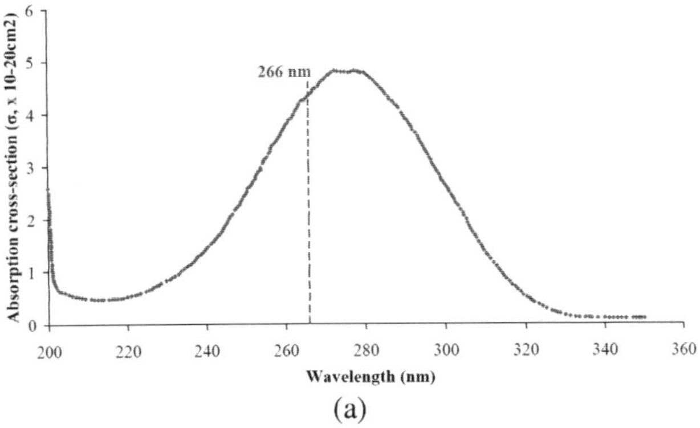

(a)

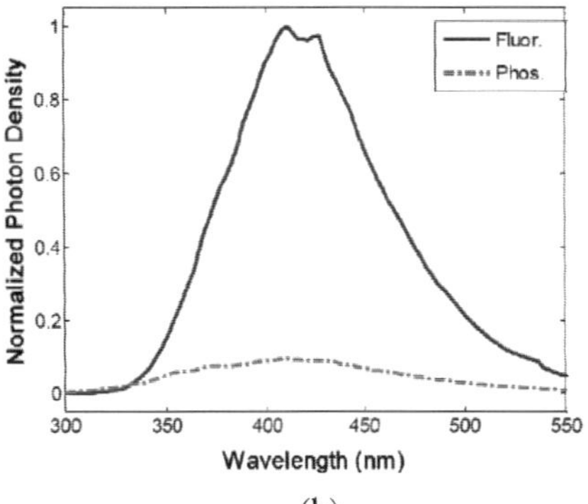

(b)

Figure 5 (a): Acetone absorption spectrum; **(b):** Acetone emission spectrum: fluorescence (solid blue line) and phosphorescence (dashed red line), after [18].

Phosphorescence is strongly quenched by oxygen. Consequently, MTV by direct phosphorescence is hardly implementable for the analysis of gases which contain oxygen [18].

3. Flow cell

The flow cell (figure 6) has a multi-layer channel design: the main part of the device is in Polyether Ether Ketone (PEEK), mechanically machined by milling and polishing steps. The channel is 200 mm in length and has a cross section of 1 mm × 5 mm.

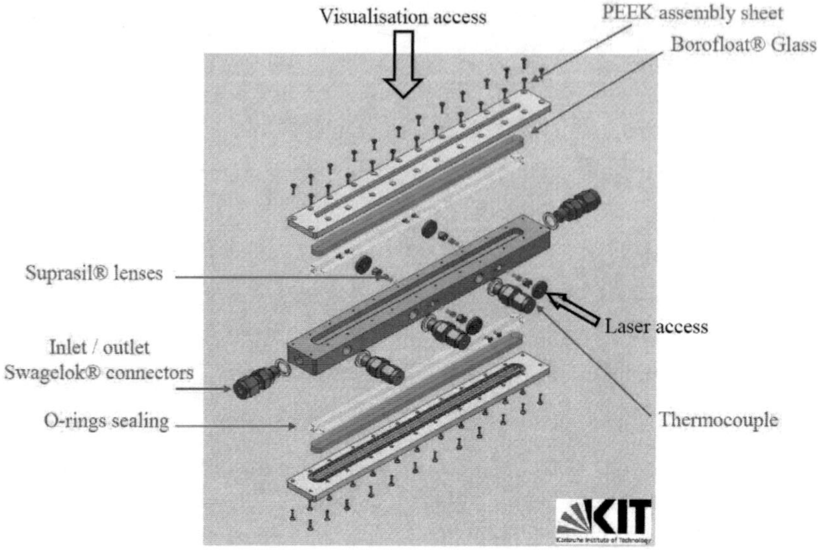

Figure 6: Channel flow.

Two Suprasil® lenses are integrated on the PEEK block for the laser beam access. Suprasil® is a fused silica grade material transparent to UV. It has been chosen among other UV transparent materials because of its low luminescence intensity. A comparative study of this intensity (I) over time is presented in figure 7 for samples of fused quartz (TSC3®), Sapphire and fused silica (Suprasil® and Dynasil® grades). It highlights the high reemission level of TSC3® and Sapphire compared to fused silica. This strong luminescence of the walls was able to mask the phosphorescence of the acetone molecules.

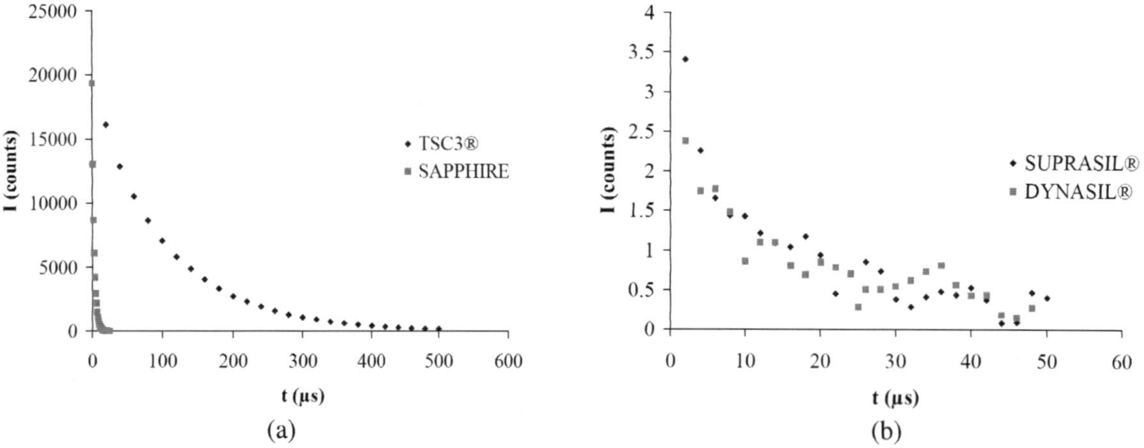

Figure 7: Reemission decay curves of different materials after excitation by 266 nm wavelength light. **(a):** TSC3® (blue diamond) and Sapphire (purple square); **(b):** Suprasil® (blue diamond) and Dynasil® (purple square).

Suprasil® lenses (A) are located at the inlet and at mid-length of the channel (figure 8). Two supplementary lenses (B) are also integrated in front of the first ones to allow transmission of the laser beam, and limit absorption and reflection at the wall. Two Borofloat® plates permit visualisation perpendicularly to the laser beam. Borofloat® is a grade of glass provided by Zell Quartz Glass industry. These plates are sealed with acetone resistant o-rings. Type K thermocouples are added along the channel wall to provide thermal information. Suprasil® lenses and thermocouples are flush mounted in order not to disturb the flow (figure 8).

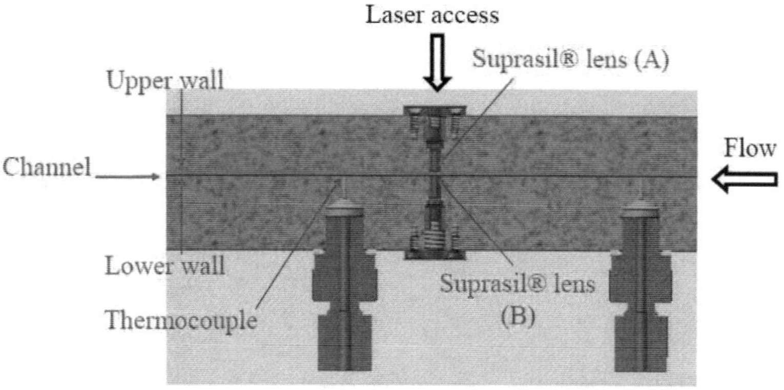

Figure 8: Thermocouples and Suprasil® lenses alignment.

The set of the flow cell layers are clamped between two PEEK sheets (figure 6). Swagelok® connectors are used for the flow inlet and outlet accesses. They are parallel to the streamwise direction and connected to the whole setup via two small chambers equipped with pressure capacitive gauges (figure 9).

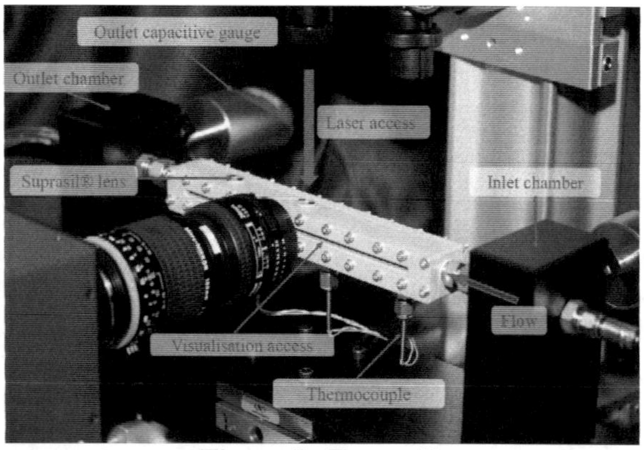

Figure 9: Flow cell.

4. Data acquisition and analysis

4.1. Image acquisition and processing

Acquisition parameters (IRO delay, gain, gate, camera exposure time…) are controlled with Davis® acquisition software and checked with a 2-channel oscilloscope. Two acquisition methods, the “standard averaging” and “on chip integration” techniques, are investigated herein.

4.1.1. Standard averaging technique: In this technique, a set of instantaneous images -one image per camera exposure time and per laser pulse- is captured and averaged. A simplified timing scheme is presented in figure 10 to illustrate this technique.

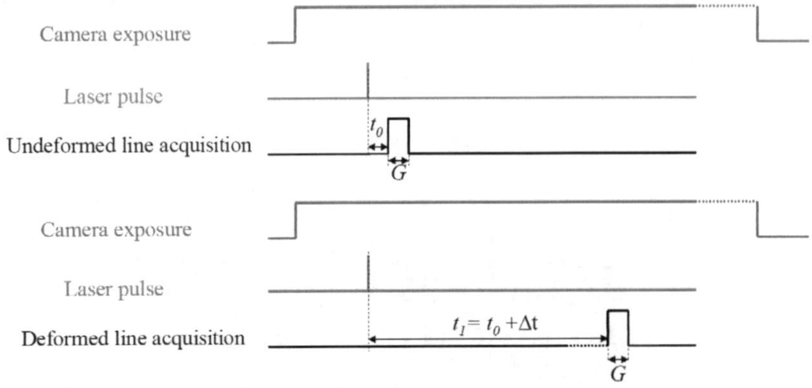

Figure 10: Timing scheme of a single image acquisition for the standard averaging technique.

In this figure, t_0 and t_1 represent the time of the initial and deformed tagged lines acquisitions, respectively. Images at t_0 and t_1 are taken separately, i.e. in two separate camera exposure times. For the images presented in figure 11, the camera exposure time is set to 2000 μs. The signal acquisition duration on the CCD camera is controlled by the intensifier activation time (IRO gate) *G* set to 1 μs and $t_1 = 100$ μs. In order to obtain the image in figure 11 (b), a set of 10000 instantaneous images is captured. These images are then post-processed with a statistical average followed by a Gaussian smoothing operation. Post processing increases the signal to noise ratio which is very poor for a single image (figure 11-a). The average of a set of images (figure 11-b) exhibits a high signal/noise ratio and therefore allows extracting velocity profiles from the deformed line acquisition. The acquisition of such a high number of images requires, however, significant time duration (about 40 min for 10000 images), during which a precise control of pressure conditions is required.

(a) (b)

Figure 11 (a): One instantaneous image; **(b):** Averaged image from a set of 10000 images.

4.1.2. On chip integration technique: In this method, the CCD exposure time is long enough to capture the emission of more than one laser pulse. This helps to increase the signal to noise ratio, as several single luminescent signals are cumulated in one CCD exposure, while being affected by read-out noise only once. The corresponding timing diagram is presented in figure 12.

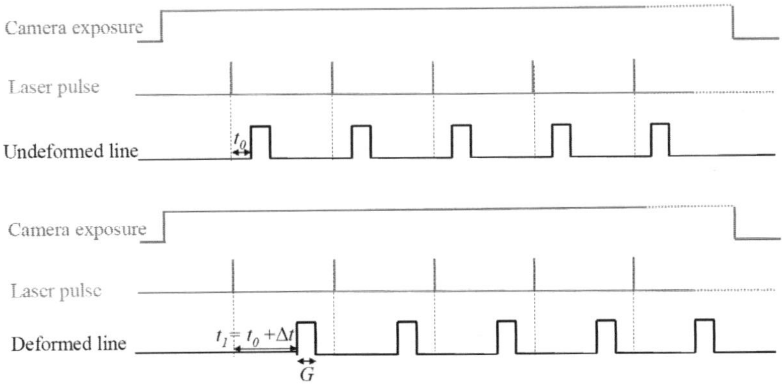

Figure 12: On chip integration timing diagram.

The signals obtained with both techniques are compared in figure 13.

(a) (b)

Figure 13 (a): On chip integration signal, averaging of 1000 images with 50 pulses integration per image; **(b):** Standard averaging technique, averaging of 10000 images with 1 laser pulse per image.

Figures 13-a and 13-b are not taken with exactly the same flow conditions neither at the same time delay, which explains the small deviations in the deformed profile. It can be observed that the on chip integration technique gives a better signal quality. In addition, the acquisition time is reduced with this technique and a constant operating pressure is then more easily controlled.

4.2. Analytical solution

First experimental velocity profiles obtained with the MTV technique are compared with the analytical solution of a laminar fully developed gas flow through rectangular channel, which can be written in a dimensionless form [21]:

$$V(x,y) = \sum_{n=1,odd}^{\infty} \sum_{m=1,odd}^{\infty} v_{n,m} \sin(n\pi x) \sin(\frac{m\pi y}{\beta}) \tag{1}$$

with

$$v_{n,m} = \frac{\pi^2}{4mn(\beta^2 n^2 + m^2)} \frac{1}{\sum\limits_{i=1,odd}^{\infty} \sum\limits_{j=1,odd}^{\infty} \frac{1}{i^2 j^2 (\beta^2 i^2 + j^2)}} = \frac{1}{A(\beta) mn(\beta^2 n^2 + m^2)} \tag{2}$$

In Eq. (2), $A(\beta)$ is a function of the aspect ratio β of the cross-section and can be accurately approximated by the polynomial $0.5059 - 0.3022\,\beta - 0.0642\,\beta^2 + 0.0747\,\beta^3$ [21]; i, j, n and m are summation integer indices; x and y are dimensionless rectangular Cartesian coordinates; $\upsilon_{n,m}$ is defined by Eq. (2).

5. First results and discussion

A typical MTV result is presented in figure 14 with a superposition of initial and deformed tagged lines. The deformed tagged purple line is taken with a time delay $\Delta t = 99.49$ µs after the initial green line. The spot observed at the bottom of the green line is due to a high energy level and a reflexion of the luminescence at the wall.

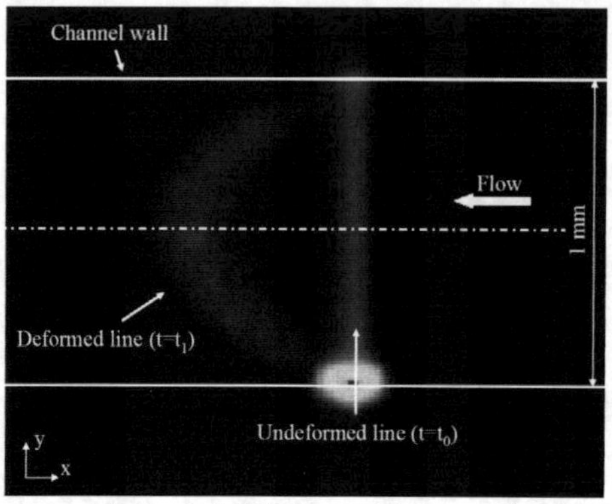

Figure 14: Superposition of initial and deformed MTV images in the median plane of the channel at equal distance from lateral walls.

The superposition of initial and deformed tagged lines is obtained with the image analysis software ImageJ®. The different colours for both lines in figure 14 underline the luminescence intensity difference with a much higher level for the initial line. This is confirmed by the intensity profiles (figure 15) measured along the dotted line (figure 14) at equal distance from the top and bottom walls.

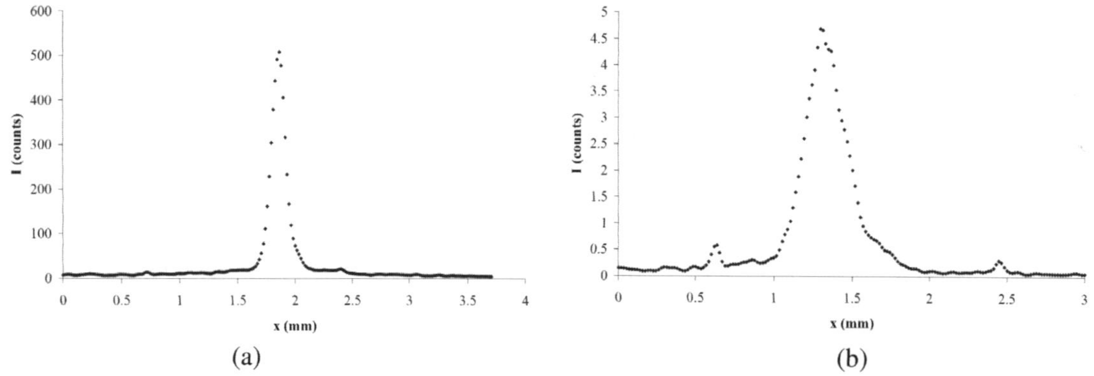

Figure 15: Intensity profiles of the signal at the channel centreline: a) initial tagged line at t_0= 0.51 µs; b) deformed tagged line at t_1 = 100 µs.

The decrease in phosphorescence intensity level with time was expected, as well as the phosphorescence lifetime duration (about 150 µs) which is comparable with data from the literature [18]. A local temperature increase can result in a phosphorescence time decrease, but this possible local effect is at the moment difficult to quantify. The temperature, which is measured at 3 different locations along the channel, is constant during the recording time (figure 16), but the presence of a hot spot is possible in the tagged region and could influence the duration of phosphorescence.

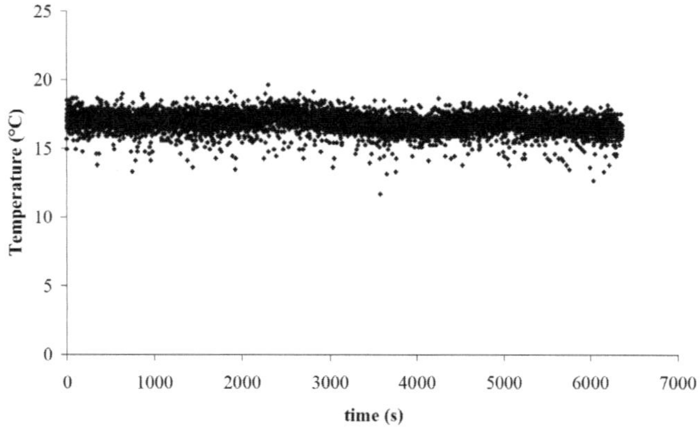

Figure 16: Temperature evolution at the channel wall.

The obtained deformed tagged line is presented in figure 17. It is successfully superposed to the theoretical displacement profile calculated from the velocity given by Eq (1) for ΔP = 190 Pa and Δt = 99.49 μs.

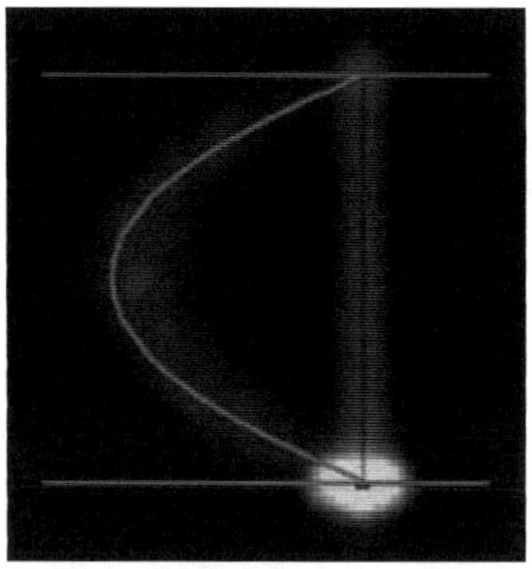

Figure 17: Superposition of deformed tagged line and theoretical displacement profile for ΔP = 190 Pa and Δt = 99.49 μs. Red edges correspond to channel walls location.

Velocity profile in the median channel plan (figure 18) is extracted from the experimental data presented in figure 17 and compared to the theoretical Poiseuille flow profile. The uncertainties on some parameters are taken into account. For the channel height, the uncertainty $\Delta h/h = \pm 3\%$. For the measured pressure, the uncertainty $\Delta P/P = \pm 0.4$ %. For the viscosity, the uncertainty $\Delta\mu/\mu = \pm 0.13$ %. Considering a plane flow between two parallel plates (in the middle of the channel, far from the walls), the total uncertainty is written as $\frac{\Delta U}{U} = 2\frac{\Delta h}{h} + \frac{\Delta P}{P} + \frac{\Delta \mu}{\mu} = \pm 6.53\%$.

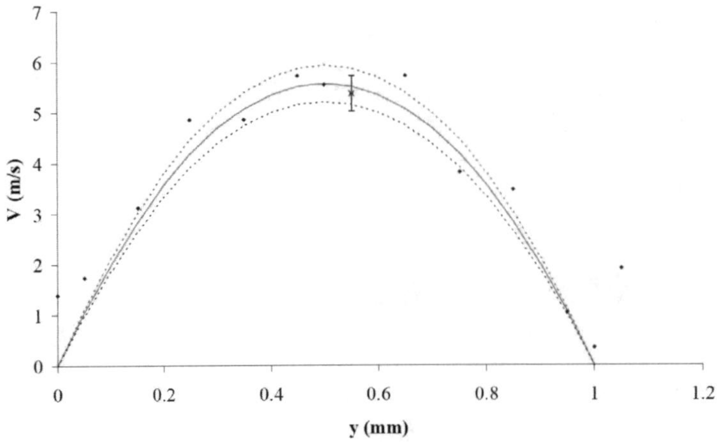

Figure 18: Velocity profile in a cross-section: experimental data (blue) and velocity profile in rectangular duct (pink, red and green) obtained for ΔP = 190 Pa.

In figure 18, the theoretical velocity profiles in a rectangular channel are plotted taking into account this total uncertainty: respectively + 6.53 % for the red and – 6.53 % for the green dashed lines. Small deviations are observed between experimental and theoretical profiles. The dispersion observed on the experimental data is mainly due to the technique used to extract these data. The displacement of the tagged molecules has been deduced from the position of the peak on a series of intensity curves like the one shown in figure 15-a, corresponding to a scan of the image along the x-direction at different y-positions, i.e. at different depths in the channel. Work is currently in progress to improve this data extraction, which maybe requires an initial smoothing of the intensity curves.

6. Conclusion and perspectives

A Molecular Tagging Velocimetry setup for gas flow analysis has been developed. An original flow cell of 1 mm x 5 mm cross-section with integrated temperature sensors has been manufactured. First promising experimental results have been obtained. The extracted velocity profiles are successfully compared to velocity profiles of a compressible Poiseuille flow in a rectangular duct. The potential of the MTV technique for the analysis of internal gas flows is therefore demonstrated. As the final goal of the work is to access velocity profiles in the slip flow regime, next tests will concern flows at lower pressures. In addition, the study of the temperature phosphorescence dependence is foreseen. This could be a first step toward Molecular Tagging Thermometry (MTT) which would be an interesting tool for the analysis of heat transfer in the slip flow regime, for which there is a crucial need in the literature [22].

Acknowledgment

This research obtained financial support from the European Community's Seventh Framework Program (FP7/2007-2013) under grant agreement no 215504 and from the ‘Fédération de Recherche Fermat’, FR 3089, Toulouse, France. Dr. Sébastien Cazin and Dr. Emmanuel Cid are acknowledged for valuable discussion.

References

[1] Pitakarnnop J, Varoutis S, Valougeorgis D, Geoffroy S, Baldas L and Colin S 2010 A novel experimental setup for gas microflows *Microfluid. Nanofluid.* **8** 57-72

[2] Szalmás L, Pitakarnnop J, Geoffroy S, Colin S and Valougeorgis D 2010 Comparative study between computational and experimental results for binary rarefied gas flows through long microchannels *Microfluid. Nanofluid.* **9** 1103-14

[3] Ewart T, Perrier P, Graur I A and Méolans J G 2007 Mass flow rate measurements in a microchannel, from hydrodynamic to near free molecular regimes *J. Fluid Mech.* **584** 337-56

[4] Maurer J, Tabeling P, Joseph P and Willaime H 2003 Second-order slip laws in microchannels for helium and nitrogen *Phys. Fluids* **15** 2613-21

[5] Wereley S T and Meinhart C D 2005 Micron Resolution Particle Image Velocimetry *Microscale Diagnostic Techniques,* ed K S Breuer (Berlin: Springer) pp 51-112

[6] Lempert W R, Boehm M, Jiang N, Gimelshein S and Levin D 2003 Comparison of molecular tagging velocimetry data and direct simulation Monte Carlo simulations in supersonic micro jet flows *Exp. Fluids* **34** 403-11

[7] Gendrich C P, Koochesfahani M M and Nocera D G 1997 Molecular tagging velocimetry and other novel applications of a new phosphorescent supramolecule *Exp. Fluids* **23** 361-72

[8] Hu H and Koochesfahani M M 2006 Molecular Tagging Technique and thermometry and its application to the wake of a heated circular cylinder *Meas. Sci. Technol.* **17** 1269-1281

[9] Hosokawa S Fukunaga T and Tomiyama A 2009 Application of photobleaching Molecular Tagging Velocimetry to turbulent bubbly flow in a square duct *Exp. Fluids* **47** 745-754

[10] Maynes D and Webb A R 2002 Velocity profile characterization in sub-millimeter diameter tubes using molecular tagging velocimetry *Exp. Fluids* **32** 3-15

[11] Thompson B R Maynes D and Webb B W 2005 Characterization of the Hydrodynamically Developing Flow in a Microtube Using MTV *J. Fluid Eng.* **127** 1003-1012

[12] Elsnab J R Maynes D Klewicki J C and Ameel T A 2010 Mean flow structure in high aspect ratio microchannel flows *J. Experimental Thermal and Fluid Science* **34** 1077-1088

[13] Miles R B, Grinstead J, Kohl R H and Diskin G 2000 The RELIEF flow tagging technique and its application in engine testing facilities and for helium-air mixing studies *Meas. Sci. Technol.* **11** 1272-81

[14] Koochesfahani M M 1999 Molecular Tagging Velocimetry (MTV): Progress and Applications *30th AIAA Fluid Dynamics Conference,* (Norfolk, VA: AIAA) pp 99-3786

[15] Stier B and Koochesfahani M M 1999 Molecular Tagging Velocimetry (MTV) measurements in gas phase flows *Exp. Fluids* **26** 297-304

[16] ElBaz A M and Pitz R W 2012 N2O molecular tagging velocimetry *Appl. Phys. B* **106** 961-969

[17] Bathel B F Danehy P M Inman J A Jones S B Ivey C B and Goyne C P 2011 Velocity Profile Measurements in Hypersonic Flows Using Sequentially Images Fluorescence Based Molecular Tagging *AIAA J.* **49** No.9

[18] Tran T T 2008 Acetone planar laser-induced fluorescence and phosphorescence for mixing studies of multiphase flows at high pressure and temperature. *PhD thesis* (Atlanta: Georgia Institute of Technology)

[19] Lozano A 1992 Laser-excited luminescent tracers for planar concentration measurements in gaseous jets. *PhD thesis* (Stanford: Stanford University)

[20] Ritchie B D 2006 Quantitative acetone PLIF measurements of jet mixing with synthetic jet actuators. *PhD thesis* (Atlanta: Georgia Institute of Technology)

[21] Morini G L 2000 Analytical determination of the temperature distribution and Nusselt numbers in rectangular ducts with constant axial heat flux *Int. J. Heat Mass Transfer* **43** 741-55

[22] Colin S 2012 Gas microflows in the slip flow regime: a critical review on convective heat transfer *J. Heat Transf.-Trans. ASME* **134** 020908:1-13

TRANSFLOW: An experimental facility for vacuum gas flows

S. Varoutis, T. Giegerich, V. Hauer, Chr. Day

Karlsruhe Institute of Technology (KIT), Institute for Technical Physics, 76021, Karlsruhe, GERMANY

Corresponding author's e-mail address: stylianos.varoutis@kit.edu

Abstract.

The TRANSFLOW experimental facility represents a reliable tool for measuring the conductance of 1:1 scale components as typically used in vacuum systems in a wide range of the Knudsen number (e.g. $10^{-4} \leq Kn \leq 10^{3}$). The main principle of this facility is the dynamic measurement of the pressure difference upstream and downstream of the duct by setting a constant mass flow rate through the test channel. Many experiments on fully developed and developing flows, based on long and short channels respectively, have been already completed and comparisons with corresponding numerical results have been successfully performed. It has been clearly proven that the TRANSFLOW experimental setup provides conductance results with overall uncertainty between 1 to 10% and it could be used as a benchmark facility for any new proposed scientific numerical method in rarefied gas dynamics and in the whole range of gas rarefaction.

1. Introduction

Gas flows under vacuum conditions consist one of the main scientific fields in rarefied gas dynamics with many engineering applications as for instance the vacuum systems of fusion reactors. The main characteristic of such flows is that in rough, medium, and high vacuum conditions, the whole range of the Knudsen number from the free molecular through the transition and slip regimes up to the hydrodynamic limit may be covered. Typical components of vacuum systems are ducts, valves and piping fittings with various lengths and cross sections, where the estimation of overall macroscopic quantities as for instance the mass flow rate, conductance and pressure are of vital importance.

In the above framework, a test facility called TRANSFLOW (test facility for Transitional Flow range experiments) was set up at the Karlsruhe Institute of Technology in 2006. The whole facility is designed for fundamental laboratory research, but sufficiently large to investigate 1:1 scale big vacuum components. It allows the measurement of the conductance for different geometries including ducts, bends and bellows with a maximum length of 1500 mm and a maximum diameter of 600 mm over a wide range of pressure and flow conditions. To the best of our knowledge, TRANSFLOW is the only facility of that kind with sufficiently large size and capable of investigating vacuum components in real scale. This allows generating experimental results, where the wall surface is of negligible influence and comparisons with results taken in micro-scale experiments, which potentially suffer from a stronger influence of the wall surface on the gas flow, may be performed.

The scope of the present work is to study on experimental basis the measured quantities of the mass flow rate and the conductance for long and short ducts of various cross sections in the whole range of the Knudsen number. In addition, the TRANSFLOW experimental facility, its limitations and the associated measurement uncertainty will be explained and finally some comparisons between numerical and experimental data will be performed.

2. Experimental set-up

2.1. General configuration

The basic principle of the TRANSFLOW test rig (Fig. 1) is the measurement of the conductance in whole range of Knudsen number at isothermal conditions. TRANSFLOW is based on the direct dynamic approach, where a constant flow Q is adjusted and the pressure difference ΔP is measured (Fig. 2). For completeness purposes the definition of conductance as $C=Q/\Delta P$ is stated. The upstream (dosing dome) and downstream (pump dome) vessels are connected with a test channel. The pump dome is equipped with two turbomolecular pumps, which are further connected to the forepumps, to maintain the vacuum conditions inside the system. For the presented experimental work, indicative results for the case of circular and triangular ducts are demonstrated. More specific, two short test tubes are used with lengths and diameters equal to L=21.2 mm, D=21.2 mm (L/D_h=L/D=1) and L=21.4 mm, D=5 mm (L/D_h=L/D=4.28) respectively (Fig. 3, a and b). In addition, a long equilateral triangular duct with length L=1277 mm and hydraulic diameter approximately D_h=16 mm (L/D_h=80) is used (Fig. 3, c). The inner surface of each duct was of standard clean technical quality without special treatment. Gas Nitrogen (N_2) was used at average ambient temperature T_0=296 K.

Figure 1: Experimental facility TRANSFLOW.

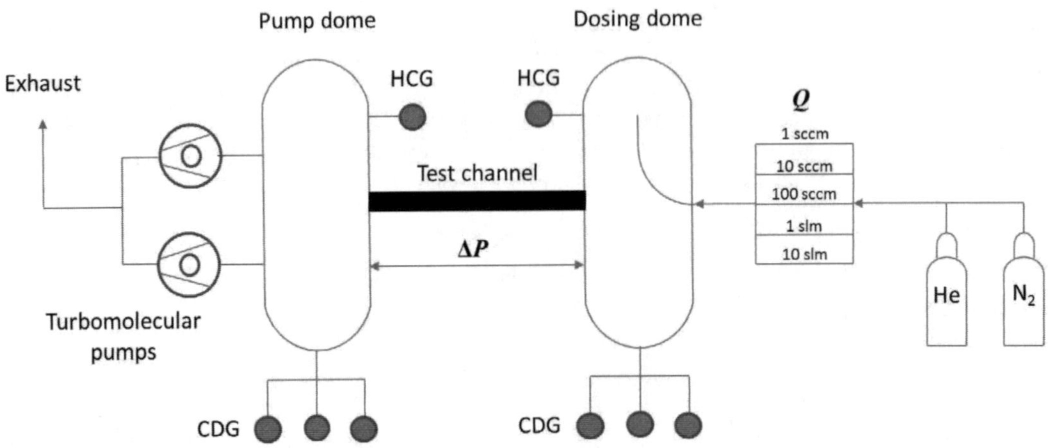

Figure 2: Simplified flow chart of TRANSFLOW experimental facility.

Figure 3: Various test channels: a) circular duct with $L=D=21.2$ mm, $L/D=1$ (left), b) circular duct with $L=21.4$ mm, $D=5$ mm, $L/D=4.28$ (middle) and c) equilateral triangular duct with $L=1277$ mm, $D_h=16$ mm, $L/D_h=80$ (right).

2.2. Dosing and pump dome

The total volume of the dosing dome including adapter flange is 0.6 m^3, while both domes consist of stainless steel. For the case of the smaller adjusted flows, which are of the order of 10^{-3} mbar.l/s, the facility allows an outgassing rate of the order of 10^{-7} mbar.l/s (e.g. this value is high for a cleaned technical stainless steel surface) and it can be assumed as negligible. All connected flanges are high vacuum CF flanges with copper sealing or Swagelok connections. The pump dome is mounted on a movable frame and it can be adjusted to the length and the geometry of the test channel under consideration.

In the pump dome, two turbomolecular pumps MAG W 2800 by Oerlikon-Leybold are attached via a VAT UHV gate valve. They are magnetically levitated and provide a maximum pumping speed of 2.8 m^3 s^{-1} for nitrogen. The pumping speed of these pumps can be adjusted by the gate valves in 1000 increments (0.1 percent steps). The pump speed can be varied by a frequency inverter between 0 and 475 Hz. The volume of the pump dome is 1.2 m^3 and is assumed quite large compared to the volume of the corresponding test channel. As roughing pump, a pumping train consisting of a Pfeiffer Roots blower, type WKP 1000, with a maximum pumping speed of 1070 m^{3}h^{-1} for nitrogen and a rotary vane pump with a maximum pumping speed of 65 m^3 h^{-1} for nitrogen is used.

Finally, it is noted that the whole facility can be heated for bake-out during the pump down phase to a temperature up to 473 Kelvin (200°C). Therefore 14 heating circuits are available, 6 on the dosing dome, 7 on the pump dome and the remaining one on the test component. The temperatures can be adjusted within 1 degree, while the temperatures of the dosing dome, pump dome and the test component can be set separately.

2.3. Dosing system

The dosing system consists of total 5 mass flow controllers (MFC, type MKS-647C) in order the injected flow to be effectively controlled. The MFCs are connected via pneumatic valves and via a pressure reducer with an outlet pressure of 1200 mbar, directly to the gas storage. The maximum flow ranges, in nitrogen equivalents, are 1 sccm, 10 sccm, 100 sccm, 1 slm and 10 slm respectively. It is noted that all MFCs are insulated to reduce the effect of changes in the ambient temperature. Finally, it is noted that calibration data is used for obtaining the final uncertainty of each device, which ranges from 0.5 to 2%.

2.4. Pressure measurement

The pressure measurement in the pump and dosing domes is performed by highly accurate pressure gauges. In each dome three capacitance diaphragm gauges (CDG) by MKS, type Baratron® 690 HA, and one hot cathode gauge (HCG) by Granville-Philips, type Stable Ion®, are installed. The operation of CDGs is independent of the used gas and they provide a maximum measurement range of 1000 Torr, 10 Torr and 1 Torr at the dosing dome or 0.1 Torr at the pump dome, respectively. On the other hand the hot cathode gauge provides a maximum measurement range of 10^{-4} mbar. For the case of CDGs the obtained uncertainty ranges between 0.003-0.1%, while for the case of HCGs the uncertainty is of the order of 5% especially in low pressure range. It is noted that for all the above pressure gauges calibration data for the estimation of the uncertainty is used as well.

2.5. Control system and data acquisition

The facility is controlled by a SIEMENS S5 programmable logic controller (PLC) including the COROS system as graphical user interface. The PLC controls the vacuum pumps, the valves and the pressure devices. For data acquisition, a Keithley multi-channel voltmeter is installed. The range of the output signal of every device is between 0 and 10 Volt for the pressure and temperature devices and between 0 and 5 Volt for the flow devices, respectively. This range is scaled to the maximum range of the physical values of every device. The sampling rate of every channel is 0.1 Hz and it can be increased to 10 Hz if necessary. The corresponding voltage values for pressure, temperature and injected flow are stored with a resolution of 16 bit to the hard disc of the measurement computer. Almost 10^2 measurements are needed for averaging, in order to receive experimental results with sufficiently good statistics.

3. Results and discussion

In this section the comparison between the available experimental and computational results of the conductance for the ducts under consideration are presented. The results cover a wide range of Knudsen number from the free molecular regime all the way up to the hydrodynamic limit. In parallel, numerical results from computational fluid dynamics (ANSYS-CFX) are compared with the present experimental data for the solution in the continuum limit.

In Fig. 4 the conductance in terms of the upstream pressure P_0 for the case of a long equilateral triangular duct (e.g. Fig. 3c) is plotted. For small values of P_0 the flow belongs to the free molecular regime, while for large values of P_0 the viscous regime is being approached. It is seen qualitatively that the two conductance limit cases - a constant value in the free molecular limit (according to [1,2]) and a direct proportionality to P_0 in the viscous limit (according to [2]) - are properly described. For intermediate pressures (P_0~1 Pa) the Knudsen minimum is observed, although it is very weak (as expected in such a long channel) and not as pronounced as with the corresponding curves for orthogonal ducts with large aspect ratios [3]. The experimental results are in very good agreement with the corresponding numerical results based on the linearized kinetic theory [1]. The obtained average relative error, which is defined as $(C^{comp}-C^{exper})/C^{exper}$, between corresponding presented experimental and numerical results, varies between 3 to 10%.

In Fig. 5, the corresponding numerical and experimental conductance, in terms of P_0, for the case of short tubes with L/D=1 (left) and L/D=4.28 (right) (e.g. Fig. 3a and 3b) respectively, is presented. For the case of short tube with L/D=1 the pressure ratio varies from P_1/P_0=1.2x10^{-2} to 8x10^{-2} with P_1 to range between 10^{-4} and 10^{-1} Pa, while for L/D=4.28 the corresponding pressure ratio varies from P_1/P_0=6.7x10^{-4} to 8.7x10^{-4} with P_1 to range between 10^{-4} and 10^{-1} Pa as well. It is noted that the numerical results for L/D=1, for simplicity purposes, are based on the assumption of P_1/P_0=0 (expansion into vacuum), while for L/D=4.28 the numerical results correspond to the exact pressure ratio given by the experiment. The influence of the assumption of expansion into vacuum (i.e. P_1/P_0=0) has been proven to result in a negligible impact on the obtained results. Comparing Fig. 5 with the corresponding Fig. 4, it is seen that qualitatively the behaviour of the conductance thoroughly differs in the hydrodynamic regime, where the conductance seems to reach asymptotically a constant value in the case of the short ducts. A good agreement is observed in specific ranges of gas rarefaction, especially in the transition and slip regime, where the relative error as defined above is within the order of 10%. Larger discrepancies between experimental and numerical results are observed in small flows in the free molecular regime, where the accuracy of the measurements is strongly affected by the offset deviation of each mass flow controller. As a result, the lower limit for the experimental facility TRANSFLOW, where reliable results can be obtained, is of the order of P_0=0.1 Pa. Also it is noted that in Fig. 5 (right), for the case of L/D=4.28, the discrepancy between experimental and DSMC results in the viscous regime [4,5] is due to the fact that DSMC results are based on insufficient number of modelled particles. Furthermore, the corresponding analytical solution for conductance in free molecular flow (dashed lines) has been calculated based on available results presented in [3]. It is seen that the numerical and experimental results tend to reach the corresponding analytical values.

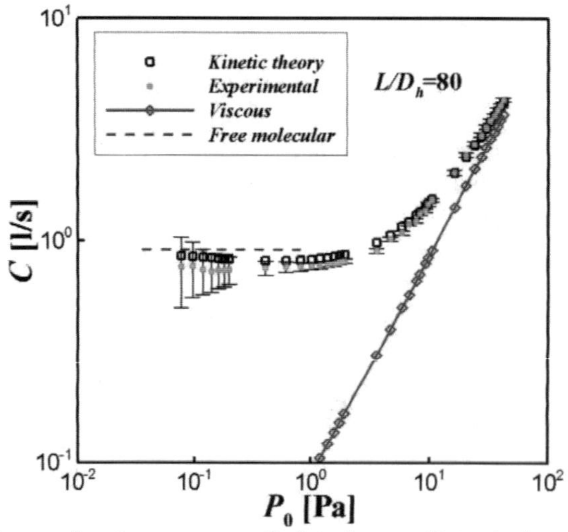

Figure 4: Conductance in terms of upstream pressure P_0 for a long equilateral triangular duct with L/D_h=80. The gas is nitrogen at ambient temperature.

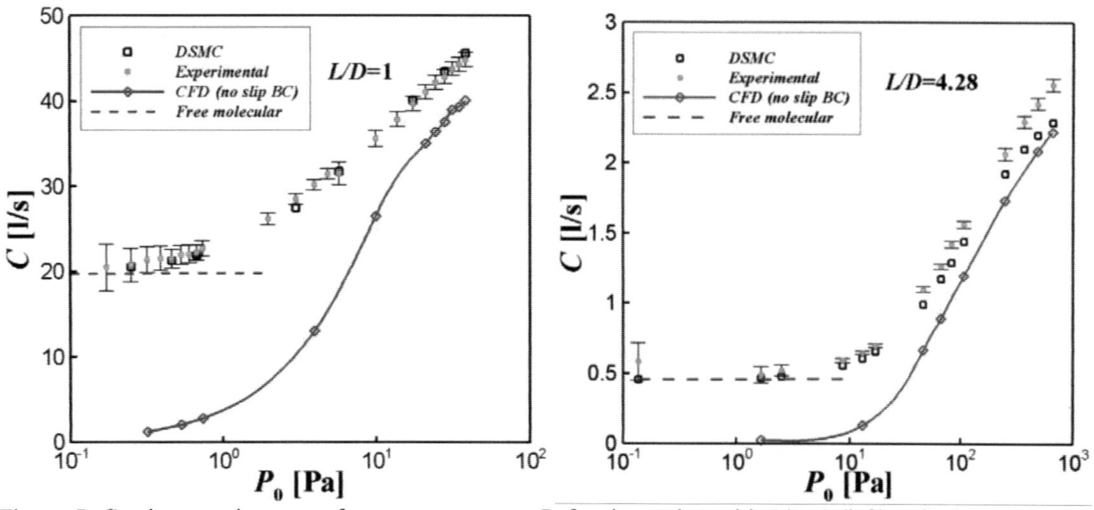

Figure 5: Conductance in terms of upstream pressure P_0 for short tubes with L/D=1 (left) and L/D=4.28 (right) respectively. The gas is nitrogen at ambient temperature.

4. Concluding remarks

From the above analysis it is clearly seen that the TRANSFLOW experimental facility is proven to be a very reliable experimental setup, where results with overall uncertainties between 1 to 10% can be obtained for test channels or industrial components. The main advantages of the facility are the flexibility in terms of the 1:1 scale of the installed components, where the influence of the wall surface can be assumed negligible, the wide range of achievable gas flows and pressures in isothermal pressure driven flows and finally the ability of measuring thermal driven flows due to the installed heating system. Such a scientific device could be used as a benchmark facility for any new proposed scientific numerical method in rarefied gas dynamics for the whole range of the Knudsen number. Currently, a research plan is under development for the extension of the applicability of the TRANSFLOW in "micro" applications and more specific in the investigation of thermal creep flows, transient flows and flows through conical micro channels.

5. Acknowledgments

This work has been supported by the European Community under the contract of Association EURATOM/KIT. The work of S. Varoutis has been carried out within the EFDA-Fusion Researcher Fellowship, while the work of V. Hauer and C. Day has been carried out within the framework of the European Fusion Development Agreement.

References

[1] Varoutis S, Naris S, Hauer V, Day C, Valougeorgis D 2009 *J. Vac. Sci. Technol. A*, **27**(1), 89.
[2] Naris S, Valougeorgis D 2008 *European Journal of Mechanics B/Fluids*, **27**, 810.
[3] Sharipov F, Seleznev V 1998 *J. Phys. Chem. Ref. Data*, **27**, 657.
[4] Varoutis S, Valougeorgis D, Sazhin O and Sharipov F 2008 *J. Vac. Sci. Technol. A*, **26**, 228.
[5] Varoutis S, Valougeorgis D and Sharipov F 2009 *J. Vac. Sci. Technol. A*, **27**(6) 1377.

Experimental analysis of heat transfer between a heated wire and a rarefied gas in an annular gap with high diameter ratio

H Chalabi[1], O Buchina[2], L Saraceno[3], M Lorenzini[1], D Valougeorgis[2], GL Morini*[1]

[1]DIENCA-Alma Mater Studiorum Università di Bologna, Bologna, IT
[2]University of Thessaly, Department of Mechanical Engineering, Volos, GR
[3]ENEA, ITFD Casaccia, Rome, IT

E-mail: gianluca.morini3@unibo.it

Abstract. In this paper a first experimental attempt is performed to measure heat conduction through rarefied air at rest contained between two concentric cylinders. The heat transfer between a heated platinum wire having a diameter (d) of 0.15 mm, disposed along the axis of a cylindrical shell in stainless steel having an inner diameter (D) of 100 mm, and a surrounded rarefied gas has been studied experimentally and numerically. The ratio between the outer and inner diameter of the annular region filled by the gas is large (D/d=667). In the annular region filled with air the pressure was varied by using a vacuum pump from atmospheric value down to 10^{-3} mbar. Temperature differences between the wire and the external stainless steel wall in the range 50-125 K were imposed and the heat power transferred from the wire to the surround was measured as a function of the gas pressure starting from air at atmospheric conditions down to 10^{-3} mbar. The experimental results obtained in these tests were compared with the numerical results obtained by using the linear and nonlinear Shakhov kinetic models.

1. Introduction

During the last decades the development of innovative miniaturised sensors has been constantly increasing due to the rapid improvement of the micro-fabrication technologies. The reduction of the dimensions of the sensors has found various interests in a large spread of technological fields. Indeed, together with the sensor reduction, this kind of new sensors make it possible to reduce their impact on the measured system, the power consumptions of the active components, their characteristic response time and, in many cases, their cost. In particular, the microsensors based on silicon with on-chip circuitry fabricated by using the integrated circuit (IC) technology have demonstrated that it becomes possible to conjugate low costs with high level of reliability and the compatibility with the modern signal processing circuitry. A variety of microsensors and microactuators fabricated by standard semiconductor technologies have been demonstrated in the last ten years: an important field of application of these miniaturized sensors is related to the measurement of the gas pressure (micro-Pirani sensors [1]), especially in the range between 1 and 10^{-5} mbar; the miniaturization of the Pirani sensors has enabled the extension of their dynamic range measurement in comparison to the classical

*[1] To whom any correspondence should be addressed.

Pirani gauges. The microPirani sensors are sometimes based on silicon sheets realizing plate geometries and sometimes based on a hot wire with a diameter of the order of hundreds of microns inserted in a system having a characteristic dimension of the order of some centimetre with a large scale difference between the wire and its surround. The heat transfer between the wire and its surround can be modelled by considering the problem of heat conduction through rarefied gases contained between two concentric cylinders. This kind of geometry has been extensively investigated in the past, for practical reasons, in order to measure thermal conductivities and thermal accommodation coefficients of gases. One of the first experimental work in this field is due to Bomelburg [2] on 1959, who measured heat conduction from 1.25, 5 and 10 μm diameters Wollaston wires placed in 10 and 25 cm inner diameter bell jars. In these experiments the author assumed thermal radiation and end losses as negligible. Dybbs and Springer [3] evidenced as the conditions of the experiments of Bomelburg had to be improved. Their experimental tests were conducted by using helium, neon and argon over a pressure range that included free molecular, transitional and temperature jump regimes. Special care was dedicated by the authors in order to design a test rig in which the effects due to end losses can be minimized and in order to eliminate the effect related to thermal radiation between the heated wire and surround. The experimental results obtained by Dybbs and Springer were in a good agreement with the numerical results obtained by using the momentum method proposed by Lees and Liu [4] over the entire Knudsen number range covered in the experiments ($0.02<Kn<50$), assuming that the thermal accommodation coefficient at the inner cylinder remained constant over this range at its free molecular value. Yu et al. [5] proposed a transient hot-wire technique in order to measure the heat flux between a fine metallic filament (platinum, nickel, tungsten, tantalum) and a rarefied gas under different heat transfer conditions ranging from a molecular to a continuum regime. Their data have been compared with the theoretical ones obtained from the solution of the BGK model by the variational method and the comparison evidenced a very good agreement between experimental and numerical values.

More recently, O'Shea and Collins [6] measured the total heat losses due to conduction heat transfer through rarefied polyatomic gas (benzene and n-hexane) in a concentric cylindrical geometry in order to study the heat transfer in temperature limited evacuated solar thermal collectors. Their experimental data have been compared with the results obtained by means of a Monte Carlo simulation in the range of Knudsen numbers between 0.1 and 100 showing a good agreement for annular gaps of the order of 2.65 mm and 2.1 mm and for an imposed temperature difference between the inner cylinder and the outer cylinder equal to 60 K.

From the analysis of the literature it is evident that a large part of the experimental results published in this area are more than 40 years old.

In addition, the more accurate experimental data appeared in literature have been obtained:

1. by imposing low temperature differences between the heated wire and the surround ($\Delta T<10$ K) [2,3,5];
2. by considering low values of the ratio between the diameter of the external cylinder (D) and the diameter of the heated wire (d) [3,5,6].

The work presented in this paper has the goal to remove these two limitations by increasing the temperature difference between the wire and the surround up to 140 K and by adopting a cylindrical geometry characterised by a large value of the D/d ratio (equal to 667).The experimental results are compared to corresponding computational ones based on linear [7] and nonlinear [8] kinetic theory in order to analyse the validity of kinetic modelling and to appreciate the deviation from the linear theory which can be evidenced experimentally when the imposed temperature difference between the wire and the external shell increases. These preliminary results will be extended and validated in the short future in order to be used as a benchmark for the test rig that will be further used in order to investigate the behaviour of this system under transient conditions.

2. The experimental test rig

In order to study the heat transfer between the wire and a surrounded rarefied gas an experimental campaign has been conducted at the Laboratory of ENEA in Rome (Italy) with the cooperation of the

DIENCA Microfluidics Lab (Bologna, Italy). A specific test rig (RAGA) has been designed and realized in order to test the heat conduction between a heated platinum wire and rarefied air within a stainless steel cylinder (Figure 1).

In Figure 1 is shown the cylindrical test section used in this experience. A stainless steel cylindrical capsule having an inner diameter (*D*) equal to 100 mm and a total length of 400 mm; it contains a platinum wire (*1*, Fig.1) fixed between two Teflon blocks (*2*, Fig. 1). The wire has a diameter (*d*) of 0.15 mm and a length (*L*) of 97 mm.

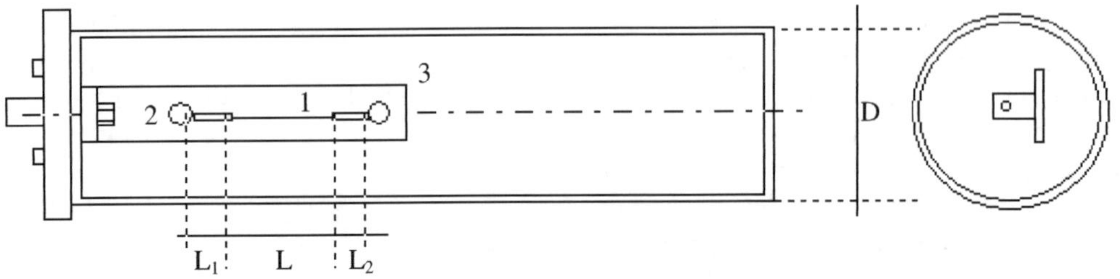

Figure 1. Sketch of the RAGA test rig: the cylindrical capsule with the centred platinum wire.

The platinum wire is fixed to a rectangular support, shown in Figure 2a, which maintains the wire on the axis of the capsule. The wire is placed in the central region of the capsule in order to minimize the effects related to the ends of the cylindrical chamber. The wire is soldered to two copper ends, having the first a diameter of 1.45 mm (d_1), a length of 10.55 mm (L_1) and the second a diameter of 1.7 mm (d_2) and a length of 9.75 mm (L_2); on these copper ends the electrodes used to heat by Joule effect the wire are attached (see Figure 2b). The Teflon blocks reduce the heat losses at the ends of the heated section. The geometry of the cylindrical test rig is characterized by a very large ratio between the outer and inner diameter of the annular region filled by the gas (*D/d*=667).

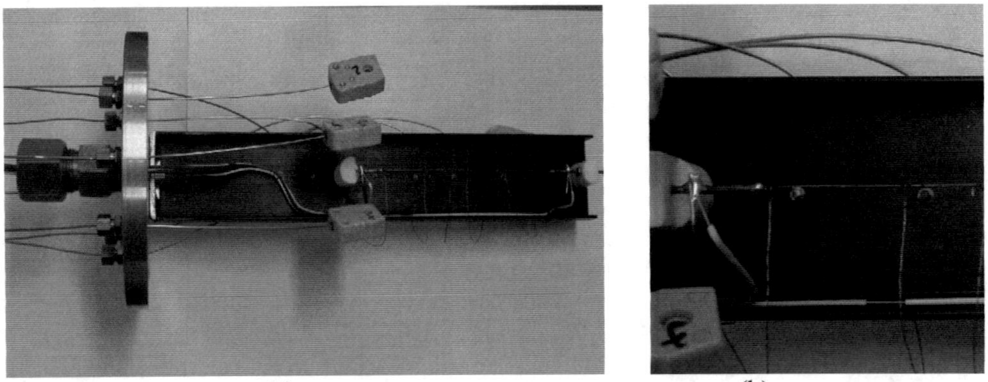

(a) (b)

Figure 2. The platinum wire fixed to the rectangular support by means of two teflon blocks (a); the copper end soldered to the platinum wire where the electrical electrode is fixed (b).

The cylindrical capsule is filled with air, initially at the atmospheric pressure. By connecting the capsule to the backing pump of a turbo-molecular pump (*Alcatel ATS-100*) it is possible to reduce the air pressure from atmospheric value down to 10^{-3} mbar. The pressure inside the capsule was monitored by means of a vacuum pressure gauge (*MKS type 925*).

The platinum wire is heated by Joule effect using a programmable DC supply (*HP6032A*) which provides a constant electrical current on the wire. The electrical resistance of the Pt wire varies with

the average wire temperature between 0.12 and 0.39 ohm; the maximum value of electrical power which has been imposed on the wire during the tests is of the order of 700 mW.
The programmable DC supply was managed by using LabView in order to impose a fixed value of the average temperature along the wire surface.
The temperature along the platinum wire was measured by means of 5 K-type thin thermocouples, having a diameter of 25 μm, glued directly on the wire surface (Figure 3a). By means of Joule heating, the wire temperature was varied up to 150°C; on the contrary the external tube is maintained at the room temperature (25-27°C). The maximum value of temperature of the wire was selected by taking into account the working range of the glue used to fix the thermocouples on the wire. The temperature of the external stainless steel tube was monitored by means of 3 K-type thermocouples (Figure 3b).
During the experience large values of the temperature difference between the wire and the external stainless steel wall of the capsule (ΔT) were imposed ranging between 50 K to 125 K.

(a) (b)

Figure 3. 5 K-type thermocouples glued on the platinum wire (a); 3 K-type thermocouples attached on the inner surface of the stainless steel external capsule.

In Table 1 the typical uncertainty of the sensors used in these tests is quoted.

Table 1. Characteristics and uncertainties of the measurement instruments.

	Range	**Uncertainty**
K-type thermocouples	0-200°C	±0.1 K
Pressure gauge	10^{-4}-10^{3} mbar	±10% R (10^{-4}-10^{3} mbar) ±5% R (10^{-3}-10^{2} mbar)
Electrical current	0-17 A	± 0.36% R + 15 mA
Voltage	0-200 V	± 0.08% R + 80 mV

Before making measurements the system was evacuated, backed out and the main filament was flashed. After that, the tests were conducted by following these steps:

1. A value of pressure is fixed inside the cylindrical shell, starting from atmospheric value.
2. By means of the DC programmable supply (*HP6032A)* a value of electrical current is imposed through the platinum wire in order to maintain the average temperature of the wire equal to the set-point (T_{sp}). A LabView code was used in order to tune the supply by using the temperature measurements on the wire.The LabView system is set in order to verify that the measured

average value of the wire temperature (calculated as average value of the 5 values of temperature measured by the K-type thermocouples ($T_{w,i}$) glued on the wire surface (Fig. 3a)), is maintained by the DC programmable supply within 2% from the imposed set-point (T_{sp}):

$$\frac{\delta T}{T_{sp}} = \frac{\sum_{i=1}^{5} T_{w,i} - 5T_{sp}}{5T_{sp}} \leq 0.02 \tag{1}$$

3. The value of the electrical power dissipated through the platinum wire for Joule heating is calculated as product between the square value of the imposed electrical current (range: 0.2-0.7 A) and the wire electrical resistance (R):

$$Q = I^2 R \tag{2}$$

In steady-state conditions the power Q is equal to the heat transferred from the wire to the surround by radiation and conduction:

$$Q = I^2 R = g(p_{gas})Q_{cond} + Q_{rad} \tag{3}$$

where $g(p_{gas})$ is a correction factor of the heat conduction term depending on the rarefaction level of the air within the vacuum chamber.

The radiative term is independent on the gas pressure within the vacuum chamber and it is evaluated by considering the formula for two concentric cylindrical surfaces [9], simplified for the limiting case where the surface of the internal body (OD of the platinum wire is 0.15 mm) is much smaller than the external, concave surface (ID of the vacuum chamber is 100 mm):

$$Q_{rad} = \sigma_0 A(T_{sp}^4 - T_{ext}^4) \tag{4}$$

where T_{sp} is the imposed wire temperature, T_{ext} is the temperature of the inner surface of the stainless steel capsule (obtained as average value of the temperature measured by the 3 K-type thermocouples of Fig.3b) and A the radiative total surface of the heated section calculated as follows:

$$A = \varepsilon_w \pi d L + \varepsilon_c (\pi d_1 L_1 + \pi d_2 L_2) + \varepsilon_c \frac{\pi}{4}(d_1^2 + d_2^2 - 2d^2) \tag{5}$$

ε_w is the emissivity of the platinum wire (ε_w =0.05) and ε_c is the emissivity of the oxidized copper ends (ε_c =0.53). The values of the emissivity of the platinum wire and of the copper ends have been checked by comparing the surface temperature of the components measured by an infrared camera (*AVIO TVS 200EX*) and by means of a thermocouple glued on the surface: the value of the emissivity has been tuned in order to obtain the same value of temperature indicated by the thermocouple. The emissivity values were checked before each experimental test. The accuracy linked to the estimation of the emissivity obtained with this method is not very high and can be estimated of the order of ±10%; however, the estimated emissivity of platinum and copper are in agreement with the literature values [10].

The heat loss due to conduction through the air is written in Eq.(3) as a product between the classical conduction term (Q_{cond}) and a correction factor $g(p_{gas})$, depending on the rarefaction level of the air, which reduces the conduction heat transfer when the gas pressure decreases:

$$g(p_{gas})Q_{cond} = g(p_{gas})\lambda_0(T_m)\frac{\Delta T}{\frac{\ln(D/d)}{2\pi L}} \tag{6}$$

where $\Delta T = T_{sp} - T_{ext}$ and λ_0 is the air thermal conductivity at atmospheric conditions and at the average temperature between the wire and the capsule ($T_m=(T_{ext}+T_{sp})/2$), calculated as a function of temperature by means of the following correlation (T_m in K):

$$\lambda_0(T_m) = -0.00039333 + 1.0184 \cdot 10^{-4} T_m - 4.8574 \cdot 10^{-8} T_m^2 + 1.5207 \cdot 10^{-11} T_m^3 \quad \left[\frac{W}{mK}\right] \tag{7}$$

Eq.(7) gives values of the air thermal conductivity in the range 0-200°C with a maximum overestimation of the values quoted in Raznjevic [11] equal to +4.2%.

Since the radiative term is independent on the gas pressure for a fixed wire temperature and capsule inner temperature, the weight of thermal radiation is not negligible especially at low gas pressure where the conduction heat transfer is strongly reduced by rarefaction.

From Eq.(3) it is evident that, for a fixed wire temperature (T_{sp}) and capsule temperature (T_{ext}), Q is a function of the air pressure within the capsule.

4. By means of the valve between the turbo-molecular pump and the stainless steel capsule the gas pressure was progressively reduced and the value of Q, evaluated by using Eq.(1), was recorded as a function of the air pressure in the chamber.
5. All the operative procedure is repeated by changing the set-point temperature of the wire.

3. The linear and non linear Shakhov kinetic model

The specific heat transfer configuration may be accurately simulated in the whole range of the Knudsen number via the Shakhov (S) kinetic model. Taking into account the axisymmetry of the problem and invariance in the axial direction (i.e. ignoring end effects) the S kinetic equation reads [7,8]

$$\xi_p \cos\theta \frac{\partial f}{\partial \hat{r}} - \frac{\xi_p \sin\theta}{\hat{r}} \frac{\partial f}{\partial \theta} = \frac{P}{\mu}(f^S - f) \tag{8}$$

where

$$f^S = f^M \left[1 + \frac{2m}{15n(k_B T)^2} q \xi_p \cos\theta \left(\frac{m\xi^2}{2k_B T} - \frac{5}{2}\right)\right] \tag{9}$$

with

$$f^M = n\left(\frac{m}{2\pi k_B T}\right)^{3/2} \exp\left(-\frac{m\xi^2}{2k_B T}\right) \tag{10}$$

being the Maxwellian.

Here, $f = f\left(\hat{r}, \xi\right)$ is the unknown distribution function, $\hat{r}$ is the radial spatial coordinate and $\xi = \left(\xi_p \cos\theta, \xi_p \sin\theta, \xi_z\right)$ is the molecular velocity vector. Also, n, T and q are the macroscopic distributions of number density, temperature and radial heat flux respectively, which may be obtained by the moments of the distribution function according to

$$n\left(\hat{r}\right) = \iiint f \xi_p d\xi_p d\theta d\xi_z \tag{11}$$

$$T\left(\hat{r}\right) = \frac{m}{3n\left(\hat{r}\right)k_B} \iiint \xi_p^2 f \xi_p d\xi_p d\theta d\xi_z \tag{12}$$

$$q(\hat{r}) = \frac{m}{2} \iiint \xi_p^3 \cos\theta f \xi_p d\xi_p d\theta d\xi_z \tag{13}$$

Furthermore, μ is the viscosity of the gas at local temperature T, while P is the local pressure of the gas given by the equation of state $P=nk_BT$, with k_B denoting the Boltzmann constant .

When the temperature difference between the wire and the external stainless steel wall, namely $\Delta T = T_{sp} - T_{ext}$ is small compared to $T_0 = T_m = \left(T_{sp} + T_{ext}\right)/2$ the unknown distribution function may be linearized according to:

$$f = f_0\left(1 + h\frac{\Delta T}{T_0}\right), \qquad \Delta T / T_0 << 1 \tag{14}$$

to yield the corresponding linearized S kinetic equation with the associated perturbed macroscopic quantities [7].

It is noted that both the linear and nonlinear heat transfer problem under consideration have been recently solved in [7] and [8] respectively deducing accurate results in the whole range of gas rarefaction for various radius ratios and temperature differences. All the details may be found in these references. Here, these kinetic algorithms are implemented to yield the linear and nonlinear heat fluxes for the specific radius ratio $D/d = 667$ and the imposed temperature differences ΔT of the experimental set-up. Based on these results a comparison between computations and measurements is performed in the next section.

4. Discussion of the results

The raw data obtained during these tests are shown in Figure 4. It is possible to distinguish 4 series of data obtained by imposing a temperature difference between the wire and the inner surface of the vacuum chamber equal to 50 K, 75 K, 100 K and 125 K.

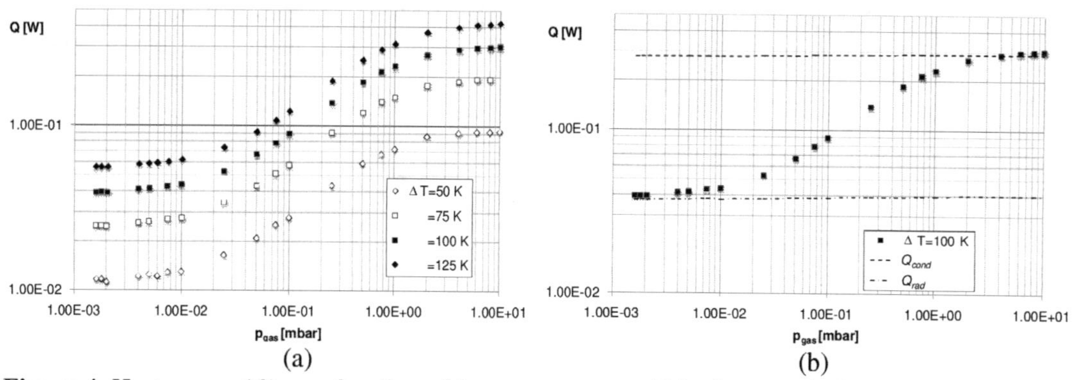

Figure 4. Heat power (Q) as a function of the gas pressure within the capsule for different values of the imposed ΔT (a); Q compared with the radiative and the conductive terms for an imposed ΔT=100 K (b).

By observing Figure 4a it is evident that Q tends to saturate both for low and large values of gas pressure. When the air pressure is larger than 1 mbar the heat power Q transferred from the heated wire to the surround becomes independent by the gas pressure: the heat is mainly transferred by conduction ($Q \cong Q_{cond}$). On the contrary, when the level of the gas rarefaction increases, the conduction becomes less efficient as heat transfer mode and it tends to become negligible; for this reason, the heat transfer for gas pressure less than 0.01 mbar is mainly due to the radiative component ($Q \cong Q_{rad}$) which is independent by the gas pressure (see Figure 4b).

Figure 4b highlights that, even in the case of a large temperature difference between the wire and the inner surface of the capsule (ΔT=100 K), the effects due to natural convection was very low within the chamber; in fact, the heat power Q is close to Q_{cond} when the gas pressure increases. In the case of Figure 4b the difference between Q and Q_{cond} is small and entirely due to the presence of the radiative heat transfer component. The same observation can be repeated for ΔT=50 K, 75 K, and 125 K. It is possible to conclude that the effects due to natural convection within the capsule in the range of temperature difference tested in this experimental campaign are very limited.

In Figure 5a the difference between the measured wire temperature and the imposed temperature set point is shown for the 4 series of tests conducted in this work. It is well evident that the system is able to guarantee a maximum difference between the measured temperature of the platinum wire and the set point less than ±2% during the whole experimental conditions. As general trend, the platinum wire was maintained to a temperature lower than the set-point when the air pressure in the chamber decreases down to 0.1 mbar; on the contrary, the wire temperature was larger than the set-point for large gas pressure.

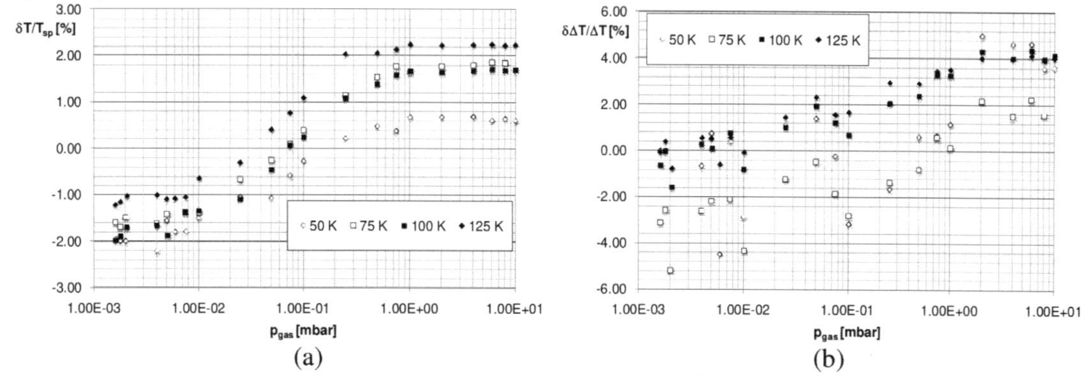

Figure 5. Stability during the experimental tests of: (a) the difference between the measured wire temperature and the imposed temperature set point (T_{sp}) (b) the temperature difference between the wire and the stainless steel shell.

In Figure 5b is shown the trend of the temperature difference between the wire and the stainless steel shell during the experimental test. It can be evidenced that the variation of ΔT is for a large part of data within 3% during the tests but, in some case a maximum variation equal to ±5% is reached. This variation is due to the choice to leave free the temperature of the external shell which is exposed to the variation of the room thermal conditions. This fact has to be considered during the discussion of the results since it determines a limitation in terms of accuracy of the results.

The numerical results obtained by the Shakhov kinetic model neglect the radiative heat transfer between the wire and its surround. In order to compare the numerical data with the experimental results it is important to subtract the radiative heat transfer component (which is independent by the gas pressure) to the value of the heat power calculated by using Eq.(1). In Figure 6 the difference Q-Q_{rad} is shown as a function of the gas pressure and compared with the predictions of the linear S model described in the previous section.

It is evident that the agreement between the numerical data and the experimental data is quite good for ΔT=50 K in the whole range of pressure investigated (see Figure 6a). When the temperature difference between the wire and surround increases (Figure 6b,c,d), the agreement between the experimental values and the numerical data tends to decrease. In particular, when the temperature difference increases the linear S method seems to underestimate the heat transfer between the wire and the gas. It is noted that the kinetic results have been obtained assuming purely diffuse reflection at the wall, i.e., the accommodation coefficient is taken equal to one (α=1), while a reduction in the value of α will further reduce the computed heat transfer.

In order to check if this difference is contributed to the implementation of linear modelling a comparison with nonlinear results is presented in Fig. 7 for ΔT=100 K. As it is seen for pressure values up to 0.2 mbar the linear and nonlinear results are identical while for the higher pressure values the nonlinear heat fluxes tend to become larger and closer to the experimental ones. This behaviour is expected since in this case the ratio $\Delta T/T_0$ is about 0.3, in which case there are no large differences between linear and nonlinear analysis. On the contrary, the experimental results put in evidence some difference with the theoretical predictions also for low values of pressure. Thus, the results shown in Figure 7 underline that the discrepancy between computations and measurements cannot be contributed solely to the differences between linear and nonlinear analysis.

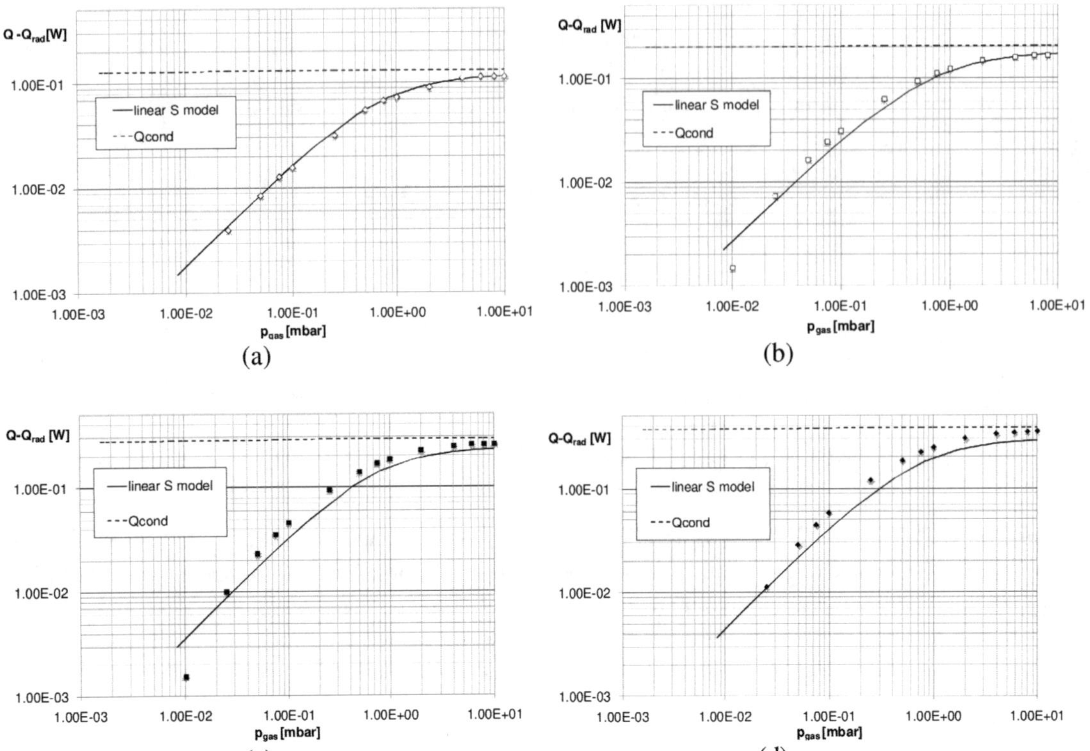

(a) (b)

(c) (d)

Figure 6. Comparison between experimental values and numerical data: (a) ΔT=50 K; (b) ΔT=75 K; (c) ΔT=100 K; (d) ΔT=125 K.

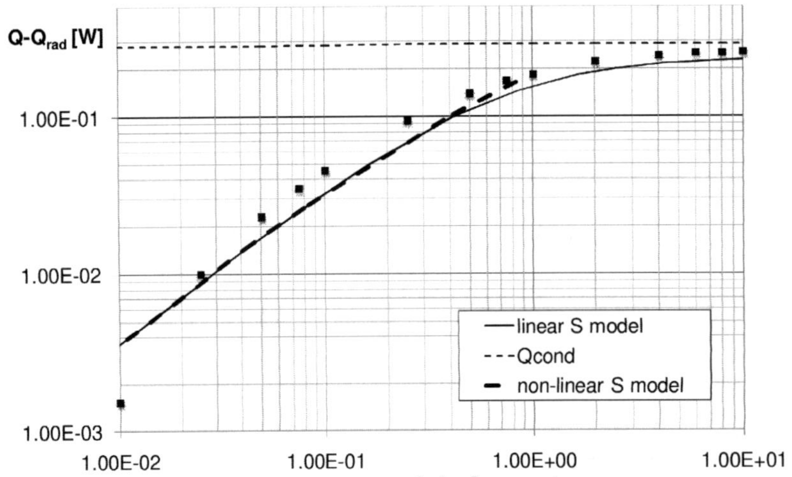

Figure 7. Comparison between experimental values and nonlinear numerical data for ΔT=100 K.

It is important to underline that in the numerical kinetic solution adopted in this work, air is considered as a pure monoatomic gas since the authors have found very good agreement with experiments with nitrogen and air in their previous works [12]. However, since in this experimental work the humidity of air was not checked and measured, the water content could be responsible of a deviation of the data

from the theoretical prediction (obtained for pure air). In addition, it is important to bear in mind that during the experience the temperature difference between the wire and the external shell can vary of ±5%, as highlighted by Figure 5b; on the contrary, the numerical results have been obtained by imposing a fixed value of ΔT. From a geometrical point of view, a possible explanation of the disagreement between the numerical predictions and the experimental data is linked to the break of symmetry introduced by the metallic holder for the wire (see Fig.2). Another sensible point of this experience is linked to the accurate determination of the emissivity of the wire and its copper supports which, as the experimental results confirm, play an important role on the accurate determination of the total heat transferred to the surround, especially when the wire temperature increases.
All these observations, put in evidence how, at the present stage, the designed test rig cannot be considered able to check in which conditions non-linear results begin to deviate from the linear ones. Further experimental tests and numerical runs have been planned in the short future in order to improve the test rig and generate new and more accurate experimental data.

5. Conclusions

In this paper an experimental and numerical analysis of the heat conduction between a heated platinum wire having a diameter (*d*) of 0.15 mm, disposed along the axis of a cylindrical shell in stainless steel having an inner diameter (*D*) of 100 mm filled with air is described. Temperature differences between the wire and the external stainless steel wall in the range 50-125 K were imposed and the heat power transferred from the wire to the surround was measured as a function of the gas pressure starting from air at atmospheric conditions down to 10^{-3} mbar. The experimental results obtained in these tests were compared with the numerical results obtained by using the linear and nonlinear Shakhov kinetic models. It has been observed a good agreement between the numerical results obtained using the linear S kinetic model and the experimental results for ΔT=50 and 75 K; on the contrary, for ΔT>75 K the experimental heat flux was larger than the numerical values. The comparison with the predictions obtained for ΔT=100 K using the non-linear Shakhov kinetic model evidenced how the non-linear model can improves the agreement with the experimental results solely for pressure larger than 0.2 mbar. On the contrary, for lower pressure the larger experimental heat fluxes cannot be justified by invoking non-linear effects. These results suggest the need of an improved accuracy in the experimental tests in order to check in which conditions non-linear results begin to deviate from the linear ones.

Acknowledgments

The research leading to these results has received funding from the European Community's Seventh Framework Programme (ITN - FP7/2007-2013) under grant agreement n° 215504.

References

[1] Górecka-Drzazga A 2009 *Vacuum* **83** 1419
[2] Bomelburg HS 1959 *Phys. Fluids* **2** 717
[3] Dybbs A, Springer GS 1965 *Phys. Fluids* **8** 1946
[4] Lees L, Liu CY 1962 *Phys. Fluids* **5** 1137
[5] Semyonov YG, Borisov SF, Suetin PE 1984 *Int. J. Heat Mass Transfer* **27** 1789
[6] O'Shea SJ, Collins RE 1992 *Int. J. Heat Mass Transfer* **35** 3431
[7] Sharipov F, Bertoldo G 2006 *J.Vac. Sci. Technol. A* **26** 2087
[8] Shakhov EM, 1968 *Fluid Dynamics* **3** 142
[9] Incropera FP, DeWitt DP, Bergman TL, Lavine AS 2007 *Introduction to Heat Transfer*, Wiley & Sons Ed.
[10] Siegel JR, Howell R, Howell. JR 2001 *Thermal radiation heat transfer* Taylor & Francis Ed.
[11] Raznjevic K, *Handbook of Thermodynamic Tables and Charts*, 1976 McGraw-Hill Ed.
[12] Pantazis S, Valougeorgis D, 2010 *Eur.J. Mech. B/Fluids* **29** 494

Implementation of SMC averaging method in a channeled molecular flow of liquids and gases

S M H Karimian[1] and S Namvar[2]
Center of Excellence in Computational Aerospace Engineering, Department of Aerospace Engineering, Amirkabir University of Technology, Tehran, Iran

E-mail: hkarim@aut.ac.ir[1], soheil_snp@aut.ac.ir[2]

Abstract. This paper proposes the application of SMC (SAM-Modified-CAM) method of averaging, formerly developed by Karimian et al, in a channelled molecular flow. While they have used periodic boundary conditions in all three directions, in this paper wall boundary condition is used in one direction and in the other two directions, periodic boundary conditions are applied. In order to verify the results obtained from SMC averaging method for the above boundary conditions, velocity profile extracted here is compared to the one obtained by Sofos et al. Results show that very good agreement is achieved. In addition, same trends reported in the work of Sofos et al for the change of velocity magnitude with the spring stiffness and also with the interaction ratio ($\frac{\varepsilon_f}{\varepsilon_w}$), have been observed in the present study as well. It is worth to mention that the velocity profile calculated by the present averaging method is much smoother than the one obtained by Sofos et al. This approves that SMC averaging method is capable of being used for cases with wall boundary conditions. To check whether the simulation has reached equilibrium, "convergence history" of maximum velocity in the channel is plotted using both SMC and CAM methods of averaging. A comparison between the two proves that SMC yields a much smoother graph which is more suitable for determination of the time step in which the equilibrium occurs. It can be concluded that SMC method of averaging is a reliable way of data sampling in molecular dynamics simulations.

1. Introduction

In MD simulation where discrete molecules are modeled rather than continuum flow, finding an appropriate way to calculate mean flow characteristics of molecules is inevitable. Two conventional ways of averaging in MD simulation are sample averaged measurement (SAM) and cumulative average measurement (CAM). Having applied these averaging to obtain molecular flow velocity, the following relations is obtained for SAM velocity of N molecules in a selected domain [2]

$$\langle \vec{V}_S \rangle_{SAM} = \frac{1}{S} \sum_{j=1}^{S} \vec{V}_{av}(t_j)$$

Where S denotes the total number of time steps, t_j indicates that the corresponding variable is a function of time at jth time step, and $\vec{V}_{av}$ is defined as

$$\vec{V}_{av}(t_j) = \frac{1}{N(t_j)} \sum_{p=1}^{N(t_j)} \vec{V}_p(t_j)$$

In which $\vec{V}_p$ is the velocity vector of a molecule in the selected domain. Moreover, the CAM velocity would be [2]

$$\langle\vec{V}_S\rangle_{CAM} = \frac{\frac{1}{S}\sum_{j=1}^{S} N(t_j)\vec{V}_{av}(t_j)}{\frac{1}{S}\sum_{j=1}^{S} N(t_j)}$$

S.M.H Karimian et al [2] have done a comparative study on SAM and CAM method of averaging for the calculation of velocity. They have implied that to obtain averaged molecular property (e.g. velocity) within a selected domain using SAM method, the molecular property is first averaged within the domain and then these average velocities are averaged over the time. This makes SAM highly sensitive to the number of molecules and their thermal velocity. As the number of molecules increase, SAM gives more realistic results. In fact, SAM is vulnerable to statistical errors for low density flows. In contrast, averaging in CAM is performed cumulatively over the molecules during past time steps which means that the number of molecules taken into account increases continuously. However obtaining flow quantities using these methods will be accompanied by large fluctuations. Thus it is difficult to distinguish whether the quantity in question is converged. To resolve this problem they proposed a more efficient method of averaging called SMC (SAM-Modified-CAM) which could be defined as follows [2]

$$\langle\vec{V}_s\rangle_{SMC} = \frac{1}{S}\sum_{q=1}^{S}\langle\vec{V}_q\rangle_{CAM}$$

The SMC method of averaging has been validated through different test cases in Ref. 2. Judging by the statistical error in all test cases, it is concluded that SMC method is superior to SAM and CAM. This is because SMC method damps out the oscillations of velocity versus time and makes the curve smoother.

In many practical cases in which molecular flow is involved, such as simulating flow in nanotubes, wall boundary conditions should be applied. For this reason, we would like to study the implementation of SMC method in a channeled molecular flow, where the flow is bounded with two layered molecular walls. Fortunately, since the results in the literature where wall boundary B.Cs are used are more than the ones in which periodic B.Cs are applied, comparison of the present results will be straightforward [3-9]. In this paper, flow of 576 molecules of liquid argon with $\rho^* = 0.642\,\sigma^{-3}$ in a channel is studied. At each region velocity is evaluated using SMC method of averaging. For the channel mid-point (in all directions), time variation of velocity is plotted to show that how SMC well affects data sampling. Although, the simulation carried out in this paper is for liquid Argon, the implementation of SMC method of averaging is also applicable to molecular flow of gases (low density flows).

2. Molecular modeling and Solution Procedure

This paper presents the simulation of 576 argon molecules which are conducted in a 36.15 $\AA$ channel. The Knudsen number of the flow is 0.04 and the pressure is 50 Mpa. The velocity direction of each molecule is determined using Maxwell-Boltzmann distribution function. The interaction between molecules is described by Lennard-Jones 12-6 potential equation:

$$U_{ij} = 4\varepsilon\left(\left(\frac{\sigma}{r_{ij}}\right)^{12} - \left(\frac{\sigma}{r_{ij}}\right)^{6}\right) \quad (1)$$

Where U_{ij} is the potential energy, r_{ij} is the distance between two interacting molecules i and j, and σ and ε are the distance characteristic length and energy scale, respectively. The intermolecular forces are then calculated from equation 1 through the following relation:

$$F_{ij} = \frac{-dU_{ij}}{dr_{ij}} \quad (2)$$

And finally the motion of the molecules is determined by Newton's second law through the Verlet's integrating scheme [10].

As mentioned before, two 2-layered walls are located at Y-constant positions. In the other directions periodic boundary conditions are applied. Each wall molecule is attached to its place with a linear spring to ensure wall solidity. The spring will consequently exert a force of $F = -K\Delta r$ to every molecule that moves away a distance equal to Δr; in this equation K is the spring stiffness. Figure 1 shows the schematic of initial placement of fluid and wall molecules.

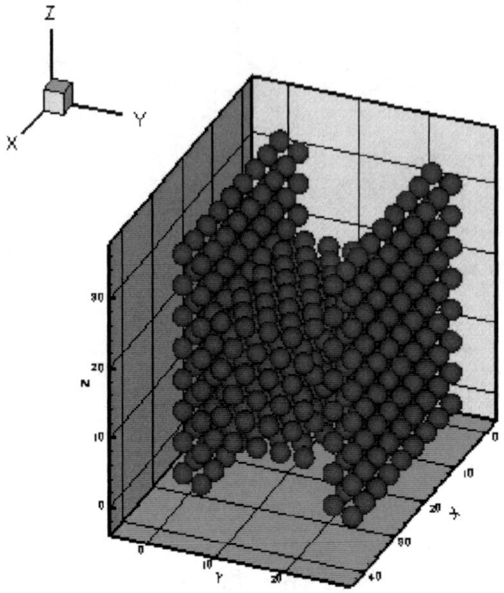

Figure 1. Schematic of initial placement of fluid and wall molecules

Wall temperature is kept constant using thermostat. In fact at each time step velocity of wall molecules are rescaled to ensure constant wall temperature. An external force is constantly applied to the fluid molecules in order to develop flow in the channel. It should be noted that all of the initial conditions of the simulation such as temperature, number of molecules, spring stiffness, number of wall layers and external force magnitude are chosen similar to a simulation performed by Sofos et al [9] in order to validate the simulation of this paper. The detailed specifications of the simulation carried out in this paper are presented in table 1.

Table 1. Specification of the system under examination.

Dimension (Angstrom)	36.15×36.15×36.15
Wall molecules	288 argon molecules
Fluid molecules	576 argon molecules
Spring stiffness ($\varepsilon\sigma^{-2}$)	57.15
External force (pN)	0.18
Thermostat (k)	Fixed on 120
Number of bins	15 and 36
Time step (s)	10^{-14}
Boundary Conditions	Periodic in X and Z directions Wall BC in Y direction

3.Results and discussion

As mentioned in Table 1, the solution domain is divided into 15 equal bins in Y direction to evaluate velocity profile. The velocities of molecules are averaged in each bin using SMC method. Based on our examination, this is the minimum number of bins that has resulted in accurate velocity profile. It should be noted that higher number of bins, for instance 36 bins, were also considered and the same results have been obtained; this is shown in figure 2. In this figure velocity profile is plotted after 2×10^6 time steps using 15 and 36 bins.

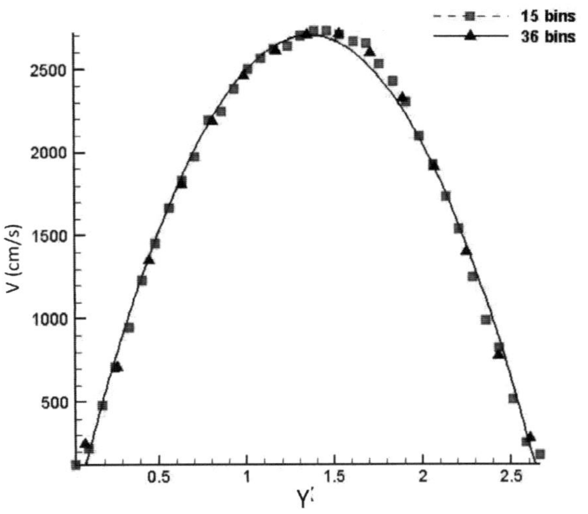

Figure 2. Velocity profile after 2×10^6 time steps using 15 bins and 36 bins

The effect of the number of time steps of simulation is shown in figure 3. As seen, result of 1.5×10^6 and 2×10^6 time steps are almost the same. The relative error between peak values of these profiles is less than 1 %.

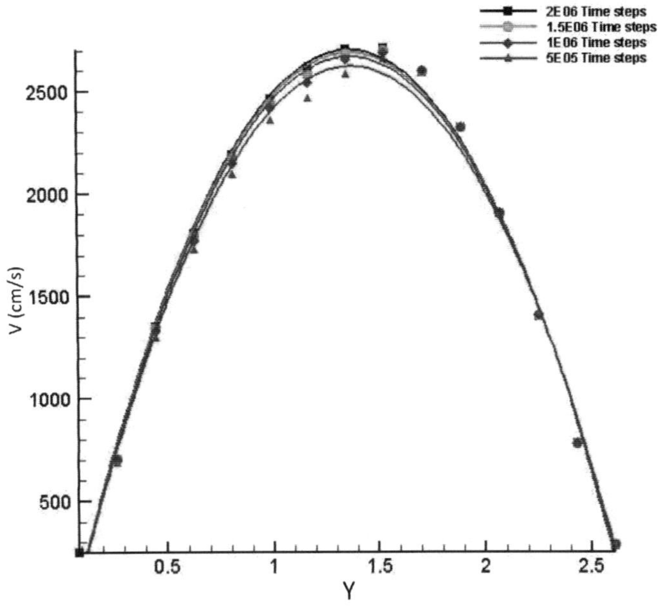

Figure 3. Effect of time step on the velocity profile

At this stage results of the present paper are compared with the results of Sofos et al [9] for the same test case in figure 4. Present results are shifted 0.7 nm to the right on the horizontal axis, i.e. in the Y direction of solution domain. Averaging method used by Sofos et al has not been mentioned in his work. In order to make a fair comparison present results are obtained using CAM method. It is worth to mention that approximately the same results are obtained if SMC method is used.

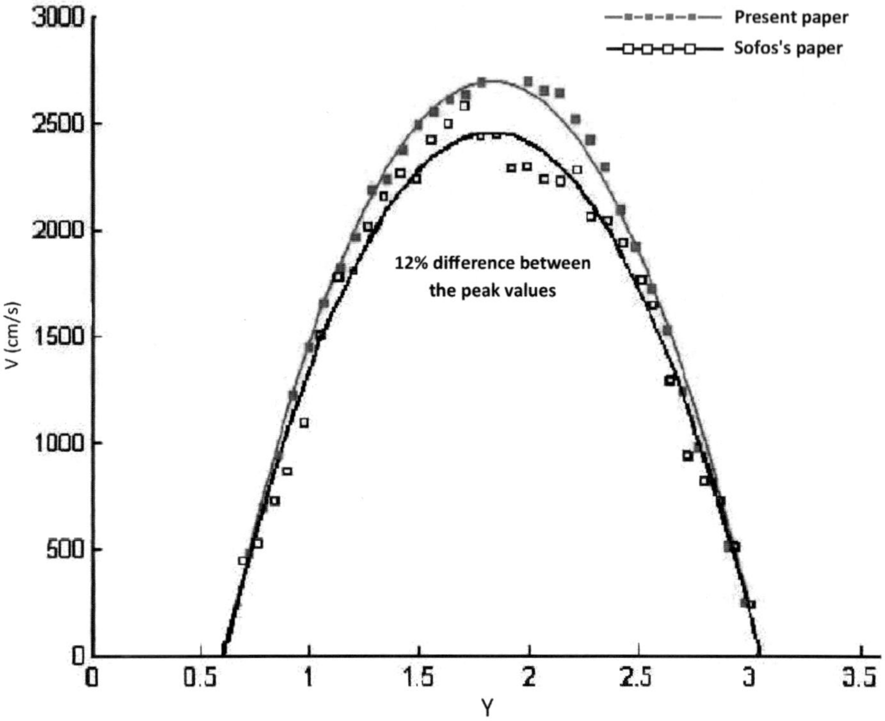

Figure 4. Comparison of velocity profiles obtained in this paper and by Sofos et al [9]

In both cases time step of 10^{-14} second is used. The bin values of velocities in the present study (denoted by solid squares) are fairly symmetric, as expected. In contrast, velocity values of Sofos et al are not symmetric at all. As seen, both velocity values are close to each other except at the right side of the peak region, where the present velocity values follow a physically meaningful path. Approximately, 12% difference is noticed in the peak value of velocity. From the above observations it can be concluded that the present velocity values are correct and more realistic than those obtained by Sofos et al.

Although both methods approximately result in the same profiles, SMC reveals its advantage when time variation of velocity (convergence history) is examined to see whether the simulation is reached equilibrium. Convergence history of velocity in the middle bin of the above test case is plotted in figure 5, having used both SMC and CAM averaging methods. These curves are plotted based on every 1000 time step results.

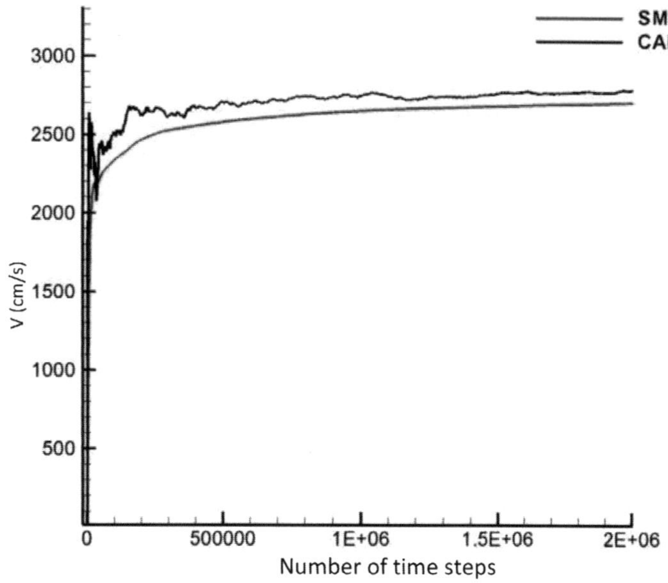

Figure 5. Velocity of the middle bin versus time obtained by both CAM and SMC method of averaging

Convergence history can be used to distinguish whether equilibrium is reached. It is obvious that convergence history using CAM method has oscillations all over its curve. Conversely, the curve obtained using SMC method has almost no oscillations. Obviously any kind of convergence criteria such as $\frac{\Delta V}{V} < \varepsilon$, where ΔV is the difference between velocity values within some successive time steps and ε is the maximum acceptable error, will work excellent on the curve obtained from SMC. Due to the excessive oscillations in the curve obtained by CAM, implementation of any kind of convergence criteria using this curve will face substantial difficulties. For a suitable convergence criterion in MD simulation see Ref. 2. Although the velocity difference between the two methods is small, based on the error analysis discussed in Ref. 2 present authors believe that velocity value predicted by SMC is more accurate.

4.Conclusion

A simulation of 576 monatomic argon molecules in a channel with two 2-layered walls is reported in this paper. All of the initial conditions have been chosen similar to [9]. The velocity profiles obtained using SMC and CAM methods are approximately the same. For validating the results the velocity profile of the present paper is compared to the one obtained by Sofos et al. For both velocity profiles the values of velocity are roughly the same except for the peak region. It is also observed that the velocity profile obtained in this paper is much smoother and symmetric. CAM and SMC methods are then used to obtain the convergence history. It is observed that SMC yields a graph with no fluctuations which is suitable for obtaining the time step in which equilibrium occurs. This is while the use of CAM results in a convergence history in which oscillations make the use of convergence condition very difficult. The privilege of SMC method is it damps the oscillations of the quantities in the domain of simulation. Being simple to implement without adding to the CPU time of simulation, the SMC method is preferred when mean flow characteristics are to be calculated.

5.References

[1] Fan X , Phan-Thien N,Teng Yong N and Diao X 2002 Molecular dynamics simulation of a liquid in a complex nano channel flow *Phys. Fluids* **14** 1146-53

[2] Karimian S M H and Izadi S 2010 A study on the measurement of mean velocity and its convergence *Int. J. Numer. Meth. Fluids* **67** 2130-40

[3] Abraham F 1978 The interfacial density profile of a Lennard-Jones fluid in contact with a (100) Lennard-Jones wall and its relationship to idealized fluid/wall systems: A Monte Carlo simulation *J.Chem. Phys.* **68** 3713-16

[4] Koplik J, Banavar J and Willemsen J 1989 Molecular dynamics of fluid flow at solid surfaces *Phys. Fluids* **1** 781-794

[5] Thompson P and Robbins M 1990 Shear flow near solids: Epitaxial order and flow boundary conditions *Phys. Rev. A* **41** 6830-37

[6] Jabbarzadeh A Atkinson J D and Tanner R I 1997 Rheological properties of thin liquid films by molecular dynamics simulations *J. Non-Newtonian Fluid Mech.* **69** 169-193

[7] Jabbarzadeh A Atkinson J D and Tanner R I 1997 Wall slip in the molecular dynamics simulation of thin films of hexadecane *J. Chem. Phys.* **110** 2612-19

[8] Bing Mi X and T. Chwang A 2003 Molecular dynamics simulations of nanochannel flows at low reynolds numbers *Molecules* **8** 193-206

[9] Sofos F, Karakasidis T E and Liakopoulos A 2009 Non-Equilibrium Molecular Dynamics Investigation of Parameters Affecting Planar Nanochannel Flows *Contem. Eng. Sci.* **2** 283-298

[10] Verlet L 1967 Computer "Experiments" on classical fluids. I. Thermodynamical properties of Lennard-Jones molecules *Phys. Rev.* **159** 98-103

Numerical simulation of micro flows with moving obstacles

G. Dechristé and L. Mieussens

Univ. Bordeaux, IMB, UMR 5251, F-33400 Talence, France. CNRS, IMB, UMR 5251, F-33400 Talence, France.

E-mail: Guillaume.Dechriste@math.u-bordeaux1.fr, Luc.Mieussens@math.u-bordeaux1.fr

Abstract. We present a numerical method for computing unsteady rarefied gas micro flows, in domains with moving boundaries, in view of applications to complex computations of moving structures in micro or vacuum systems. The flow is described by the Bhatnagar-Gross-Krook (BGK) model of the Boltzman equation. A standard approach to simulate incompressible viscous flows with moving boundaries is the immersed boundary method. In this work, we propose an extension of this approach to a deterministic simulation of rarefied flows. The immersed boundary approach consists in keeping the same mesh all along the calculation: every cell of the mesh remains fixed for all time steps while the domain occupied by the gas changes. This strategy avoids to use moving meshes and remeshing approaches, and should be easily applied to problems with complex geometries. The method has been tested with both specular and diffuse boundary conditions and has been validated on several 1D problems (moving piston, actuator, etc.). Up to our knowledge, this is the first time that a deterministic simulation method for rarefied flows with moving obstacles based on the immersed boundary method is presented.

1. Introduction

The standard approach used for the computation of rarefied gas flows is the Direct Simulation Monte Carlo [1]. For low-speed flows, the results obtained with this method are often noisy, which is a major drawback for the simulation of Microelectromechanical systems (MEMS). However when the Navier-Stokes solution is known, low-variance approaches have been recently proposed [2, 3] and have been proved to be very efficient. Another approach to get accurate results are the deterministic methods [4]. In order to reduce the numerical cost required by the computation of binary collisions, the Boltzmann equation is often replaced by the simpler BGK model [5].

Generally, problems involving moving obstacles are treated with remeshing methods. However, this approach needs high computational time and high numerical cost, especially for kinetic equations. Recently, Russo and Filbet [6] proposed a semi-langrangian scheme for one dimensional problems to avoid the remeshing approach.

A well-known Eulerian approach to simulate incompressible viscous flows with moving boundaries is the immersed boundary method [7]. This approach consists in solving the equation on a Cartesian grid. Even the solid part of the domain is meshed, but the equation is not solved in the solid cells. These cells are instead filled with a boundary condition value. The main difficulty is then to properly choose this condition. The advantages of this strategy are the low computational cost and its natural extension to higher dimensions. Here, we propose an extension of this method to a deterministic simulation of rarefied flows based on the BGK model.

The outline of this article is the following. In section 2, we briefly present the kinetic description of a rarefied gas. Then, the application of the immersed boundary method is described in section 3. Finally, in section 4 the method is validated on two different test cases: a moving piston and an actuator.

2. Kinetic description of rarefied gas dynamics

Gas flows can be modelled either at a macroscopic or microscopic scale. Euler and Navier-Stokes equations describe the macroscopic behaviour of the flow, while the Boltzmann equation describes the flow at a microscopic scale. The level of rarefaction of a gas can be estimated through the Knudsen number (Kn), which is defined by the ratio between the mean free path (l) of the molecules and a characteristic macroscopic length of the system (L):

$$Kn = \frac{l}{L}. \tag{1}$$

Large Knudsen numbers correspond to a rarefied flow, while small Knudsen numbers indicate that the flow is close to the equilibrium and can be described by Euler or Navier-Stokes equations. It is generally admitted [1] that the macroscopic equations are no longer sufficient when the Knudsen number is greater than 0.1.

2.1. Boltzmann equation and BGK model

In kinetic theory, the basic model used for the description of the gas is the Boltzmann equation:

$$\partial_t f + \mathbf{v} \cdot \nabla_{\mathbf{x}} \cdot f = Q(f), \tag{2}$$

where f is the distribution function defined such that $f(t, \mathbf{x}, \mathbf{v})\mathbf{dxdv}$ is the mass of molecules located in the space volume $\mathbf{dx}$ with a velocity contained in an elementary volume $\mathbf{dv}$ at time t. The macroscopic quantities are therefore defined by the first moments of f:

$$\begin{pmatrix} \rho(t, \mathbf{x}) \\ \rho\mathbf{u}(t, \mathbf{x}) \\ E(t, \mathbf{x}) \end{pmatrix} = \int_{\mathbb{R}^d} \begin{pmatrix} 1 \\ \mathbf{v} \\ \frac{1}{2}|\mathbf{v}|^2 \end{pmatrix} f(t, \mathbf{x}, \mathbf{v})\mathbf{dv}. \tag{3}$$

The temperature of the gas may be deduced from the conservative quantities though the relation $E = \frac{1}{2}\rho|\mathbf{u}|^2 + \frac{d}{2}\rho RT$ where d is the dimension of the velocity space and R is the gas constant, which is defined by the ratio between the Boltzmann constant and the molecular mass of the gas.

It is well known that in a thermodynamical equilibrium state, the distribution function f is equal to a Maxwellian distribution $\mathcal{M}(\rho, \mathbf{u}, T)$:

$$\mathcal{M}(\rho, \mathbf{u}, T) = \frac{\rho}{(2\pi RT)^{\frac{d}{2}}} \exp\left(-\frac{|\mathbf{v} - \mathbf{u}|^2}{2RT}\right). \tag{4}$$

The collision integral $Q(f)$ given by the Boltzmann operator [8] requires a very expensive computational cost. In order to avoid this problem, it is interesting to consider the simpler BGK model, proposed by Bhatnagar, Gross and Krook [5] where the collision term is modelled by a relaxation of the distribution function towards its corresponding Maxwellian equilibrium:

$$Q(f) = \frac{1}{\tau}\left(\mathcal{M}(\rho, \mathbf{u}, T) - f\right), \tag{5}$$

where τ is the relaxation time, defined by $\tau = \frac{\mu}{\rho RT}$. According to [1], $\mu = \mu_{ref}\left(\frac{T}{T_{ref}}\right)^{\omega}$ where μ_{ref}, ω and T_{ref} are constants that depend on the gas.

2.2. Boundary conditions

Boundary conditions may be modelled by two standard approaches: either specular or diffuse reflection. In the former model, incident molecules of velocity $\mathbf{v}$ are remitted with a velocity symmetric to $\mathbf{v}$ with respect to the wall. Consequently, if the wall moves at a velocity $\mathbf{u}_w$, the specular reflection condition will be given by the following relation:

$$f(t, \mathbf{x}, \mathbf{v}) = f(t, \mathbf{x}, \mathbf{v} - 2(\mathbf{v} - \mathbf{u}_w) \cdot \mathbf{n_x} \mathbf{n_x}) \quad \text{if} \quad (\mathbf{v} - \mathbf{u}_w) \cdot \mathbf{n_x} > 0, \tag{6}$$

where $\mathbf{n_x}$ is the unit normal vector to the boundary, pointed to the gas.

In the diffuse reflection model, the molecules colliding with the boundary are remitted with a temperature T_w equal to the wall temperature and with a random velocity normally distributed around $\mathbf{u}_w$. This reads:

$$f(t, \mathbf{x}, \mathbf{v}) = \phi \mathcal{M}(1, \mathbf{u}_w, T_w) \quad \text{if} \quad (\mathbf{v} - \mathbf{u}_w) \cdot \mathbf{n_x} > 0, \tag{7}$$

where ϕ is defined such that the normal mass flux across the wall is zero:

$$\phi = -\frac{\int_{(\mathbf{v}-\mathbf{u}_w)\cdot\mathbf{n_x}<0} f(t, \mathbf{x}, \mathbf{v})(\mathbf{v} - \mathbf{u}_w) \cdot \mathbf{n_x}\, \mathbf{dv}}{\int_{(\mathbf{v}-\mathbf{u}_w)\cdot\mathbf{n_x}>0} \mathcal{M}(1, \mathbf{u}_w, T_w)(\mathbf{v} - \mathbf{u}_w) \cdot \mathbf{n_x}\, \mathbf{dv}}. \tag{8}$$

2.3. Fluid limits of kinetic equation

Multiplying the Boltzmann equation (2) by $(1, \mathbf{v}, \frac{1}{2}|\mathbf{v}|^2)$ and integrating it with respect to the velocity variable give the conservation laws:

$$\begin{aligned} \partial_t \rho + \nabla \cdot (\rho \mathbf{u}) &= 0, \\ \partial_t (\rho \mathbf{u}) + \nabla \cdot (\rho \mathbf{u}^2 + \overline{\overline{p}}) &= 0, \\ \partial_t E + \nabla \cdot (E\mathbf{u} + \overline{\overline{p}}\mathbf{u} + \mathbf{q}) &= 0. \end{aligned} \tag{9}$$

This system can be closed if the gas is in equilibrium and in that case the stress tensor and the heat flux vector are

$$\overline{\overline{p}} = pI, \qquad \mathbf{q} = 0. \tag{10}$$

Equations (9) and (10) give the Euler equations of gas dynamics. For a regime close to the equilibrium, the Chapman-Enskog expansion leads to the Navier-Stokes equations with

$$\overline{\overline{p}} = pI - \mu[\nabla \mathbf{u} + (\nabla \mathbf{u})^T] - \mu_B (\nabla \cdot \mathbf{u}) I, \qquad \mathbf{q} = -\kappa \nabla T, \tag{11}$$

where μ, μ_B and κ are the viscosity, the bulk viscosity and the thermal conductivity.

In order to make relatively fast numerical tests, this study is restricted to a simpler BGK model with one dimensional space variable x and one dimensional velocity variable v. In this case, the corresponding Navier-Stokes equations are easily calculated. Indeed, computing the moments of f naturally gives $\mu = 0$ and $\mu_B = 0$. Moreover, the heat flux vector can be calculated by a Chapman-Enskog expansion [9], and we get:

$$q = \frac{3}{2} \tau \rho R^2 T \partial_x T. \tag{12}$$

3. Immersed boundary method

Because the numerical tests are restricted to one dimensional cases, the following method will be described for a one dimensional flow. The computational domain is defined by $\Omega = [a, b]$ such that $\Omega = \Omega_{gas}(t) \cup \Omega_{solid}(t)$ where $\Omega_{gas}(t)$ represents the intervals occupied by the gas and $\Omega_{solid}(t)$ are the intervals filled by the walls at time t. Naturally, although Ω remains fixed for every times, $\Omega_{gas}(t)$ as well as $\Omega_{solid}(t)$ may vary in time when the obstacles are moving. With these definitions, equation (2) gives:

$$\begin{aligned} \partial_t f + v\partial_x f &= \frac{1}{\tau}(\mathcal{M}(\rho, u, T) - f) \qquad && x \in \Omega_{gaz}(t), \\ f(t, x \in \partial\Omega_{gas}(t), v) &= f_w(t, x, v) \qquad && (v - u_w) \times n_x > 0, \end{aligned} \tag{13}$$

where f_w is defined by equation (6) or (7).

3.1. Numerical method

The domain is discretized by a Cartesian grid of $(N + 1)$ nodes $x_{i+1/2} = i\triangle x$ with step $\triangle x = (b - a)/N$. The cells of the grid are denoted by $C_i = [x_{i-1/2}, x_{i+1/2}[$ and their center are $x_i = (i + \frac{1}{2})\triangle x$. At each discrete time step $t_n = n\triangle t$, we denote by $f_i^n(v)$ an approximation of the average of $f(t_n, x, v)$ over the cell C_i. Similarly, Ω^n_{gas} (resp. Ω^n_{solid}) denotes an approximation of $\Omega_{gas}(t)$ (resp. $\Omega_{solid}(t)$) and is defined by the union of the cells C_i whose center x_i are in $\Omega_{gas}(t)$ (resp. $\Omega_{solid}(t)$). Finally, we define the normal vector to the interface between a gas cell C_i and its neighbouring solid cell by:

$$\nu_i = \begin{cases} 1 \text{ if } C_{i+1} \in \Omega^n_{gas} \text{ and } C_{i-1} \in \Omega^n_{solid} \\ -1 \text{ if } C_{i-1} \in \Omega^n_{gas} \text{ and } C_{i+1} \in \Omega^n_{solid} \end{cases}. \tag{14}$$

For every cell in Ω^n_{gas}, we use the following standard first order upwind scheme:

$$f_i^{n+1} = f_i^n - \frac{\Delta t}{\Delta x}(\mathcal{F}^n_{i+\frac{1}{2}} - \mathcal{F}^n_{i-\frac{1}{2}}) + \frac{\Delta t}{\tau_i^n}(\mathcal{M}_i^n - f_i^n), \tag{15}$$

where the numerical flux at interface $x_{i+1/2}$ is defined by $\mathcal{F}^n_{i+1/2} = v^+ f_i^n + v^- f^n_{i+1}$, where $v^+ = \max(v, 0)$ is the positive part of v and $v^- = \min(v, 0)$ is the negative part of v. The calculation of this numerical flux at an interface between a gas cell C_i and a solid cell $C_{i-\nu_i}$ requires the value of $f^n_{i-\nu_i}$ for all $v \times \nu_i > 0$. For such velocities, there are two cases. First, if $(v - u_w) \times \nu_i > 0$, then we define $f^n_{i-\nu_i}$ by using the boundary condition (like in the ghost cell technique) and hence set $f^n_{i-\nu_i}$ to $f^n_{w,i-\nu_i}$ defined by:

- Specular reflection:

$$f^n_{w,i-\nu_i}(v) = f_i^n(v - 2(u_w - v)). \tag{16}$$

- Diffuse reflection:

$$\begin{aligned} & f^n_{w,i-\nu_i}(v) = \phi\mathcal{M}(1, u_w, T_w), \\ & \text{with} \quad \phi = -\frac{\int_{(v-u_w)\nu_i<0} \left(f_i^n(v)(v - u_x)\nu_i\right) dv}{\int_{(v-u_w)\nu_i>0} \left(\mathcal{M}(1, u_w, T_w)(v - u_w)\nu_i\right) dv}. \end{aligned} \tag{17}$$

At the contrary, if $(v - u_w) \times \nu_i < 0$, we propose to use the extrapolated value computed with the neighbouring gas cell C_i, that is to say we set $f^n_{i-\nu_i}(v) = f_i^n(v)$.

The velocity variable is approximated by using a discrete velocity method [10]. We define $f^n_{i,k}$ as an approximation of $f(t_n, x_i, v_k)$ where v_k are the nodes of a bounded Cartesian grid with step $\triangle v$. The Maxwellian is approximated on this grid by a discrete Maxwellian which has the same moments as f. Then, equation (15) can be readily applied to the discrete velocity model where the flux ϕ for the diffuse boundary condition in equation (17) becomes:

$$\phi = -\frac{\sum\limits_{(v_k-u_w)\nu_i<0} f^n_{i-n_i,k}(v_k-u_w)\nu_i\triangle v}{\sum\limits_{(v_k-u_w)\nu_i>0} \mathcal{M}(1,u_w,T_w)(v_k-u_w)\nu_i\triangle v}. \tag{18}$$

If $v_k - 2(u_w - v_k)$ is not contained in the velocity grid, the value of $f_i(v_k - 2(u_w - v_k))$ for the specular reflection condition in equation (16) is calculated by a linear interpolation from $f_{i,K}$ and $f_{i,K+1}$ where K is defined by: $v_K < v_k - 2(u_w - v_k) \leq v_{K+1}$.

Positivity and stability properties of the scheme hold in a unbounded domain under the CFL condition:

$$\Delta t < \min(\frac{\Delta x}{v_k}, \tau) \tag{19}$$

It would be desirable to get the stability of the scheme in a bounded domain. This property is directly linked to the mass conservation. For the moment, the total mass is not exactly preserved by the method (we are currently working to solved the issue). Consequently, it is difficult to prove a stability property. However we did not observed any effects in our numerical tests.

3.2. Transition solid cell/gas cell

In order to take into account the time variation of $\Omega_{gas}(t)$ and $\Omega_{solid}(t)$, a backward Euler scheme is applied for computing the movement of the boundary. This reads:

$$x_w^{n+1} = x_w^n + \triangle t \times u_w^n, \tag{20}$$

where x_w^n is an approximation of the boundary of $\Omega_{solid}(t^n)$. After the computation of f_i^{n+1} in $C_i \in \Omega^n_{gas}$ by equation (15), x_w^{n+1}, Ω^{n+1}_{gas} and Ω^{n+1}_{solid} are calculated. For simplicity, assume that $\nu_i = 1$. It may happen that a gas cell C_i becomes a solid cell if $x_w^{n+1} > x_i > x_w^n$. Even if f_i^{n+1} is calculated there, its value becomes useless for the next time step. Similarly, a solid cell C_i may become a gas cell if $x_w^{n+1} < x_i < x_w^n$. In that case, the value of f_i^{n+1} is required for computing the next time step. We propose to use the extrapolated value computed with the neighbouring gas cell C_{i+1}, that is to say we set $f_i^{n+1} = f_{i+1}^{n+1}$.

The method can be summarized in four steps:

- Calculation of the numerical flux at all interfaces between gas cells and solid cells.
- Calculation of f_i^{n+1} in all cells $C_i \in \Omega^n_{gas}$.
- Computation of Ω^{n+1}_{gas} by the calculation of x_w^{n+1}.
- Computation of f_i^{n+1} in the new gas cells.

We emphasize that the present method reduces to the standard scheme described in [10] for non-moving boundaries. Moreover, the principal advantage of the method is that the only difference between moving and non-moving boundaries is the implementation of the tracking of the moving boundary. This implementation should be easily extended to higher dimension by using level set methods.

4. Numerical comparisons

The method described above has been validated with two test cases: a moving piston and an actuator. A convergence study has been led for each case, and it has been shown that the immersed boundary method fully preserves the first order accuracy of scheme (15) used to discretize the transport equation.

4.1. Moving piston

We consider a domain of total length $L = 20$ with a piston moving at velocity $\sin(t)$ at the left edge of the domain. The variables are initialized as follows:

$$u = 0, \qquad \rho = 1, \qquad RT = 1, \qquad x_{piston}(t = 0) = 2. \tag{21}$$

For all the following simulations, the domain has been discretized by 2000 cells ($\Delta x = 2 \cdot 10^{-2}$). The time step is chosen according to (19) and is around 10^{-3}. The computation lasts less than 5 minutes for each case on an Intel Xeon CPU X5560, 2.80GHz.

First, the velocity of the gas near the piston is the same as the velocity of the piston. The specular boundary condition (16) is implemented with $u_w = \sin(t)$. The solution of the BGK equation (2) with specular boundary conditions (6), is compared to the solution of Euler equations (9) and (10). In order to ensure the compatibility of the two models, the simulations are performed in the context of small Knudsen number ($\tau = 10^{-3}$). Euler equations are written in Lagrangian coordinates [11] and solved by using the central scheme of Nessayhu-Tadmor [12]. A shock wave is created by the motion of the piston and propagates inside the gas. A rapid change of the density and temperature can be observed when the sign of the velocity of the piston changes. The profiles of density and temperature are plotted on figure 1 after the piston has moved forward and backward.

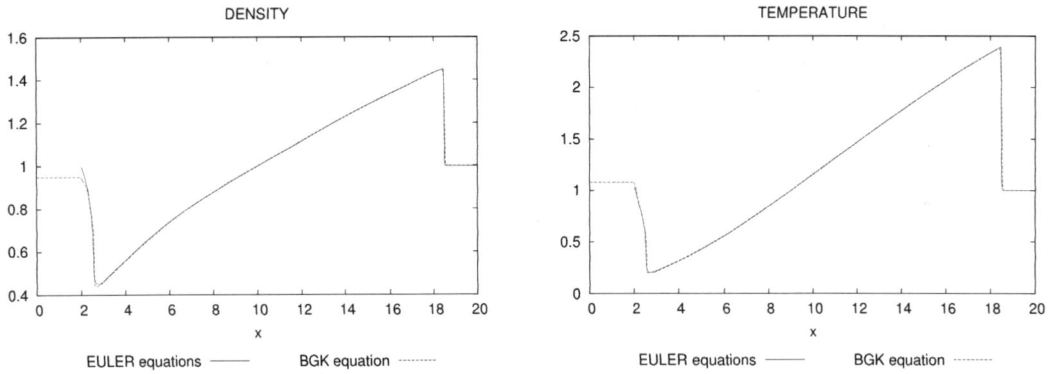

Figure 1. Comparison between Euler and BGK equations. Density and temperature profiles at $t = 2\pi$.

Second, the temperature of the gas close to the piston is set to the temperature of the piston by using diffuse boundary condition (17), with $u_w = \sin(t)$ and $T_w = 1$. Results obtained with the BGK equation (2) associated now with diffuse reflection boundary conditions (7), are compared with results obtained with Navier-Stokes equations, eq. (9) and (11). Here, again, the Navier-Stokes equations are written in Lagrangian coordinates and solved with the central scheme. The results are quite similar to the previous case, except that the density near to the piston increases a lot. Density and temperature are plotted on figure 2 after the first compression.

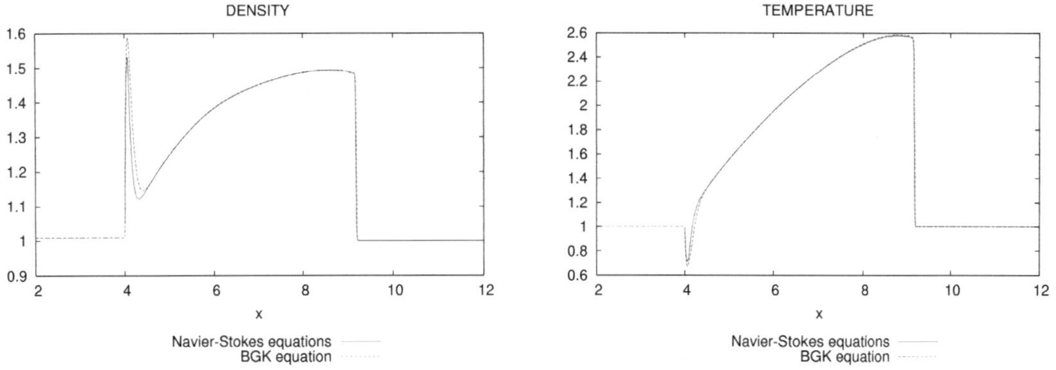

Figure 2. Comparison between Navier-Stokes and BGK equations. Zoom of density and temperature profiles at $t = \pi$.

Note that the computational domain of the BGK computation differs from the computational domain of Euler equations. Indeed, the former equation has been computed on a fixed domain (immersed boundary method) while the computation of the latest set of equations is done on a domain moving with the gas (equations written in Lagrangian coordinates). The same remark can be done for the comparison between BGK and Navier-Stokes equations.

The good agreement between the solution of the BGK equation and the predictions of Euler and Navier-Stokes equations ensure that the immersed boundary method is accurate when the velocity of the moving object is independent of the behaviour of the gas. Indeed, in this test case the piston moves whatever the state of the gas is.

In the following, we study a case where the interaction between the gas and the wall is more complex.

4.2. Actuator

A flat plate is contained in a slab. At initial time, the temperature, the pressure and the density are the same in the whole domain and are denoted by T_0, P_0 and ρ_0. The right side of the plate is heated to T_w whereas the left side is maintained at the initial temperature T_0 of the gas. In the same way, the right wall of the slab is heated and the left wall is kept to the initial temperature. The experiment is illustrated in figure 3. Because the gas is heated close to the right side of the plate, the temperature and the pressure increase there and the movement of the plate is induced by the pressure difference between the two sides.

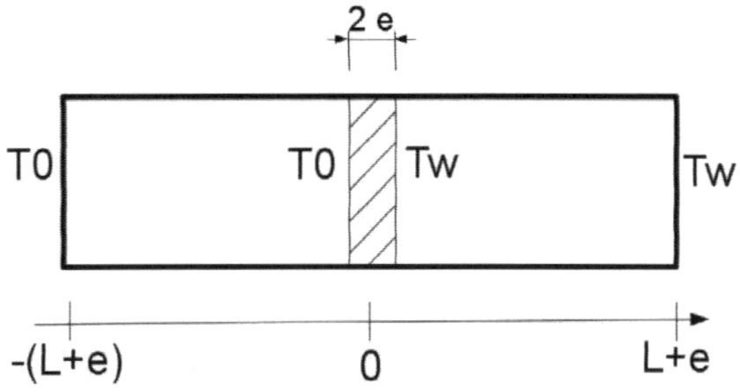

Figure 3. Schematic view of the actuator.

The equilibrium state of the gas and the equilibrium position of the plate can be easily calculated. Once the equilibrium state is reached, temperature, pressure and density are constant in each part of the slab. The state of the gas can be fully described with the perfect gas relation and the mass conservation. Indeed, denoting by P_{equi} and ρ_{left} (resp. ρ_{right}) the equilibrium pressure and left (resp. right) density and by x_{equi} the coordinate of the center of the plate once equilibrium is reached, we get:

- on the left side:

$$\begin{cases} \rho_0 R T_0 = P_0 \\ \rho_{left} R T_0 = P_{equi} \\ \rho_0 L = \rho_{left}(L + x_{equi}) \end{cases} \Longrightarrow P_{equi} = \frac{L}{L + x_{equi}} P_0. \tag{22}$$

- on the right side:

$$\begin{cases} \rho_0 R T_0 = P_0 \\ \rho_{right} R T_w = P_{equi} \\ \rho_0 L = \rho_{right}(L - x_{equi}) \end{cases} \Longrightarrow P_{equi} = \frac{T_w}{T_0} \frac{L}{L - x_{equi}} P_0. \tag{23}$$

From equations (22) and (23), it comes:

$$x_{equi} = L \frac{1 - T_w/T_0}{1 + T_w/T_0}, \qquad \rho_{left} = \frac{P_{equi}}{RT_0}, \qquad \rho_{right} = \frac{P_{equi}}{RT_w}. \tag{24}$$

The immersed boundary method is used for the simulation of this experiment. The domain is discretized by 800 cells ($\Delta x = 2.5 \cdot 10^{-2}$). The BGK equation (2) is solved with diffuse boundary conditions (7). The velocity of the plate is calculated through the fundamental principle of dynamics with an external force corresponding to the difference of pressure at each side of plate:

$$m \frac{du_w}{dt} = (P_{left} - P_{right})S, \tag{25}$$

where m and S are the mass and the area of plate.

At initial time, the domain is filled with argon. $P_{left} = P_{right} = P_0$ and the velocity of the plate is $u_w = 0$. With these parameters, the Knudsen number, based on the length of the computational domain ($2L = 2m$), is equal to $8 \cdot 10^{-2}$. Since the temperature at the right side of the slab increases, the pressure increases too, and the piston moves from the right side to the left side with a positive acceleration. There is a time at which $P_{left} = P_{right}$, and hence $\frac{du_w}{dt} = 0$. But since $u_w \neq 0$, the plate keeps on moving. As a result, it will evolve toward the equilibrium point by oscillating around it. These oscillations are observed on figure 4 in which the velocity of the center of the plate is plotted. The results are obtained after 24 hours of computation on an Intel Xeon CPU X5560, 2.80GHz. This long computational time is due to the fact than the equilibrium time is very large compared to the computational time step ($\Delta t \approx 10^{-6}$). Computations have been launched for several Knudsen numbers (0.04, 0.8 and 10) and the oscillatory behaviour of the plate has been observed in each case.

Taking $T_0 = 270$ K, $T_w = 330$ K and $P_0 = 10$ Pa, the error of the simulated equilibrium position with respect to the exact equilibrium position is equal to 5% . Indeed, the value of the exact equilibrium position comes from equations (24): $x_{equi}^{exact} = -0.1m$. The predicted equilibrium position is calculated by the average between the center of the leftmost and the rightmost solid cells at time $t = 7s$: $x_{equi}^{predicted} = -0.105m$ (Fig. 5).

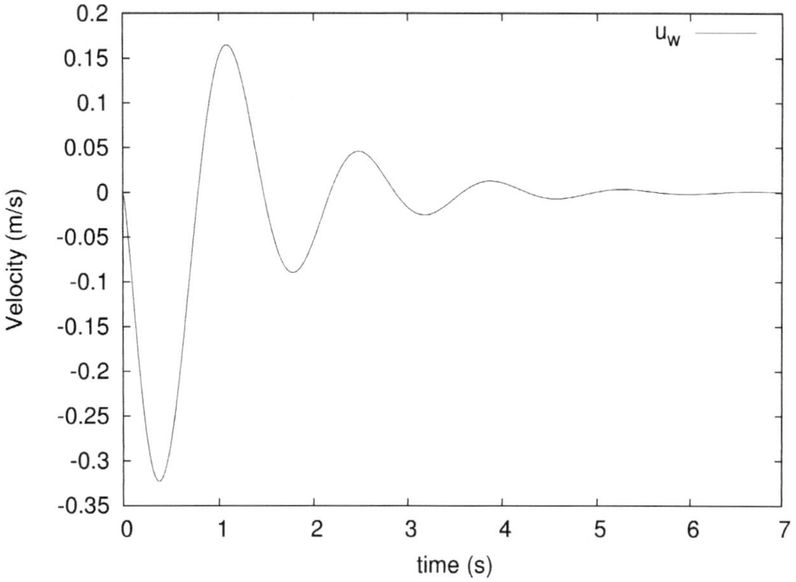

Figure 4. Velocity of the moving plate.

Moreover, the error of the predicted pressure ($P_{equi}^{predicted} = 11.15Pa$) with respect to the exact pressure ($P_{equi}^{exact} = 11.11Pa$) is 3.6%. The errors for the right and left density are in the same order of magnitude (3% and 3.1%).

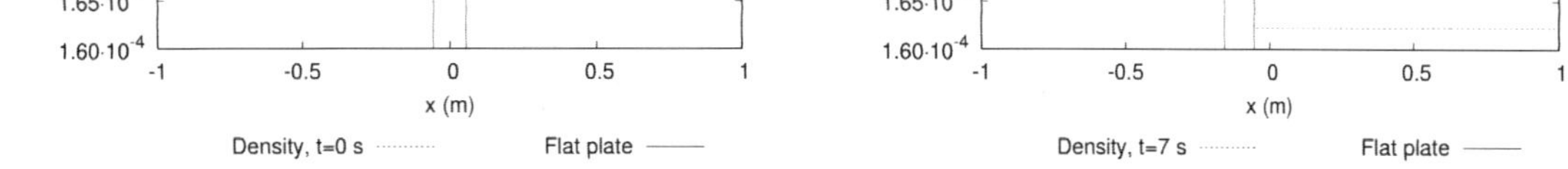

Figure 5. Density profiles and position of the flat plate at initial time ($x = 0$) and equilibrium time ($x = -0.101m$).

5. Conclusion

We proposed an extension of the immersed boundary method, already used for the simulation of incompressible viscous flows, to the simulation of the BGK model of gas dynamics. The method has been validated for one dimensional problems and we showed that the method accurately predicts the behaviour of the gas near the boundary. The interaction between gas and solid boundaries is well predicted. The application of our method to higher dimensional problems will be presented in a forth coming work. In particular, our project is to use this approach to study the radiometric effect.

References

[1] Bird G A 1994 *Molecular Gas Dynamics and the Direct Simulation of Gas Flows* (*Oxford Engineering Science Series* vol 42) (Oxford Science Publications)
[2] Baker L and Hadjiconstantinou N 2005 *Physics of Fluids* **17** 1–4
[3] Homolle T and Hadjiconstantinou N 2007 *Journal of Computational Physics* **226** 2341–2358
[4] Aristov V V 2001 *Direct Methods for Solving the Boltzmann Equation and Study of Nonequilibrium Flows* (Kluwer Academic Publishers)
[5] Bhatnagar P L, Grosss E P and Krook M 1954 *Physical Reeview* **94** 511–525
[6] Russo G and Filbet F 2009 *Kinetic and related models* **2** 231–250
[7] Dong H, Mittal R and Najjar F 2006 *Journal of Fluid Mechanics* **566** 309–343
[8] Sone Y 2007 *Molecular Gas Dynamics. Theory, Techniques and Applications.* Modeling and Simulation in Science, Engineering and Technology (Birkhause)
[9] Chapman S and Cowling T G 1970 *The mathematical theory of non-uniform gases* (Cambridge University press)
[10] Mieussens L 2000 *Mathematical Models and Methods in Applied Sciences* **10** 1121–1149
[11] Godlewski E and Raviart P A 1996 *Numerical Approximation of Hyperbolic System of Conservation Laws* (*Applied Mathematical Sciences* vol 118) (Springer)
[12] Nessyahu H and Tadmor E 1990 *Journal of Computational Physics* **87** 408–463

Simulation of gas flows in micro/nano systems using the Burnett equations

Fubing Bao[1], Xia Yu[1], Jianzhong Lin[1,2]

[1] Institute of Flow Measurement and Simulation, China Jiliang University, Hangzhou 310018 China
[2] Department of Mechanics, State Key Laboratory of Fluid Power Transmission and Control, Zhejiang University, Hangzhou 310027 China

E-mail addresses: dingobao@zju.edu.cn

Abstract. The gaseous flow characteristics in micro/nano systems are studied using the Burnett equations. The Burnett equations are the second-order approximation of the Chapman–Enskog solution to the Boltzmann equation. This set of equation is appropriate to describe the rarefied gas flows. Velocity slip and temperature jump boundary conditions are applied on solid surface. Several techniques, including the usage of relaxation method on slip values and Burnett terms, are introduced to increase the stability of the Burnett equations. Hence, convergent results at Knudsen number up to 0.5 are achieved for the first time. The results of the Burnett equations are first verified using the corresponding experimental and the DSMC data. The Navier-Stokes and the Burnett equations give almost the same result when the flow is in slip regime (Kn<0.01). But when the flow is in transition regime (0.1<Kn<0.5), the results of Navier-Stokes equations deviate from DSMC data while those of Burnett equations still agree very well. The Burnett equations are then used to study the gaseous flows in micro/nano systems. The pressure driven plane Poiseuille and the backward-facing step flows, the flows through micro filters are then studied. Different inlet to outlet pressure ratios, size effects, shape effects and boundary conditions are analyzed. The competitive relations of rarefied and compressible effects in micro gas flows are discussed.

1. Introduction

Micro/nano-scale gaseous flows are always encountered in microfluidic and nanofluidic systems. Flow and heat transfer characteristics in these systems have attracted considerable investigations during the past decade[1-2]. In micro/nano-scale flows, small length scale gives rise to high Knudsen number, Kn, which is defined as the ratio of the mean free path to the characteristic length scale. The Knudsen number is usually used to describe the departure from the continuum regime. As the value of Kn increases, rarefaction effect becomes more important. The appropriate flow and heat transfer models are based on the range of Kn. A classification of different flow regimes is proposed by Tsien[3]: continuum flow (ordinary density) for $Kn \leq 0.001$, slip flow regime (slightly rarefied) for $0.001 < Kn \leq 0.1$, transition regime (moderately rarefied) for $0.1 < Kn \leq 10$ and free molecule flow (highly rarefied) for $Kn > 10$. The Navier-Stokes equations together with nonslip boundary conditions are appropriate to model the continuum flow. Deviation from the state of continuum is relatively small in slip flow regime and the flow is still governed by the Navier-Stokes equations. The rarefaction effect is modelled through the partial slip at the wall using slip boundary conditions. In free molecule regime, the collisionless Boltzmann equation can be used. While in transition regime, the flow can neither be considered as an absolutely continuous medium nor a free molecule flow and is hard to describe.

In micro/nano-scale gaseous flows, the characteristic length scale is comparable to mean free path of gas. The rarefaction effect becomes important and cannot be neglected any more[4-5]. Experiments conducted by Harley *et al.*[6] and Arkilic *et al.*[7] on low Reynolds number gas flows in microchannels showed that conventional analyses are unable to predict the observed flow rates with any degree of accuracy. The Navier-Stokes equations, which are based on linear constitute relationship between shear and stress, cannot describe the gaseous flow in this regime any more[1, 4]. The direct simulation Monte Carlo (DSMC) method is widely used nowadays to investigate the flow and heat transfer behaviours in microscale gaseous flows[8-9]. But this method requires a large number of particles for accurate simulation and is very expensive both in computational time and memory requirements, especially for the low-speed flow in MEMS and nanodevices[2].

As an alternative, higher-order extended or generalized hydrodynamic equations have been proposed which can perform equally well in both the continuum and slip/transition regimes[2, 4]. The Burnett equations, which add additional nonlinear second-order terms to the stress and heat flux terms, represent a second order correction to the equilibrium distribution function of gas flow[10]. These equations have been proposed and proved to be suitable to model the flows near equilibrium in slip and early transition regime[11].

The Burnett equations have been a subject of considerable investigation in recent years. They were firstly used to model the shock wave in the transition regime. Fiscko and Chapman[12] showed that the Burnett equations provide much more accurate numerical solutions than the Navier-Stokes equations for one-dimensional hypersonic shock structure in monatomic gases. Agarwal *et al.*[2] investigated the hypersonic shock structure, hypersonic blunt body flow and axisymmetric hemispherical nose flow using the Burnett equations, and also found that the Burnett equations can present better results than the Navier-Stokes equations in transition regime. Xu[13] showed that the discrepancy between the Navier–Stokes and the DSMC results can be resolved based on the simulation results of higher-order equations, such as Burnett and super-Burnett ones. Ohwada and Xu[14] theoretically showed the consistency between the traditional Chapman–Enskog expansion and the successive approximation for the BGK equation up to the super-Burnett order and they designed a numerical scheme to efficiently solve the Burnett equations.

The necessity of using the Burnett equations for micro/nano flows has been demonstrated by many researcher[15]. The Couette flows between two parallel plates were simulated using the Burnett equations and the results were compared with Navier-Stokes results[16-17]. Agarwal *et al.*[2] and Fang[18] studied plane Poiseuille flow numerically and found that the velocity profiles agree well with those from μFlow calculations by Beskok and Karniadakis[19].

However, because of the numerical instability associated with the Burnett equations[20-21], convergent results in Poiseuille flow can only be obtained at small Kn. Agarwal *et al.*[2] got convergent solutions of the Burnett equations to Poiseuille flow when Kn=0.2. But when Knudsen number becomes higher than 0.2, no convergent results could be achieved in the past. However, many gas flows in micro/nano devices fall in the range of $0.2<Kn<0.5$.

In present paper, the Burnett equations were adopted to indicate the gaseous flow characteristics in micro/nano-scale channels. The pressure driven plane Poiseuille flows[22], the backward-facing step flow and the flows in filters[23] were modelled and analyzed.

2. The Burnett equations

The Burnett equations are the second-order approximation of the Chapman-Enskog solution to the Boltzmann equation[10]. The augmented Burnett equations[20] are adopted in present paper. The governing equations of two dimensional unsteady compressible viscous flow can be expressed in Cartesian coordinates as follows,

$$\frac{\partial Q}{\partial t}+\frac{\partial E}{\partial x}+\frac{\partial F}{\partial y}=0 \tag{1}$$

where

$$Q=\begin{bmatrix}\rho\\ \rho u\\ \rho v\\ e_t\end{bmatrix},\quad E=\begin{bmatrix}\rho u\\ \rho u^2+p+\sigma_{11}\\ \rho uv+\sigma_{12}\\ (e_t+p)u+\sigma_{11}u+\sigma_{12}v+q_1\end{bmatrix},\quad F=\begin{bmatrix}\rho v\\ \rho uv+\sigma_{21}\\ \rho v^2+p+\sigma_{22}\\ (e_t+p)v+\sigma_{21}u+\sigma_{22}v+q_2\end{bmatrix} \tag{2}$$

where e_t is the total energy, $e_t=\rho C_V T+\frac{1}{2}\rho(u^2+v^2)$. The viscous stress tensor $\boldsymbol{\sigma}$ and heat flux vector $\boldsymbol{q}$ can be derived as approximate solutions of the Boltzmann equation using the Chapman-Enskog expansion. In the Burnett equations,

$$\begin{aligned}\boldsymbol{\sigma}&=-2\mu\boldsymbol{e}\\ &+\frac{\mu^2}{p}\left[\omega_1\nabla\cdot\boldsymbol{v}\boldsymbol{e}+\omega_2(\mathrm{D}\boldsymbol{e}-2\overline{\overline{\nabla\boldsymbol{v}\cdot\boldsymbol{e}}})+\omega_3 R\overline{\overline{\nabla\nabla T}}+\frac{\omega_4}{\rho T}\overline{\overline{\nabla p\nabla T}}+\omega_5\frac{R}{T}\overline{\overline{\nabla T\nabla T}}+\omega_6\overline{\overline{\boldsymbol{e}\cdot\boldsymbol{e}}}\right]\\ &+\frac{\mu^3}{p^2}\left[\frac{3}{2}\omega_7 RT\overline{\overline{\nabla(\nabla\cdot\nabla\boldsymbol{v})}}\right]\end{aligned} \tag{3}$$

$$\begin{aligned}\boldsymbol{q}&=-\kappa\nabla T\\ &+R\frac{\mu^2}{p}\left[\theta_1\nabla\cdot\boldsymbol{v}\nabla T+\theta_2(\mathrm{D}\nabla T-\nabla\boldsymbol{v}\cdot\nabla T)+\theta_3\frac{T}{p}\nabla p\cdot\boldsymbol{e}+\theta_4 T\nabla\cdot\boldsymbol{e}+3\theta_5\nabla T\cdot\boldsymbol{e}\right]\\ &+\frac{\mu^3}{p\rho}\left[\theta_6\frac{RT}{\rho}\nabla(\nabla\cdot\nabla\rho)+\theta_7 R\nabla(\nabla\cdot\nabla T)\right]\end{aligned} \tag{4}$$

where $\boldsymbol{e}=\overline{\overline{\nabla\boldsymbol{v}}}$. The double line over a tensor represents a divergence-free symmetric tensor,

$$\overline{\overline{f_{ij}}}=\frac{f_{ij}+f_{ji}}{2}-\frac{\delta_{ij}f_{mm}}{3}.$$

In equations (3) and (4), D is the material derivative. The coefficients in equations (3) and (4) for a hard sphere gas are: $w_1=10/3$, $w_2=2$, $w_3=3$, $w_4=0$, $w_5=3$, $w_6=8$, $w_7=2/9$, $\theta_1=75/8$, $\theta_2=-45/8$, $\theta_3=-3$, $\theta_4=3$, $\theta_5=117/4$, $\theta_6=-5/8$, $\theta_7=11/16$.

The pressure in above equations obeys the ideal gas law,

$$p=\rho RT. \tag{5}$$

Equations (1) to (5) form a closed set of equations, and ρ, u, v and T can be solved with appropriate boundary conditions.

3. Slip boundary conditions

Non-slip boundary conditions are generally unrealistic in slip and transition regime flows as there are not enough collisions near the wall to equilibrate the flow field[24]. So, slip boundary conditions are specified at the wall. The classical slip boundary conditions are the Maxwell-Smoluchowski first-order slip conditions:

$$u_s - u_w = \frac{2-\sigma_v}{\sigma_v}\lambda \left.\frac{\partial u}{\partial n}\right|_w + \frac{3}{4}\frac{\mu}{\rho T}\left.\frac{\partial T}{\partial \tau}\right|_w \tag{6}$$

$$T_s - T_w = \frac{2-\sigma_T}{\sigma_T}\frac{2\gamma}{Pr(\gamma+1)}\lambda \left.\frac{\partial T}{\partial n}\right|_w \tag{7}$$

where the subscripts s and w denote the slip and wall values. σ_v and σ_T are, respectively, the tangential momentum and thermal accommodation coefficients. The second term on the right of equation (6) is thermal creep contribution to slip velocity.

Analytical models derived using the first-order slip boundary condition have been shown to be relatively accurate up to Knudsen numbers of approximately 0.1[25]. For $Kn>0.1$, however, experimental studies have shown that models based on the first-order slip condition show considerable discrepancies against observed data[26]. In the present study, the following general slip boundary condition proposed by Beskok[24] is adopted,

$$u_s = \frac{1}{2}\left[(2-\sigma_v)u_\lambda + \sigma_v u_w\right] \tag{8}$$

$$T_s = \frac{\frac{2-\sigma_T}{Pr}\frac{2\gamma}{\gamma+1}T_\lambda + \sigma_T T_w}{\sigma_T + \frac{2-\sigma_T}{Pr}\frac{2\gamma}{\gamma+1}} \tag{9}$$

where, the subscript λ represents the value calculated one mean free path away from the wall. This kind of slip boundary condition corresponds to a high-order slip boundary condition by simply expanding u_λ in terms of u_s, using Taylor series expansion. The tangential momentum and thermal accommodation coefficient are both set to 1.0 in present study.

4. Numerical Algorithm

The present computation is based on the SIMPLE algorithm. The Burnett terms are treated as source terms. Slip boundary conditions are adopted to describe the velocity slip and temperature jump phenomena at wall. Generally, the Reynolds number and Mach number in microscale gas flows are small. But because of the large pressure drop along the channel, the density varies greatly and the effect of compressibility must be taken into consideration. The convergent Navier-Stokes results are used as initial values to increase the convergence.

As mentioned by many researchers[2, 18], the computation of Burnett equations is unstable at fine grids. Convergent solutions can only be achieved on very coarse grids in their simulations. In rarefied gas flows, the largest departure from thermal dynamics equilibrium takes place near the boundary. As to the proposal of Lockerby and Reese[16], a relaxation method for the boundary values is first adopted in the present computation. The relaxation factor, R_f, is introduced to modify the boundary value of the new iteration as follows:

$$B_{new} = B_{old} + R_f\left(B - B_{old}\right) \tag{10}$$

where B_{old} represents for the values of previous iteration. B represents for the value calculated from present values.

High-order derivatives will bring large variations. The third-order and fourth-order derivatives in the

Burnett equations result in large change of Burnett terms and lead to the divergence of computation. So the relaxation method is also applied to the Burnett terms in present study. By applying the relaxation method to boundary values and Burnett terms, the convergent results of the Burnett equations can be achieved up to Kn=0.5. The general slip boundary condition in equation (8) and (9) are not appropriate any more when Kn is larger than 0.5 since the slip value are calculated from the location one mean free path away from the wall. This kind of boundary condition is not suitable if the channel width is less than one mean free path.

5. Results and Discussions

5.1. Poiseuille flow

The plane Poiseuille flow is driven by the pressure difference between the inlet pressure and outlet pressure. The ratio of these two pressures is defined as Π. Pong *et al.*[27] studied the pressure distributions along planar microchannels using an array of surface micromachined piezoresistive pressure sensors. In the experiments they found that pressure distributions are nonlinear in the microchannels due to compressibility. The present computed pressure distributions along the stream-wise direction for nitrogen flow are compared with experimental results of Pong *et al.* in figure 1. Five inlet pressures are compared (p_i=135, 170, 205, 240 and 275kPa). Our present numerical results agree very well with experimental data. When the pressure ratio is small, the effect of compressibility is small and the pressure gradient is almost constant. The trend towards a nonlinear distribution is clear when the pressure ratio becomes large.

The detailed flow properties in microchannels in transition flow regime are beyond the present experiment technology. As a result, the molecule-based DSMC method is usually considered as an accurate method. Velocity profiles at Π =2.28, non-dimensionalized with the inlet velocity, u_i, at three different locations, are shown in figure 2, compared with DSMC results of Beskok[19]. The results of the Navier-Stokes with the same boundary conditions are also presented in the figure. The Knudsen number based on the outlet pressure is 0.20. It is found that the centerline velocities of the Navier-Stokes equations are slightly under predicted compared to DSMC results, while the Burnett results agree very well with DSMC, at all these three locations.

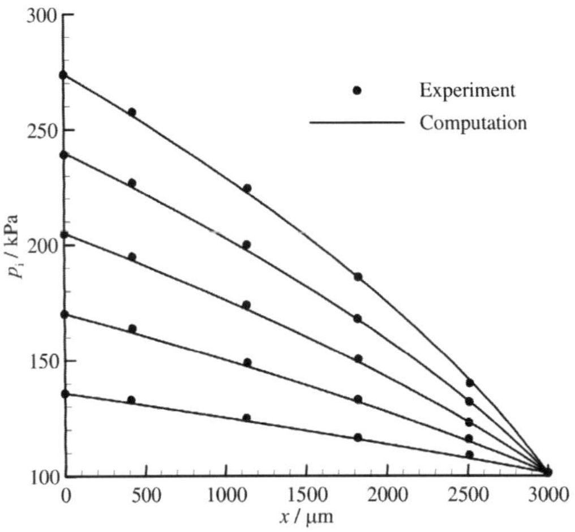

Figure 1. Pressure distribution along the channel at five inlet pressures.

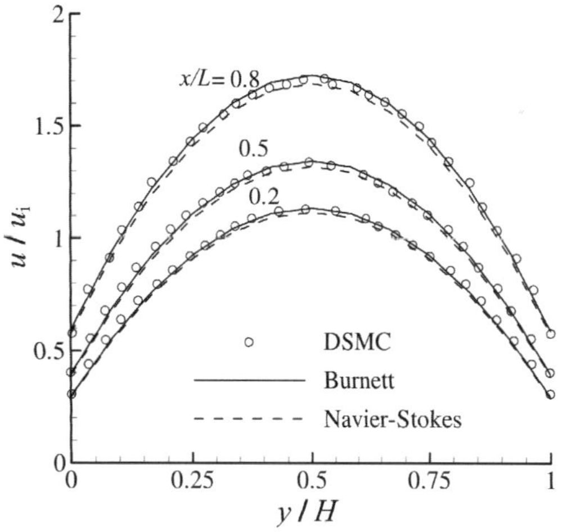

Figure 2. Velocity profiles at three locations, x/L =0.2, 0.5 and 0.8.

The velocity profiles at x/L=0.5 at five Knudsen numbers (Kn=0.01, 0.1, 0.2, 0.3, 0.4) are shown in figure 3 to give a detailed description of slip velocity. These velocities are non-dimensionalized with corresponding average velocities, u_{ave} at x/L=0.5. When Knudsen number is small, the maximum velocity is about 1.5 time the averaged value. But when Kn=0.4, this value falls to only 1.1. The velocity profile is flatter at higher Knudsen number. Also, we can found from the figure that the results of these two methods agree with each other at small *Kn*, e.g. *Kn*=0.01. But when *Kn* is larger, e.g. *Kn*=0.4, the results are different.

The axial variation of local Poiseuille number at different *Kn* is showed in figure 4. The Reynolds numbers in these cases are 2.2. Because of the difference in dimension, the inlet pressure is chosen to keep *Re* constant. The local Poiseuille number decreases when proceeding along the channel. At small *Kn* (e.g. *Kn*=0.01), *Po* decrease rapidly first near the inlet and then tends to a constant. But at large *Kn* (e.g. *Kn*=0.3), *Po* decreases along the channel, not only in the entrance region, but in the whole channel. The fluid accelerates along the channel because of the density reduction brought about by the pressure drop. The normal velocity gradient at the wall also increases along the channel. The numerical results in figure 4 reveal that the increase in u outweighs the increase in normal velocity gradient when *Kn* is large.

Figure 4 also demonstrates that the *Po* decreases with the increase of *Kn*. The velocity profile at larger *Kn* is flatter than that at smaller *Kn*. The normal velocity gradient at the wall decreases with the increase of *Kn* number. So the friction factor or the shear stress decreases with the increase of *Kn*.

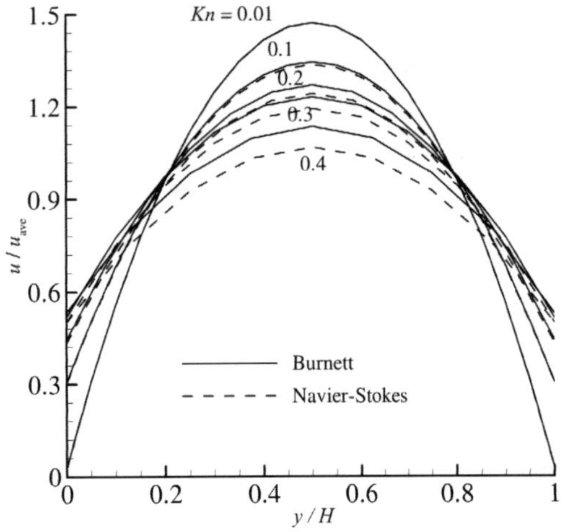

Figure 3. Velocity profiles of the Burnett and Navier-Stokes equations at different *Kn* at *x/L*=0.5.

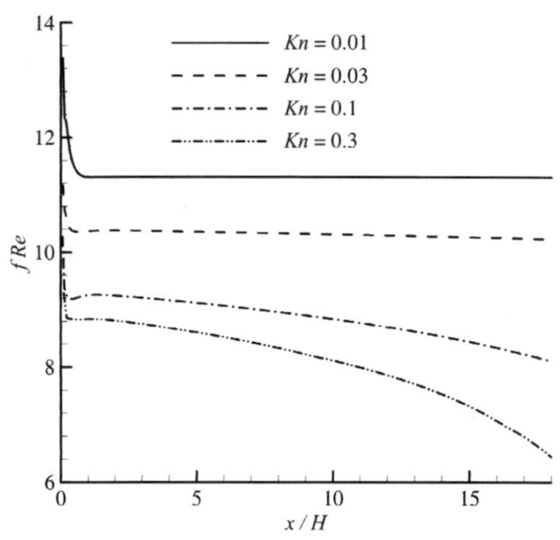

Figure 4. Variation of *Po* along the channel at different *Kn*.

5.2. Micro-filter flow

The flows in and around microfilter are then studied. Usually, the width of filter hole is much larger than the height, *h*. So, only two-dimensional flow is considered here. Considering that the filter holes repeat periodically, the single micro/nano-filter shown in figure 5 is studied in present paper. The opening factor of filter is defined as $\beta = h / H$. The flow is driven by the pressure difference between inlet and outlet. The outlet pressure is 10^5Pa throughout this section.

The present work first compares the velocity and pressure profiles obtained from the Burnett simulations with those of DSMC method[28]. The filter geometry and boundary conditions are all set the same in these two methods. The *Kn* of the flow in the simulation is 0.057. The distributions of centerline velocity and pressure are shown in figure 6. As can be seen from the figure, the velocity and pressure calculated from the Burnett equations agree well with DSMC results, which indicates that the Burnett equations are able to describe the gas flow characteristics in micro/nano-filters.

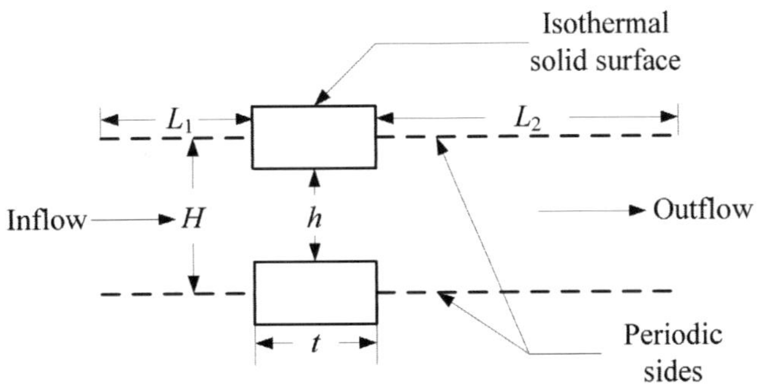

Figure 5. Schematic view of the micro/nano-filter

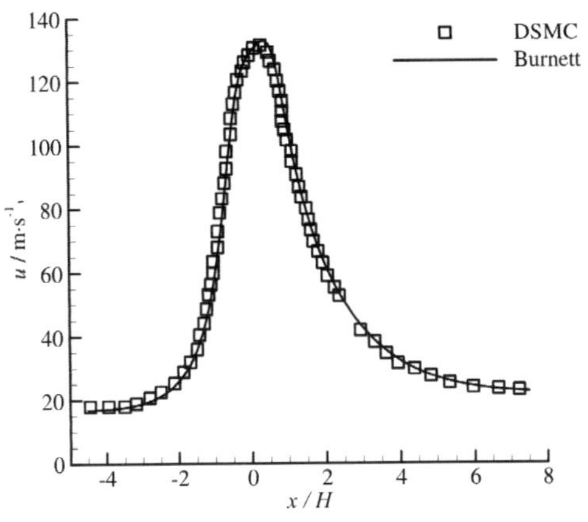

Figure 6(a). Centerline velocity distribution.

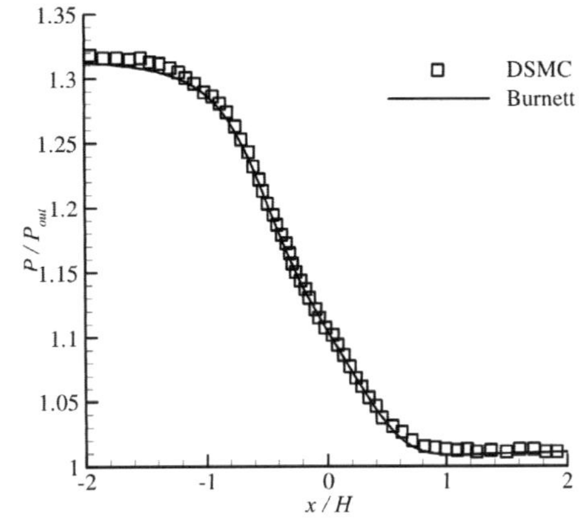

Figure 6(b). Centerline pressure distribution.

The variation of non-dimensional pressure drop, K, with Reynolds number, Re, is then analyzed, where K is defined as $K = \Delta P \Big/ \frac{1}{2}\rho U_{in}^2$. The opening factor, β, and the ratio of the filter thickness to height, t/h, are kept the same in the calculation. Three kinds of micro/nano-filter are investigated. The variation of K with Re are shown in figure 7, at different Kn numbers. As can be find from the simulation results, K is proportional to $1/Re$. The relation between them can be fitted according to the simulation as follows:

$$K = C_1\left(\frac{11}{Re}+1\right) \tag{11}$$

where C_1 is a function of factor β and the ratio t/h. It can also be found that the non-dimensional pressure drop is independent of Kn. This conclusion is different from that of Mott[29], in which K varies at different Kn.

The variation of K with opening factor, β, are then analyzed, with Re and t/h kept the same. Non-dimensional pressure drops at different β are calculated and the variations of pressure drop with factor β are shown in figure 8. As can be found from the figure, the pressure drop K decreases with the increase of factor β. According to the numerical results, the following relation between K and β can be achieved at different Re and t/h

$$C_1 = K\Big/\left(\frac{11}{Re}+1\right) = C_2\beta^{-1.78} \tag{12}$$

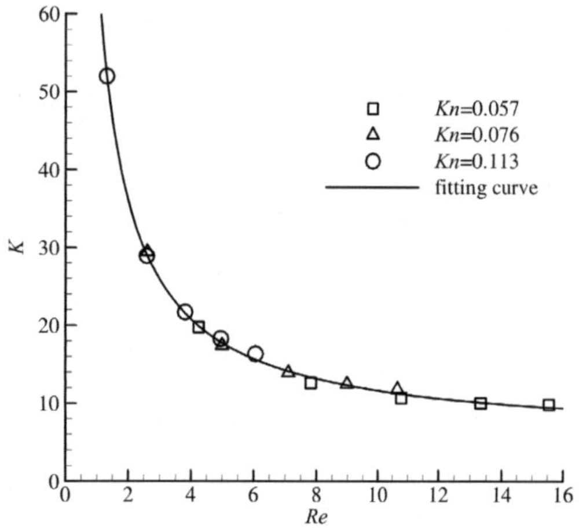

Figure 7. Variations of K with Re.

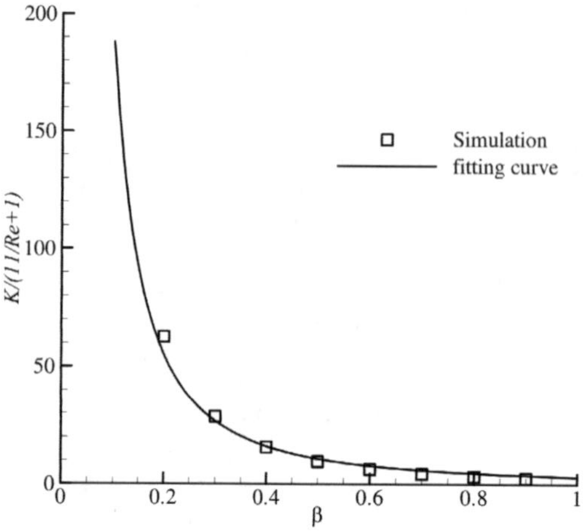

Figure 8. Variations of K with opening factor β.

The effect of the ratio of filter thickness to height on pressure drop are analyzed at last. Different ratios t/h are calculated and we can find from the results that the non-dimensional pressure drop increases proportionally with t/h. According to the numerical results, the following relation between K and t/h can be achieved at different Re and β.

$$C_2 = K\beta^{1.78} \Big/ \left(\frac{11}{Re}+1\right) = 1.91\, {t}/{h} + 1.25 \tag{13}$$

The variations of K with ratio t/h are shown in figure 9, with above fitting curve also included.

In conclusion, according to the numerical results, the variations of non-dimensional pressure drop K with Re, opening factor β and the ratio t/h can be characterized as follows,

$$K = \beta^{-1.78}\left(1.91\, {t}/{h} + 1.25\right)\left(\frac{11}{Re}+1\right). \tag{14}$$

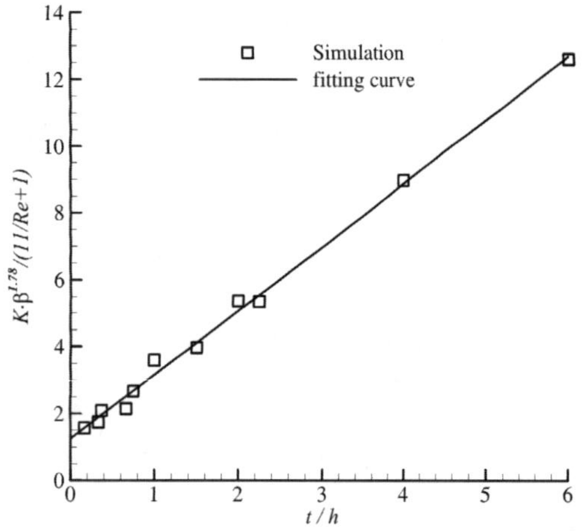

Figure 9. Variations of K with opening factor t/h.

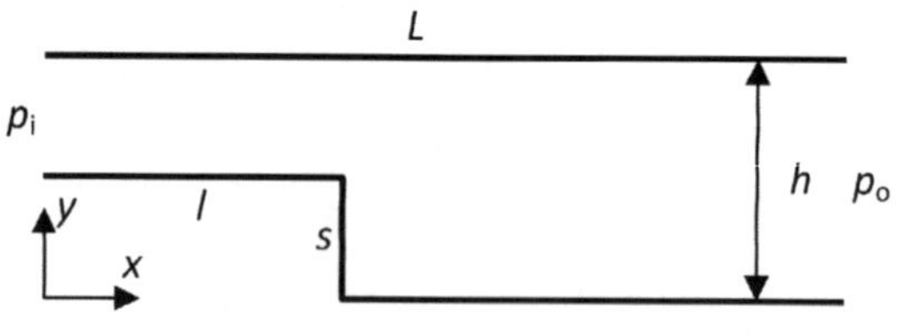

Figure 10. Schematic of backward-facing step flow.

5.3. Backward-facing step flow

The flow characteristics of backward-facing step flow are then studied. As shown in figure 10, the channel height is 0.5μm and the step ratio(*s/h*) is 0.5. The outlet pressure is 100KPa and the pressure ratio is 2. The *Kn* at inlet and outlet are both 0.136. The *Re* number based on the mass flow rate per unit width is 0.74.

The streamwise velocities at five transverse locations are shown in figure 11(a). It can be found from the figure that the streamwise velocity is always positive, which indicates that no reverse flow exits after the step. The adverse pressure gradient after the step is very small, too small to stagnate the flow. This is different from macro backward-facing step flow.

The slip velocities at top and bottom wall are coincidental in most regions, as illustrated in the figure. The effect of step can be neglected when *x/h* is large enough. The transverse velocities at different *y/h* locations are shown in figure 11b. The transverse velocity is negative and asymmetric toward the bottom wall, which results from the momentum transfer downward owing to the appearance of downward expansion. The max nondimensional transverse velocity is about 0.9. The transverse velocity becomes zero at about *x/h*=4.5. The streamwise velocities at different streamwise locations are shown in figure 11c. The velocity profile across the channel are symmetrical when *x/h*=5.

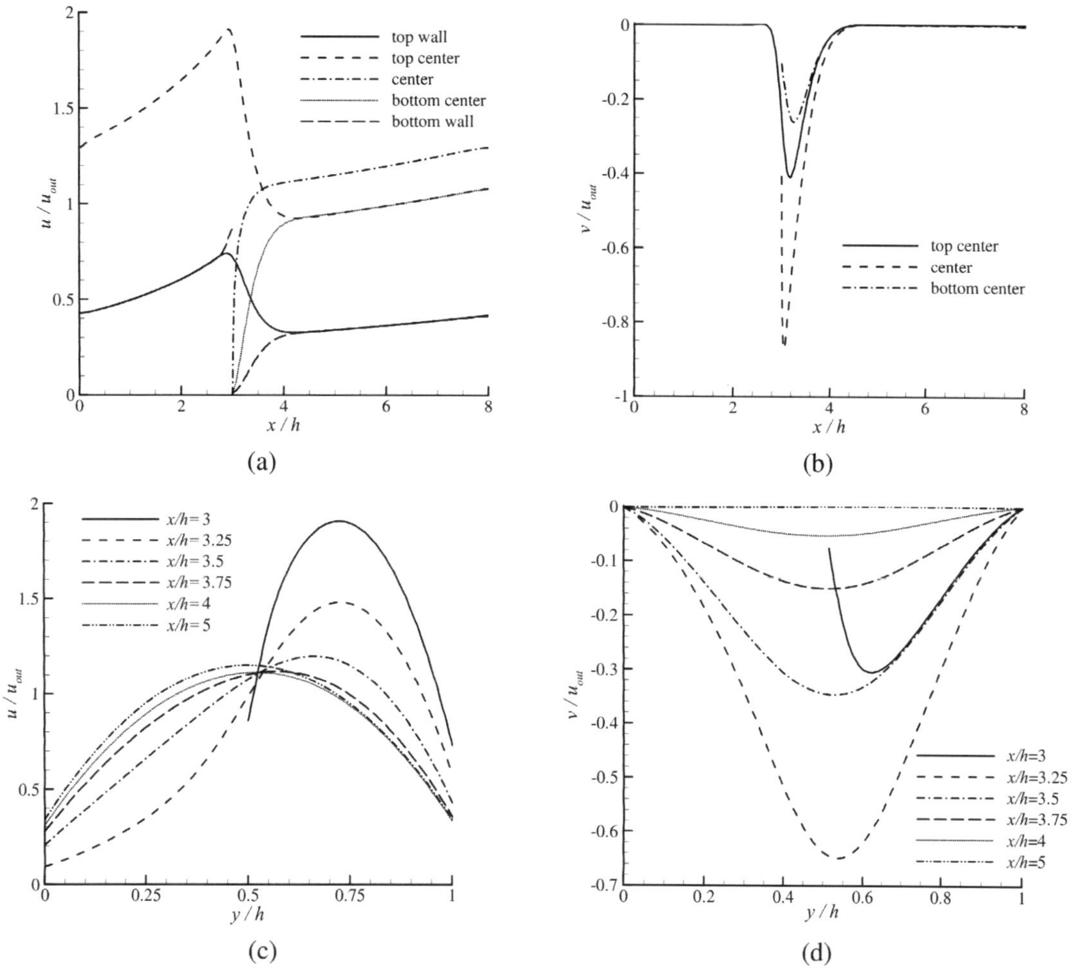

Figure 11. Velocity profiles in backward-facing step flow: (a) streamwise velocities at different transverse locations, (b) transverse velocities at different transverse locations, (c) streamwise velocities at different streamwise locations, (d) transverse velocities at different streamwise locations.

The transverse velocities at different streamwise locations are shown in figure 11d. When *x/h*=5, the transverse velocities is zero. The effect of step can be neglected when *x/h*=4.5 in present case. This location changes with pressure ratio, *Kn* number and step ratio *s/h*. In present simulation, *x/h*=8 is used to ensure that the characteristics of backward-facing step flow can be fully described.

The backward-facing step flows at different pressure ratios are then analyzed. The channel height is 0.5μm and the step ratio(*s/h*) is 0.5. Different pressure ratios are studied, from 1.12 to 5. The *Kn* at outlet is 0.136 and the *Kn* at inlet is a function of inlet pressure. *Kn* decreases with the increase of pressure. The variation of mass flow rate with pressure ratio is shown in figure 12. The mass flow rate increases with the increase of pressure ratio. The relation is not linear as in traditional flow. The gradient increases with the pressure ratio.

The centerline streamwise velocities at different pressure ratios are shown in figure 14, nondimensionalized with each average outlet velocity. The average outlet velocities at four pressure ratios, *Π*=1.5, 2, 2.5 and 3 are 9.85m/s, 21.68 m/s, 35.18 m/s and 50.07m/s, respectively. The velocity profiles at outlet are parabolic, so when nondimensionalized with average velocity, the center velocities at outlet are the same, as can be seen from the figure. The centre velocities are not 1.5, as in non-slip traditional flow. That's because of the existence of the slip velocity at the wall. The velocity profile across the channel is much flatter. When *Π*=1.5, the centerline velocity varies almost linearly with *x*, after *x/h*=4.5. But when *Π*=3, the centerline velocity are not linear with *x* because of the compressibility. This phenomenon is more obvious when the pressure ratio is larger.

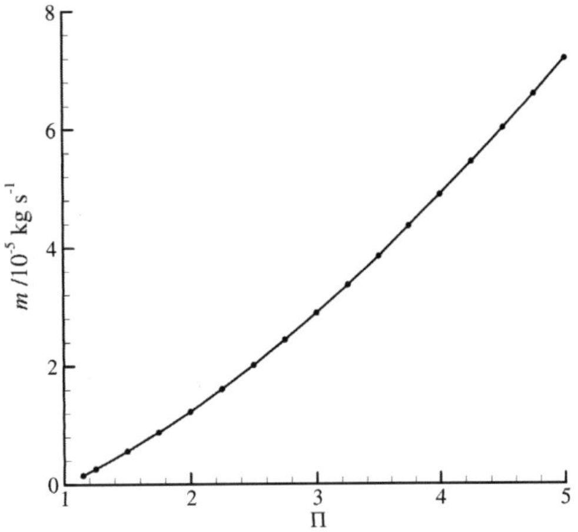

Figure 12. Variation of mass flow rate with pressure ratio.

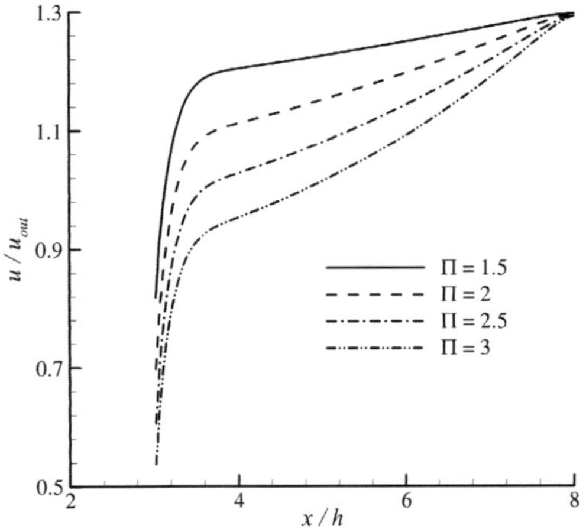

Figure 13. Centerline velocities along the channel at different pressure ratios.

The variations of slip velocity at top wall with pressure ratio are shown in figure 14, nondimensionalized with outlet velocity. The *Kn* at outlet is the same at different pressure ratios, so the slip velocities are calculated from the same location. The velocity profiles are the same, so the slip velocity are also the same, as illustrated from figure 14. At inlet, the mean free paths are different at different inlet pressures. The slip velocities are calculated from different locations. So the slip velocities nondimensionalized with average inlet velocities are different. The larger pressure ratio is, the smaller slip velocity will be.

The backward-facing step flows at different step ratios are then studied. The channel height is 0.5μm and the pressure ratio is 2. The *Kn* at outlet is 0.136. The variations of slip velocities at top wall are first illustrated in figure 15, nondimensionalized with average outlet velocity. The average outlet velocities at seven step ratios, *s/h*=0.2, 0.3, 0.4, 0.5, 0.6, 0.7 and 0.8 are 49.80m/s, 40.63m/s, 30.01m/s,

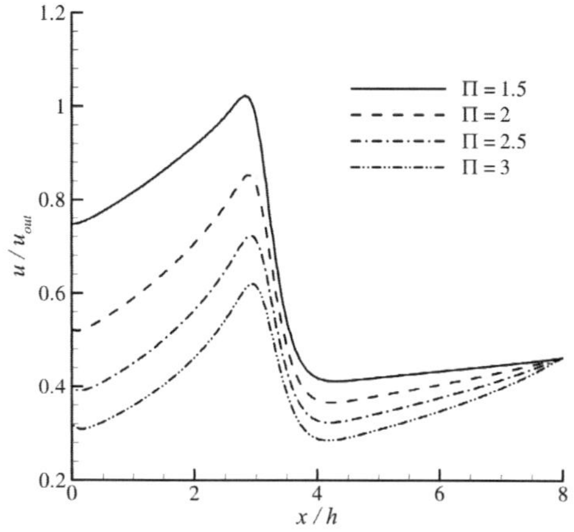

Figure 14. Slip velocities at different pressure ratios.

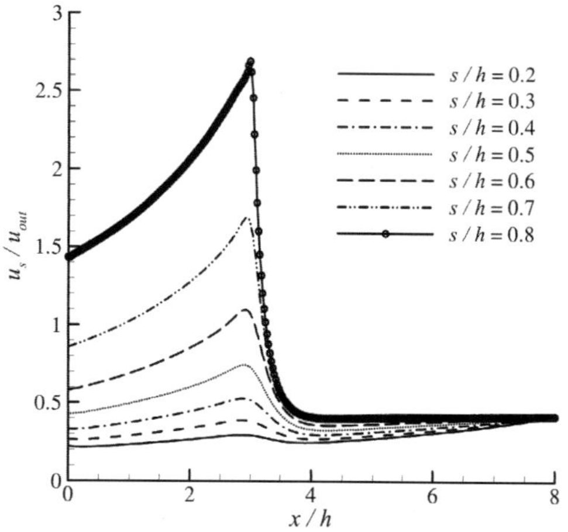

Figure 15. Slip velocities along the channel at different step ratios.

21.68m/s, 13.38m/s, 6.83m/s and 2.42m/s, respectively. The velocity profiles are parabolic and the *Kn* is the same at outlet. The nondimensional slip velocity at outlet are the same, u_s=0.42, as can be seen in the figure. The slip velocity before the step increases with the increase of step ratio.

The velocity gradient is larger in the front part of the backward-facing step flow. That's because the pressure are mainly drop in the front part of the step flow at large step ratio. The pressure gradient is much larger in the front part of the backward-facing step flow.

6. Conclusions

The gaseous flow characteristics in two dimensional micro/nano systems are investigated using the Burnett equations. Velocity slip and temperature jump boundary conditions are applied on solid surface. The relaxation method on slip values and Burnett terms are introduced to increase the stability of the Burnett equations. Convergent results at Knudsen number up to 0.5 are achieved for the first time in the present study.

The pressure driven plane Poiseuille flow, the backward-facing step flows and the flows in and around micro filters are studied. Different inlet to outlet pressure ratios, size effects, shape effects and boundary conditions are analyzed. The competitive relations of rarefied and compressible effect in micro gas flows are discussed. The Burnett equations are high efficiency method to model the gaseous flow in slip and early transition flow regimes in micro/nano-scale devices.

Acknowledgement

This work was supported by the National Natural Science Foundation of China (Grant No. 10902104), the Natural Science Foundation of Zhejiang Province(Grant No. Y6090589) and a grant from the Major State Basic Research Development Program of China (973 Program) (Grant No. 2011CB706501).

Reference

[1] Ho C M and Tai Y C Micro-Electro-Mechanical-Systems(MEMS) and fluid Flows 1998 *Annu. Rev.Fluid Mech.* **30** 579-612

[2] Agarwal R K, Yun K Y and Balakrishnan R Beyond Navier-Stokes: Burnett equations for flows in the continuum-transition regime 2001 *Phys.Fluids* **13** 3061-85

[3] Tsien H S Superaerodynamics, mechanics of rarefied gases 1946 *J. Aeronautical Sci.* **13** 653-664

[4] Gad-el-Hak M The fluid mechanics of microdevices - The Freeman Scholar Lecture 1999 *J.*

Fluids Eng.-T. ASME **121** 5-33
[5] Karniadakis G, Beskok A and Aluru N 2005 *Microflows and Nanoflows, fundamentals and Simulation* (Berlin: Springer)
[6] Harley J C, Huang Y, Bau H H Gas flow in micro-channels 1995 *J. Fluid Mech.* **284** 257-74
[7] Arkilic E B, BREUER K S and SCHMIDT M A Measurement of the mass flow and tangential momentum accommodation coefficient in silicon micromachined channels 2001 *J. Fluid Mech.* **437** 29-43
[8] Alexander F J and Garcia A L The Direct Simulation Monte Carlo Method 1997 *Comput. Phys.* **11** 588-593
[9] Oran E S, Oh C K and Cybyk B Z Direct simulation Monte Carlo: recent advances and applications 1998 *Annu. Rev. Fluid Mech.* **30** 403-441
[10] Chapman S and Cowling T G 1970 *The Mathematical Theory of Non-uniform Gases* (Cambridge: Cambridge University Press)
[11] Zhong X L On numerical solutions of Burnett equations for hypersonic flow past 2-D circular blunt leading edges in continuum transition regime 1993 *AIAA paper* 93-3092
[12] Fiscko K A and Chapman D R 1988 Comparison of Burnett, super-Burnett and Monte Carlo solutions for hypersonic shock structure in *Proc. of the 16th International Symposium on Rarefied Gas Dynamics*
[13] Xu K Super-Burnett solutions for Poiseuille flow 2003 Phys. Fluids **15** 2077-80
[14] Ohwada T and Xu K The kinetic scheme for the full-Burnett equations 2004 *J. Comput.l Phys.* **201** 315-32
[15] Balakrishnan R.An approach to entropy consistency in second-order hydrodynamic equations 2004 *J. Fluid Mech.* **503** 201-45
[16] Lockerby D A and Reese J M High-resolution Burnett simulations of micro Couette flow and heat transfer 2003 *J. Comput. Phys.* **188** 333-47
[17] Bao F B, Lin J Z and Shi X Burnett simulation of flow and heat transfer in micro Couette flow using second-order slip conditions 2007 *Heat Mass Transfer* **43** 559-66
[18] Fang Y C 2003 Parallel simulation of microflows by DSMC and Burnett equations (Western Michigan University)
[19] Beskok A and Karniadakis G E A model for flows in channels, pipes, and ducts at micro and nano scales 1999 *Microscale Thermophys. Eng.* **3** 43-77
[20] Zhong X L, MacCormack R W and Chapman D R Stabilization of the Burnett equations and application to high-altitude hypersonic flows 1991 *AIAA paper* 91-0770
[21] Bao FB and Lin JZ. Linear stability analysis for various forms of one-dimensional burnett equations 2005 *Int. J. Nonlinear Sci.* **6** 295-303
[22] Bao F B and Lin J Z Burnett simulation of gas flow in microchannels 2008 *Fluid Dyn. Res.* **40** 679-94
[23] Bao F B, Wang J X and Lin J Z Flow Characteristics of Microfilter in Slip/transition Regime 2011 *Mater. Sci. Forum* **694** 831-7
[24] Beskok A, Karniadakis G E and Trimmer W Rarefaction and compressibility effects in gas microflows 1996 *J. Fluids Eng.-T. ASME* **118** 448-56
[25] Arkilic E B, Schmidt M A and Breuer K S Gaseous slip flow in long microchannels 1997 *J. Microelectromechanical Systems* **6** 167-78
[26] Colin S, Lalonde P and Caen R Validation of a second-order slip flow model in rectangular microchannels 2004 *Heat Transfer Eng.* **25** 23-30
[27] Pong K C, Ho C M and Liu J Q Non-linear pressure distribution in uniform micro-channels 1994 *ASME FED* **197** 51-6
[28] AKtas O and Aluru N R A combined continuum DSMC technique for multiscale analysis of microfluidic filters 2002 *J. Comput. Phys.* **178** 342-72
[29] Mott D R, Oran E S and Kaplan C R Micromachined pariticlefilter simulations and scaling laws 2001 *J. Thermophys Heat Transfer* **15** 473-7

Quantitative measurement of gas pressure drop along T-shaped micro channels by interferometry

Y. Li[1,4], S. Joseph[1], S. Colin[2], L. Baldas[2], C. Barrot[2], S. Orieux[2], D. Newport[1] and J.J. Brandner[3]
[1]Stokes Institute, University of Limerick, Limerick, Ireland
[2]Université de Toulouse, INSA, UPS; ICA, Toulouse, France
[3]Institut für Mikroverfahrenstechnik (IMVT), Karlsruher Institut für Technologie (KIT), Eggenstein-Leopoldshafen, Germany

E-mail address: yongli.li@ul.ie

Abstract. The study of gas flows in microchannels has received considerably more attention in the literature from a simulation perspective than an experimental. The majority of the experimental work has emphasis on the global measurements at the inlet or exit of the microchannel instead locally along it. In this paper some efforts were made to measure the pressure drop along T-shaped micro channel by using interferometry. The two side channels were served as gas entrances and they were both open to air and the channel outlet was being vacuumed during experiments. A Mach-Zehnder interference microscopy was built for the measurement of gas pressure drop along the mixing channel. Some points along the mixing channel were selected for interferometric measurements. Simulations were first developed in unsteady condition by using Ansys Fluent to verify the nonexistence of transient phenomena of gas flow in the defined condition and then run again in steady condition to get the theoretical pressure drop that was would be used for comparison with experimental results.

1. Introduction

The experimental study on gas flow in microchannels was started 100 years ago, but the indirect measurement such as the pressure or temperature measurement near channel inlets and outlets still remains the most popular methods. The flow visualization in micro scale develops mainly in liquid phase microchannel flows rather than in gas phase [1], for example, the particle-based techniques PIV and LDV well developed in liquid flow has limitation in gaseous flow measurement because of the challenges associated with seeding and particle inertia [1, 2]. Molecular tagging approaches have advantages for velocimetry in gas-phase as it uses molecular tracers instead of particles. The flowing medium is premixed with molecule-seeds that can be turned into long-lifetime tracers upon excitation by photons of an appropriate wavelength [2]. But Molecular tagging methods involving seeding are often undesirable. The molecular seed can be toxic or environmentally hazardous, and it is often difficult to distribute the seeding throughout the flow [3]. So except the flow visualization techniques mentioned above, some totally non-intrusive techniques such as interferometry [4] have still been being developed in microfluidics. Based on some previous work [4, 5], this paper will study gas flow by using micro interferometry.

[4] To whom any correspondence should be addressed.

Published under licence by IOP Publishing Ltd

Interferometry can be applied to the measurement or visualization of changes in physical properties of transparent objects and such properties include mass density of fluids, electron density of plasma, temperature of fluids, and chemical species concentration in reacting gases and state of stress in solids [6]. The change of the refractive index is due to the change in density of the medium if it is homogeneous. For gases, the phase information is related to many physical properties, for example, temperature, pressure or density. In this paper, the local pressure measurement along microchannels by interferometry will be studied.

2. Experimental

2.1 Principles

If the microchannel together with gases is treated as a phase object and the probing optical rays are assumed to be straight lines. The optical path length difference $\Delta\Phi\,(x,y)$ of a ray traveling in z direction (the direction of depth) through a phase object is given by: [6]

$$\Delta\Phi\,(x,y) = \int [n(x,y,z) - n_0]\, dz \tag{1}$$

n_0 is the refractive index of phase object of the initial holographic exposure, n(x,y,z) is the refractive index distribution at the time of the second exposure.

The phase change $\Delta\delta$ is related to the optical length change by:

$$\Delta\delta = \frac{2\pi}{\lambda}\,\Delta\Phi \tag{2}$$

The equation (1) and (2) show that for one phase change measurement, two refractive index, n_0 and n(x,y,z), are necessary to be determined experimentally, the former is for the reference condition, and the latter is for the real gas flow condition.

2.2 Interferometry setup

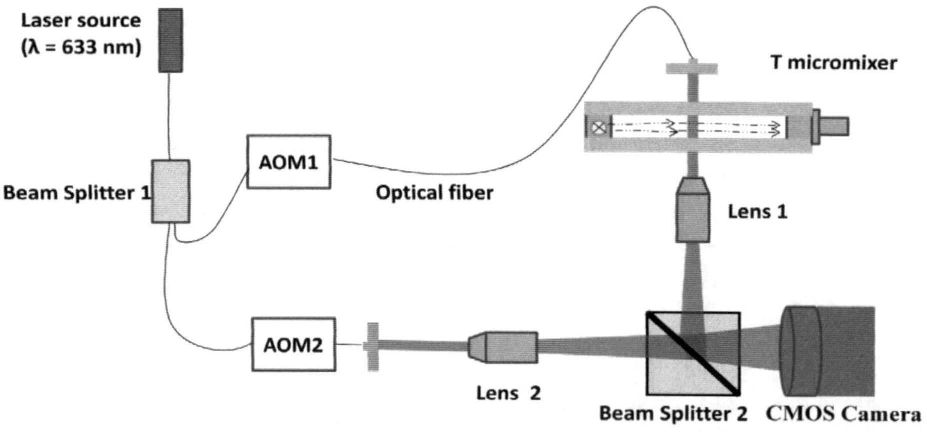

Figure 1. Schematic of the Mach-Zehnder interferometric microscope

Figure 1 shows the Mach-Zehnder interferometric microscope used in this study. A He-Ne laser with a wavelength of 633 nm is connected to the beam splitter 1 (BS1) by optical fibres. Then BS1 splits the laser beam into two laser beams with equal intensity: reference beam and measurement beam. Both beams travel through the acoustic optic modulators (AOM1 or 2) separately. The reference beam goes across the channel sample, and the gas property change can be detected by the phase information at the local measurement point. After magnification by tube lens (Lens1 and Lens2), the two beams recombine again at the beam splitter 2 (BS2). A CMOS camera is put after the recombined beam to acquire the fringes images. The speed of camera is set as 336 f/s that is 4 times of the frequency shift introduced by AOMs.

2.3 Microchannel

The channel used in this work was manufactured in a previous work (patent pending). The channel comprises of three layers: a polymer layer (detailed in Figure1) is sandwiched by two quartz glass plates that give optical access.

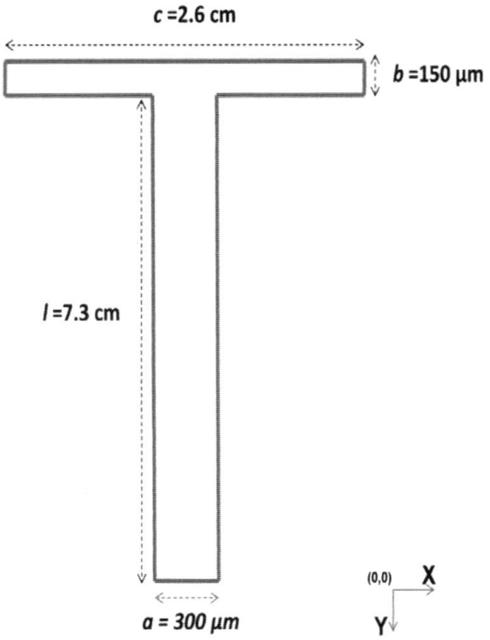

Figure 2. Demonstration of channel polymer layer

From the physical point of view, the local heating of the wall due to the continuous laser could be an issue as it introduces undesirable heat source to the gas flow. It is important to make this influence as small as possible through ensuring: a) avoiding significant laser absorbance by using quartz glasses. The two quartz glass plates with1 mm each have about 90% transmission ratio in the wavelength of 633 nm [7]; b) using lower laser intensity, the laser power used in this work is 35mv, and it is split into two beams, then both beams pass though the optical fibres, attenuating the intensity in the optical fibres. Especially the power suppliers of AOMs can also be used to reduce the output laser intensity; c) Fast measurement: it takes about 1s for 10 measurements. d) Studying pressure drop instead of temperature change. The later simulation results show that in the experimental condition, phase change is not sensitive to temperature change but only to pressure drop. So the laser induced heat is negligible in this work.

The T-mixer has a square section of 300 x 300 μm^2 and a length of 73 mm for the mixing channel and a rectangular section of 150 μm in width x 300 μm in height and a length of 12 mm for each side channel (see Figure 2). Each ending of the channel has a connector that does not have good optical transparency and it occupies roughly 1.3 cm. The analysis near the channel endings is limited.

2.4 Experimental process

Several points along the microchannel were selected for phase change analyse, but each time the interferometric system can only measure the phase information at one location, so it is important to develop a system to keep the gas flow constant, which allows the pressure change measurement in different locations sequentially without changing the experimental condition. The system shown in Figure 3 was developed for this purpose. The two entrance channels are open to air to make the inlet pressure constant at 1atm. A vacuum pump is running during the experiments to keep the outlet pressure stable at around 4000 Pa. A pressure gauge is used to monitor the outlet pressure.

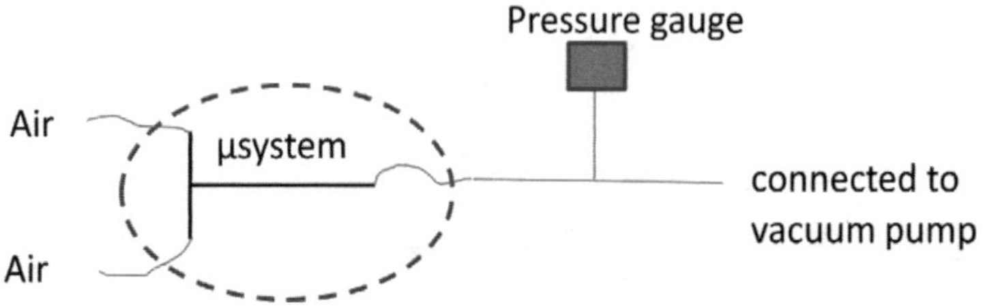

Figure 3. Schema of the gas mass flow control system

2.5 Phase extraction process

There are two methods for phase extraction; one is known as phase stepping where an interferogram is recorded after each known phase step is introduced to the system. The second is the integrating-bucket method, in which the intensity is detected as the phase is shifted continuously [8]. The former is used in this study.

The phase difference between the beams of the interferometer is determined by a measure of the intensity at the interference point for a range of phase shifts, which is introduced by using AOMs in this work. The mathematical expression of the interference of two light waves has already been derived. This can be expressed in terms of the average intensity and the modulation level [9].

$$I_i = I_0(1 + \gamma \cos \delta) \tag{3}$$

From Equation (3) it can be seen that there are three unknowns, hence a minimum of three equations is required to solve for the phase data. By the introduction of a known phase step, α_i, the interference expression becomes:

$$I_i = I_0(1 + \gamma \cos(\delta + \alpha_i) \tag{4}$$

A four-step method is adopted. By repetition of this process with varied phase steps of multiples of $\pi/2$ Equation (4) may be solved. The four steps introduced are $\alpha = 0$, $\pi/2$, π and $3\pi/2$ and result in the three intensity equations (5-8).

$$I_1 = I_0[1 + \gamma \cos(\delta)] \tag{5}$$

$$I_1 = I_0[1 + \gamma \cos(\delta + \pi/2)] \tag{6}$$

$$I_1 = I_0[1 + \gamma \cos(\delta + \pi)] \tag{7}$$

$$I_1 = I_0[1 + \gamma \cos(\delta + 3\pi/2)] \tag{8}$$

Hence, the unknown phase difference in the interference pattern can be retrieved by: [10]

$$\delta = tan^{-1}\frac{(I_4-I_2)}{(I_1-I_3)} \tag{9}$$

3. Simulation and calculation

3.1 Simulation with Ansys Fluent

The computational volume was build with GAMBIT. The domain was represented with a grid constructed from hexagonal cells. Along the X - axial direction, the grid had 20 points evenly distributed in the T – junction and 150 points were distributed in each entrance channel with a slightly less clustered tendency towards the inlet. Along the Y - axial direction, the grid had 10 points evenly distributed in the T-junction and 800 points were distributed in the main channel with a slightly less clustered tendency towards the outlet. Along the Z-axial direction (vertical to screen), the grid had 10 points evenly distributed. Then the simulation was performed with Ansys Fluent. The inlet pressures of 100 kPa and the outlet pressure of 4 kPa were taken as the boundary condition.

The possibility of existence of transient flow phenomena for symmetrical 1:1 mixing in T-shaped micro channels with rectangular cross-section was approved in Dreher's work [11] depending on Reynolds numbers. In this case of study, the transient phenomena may introduce pressure or temperature fluctuations that could contribute to significant phase fluctuations at one location, which could make the extracted phase values more scattered even failed. So it was necessary to avoid this happening in this work. Because the T-mixer dimension and the fluids used in Dreher's work were different from ours, the range of Reynolds numbers to get steady flow could be different. For this purpose a simulation with Ansys Fluent was run and the transient calculations were first performed. Some flow points with a distance of 100 μm to the mixing channel wall were selected and their velocities were monitored over time. The time step was set as 1 x 10^{-5} s and the duration correspondent to real experiment was 50 ms. The results show that the velocity fluctuation is less than 0.03% (see Figure 4), which proves the flow in the experiment would be steady flow. Then the simulation was run again under steady condition. The pressure and temperature distribution inside the T micro channel were simulated by using laminar model.

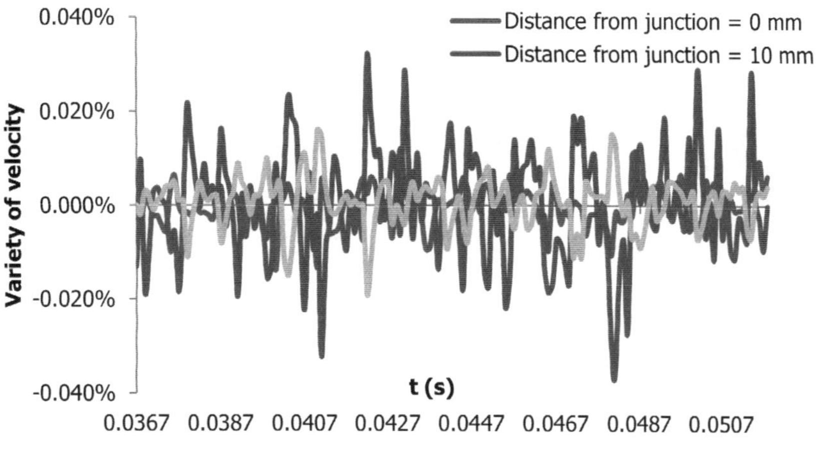

Figure 4. Velocity fluctuation

3.2 Relationship between the refractive index of air and pressure / temperature

The refractive index of air was well studied in Bengte's work [12] and then updated by Birch [13, 14]. It can be expressed as a function of air pressure and temperature by the equations (10) and (11). The agreement between calculation and experiment within an experimental uncertainty can be reduced to 3 x 10^{-8}. In reality, due to the carbon dioxide and humidity content of laboratory air can be very different, so the prediction of air pressure and temperature from the equations (10) and (11) can be expected to have a higher uncertainty than reported in literature.

The value of $(n-1)_s$, the refractivity of standard air at 1 atmosphere and 15°C, is calculated from the dispersion formula which was given as

$$(n-1)_s \times 10^8 = 8342.54 + 2406147(130 - \sigma^2)^{-1} + 15998(38.9 - \sigma^2)^{-1} \tag{10}$$

Where σ is the vacuum wavenumber and is expressed in μm^{-1}.

$$(n-1)_{t,p} = \frac{p(n-1)_s}{96095.43} \times \frac{1+10^{-8}(0.601-0.00972\,t)p}{1+0.0036610\,t} \tag{11}$$

Where $(n-1)_{tp}$ is the refractivity of standard air, the temperature t is expressed in degrees Celsius and the air pressure p in Pascal.

4. Results and discussion

The channel materials were quratz glass and polymers that had low thermal conductivity, but as the temperature could influence the phase change, the simulations with both adiabatic and isothermal boundary conditions were carried out separately. It is believed that the real temperature distribution should be between those found in the two simulations. As the goal of this work was to find the pressure drop along the channel by analyzing phase information obtained, it was expected that the temperature influence could be neglected. Figures 5 and 6 show that in this experimental condition, thermal boundary condition settings have much influence in temperature distribution but not in pressure distribution.

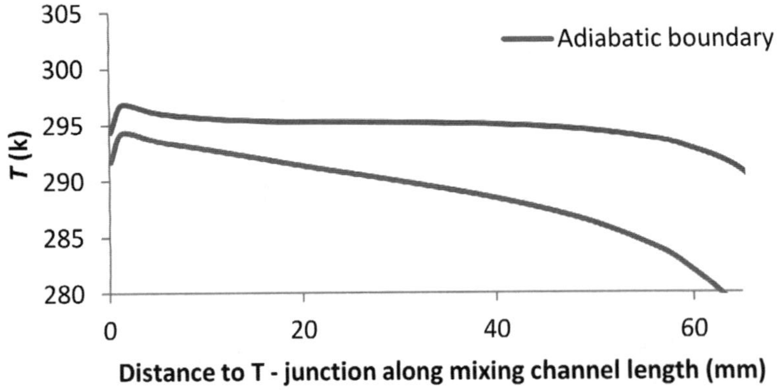

Figure 5. Temperature distributions along mixing channel

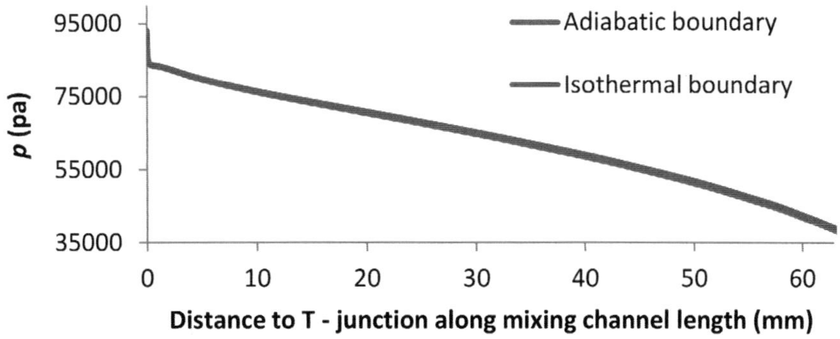

Figure 6. Pressure distributions along mixing channel

From the pressure and temperature distribution curves, the phase change information were calculated and presented in Figure 7. It is noted that for the phase change curve along the mixing channel is almost independent to thermal boundary settings, the same as pressure distribution (see Figure.5) but different from temperature distribution (see Figure. 6). That is, the temperature difference in the two cases does not cause significant phase difference, which indicates the phase analysis will be only the function of pressure in the defined experimental condition, as the contribution to phase change in gas flow are mainly by temperature and pressure change.

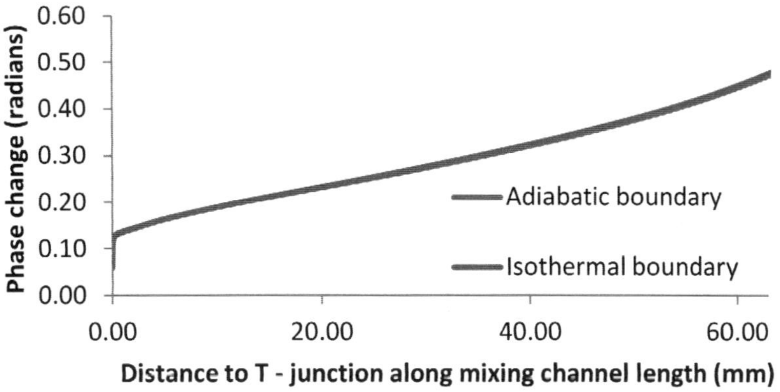

Figure 7. Expected phase change along the mixing channel

In experiments, the phase difference between the reference condition and real flow condition will be measured by interferometry microscope shown in Figure 1, then extracted by the 4 steps methods phase extraction process and finally translated into pressure difference. For reference condition, the pump is shut down and there is no flow inside the micro system, so the air inside the channel has the pressure of 1 atm. The phase change curve (see Figure 7) has a converse tend compared to the pressure drop curve (see figures 5 and 6), for the reason that the air at atmosphere pressure is taken as reference condition, and the pressure in the defined gas flow condition is always below 1atm.

Figure 7 shows that the phase change happening in this work will be between 0.1 and 0.5 radians, compared to the temporal noise of this setup, 0.01 radians, in some areas of the micro channel the phase measurement could be good while it does exist some areas where the phase measurement could have a high deviation.

The noise level is an important factor to limit the use of this technique for gas pressure drop analyse. There exists vibration during the experiments, which introduces temporal noise to the measurement. The optical length fluctuation due to the temporal noise in the present setup is roughly 1 nm. Spatial noise is not very sensitive in this case, because the average value for the measurement at one point is taken instead of analyzing the phase pixel by pixel, the special noise could be averaged out. To reduce noise level in these experiments, there are some solutions, for example, using high speed detectors, increasing gas density or channel dimension, or even performing the experiments in midnight when the laboratory and its surroundings are quiet, etc.

5. Conclusions

The flow studied in this work is a steady flow and there will be no influences on phase measurement due to flow fluctuations. Pressure drop measurement by interferometry in this selected experimental condition is feasible as the phase change is not sensitive to temperature change. But the interferometry technique is sensitive to noise, and the good experimental results can only be expected in some locations of the mixing channel under the defined flow condition.

6. References

[1] Sinton D 2004 *Microfluid Nanofluid* **1** 2–21
[2] Stier B, Koochesfahani M M 1999 *Experiments in Fluids* **26** 297-304
[3] Miles R B, Grinstead J, Kohl R H, Diskin G 2000 *Meas Sci Technol* **11** 1272–1281
[4] Newport D, Garvey J , Dalton T, Egan V and Whelan M 2004 *Microscale Therm Eng*, **8** 141–154
[5] Li Y, Newport D, Joseph S, and Carreras, M P 2010 *Proc. of 2nd International GASMEMS Workshop* Vol(5) pp1-8
[6] Charles M V 1979 *Holographic Interferometry* (NY: John Wiley & Sons)
[7] Technical Information Sheets from suss-microoptics *http://www.suss-microoptics.com/downloads/SMO_TechInfo_Sheet_09.pdf*
[8] Gravey J 2005 *the application of interferometry to micro channel phase measurement* (PhD thesis of University of Limerick)
[9] Cloud G L 1995 *Optical Methods of Engineering Analysis* (Cambridge, Cambridge University Press)
[10] Wyant J C 1982 *Laser Focus* **18** 65-71
[11] Dreher S, Kockmann N, and Woias P 2009 *heat transfer engineering* **30** (1–2) 91-100
[12] Edlen B 1966 *Metrologia* **2 (2)** 71-80
[13] Birch K P and Downs M J 1993 *Metrologia* **30** (155) 155-162
[14] Birch K P and Downs M J 1994 *Metrologia* **31** (315) 315-316

Acknowledgements

The research leading to these results has received funding from the European Community's Seventh Framework Programme (ITN - FP7/2007-2013) under grant agreement n° 215504.

Pressure dependence of the quality factor of a micromachined cantilever in rarefied gases

Michael Stifter, Matthias Sachse, Thilo Sauter and Wilfried Hortschitz
Institute for Integrated Sensor Systems, Austrian Academy of Sciences,
Viktor Kaplan Straße 2, A-2700 Wr. Neustadt

Franz Keplinger
Institute of Sensor and Actuator Systems, Vienna University of Technology,
Gusshausstraße 27-29, A-1040 Wien

E-mail: Michael.Stifter@oeaw.ac.at

Abstract. We present a study of the damping behavior of monocrystalline silicon cantilevers in different rarefied gas regimes. Mechanical quality factors Q were analyzed at controlled ambient pressures in the range of 0.01 Pa to 100 Pa. Emphasis was laid on the investigation of the fundamental vibration mode. Hence, the test structures were harmonically excited by the Lorentz force acting on the current carrying lead attached to the top surface of the cantilever. The micromachined clamped-free cantilevers featuring a length of 2 mm, a width of 1.5 mm and a thickness of 20 μm, were manufactured in SOI technology. The experimental results were compared with existing theories revealing an underestimate of the damping parameter for the Knudsen range $Kn = 0.1$ to 10. So far, squeeze-film damping by free molecular flow and kinetic damping were taken into account in damping models for the quasi-molecular regime. However, our measurements indicate that also the ongoing molecular flow around the test structures has to be considered. Hence the damping coefficient has to be calculated with methods of the free molecular aerodynamics. Thus, we used an algorithm based on the random walk model that allows the usage of already available knowledge in the field of Direct Simulation Monte Carlo. With this approach the quality factor of a squeezed-film damped cantilever in the quasi-molecular regime was derived. The results were compared with the most recent stochastic model, where the theoretical predictions and the experimental investigations indicate significant squeezing up to a Knudsen number of 10. In a superposition of both damping mechanisms, kinetic and squeeze-film damping, a satisfactory characterization of the damping behavior of an oscillating cantilever in the quasi-molecular regime with Knudsen numbers in the range of 10 down to 0.02 was achieved.

1. Introduction

For many years, micromechanical structures with all kinds of geometries have been used in various ways in microelectromechanical systems (MEMS-devices). Single-crystalline silicon has superior mechanical properties with regard to elastic parameters which permit reproducibility and long-term stability better than at other materials [1]. Furthermore, silicon-based production enables an integration of microelectronic and micromechanical components.

The resolution of systems with resonant vibrating structures is determined by the mechanical quality factor. With increasing quality factor, the resonance amplification becomes more and more pronounced. As a result, the resolution of a sensor that evaluates the resonant frequency is improved, too.

If the mechanical quality factor is derived theoretically, several kinds of damping mechanisms, like thermo-elastic damping (TED), gas damping or structural damping have to be considered which are adequately described in literature [2–9].

This work emphasizes the dependency of the quality factor on the ambient pressure, since the related extrinsic damping mechanism dominates for typical MEMS structures in the pressure range of 1 up to 100 Pa. Due to the gas-solid interactions this range is further divided into molecular and transition regime. In these regimes, a part of the extrinsic damping is caused by collisions of impinging molecules with the structure. The regime names indicate that the conventional equations of continuum theory are inappropriate to describe rarefied gas flows. Continuum (gas) damping requires a sufficient particle density in the flow, which is not guaranteed in the transition regime. The calculation of transition flows can only be carried out on the basis of the kinetic theory, where the characteristic equation is the Boltzmann-equation. This equation ensures not only a complete description of the free molecular flow, but implies also continuum flows as an approximation. There are three dimensionless parameters which determine the flow regime: the Mach number M, the Knudsen number Kn and the Reynolds number Re [10]. For typical MEMS applications two of them, M and $Re \ll 1$, are negligible because of subsonic vibration velocities. This leaves the Knudsen number as the most important fundamental parameter,

$$Kn = \frac{\lambda}{b} = \frac{k_B T}{\sqrt{2}\pi p \sigma^2 b}, \tag{1}$$

where λ is the mean free path, k_B the Boltzmann constant, T the ambient temperature, p the ambient pressure, $\sigma = 4.19 \cdot 10^{-10}$ m the diameter of air molecules and b the width of the cantilever [11]. Both in the theoretical and experimental description, the overall pressure range is divided into an intrinsic (0.01 Pa to 1 Pa) and a quasi-molecular (1 Pa to 100 Pa) regime, depending on the dominant damping mechanism [2]. For cantilevers with a characteristic width b of 1.5 mm the Knudsen number in the intrinisic regime is always greater than or equal to 10. This is therefore a sufficient condition to separate the molecular (Knudsen) from the transition flow.

The dominant damping mechanism in the intrinsic regime is structural damping caused by the mounting of the vibrating structure to the chip carrier [4, 8]. Another damping mechanism in the intrinsic regime is the thermo-elastic damping (TED), where the flexural oscillations are modeled as acoustic modes interacting with the thermally excited phonons of the vibrating structure. Through the interaction between the field of the mechanical stress and strain and the temperature field of thermal phonons, energy can be dissipated.

The quasi-molecular regime coincides with Knudsen numbers in the range of 0.1 to 10, whereas at the lower boundary the molecular flow enters into the slip flow. The dominant damping mechanisms in this regime are structure-molecular and intermolecular collisions. Analytical solutions of the Boltzmann equation in the quasi-molecular regime do not exist, but several approximations like the Burnett equation are available [12]. One possibility to get a better insight into this transition regime is to use the assumptions of the free molecular flow and to compare the systematic deviations of the measurement results with the analytical solutions of the free molecular flow. As an important precondition for the experimental evaluation surrounding surfaces from the bulk material, the mounting and the measurement chamber are in sufficiently large distance to the cantilever so that their influence on the damping of the fundamental mode of the device can be neglected.

2. Device Fabrication

The fabrication process for the device is based on the silicon-on-insulator (SOI) technology. The handle layer of the SOI wafer is passivated with silicon dioxide (SiO_2). First of all, metal lines are evaporated and patterned with lift-off technique using an image reversal photoresist (AZ 5214). Both the current-carrying leads and the bond pads consists of a 300 nm thick gold layer and a 70 nm chromium layer acting as bonding agent. In the resulting multi-layered structure the stresses of the chromium and the gold layer compensate each other and the structure remains undeflected. In a second step, the silicon wafer is etched from the back side anisotropically with 30 wt % KOH solution at 75 °C. Afterwards, the membrane is dry etched from the top side using a DRIE (Deep-Reactive-Ion-Etching) process, which finally forms the cantilever.

The structure, schematically shown in figure 1, consists of a clamped-free cantilever with a length l of 2 mm, a width b of 1.5 mm, and a thickness h of the silicon device layer of 20 μm. The thickness of the handle wafer is 350 μm.

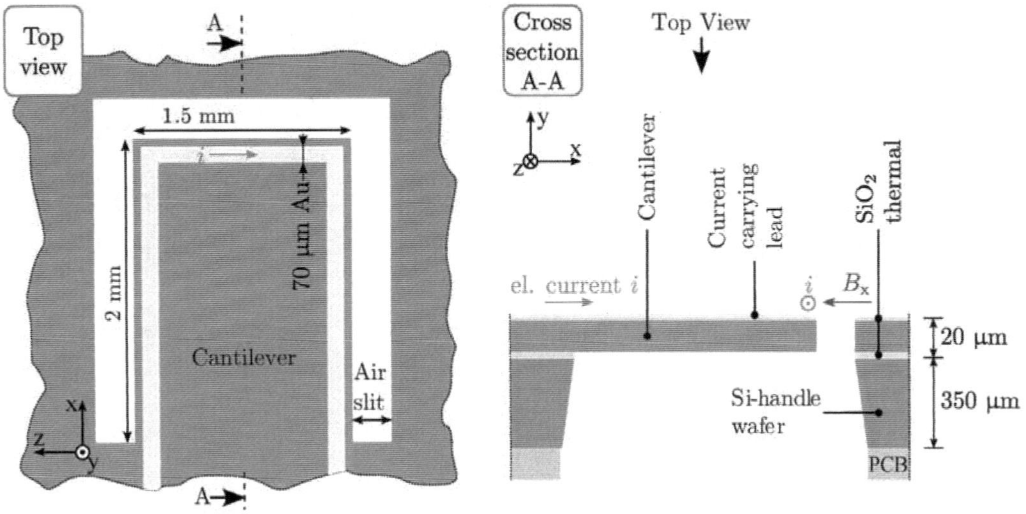

(a) Enlarged top view of the cantilever.

(b) Schematic cross section of the test device.

Figure 1: Schematic top view and cross-section of the test device with the Au-lead on the top of the Lorentz-force actuated cantilever.

3. Experimental Setup

A scanning laser Doppler-vibrometer (Polytec MSA400 Micromotion Analyzer, shown in figure 2) comprising displacement and velocity decoders is used to characterize the out-of-plane vibrations of the cantilever structures. Because of its considerable broadband noise, the vibrometer output signal is measured with a lock-in amplifier (SR830, Stanford Research Instruments). A waveform generator excites the vibrating structure around its natural frequency in a static magnetic field. The sinusoidal current in the lead on the top of the cantilever causes a harmonic oscillation due to the Lorentz force. The waveform generator sweeps stepwise ($\Delta f = 0.1$ Hz) around the natural frequency and synchronizes the vibrometer DSP and the Lock-in amplifier measuring the vibrometer's output signal. Owing to high intrinsic quality factors, the cantilever vibration may take several seconds to settle. The amplitude of the cantilever vibration due to the excitation is always much smaller than its length l to ensure linearity.

The test device shown in figure 3 is placed in a vacuum chamber surrounded by a Helmholtz coil that generates the static magnetic field. The nitrogen pressure in the vacuum-chamber is measured with a high vacuum gauge (Pfeiffer IMR 265) and maintained by a flow controller (MKS 50 sccm) at the high vacuum port of a turbomolecular pump, enabling a dynamic equilibrium in the pressure range of 20 mPa to 200 Pa. As a result a fully automated control of the nitrogen atmosphere is established for interactions with the vibrating test structure.

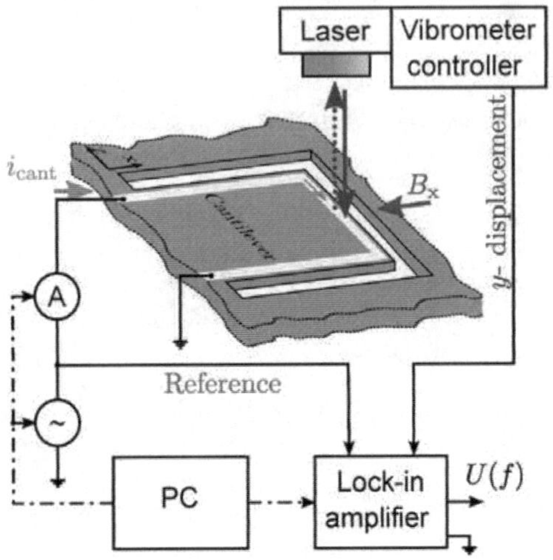

Figure 2
Schematic setup for the characterization of the test structures.

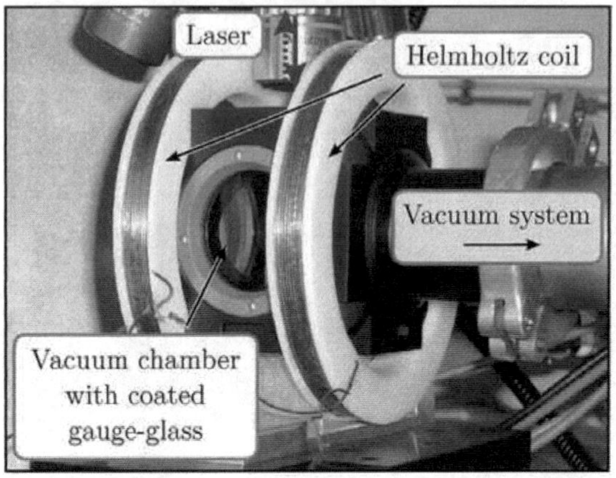

Figure 3
Vacuum chamber (black cube) with the surrounding Helmholtz coil.

4. Mechanical Quality Factor of Vibrating Cantilevers

One possible and widely used definition of the quality factor of a damped mechanical system is based on energy considerations

$$Q = \frac{2\pi E_\mathrm{i}}{E_\mathrm{d}}, \tag{2}$$

where E_i is the stored vibrational energy and E_d is the dissipated energy per oscillation period. In the case of a weakly damped system it is feasible to determine the quality factor using the linewidth of the resonance. At the lower boundary of the investigated pressure range intrinsic and quasi-molecular damping governs the quality factor. According to equation (2) the resulting quality factor

$$Q = \left(\frac{1}{Q_\mathrm{intrinsic}} + \frac{1}{Q_\mathrm{quasi-molecular}} \right)^{-1} \tag{3}$$

is the reciprocal sum of all inverse individual quality factors. The intrinsic quality factor is independent of the ambient pressure and can therefore be found experimentally at the ultimate pressure of the vacuum system (20 mPa). In 1992 Blom et al. first published the quasi-molecular quality factor of a cantilever,

$$Q_{\text{Cantilever}} = \frac{\rho A \omega}{f_1}, \tag{4}$$

where ρ is the density of the beam material, A is the cross-sectional area, ω and f_1 denote the resonance angular frequency and the dissipative drag parameter per unit length [2]. The drag parameter

$$f_1 = k_{\text{m}} b p \tag{5}$$

was first published by Drawin in 1962, where k_{m} is the damping coefficient, b is the width of the cantilever and p the ambient pressure [13]. The idea of his derivation is based on a net difference of the particle density between the top and the bottom side of the vibrating cantilever. In consequence, a nonzero molecular pressure difference acts on the cantilever, which counteracts the propulsive forces. Several publications have laid emphasis on the determination of the damping coefficient, but Martin et al. in 2008 has given an upper limit

$$k_{\text{m}} = \frac{C_{\text{D}}}{c} = 3.9008/c, \tag{6}$$

with full momentum accommodation, where the damping coefficient k_{m} is expressed in terms of the most probable speed c (=thermal speed) and a damping parameter C_{D} [2, 14, 15]. Thermal effects are completely neglected, but the result of equation (6) is valid for a wide range of MEMS cantilevers, where the height h is much smaller than the width b of the beam. The minimum quality factor of a cantilever can be retrieved by combining equations (4), (5) and (6)

$$Q_{\text{kinetic}} = \frac{\rho h \omega}{p} \frac{c}{C_{\text{D}}}, \tag{7}$$

with a dominant damping mechanism due to structure-molecular collisions in the quasi-molecular regime.

5. Squeeze-film Damping in the Quasi-Molecular Regime

In the range of Knudsen numbers $Kn \ll 1$ viscous damping mechanisms like squeeze-film damping dominate the gas-surface interactions. Viscous squeeze-film damping is well known and exactly analyzed [16, 17]. Due to flexural vibrations of the cantilever perpendicular to a fixed wall the gas in between is compressed [16]. The gas can partially escape through air slits that surround the moving cantilever opposite to the chip carrier, which induces friction losses since compression work is not fully recovered. Rarefaction effects were considered by Veijola et al. into the linearized Reynolds equation using an effective viscosity model, which permits the prediction of the flow in rarefied gases [18]. The Reynolds equation is a simplification of the Navier-Stokes equation. Hence, it stands for a continuum approach, that is adapted to the transition between the macroscopic and microscopic description of rarefied gas flows [19]. But the Navier-Stokes equation in turn applies only to a subset of problems covered by the Boltzmann equation.

A numerical solution of the latter based on the DSMC method was chosen by Alexeenko et al., and a promising analytical approach was introduced by Suijlen et al. in 2009 [20, 21]. This work presents a model of squeezed-film damped beams in the molecular regime. The basic idea of this derivation is a variation of the particle density between the vibrating cantilever and the fixed wall. The dynamic behavior of the beam counteracts the random walk of the gas molecules which are forced to sustain a steady-state particle density in the gap. For harmonic motion, this results in a complex-valued force, where the diffusion time

$$\tau = \frac{8}{\pi^3} \cdot \frac{A}{(d\bar{c})} \tag{8}$$

of the gas molecules is calculated from the Brownian motion. In equation (8) A is the surface area of the cantilever, d is the averaged distance between the moving cantilever and the fixed wall and $\bar{c}$ denotes the mean speed of the gas molecules (Maxwell–Boltzmann distribution). In an earlier publication by Legtenberg et al. in 1995 it was already speculated that in the quasi-molecular regime not only kinetic, but also squeeze-film damping lowers the quality factor Q [22]. According to the calculations of [2, 21, 23] the quality factor of a squeeze-film damped cantilever in the free molecular regime was derived

$$Q_{\text{squeeze}} = \frac{\rho h d \omega}{p} \cdot \frac{1 + (\omega\tau)^2}{\tau}. \tag{9}$$

Adding the losses is equal to summing up all contributions to the inverse of the quality factor, and the pressure-dependent quality factors can be summarized to

$$Q_{\text{quasi-molecular}} = \left(\frac{1}{Q_{\text{kinetic}}} + \frac{1}{Q_{\text{squeeze}}} \right)^{-1}. \tag{10}$$

6. Results and Discussion

In order to obtain data with significant as well as negligible squeeze-film damping several identical structures like the micrograph depicted in figure 4 are produced. In the case of testing exclusively kinetic damping the PCB underneath the vibrating structure is partially removed (see figure 5).

Otherwise a fixed wall is placed 350 µm (which conforms to the thickness of the handle wafer) below the test-structure. Usually the lateral air slit is 100 µm, whereat the influence of different slit widths on the quality factor is tested in the transition regime. The experimental data shown in figure 6 are compared with analytical solutions of the equations (3) and (7) on the one hand and with the most recent stochastic model by Okada et al. on the other hand [24].

From figure 6 it is obvious that the assumptions of the free molecular flow are no longer valid for Knudsen numbers in the range of 0.1 or smaller. The intermolecular collision frequency increases with decreasing Knudsen numbers, where in figure 6(a) the Knudsen flow evolves into the pressure-independent viscous flow [2]. Compared to that, there is a stronger structure-molecular dependence observable in the results in figure 6(b) due to the smaller air slit.

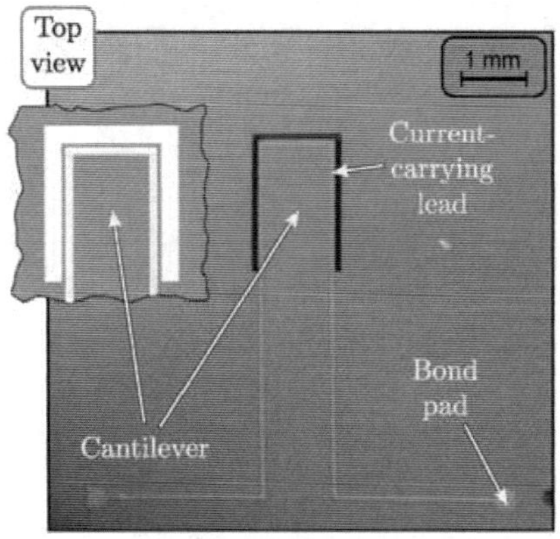

Figure 4: Micrograph of the top view of a cantilever with a typical air slit width of 100 μm.

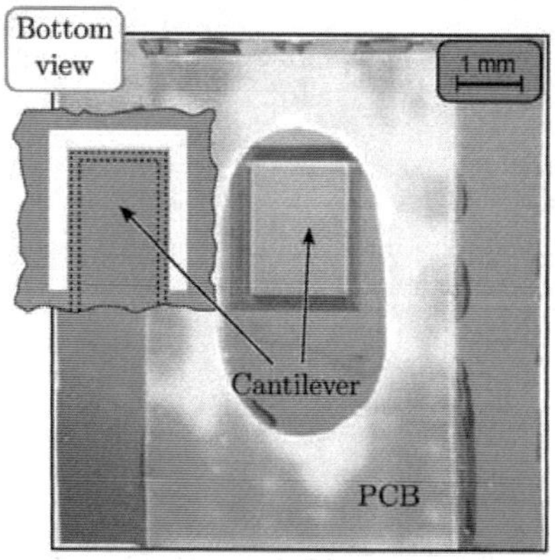

Figure 5: Micrograph of a bottom view with removed fixed wall below the test structure.

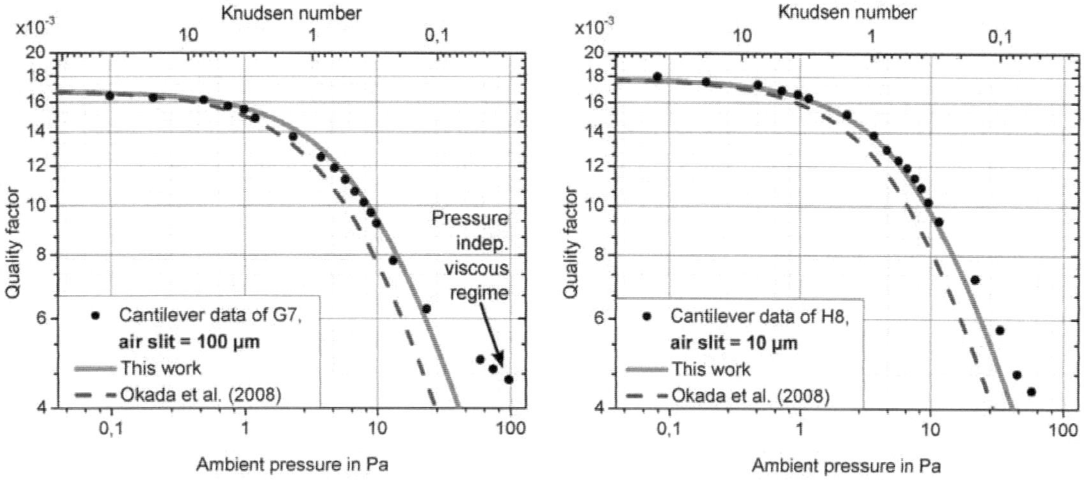

(a) Air slit width of 100 μm.

(b) Air slit width of 10 μm.

Figure 6: Experimental and theoretical results of the pressure dependence of the quality factor due to intrinsic and kinetic damping mechanisms of a cantilever with a length of 2 mm and a width of 1.5 mm, and with two different widths of the air slit.

This structure-molecular dependence increases with a fixed wall below the vibrating structure. The experimental data shown in figure 7 are similarly compared with analytical solutions of the equations (3) and (7), but additionally with equations (9) and (10), because of the fixed wall below the cantilever. In figure 7 the uncertainty is mainly determined by the lack of knowledge of the equilibrium distance d between the vibrating structure and the fixed wall. It is assumed that this distance corresponds to the thickness of the handle wafer, which is 350 μm with an estimated uncertainty of 10 %. The red dotted curve depticted in figure 7 represents exclusively kinetic damping. Evidently, squeeze-film damping according to equations (9) is the second dominant damping mechanism in the transition regime, where the modeled quality factor Q perfectly matches the experimental data.

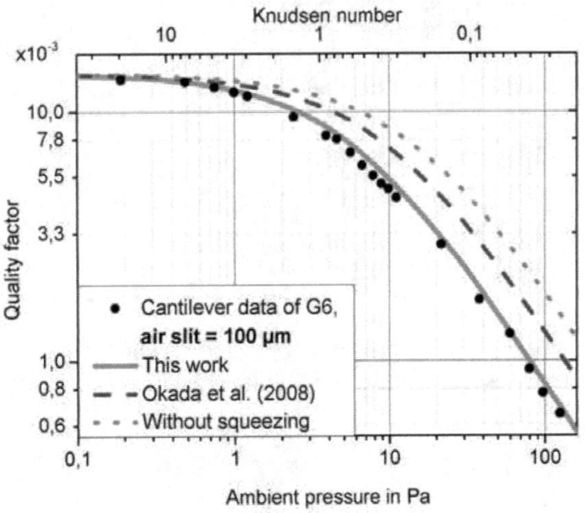

Figure 7: Comparison of analytical models with measurement results for a cantilever featuring a length of 2 mm, a width of 1.5 mm, an air slit of 100 µm, and a fixed wall about 350 µm below the cantilever.

7. Conclusion and Outlook

A comprehensive study of the damping behavior of monocrystalline silicon cantilevers in different rarefied gas regimes is presented and experimentally verified with the use of a micromotion analyzer. It was demonstrated that two damping mechanisms, namely kinetic and squeeze-film damping, dominate in the pressure-dependent transition regime with low Knudsen numbers in the range of 0.02 to 10. Both damping mechanisms are derived from the assumptions of a free molecular flow, avoiding a priori phenomenological presumptions.

Further analytical models in the transition regime have to be carried out and analyzed with MEMS devices in order to test which kinetic equation would describe the interaction of diatomic gases in a non-thermal equilibrium with the structure. Similarly, the dependency of squeeze-film damping on different gaps d between vibrating structure and fixed wall has to be studied for a thorough verification of the models. These considerations are topics of further examinations.

Acknowledgments

This work was supported by the fund of "Förderung der wissenschaftlichen Forschung (FWF)" with the project No. L477-N14.

References

[1] Bao M 2000 *Micro Mechanical Transducers* (Handbook of Sensors and Actuators)
[2] Blom F R, Bouwstra S, Elwenspoek M, Fluitman J H J 1992 *J. Vac. Sci. Technol. B* **10** 19–26
[3] Hosaka H, Itao K, Kuroda S 1995 *Sens. Actuators* **A49** 87–95
[4] Yang J, Ono T, Esashi M 2002 *Journal of Microelectromechanical Systems* **11 (6)** 775–783
[5] Lifshitz R, Roukes M L 2000 *Phys. Rev. B* **61** 5600–5609
[6] Roszhart T 1990 *Proc. IEEE Solid-State Sensor and Actuator Workshop* 13–16
[7] Prabhakar S, Vengallatore S 2007 *Journal of Micromechanics and Microengineering* **17** 532–538
[8] Hao Z, Erbil A, Ayazi F 2003 *Sensors and Actuators A* **109** 156–164
[9] Zener C 1937 *Phys. Rev.* **52** 230–235
[10] Gombosi TI 1994 *Gaskinetic theory* (Cambridge Univ. Press)
[11] Bird G A 1994 *Molecular Gas Dynamics and the Simulation of Gas Flows* (Oxford, U.K.)
[12] Barber R W and Emerson D R 2006 *Heat Transfer Engineering* **27, issue 4** 3–12
[13] Drawin H W 1962 *Vakuum-Technik* **11** 45–49
[14] Christian R G 1966 *Vacuum* **16** 175–178
[15] Martin M J, Houston B H, Baldwin J W, Zalalutdinov M K 2008 *Journal of Microelectromechanical Systems* **17** 503–511
[16] Bao M, Yang H 2007 *Sensors and Actuators A* **136** 3–27
[17] Veijola T 2004 *Journal of Micromechanics and Microengineering* **14** 1109–1118
[18] Veijola T , Kuisma H, Lahdenperä J, Ryhänen T 1995 *Sensors and Actuators A* **48** 239–248
[19] Leung R C W, Thurber T, and Ye W 2011 *Microfluid Nanofluid* **11** 753–762
[20] Alexeenko A A , Gimelshein S F, Muntz E P, Ketsdever A D 2006 *Kinetic modeling of temperature driven flows in short microchannels* **45, Issue 11** 1045–1051
[21] Suijlen M, Koning JJ, Gils MAJ van, Beijerinck HCW 2009 *Sensors and Actuators A: Phys* **156 (1)** 171–179
[22] Gui C, Legtenberg R, Elwenspoek M, Fluitman JH 1995 *J. Micromech. Microeng.* **5** 183–185
[23] Landau LD, Lifshitz EM 1987 *Fluid Mechanics* (Pergamon, Oxford)
[24] Okada H, Itoha T, Suga T 2008 *Sensors and Actuators A: Physical* **147** 359–364

Heat transfer in micro devices packaged in partial vacuum

Anirudh Rana[1], **Manuel Torrilhon**[2], **Henning Struchtrup**[1]

[1]Department of Mechanical Engineering,
University of Victoria, Victoria BC, Canada
[2]Department of Mathematics,
RWTH Aachen University, Aachen, Germany

E-mail: `anirudh@uvic.ca`

Abstract. The influence of rarefaction effects on technical processes is studied numerically for a heat transfer problem in a rarefied gas, a box with bottom heated plate. Solutions obtained from several macroscopic models, in particular the classical Navier-Stokes-Fourier equations with jump and slip boundary conditions, and the regularized 13 moment (R13) equations [Struchtrup & Torrilhon, *Phys. Fluids* 15, 2003] are compared. The R13 results show significant flow patterns which are not present in the classical hydrodynamic description.

1. Introduction

Accurate prediction of gas flow through microfluidic devices is important to enable the optimum performance, design and fabrication of micro-electro-mechanical systems (MEMS). Simulation of the gas flow through these micro devices is, however, more challenging than in classical flow regimes, since gas rarefaction leads to the breakdown of the underlying assumptions of the classical continuum theory, i.e., the Navier-Stokes-Fourier (NSF) equations.

The state of a gas in such microsystems is described by the Knudsen number which is defined as $Kn = \frac{\lambda}{L}$, where L is the length scale of device and λ is the mean free path of gas molecules, given by $\frac{\mu}{\rho\sqrt{RT}}$. Here, μ is the dynamic viscosity, R is the gas constant and T and ρ are the temperature and density of the gas, respectively.

Based upon the value of Knudsen number, gas flows in micro devices can be categorized into three regimes. For $Kn \leq 10^{-3}$, the gas is considered as a continuum, while for $Kn \geq 10$ it is considered to be in the free molecular flow regime. For Knudsen numbers ranging between $10^{-3} \leq Kn \leq 10$, known as the transition regime, the gas exhibits rarefaction effects such as Knudsen layers, Knudsen minimum, heat flux without temperature gradient, thermal creep, etc. [2]-[5].

The transition regime is a challenging regime to model, and perhaps the one most frequently encountered in MEMS devices [6]. Many MEMS applications involve the analysis of gas flow through microdevices such as micro-Pirani gauges [7][8], microcantilever heaters [9][10], etc. In these microdevices, where the characteristic length scale of the device is comparable with the mean free path, the Knudsen number ceases to be sufficiently small for the NSF equations to hold.

Direct numerical solution of the Boltzmann equation gives an accurate microscopic description of gas flows at all Knudsen numbers, but requires detailed information on microscopic phase space, and thus typically huge computational times [11]. The direct simulation Monte Carlo (DSMC) method, proposed by Bird [12], is another commonly used numerical method for simulation of high Knudsen number gas flows. In MEMS, however, where the mean streamwise velocity is several orders of magnitude smaller than the mean molecular velocity $\left(\sqrt{2RT}\right)$, the DSMC method produces significant statistical noise. Normally, a large number of samples, and therefore huge computational times, are required to reduce the noise. Recently, several variants of the DSMC method were presented which have, for some flow problems, reduced computational time [13][14].

For the purpose of design optimization, which requires numerous repetition of simulations with different parameters, microscopic solvers are impractical due to the large computational time involved. Moreover, since the solution is produced in the microscopic picture, the underlying physics for a typical thermodynamic process remains somewhat hidden. Indeed, microscopic solvers consider the behavior of individual particles, but not their collective behavior. This makes it impossible to deduce, a priory, the relative size of relevant macroscopic physical quantities, e.g., density, temperature, heat flux, stress, and their relative effect on the particular process under consideration.

Beside microscopic approaches, a thermodynamic process can also be described by macroscopic models, derived from the Boltzmann kinetic equation either by asymptotic expansion, such as in the Chapman-Enskog expansion [15], or by reconstruction of the distribution function, such as in Grad's moment method [16]. In macroscopic theories, the behavior of a gas is described through physical quantities such as mass density, temperature, velocity, heat flux, stress tensor, and so on. The goal of these macroscopic models is to reduce the high dimensional phase space of the particle description to a low-dimensional continuum model by relating the physical quantities as moments of the probability density function $f(c, x, t)$. Macroscopic models, despite some limitation on the their accuracy, give an easy and detailed access to physical quantities and their relative effect on the process. In recent years, we developed the regularized 13 moment equations (R13), which give good predictions for rarefied flows with Knudsen numbers up to 0.5 [1][18]-[23].

In the present contribution we study the gas flow behaviour and heat transfer characteristics of a rarefied gas confined in a bottom heated square domain. A similar problem has been investigated using DSMC or another kinetic approaches in [9][24]-[30]. In the following we shall consider the effects of Knudsen number and convection on heat transfer from the heated plate using finite difference numerical solutions of macroscopic methods. The developed numerical method is applied to the R13 equations, and to the Navier–Stokes equations supplemented by the first and second order velocity-slip and temperature-jump conditions. The main objective here is the implementation of a computationally efficient yet accurate, macroscopic description of such flows, so as to anticipate a better understanding of heat transfer process in micro devices.

2. Statement of the Problem

We consider the steady state heat transfer process through a monatomic gas, trapped inside a square enclosure of width L with a uniformly heated surface at the bottom ($y = 0$), as shown in Fig. 1. It is assumed that the temperature of the heated surface of the cavity is T_H and the other sides are considered isothermal at an environment temperature $T_0 = 273K$. The ratio between hot chip surface temperature, T_H, and T_0 was taken to be 1.1. Results will be presented for Knudsen numbers ranging from the continuum limit to the early transition regime, i.e., $10^{-3} \leq Kn \leq 1$. Effects due to radiation and gravity are assumed to be negligible.

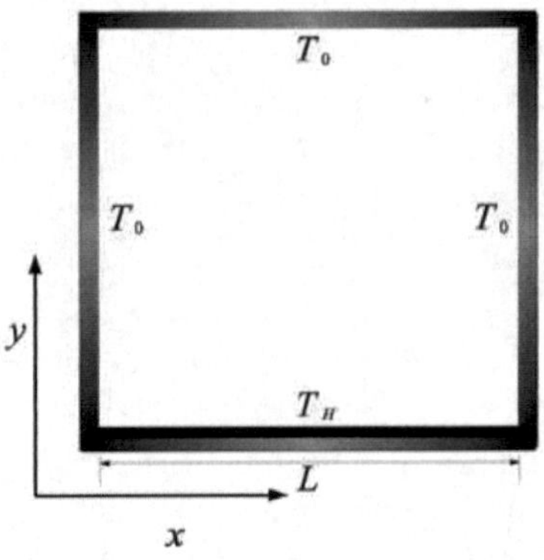

Figure 1. Schematic diagram for bottom heated plate.

3. Macroscopic Transport Equations

Macroscopic equations for rarefied gases are obtained by multiplying the Boltzmann equation with weight functions and integrating it over the velocity space. By taking appropriate weight functions, we obtain the conservation laws for mass, momentum, and energy [31],

$$\frac{\partial \rho}{\partial t} + \frac{\partial \rho v_i}{\partial x_i} = 0, \tag{1a}$$

$$\frac{\partial \rho v_i}{\partial t} + \frac{\partial \left(p_{ik} + \rho v_i v_k\right)}{\partial x_k} = 0, \tag{1b}$$

$$\frac{\partial \rho \left(u + \frac{1}{2} v^2\right)}{\partial t} + \frac{\partial \left(\rho e v_k + p_{ik} v_i + q_k\right)}{\partial x_k} = 0, \tag{1c}$$

where t and x_i are temporal and spatial coordinates, respectively. Moreover, ρ, v_k, p_{ik}, u, and q_k denote mass density, velocity, pressure tensor, internal energy density and the heat flux, along with

$$p_{ij} = p\delta_{ij} + \sigma_{ij} \text{ and } u = \frac{3\theta}{2}. \tag{2}$$

Here, p and σ_{ij} are the pressure and the stress tensor, respectively. Also, for ideal monatomic gases the pressure is given by the ideal gas law, $p = \rho\theta = \rho RT$, where θ is the temperature in energy units, R is the gas constant and T is the thermodynamic temperature.

Closure of the conservation laws (1a-1c) requires to specify the pressure tensor, p_{ij} and the heat flux vector, q_k. Several methods to this end can be found in the literature.

One approach to the closure of the macroscopic equations is given by a series expansion of the distribution function in terms of Knudsen number, the Chapman-Enskog (CE) expansion [15]. The Euler and the Navier-Stokes-Fourier equations result from the zeroth and first order expansions, respectively, while higher order expansions lead to the Burnett and Super-Burnett equations [32]. However, in many cases, the higher order CE expansion leads to unstable or unphysical results [33]. Moreover, there is no complete theory for boundary conditions of the higher order equations.

Another approach is due to Grad [16], known as the Grad's moment methods, in which the velocity distribution function is reconstructed by an expansion of the distribution function into Hermite polynomials. The method yields extended sets of transport equations, the best-known of which is the 13 moment system. The Grad method is not linked to the Knudsen number, and thus it provides no information on which, and how many, moments one should consider for a given process.

The regularized-13 (R13) equations [1][31] were obtained by combining elements of the CE and Grad methods, they are a regularized version of the classical Grad's 13-moment equations. The R13 equations are of third order in the Knudsen number in the CE sense, and for small enough Knudsen numbers ($Kn \lesssim 0.5$) the equations describe all known rarefaction effects with good accuracy [19]-[22].

The equations for heat-flux vector and stress tensor in the R13 system are their respective moment equations

$$\frac{\partial q_i}{\partial t} + v_k\frac{\partial q_i}{\partial x_k} + \frac{7}{5}q_k\frac{\partial v_i}{\partial x_k} - \sigma_{ik}\frac{\theta}{\rho}\frac{\partial \rho}{\partial x_k} - \frac{\sigma_{ik}}{\rho}\frac{\partial \sigma_{kl}}{\partial x_l} + \frac{7}{2}\frac{\partial \theta\sigma_{ik}}{\partial x_k} + \frac{5}{2}\sigma_{ik}\frac{\partial \theta}{\partial x_k} + \theta\frac{\partial \sigma_{ik}}{\partial x_k}$$
$$+\frac{7}{5}q_i\frac{\partial v_k}{\partial x_k} + \frac{2}{5}q_k\frac{\partial v_k}{\partial x_i} + \frac{1}{2}\frac{\partial R_{ik}}{\partial x_k} + \frac{1}{6}\frac{\partial \Delta}{\partial x_i} + m_{ikl}\frac{\partial v_k}{\partial x_l} + \underline{\frac{5}{2}p\frac{\partial \theta}{\partial x_i}} = \underline{-\frac{2}{3}\frac{p}{\mu}q_i}, \quad (3)$$

$$\frac{\partial \sigma_{ij}}{\partial t} + v_k\frac{\partial \sigma_{ij}}{\partial x_k} + \frac{\partial m_{ijk}}{\partial x_k} + \frac{4}{5}\frac{\partial q_{\langle i}}{\partial x_{j\rangle}} + 2\sigma_{k\langle i}\frac{\partial v_{j\rangle}}{\partial x_k} + \sigma_{ij}\frac{\partial v_k}{\partial x_k} + \underline{2p\frac{\partial v_{\langle i}}{\partial x_{j\rangle}}} = \underline{-\frac{p}{\mu}\sigma_{ij}}, \quad (4)$$

For $\Delta = R_{ij} = m_{ijk} = 0$, the above equations reduce to Grad's 13-moment equations [16]. The R13 closure for Δ, R_{ij} and m_{ijk} results from taking into account the appropriate terms of higher order moment equations, so that

$$\begin{aligned}
\Delta &= 5\frac{\sigma_{kl}\sigma_{kl}}{\rho} + \frac{56}{5}\frac{q_k q_k}{p} - 12\frac{\mu}{p}\left(\theta\frac{\partial q_k}{\partial x_k} - \theta q_k\frac{\partial \ln p}{\partial x_k}\right), \quad (5)\\
R_{ij} &= \frac{20}{7}\frac{\sigma_{k\langle i}\sigma_{j\rangle k}}{\rho} + \frac{192}{75}\frac{q_{\langle i}q_{j\rangle}}{p} - \frac{24}{5}\frac{\mu}{p}\left(\theta\frac{\partial q_{\langle i}}{\partial x_{j\rangle}} - \theta q_{\langle i}\frac{\partial \ln p}{\partial x_{j\rangle}}\right),\\
m_{ijk} &= \frac{20}{15}\frac{q_{\langle i}\sigma_{jk\rangle}}{p} - 2\frac{\mu}{p}\left(\theta\frac{\partial \sigma_{\langle ij}}{\partial x_{k\rangle}} - \theta\sigma_{\langle ij}\frac{\partial \ln p}{\partial x_{k\rangle}}\right).
\end{aligned}$$

The indices inside angular brackets denote the symmetric trace-free part of tensors [31]. The R13 constitutive relations given by eqs. (5) add second order derivatives to stress tensor, σ_{ij} and heat flux, q_i, in eqs. (3 and 4).

The R13 equations are an extension to the classical Navier-Stokes- Fourier (NSF) equations of hydrodynamics. An analysis of the order of Knudsen number in (3-4) shows that only the underlined terms are of first order in Knudsen number, while all others are of higher order [34]. Thus, the underlined terms are the NSF contributions,

$$q_i^{NSF} = -\frac{15}{4}\mu\frac{\partial \theta}{\partial x_i}, \text{ and } \sigma_{ij}^{NSF} = -2\mu\frac{\partial v_{\langle i}}{\partial x_{j\rangle}}. \quad (6)$$

From now on, all equations and results will be given in dimensionless quantities, defined as

$$\widehat{x_i} = \frac{x_i}{L}, \widehat{\theta} = \frac{\theta}{\theta_0}, \widehat{\rho} = \frac{\rho}{\rho_0}, \widehat{v_i} = \frac{v_i}{\sqrt{\theta_0}}, \widehat{q_i} = \frac{v_i}{\rho_0\sqrt{\theta_0}^3},$$
$$\widehat{\sigma_{ij}} = \frac{\sigma_{ij}}{\rho_0\theta_0}, \hat{\Delta} = \frac{\Delta}{\varrho_0\,\theta_0^2}, \hat{R_{ij}} = \frac{R_{ij}}{\varrho_0\,\theta_0^2}, \text{ and } \hat{m}_{ijk} = \frac{m_{ijk}}{\varrho_0\sqrt{\theta_0}^3}.$$

Here, ρ_0 is the average mass density. As a result the Knudsen number arises, which is given by

$$Kn = \frac{\mu_0}{\rho_0\sqrt{\theta_0}}\frac{1}{L},$$

where L is the length of the bottom wall. Therefore, the flow considered is fully defined by setting the temperature ratio T_H/T_0 and the Knudsen number.

3.1. Boundary conditions

Maxwell's accommodation model for the boundary conditions in kinetic theory was used to derive boundary conditions for the R13 system, which link the moments of the gas near the wall to the tangential wall velocity v_τ^W and wall temperature θ^W. The details of the construction procedure for the boundary conditions (BCs) for the R13 equations can be found in [17], the final result reads

$$v_n = 0 \tag{7a}$$

$$\sigma_{\tau n} = \frac{-\chi}{2-\chi}\sqrt{\frac{2}{\pi\theta}}\left(\mathcal{P}\mathcal{V}_\tau + \frac{1}{5}q_\tau + \frac{1}{2}m_{\tau nn}\right), \tag{7b}$$

$$q_n = \frac{-\chi}{2-\chi}\sqrt{\frac{2}{\pi\theta}}\left(2\mathcal{P}\mathcal{T} - \frac{1}{2}\mathcal{P}\mathcal{V}_\tau^2 + \frac{1}{2}\theta\sigma_{nn} + \frac{1}{15}\Delta + \frac{5}{28}R_{nn}\right), \tag{7c}$$

$$R_{\tau n} = \frac{\chi}{2-\chi}\sqrt{\frac{2}{\pi\theta}}\left(6\mathcal{P}\mathcal{T}\mathcal{V}_\tau + \mathcal{P}\theta\mathcal{V}_\tau - \mathcal{P}\mathcal{V}_\tau^3 - \frac{11}{5}\theta q_\tau - \frac{1}{2}\theta m_{\tau nn}\right), \tag{7d}$$

$$m_{nnn} = \frac{\chi}{2-\chi}\sqrt{\frac{2}{\pi\theta}}\left(\frac{2}{5}\mathcal{P}\mathcal{T} - \frac{3}{5}\mathcal{P}\mathcal{V}_\tau^2 - \frac{7}{5}\theta\sigma_{nn} + \frac{1}{75}\Delta - \frac{1}{14}R_{nn}\right), \tag{7e}$$

$$m_{\tau\tau n} = \frac{-\chi}{2-\chi}\sqrt{\frac{2}{\pi\theta}}\left(\frac{1}{5}\mathcal{P}\mathcal{T} - \frac{4}{5}\mathcal{P}\mathcal{V}_\tau^2 + \frac{1}{14}R_{\tau\tau} + \theta\sigma_{\tau\tau} - \frac{1}{5}\theta\sigma_{nn} + \frac{1}{150}\Delta\right), \tag{7f}$$

where

$$\mathcal{P} = \rho_W\sqrt{\theta_W}\sqrt{\theta} = \left(\rho\theta + \frac{1}{2}\sigma_{\tau\tau} - \frac{1}{120}\frac{\Delta}{\theta} - \frac{1}{28}\frac{R_{\tau\tau}}{\theta}\right). \tag{8}$$

Here, $\mathcal{V}_\tau = v_\tau - v_\tau^W$ and $\mathcal{T} = \theta - \theta_W$ are slip velocity and temperature jump, respectively, and subscripts n and τ stand for the normal and tangential directions with respect to the surface. The accommodation coefficient in the Maxwell model is denoted by χ, representing the fraction of molecules undergoing diffuse reflection. For all numerical results presented in this paper accommodation factor is set to be unity [2].

Equation (7b) relates the velocity slip $\mathcal{V}_\tau$, to the shear stress $\sigma_{\tau n}$, associated with the gradient of tangential velocity in the normal direction, and the heat flux q_τ, associated with the gradient of temperature parallel to the boundary. Therefore, in contrast to continuum flows, a tangential temperature gradient along the surface can also induce a flow from the colder part to the hotter part, usually referred to as thermal creep [2].

In the slip flow regime, the NSF equations must be furnished with slip and jump boundary conditions. Slip and jump boundary conditions for the Navier-Stokes-Fourier equations are obtained by replacing, $\sigma_{ij} = \sigma_{ij}^{NSF}$ and $q_i = q_i^{NSF}$ from Eqs. (6), in Eqs.(7a-7c) and ignoring higher order moments, i.e., $m_{\tau nn}$, Δ and R_{nn}. As a result we obtain [5]

$$v_n = 0 \tag{9a}$$

$$\sigma_{\tau n}^{NSF} = \frac{-\chi}{2-\chi}\sqrt{\frac{2}{\pi\theta}}\left(\mathcal{P}\mathcal{V}_\tau - \frac{3}{4}\mu\frac{\partial\theta}{\partial x_\tau}\right), \tag{9b}$$

$$q_n^{NSF} = \frac{-\chi}{2-\chi}\sqrt{\frac{2}{\pi\theta}}\left(2\mathcal{P}\mathcal{T} - \frac{1}{2}\mathcal{P}\mathcal{V}_\tau^2 - \mu\theta\frac{\partial v_{\langle n}}{\partial x_{n\rangle}}\right), \tag{9c}$$

where $\mathcal{P} = \left(\rho\theta + \frac{1}{2}\sigma_{\tau\tau}\right)$.

Second order slip and jump boundary conditions can also be derived from (7a-7c), by replacing the heat flux and stress using Navier-Stokes-Fourier constitutive relations (6) in higher order moments, i.e., $m_{\tau nn}$, Δ and R_{nn} [5]

$$v_n = 0 \tag{10a}$$

$$\sigma_{\tau n}^{NSF} = \frac{-\chi}{2-\chi}\sqrt{\frac{2}{\pi\theta}}\left(\mathcal{P}\mathcal{V}_\tau + \frac{1}{5}q_\tau^{NSF} + \frac{1}{2}m_{\tau nn}^{NSF}\right), \tag{10b}$$

$$q_n^{NSF} = \frac{-\chi}{2-\chi}\sqrt{\frac{2}{\pi\theta}}\left(2\mathcal{P}\mathcal{T} - \frac{1}{2}\mathcal{P}\mathcal{V}_\tau^2 + \frac{1}{2}\theta\sigma_{nn}^{NSF} + \frac{1}{15}\Delta^{NSF} + \frac{5}{28}R_{nn}^{NSF}\right). \tag{10c}$$

Here, the superscript NSF denotes that the actual heat flux and stress are replaced using the Navier-Stokes-Fourier constitutive relations.

4. Results and discussion

To obtain the numerical solution a central finite-difference scheme was used and the system of discretized equations has been solved using a quasi-minimal residual method. This numerical algorithm has been described and validated against the existing results for the case of the lid driven cavity in [35]. The numerical algorithm converges with approximately second-order numerical accuracy, and a grid independent solution (with less then 1% deviation) was obtained approximately at 70×70 mesh size.

4.1. Rarefaction effects on flow field characteristics

Fig. 2(a)-(d) show the results for Knudsen numbers 0.06, 0.12, 0.335 and 0.5, respectively, for the temperature ratio $T_H/T_0 = 1.1$. The flow fields are symmetric to the centerline ($x = 0.5$); for compact presentation, the figures show NSF results (using 1st order BC) in the left part, and R13 results on the right. The figures show temperature contours and velocity streamlines.

As we see from Fig. 2, for all Kn, two primary circulating vortices are induced near the heated surface, along the vertical axes. The formation of primary vortices, is induced by the tangential temperature gradient in the gas near the vertical surface that forces a creep-driven fluid into a circulatory motion.

Interestingly, in addition of the primary vortices, the R13 equations predict a secondary counter-circulating flow induced along the corners of the top plate. These predictions are qualitatively similar to those seen in Monte Carlo simulations for vacuum packaged MEMS devices [27] and [38]. This flow behavior results form the interplay of the variables as given by the transport equations, and from the slip equation (7b), which relates the tangential velocity v_τ to shear stress, $\sigma_{\tau n}$, tangential heat flux, q_τ, and higher order contribution, $m_{\tau nn}$. The latter can be rearranged to give an expression for the tangential velocity at the right vertical surface as,

$$v_x \propto \left(\frac{2-\chi}{\chi}\sqrt{\frac{\pi\theta}{2}}\sigma_{xy} - \frac{1}{5}q_y - \frac{1}{2}m_{yxx}\right). \tag{11}$$

At moderate Knudsen numbers, the tangential velocity has two opposing contributions, one from the tangential heat flux, q_y and the second from the shear stress, σ_{xy}.

For example, on the right vertical surface, along $x = 1$ (see Fig. 1), the tangential heat flux, q_τ, is positive mainly due to a negative temperature gradient (Fourier law). The magnitude of q_τ increases with the Knudsen number (also with the temperature ratio), whereas it decreases as the gas moves further away from the bottom surface. Along the vertical surface, both the Navier-Stokes-Fourier (NSF) and the R13 equations predict positive shear stress. However, the Navier-Stokes equations under-predict the shear stress, $\sigma_{\tau n}$, in comparison to the R13 equations. In particular, the NSF equations do not take into account the thermal stress, i.e., $\sigma_{ij}^T \propto -\frac{\mu}{p}\frac{\partial q_{\langle i}}{\partial x_{j\rangle}}$, which appears in the stress balance equation (4).

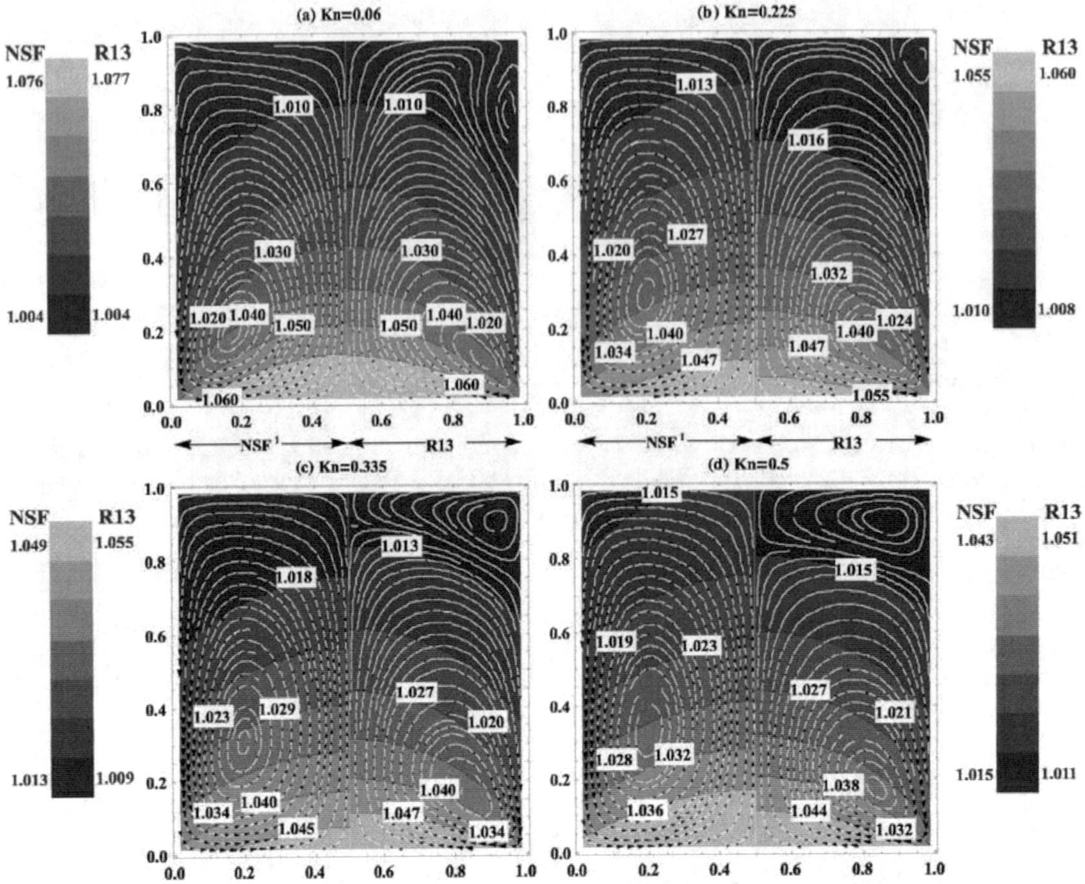

Figure 2. Velocity streamlines and temperature contours for the bottom heated cavity. Temperature ratio is $T_H/T_0 = 1.1$, and Knudsen numbers are 0.06, 0.225, 0.335, and 0.5, in sub-figures (a), (b), (c), and (d), respectively. Results computed with Navier -Stokes-Fourier equations with first order boundary conditions are depicted on the left-half domain, while the results computed with R13 are shown on the right-half portion of the domain.

For the lower portion of the vertical surface ($0 \leq y \lesssim 0.6$), q_y is greater than σ_{xy}, which, according to (11), induces a downward motion in the fluid and thus the primary eddies. However, close to the top plate shear stress dominates over heat flux which produces an upward motion in gas, hence the formation of the secondary vortices. Since the NSF equations overestimate the heat-flux and underestimate the shear stress near the boundary, these secondary eddies do not appear in the Navier-Stokes-Fourier solutions.

For a temperature ratio of 1.1, the magnitude of the dimensionless velocity in the primary and secondary eddies is of the order of 10^{-2} and 10^{-4}, respectively. Both velocities increase slightly with the Knudsen number.

With increasing Knudsen number, all gradients inside the cavity become steeper. Consequently, the secondary eddies becomes more and more pronounced. For Kn$\simeq$ 0.8, the secondary eddies span almost 40% of the cavity area. Although at Kn ≥ 0.5 the results computed with both NSF or R13 equation will be quantitatively different then the more accurate solvers like DSMC, the R13 equations are able to qualitatively predict the behavior of the underlying process. Moreover, the overall CPU time required for solution of NSF or R13 is far less then for DSMC, especially for slow flows. For example, for a typical case considered in this study, R13 and NSF equations simulation takes about 20 and 12 minutes with an un-optimized matlab code, respectively. On the other hand, a typical DSMC simulation takes about 150 hours [27].

4.2. Heat transfer

Now we will study the effects of rarefaction on the heat transfer from the hot surface, by comparing the results of the NSF and R13 equations, for various Knudsen numbers.

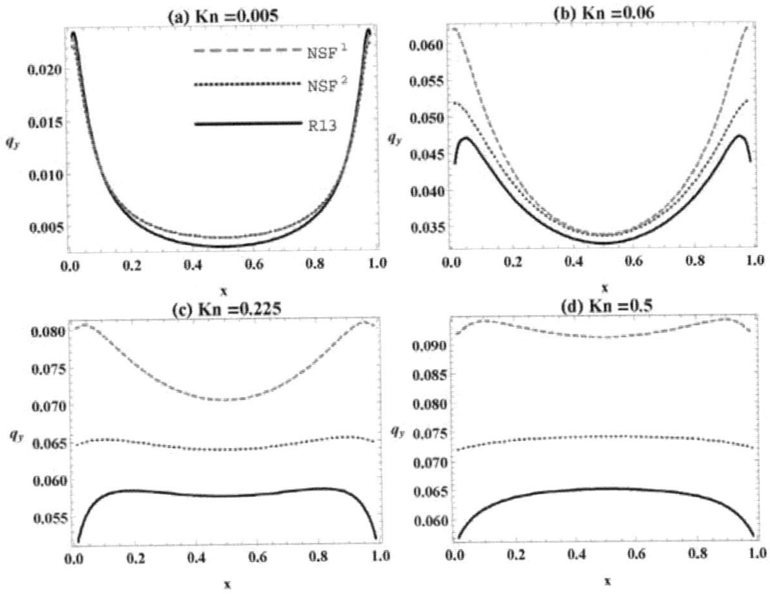

Figure 3. Normal heat flux profiles along the bottom plate, obtained by solving R13 (continuous, black), NSF with first (dashed, red) and second order (dotted, blue) boundary conditions for $Kn = 0.005, 0.06, 0.225$ and 0.5, at tempetature ratio 1.1.

Figure 3 illustrates the variation of the normal heat flux, q_y, along the bottom plate, obtained by solving R13 (continuous, black), NSF (dashed, red) with first order boundary conditions (9) and second order (dotted, blue) boundary conditions (10). At the relatively small Knudsen number 0.005, in Fig. 3(a), all models show good agreement. However at larger Knudsen numbers, in Fig 3(b)-(d), the heat transfer predictions by the classical NSF equations lead to a significant overestimation. This result confirms the conclusion obtained by Liu et al [27] from DSMC computations.

Next, in Fig. 4, we show the net dimensionless heat transfer from the bottom surface, defined as

$$Q_y = \int_0^1 q_y(x,0)dx \, .$$

This is an important quantity for device design, since it describes the thermal performance. Results computed with $T_H/T_0 = 1.1$, using R13 (continuous, black) are compared with NSF with first (dashed, red) and second order (dotted, blue) boundary conditions, are presented in Fig. 4(a). As shown, NSF over-calculates the net heat transfer, however, second order boundary conditions shows a significant improvement over first order boundary conditions.

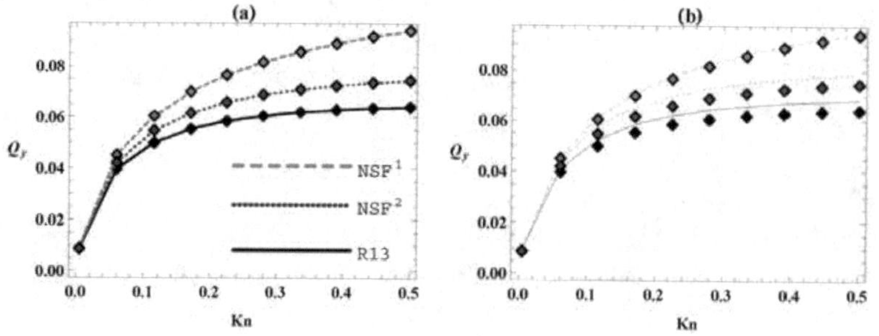

Figure 4. Navier-Stokes results with first (dashed, red) and second order (dotted, blue) boundary conditions are compared with R13 (continuous, black) for various Kn, at temperature ration T_H/T_0 =1.1. In (a), the net heat transfer from the bottom plate is compared between NSF and R13. In (b), the heat transfer through a stationary gas (denoted by curves) is compare against the heat transfer in the moving gas (denoted by symbols, $\diamondsuit$).

To understand the extent to which the convection affects the heat transfer we compute the heat transfer through a stationary gas, where all velocity components vanish, $v_i = 0$, and compare it against the heat transfer in the moving gas, $v_i \neq 0$. Note that this implies a change in boundary conditions, so that no transpiration flow occurs. In an experiment, one will observe the moving gas. As we see from Fig. 4(b), for all continuum models the heat flux from the pure heat conduction is slightly higher than that from the convective heat transfer, the heat transfer is slightly weakened by convection.

5. Conclusions

A numerical investigation of convective heat transfer in a thermal cavity is presented and the influence of the rarefaction effects on the heat transfer in a thermal cavity are studied using Navier-Stokes-Fourier equations with first and second order jump and slip boundary conditions and the R13 equations. It is shown that the NSF equations, with first order boundary conditions, over-estimate the heat transfer, however, second order boundary conditions show an improvement over first order boundary conditions. The R13 results show significant flow patterns which are not present in the NSF equations. This convective mechanism is described in detailed through relative terms in the boundary conditions and their size.

References

[1] Struchtrup H and Torrilhon M 2003 Regularization of Grad's 13 moment equations: Derivation and linear analysis *Phys. Fluids* **15** 2668
[2] Sone Y 2002 *Kinetic Theory and Fluid Dynamics Birkhäuser* (Boston)
[3] Aoki K, Sone Y and Waniguchi Y 1998 A rarefied gas flow induced by a temperature field: Numerical analysis of the flow between two coaxial elliptic cylinders with different uniform temperatures *Computers Math. Applic.* **35**, No. 1/2, 15-28
[4] Sone Y and Yoshimoto M 1997 Demonstration of a rarefied gas flow induced near the edge of a uniformly heated plate *Phys. Fluids* **9** (11)
[5] Struchtrup H and Torrilhon M 2008 Higher-order effects in rarefied channel flows *Phys. Rev.* E **78** 046301
[6] Gad-el-Hak M 2005 *The MEMS Handbook* second edition volumes I–III, (Boca Raton: CRC Press)
[7] Zhang F T, Tang Z, Yu J, and Jin R C 2006 A micro-Pirani vacuum gauge based on micro-hotplate technology *Sensors and actuators A* **126** 300-305
[8] Turnbull H, Barton R C, and Rivire J C 1962 *An Introduction to Vacuum Technique* (Grotge Newnes Limited)
[9] Lee J, Wright T L, Abel M R, Sunden E O and Marchenkov A 2007 Thermal conduction from microcantilever heaters in partial vacuum *J. Appl. Phys.* **101** 014906

[10] Roesle M, Coskun V and Steinfeld A 2011 Numerical Analysis of heat loss from a parabolic trough absorber tube with active vacuum system *J. of Solar Energy Eng.* **133** 031015-1

[11] Hadjiconstantinou N G and Garcia A L 2001 Statistical Error in Particle Simulations of Low Mach Number Flows *Proc. of the First MIT Conf. on Computational Fluid and Solid Mechanics,* Elsevier

[12] Bird G A1994 *Molecular Gas Dynamics and the Direct Simulation of Gas Flows* (Oxford University Press, Oxford)

[13] Radtke G A and Hadjiconstantinou N G 2009 Variance-reduced Particle Simulation of the Boltzmann Transport Equation in the Relaxation-time Approximation *Physical Review* E **79** 056711

[14] Fan J and Shen C 2001 Statistical simulation of low-speed rarefied gas flows *J. Comput. Phys.* **167** 393

[15] Chapman S and Cowling T G 1970 *The Mathematical Theory of Nonuniform Gases* (Cambridge University Press)

[16] Grad H 1949 On the kinetic theory of rarefied gases *Comm.Pure Appl.Math.* 2 p.331-407

[17] Struchtrup H and Torrilhon M 2008 Boundary conditions for regularized 13-moment-equations for micro-channel-flows *J. Comp. Phys.* **227** 1982-2011

[18] Struchtrup H and Taheri P 2011 Macroscopic transport models for rarefied gas flows: A brief review *IMA J. Apppl. Math.* 1-26

[19] Taheri P, Rana A S, Torrilhon M and Struchtrup H 2009 Macroscopic description of steady and unsteady rarefaction effects in boundary value problems of gas dynamics *Continuum Mech. Thermodyn.* **21** 423–443

[20] Taheri P and Struchtrup H 2010 An extended macroscopic transport model for rarefied gas flows in long capillaries with circular cross section *Phys. Fluids* **22** 112004

[21] Taheri P, Torrilhon M and Struchtrup H 2009 Couette and Poiseuille microflows: Analytical solutions for regularized 13-moment equations *Phys. Fluids* **21** 017102

[22] Taheri P and Struchtrup H 2010 Rarefaction effects in thermally–driven microflows *Physica* A **389** 3069–3080

[23] Gu X and Emerson D R 2009 A high-order moment approach for capturing non-equilibrium phenomena in the transition regime *J. Fluid Mech.* **636** 177

[24] Cai C 2008 Heat transfer in vacuum packaged microelectromechanical system devices *Phys. of Fluids* **20** 017103

[25] Sone Y 2009 Comment on "Heat transfer in vacuum packaged microelectromechanical system devices" [Phys. Fluids 20, 017103 (2008)]*Phys. of Fluids* **21** 119101

[26] Sone Y 1985 Boundary temperature effect in a highly rarefied gas *Phys. of Fluids* **28** 419

[27] Liu H, Wang M, Wang J, Zhang G, Liao H, Huang R and Zhang X 2007 Monte Carlo simulations of gas flow and heat transfer in vacuum packaged MEMS devices *Applied Thermal Engineering* **27** 323–329

[28] Papadopoulos D H and Rosne D E 1995 Enclosure gas flows driven by non-isothermal walls *Phys. of Fluids* **7** (11)

[29] Baker J, Calvert M E, Power D J, Chen E T, Ramalingam M L and Lamp T R 1996 On the role of the Knudsen number with respect to heat transfer in micro-scale flows *Proc. of the 31st Intersociety Energy Conversion Eng. Conf.* Vol. **2**, 11-16, pp. 1396-1401

[30] Passian A and Wig A 2002 Knudsen forces on microcantilevers *J. Appl. Phys.* **92** No. 10

[31] Struchtrup H 2005 *Macroscopic Transport Equations for Rarefied Gas Flows* (Springer, New York)

[32] Garcia L, Velasco R M and Uribe F J 2008 Beyond the Navier–Stokes equations: Burnett hydrodynamics *Phys. Rep.* **465** 149

[33] Bobylev A V 2006 Instabilities in the Chapman-Enskog expansion and hyperbolic Burnett equations *J. Stat. Phys.* **124** (2-4) p.371-399

[34] Struchtrup H and Torrilhon M 2008 Higher-order effects in rarefied channel flows *Phys. Rev.* E **78** 046301

[35] Rana A S, Torrilhon M and Struchtrup H 2011A robust numerical method for the R13 equations of rarefied gas dynamics: Application to lid driven cavity (Submitted) *J. Comp. Phys.*

[36] Karniadakis G E and Beskok A 2002 *Micro Flows: Fundamentals and Simulation* (Springer, New York)

[37] Struchtrup H and Torrilhon M 2004 Regularized 13-moment-equations: Shock structure calculations and comparison to Burnett models *J. Fluid Mech.* **513** 171-198

[38] Masters N D and Ye W 2007 Octant flux splitting information preserving DSMC method for thermally driven flows *J Comput. Phys.* **226** pp. 2044-2062

Tangential Momentum Accommodation Coefficient measurements for various materials and gas species

H Yamaguchi*, Y Matsuda and T Niimi
Department of Micro-Nano Systems Engineering, Nagoya University, Furo-cho, Chikusa, Nagoya, Aichi 464-8603, JAPAN

* E-mail: hiroki@nagoya-u.jp

Abstract. The tangential momentum accommodation coefficient was measured from the gaseous flow through a single microtube. The mass flow rate was measured by the constant volume technique, where the rate is related to time rate of pressure variation in a fixed volume tank. The measured mass flow rate was fitted by the theoretical mass flow rate expressed by the slip velocity at the surface to deduce the slip coefficient, which can be related to the tangential accommodation coefficient. The mean Knudsen number, which is determined as the mean pressure of the inlet and the outlet, was set to be below 0.32, where the second-order slip boundary condition was suggested to be valid. The measurement system was designed to allow using a microtube with large diameter of several hundred micrometers. Since low pressure environment was essential for large Knudsen number condition for the flow through such large microtube, a low leakage measurement system realized by applying the UHV technology is needed. Applicability of sub millimeters size microtubes allowed us to measure the tangential momentum accommodation coefficients on various materials. In this study, we measured the tangential momentum accommodation coefficients on an engineering metal surface for various gas species.

1. Introduction

In micro gaseous flows, Knudsen number *Kn*, which is defined as a ratio of the molecular mean free path λ to the characteristic length of the system *L*, becomes large. In such so-called "high Knudsen number flows", the collision number of gas molecules to surfaces can not be neglected compared to that to other gas molecules. The ratio of a surface area to a volume is also large. Those characteristics lead to the importance of gas-surface interaction in the micro thermal and flow fields. For precise numerical simulations, accurate gas-surface interaction model for a boundary condition is essential.

To illustrate overall gas-surface interaction without any detailed processes, the accommodation coefficient α is the most widely used empirical parameter, which is also often used as a parameter for a gas-surface interaction models for numerical simulations. Related to flow properties, like a pressure loss of a fluid system, the tangential momentum accommodation coefficient α_t is important, which is defined as,

$$\alpha_{\mathrm{t}} = \frac{\tau_{\mathrm{i}} - \tau_{\mathrm{r}}}{\tau_{\mathrm{i}}} \qquad (1)$$

where τ is tangential momentum of gas molecules and indices i and r represent incident and reflected molecules, respectively. The tangential momentum accommodation coefficient represents the

Published under licence by IOP Publishing Ltd

statistical characteristics of the mean exchange rate, efficiency, or probability of the tangential momentum. The energy (=thermal) accommodation coefficient, which describes the energy exchange by the similar definition, is also often used.

Especially for the machined and non-treated "as is" metal surfaces, often called engineering surfaces, the complete accommodation has been assumed unquestionably. However, the recent precise measurements on the energy accommodation coefficient suggested an incomplete thermal accommodation on metal surfaces. This result indicates the incomplete accommodation of gas molecules on the engineering surfaces, which leads to doubt the approximation of the tangential momentum accommodation coefficient as unity.

However, it is still not easy to measure the tangential momentum accommodation coefficients on an engineering metal surface of "bulk" materials, and few data were reported [1]. The gas flow through the stainless steel tube was studied by several researchers: Reference [2] measured by a single cylindrical tube with several inert gases, but the tangential momentum accommodation coefficient was not obtained. Reference [3] evaluated the conductance with a bundle of tubes. References [4] and [5] were studied under rarefied gas flow conditions, which were different in the ratio of the boundary layer thickness to the diameter of the tube compared to other researches.

In this paper, the mass flow rate through a microtube under the slip flow regime was measured by the constant volume method, by which the tangential momentum accommodation coefficient was calculated. The second-order effect of the velocity slip condition was considered in the analysis. A microtube with an engineering metal surface, which was manufactured from a bulk material, was employed by using the low pressure environment.

2. Theory

The tangential momentum accommodation was measured from the mass flow rate through a single microtube [6].

The constant volume method [7-10], in which the mass flow rate was related to the pressure change of the outlet tank through the equation of state under an ideal gas approximation, was employed. The mass flow rate was obtained by

$$Q_{\mathrm{m}} = \frac{V}{RT}\frac{\delta P}{\tau} \tag{2}$$

where V, R, T, δP, τ are volume, the specific gas constant, temperature, pressure change, measurement time, respectively. The mass flow rate through a microtube is theoretically obtained in frame of the Navier-Stokes equation with the second-order velocity slip boundary condition as [11, 12],

$$Q_{\mathrm{m}} = \frac{\pi D^4 \Delta P P_{\mathrm{m}}}{128\mu RTL}\left(1 + 8A_1 Kn_{\mathrm{m}} + 16A_2 Kn_{\mathrm{m}}^2 \frac{\mathcal{P}+1}{\mathcal{P}-1}\ln\mathcal{P}\right) \tag{3}$$

where D, ΔP, P_{m}, μ, L, Kn_{m}, $\mathcal{P}$ are the diameter of a microtube, the pressure difference from the inlet to the outlet, the mean pressure of the inlet and the outlet, the viscosity coefficient, the length of a microtube, the mean Knudsen number based on the mean pressure of the inlet and the outlet, and the pressure ratio of the inlet to the outlet, respectively. Experimentally measured mass flow rates are normalized to S as follows:

$$S = Q_{\mathrm{m}} \Big/ \left(\frac{\pi D^4 \Delta P P_{\mathrm{m}}}{128\mu RTL}\right)$$

$$= 1 + 8A_1 Kn_{\mathrm{m}} + 16A_2 Kn_{\mathrm{m}}^2 \frac{\mathcal{P}+1}{\mathcal{P}-1}\ln\mathcal{P} \tag{4}$$

The normalized mass flow rates S are plotted as a function of the mean Knudsen number Kn_m and fitted with theoretical expression of equation (4) by the least-square method. The coefficient A_1 is function of the slip coefficient and also of the tangential momentum accommodation coefficient. Therefore, the tangential accommodation coefficient can be deduced from the procedure. A scattering model is used to relate the slip coefficient and the tangential momentum accommodation coefficient. Following the Maxwell scattering model, the tangential momentum accommodation coefficient α_t^M becomes as follows:

$$\sigma_p = \frac{\sqrt{\pi}}{2}\frac{2-\alpha_t^M}{\alpha_t^M} \tag{5}$$

While, it was suggested [13, 14] that this model gave inconsistent slip velocity coefficient for the diffuse reflection with the analytical result. The model by Sharipov [15] based on the correct result for full accommodation by the variational approach to the BGK kinetic equation was proposed as follows:

$$\sigma_p = \frac{2-\alpha_t^S}{\alpha_t^S}\left(\sigma_p(\alpha_t^S = 1) - 0.1211(1-\alpha_t^S)\right) \tag{6}$$

where $\sigma_p(\alpha_t^S=1)=1.016$. In this study, the Maxwell diffuse-specular scattering model and the model by Sharipov were employed, which gave quantitatively different results but not qualitatively.

For the measurement accuracy of the obtained tangential momentum accommodation coefficient, the uncertainties in the mass flow measurement by the constant volume method caused from the non-isothermal effect, the outlet tank volume, temperature and the linear fitting process, those in the theoretical mass flow rate from the inner diameter the length of the microtube, temperature, and pressure, and that from the fitting procedure of the measured mass flow rate by that from the theory were considered [6]. The outlet tank volume was measured from the pressure change induced by changing the volume in a syringe connected to the tank [6] for the precise measurement. Since the error arose mainly from the inner diameter of the microtube, it was estimated from the mass flow rate in the hydrodynamic regime using the regression analysis. The total measurement uncertainty was evaluated at most 2.3%. The uncertainties were evaluated on each experiment, and were reflected in the results.

3. Experimental setup

The experimental setup was explained elsewhere [6]. The schematic image of the experimental setup is shown in figure 1. Two tanks connected with a single microtube were set to slightly different pressures, which caused a pressure-driven flow through the microtube.

The microtubes of SUS304 with the nominal inner diameter of D=0.29mm (Ohbakiko) were employed. The inner surface of the microtube was estimated as an engineering surface. Three different tubes were applied to eliminate individual differences; S1, S2, and S3. The microtubes were evacuated for a long time in advance of measurement without any special surface treatment; i.e. "as is" condition of an engineering surface. The length of the microtubes were L=50.0mm, which were much greater than the inner diameter allowing to neglect the inlet and the outlet effects.

Argon, nitrogen and oxygen were employed as test gases. Pressures were measured by temperature controlled capacitance manometers (MKS Baratron 627B Series; accuracy ±0.15%). The leakage was less than 0.1% of the mass flow rate through a microtube. The pressure ratio was set from 3 to 5, and the mean Knudsen number Kn_m was less than 0.32, where the velocity slip condition of the second-order was valid [16, 17]. The highest pressure condition in the inlet tank among all measurement was about 500Pa.

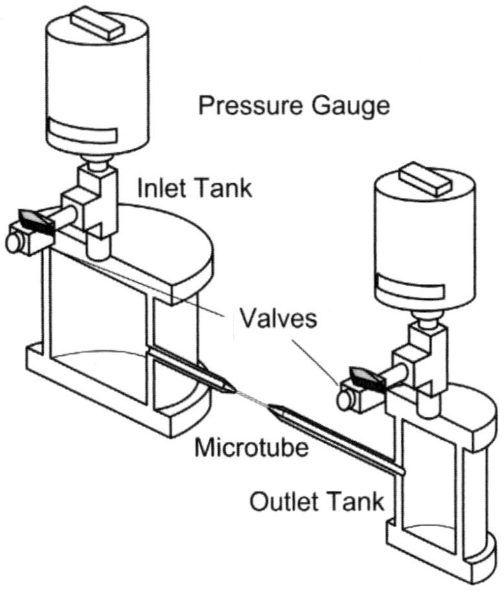

Figure 1. Schematic image of experimental setup.

4. Results and Discussions

The inner diameters of the microtubes were estimated from the mass flow rate in the hydrodynamic regime. The obtained values are listed in table 1. They slightly differ among gas species due to the difference in the adsorption coefficients. For evaluation of the mass flow rate, the estimated inner diameter for each microtube and each gas species was used. The difference among the gas species might be attributed to the adsorption on the inner surface; however, it is still under investigation.

Table 1. Estimated inner diameter of the microtubes in [μm].

microtube	Ar	N_2	O_2
S1	298.2±0.2	297.7±0.2	297.6±0.2
S2	298.3±0.2	298.1±0.2	297.8±0.2
S3	296.6±0.2	296.0±0.2	296.1±0.2

The obtained mass flow rates was normalized and plotted against the mean Knudsen number. The normalized data were well fitted by quadratic functions; equation (4). The typical example of the result for S1 microtube with argon is shown in figure 2. The determination coefficients of the fitted curves were larger than 0.99 for all conditions. The second-order effect was confirmed to be small for all cases, which indicates the mean Knudsen number range was appropriate for the slip flow assumption.

The tangential momentum accommodation coefficients were calculated for two models, and plotted in figure 3. The obtained values seem to show some tendency between the microtubes of S1, S2, and S3, which could be an individual difference of engineering surfaces. However, these discrepancies are small. Considering the measurement error, it is concluded that the difference is not significant and negligible. The obtained values of the tangential momentum accommodation coefficients were quite close to unity, but clearly less than unity. The accurate measurement allowed us to measure the small discrepancy of the tangential momentum accommodation coefficient from the unity. Thus, the accommodation even for tangential momentum is incomplete on the non-treated "as is" engineering surface of SUS304.

This result is consistent with the report on the thermal accommodation coefficient α_E [18]; α_E=0.95 for argon and α_E =0.87 for nitrogen. Thus, the common assumption of complete accommodation on engineering surfaces must be inappropriate. Comparing the accommodation coefficients for the tangential momentum with that for the energy, there are some discrepancies. Therefore, it is also inappropriate to employ the Maxwell diffuse-specular scattering model, where all the accommodation coefficients are assumed to be the same.

The effect of gas species on the tangential momentum accommodation coefficient was negligibly small on the engineering surface, which is different from the tendency for non-metal surfaces already shown in the literatures [13]. However, this might be because the difference in the molecular weight was too small among argon, nitrogen, and oxygen molecules. Further measurement is required to investigate on the effect.

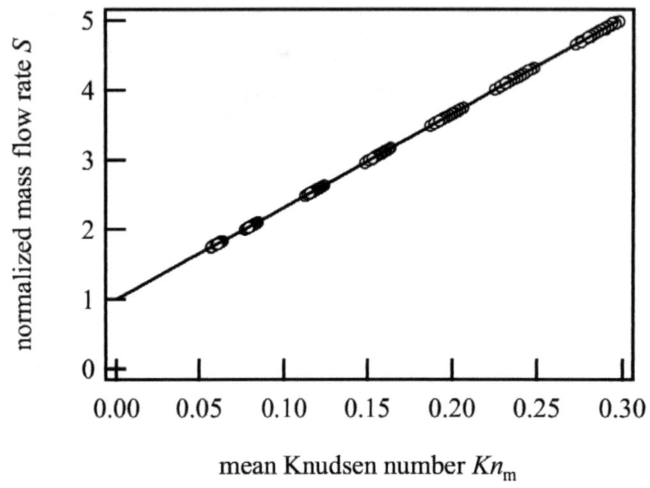

Figure 2. Typical example of the obtained normalized mass flow rate against the mean Knudsen number (Argon, S1 microtube).

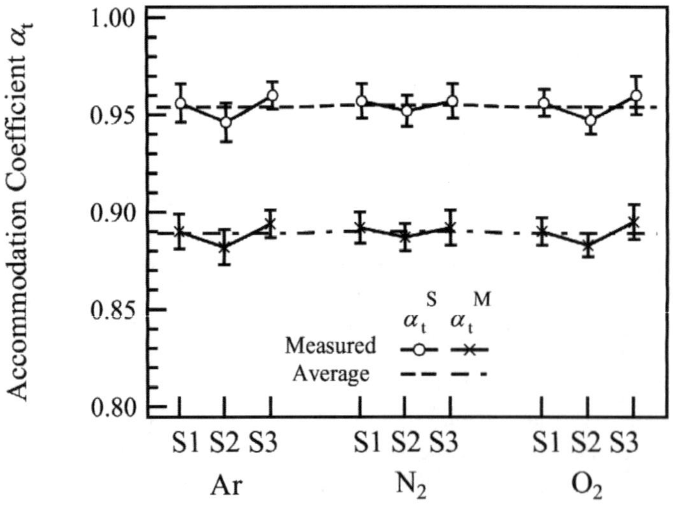

Figure 3. Obtained tangential momentum accommodation coefficients.

5. Conclusion

The tangential momentum accommodation coefficient was successfully obtained from the mass flow rate measurement through a microtube with an engineering metal surface of SUS304, which was manufactured from a bulk material, in the slip flow regime. The obtained tangential momentum accommodation coefficients for argon, nitrogen, and oxygen molecules were less than unity, which is inconsistent with the assumption of full accommodation on metal surfaces often used in the numerical simulations.

The measurement on other metal, polymer, glass and quartz surfaces with gases of much larger range in the molecular weight, like He and Xe, would be the future work.

References

[1] Agrawal A and Prabhu S V 2008 *J. Vac. Sci. Technol.* **A 26** 634
[2] Tison S A 1993 *Vacuum* **44** 1171
[3] Marino L 2009 *Microfluid. Nanofluid.* **6** 109
[4] Varoutis S, Naris S, Hauer V, Day C and Valougeorgis D 2009 *J. Vac. Sci. Technol.* **A 27** 89
[5] Demsis A, Prabhu S and Agrawal A 2010 *Exp. Therm. Fluid Sci.* **34** 1448
[6] Yamaguchi H, Hanawa T, Yamamoto O, Matsuda Y, Egami Y and Niimi T 2011 *Microfluid. Nanofluid.* **11** 57
[7] Ewart T, Perrier P, Graur I and Méolans J G 2007 *Microfluid. Nanofluid.* **3** 689
[8] Arkilic E B, Breuer K S and Schmidt M A 2001 *J. Fluid Mech.* **437** 29
[9] Ewart T, Perrier P, Graur I A and Méolans J G 2006 *Exp. Fluids* **41** 487
[10] Ewart T, Perrier P, Graur I A and Méolans J G 2007 *J. Fluid Mech.* **584** 337
[11] Graur I A, Méolans J G and Zeitoun D E 2006 *Microfluid. Nanofluid.* **2** 64
[12] Arkilic E B, Schmidt M A and Breuer K S 1997 *J. MEMS* **6** 167
[13] Perrier P, Graur I A, Ewart T and Méolans J G 2011 *Phys. Fluids* **23** 042004
[14] Graur I A, Perrier P, Ghozlani W and Méolans J G 2009 *Phys. Fluids* **21** 102004
[15] Sharipov F 2004 *Proc. 5th Int. Conf. on Thermal and Mechanical Simulation and Experiments in Microelectronics and Microsystems* (Belgium: Shaker) p 243
[16] Maurer J, Tabeling P, Joseph P and Willaime H 2003 *Phys. Fluids* **15** 2613
[17] Colin S, Lalonde P and Caen R 2004 *Heat Transfer Eng.* **25** 23
[18] Trott W M, Castañeda J N, Torczynski J R, Gallis M A and Rader D J 2011 *Rev. Sci. Inst.* **82** 035120

Development of Pressure-Sensitive Channel Chip for Micro Gas Flows

Yu Matsuda[1], Hiroki Yamaguchi[1] and Tomohide Niimi[1]

[1] Department of Micro-Nano Systems Engineering, Nagoya University, Furo, Chikusa, Nagoya, Japan

E-mail: y.matsuda@nagoya-u.jp

Abstract. Optical measurement techniques are useful for experimental studies on micro gas flows, which enable us to non-intrusively measure the flows with a high spatial resolution. The pressure-sensitive paint (PSP) technique, which is based on the emission of photons from luminophore, is a potential diagnostic tool for pressure measurement of micro gas flows. However, measurements by conventional PSPs are limited to the sub-millimeter order spatial resolution of ca. 200 μm, indicating the difficulty of the micro scale measurements. The present study proposes pressure-sensitive channel chip (PSCC) which is a micro channel with the capability of measuring pressure. We focused on the poly(dimethylsiloxane) (PDMS) micro-molding technique, which is one of the most popular techniques to fabricate a micro channel easily. Moreover, PDMS is a polymer used as a binder in PSP because of high optical transparency, gas permeability, and gas diffusivity. Thus, we developed a micro channel by the PDMS micro-molding technique with mixing a pressure-sensitive luminophore into PDMS: i.e. a micro channel fabricated by PSP, which is named PSCC. A flow through a micro converging-diverging nozzle with the throat width of 120 μm was demonstrated. The pressure distribution on the nozzle surface was successfully obtained by PSCC.

1. Introduction

It has been strongly desired to understand thermo-fluid phenomena around micro- and nano-systems such as a magnetic head slider of hard disk drive, a micro thruster for a micro satellite, and a semiconductor thin film fabrication system and so on. Flows through a micro system should be treated as high Knudsen number flows ($Kn = \lambda/L$: λ is the mean free path for gas molecules and L is the characteristic length of the flow). There are no appropriate techniques for the measurement of gas pressure on solid surfaces inside micro-systems. To measure pressure distributions in high Knudsen number regimes, we have focused on a pressure sensitive paint (PSP) technique [1, 2]. Because PSP works as a so-called "molecular sensor", it seems suitable for the analyses of high Knudsen number flows, which require diagnostic tools in molecular level. The pressure measurements of the flows in a small object were carried out by Huang et al. [3] and Nagai et al.[4]. However, the spatial resolutions of their measurements are limited to sub-millimeter order of ca. 200 μm due to the aggregation of luminescent molecules in polymer binders [5]. Moreover, conventional PSPs are very thick [2] (> 5 μm) compared to the dimension of micro-devices owing to the use of polymer binders. In the previous study, we developed pressure-sensitive molecular film (PSMF) by the Langmuir-Blodgett (LB) technique to construct ordered molecular assemblies, and tested PSMF to clarify the feasibility of the pressure measurement around micro-devices [6, 7, 8]. Because the fabrication of PSMF takes a long time and the application of PSMF is restricted only to a flat chemically-modified surface, the measurement by PSMF is not easy to be applied to practical measurements. We proposed pressure-sensitive channel chip (PSCC) which is a micro channel with the capability of measuring pressure [9]. We focused on the poly(dimethylsiloxane) (PDMS) micro-molding technique, which is one of the most popular techniques to fabricate a micro channel easily. Moreover, PDMS is a polymer used as a binder in PSP because of high optical transparency, gas permeability, and gas diffusivity. Thus, we developed a micro channel by the PDMS micro-molding technique with mixing a pressure-sensitive luminophore into PDMS: i.e. a micro channel fabricated by PSP, which is named PSCC.

2. Fabrication of PSCC

The fabrication process of PSCC is very simple, because most of the processes are the same as the PDMS molding technique. PSCC consists of three parts: (i) a sensor layer; (ii) a gas barrier layer; (iii) a supporting PDMS block. A sensor layer is fabricated by a pressure-sensitive luminophore and PDMS with high gas permeability. The high gas permeability of a binder is required to achieve high pressure sensitivity. A gas barrier layer is necessary to prevent the permeation of oxygen molecule from a sensor layer to a supporting block and vice versa, because oxygen molecules through a supporting block to a sensor layer changes the luminescent intensity of PSCC regardless of the pressure in the channel and will cause measurement error. The high gas permeability and diffusivity of PDMS are drawbacks from this standpoint. A supporting block is fabricated by pure PDMS to fix these layers.

The actual fabrication procedure is as follows (see Figure 1): firstly, a master of SU-8 photoresist (MicroChem, SU-8 3050) on a silicon wafer was fabricated using standard photolithography procedures (Figure 1(1)). Pt(II) tetrakis(pentafluorophenyl)porphyrin (PtTFPP) was adopted as a pressure-sensitive luminophore, and a PDMS prepolymer (Dow corning TORAY, Silpot 184) was prepared by mixing PtTFPP in the mass ratio of PtTFPP:PDMS= 0.1 mg : 1.0 g. Then the degassed PDMS prepolymer containing PtTFPP was poured over the master, and was spun at 1000 rpm to make thin layer of the PDMS with luminophore (sensor layer, see Figure 1(2)). The parylene (KISCO, dix C) layer was deposited on the sensor layer using parylene coater (KISCO, DACS-LAB) as a gas barrier layer [10] with the thickness of ca. 0.5 μm (Figure 1(3)). A pure PDMS prepolymer (without PtTFPP) was poured over the sample and was cured at 70°C for 60 min to form a supporting block (Figure 1(4)). This three-layer structure was peeled off from the master, and the PSCC was obtained. For the measurement, the PSCC was holed for tube connections and was placed on a glass slide (Figure 1(5)).

3. Stern-Volmer relation

The pressure measurement technique using PSCC is based on the oxygen quenching of luminescence. When PSCC is illuminated by UV/blue light (350 – 420 nm), the luminescent molecules are excited by absorption of photon energy, and then the molecules emit phosphorescence. On the other hand, oxygen molecule with triplet ground state acts as a quencher of the phosphorescence (630 – 700 nm for PtTFPP[1, 4]). As a result, the phosphorescence intensity decreases as an increase in the partial pressure of oxygen gas. Pressure on the PSCC surface can be deduced from the relationship between pressure p and the luminescence intensity I (Stern-Volmer plot [1, 2]),

$$\frac{I_{\rm ref}}{I} = A + B\frac{p}{p_{\rm ref}}, \tag{1}$$

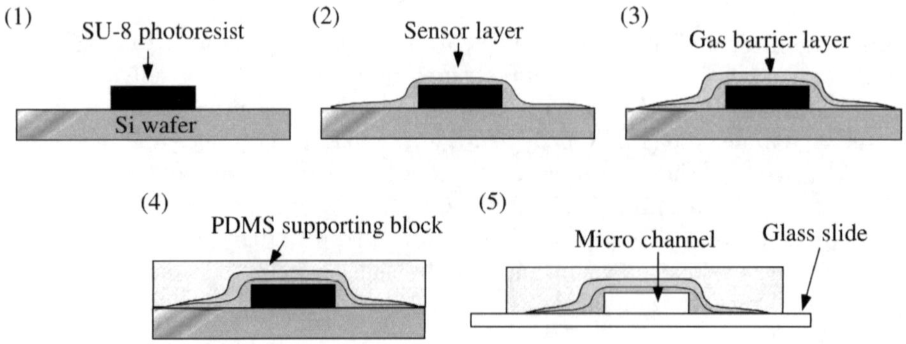

Figure 1. A schematic process of PSCC.

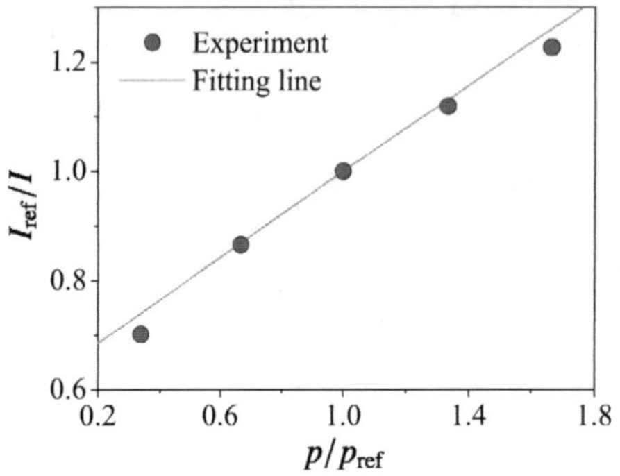

Figure 2. Stern-Volmer plot for the PSCC used in this study, where the reference pressure $p_{\mathrm{ref}} = 15$ kPa. Fitting line: $I_{\mathrm{ref}}/I = 0.61 + 0.39p/p_{\mathrm{ref}}$

where the subscript "ref" describes the known (reference) condition. The coefficients A and B are the Stern-Volmer coefficients which are determined by a calibration test, and show temperature dependencies. In this study, the temperature dependencies of the Stern-Volmer coefficients was neglected, because our experiment was carried out in a constant temperature condition.

The Stern-Volmer plot of our PSCC is shown in Figure 2, where the reference pressure $p_{\mathrm{ref}} = 15$ kPa. It is shown that the experimental data was well fitted by Equation 1 with the constraint condition of $A + B = 1$ at $(I_{\mathrm{ref}}/I, p/p_{\mathrm{ref}}) = (1, 1)$. The pressure sensitivity of PSCC of 0.39 is similar to that of conventional PSPs [1], and is applicable for pressure measurements.

4. Experimental Setup

The PSCC employed in this study was a converging-diverging nozzle. As shown in Figure 3, the converging and the diverging angle were 90° and 24°, respectively. The throat width and the diverging length were 120 and 485 μm, respectively. The depth of the nozzle was 59 μm. The micro-nozzle was put on a holder with an inlet and an outlet port (see Figure 4). Oxygen (21.0 %) gas balanced with nitrogen gas was supplied from the inlet port, and the outlet port was evacuated by a scroll pump (Ulvac, DVS-631). The pressures of the inlet and the outlet were monitored by capacitance manometers (Keyence, AP-44) connected to pressure taps. Since both the inlet and the outlet port were attached far from the nozzle throat (1.2 cm), the effects of inlet and outlet conditions can be negligible.

The images of the PSCC emission were recorded with an EM-CCD camera (Princeton

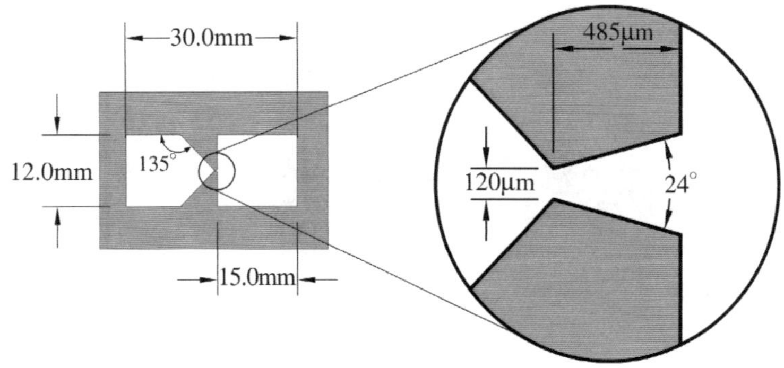

Figure 3. Geometry of PSCC (converging-diverging nozzle).

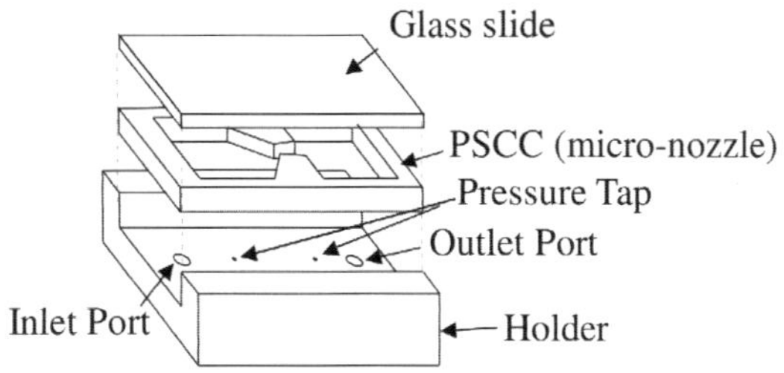

Figure 4. Schematic image of micro-nozzle assemblies.

Instruments, ProEM:1024B) mounted on a fluorescent microscope (Olympus, BX61). We adopted a band-pass filter (400 ± 20 nm), a long-pass filter (600 nm), and a dichroic mirror (reflection at $400 - 450$ nm, transmission at $535 - 700$ nm) in the microscope. An objective lens, Olympus MPlanFLN 10× with a numerical aperture of 0.30, was used for the measurement. The images were taken by a camera with a binning of 2×2pixel, 1pixel of the image corresponds to 2.6 μm.

5. Results and discussions

Figure 5 shows the pressure distribution of the nozzle with the inlet port pressure $p_{\text{in}} = 30.1$ kPa and the outlet port pressure $p_{\text{out}} = 2.9$ kPa. The images were taken with exposure time of 0.05s, and 32 images were averaged to reduce the noise for wind-on and wind-off (reference) images. The pressure distribution on the nozzle surface was successfully obtained without any aggregations of luminophore observed in conventional sprayed PSPs[5, 8]. The PSCC captured the sharp pressure drop near the throat $160 < x < 250$ μm as shown in Figure 6. As shown in Figure 7, the pressure distribution at the nozzle exit was a parabolic curve. The pressure near the wall was measured, while the pressures at the edge of the wall contain considerable errors due to the scattered light. The fluctuation of the measured pressure values were ± 0.5 kPa at the lowest SN ratio area (ex. $x = 10\mu$m) and this is much smaller than that by PSMF [8]. The pressures at $x = 0$ and 850 μm were 20.5 ± 0.3 kPa and 11.7 ± 0.5 kPa, respectively. These measured pressures were considerably different from $p_{\text{in}} = 30.1$ kPa and $p_{\text{out}} = 2.9$ kPa at the pressure taps, which were attached far from the channel (ca. 1 cm). These pressure drops would be caused by wall friction and the acceleration at the long converging area of the nozzle (see Figure 3).

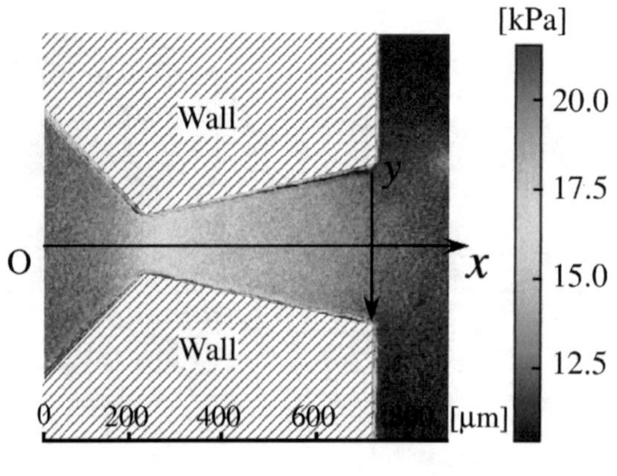

Figure 5. Pressure distribution of the nozzle.

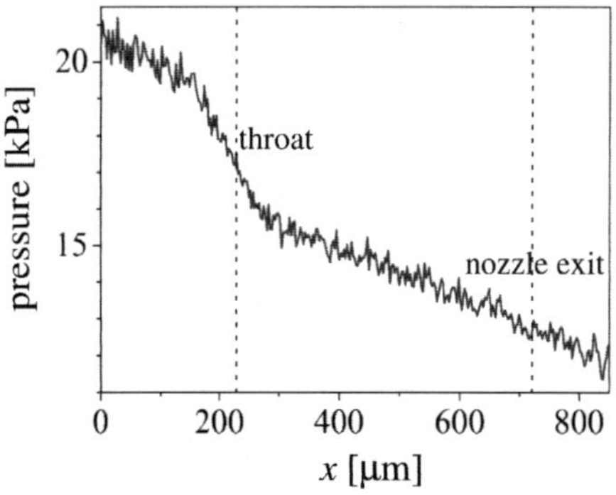

Figure 6. Pressure distribution along the centerline of the nozzle.

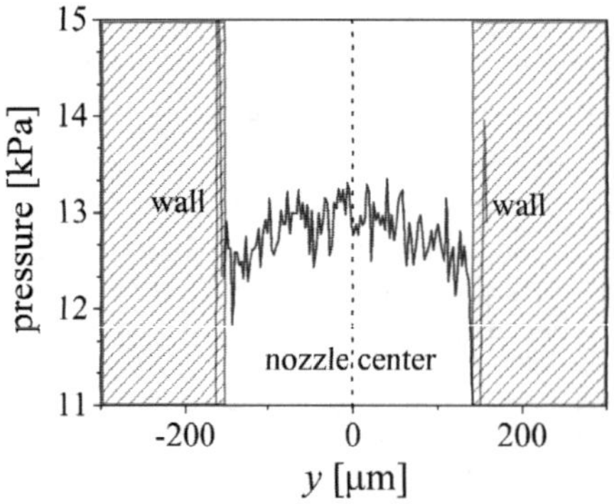

Figure 7. Pressure distribution along the crossline of the nozzle exit.

6. Conclusion

We have developed a new pressure sensing tool named pressure-sensitive channel chip (PSCC) by combining the pressure-sensitive paint (PSP) technique with the poly(dimethylsiloxane) (PDMS) micro-molding technique. PSCC is a micro channel which itself works as a pressure "distribution'' sensor. In this study, a micro converging-diverging nozzle with the throat width of 120 μm was prepared and the pressure distribution on the nozzle surface was successfully obtained by PSCC. It is shown that PSCC is feasible to measure the pressure in micro gas flows.

As a future work, we are going to apply PSCC to dissolved oxygen (DO) measurement. The measurement of DO in water is very important in many research fields, because oxygen plays an important role in biochemical reactions, chemical reactions and so on. The DO distribution measurement will facilitate understanding of the mechanism of such reactions. We have proposed a novel utilization of PSP as a DO sensor film for macro scale unsteady DO distribution measurements [11]; thus it is considered that PSCC can be applied to "micro scale" DO measurement. For DO measurement, Equation 1 is written as follows:

$$\frac{I_{\mathrm{ref}}}{I} = A_c + B_c \frac{[\mathrm{O}_2]}{[\mathrm{O}_2]_{\mathrm{ref}}}, \tag{2}$$

where $[\mathrm{O}_2]$ is DO concentraion. Since the measurement system is the same as presented this paper, this new method easy to be developed. PSCC will be a promising measurement method for micro DO measurement method.

References

[1] Liu T and Sullivan J P 2005 *Pressure and temperature sensitive paints* (Springer)
[2] Bell J H, Schairer E T, Hand L A, Mehta R D 2001 *Annu. Rev. Fluid. Mech.* **33** 155
[3] Huang C Y, Sakaue H, Gregory J W, Sullivan J P 2002 *Proc. 40th aerospace sciences meeting and exhibit* AIAA2002-0256
[4] Nagai H, Naraoka R, Sawada K, Asai K 2008 *AIAA Journal* **46** 215
[5] Mori H, Niimi T, Uenishi H, Hirako M 2005 *Phys. Fluids.* **17** 100610
[6] Matsuda Y, Mori H, Niimi T, Uenishi H, Hirako M 2007 *Exp. Fluids.* **42** 543
[7] Matsuda Y, Mori H, Sakazaki Y, Uchida T, Suzuki S, Yamaguchi H, Niimi T 2009 *Exp. Fluids.* **47** 1025
[8] Matsuda Y, Uchida T, Suzuki S, Misaki R, Yamaguchi H, Niimi T 2011 *Microfluid. Nanofluid.* **10** 165
[9] Matsuda Y, Misaki R, Yamaguchi H, Niimi T 2011 *Microfluid. Nanofluid.* **11** 507
[10] Heo Y S, Cabrera L M, Song J W, Futai N, Tung Y C, Smith G D, Takayama S 2007 *Anal. Chem.* **73** 1126
[11] Matsuda Y, Nagashima F, Yamaguchi H, Niimi T 2011 *Sens. Actuator B-Chem.* **160** 1464

Numerical modeling of Knudsen's thermal creep experiment

Yu.A. Anikin[1,2], O.I. Dodulad[1,2], Yu. Yu. Kloss[1,2], D.V. Martynov[1,2], P.V. Shuvalov[1,2], F.G Tcheremissine[1,3]

[1] Moscow Institute of Physics and Technology, Moscow, Russia
[2] National Research Centre "Kurchatov institute", Moscow, Russia
[3] Dorodnicyn Computing Centre of RAS, Moscow, Russia

E-mail: tcherem@ccas.ru

Abstract. In his study of thermal creep phenomena in 1910, M. Knudsen build a construction that was composed of 10 pairs of cylindrical tubes of different diameters heated at its junctions. He observed that at some conditions the construction worked as a gas compressor and he was able to obtain the pressure ratio at the ends of the device up to 10. The aim of the present work was to model the experiment using the power of modern computers and the efficient method for solving the Boltzmann kinetic equation. The computations were made for the same set of temperatures, gas pressures and geometrical parameters of the installation as in the experiment, and for molecular hydrogen with real Lenard-Jones potential and realistic rotational energy relaxation time. The Boltzmann equation was solved by the conservative projection method. The computations were carried out with the use of specially developed by the authors program-solving environment for modeling of rarefied gas flows with arbitrary geometry and real gas parameters. The obtained results are in reasonable good agreement with the experimental data.

1. Introduction

In 1910 Martin Knudsen has studied thermal creep phenomena in a cylindrical tube with the temperature gradient applied on the walls and constructed a device operating as a pump. In contrast to classical vacuum pumps, the device worked without moving parts and doesnt use oil, hence it was environmentally friendly. The last feature was not much considered in that time, but became very important now.

Knudsen has measured pressure difference at the ends of the device built from the capillary and large-scale cylindrical tubes connected in series [1]. The diameter of the capillary tubes was about 0.5 mm and the temperature grows linearly along the tubes from T_1 to $T_2 > T_1$. The diameter of the large-scale tubes was about $10 - 15$ mm and the temperature decreases there from T_2 to T_1. It was noticed that under certain conditions pressure rate on the device ends rises up to 10 times, i.e. it could operate as a pump. In the beginning of the 20th century such pumps dont obtain practical applications because the diameter of the capillary tubes limited maximal operating pressure by approximately $1.33 \cdot 10^4$ Pa.

The aim of the present work consists in modeling of the Knudsen experiment on the base of solution of the Boltzmann kinetic equation for a diatomic gas. The rotational degrees of freedom are taken into account by the application of a 2-levels relaxation model [2].The Boltzmann

Published under licence by IOP Publishing Ltd

equation is solved by a splitting scheme with three stages: transport stage, elastic relaxation and rotational-translational relaxation. The transport equation is approximated by the finite volumes method. Tetrahedral unstructured grids generated with the help of GMSH software [3] are used. The collision integral is evaluated by the conservative projection method [4]. The method ensures that the laws of conservation of mass, momentum and energy are strictly observed and the collision integral of the Maxwellian distribution function is equal to zero.

Rarefied gas flows induced by thermal creep in microchannels have been studied in [5, 6] with the help of DSMC and BGK.

Using aerogel membranes make it possible to create Knudsen pump prototypes performing operations at atmospheric pressure [7].

2. Mathematical framework

The Boltzmann equation is solved

$$\partial f/\partial t + \xi \cdot \partial f/\partial \mathbf{x} = I(f, f)$$

$$I(f, f) = \int_{-\infty}^{\infty} \int_{0}^{2\pi} \int_{0}^{b_m} (f' f'_* - f f_*) g b db d\varphi d\xi_*$$

A spherical domain on a uniform grid with a spacing h is chosen as the discrete velocity space. The maximum velocity is taken equal to $4.8\sqrt{kT_{max}/m}$ because the values of the distribution function for the greater velocities are negligible small.

For convenience of data input and analysis, all values are represented in the dimensionless form. Reference quantities are selected, and all parameters are normalized to those quantities. Free path is chosen as the length unit, characteristic free time as the time unit, and thermal velocity as the velocity unit.

To ensure high efficiency of complex calculations, there was developed a problem-solving environment (PSE) [8] enabling parallel computations. The PSE is capable effectively solve a large class of rarefied gas dynamics problems. It includes solvers which enable quick calculations of 2D and 3D gas flows on rectangular structured grids and calculations of flows on non-structured tetrahedral grids allowing description of arbitrary 3D geometries. This makes the PSE a universal tool not just for academic purposes but for engineering analysis as well.

High modularity makes the PSE flexible and scalable which is important for its future development. Due to the cross-platform design of the code and the possibility to use graphic processors that are rapidly developing at present time, the problems can be solved in a wide range of architectures.

Fig.1 shows the diagram of data flow from user-supplied input parameters and computational grid generation to visualization of results with the help of third parties software. In the centre of the figure is the solver – a program that carries out all calculations required for gas modeling.

The efficient paralleling, on the one hand, allowed a substantial reduction in the computation time (from weeks to hours), and, on the other hand, enabled precision calculations on fine grids that require large memory. Today solvers use two paralleling technologies – the MPI (Message Parallel Interface) that is used for computations on clusters of ordinary processors, and the Nvidia CUDA (Computer Unified Device Architecture) designed for general purposes computations on the GPUs.

3. Modeling results

Fig. 2 shows the design of i-th cascade of the Knudsen pump. Capillar and large-scale tubes with the same length L are connected successively.

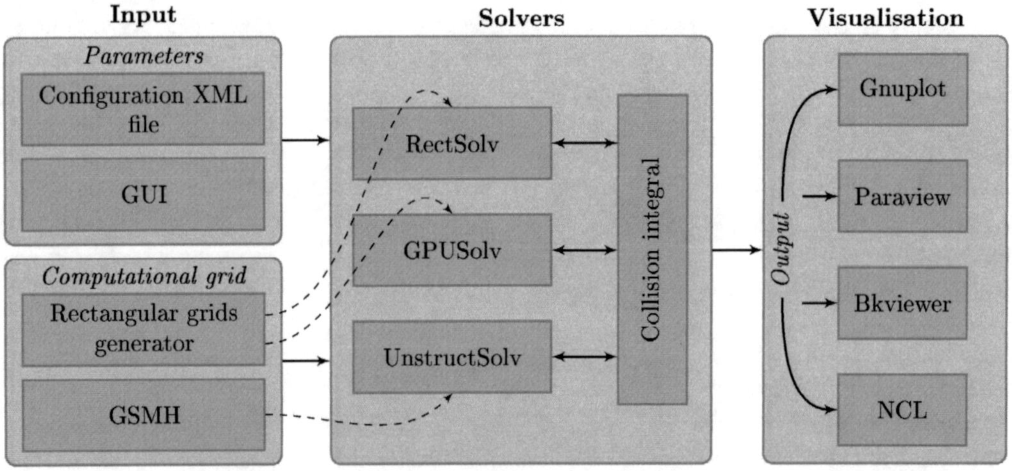

Figure 1. Diagram of the PSE

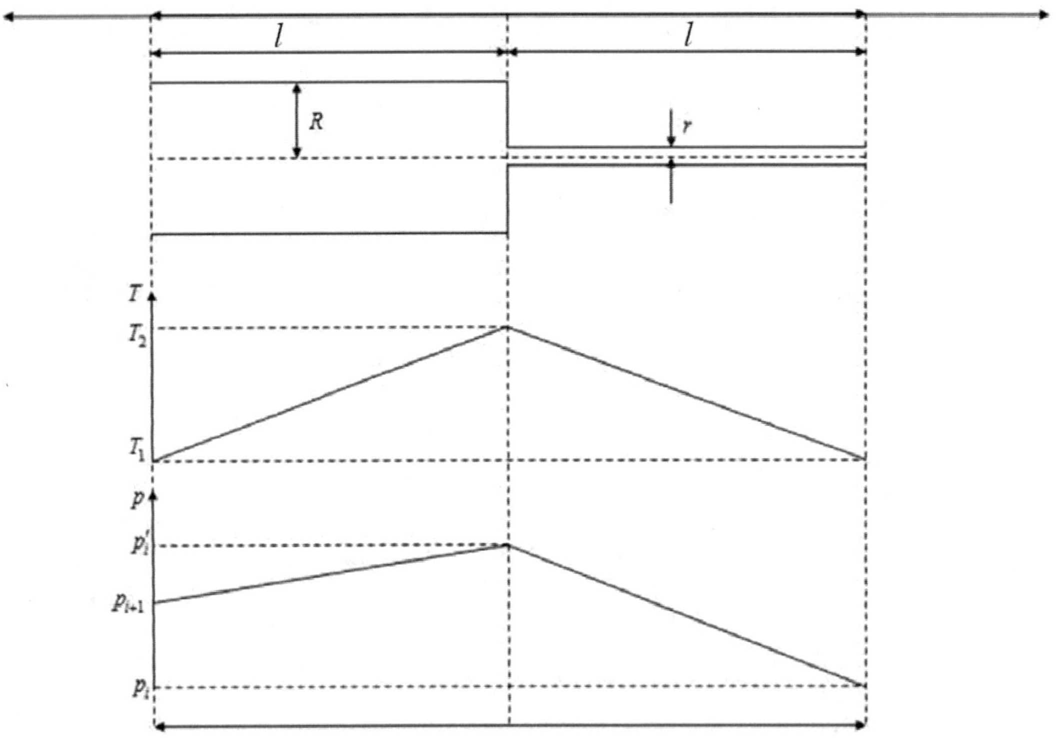

Figure 2. The i-th cascade of the Knudsen pump

In the first series of experiments Knudsen takes $T_1 = 373K$ and $T_2 = 623K$. In the present paper Lenard-Jones potential for Hydrogen molecules was used with parameters $\varepsilon = 38K$ and $\sigma = 2.915$ Å. The ratio of rotational and translational relaxation times was taken equal to 100 and rotational energy accommodation coefficient at the wall was taken equal to 0.5.

Tetrahedral mesh for one cascade of the device is shown in Fig. 3. The edges of tetrahedrons near the symmetry line were taken much larger then the edges near the walls.

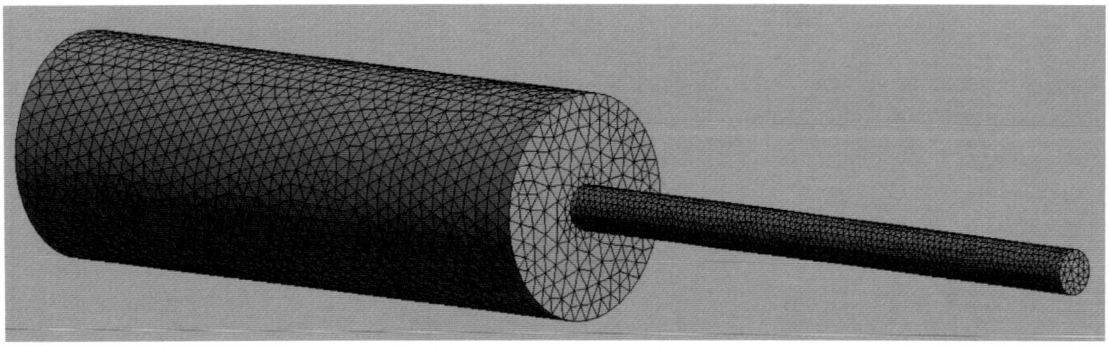

Figure 3. Tetrahedral mesh

Boundary conditions of perfect accommodation with outgoing Maxwellian function were posed at the walls and external flanks of the tubes. The computation was continued until the achievement of a steady distribution of gas parameters in the system.

Fig. 4 presents a plot of pressure ratio $a = \frac{p_2}{p_1}$ versus Knudsen number Kn for $l/r = 10$.

Table 1 shows pressure ratio on the ends of the tube for various ratios l/r.

Table 1. Pressure ratio

Kn	$l/r = 1$	$l/r = 5$	$l/r = 10$	$l/r = 50$	$l/r = 100$
0.1	1.037	1.035	1.035	1.036	1.035
0.5	1.099	1.130	1.135	1.134	1.135
1.0	1.115	1.165	1.183	1.186	1.184
2.5	1.127	1.203	1.219	1.220	1.221
5.0	1.136	1.216	1.235	1.236	1.237
10.0	1.150	1.227	1.248	1.248	1.249

One could make a conclusion based on the data provided that the steady-state pressure ratio $a = \frac{p_2}{p_1}$ does not depend on the tube length if the condition $\frac{l}{\chi} > 5$ is met, where $\chi = min(r, \lambda)$, λ being the molecular mean free path.

The results of the numerical modeling are compared with the experimental data of Knudsen [1] in Table 2.

Fig. 5, 6, 7 show steady-state distribution of temperature, density and pressure for one-cascade Knudsen pump with $\frac{p_1+p_2}{2} = 34.931$ Pa. The values on the plot were normalized by the following reference values: $373K$ for temperature, 28.7 Pa for pressure and $5.76 \cdot 10^{23} m^{-3}$ for density.

4. Conclusion

Numerical modeling of the Knudsen pump was carried out. The Boltzmann equation was solved by the conservative projection method. The rotational degrees of freedom were taken into

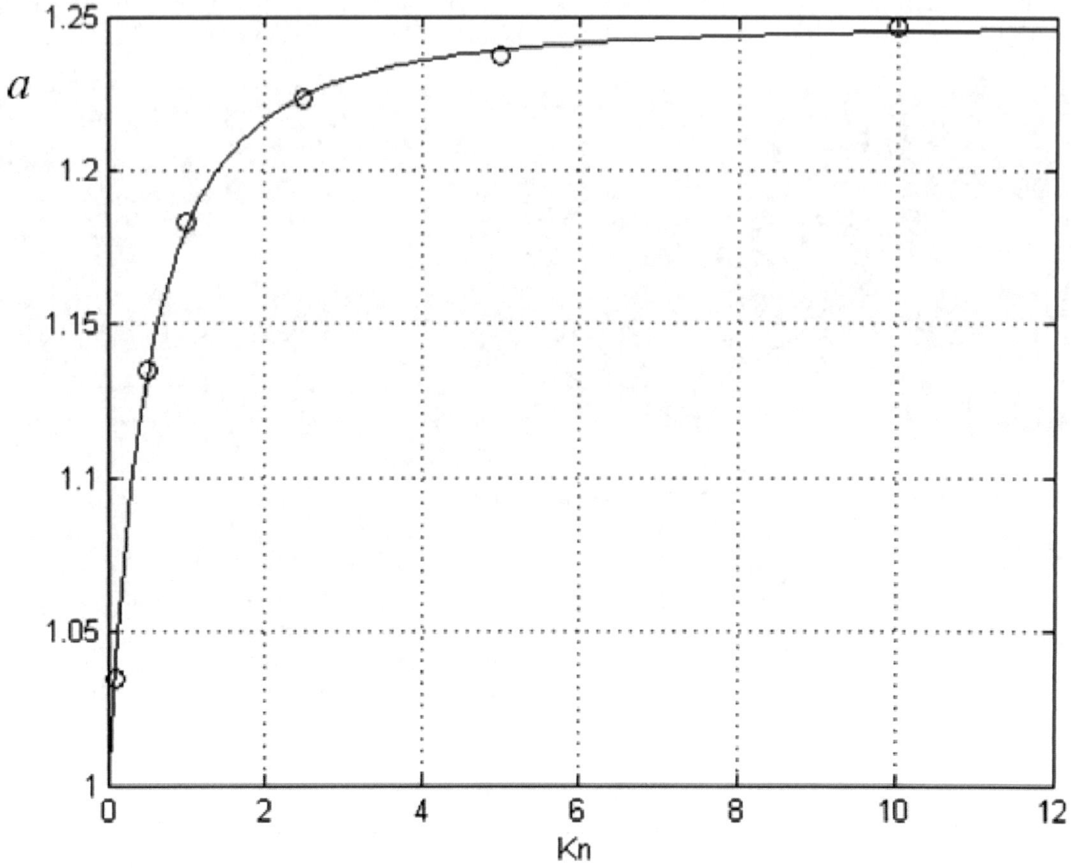

Figure 4. Pressure ratio versus Knudsen number

Table 2. The numerical results compared with the experimental data

Data source	p_2	p_1	$\frac{p_1+p_2}{2}$	$p_2 - p_1$	$\frac{p_2}{p_1}$
Experimental data	2.973	2.813	2.893	0.160	1.057
Numerical modeling	2.973	2.813	2.893	0.160	1.057
Experimental data	37.597	32.131	34.931	5.466	1.170
Numerical modeling	37.264	32.597	34.931	4.666	1.170
Experimental data	189.851	175.319	182.652	14.532	1.083
Numerical modeling	189.051	176.252	182.652	12.799	1.075
Experimental data	702.342	689.943	696.609	12.399	1.018
Numerical modeling	703.542	689.677	696.609	13.866	1.018

account by the application of the 2-levels relaxation model. The obtained results are in good agreement with the experimental data of Knudsen.

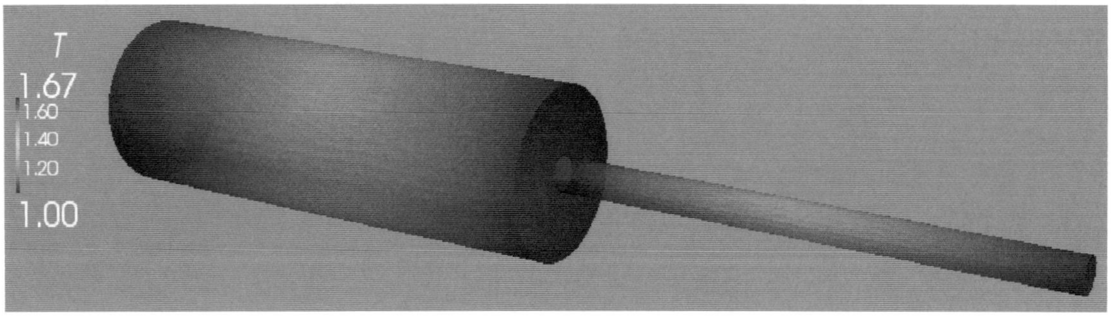

Figure 5. Steady-state distribution of temperature

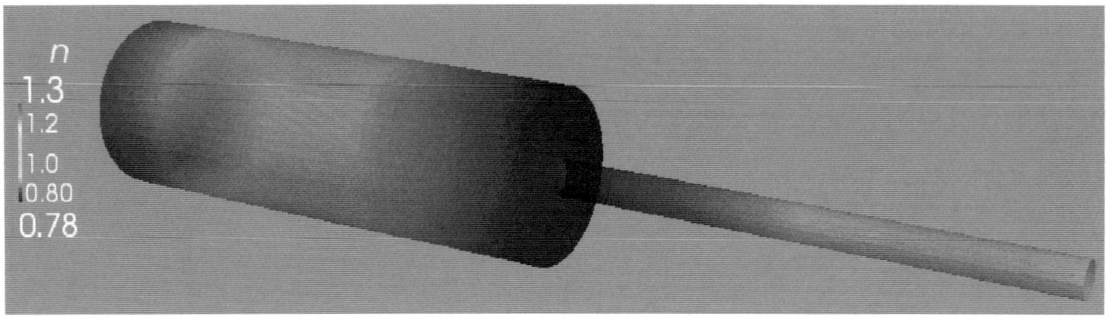

Figure 6. Steady-state distribution of density

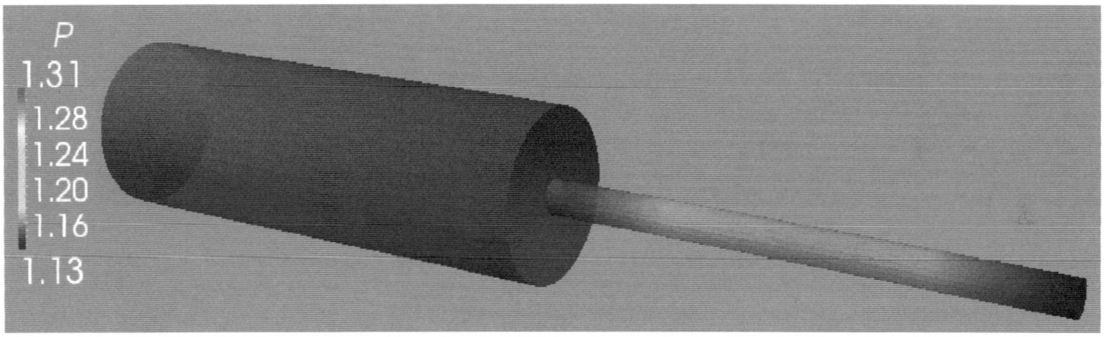

Figure 7. Steady-state distribution of pressure

References

[1] Knudsen M 1910 Eine Revision der Gleichgewichtsbedingung der Gase. Thermische Molekularströmung. *Ann. der Phys* **31** 9 pp 205-229

[2] Tcheremissine F G 2007 Two level kinetic model for rotational-translational transfers in a rarefied gas. URL http://www.chemphys.edu.ru/pdf/2007-10-22-001.pdf

[3] www.geuz.org/gmsh

[4] Tcheremissine F G 2008 Solution of the Boltzmann Kinetic Equation for Low Speed Flows *Transport Theory and Statistical Physics* **37** 5 pp 564-575

[5] Sone Y, Waniguchi Y, Aoki K 1996 One-way flow of a rarefied gas induced in a channel with a periodic temperature distribution *Phys. Fluids* **8** 8 pp 2227-2235

[6] Muntz E P, Sone Y, Aoki K, Vargo S, Young M 2002 Performance Analysis and Optimization Considerations

for a Knudsen Compressor in Transitional Flow *J. Vac. Sci. Technol. A* **20** 1 pp 214-224

[7] Han Y L, Muntz E P, Alexeenko A, Young M 2007 Experimental and Computational Studies of Temperature Gradient-Driven Molecular Transport in Gas Flows through Nano/Microscale Channels *Nanoscale and Microscale Thermophysical Engineering* **11** pp 151-175

[8] Anikin Yu A, Derbakova E P, Dodulad O I, Kloss Yu Yu, Martynov D V, Rogozin O A, Shuvalov P V and Tcheremissine F G 2010 Computing of gas flows in micro- and nanoscale channels on the base of the Boltzmann Kinetic equation *Procedia Computer Science* **1** 1 pp 735-744

Effect of surface roughness: comparison between continuum and kinetic approaches

O Rovenskaya and G Croce
University of Udine, via delle Scienze 208, 33100 Udine, Italy

E-mail: olga.rovenskaya@uniud.it, giulio.croce@uniud.it

Abstract. In the present work a numerical analysis of the flow field in rough microchannels is carried out using two approaches: Navier-Stokes equations provided with first order slip-boundary condition and kinetic S-model equation with Maxwell diffuse reflecting boundary condition. An implicit scheme is used for the solution of S-model equation and an algorithm allowing massive parallelization in both physical and velocity spaces has been developed. The roughness geometry is modelled as a series of triangular obstructions with relative roughness ε equals to 1.25%, 2.5% and 5%. A wide range of Mach numbers is considered, from nearly incompressible to chocked flow conditions and a Reynolds number up to 170. To estimate rarefaction effect the flow at Knudsen number ranging from 0.01 to 0.08 and fixed pressure ratio has been considered. Accuracy and discrepancies between full Navier - Stokes and S-model solutions are discussed, assessing the range of applicability of first order slip condition in rough geometries. The effect of the roughness is discussed via Poiseuille number as a function of local Knudsen and Mach numbers.

1. Introduction

In MEMS fabrication, due to the small scale, it is nearly impossible to create an actual smooth surface. On the other hand, surface roughness may have a significant impact on microchannel performances, both in terms of pressure drop and heat transfer. In the recent past several studies have been devoted to the estimation of roughness effect in these systems. Since roughness may assume different shapes depending on the manufacturing processes, a popular approach is to analyse artificial, ad hoc designed rough textures. This is necessary for numerical analysis, but allows easier parametrization also in experimental work; as an example, an experimental analysis of artificially generated roughness has been carried out as by Kandlikar et al. [1]. In most cases, experimental investigation detected relatively small roughness effect [2, 3], in the sense that its magnitude may be masked by the relatively large uncertainty bar. Nonetheless, even recent experimental works strongly suggest that roughness does affect Poiseuille number in microchannel rarefied flow, although the magnitude of experimental uncertainty does not allow a definitive answer [4]. Thus, the numerical modelling offers an appealing alternative to an experiment, allowing a detailed description of the surface imperfections, where we can easily separate roughness from other microscale effects.

The roughness problem has been first addressed, both via direct numerical analysis and simplified modelling, mainly in the incompressible, no-slip flow regime. The numerical simulation of pressure drop in the presence of three dimensional square box-shaped rough elements was studied in [5], while other authors [6] investigated the effect of several geometries based on 2D squared and saw-tooth obstruction. Both pressure drop and heat transfer in tube and plane channels were considered.

There are few works devoted to the rarefied gas flow through rough channels using direct Monte Carlo simulation [7 -9]. Knudsen numbers from 0.02 to 0.12 were considered, in the presence of very rough geometries, from 5% up to 12%. In both papers, the low Mach number level prevented any compressibility effect. In [10] coupling effects of rarefaction and roughness without the effect of compressibility has been considered using the Lattice Boltzmann equation. Numerical results show that the roughness effect increases the friction factor near the wall, while the rarefaction effect decreases it.

Thus, in the literature the interaction with compressibility effect is generally not taken into account. Only recently Ji et al. [11] studied the influence of roughness in slip flow regime with second order slip boundary conditions up to a maximum exit Mach number around 0.5. They simulated the roughness with rectangular elements on two parallel plates and showed that the effect of wall roughness is reduced with increasing Knudsen number. The prediction of roughness effect for gaseous flow, in the presence of compressibility up to chocked flow conditions and/or rarefaction effects, has been analyzed by Croce [12] and Hakak Khadem [13] using Navier-Stokes equations.

While for incompressible flow the numerical tools for roughness analysis are just the usual, well established Navier-Stokes equations, for gaseous flow the rarefaction effect may require different models. The applicability of continuum numerical approach based on the solution of Navier-Stokes equations should be, in fact, limited to flows where the roughness size is larger than the mean free path of the gas. This condition is obviously stricter than the usual condition of hydraulic diameter larger than the mean free path, and may not be fulfilled in actual application of practical interest. Thus, even if the use of appropriate slip boundary conditions allows to partially overcome this limitation, a correct description of the gas-roughness interaction should require an expensive kinetic numerical approach, at least near the roughness surface.

The aim of this paper is, thus, to investigate the effects of rarefaction, compressibility and roughness using two approaches: Navier-Stokes equations provided with first order slip-boundary condition and kinetic S-model equation with Maxwell diffuse reflecting boundary condition. The analysis is limited to first order slip, since second order models are still geometry dependent and not so developed for reliable, arbitrary geometry simulations. Accuracy and discrepancy between full Navier - Stokes and S-model solutions are discussed, assessing the range of applicability of first order slip condition in rough geometries. The effect of the roughness is discussed via Poiseuille number as a function of local Knudsen and Mach numbers.

2. Numerical method for kinetic approach

In the most general case the gas flow is modelled by the Boltzmann transport equation (BTE), describing the evolution of a particle distribution function $f = f(t, \mathbf{x}, \xi)$ in a six-dimensional phase space:

$$\frac{\partial f}{\partial t} + \xi \frac{\partial f}{\partial \mathbf{x}} = J(f, f) \tag{1}$$

where t is the time, $\mathbf{x} = (x, y, z)$ is the position vector, $\xi = (\xi_x, \xi_y, \xi_z)$ is the particle velocity vector, while the operator $J(f, f)$ takes into account the binary collisions between particles. For the sake of simplicity and computational efficiency, the collision integral $J(f, f)$ is replaced by the S- model [14], which can be written as:

$$J_s(f, f) = \frac{p}{\mu}(S - f) \tag{2}$$

$$S = M\left(1 + \frac{2m\mathbf{qc}}{15n(kT)^2}\left(\frac{mc^2}{2kT} - \frac{5}{2}\right)\right) \tag{3}$$

$$M(n,\mathbf{V},T)=n\left(\frac{m}{2\pi kT}\right)^{3/2}\exp\left[-\frac{m\mathbf{c}^2}{2kT}\right] \tag{4}$$

where M is the local Maxwellian distribution function, $\mathbf{c}=\xi-\mathbf{V}$ is the relative speed of a single particle against a background gas, $\mathbf{q}$ is the heat flux vector, p is the local pressure and μ and λ are the gas viscosity and conductivity at local temperature T, $\Pr = 2/3$ is the Prandtl number, k is the Boltzmann constant.

The inlet and outlet boundary conditions for the kinetic equation (1) impose a Maxwellian distribution function for the incoming molecules:

$$f(\xi)\big|_{in}=M(p_{0i}/T_{0i},0,T_{0i}) \text{ if } \xi\cdot\vartheta(\mathbf{x})>0 \tag{5}$$

$$f(\xi)\big|_{exit}=M(p_e/T_{0i},0,T_{0i}) \text{ if } \xi\cdot\vartheta(\mathbf{x})>0 \tag{6}$$

where $\vartheta(\mathbf{x})$ is the unit vector normal to the boundary directed toward the gas. Note that such boundary condition, assuming Maxwellian distributions, requires equilibrium conditions, and thus the boundaries must be pushed far away from solid boundaries and viscous regions.

Maxwell diffuse reflecting boundary conditions with the full accommodation are applied on solid walls:

$$f\big|_w(\mathbf{x})=\omega(\mathbf{x})M(1,\mathbf{V}_w,T_w) \qquad \xi\cdot\vartheta(\mathbf{x})>0 \tag{7}$$

$$\omega(\mathbf{x})=-\frac{\sum_{\xi\cdot\vartheta(\mathbf{x})<0}\xi\cdot\vartheta(\mathbf{x})f(t,\mathbf{x},\xi)d\xi}{\sum_{\xi\cdot\vartheta(\mathbf{x})>0}\xi\cdot\vartheta(\mathbf{x})M(1,\mathbf{V}_w,T_w)d\xi} \tag{8}$$

The parameter $\omega(\mathbf{x})$ is determined so as to avoid a mass flux across the wall. All the particles coming off the surface are emitted with the Maxwell distribution functions corresponding the zero mean flow velocity, the temperature is equal to the wall temperature T_w, and the density is calculated from the condition of equality of the fluxes of particles coming on and off the wall. At the symmetry line the specular boundary condition is imposed.

The problem is recast in terms of dimensionless variables introducing the following reference values: the number density n_∞, is the temperature T_∞, the most probable velocity $v_\infty=\sqrt{2RT_\infty}$. The height of the channel H is taken as thereference length scale. For the hard-sphere molecular model the dimensionless viscosity coefficient is $\mu/\mu_0=(T/T_0)^{0.5}$. Thus, non-dimensional variables are ξ/v_∞, $\mathbf{x}/H$, $f/(n_\infty v_\infty^{-3})$, μ/μ_∞, n/n_∞, T/T_∞ and p/p_∞. For sake of simplicity, in the following the dimensionless quantities keep the same designations as the dimensional ones. In the non-dimensional form the S - model equation is thus written as follows:

$$\frac{\partial f}{\partial t}+\xi\frac{\partial f}{\partial \mathbf{x}}=\frac{8\rho\sqrt{T}}{5Kn\sqrt{\pi}}(S-f)=\nu(S-f) \tag{9}$$

where ν is the collision frequency, Kn is the reference Knudsen number based on the height between plate H and the mean free path λ_∞ at reference (isoentropic) state which is defined as:

$$\lambda_\infty=\frac{16\mu_\infty}{5mn_\infty\sqrt{2\pi RT_\infty}} \tag{10}$$

The macroscopic (bulk) distributions of number density, velocity vector, temperature and heat flux respectively are defined as:

$$n=\int fd\xi,\ n(u,v)^T=\int(\xi_x,\xi_y)^T fd\xi,\ nT=\int\mathbf{c}^2 fd\xi,\ (q_x,q_y)^T=\int(c_x,c_y)^T\mathbf{c}^2 fd\xi \tag{11}$$

Taking advantage of the two-dimensionality of the flow, the z component of the molecular velocity may be eliminated by the projection procedure. This is performed by introducing the functions:

$$g = \int_{-\infty}^{\infty} f d\xi_z \qquad h = \int_{-\infty}^{\infty} \xi_z^2 f d\xi_z \tag{12}$$

For the sake of simplicity all schemes will be written for general S- model equation (9).

To discretize the S-model equation we construct a two-dimensional grid $\{\boldsymbol{\xi}_\Gamma\} = (\xi_{x\alpha}, \xi_{y\beta})$, $\alpha, \beta = 1, \ldots, N_\xi$ bounded by $v_{\max}$, where $\boldsymbol{\Gamma} = (\alpha, \beta)$ defines velocity grid points indices. Velocity space nodes are located at Gaussian abscissas, and Gaussian integration is performed via proper Gaussian weight function. In the discrete form the S -model equation (9) is the following:

$$\partial_t f_\Gamma + \xi_{x\Gamma} \frac{\partial f_\Gamma}{\partial x} + \xi_{y\Gamma} \frac{\partial f_\Gamma}{\partial y} = \nu(S_\Gamma - f_\Gamma) \tag{13}$$

In the physical space a 2D, non uniform grid is defined by nodes (x_i, y_j), for $i = 1, i_{max}$ and $j = 1, j_{max}$ and $N_c = i_{\max} \cdot j_{\max}$ cells centred around each node. Denoting $f^n_{\Gamma,i,j}$ as the approximations of $f(t^n, \xi_\Gamma, x_i, y_j)$ the numerical scheme for the equation (13) can be written as:

$$f^{n+1}_{\Gamma,i,j} = f^n_{\Gamma,i,j} - \frac{\Delta t}{\Delta x}\left(F^n_{\Gamma,i+1/2,j} - F^n_{\Gamma,i-1/2,j}\right) - \frac{\Delta t}{\Delta y}\left(F^n_{\Gamma,i,j+1/2} - F^n_{\Gamma,i,j-1/2}\right) + \Delta t \nu^n_{i,j}\left(S^n_{\Gamma,i,j} - f^n_{\Gamma,i,j}\right) \tag{14}$$

where the transport term is approximated by a standard finite volume scheme, while the nonlinear relaxation term is discretized via a standard centred approximation. In particular, the numerical fluxes are defined as:

$$F^n_{\Gamma,i+1/2,j} = \frac{1}{2}\left(\xi_{x\alpha}\left(f^n_{\Gamma,i+1,j} + f^n_{\Gamma,i,j}\right) - \left|\xi_{x\alpha}\right|\left(\Delta f^n_{\Gamma,i+1/2,j} - \Phi^n_{\Gamma,i+1/2,j}\right)\right) \tag{15}$$

$$F^n_{\Gamma,j+1/2,j} = \frac{1}{2}\left(\xi_{y\beta}\left(f^n_{\Gamma,i,j+1} + f^n_{\Gamma,i,j}\right) - \left|\xi_{y\beta}\right|\left(\Delta f^n_{\Gamma,i,j+1/2} - \Phi^n_{\Gamma,i,j+1/2}\right)\right) \tag{16}$$

$$\Delta f^n_{\Gamma,i+1/2,j} = f^n_{\Gamma,i+1,j} - f^n_{\Gamma,i,j}$$

where $\Phi^n_{\Gamma,i+1/2,j} = \min\bmod\left(\Delta f^n_{\Gamma,i-1/2,j}, \Delta f^n_{\Gamma,i+1/2,j}, \Delta f^n_{\Gamma,i+3/2,j}\right)$ is the flux limiter function, allowing for a second order of the scheme. Since the scheme is explicit the time step should be limited by the condition:

$$\Delta t_K = CFL / \max_{i,j}\left(\nu + v_{\max}/\Delta x + v_{\max}/\Delta y\right) \tag{17}$$

The equation (13) is solved on a curvilinear, structured mesh. If $\zeta(x, y)$ and $\eta(x, y)$ denote curvilinear coordinates, equation (13) may be recast in terms of the new coordinates as in the following:

$$\frac{1}{J}\partial_t f_\Gamma + \partial_\zeta\left(\xi_\Gamma \frac{\nabla\zeta}{J} f_\Gamma\right) + \partial_\eta\left(\xi_\Gamma \frac{\nabla\eta}{J} f_\Gamma\right) = \frac{\nu}{J}(S_\Gamma - f_\Gamma) \tag{18}$$

Where $\nabla\zeta = (\partial_x\zeta, \partial_y\zeta)$, $\nabla\eta = (\partial_x\eta, \partial_y\eta)$ and $J = \partial_x\zeta\partial_y\eta - \partial_y\zeta\partial_x\eta$. If we define a uniform grid $\zeta_i = i\Delta\zeta, \eta_j = j\Delta\eta$ a scheme very similar (14) may be used.

In the implicit scheme the gain term S_Γ is positive and, thus, can be treated explicitly. Then the implicit first order scheme can be written as:

$$f^{n+1}_{\Gamma,i,j} = f^{n}_{\Gamma,i,j} - \frac{\Delta t}{\Delta x}\left(F^{n+1}_{\Gamma,i+1/2,j} - F^{n+1}_{\Gamma,i-1/2,j}\right) - \frac{\Delta t}{\Delta y}\left(F^{n+1}_{\Gamma,i,j+1/2} - F^{n+1}_{\Gamma,i,j-1/2}\right) + \Delta t v^{n}_{i,j}\left(S^{n}_{\Gamma,i,j} - f^{n+1}_{\Gamma,i,j}\right) \quad (19)$$

or

$$\left(\frac{I}{\Delta t} + T + v^{n}\right)\delta f = RHS^{n} \quad (20)$$

$$(Tf^{n})_{\Gamma,i,j} = \frac{1}{\Delta x}\left(F^{n}_{\Gamma,i+1/2,j} - F^{n}_{\Gamma,i-1/2,j}\right) + \frac{1}{\Delta y}\left(F^{n}_{\Gamma,i,j+1/2} - F^{n}_{\Gamma,i,j-1/2}\right) \quad (21)$$

where $\delta f = f^{n+1} - f^{n}$, I is the identity matrix, $(Tf^{n})_{\Gamma,i,j}$ is a matrix with the only first order fluxes, and right –hand side:

$$RHS^{n}_{\Gamma,i,j} = -\frac{1}{\Delta x}\left(F^{n}_{\Gamma,i+1/2,j} - F^{n}_{\Gamma,i-1/2,j}\right) - \frac{1}{\Delta y}\left(F^{n}_{\Gamma,i,j+1/2} - F^{n}_{\Gamma,i,j-1/2}\right) + v^{n}_{i,j}\left(S^{n}_{\Gamma,i,j} - f^{n}_{\Gamma,i,j}\right) \quad (22)$$

which contains the limiters for the second order scheme. In particularly, If quantities $f^{n}_{\Gamma,i,j}$ are stored as $f^{n}_{\Gamma} = (f^{n}_{i,j})_{\Gamma}$, it can be seen that T is a $(N_c \times N_\xi)\times(N_c \times N_\xi)$ block diagonal matrix with $N_c \times N_c$ pentadiagonal blocks T_Γ:

$$T_{\Gamma,i-1,j} = -\frac{1}{\Delta x}\xi^{+}_{x\alpha},\ T_{\Gamma,i,j-1} = -\frac{1}{\Delta y}\xi^{+}_{y\beta},\ T_{\Gamma,i,j} = \frac{1}{\Delta x}|\xi_{x\alpha}| + \frac{1}{\Delta y}|\xi_{y\beta}|$$

$$T_{\Gamma,i,j+1} = \frac{1}{\Delta y}\xi^{-}_{y\beta},\ T_{\Gamma,i+1,j} = \frac{1}{\Delta x}\xi^{-}_{x\alpha}$$

The boundary conditions are treated explicitly and the right –hand side contains the boundary terms and the coefficients of T corresponding to the boundary conditions are set to 0.

Thus, equation (20) can be symbolically expressed as for each $\mathbf{\Gamma}$:

$$\left[D + L + U\right]\delta f = RHS \quad (23)$$

where D, L, U are the matrices of size $N_c \times N_c$, δf and RHS are vectors of size N_c and D consists of only diagonal terms, L only the terms in the lower triangular matrix, and U the upper triangular matrix. The left –hand side of equation (23) can be approximately factored into the product of two operators:

$$(D+L)D^{-1}(D+U)\delta f = RHS \quad (24)$$

This scheme can be implemented in the following sequence:

$$(D+L)\delta f^{*} = RHS \quad (25)$$

$$(D+U)\delta f = D\delta f^{*} \quad (26)$$

The distribution function at the next time step is:

$$f^{n+1} = f^{n} + \delta f$$

The solution of linear systems (23) is local in **Γ**, and therefore completely parallelizable. The computation of macroparameters for gain term S and collision frequency ν is local in *i*, *j*, thus it is parallelized in physical space. The software code was written in C++ with the use of MPI (Message Passing Interface). The kinetic code ran on double processors, quad core systems, using 8 parallel processes. Running in parallel on 8 cores, on the grid 720 × 50 in the physical space the CPU time per time step is 36 × 8 = 288 sec. The maximum speed up achieved by 8 processors was 94 %.

3. Numerical solution of the Navier–Stokes equation

The viscous, compressible Navier-Stokes equations for 2D laminar flow can be written in terms of conservative variables:

$$\frac{\partial \mathbf{U}}{\partial t}+\frac{\partial \mathbf{F}(\mathbf{U})}{\partial \mathbf{x}}=0 \tag{27}$$

$$\mathbf{U}=\left(\rho,\rho\mathbf{V},e_{tot}\right) \tag{28}$$

where the flux vector **F(U)** may be decomposed into the convective (inviscid) and diffusive (viscous) components:

$$\mathbf{F}=\mathbf{F}^{\mathbf{inv}}-\mathbf{F}^{\mathbf{v}} \tag{29}$$

$$\mathbf{F}^{\mathbf{inv}}=\left(\rho\mathbf{V},\rho\mathbf{V}\mathbf{V}+p\mathbf{I},\mathbf{V}(e_{tot}+p)\right)^{T} \tag{30}$$

$$\mathbf{F}^{\mathbf{v}}=\left(0,\boldsymbol{\tau},\boldsymbol{\tau}\cdot\mathbf{V}+\mathbf{q}\right)^{T} \tag{31}$$

where **I** is the identity matrix and **q** is the heat flux vector. Pressure *p* and total energy per unit volume e_{tot} are linked by the equation of state for ideal gas:

$$p=(\gamma-1)\left(e_{tot}-\frac{\rho\mathbf{V}^{2}}{2}\right) \tag{32}$$

Navier-Stokes solver is based on a hybrid finite difference-finite volume method. The in-house FORTRAN code was developed by Croce, has formal second order accuracy in space and time, has already been applied and validated for the microflow simulations of (Croce et al. [12]) and, thus, is here only briefly outlined. Adopting a curvilinear, structured mesh we may define a computational cell of volume Λ_{ij} with faces A_k, k = 1, 4 around each node. At each of these computational cells the flux balance offers:

$$\Lambda_{ij}\frac{\partial \mathbf{U}_{ij}}{\partial t}=\sum_{k=1}^{4}\mathbf{F}(\mathbf{U})\cdot A_k\eta(\mathbf{x})_k \tag{33}$$

Fluxes are defined via neighbouring value averaging, and an artificial dissipation term *d* is added to prevent checker boarding and numerical instabilities. Thus, considering a face located at *i*+1/2 we get:

$$\mathbf{F}^{inv}(\mathbf{U})_{i+\frac{1}{2}}=\frac{\mathbf{F}^{inv}(\mathbf{U})_{i}+\mathbf{F}^{inv}(\mathbf{U})_{i}}{2}+d_{i+\frac{1}{2}} \tag{34}$$

Artificial dissipation term *d* are given by a blend of second and fourth order differences, scaled by the maximum eigenvalue of jacobian matrix of vectors $\mathbf{F}^{\mathbf{inv}}$, as suggested in (Pulliam [15]). Second order terms are switched on near discontinuities, as in Pulliam [15]. Viscous flux vectors are evaluated with second order finite differences at *i*+1/2. The solution is advanced in time via Crank Nicolson integration scheme. The use of the spatially factored ADI scheme originally proposed by Beam and Warming (Hirsh [16]) leads, at each time step, to the resolution of two series of block tridiagonal algebraic systems, rather than the original pentadiagonal block system arising from flux discretization.

So far, no general consensus has yet been reached on the expression of second order slip conditions for arbitrary geometries and most of the proposed models are related to a specific reference case and geometry, and are, thus, not easily extended to arbitrary geometry. For these reasons, the Maxwell first order slip boundary condition is here used. Thus, at solid walls we have:

$$u_{gas} - u_w = s_p Kn \left(\frac{\partial u_{\iota(\mathbf{x})}}{\partial \vartheta(\mathbf{x})} + \frac{\partial u_{\vartheta(\mathbf{x})}}{\partial \iota(\mathbf{x})} \right) \tag{35}$$

$$s_p = \frac{\sqrt{\pi}}{2} \frac{2-\sigma_v}{\sigma_v} (1 + 0.1366\sigma_v) \tag{36}$$

where $\vartheta(\mathbf{x})$ and $\iota(\mathbf{x})$ are the unit vectors respectively normal and tangential to the solid wall. σ_v is the tangential momentum accommodation coefficient, which is not only a function of the material property of gas–solid interface, but depends also on Knudsen number. In the computations s_p is close to one (since the portion of incident particles scattered diffusely σ_v is equal to one). The additional derivative along the tangential direction is essential in capturing even the qualitative behaviour of slip flow along curved walls.

A Dirichlet temperature boundary condition is imposed at the wall. The wall temperature is fixed at the inlet total temperature value to minimize the effect of viscous dissipation. In the energy equation, the Smoluchowski temperature jump is used:

$$T_{gas} - T_w = s_T Kn \frac{\partial T}{\partial \vartheta(\mathbf{x})} \tag{37}$$

$$s_T = \frac{2-\sigma_T}{\sigma_T} \frac{2\gamma}{\gamma+1} \frac{1}{\mathrm{Pr}} \tag{38}$$

where s_T is the temperature jump coefficient,. Since the isothermal flow is considered we assume the temperature accommodation coefficient σ_T equals to one.

4. Results and discussion

The two-dimensional problem geometry is illustrated in figures 1 and 2. A plane channel of length L and height H connects two large reservoirs of size $L_x \times L_y = 30H \times 10H$. The reservoir size was chosen as to be sufficiently large, so that their further increase by 50% did not change the mass flow rate beyond 0.5%. The flow is driven by the pressure difference between the two reservoirs and is examined using both kinetic and continuum approaches. Due to the symmetry of the flow only one half of the domain was modelled. The inlet and outlet reservoirs are included in the computational domain. This allows a realistic description of inlet and outlet losses and is required, as noted above, for the kinetic solution. In order to save computational resources, we considered a channel with an inlet smooth section of length $L_s = 9H$ followed by a rough one of length $L_r = 3.15H$ (see figure 1). The roughness geometry is modelled as a series of triangular obstructions with sharp angle of 45°, as shown in figure 2. The details of the roughness element shape, as well as their pitch, have certainly an impact on the quantitative evaluation of the channel performances, as shown in [17]. However, the characterization of the actual surface texture (which is a function of the manufacturing process and parameters) with the level of detail required for the CFD discretization is a demanding task and beyond the scope of the present paper. As such, any choice of range of pitch, height or random distribution of peaks would be somewhat arbitrary. Thus, we focus on the qualitative interaction

between compressibility, rarefaction and representative length scales, rather than on the detailed geometrical effect. Due to this choice, the computations are limited to a single geometry, properly scaled in order to get different relative roughness.
The height of a single peak is $h = \varepsilon H$ and the peaks spacing is s = $5h$. We consider relative roughness ε of 1.25%, 2.5% and 5%. Thus, the rough section L_r includes 12 roughness ridges at ε = 5%, 24 ridges at ε = 2.5% and 48 ridges at ε = 1.25%.

The results have been obtained on non uniform, structured meshes in physical space; in particular, we used of 672 × 50 nodes for ε =5%, 720 × 50 nodes for ε = 2.5% and 1.25%, and 440 × 50 nodes for the smooth channel. For all of these grids the minimum grid spacing close to the wall is of the order of 0.001 H, with around 21 nodes within the rough layer. The velocity space $-v_{max} \leq \xi_x, \xi_y \leq v_{max}$, ($v_{\max} = 4.68$), is discretized on a non uniform 16×16 mesh; the node spacing is based on the proper Gaussian abscissas, allowing for accurate integration. Grid independence analysis was carried out for both space and velocity domain discretization: doubling velocity mesh points gives a 1% variations in mass flow rate.

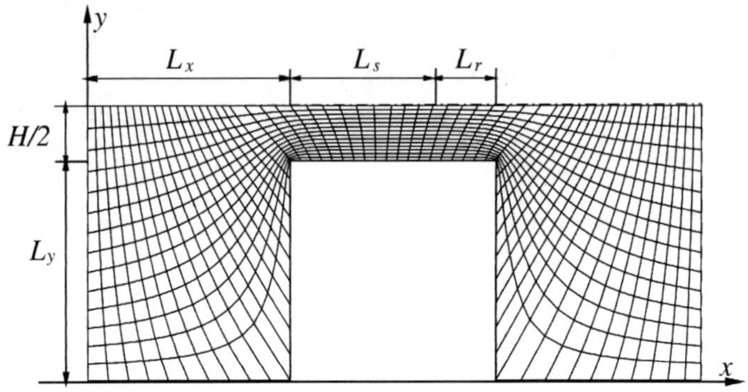

Figure 1. Schematic computational domain.

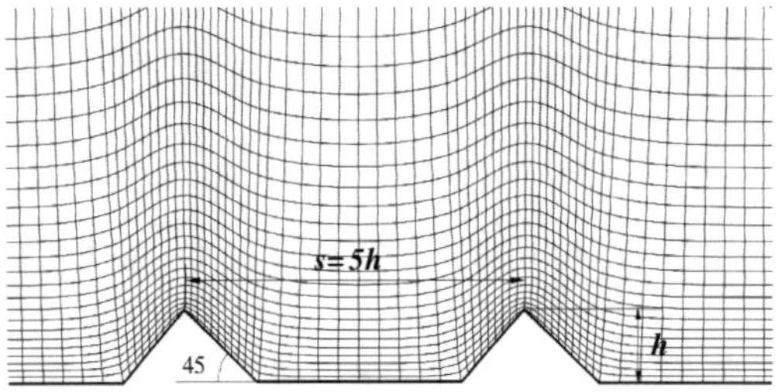

Figure 2. Rough elements parameters.

The gas in the reservoirs is assumed in equilibrium at constant inlet pressures p_{0i} or outlet pressure p_e, with $p_{0i} > p_e$, and temperature T_{0i}. The compressibility effect is monitored via the local value of Ma and the isentropic exit Mach number Ma_{is} (i.e., the Ma that would arise from an isentropic flow with the same pressure ratio as the real one):

$$\frac{p_{0i}}{p_e} = \left(1 + \frac{\gamma - 1}{2} Ma_{is}^2\right)^{\gamma/(\gamma-1)} \tag{39}$$

The temperature of the walls is equal to the temperature in reservoirs T_{0i}.
We define also the isentropic Reynolds number as following:

$$\mathrm{Re}_{is} = \frac{\rho_{is} u_{is} H}{\mu_{is}} \tag{40}$$

Thus, the corresponding Knudsen number based on the isentropic values is:

$$Kn_{is} = \frac{16}{5}\sqrt{\frac{\gamma}{2\pi}}\frac{Ma_{is}}{\mathrm{Re}_{is}} \tag{41}$$

The following computations have been carried out for pressure ratios $p_{0i}/p_e = 1.1$, 2 and 3, allowing for a Kn_{is} range from 0.01 to 0.08. Maximum local Knudsen number Kn_e is reached after the rough region and ranges from 0.01 to 0.055.

Global and local channel performances are defined in terms of the Poiseuille number *Po*, where the friction factor *f* is computed as an average value either over a portion of rough section of the channel or over a single geometrically periodic roughness element of length *s*. In both cases, *f* is defined as:

$$f = 2\frac{\Delta p}{\Delta L}\frac{D_h}{<\rho><u>^2} \tag{42}$$

where $D_h = 2H$ is the hydraulic diameter, the variations ΔL and Δp are computed between the inlet and outlet section, either of the whole rough portion of the channel or of the single roughness element, <> means average over cross section.

The Poiseuille number *Po* is then computed as:

$$Po = f \times \mathrm{Re} = 2\frac{\Delta p}{\Delta L}\frac{D_h^2}{\mu<u>} \tag{43}$$

For uncompressible flow through smooth channel with first order slip boundary condition it's possible to compute an analytical expression for the Poiseuille number:

$$Po_{smooth} = \frac{96}{1+12\cdot s_p \cdot Kn} \tag{44}$$

For either compressible or rough surface flows no analytical solution is available. However, a few correlations have been proposed. In particular, an expression for rough surfaces, based on the results obtained by the Lattice Boltzmann computations, was proposed in [10]:

$$Po_{rough} = Po_{smooth} \times (1 + 0.8\varepsilon + 24\varepsilon^2) \tag{45}$$

while Asako et al. [18] suggested a local Po correlation for a developing compressible flow:

$$Po_{comp} = 96 - 4.55 \times Ma + 274.8 \times Ma^2 \tag{46}$$

Equation (44) allows for a first assessment of numerical results. Figure 3 compares Poiseuille numbers from kinetic and NS computations in a smooth channel with the first order expression equation (44) and numerical results of Varoutis et al. [19] obtained by BGK-model equation. We consider a small pressure ratio $p_{0i}/p_e = 1.1$ (small compressibility effect) and an exit Knudsen number Kn_e range from 0.01 to 0.81. It can be seen that S- model results are close to the BGK-model ones from [19] in the whole range of *Kn*, thus the numerical scheme for the solution of S-model can accurately predict rarefied flow. We should expect that first order expression is accurate at lower *Kn*. The NS results well correlate with both analytical expression and kinetic computations. As expected, kinetic results deviation from slip solutions become apparent above $Kn = 0.05$.

Figure 4 shows the averaged Poiseuille *Po*, defined by equation (43), between the inlet and outlet section of a rough sector of length $L^* = 2.3\,H$ at the end of the channel, i.e. including 9, 18 or 36 rough modules, depending on ε. The length and the position of L^* was chosen in order to exclude the developing flow region at the beginning of the rough sector. The Poiseuille number increases with *Re*; due to equation (41), this means also an increase in *Po* related to a decrease in Kn. The picture for a higher pressure ratio (higher actual *Ma*), but roughly comparable width of *Re* range, shows a much steeper slope. This clearly suggests that the dominant effect is the compressibility.

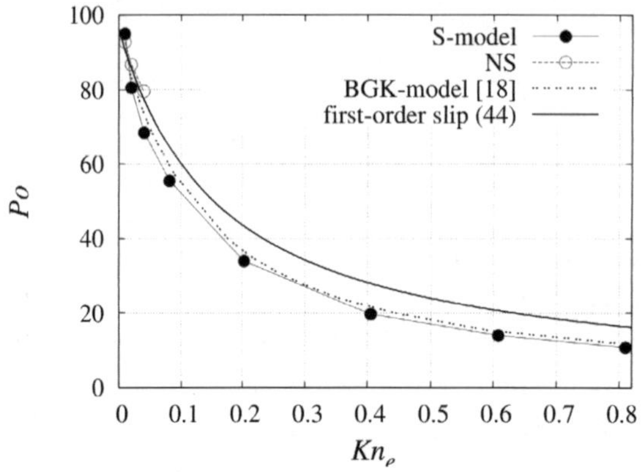

Figure 3. The rarefaction effect on the Poiseuille number in a smooth channel: $p_{0i}/p_e = 1.1$.

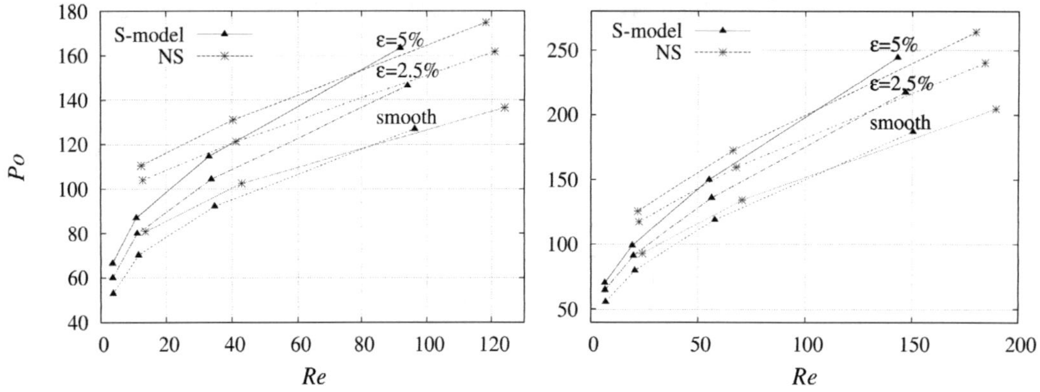

Figure 4. Averaged Poiseuille number vs. averaged Re: (left) $p_{0i}/p_e = 2$, (right) $p_{0i}/p_e = 3$.

Figure 4 shows significant discrepancies between NS and kinetic results at low level of *Re* and high roughness. These conditions correspond to the higher rarefaction levels: as an example, at $\varepsilon = 5\%$, $p_{0i}/p_e = 3$ and $Re = 21$ we have an exit Knudsen number equals to 0.055. In particular, NS over-estimates the rough effect in comparison with S-model results, at higher rarefaction level, up to nearly 30%, while for $Re \approx 100$ the discrepancies are negligible.

In figures 5, the local Poiseuille number for $\varepsilon = 0$, 2.5% and 5% and pressure ratios $p_{0i}/p_e = 3$ and 2 respectively is plotted as a function of the cross section averaged Mach number. The Reynolds number is nearly constant along the channel, but the Mach number increases due to the acceleration of the gas flow. For rough channels the local *Po* is computed between the inlet and outlet sections of each rough module. In figures 5 we can easily identify the transition region between the smooth and rough sectors. The developing flow in this area yields a quick decrease in Poiseuille. As soon as this transition is completed (i.e., at the minimum in the *Po* curve), Poiseuille becomes a monotone function of *Ma*. Figure 5 shows, at the $p_{0i}/p_e = 2$ and lower Kn_{is}, that for the same value of local Ma both kinetic and NS solutions offer the same value of *Po*, for both smooth and rough channels. For rough geometries and higher rarefaction, on the other hand, NS over estimates the roughness effect.

Moreover, the impact of compressibility, especially at higher pressure ratio is much more significant than the variation due to roughness: *Po* at Ma ≈ 0.6 is three times higher than at Ma ≈ 0.3, while the increase between rough and smooth channel at the same Ma ≈ 0.6 is at most 40%.

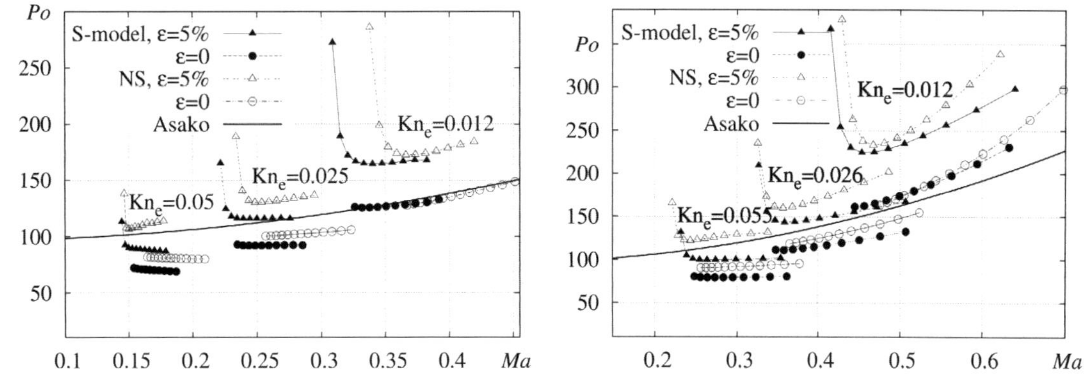

Figure 5. Local Poiseuille number at $p_{0i}/p_e = 2$ (left); $p_{0i}/p_e = 3$ (right).

Furthermore, in figure 5 the comparison with correlation equation (46) for smooth channel of Asako [19] is presented. For lower Kn_{is} equals to 0.01 and $p_{0i}/p_e = 2$ there is a small difference between equation (46) and numerical solution. When Knudsen number increases the difference becomes larger and local Po are lower than that obtained from the Asako's expression. This is due to the rarefaction effect, which is not included in the model of Asako et al. [19]. The rarefaction alleviates the friction effect of the wall on the gas flow. It should be mentioned that for higher pressure ratio $p_{0i}/p_e = 3$ (see figure 5) the deviation from equation (46) is large both for lower Kn, due to high Ma numbers exceeding the correlation range of validity, and for higher Kn, due to the rarefaction effect.

The increase in Poiseuille number due to the roughness, and the behavior of kinetic and continuum solvers are different at different Mach numbers. Looking at figure 5, for $Kn_{is} = 0.01$, at $\varepsilon = 5\%$ and local $Ma \approx 0.6$ the Poiseuille number increase, with respect to the smooth channel, is 30% for kinetic results and 41% for NS ones. At a lower Ma, namely $Ma \approx 0.37$, and the same roughness the increase is smaller, around 30%, and is nearly identical in both kinetic and NS results. Accordingly, for pressure ratio $p_{0i}/p_e = 3$ the difference between NS and kinetic results increases towards to the outlet reservoir (i.e. at higher Ma) in the rough channel, while for smooth channel the difference is small. For smaller pressure ratio $p_{0i}/p_e = 2$ there is no difference between kinetic and NS results for smooth and rough channels at lower Kn.

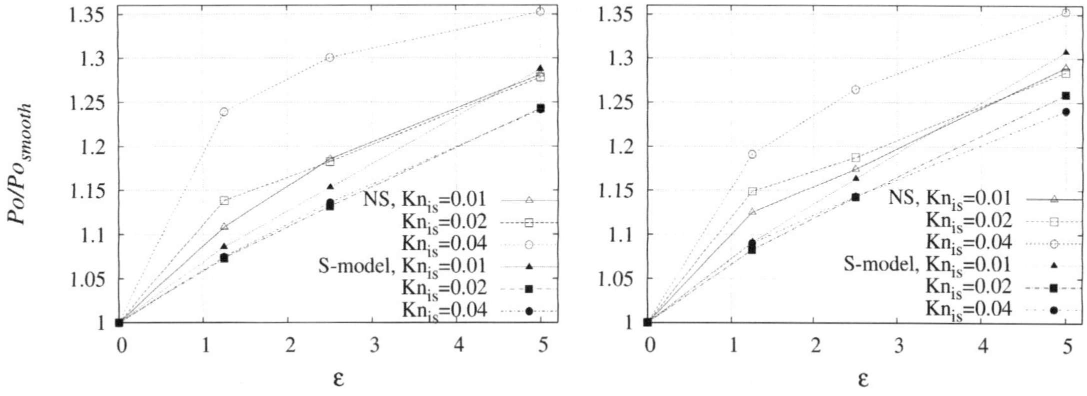

Figure 6. Po/Po_{smooth} via ε: (left) $p_{0i}/p_e = 2$, (right) $p_{0i}/p_e = 3$.

In figure 6 we plot the increase of averaged rough Poiseuille number with respect to smooth value, based on the non-entrance region, as a function of roughness height ε for different Kn_{is} from 0.01 to 0.08. At higher *Kn* kinetic and continuum solutions show quite different behaviours. For kinetic results the maximum deviation is reached at the lowest Kn_{is} for both pressure ratios: thus, the surface roughness has a more significant effect on a weaker rarefied gas flow. Similar behavior has also been observed in the DSMC results of Sun [7], whose calculations indicated *Po* significantly decreased when the Knudsen number increased from 0.02 to 0.08. On the opposite, for NS results maximum deviation of *Po* reaches at higher rarefaction $Kn_{is} = 0.04$ ($Kn_e = 0.055$).

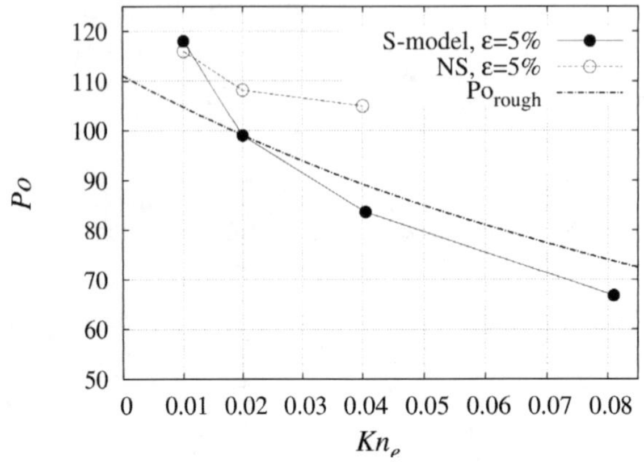

Figure. 7. The rarefaction effect on the Poiseuille numbers in a rough channel.

In order to concentrate on the coupling effect of rarefaction and roughness the flow for small pressure ratio $p_{0i}/p_e = 1.1$ and $0.01 < Kn_e < 0.08$ has been considered. In figure 7 the Poiseuille numbers for rough surface (ε = 5%) obtained by NS and S- model equations are compared with the correlation equation (45) suggested in [7]. Such picture indicates that kinetic computed *Po* correlates well enough with expression (45) for high Knudsen number, while at lower ones the correlation fails and NS results offer good agreement with kinetic data. It is worth to notice that, while in a smooth channel continuum and kinetic approaches show good agreement up to $Kn_e = 0.05$, for the rough geometries the discrepancies appear at lower even at $Kn_e = 0.02$. This is due to the fact that the effect is related to geometrical details (peak height) of a much smaller scale than the channel height; thus, the significant length scale is smaller than *H*, and, as a consequence, a representative *Kn* should actually be higher than the *H* based one.

5. Conclusion

The numerical analysis of the flow field in rough microchannel with relative roughness ranging from 0% to 5% has been carried out using two approaches: Navier-Stokes equations provided with the first order slip-boundary condition and kinetic S-model equation with Maxwell diffuse reflecting boundary condition. The use of implicit scheme for the solution of S-model equations and the optimization of the algorithm using parallelization in both physical and velocities spaces allow for efficient computations.

It was found that compressibility is the dominant effect increasing the value of Poiseuille number. Such effect often overcomes the Poiseuille number significant decrease due to rarefaction effects. As expected at a fixed pressure ratio and rarefaction level the increase in roughness gives an increase in Poiseuille number which may be significant.

The difference between NS and S-model results is smaller at lower *Kn* and increase with *Kn*, as expected; however, while for smooth channels the two approaches agrees well up to $Kn = 0.05$, in the rough case the discrepancies arise earlier. This is consistent with the idea that the representative length scale for the flow around the roughness peaks is their height, which is a few percent of *H*. Thus, locally rarefaction has a greater impact on the flow and first order slip may become inadequate. This interpretation is supported by the fact that NS consistently over-estimates the rough effect in comparison with S-model results.

6. References

[1] Kandlikar S, Schmitt D, Carrano A and Taylor J 2005 *Phys. of Fluids* **17** 1–11
[2] Turner S E, Lam L C, Faghri M and Gregory O J 2004 ASME *J. Heat Transfer* **126** 753–762
[3] Lorenzini G, Morini G L, Salvigni S, 2010, Int. J. Thermal Sciences **49** 248-255
[4] Demsis A, Prabhu S V and Agrawal A 2010 *Exper. Ther. Fluid Science* **34** 1448-1455
[5] Hu Y, Werner C and Li D 2003 ASME J. Fluid Eng. **125** 871–879
[6] Croce G and D'Agaro P 2004 *Superlattices and Microstructure* **35** 601–616
[7] Mala G M and Li D 1998 *Int. J. Heat Fluid Flow* **20** 142–8
[8] Cao B, Chen M and Guo Z 2004 *Chinese Phys. Lett.* **21** 1777–1779
[9] Sun H and Faghri M 2003 Numer. Heat Transfer Part A **43** 1–8
[10] Liu C, Yanga J and Ni Y 2011 *Comp. and Math. with Applic.* **61** 3528–3536
[11] Ji Y, Yuan K and Chung J 2006 *Int. J. Heat Mass Transfer* **49** 1329–1339
[12] Croce G and D'Agaro P 2007 *Heat Transfer Eng.* **28** 688-695
[13] Hakak Khadem M, Shams M and Hossainpour S 2009 *Int. J. of Aerospace and Mech. Eng.* **3** 4
[14] Shakhov EM 1974 *A method for calculating rarefied gas flows* (Moscow: Nauka)
[15] Pulliam TH 1986 *AIAA Journal* **24** 1931-1940
[16] Hirsh C 1990 *Numerical computation of internal and external flows* (Chichester: Wiley)
[17] Croce G, D'Agaro P, Nonino C 2007 *Int. J. Heat Mass Transfer* **50** 5249-5259
[18] Asako Y, Pi T, Turner S E and Faghri M 2002 *Int. J. Heat Mass Transfer* **46** 3041–3050
[19] Varoutis S, Lihnaropoulos J, Valougeorgis D *Proc. of the 1st European Conf. on Microfluidics* (Electronic Materials) ed S Colin, G L Morini (Bologna)

Acknowledgments

The research leading to these results has received funding from the European Community's Seventh Framework Programme FP7/2007-2013 under grant agreement ITN GASMEMS n°215504.

Detailed investigation on the effect of wall spring stiffness on velocity profile in molecular dynamics simulation

S Namvar[1] and S M H Karimian[2]
Center of Excellence in Computational Aerospace Engineering, Department of Aerospace Engineering, Amirkabir University of Technology, Tehran, Iran

E-mail: soheil_snp@aut.ac.ir[1], hkarim@aut.ac.ir[2]

Abstract. In this paper, motion of 576 monatomic argon molecules is studied in a channel with two 2-layered wall molecules. The effect of wall spring stiffness (K) on maximum value of velocity profile is investigated in the channel. It was observed that for K<500 $\varepsilon\sigma^{-2}$, there is a decrease in the maximum value of velocity profile with an increase in K. This observation has been already reported by Sofos et al. To investigate a wider range of spring stiffness, in this paper the value of K was increased to more than $500\varepsilon\sigma^{-2}$. In this range of wall spring stiffness the behavior of maximum value of velocity profile changed; it increased with an increase in K. In a separate simulation the external force applied to the molecules was also increased and the same non-monotonic behavior of maximum value of velocity was observed. To clarify the reason of this behavior, the concepts of original and effective wall are introduced and through several test it is inferred that the mentioned concepts are not successful to demonstrate the reason of such behavior. It is suggested to obtain non-dimensional parameters governing the simulation in order to investigate the effect of every involved parameter on such a behavior. It is finally concluded that while wall spring stiffness affects the maximum velocity magnitude within the flow, the interaction of the two has not been clearly shown yet. The behavior of the maximum velocity is non-monotonic with the change of K. This is why no specific criterion has been reported for suitable value of wall spring stiffness in molecular dynamics simulation.

1. Introduction

The progress of human science has resulted in smaller devices and systems like nanotubes and Micro Electro Mechanical Systems (MEMS). Microscopic fluid flow like heat transfer in a thin film of solids or flow in a nano-channel has consequently arisen. Since conventional fluid mechanics fails to predict the flow characteristics in micro-nano scales [1], molecular dynamics simulation (MD) is used. MD simulation calculates the molecules' velocity and position in each time step and the flow characteristics like mean velocity, density, pressure and temperature is obtained using different methods of sampling[2].

In molecular dynamics one has to consider suitable boundary conditions as well as a proper model for a successful simulation. Wall boundary condition is one of the most important BCs which describes the molecules' behavior at a solid boundary. Needless to say, it is not possible to use no-slip boundary condition in MD simulation since the reflection of molecules after collision with the wall should be calculated precisly based on molecular dynamics formulations and the modeling of wall molecules which is very important. Note that modeling of wall molecules could affect the fluid behavior both near the wall and far from it.

In 1977 Abraham used a function to introduce a continuous molecular wall [3]. He also modeled a molecular wall with separate solid molecules. Later in 1989 Joel Kuplik studied fluid flow at solid surfaces with use of MD simulation [4]. He used fixed molecules to model the molecular wall.

One of the most efficient methods for molecular wall modeling is to attach each molecule to its place with a spring. In this way the molecules will be forced back to their initial place when they are moved. Many simulations have utilized springs, whether linear or non-linear, in their molecular wall [1,6-9]. Although a wide range of spring stiffness (57-6000$\varepsilon\sigma^{-2}$) has been used in different simulations yet there is not a specific criterion available for choosing proper spring stiffness. It is only assumed that spring stiffness should be high enough to prevent the fluid molecules from passing through the wall.

In this paper an investigation on spring stiffness is done to emphasize that the choice of spring stiffness is important in molecular dynamics simulation since different values of spring stiffness will result in different values of maximum velocity in a channel. However no method has been reported for choosing proper spring stiffness. The intention of authors has been to report some observations concerning the effect of wall spring stiffness on maximum velocity and channel's effective width.

2. Molecular modeling and Solution Procedure

This paper presents the simulation of 576 liquid argon molecules ($\rho^* = 0.642\,\sigma^{-3}$) which are conducted in a 36.15 $\mathring{A}$ channel. Although, the simulation carried out in this paper is for a liquid, the same procedure is applicable to gases.

The Knudsen number of the flow is 0.04 and the pressure is 50 Mpa. The velocity direction of each molecule is determined using Maxwell-Boltzmann distribution function. The interaction between molecules is described by Lennard-Jones 12-6 potential equation:

$$U_{ij} = 4\varepsilon\left(\left(\frac{\sigma}{r_{ij}}\right)^{12} - \left(\frac{\sigma}{r_{ij}}\right)^{6}\right) \tag{1}$$

Where U_{ij} is the potential energy, r_{ij} is the distance between two interacting molecules i and j, and σ and ε are the distance characteristic length and energy scale, respectively. The intermolecular forces are then calculated from equation 1 through the following relation:

$$F_{ij} = \frac{-dU_{ij}}{dr_{ij}} \tag{2}$$

And finally the motion of the molecules is determined by Newton's second law through the Verlet's integrating scheme [10].

As mentioned before, two 2-layered walls are located at Y-constant positions. In the other directions periodic boundary conditions are applied. Each wall molecule is attached to its place with a linear spring to ensure wall solidity. The spring will consequently exert a force of $F = -K\Delta r$ to every molecule that moves away a distance equal to Δr; in this equation K is the spring stiffness. Figure 1 shows the schematic of initial placement of fluid and wall molecules.

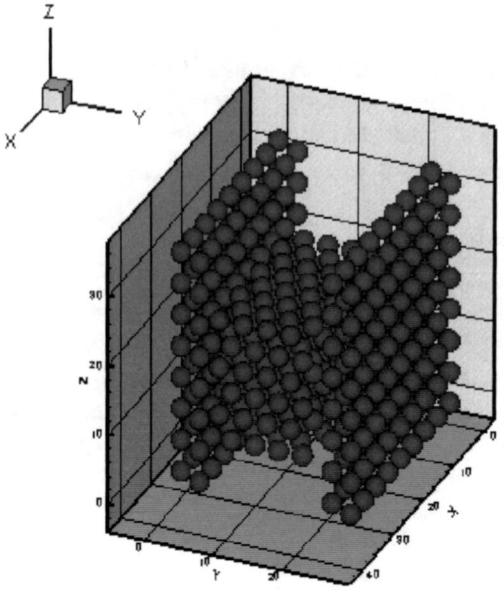

Figure 1. Schematic of initial placement of fluid and wall molecules

Wall temperature is kept constant using thermostat. In fact at each time step velocity of wall molecules are rescaled to ensure constant wall temperature. An external force is constantly applied to the fluid molecules in order to develop flow in the channel. It should be noted that all of the initial conditions of the simulation such as temperature, number of molecules, spring stiffness, number of wall layers and external force magnitude are chosen similar to a simulation performed by Sofos et al [9] in order to validate the simulation of this paper. The detailed specifications of the simulation carried out in this paper are presented in table 1.

Table 1. Specification of the system under examination.

Dimension (Angstrom)	36.15×36.15×36.15
Wall molecules	288 argon molecules
Fluid molecules	576 argon molecules
Spring stiffness ($\varepsilon\sigma^{-2}$)	Different values
External force (pN)	0.18 and 1
Thermostat (k)	Fixed on 120
Number of bins	15
Time step (s)	10^{-14}
Boundary Conditions	Periodic in X and Z directions Wall BC in Y direction

3. Results and discussion

Having done several simulations with different values of spring stiffness, 3 major observations could be reported as below:

1. The maximum value of the velocity profile is plotted against spring stiffness. To manage this, the domain of flow is divided into 15 bins and the velocity profile is obtained by averaging the magnitude of molecules' velocity in every bin using SMC method of averaging [2]. The result is shown in figure 2. Velocity profile is a proper variable to be observed with the change of spring stiffness since it well describes the flow throughout the domain and its convergence is relatively rapid.

Figure 2 shows that as wall spring stiffness (k) increases the value of maximum velocity in the channel decreases for K<500 $\varepsilon\sigma^{-2}$. This behavior of maximum velocity has been already reported by Sofos et al [9]. Note that all conditions of the present simulations including initial conditions, boundary conditions, wall molecule arrangement and its spring stiffness, number of molecules, channel width, and magnitude of external force are the same as those set by Sofos et al. For K>500 $\varepsilon\sigma^{-2}$ for which the simulation is not conducted by Ref. 9, calculations performed in the present study shows interestingly that the maximum velocity increases with the increase of K.

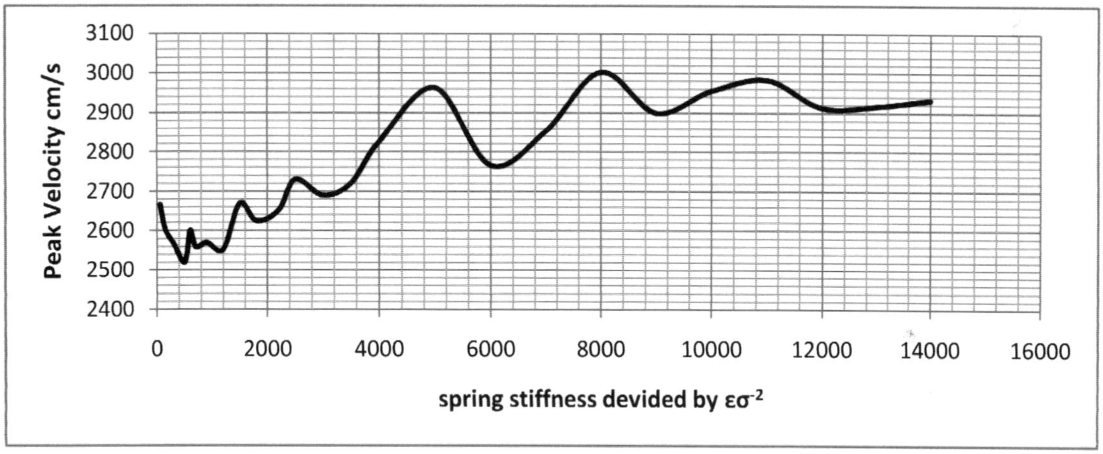

Figure 2. Maximum velocity of channel vs. spring stiffness for $F_{ext} = 0.18\,pN$

Since the spring stiffness affects maximum value of the velocity profile, it can be inferred that for a channel with dimensions mentioned in Table 1, spring stiffness affects the flow even far from the walls. That's why the spring stiffness value has to be carefully chosen for such a channel.

2. In order to investigate the effect of the external force on the above issue external force is changed from $0.18pN$ to $1pN$, in the above test case. Variation maximum velocity versus spring stiffness is plotted in Figure 3 for $F_{ext} = 1\,pN$.

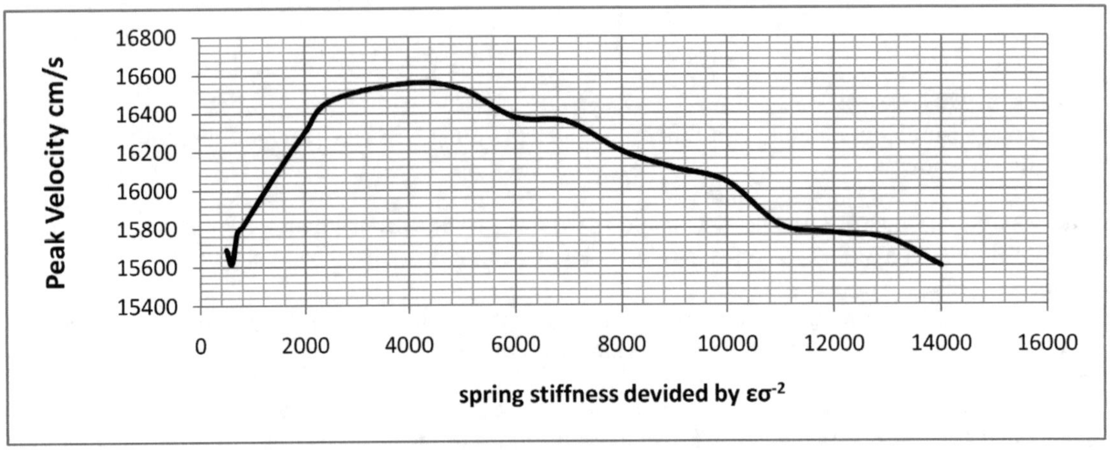

Figure 3. Maximum velocity of channel vs. spring stiffness for $F_{ext} = 1\ pN$

In this figure, after a small decrease in the maximum velocity for $K < 600\varepsilon\sigma^{-2}$, it starts to increase up to about K=4000 $\varepsilon\sigma^{-2}$, and then it again decreases. From these observations one can conclude that the maximum velocity do not obey a monotonic behavior.

3. Different scenarios were considered to clarify the reason of such behavior. Although we were not successful in this regard, one of the scenarios worth to be mentioned in this paper.

If the original width of channel is defined as the distance between the two walls at the beginning of the simulation, then an effective width of channel can be defined as the average distance between the displaced molecules of walls at the two sides of the channel, as shown in figure 4.

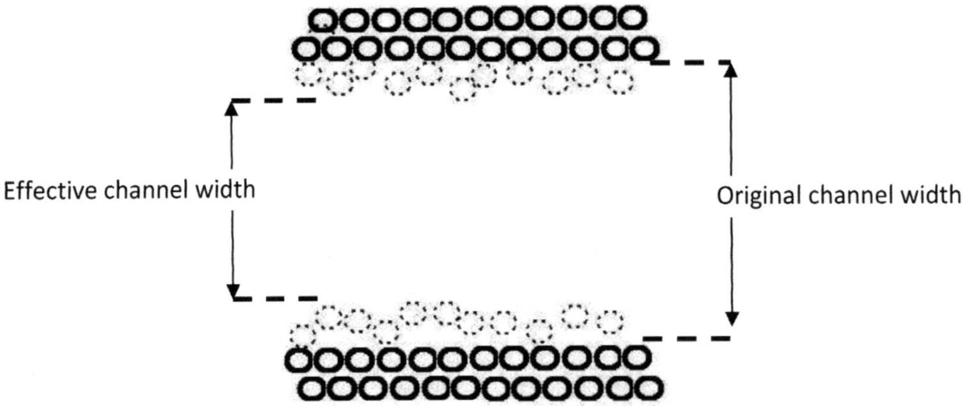

Figure 4. Large movement of wall molecules result in narrower channel

To explain the behavior of maximum velocity with spring stiffness we may assume that as the value of spring stiffness increases the oscillations of wall molecules are limited. This will result in a wider effective channel width and therefore a higher maximum velocity, as mentioned in [9] as well. Therefore maximum velocity should vary directly by spring stiffness.

Several simulations have been carried out for different values of spring stiffness having external force of 0.18 pN to verify the above theory. Since the effective width of channel is a function of molecular behavior of simulation, its value should be time-averaged. For this purpose we have used SMC method of averaging. The results of the simulations are shown in table 2 and figure 5.

Table 2. Channel's effective width with change of spring stiffness.

spring stiffness $\varepsilon\sigma^{-2}$	effective channel width (m)
10	2.964830E-09
30	2.955200E-09
50	2.956044E-09
70	2.959049E-09
100	2.963865E-09
150	2.970468E-09
300	2.984373E-09
500	2.994998E-09
700	3.001693E-09
1000	3.008557E-09
Standard deviation	1.889943E-11

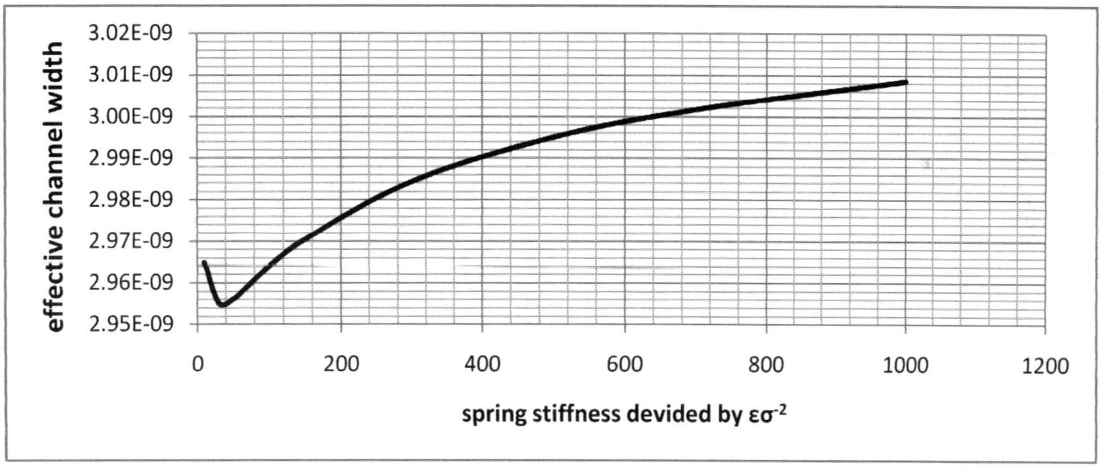

Figure 5. Different effective channel width with change of spring stiffness

Except for the first part of the figure 5, the rest of it can be explained based on the above-mentioned theory. Having compared figures 5 and 2 with each other, it can be seen that the trend of results are the same. However the turning points in these figures are not that close to each other. Although we put lots of efforts to offer better explanation for the behavior of maximum velocity with spring stiffness, it seems that we were not successful.

4. Conclusions and suggestions

Since the spring stiffness is able to affect the flow even far from the walls, suitable spring stiffness should be used for molecular dynamics simulations otherwise inaccurate results may be obtained. It is noticed that the behavior of maximum velocity magnitude is non-monotonic when the spring stiffness is changed for the both external forces applied in this paper. It first decreases and after specific spring stiffness it increases. The idea of original and effective width has been proposed to clarify the reason of such behavior; however it was not successful.

It is suggested that the tests done on the effective width be repeated with different external force magnitudes. To investigate all of the parameters affecting the velocity profile, it is proposed that the dimensionless parameters of this molecular dynamics simulation are obtained. In this way it is possible to study the reason of maximum velocity behavior considering all of the dominant parameters including spring stiffness, external force, channel width and etc.

5.References

[1] Fan X , Phan-Thien N,Teng Yong N and Diao X 2002 Molecular dynamics simulation of a liquid in a complex nano channel flow *Phys. Fluids* **14** 1146-53

[2] Karimian S M H and Izadi S 2010 A study on the measurement of mean velocity and its convergence *Int. J. Numer. Meth. Fluids* **67** 2130-40

[3] Abraham F 1978 The interfacial density profile of a Lennard-Jones fluid in contact with a (100) Lennard-Jones wall and its relationship to idealized fluid/wall systems: A Monte Carlo simulation *J.Chem. Phys.* **68** 3713-16

[4] Koplik J, R.Banavar J and F.Willemsen J 1989 Molecular dynamics of fluid flow at solid surfaces *Phys. Fluids* **1** 781-794

[5] Thompson P and Robbins M 1990 Shear flow near solids: Epitaxial order and flow boundary conditions *Phys. Rev. A* **41** 6830-37

[6] Jabbarzadeh A Atkinson J D and Tanner R I 1997 Rheological properties of thin liquid films by molecular dynamics simulations *J. Non-Newtonian Fluid Mech.* **69** 169-193

[7] Jabbarzadeh A Atkinson J D and Tanner R I 1997 Wall slip in the molecular dynamics simulation of thin films of hexadecane *J. Chem. Phys.* **110** 2612-19

[8] Bing Mi X and T. Chwang A 2003 Molecular dynamics simulations of nanochannel flows at low reynolds numbers *Molecules* **8** 193-206

[9] Sofos F, Karakasidis T E and Liakopoulos A 2009 Non-Equilibrium Molecular Dynamics Investigation of Parameters Affecting Planar Nanochannel Flows *Contem. Eng. Sci.* **2** 283-298

[10] Verlet L 1967 Computer "Experiments" on classical fluids. I. Thermodynamical properties of Lennard-Jones molecules *Phys. Rev.* **159** 98-103

1st European Conference on Gas Micro Flows (GasMems 2012) IOP Publishing
Journal of Physics: Conference Series **362** (2012) 012040 doi:10.1088/1742-6596/362/1/012040

Rebuilding of Rothe's nozzle measurements with OpenFOAM software

Erik Arlemark[1], Gennady Markelov[2] and Silvia Nedea[3]

[1,3]Eindhoven Univeristy of Technology, Eindhoven, Netherlands
[2]AOES Netherlands BV, Noordwijk, Netherlands

E-mail: [1] E.J.Arlemark@tue.nl,[2] Gennady.Mardelov@aoes.com,[3] S.V.Nedea@tue.nl

Abstract. In this paper the dsmcFoam solver is tested and validated for the the three main solver functionalities of 1) free-stream boundary conditions, 2) kinetic intermolecular collision including internal degrees of freedom and 3) gas/surface interactions. The free-stream utility was improved such that a spatially uniform field of particles gets inserted now yielding reliable results for the cells located close to these patches. Implementation of the collision models were validated for two test cases (monatomic gas mixtures and diatomic gas) by observing the equilibration of both the kinetic and internal energies. It was found that the present code had good agreement to the independent codes of HAWK and SMILE as well as to results by G. Bird. The validation of the present codes treatment for the gas/surface interactions was evaluated using the benchmark case of Rothe's nozzle measurements. Results show that the present version of dsmcFoam obtained good agreements for this case compared to the measurements of Rothe for density and temperature. It was also found that the Navier-Stokes solver of OpenFOAM produced reasonable results, even though the local Knudsen number of the flow exceeds the range of applicability for this method, Kn=0.1.

1. Introduction

Micro-electro mechanical systems (MEMS) become more common in our daily life and the area of their application is growing very fast. Design of MEMS and evaluation of their performance are mainly done using experimental methods. Obviously, an application of numerical methods can simplify and speed up development of new MEMS. Due to the small size of MEMS an inherent gas-flow can be in both continuum and transitional regimes despite a rather high working pressure. The continuum regime of gas flows are commonly and preferably simulated by the Navier-Stokes-Fourier (NSF) equations. However, experiments such as those performed by Arkilic *et al.* [7] and Colin [8] have shown that the conventional NSF equations may not produce accurate results for rarefied gas flows. It is normal practice to determine the rarefaction degree of gas flows by the Knudsen number (Kn). The Kn is the ratio of the average traveling distance of a molecule between successive collisions and the characteristic length scale of the flow. The NSF equations applied with continuous boundary conditions of velocity and temperature are commonly known to be valid up to a Knudsen number of 0.001 if no discontinuous boundary conditions are applied [9]. However, the applicability of the NSF equations can be extended to about 0.1 in case the discontinuous boundary conditions of velocity slip and temperature jump are applied [3]. A kinetic approach may therefore be necessary in order to simulate gas flows with Kn numbers higher than 0.1, for example, the direct simulation Monte Carlo (DSMC)

method [3]. The computational cost for this method, however, rapidly increases as the density increases. A suitable approach to obtain a relatively cheap method would be to combine the continuum and kinetic solvers to a hybrid method. However, it is important to obtain a clear understanding of the limits of either solver before the hybrid method can be applied.
According to G. Bird most DSMC programs are difficult to use and/or are not generally available. The main focus of this paper is to validate the DSMC solver (dsmcFoam) of the free and open CFD toolbox OpenFOAM version 2.0.1 by OpenCFD Ltd [2]. Some of the main attractive features of the dsmcFoam solver are that it is applicable for arbitrary 2D/3D geometries, it has parallel processing capacity, it utilizes a Maxellian distribution function for a free-stream flow patch, it incorporates the treatment of diffusive and specular wall reflection models and it applies the variable hard sphere model and a Larsen-Borgnakke internal energy redistribution model [12]. G. Bird considers dsmcFoam being an interesting new program initiative, but states definitive benchmark results are not yet available [13]. This work extends the validation of the dsmcFoam performed by Scanlon et al. [12] as the dsmcFoam solver is here validated for independent cases consisting of collision treatment, free-stream functionality and gas/wall interactions. The cell sizes of each case has been designed to approximately have five molecules in respect to time average in every cell. The DSMC test case of gas/wall interaction consists of the benchmark case of the Rothe nozzle [1] for which the compressible NSF solver of OpenFOAM (rhoCentralFoam) is also applied. Our aim is thus to validate both the NSF and dsmcFoam solvers of OpenFOAM against this well studied problem in order to further use it for MEMS applications.

2. Free-stream improvement of dsmcFoam

The free-stream utility is used to inject particles into the computational domain. This utility applies a Maxwellian velocity distribution function with prescribed values of density, velocity vector and temperature to inject a correct amount of particles at correct velocities. The injected particles move with their velocity and simulation time step. The validation of the free-stream utility was first considered when the Rothe nozzle case was simulated, where the default free-stream utility caused inaccurate pressure values for the cells closest to the inlet patch (about 10% lower than expected and varying with the time step size). A test case for the free-stream utility was thereafter designed which could clearly point out if molecules where inserted in an erroneous manner.
The test case for the free stream consists of a box with inlet on the left hand side and a vacuum on the opposing right hand side, see figure 1. The four remaining sides are set to reflect the particles specularly. The inlet patch is set to produce a free particle stream having a number density of 2.4513×10^{24} m^{-3}, a bulk velocity of 3000 m/s and a bulk temperature of 10 K. For this case every simulated particle represents 4×10^{10} molecules. The temperature is set low so that molecules obtain a low relative velocity to each other, which allow for the distribution of particles to be isolated and studied, only involve effects arising from the free-stream utility. The simulation box geometry is 0.1 m from inlet to outlet and have a square cross-section with side-width of 0.08 m. The box consists of 20 cells in the flow direction and 16 cells in the two other directions. The time step for this case is chosen to be 10 μs. The case is set not to treat collisions, since it is desired to isolate the phenomena of the insertion of DSMC particles.
In figure 1 (a) it is seen that particles of the conventional free stream utility are released in batches which flow down stream as blocks of particles moving towards the outlet. The free stream function is modified here such that the newly inserted molecules are moved with a step-fraction randomly chosen between zero and unity while all other molecules are moved with an initially full step fraction. This change of procedure causes a spatially correct uniformly distributed insertion of molecules at the inlet patches of the flow geometry. The result for the same test case using the modified free stream utility is shown in figure 1 (b). Here it is seen

that a more uniform distribution of molecules is present.

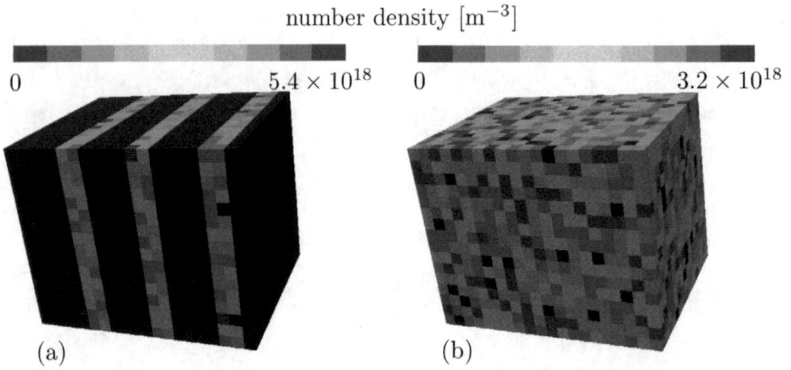

Figure 1. number density of particles in the periodic box. (a) Default free stream utility. (b) improved free stream utility.

3. Collision treatment of dsmcFoam

Two validation cases of G. Bird [3] are chosen in order to validate collision treatment of the variable-hard-sphere model and the Larsen-Borgnakke model of dsmcFoam. The latter of the two models considers polyatomic molecules and thereby treats the interchange of translational energy and rotational energy (vibrational energy can not be treated for dsmcFoam). In the first validation case only the variable hard sphere model is tested for a gas mixture consisting of 90% of argon molecules (number density=1.647×10^{19} m^{-3}) and 10% of helium molecules (number density=0.183×10^{19} m^{-3}). For these cases every simulated particle represents 5×10^{10} molecules. The helium molecules are initialized at a temperature of 500 K and the argon molecules are initiated at a temperature of 100 K. The temperature convergence is studied as a function of simulation time. The molecules are simulated in a cubic domain with specular reflecting walls. A comparison of these results is also made with the DSMC codes Hawk [4] and SMILE [5]. The results of this case are shown in figure 2.

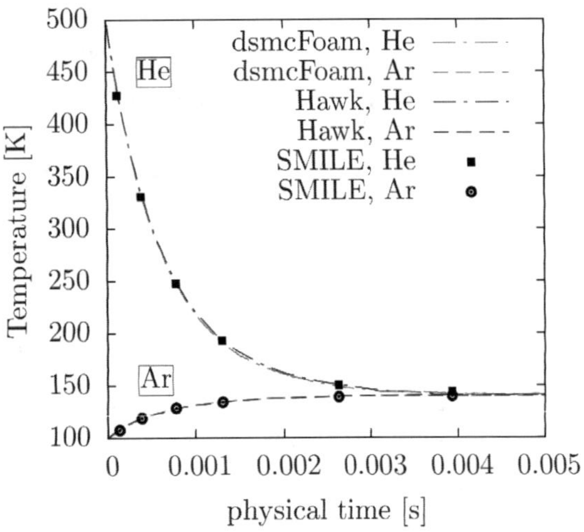

Figure 2. Temperature equilibrium of helium and argon gas species.

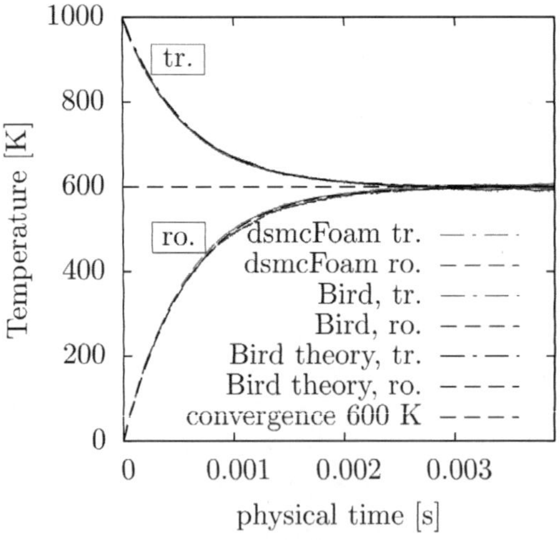

Figure 3. Equilibration of translational (tr) and rotational (ro) temperatures.

It is seen in figure 2 that the three models have the same rate of convergence to the temperature of about 140 K which occurs approximately at a physical time of 0.004 seconds. In figure 3 it is seen that the translational and rotational temperatures of the dsmcFoam model converges to a temperature of 600 K at the physical time of about 0.004 seconds, in similarity to theoretical and DSMC results of G. Bird.

4. Rothe nozzle benchmark validation

The treatment of gas-wall interactions for dsmcFoam is validated using experimental nozzle flow data (Exp) obtained by Rothe [1]. The inlet radius of the converging cone has a height of about 8.3 mm and a half angle of 30 degrees. This is followed by a diverging cone with an exit radius of about 21 mm and a half angle of 20 degrees. The two cones are connected via a throat section with a radius, $r = 2.55$ mm. The longitudinal radii of curvature at the throat equals half of the throat radius. For this nozzle an air flow is present at the inlet and a vacuum condition at the outlet. The Reynolds number notation by Rothe [1], $B = \rho_0(2H_0)^{1/2}r/\mu_0$, is adopted in this investigation, where H_0, μ_0 are the specific enthalpy and dynamic viscosity respectively in the stagnation chamber. In this paper the $B = 590$ case of Rothe is chosen to be simulated, with an applied inlet pressure of 473.86 Pa and a temperature of 300 K in agreement to [5].
The simulation geometry and mesh for the dsmcFoam solver are illustrated in figure 4, for which the results are compared with results obtained by the DSMC solver SMILE [5] and the NSF solver of OpenFOAM (NSF-OF). The geometry consists of an axisymmetric two degree wedge of the real nozzle. Although the nozzle geometry is given by Rothe [1] it is important to apply inlet and outlet domains to include certain upstream and downstream effects of the nozzle domain. Upstream of the converging cone a 10 mm long inlet domain is introduced with the free-stream path applied on the left hand side, see figure 4. Downstream of the diverging cone a vacuum chamber is applied with a height of about 29 mm and a length of about 25 mm. Investigations has been performed by Ivanov et al. [5] concerning sensitivity of these inlet and outlet conditions, based on which the dimensions of the inlet domain and the outlet domains are chosen similarly here. With a similar shape of inlet domain Ivanov et al. found that there was no significant affect dependent on whether the insertion of particles was performed having an (on average) radially uniform velocity profile or a parabolic velocity profile. Due to this reasoning molecules are inserted in this investigation with a flat velocity profile. The mesh is structured except for the converging cone section which is unstructured. The standard mesh of this case consists of 68 017 cells, where only one cell layer is placed in the symmetry direction. The side planes of the wedge are simulated as symmetry-planes by applying specular reflecting surfaces. The dsmcFoam simulation of the Rothe nozzle uses the free stream utility at the inlet and outlet of the nozzle. Collisions are treated by a variable hard sphere model with the extension of Larssen-Borgnakke method for exchanging translational and rotational energy. The temperature of the Rothe nozzle is sufficiently low for vibrational energy to be negligible. Hence the variable hard sphere model with Larsen-Borgnakke translational/rotational energy exchange treatment is considered adequate. The walls of the nozzle scatter reflected molecules according the the fully diffuse Maxwellian distribution of a temperature of 300 K. It should be noted that the SMILE solver treating wall collisions as adiabatic. The NSF was applied with the Maxwellian velocity slip [10] (slip) or both Maxwellian velocity slip and the Smoluchowski temperature jump [11], here referred to as discontinuous boundary conditions (D-BC). The case is run in parallel on four processors divided as illustrated in figure 4. The hardware used is a cluster computer with 16 cpu nodes of 2.34 Ghz AMD opteron processors with 62.95GB RAM memory. For the steady state case to equilibrate takes about a week and averaging of the properties takes also about one week. The NSF solver is run on a single machine.

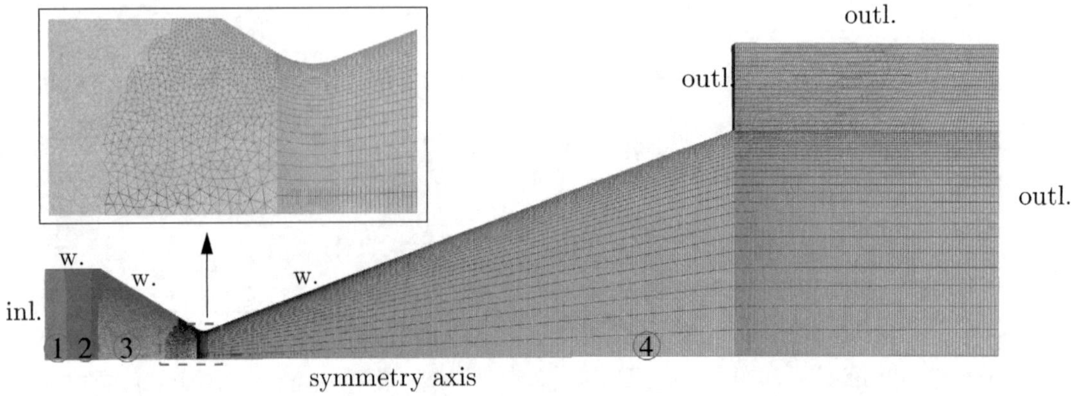

Figure 4. Nozzle mesh for dsmcFoam, where inl., outl. and w. refer to inlet, outlet and wall respectively. Mesh is a two degree axisymmetric wedge of real nozzle. The processors domains are indicated by numbers in red circles.

The dsmcFoam solver simulates nitrogen gas according to the variable hard sphere model having the parameters shown it table 1. In the simulation for the B=590 case every dsmc particle represents 5 $\times 10^8$ real molecules.

Table 1. Nitrogen data for variable hard sphere model with Larsen-Borgnakke translational/rotational energy exchange treatment [3].

property	value
mass	46.5×10^{-27} [kg/m^{-3}]
d$_{ref}$	4.17×10^{-10} [m]
internal degrees of freedom	2 [-]
ω	0.74 [-]
T$_{ref}$	273 [K]
relaxation collision number	5.0 [-]

4.1. results

The Knudsen number is calculated for the B=590 case using the dsmcFoam solver. The Kn number is calculated as the ratio of the mean free path, λ, to the characteristic length scale of the case, L. The mean free path for the nitrogen gas is calculated for the variable hard sphere model as follows:

$$\lambda = \frac{1}{\sqrt{2} n \pi d_{ref}^2 \left(\frac{T_{ref}}{T}\right)^{\omega - 0.5}}, \tag{1}$$

where n, d_{ref} and T_{ref} are the number density, reference diameter and temperature respectively for the gas and ω is the exponent that relates viscosity to temperature i.e. $\mu \propto T^{\omega}$ [3]. The length scale is calculated as follows:

$$L = \frac{\Phi}{\left|\frac{\partial \Phi}{\partial x_i}\right|}. \tag{2}$$

In (2) the parameter Φ is a macroscopic property of the flow e.g. density, bulk velocity or temperature. The Knudsen number is mapped for the B=590 case by modifying the post processing utility dsmcFieldsCalc of OpenFOAM. In figure 5 the local Knudsen number is illustrated by using the calculated number density in (1) and temperature in (2).

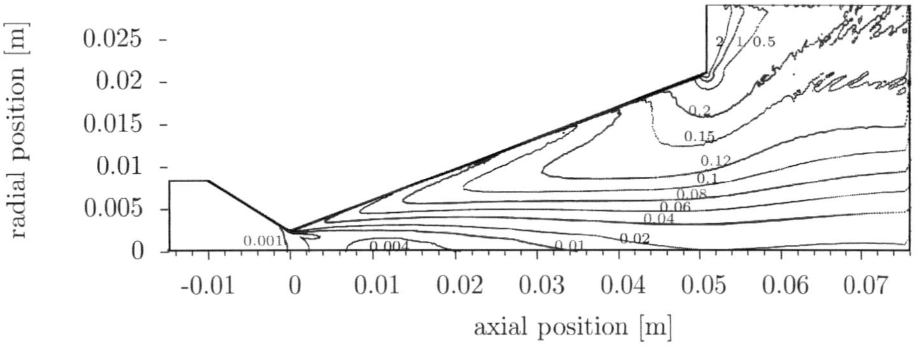

Figure 5. Kn number contours for B=590 case.

From figure 5 the applicability of the NSF equation can be prematurely evaluated. Since the NSF equations should be applicable for Kn numbers up to 10^{-3} it is reasonable to assume this method is applicable from the inlet to the throat of the the nozzle, but erroneous further downstream. With discontinuous boundary conditions the NSF method should be applicable up to Knudsen numbers of about 0.1. For the nozzle case it is not clear how far downstream this method is applicable since the wall region and axial part of the domain has higher Kn than the rest of the cross sectional domain.

In the left part of figure 6 the centerline profile of mass density is displayed throughout the nozzle. It can be seen that slightly downstream from the throat the NSF-OF solver deviates slightly from the experimental data. Downstream of the throat all of the solvers are considered to produce reasonable results. In the right part of figure 6 the radial profile of the density is displayed scaled by the axial value. It can be seen that the results of the dsmcFoam and the SMILE solvers agree well with the experimental data. It should also be noted that for the density profiles the adiabatic wall method produced similar results as for the thermal wall.

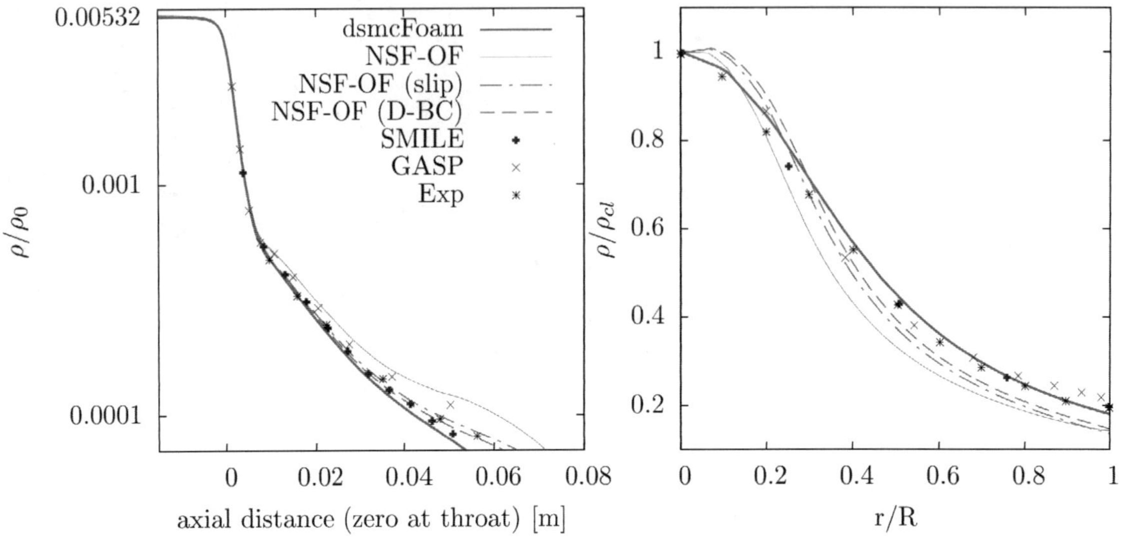

Figure 6. Left, density profile in axial direction. Right, density profile in radial direction at axial position 0.0477 m.

In figure 7 the rotational temperature is displayed. In the left part of figure 7 the centerline profile of temperature is displayed throughout the nozzle. It can be seen here that all the solvers produce reasonable results till an axial position of about 0.03 m after which the NSF-OF with slip and NSF-OF with D-BC has significantly lower temperature that the experimental data. In the right part of figure 7 the radial profile of the density is displayed scaled by the axial value. Here it can be seen that all the solvers are scattered from the experimental data. It is of particular interest to note that the dsmcFoam with thermal wall wall does not manage to capture the correct temperature jump at the wall. However the NSF-OF with D-BC manages to capture the validation temperature at the wall the best being about 15 K above.

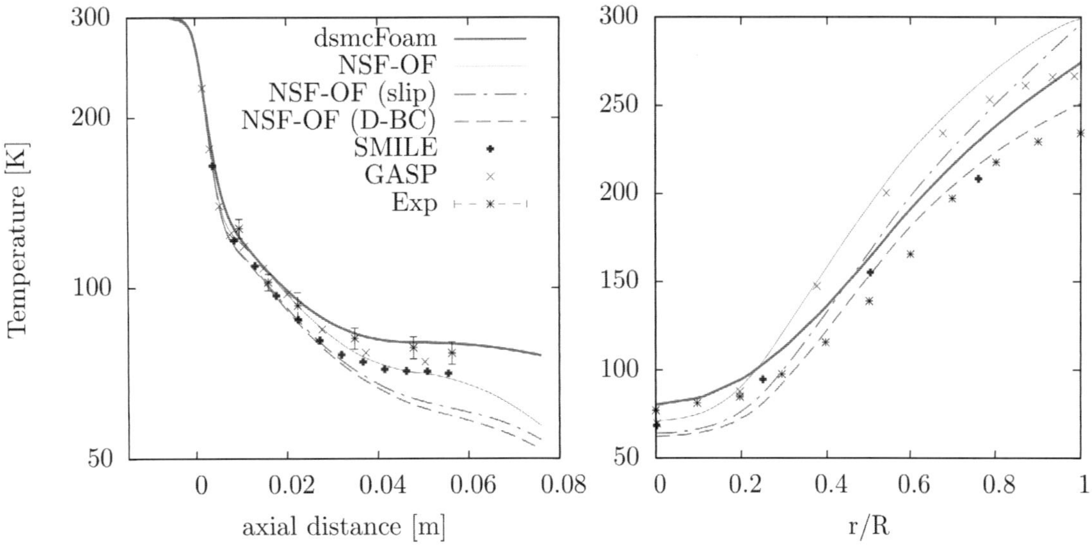

Figure 7. Left, temperature profile in axial direction, experimental error-bars from [1]. Right, temperature profile in radial direction at axial position 0.0477 m.

In figure 8 the Mach number profiles are displayed for the dsmcFoam, NSF-OF with D-BC and SMILE solvers for the B=590 case. It can be seen that all the Mach number contours are in reasonable agreement, with the largest discrepancy for the NSF-OF solver with D-BC which has a slightly lower Ma number throughout the domain compared to the DSMC solvers.

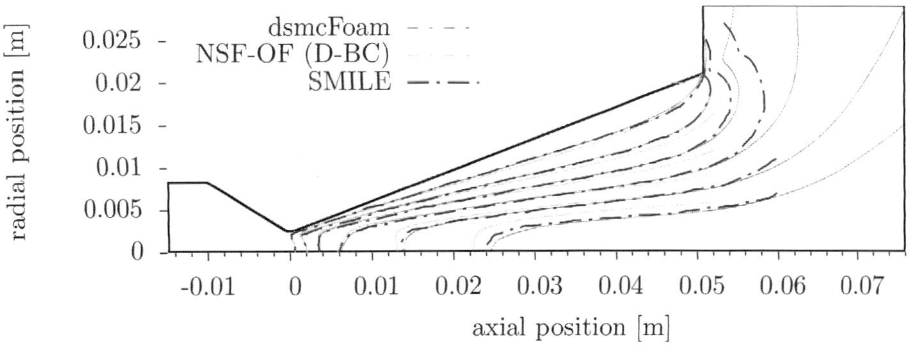

Figure 8. Ma number contours for B=590 case. Contours from left to right are: 1, 1.5, 2, 2.5, 3 and 3.5

In figure 9 the density profiles are displayed for the dsmcFoam, NSF-OF with D-BC and SMILE solvers for the B=590 case. It can be seen that all the solvers agree reasonably well except for the for the 0.005 profile of the NSF-OF with D-BC at the axial position.

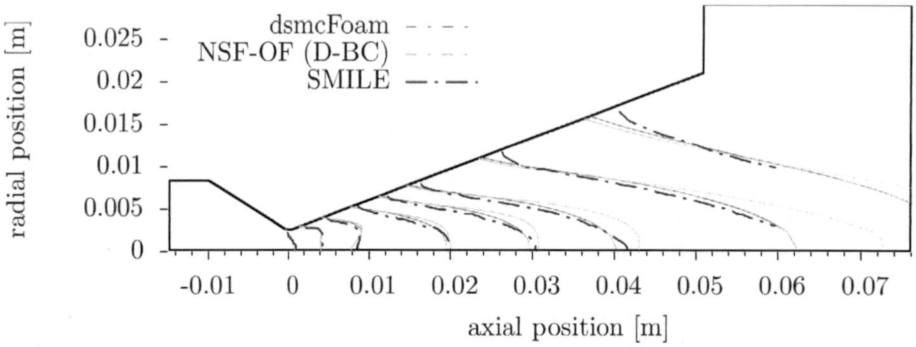

Figure 9. density contours scaled by inlet density of 0.0053223 kg/m^3 for B=590 case. Density contours from left to right are: 0.6, 0.2, 0.1, 0.05, 0.03, 0.02, 0.01, 0.005.

In figure 10 the temperature profiles are displayed for the dsmcFoam, NSF-OF with D-BC and SMILE solvers for the B=590 case. Here it can be seen that both DSMC solvers produce similar results for the 0.8, 0.6, 0.4 and 0.3 contours within the nozzle. However the dsmcFoam model produces a higher temperature than SMILE in the axial domain as can be seen for the 0.25 contour. It can also be seen that the NSF-OF with D-BC only agrees with the results of the DSMC solvers for the 0.8, 0.6 and 0.4 contours.

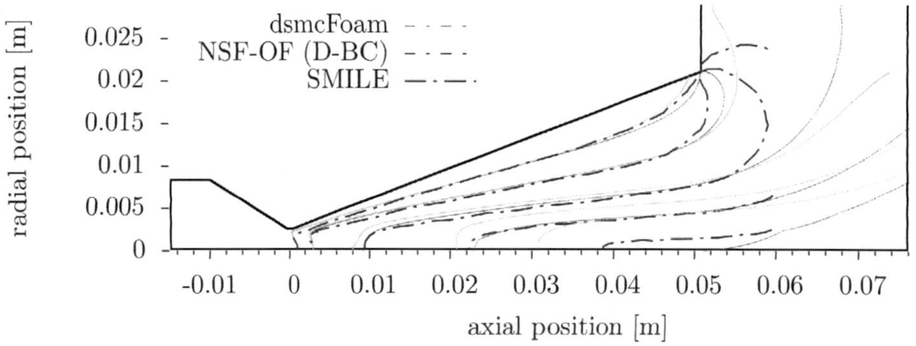

Figure 10. temperature contours scaled by inlet temparature of 300 K for B=590 case. Temperature contours from left to right are: 0.8, 0.6, 0.4, 0.3 and 0.25.

5. Discussion and conclusions

The free stream utility of dsmcFoam is improved in such a way that a uniform molecular field gets inserted at the free-stream patches. The collision treatment of dsmcFoam is validated where the variable hard sphere model show very good agreement with results of the alternative DSMC solvers HAWK [4] and SMILE [5]. The Larsen-Borgnakke treatment method for exchanging translational and rotational energy is validated with very good agreement to both theoretical and numerical results of G. Bird [3]. The dsmcFoam solver with corrected of free stream utility gave Rothe nozzle case good cell value results close to the inlet patch, which was not the case for the default free-stream utility. The dsmcFoam solver also in general showed to produce good results for the benchmark case of the Rothe nozzle [1] in comparison to the experimental results of Rothe and the numerical results of SMILE. Aslo the NSF solver of OpenFOAM is validated here both with and without discontinuous boundary conditions and show good results for large sections of the diverging cone of the Rothe's nozzle case. However, close to the vacuum patch the rarefaction degree of the flow becomes too great for the NSF solver to produce accurate results for which other solvers are necessary. Further work will be devoted to implement adiabatic wall conditions for the dsmcFoam solver when applied to Rothe's nozzle case. Further work will also be devoted to investigate the dsmcFoam and NSF solvers applicability for lower gas pressures of Rothe's nozzle case such as the measurment results having Reynolds numbers of 260 and 110.

6. Acknowledgement

The research leading to these results has received funding from the European Community's Seventh Framework Programme (ITN-FP7/2007-2013) under grant agreement n° 215504.

References

[1] Rothe D E 1971 *AIAA Journal* **9** 809-811
[2] OpenFOAM, the Open Source CFD Toolbox, www.openfoam.com/, visited 20th March 2012.
[3] Bird G A 1994 *Molecular Gas Dynamics and the Direct Simulation of Gas Flows* (Claredon: Oxford).
[4] Rieffel M, Taylor S and Watts J 1998. Electrical Engineering and Computer Science. Paper 155.
[5] Ivanov M S, Markelov G N, Ketsdever A D, Wadsworth D C. 1999 AIAA Paper 99-0166.
[6] Saleh J 2002 *Fluid flow handbook* (McGraw-Hill Professional)
[7] Arkilic A, Schmidt M and Breuer K 1997 *Journal of Microelectromenchanical Systemcs* **6(2)** 167-178.
[8] Colin S 2005 *Microfluidics and Nanofluidics* **1(3)** 268-279.
[9] Gad-el-Hak 1999 *Journal of Fluids Engineering, Transactions of the ASME* **121(1)** 5-33.
[10] Maxwell J 1879 *Philosophical Transactions of the Royal Society of London* **170** 231-256.
[11] von Smoluchowski M 1879 *Annalen der Physik und Chemie* **64** 101-130.
[12] Scanlon T J, Roohi E, White C, Darbandi M, Reese J M 2010 Computers & Fluids **39** 2078-2089
[13] DSMC resources from Graem Bird, www.gab.com.au, last updated 11th January 2012.

Gas mixing enhancement in minichannels using a rotationally oscillatory circular cylinder

M Raisee[1] and M Cheraghi

School of Mechanical Engineering, College of Engineering, University of Tehran, Tehran, Iran

[1]Email: mraisee@ut.ac.ir

Abstract. Oscillating structures and actuators can induce flow kinematics that enhances mixing. This approach is specifically effective for mixing enhancement in meso-scale channels, where the flow kinematics can be actively controlled using micro-electro-mechanical-systems (MEMS). In this paper, numerical results for mixing of two incompressible ideal gas (Schmidt number of 1.0) streams through a 2D minichannel via a rotationally oscillating circular cylinder are presented and discussed. Simulations are performed for blockage ratio of D/H=1/3 and Reynolds number of 100 and oscillation amplitudes of $\pi/3$, $2\pi/3, \pi/2$ and π for subharmonic ($F < 1$), harmonic ($F = 1$) and superharmonic ($F > 1$) regimes. Numerical results indicate that mixing performance is improved by about 70% compare to the plane channel at oscillation amplitude of π and excitation frequency of 25% higher than the natural frequency of vortex shedding of a stationary cylinder. It is shown that the mixing efficiency is increased by increasing of amplitude in all the cases except at very low excitation frequencies. This study also shows that when the excitation frequency is equal to the vortex shedding frequency the maximum power is required for mixing of two gases.

1. Introduction

Effective mixing is a major problem in many MEMS devices where mixing of two or more fluids is the main goal. Because of small dimensions and low velocities, flow regime in MEMS devices is often laminar and thus effective mixing may require long length and time which is not of practical interest. Therefore, innovative methods for fluid mixing are required. Pulsed-flow mixing, electrokinetic instability, hydrodynamic instability effects due to vortex shedding and boundary modification are methods that may be used for mixing enhancement in MEMS devices [1-7]. In the present investigation, attention is focused on the fluid mixing in a straight 2D channel via vortex shedding from a rotationally oscillating circular obstacle.

A number of researchers have investigated fluid flow structure around a circular cylinder in external flow because of its numerous engineering applications. However, only a few studies in the literature have studied fluid flow past a stationary or oscillating confined cylinder with the objective of either heat transfer or fluid mixing augmentation. One of the earliest studies on this subject was reported by Fu and Tong [8]. They performed numerical simulations to examine the effects of transversely oscillating cylinder on the cooling of an array of heated blocks. Their results showed that heat transfer markedly increases when the excitation frequency is in the lock-in regime. Recently, Celik et al. [9] have studied laminar flow past a transversely oscillating circular cylinder in a 2D channel. They found that the shedding frequency of laminar flow past a confined stationary cylinder is two times higher than the vortex shedding frequency of external flow past a cylinder. More recently, Celik et al. [10] have examined the effects of transversely oscillating cylinder on the wall heat transfer. According to their results, transversely oscillating cylinder significantly enhances the wall heat transfer. Maximum heat transfer enhancement was observed for the frequency ratio of $F = 0.75$ and faster oscillations was found to be not beneficial. In another study, Celik et al. [11] have investigated mixing enhancement in

the channel via a transversely oscillating cylinder. They found that the most favorable mixing performance occurs when the oscillation frequency is in lock-in regime.
In practice, the transverse oscillation occupies the channel height and demands more complicated movement mechanism, control and set-up than a rotationally oscillating cylinder. In the present paper, we investigated the utilization of a rotationally oscillating cylinder with various amplitudes and frequencies for mixing of two gases. For each frequency and amplitude, the mixing efficiency and the required power are calculated.

2. Problem Definition and Governing Equations

In this section, the geometry, the governing equations and the numerical method are described.

2.1. Geometry

Figure 1 demonstrates the schematic of the computational domain of the present investigation. Two fully-developed laminar ideal gas flows with a parabolic velocity profile and mean velocity of U_m enter the channel from the upper and the lower parts of the channel. Two streams are initially separated by a plate of infinitesimal thickness. The separation plate and channel lengths are 2D and 31D, respectively. The circular obstacle is positioned at 4D away from the channel inlet. The circular cylinder is subjected to forced rotational oscillations that result in time periodic tangential velocity (V_θ) on the cylinder according to following formula:

$$V_\theta = \theta_{max} \pi f_e D \sin(2\pi f_e t) \tag{1}$$

where f_e and $\theta_{\max}$ are the excitation frequency and the maximum rotation angle, respectively.

In this study, computations are carried out for a blockage ratio of 1/3, Reynolds number of 100 and Schmidt number of 1. The maximum rotation angle varies between 0 and π, whilst frequency ratio (F) is ranged between 0.25 and 2.0. The frequency ratio (F) is defined as the ratio of the excitation frequency (f_e) to the vortex shedding frequency of flow around a stationary confined cylinder (f).

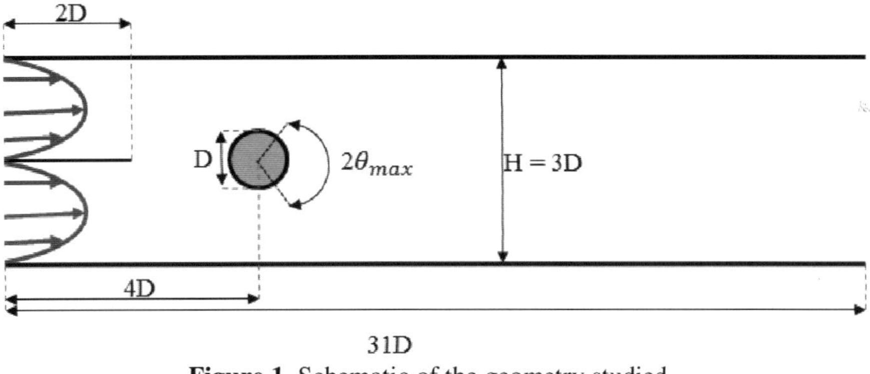

Figure 1. Schematic of the geometry studied.

2.2. Governing Equations and Numerical Method

The governing equations of flow and scalar for an unsteady incompressible flow are written as:

Continuity:

$$\frac{\partial U_j}{\partial x_j} = 0 \tag{2}$$

Momentum:

$$\frac{\partial U_i}{\partial t}+\frac{\partial\left(U_iU_j\right)}{\partial x_j}=-\frac{1}{\rho}\frac{\partial P}{\partial x_i}+\nu\frac{\partial^2 U_i}{\partial x_j\partial x_j} \tag{3}$$

Scalar:

$$\frac{\partial c}{\partial t}+\frac{\partial\left(U_jc\right)}{\partial x_j}=d\frac{\partial^2 c}{\partial x_j\partial x_j} \tag{4}$$

where ρ, ν , c and d are the density, the kinematic viscosity, concentration and the mass diffusivity of the fluid, respectively.

Computations of the present work are obtained using OpenFOAM code. The pressure field is linked to the velocity field using SIMPLE algorithm. Second order upwind scheme is employed for approximation of nonlinear convective terms in momentum and scalar equations, while temporal derivatives in equations (3) and (4) are discretized using the second order Crank-Nicolson scheme. In the simulations, uniform inlet concentrations are assumed for the upper and lower streams. No-slip boundary condition is considered for all solid boundaries. Zero gradient concentration boundary condition is applied on the cylinder surface and on the channel walls. At the channel outlet, zero Neumann boundary condition is applied for both velocity and concentration. The computational domain is meshed into 21600 quadrilateral structured cells. A series of grid-independency tests have been carried out to assess the accuracy of the results of the present investigation. The outcome of these studies has shown that the grid used here is sufficiently fine to produce grid-independent results.

3. Results and Discussion

In this section, the numerical results of present investigation are presented and discussed to study the effects of excitation frequency and amplitude of oscillations on the mixing performance.
The criterion used here to evaluate the mixing performance is mixing efficiency defined by Erickson and Li [12] as:

$$\eta(x)=\left(1-\frac{\int_0^H\left|c(x,y)-c_m\right|dy}{\int_0^H\left|c_0(x,y)-c_m\right|dy}\right)\times 100 \tag{5}$$

where c_m is the concentration in perfect mixing condition ($c_m=0.50$) , c_0 is the concentration for an unmixed condition (c_0 = 1.0 and 0 for the lower and upper portion of the channel, respectively).

3.1. Effects of frequency and amplitude

The effect of oscillating cylinder on the mixing of two gases is a function of the excitation frequency and the amplitude. Figure 2 demonstrates the variation of mixing efficiency along the channel for various non-dimensional excitation frequencies (F) and amplitudes ($\theta_{\max}$). It is seen that a considerable mixing is achieved compare to the plane channel (P. Ch.) by the insertion of the cylinder in the channel. For the stationary cylinder, as expected the maximum mixing efficiency ($\approx$ 80%) occurs at the channel outlet. A relatively long length (≈20D) is however required to reach this maximum mixing performance. Depending on the excitation frequency, oscillation of the cylinder with the amplitude of $\theta_{max}=\pi/3$ can either enhance or diminish the mixing performance. Slow oscillation of the cylinder with frequency ratio of 0.25 is seen to have insignificant influence on the efficiency compare to the stationary cylinder. By increasing the frequency ratio to 0.50 the mixing efficiency becomes lower than those obtained for $F=0.0$ and 0.25. As the frequency ratio, approaches to the lock-in regime ($F=0.75$), the beneficiary of the rotation starts to appear. For the frequency ratio of 0.75, the mixing efficiency substantially increases to more than 90% at the channel outlet. For higher frequency ratios of 1.00 and 1.25, the efficiency is not changed much at the channel outlet, but fully-mixed condition is reached at the shorter distances from the channel inlet. More specifically, for $F=1.25$, the mixing efficiency reaches its maximum value at $x/D \approx 11$, whilst for $F=0.75$ this condition occurs at $x/D \approx 22$. For super-harmonic oscillation ($F>1$), the mixing efficiency drops with increase in excitation frequency. Note that for maximum excitation frequency (F = 2), the mixing efficiency distribution approaches to that of the plane channel. For amplitude of $\pi/2$, the stationary cylinder appears to be more effective for fluid mixing than the rotating cylinder in subharmonic regime under ($F=0.25$ and 0.5). Similar to the previous case, the frequency ratio of 0.75 substantially enhances the mixing efficiency, giving the peak mixing efficiency at the outlet. Further increase of frequency ratio to $F=1.0$ and 1.25 gives lower efficiencies but shorter mixing lengths. At this amplitude, the frequency ratio of 1.5 provides mixing levels similar to those obtained for $F=0.75$. Once again, further increase of F to 1.75 and 2 reduces the mixing efficiency in a way that least overall mixing efficiency is reached for $F=2.0$.

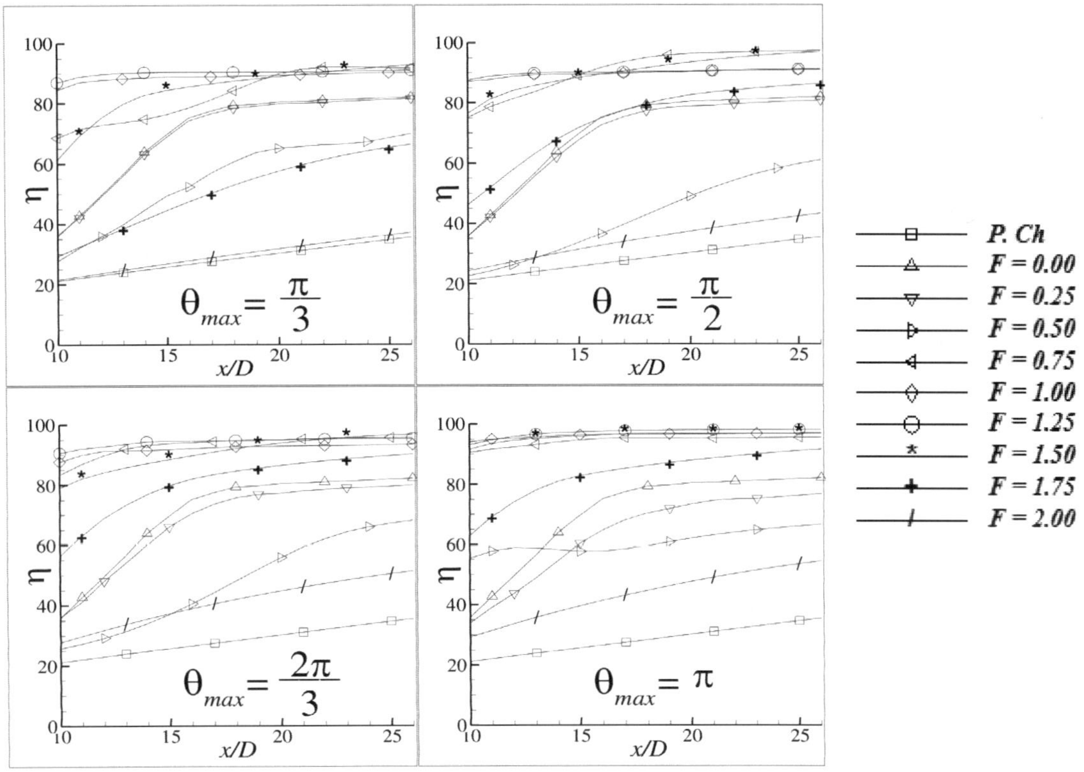

Figure 2. Effect of frequency and amplitude on the mixing efficiency Re = 100 and Sc =1.0.

The overall trend of efficiency curves for higher amplitudes of $2\pi/3$ and π are similar to those obtained for the lower amplitudes of $\pi/3$ and $\pi/2$. It is clearly seen that by increasing the oscillation amplitude the mixing efficiency increases. For $\theta_{max} = \pi$, the maximum mixing efficiency is obtained in lock-in regime $(0.75 < F < 1.25)$. In this regime, the mixing efficiency is insensitive to the excitation frequency.

The above results can be interpreted by examining the effects of vortex shedding on the mixing process. To this end, figure 3 demonstrates the vorticity and the concentration contours in the channel for amplitude of π and frequency ratios of 0.50, 1.25 and 2.00. These contours correspond to the instant that cylinder starts to rotate anticlockwise. At the lowest excitation frequency, the generated vortices are too few to effectively mix two streams. As a result, a little mixing occurs in the bulk fluid which is consistent with the efficiency distribution shown in figure 2. At the highest excitation frequency (F = 2.0), while more vortices are generated, but too small in size to vigorously mix the two fluids. Consequently, unmixed fluids are clearly visible in the end of the channel. At frequency ratio of F = 1.25, the optimum condition for mixing seems to be established. This condition creates more strong vortices and thus well mixing condition is observed for $x/D > 11$.

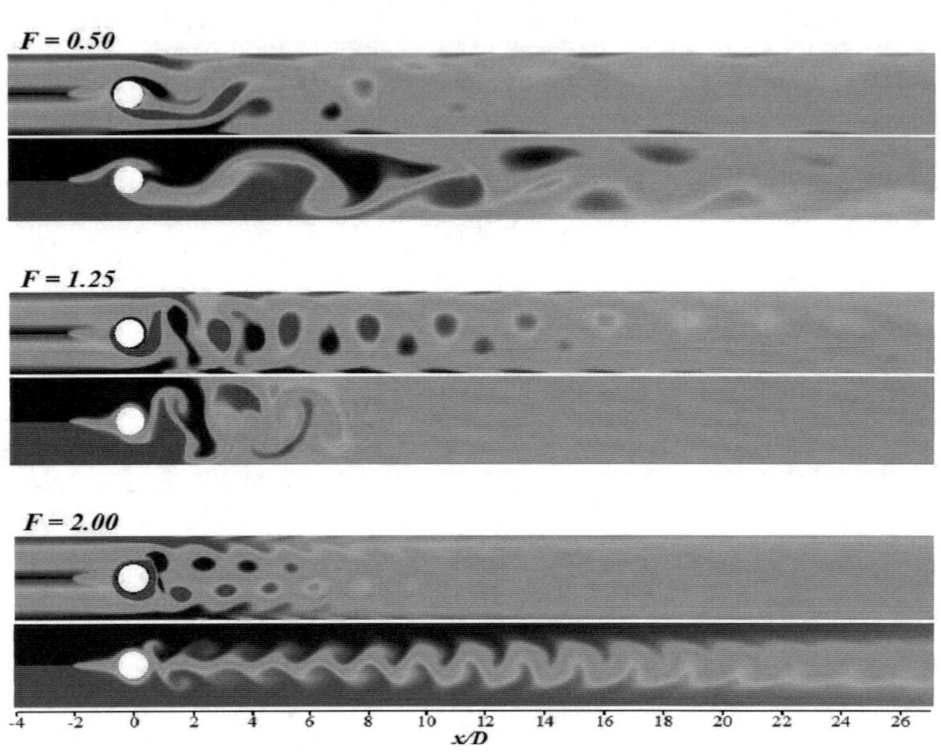

Figure 3. Vorticity and concentration contours along the channel for $\theta_{max} = \pi$, Re = 100 and Sc =1.0.

For fixed frequency ratio of F=1.25, figure 4 illustrates the effect of the oscillation amplitude on the mixing. Similar to figure 3, these contours correspond to the instant that the cylinder starts to rotate anticlockwise. Zones represented by A, B and C are selected to aid the interpretation of the results. Zone A shows the region of lower flow immediately after the cylinder. This figure shows that the height of this region increases with increase of the amplitude. Zone B represents the portion of the lower fluid where is affected by the vortex motion. As shown, for $\theta_{max} = \pi/3$, this zone is attached to the lower stream. By increasing the amplitude to $2\pi/3$, this zone displaces to higher altitudes but still does not detach. It is seen that zone B is separated from the lower stream at $\theta_{max} = \pi$. The region on the lower wall, where there is continuous unmixed fluid, is denoted by C. While this zone starts at $x/D \approx 6$ for $\theta_{max} = \pi/3$, it approaches the cylinder as the oscillation amplitude increases. This shows that mixing enhancement increases with increase of the oscillation amplitude.

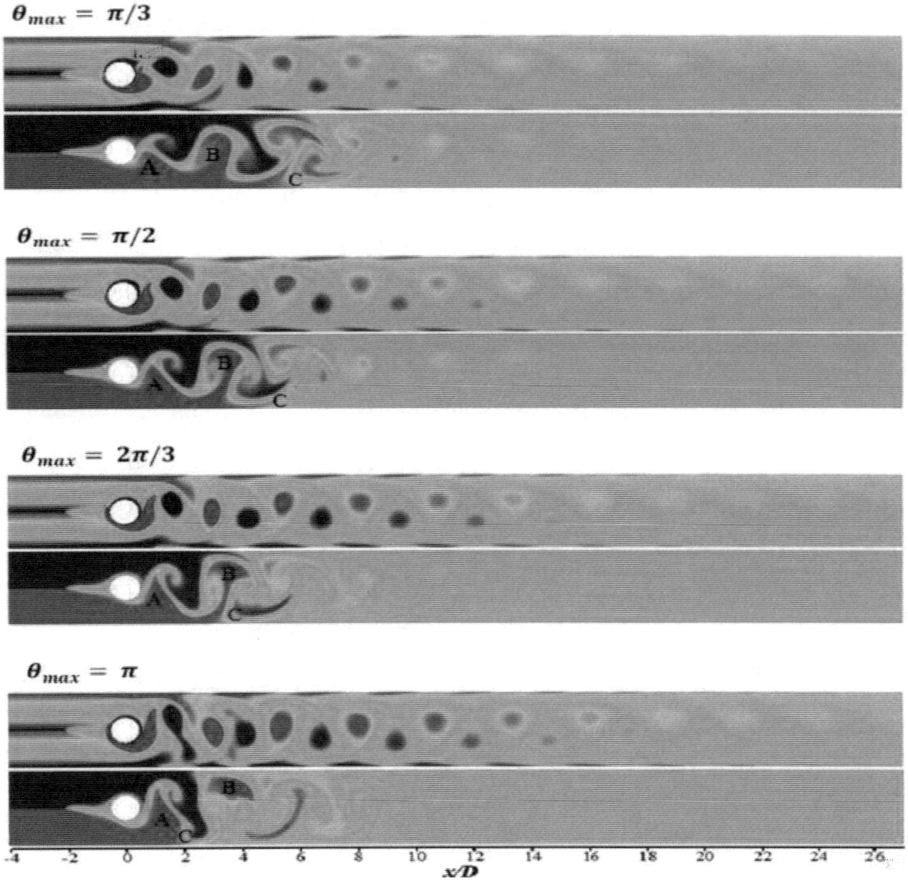

Figure 4. Vorticity and concentration contours along the channel for F = 1.25, Re = 100 and Sc = 1.0.

3.2. Power requirement

This section investigates the total power requirement for the proposed mixing mechanism. In order to calculate the mechanical power for oscillation, the average torque required for the oscillation is first computed by

$$\bar{T} = \frac{1}{4}\int_0^{2\pi} \mu \frac{\partial U}{\partial n} D^2 d\theta \tag{6}$$

where n is the direction normal to the surface.

Subsequently, the required mechanical power for one cycle of oscillation is obtained using

$$\text{Mechanical power} = \frac{1}{\tau}\int_0^{\tau} \bar{T}(t)\dot{\theta}(t)dt \tag{7}$$

Considering $\bar{P}$ as the time average pressure in one cycle of oscillation

$$\bar{P} = \frac{1}{\tau}\int_0^{\tau} P dt \tag{8}$$

pressure drop along the channel and pumping power are obtained as follows

$$\Delta P = \left(\frac{1}{H}\int_0^{H} \bar{P} dy\right)_{\text{inlet}} - \left(\frac{1}{H}\int_0^{H} \bar{P} dy\right)_{\text{outlet}} \tag{9}$$

$$\text{Pumping power} = A U_m \Delta P \tag{10}$$

where A and U_m are channel cross-sectional area and mean velocity, respectively.

Subsequently, we calculated the power ratio, Po, which is the ratio of the required power for oscillating cylinder in a channel (i.e., sum of the oscillation and pumping powers) to the required pumping power of a plane channel flow (obtained by multiplying the pressure drop in the channel with the sustained volumetric flow rate). As shown in figure 5, for each excitation frequency, the required power increases with increase of the oscillation amplitude. It is interesting to note that the total required power increases with the frequency ratio up to F = 1.00 and then decreases by increasing the frequency in the superharmonic region. It is worth mentioning that the same trend is observed for the pumping power. The mechanical power increases as the frequency increases up to F = 2.00. Another point about mechanical power is that its increase is mainly due to increase in amplitude and frequency as the average shear stress on the cylinder dose not change considerably in the investigated range of amplitudes and frequencies.

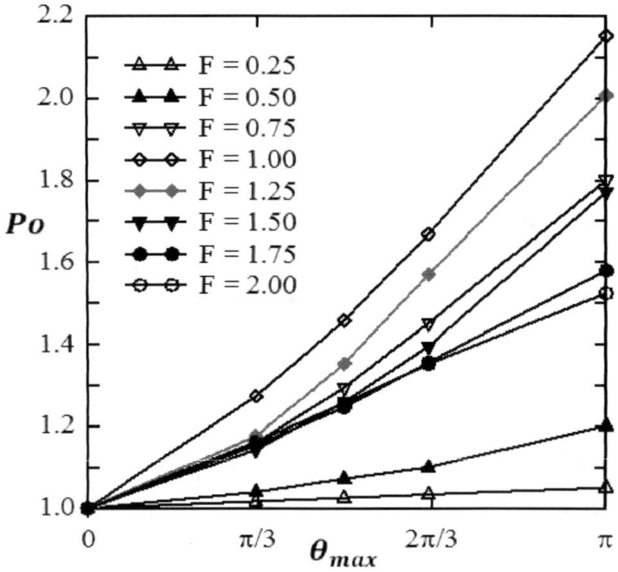

Figure 5. Total required power of the system versus frequency, Re = 100 and Sc =1.0.

Conclusions

Utilization of a rotationally oscillating cylinder as a gas mixer in a 2D minichannel investigated numerically at a constant Reynolds number of 100 and a wide range of amplitude and frequencies. In many applications, considering the cost of instruments for rotating the cylinder, use of a stationary cylinder can provide a relatively good mixing at the end of the channel (up to 80% for Pe = 100). Oscillation of cylinder with subharmonic frequencies (F = 0.25 and F = 0.50) and superharmonic frequencies (F = 2.00) resulted in less mixing performance than dose a stationary cylinder. In general, for all conditions investigated, considering the mixing length and efficiency, frequency ratio of 1.25 can be considered as the optimum frequency ratio. Except at very low frequency (F = 0.25), mixing performance increases as the oscillation amplitude increases. The total power required for the system increases as the oscillation amplitude increases. At a fixed amplitude, in subharmonic frequencies (F < 1), the required power increases as the excitation frequency increases. In contrast, at superharmonic region (F > 1), less power is required. Further work is in progress to consider flow compressibility and rarefaction effects.

References:

[1] Aref H 1984 Stirring by chaotic advection *J. Fluid Mech.* **143** 1-21

[2] Hessel V, Lowe H and Schonfeld F 2005 Micromixers_A review on passive and active mixing principles *Chem. Eng. Sci.* **60** 2479-501

[3] Nguyen N.T and Wu Z 2005 Micromixers_A review *J. Micromech. Microeng* **15** R1-R16.

[4] Jayaraj S, Kang S and Suh Y.K 2007 A review on the analysis and experiment of fluid flow and mixing in micro-channels *J. Mech. Sci. Technol.* **21** 536-48.

[5] Suh Y.K and Kang S 2010 A Review on mixing in microfluidics *Micromachines.* **1** 82-111

[6] Mansur EA, Mingxing YE, Yundong W and Youyuan D 2008 A State-of-the-Art Review of mixing in microfluidic mixers *Chinese Journal of Chemical Engineering.* **16** 503-16

[7] Briscoe B.J, Lawrence C.J and Mietus W.J.P 1999 A review of immiscible fluid mixing advances *Colloid and Interface Science.* **81** 1-17

[8] Fu W.S and Tong B.H 2004 Numerical investigation of heat transfer characteristics of the heated blocks in the channel with a transversely oscillating cylinder *Int. J. Heat Mass Transfer.* **47** 341_51

[9] Celik B, Akdag U, Gunes S and Beskok A *2008 Flow* past an oscillating circular cylinder in a channel with an upstream splitter plate *Physics of Fluids.* **20** 103603.

[10] Celik B, Raisee M and Beskok A 2010 Heat transfer enhancement in a slot channel via a transversely oscillating adiabatic circular cylinder *Int. J. Heat Mass Transfer* 626–34.

[11] Celik B and Beskok A 2009 Mixing induced by a transversely oscillating circular cylinder in a straight channel *Physics of Fluids.* **21** 073601

[12] Erickson D and Li D 2002 Influence of surface heterogeneity on electrokinetically driven microfluidic mixing *Langmuir.* **18** 188392.

A Kinetic Model for Gas Mixtures Based on a Fokker–Planck Equation

Hossein Gorji and Patrick Jenny
Institute of Fluid Dynamics, Sonneggstrasse 3, 8092 Zürich, Switzerland
E-mail: gorjih@ifd.mavt.ethz.ch

Abstract. We present a generalized nonlinear Fokker–Planck equation, which describes the dynamics of rarefied monatomic gas mixture flows. The devised kinetic model leads to correct transfer of energy and momentum between gas species and to consistent evolution of molecular stresses and heat fluxes with respect to the generalized Boltzmann equation. Thus, the correct diffusion coefficient together with the mixture viscosity and mixture heat conductivity coefficients are obtained. The strength of the presented model lies on the computational efficiency, which is due to the fact that the resulting stochastic processes are continuous in time. Therefore, unlike in Direct Simulation Monte Carlo (DSMC), here simulated particles do not collide with each other, but move along independent continuous stochastic paths. Another aspect of the new Fokker–Planck model equation is that here the effect of collisions is described via drift and diffusion type processes. Accordingly, a scheme can be derived for which the time step size limitation of the corresponding numerical simulation becomes independent of the Knudsen number. Consequently, this leads to more efficient simulations, especially in low or intermediate Knudsen numbers. Results are presented for helium-argon mixture in a one dimensional geometry. The calculated mixture viscosity is found to be in accordance with experimental data, which reveals the accuracy and relevance of the approach.

1. Introduction

Complexity of the Boltzmann collision operator together with the high dimensionality of its solution domain are the main motivations for devising simpler yet accurate approximations for rarefied gas flows. Among different approaches, Fokker–Planck equations provide computationally attractive kinetic models, which at the same time fulfill the basic properties of the Boltzmann equation, e.g. see [2]. Recently, a Fokker–Planck model for simple monatomic gas flows was proposed in [8], where accurate results were obtained up to intermediate Knudsen numbers. Extensions of the Fokker–Planck model to a nonlinear model equation in [5] resulted in the correct Prandtl number for monatomic molecules. The model was also generalized for simple diatomic gas flows with rotational and vibrational degrees of freedom [6].

In this study, the Fokker–Planck kinetic model is extended in order to include the kinetics of monatomic gas mixtures. The paper starts with the derivation of the model, where appropriate drift and diffusion coefficients are obtained. Then, an energy conserving numerical scheme based on the particle method is derived in §3 . Afterwards, the results are presented for an argon-helium mixture in the hydrodynamic limit considering different mixture fractions. The paper is closed by some concluding remarks.

2. Kinetic Model

Consider a mixture of two gas species α and β with molecular masses m^α and m^β. The statistical state of the mixture is determined by the molecular velocity distribution of each species. Here the probability density function (PDF) $f(\boldsymbol{V};\boldsymbol{X},t)$ defined over the velocity sample space $\boldsymbol{V}$, position $\boldsymbol{X}$ and time t, is employed. In the absence of an external force, the Fokker–Planck system of equations

$$\frac{\partial f^\alpha}{\partial t} + V_i \frac{\partial f^\alpha}{\partial X_i} = -\frac{\partial}{\partial V_i}\left(A_i^\alpha f^\alpha\right) + \frac{\partial^2}{\partial V_i \partial V_i}\left(\frac{D^{\alpha 2}}{2} f^\alpha\right) \tag{1}$$

$$\frac{\partial f^\beta}{\partial t} + V_i \frac{\partial f^\beta}{\partial X_i} = -\frac{\partial}{\partial V_i}\left(A_i^\beta f^\beta\right) + \frac{\partial^2}{\partial V_i \partial V_i}\left(\frac{D^{\beta 2}}{2} f^\beta\right) \tag{2}$$

describe the evolution of species PDFs, i.e. f^α and f^β in time. Note that superscripts α and β relate to gas species α and β, respectively. Also the index summation convention was used throughout this paper. The idea here is to choose the drift coefficients A_i^α and A_i^β and the diffusion coefficients D^α and D^β in a way that the above system become consistent with Boltzmann up to a certain order of velocity moments. First, we propose a linear model for A_i^α and A_i^β and second, a cubic function will be studied.

2.1. Linear Model

The relaxation of the mean species velocities $\boldsymbol{U}^{\alpha,\beta}$, species temperatures $T^{\alpha,\beta}$ and densities $\rho^{\alpha,\beta}$ can be calculated from the Boltzmann equation. Before we proceed note that for the rest of this study we only discuss the equations related to species α. The corresponding equations for the other species β can be obtained by switching the indices α and β in all terms. Given the intermolecular force constant $\chi^{\alpha\beta}$, the Boltzmann collision operator for Maxwell molecules leads to [7]

$$\frac{dU_i^\alpha}{dt} = -A^{\alpha\beta}\rho^\beta\left(U_i^\alpha - U_i^\beta\right) \tag{3}$$

and

$$\frac{dT^\alpha}{dt} = -2\frac{A^{\alpha\beta}\rho^\beta}{1+\sigma^{\alpha\beta}}\left[\sigma^{\alpha\beta}(T^\alpha - T^\beta) - \frac{m^\alpha}{3k}(\boldsymbol{U}^\alpha - \boldsymbol{U}^\beta)^2\right], \tag{4}$$

where the constant $A^{\alpha\beta}$ is given by

$$A^{\alpha\beta} = \frac{8\sqrt{\pi}}{3}\,\Gamma(\frac{5}{2})\,\mathscr{A}_1\left(\frac{(m^\alpha + m^\beta)\chi^{\alpha\beta}}{m^\alpha m^\beta}\right). \tag{5}$$

Here the Boltzmann constant is denoted by k and the mass ratio between two species is given by $\sigma^{\alpha\beta} = m^\alpha/m^\beta$. Also note that $\mathscr{A}_1 \approx 0.422$ is the collision cross section for Maxwel type molecules [3]. In order to construct a consistent Fokker–Planck model for relaxation of mean velocities and temperature, the drift coefficient reads

$$A_i^\alpha = -\overbrace{A^{\alpha\beta}\rho^\beta\left(V_i - U_i^\beta\right)}^{\text{collisions between unlike molecules}} - \overbrace{A^{\alpha\alpha}\rho^\alpha\left(V_i - U_i^\alpha\right)}^{\text{collisions between same molecules}}. \tag{6}$$

Further, based on the energy equation (4) the diffusion coefficient becomes

$$\begin{aligned} D^{\alpha 2} &= \overbrace{2\frac{A^{\alpha\beta}\rho^{\beta}}{1+\sigma^{\alpha\beta}}\left(\frac{kT^{\alpha}}{m^{\alpha}}+\frac{kT^{\beta}}{m^{\beta}}+\frac{1}{3}\left|\boldsymbol{U}^{\alpha}-\boldsymbol{U}^{\beta}\right|^{2}\right)}^{\text{collisions between unlike molecules}} \\ &+ \overbrace{2A^{\alpha\alpha}\frac{\rho^{\alpha}kT^{\alpha}}{m^{\alpha}}}^{\text{collisions between same molecules}} . \end{aligned} \tag{7}$$

Using the above discussed closures for A_i^{α} and D^{α}, consistency between the Fokker–Planck system and the Boltzmann equation can be achieved for the relaxation of mean velocities and energies. Thus e.g. the diffusion property of a binary gas mixture can be calculated accurately using the Fokker–Planck model. However, still the viscosity and heat conductivity of the mixture calculated from the model may deviate from those obtained from Boltzmann. This is the main intention for the cubic model, which is discussed in the following sub-section.

2.2. Cubic Model

The idea of this part is to employ a cubic ansatz for the drift coefficient in order to match the evolution of the pressure tensor and the heat flux vector resulting from the Fokker-Planck model with those resulting from the Boltzmann equation. Therefore, similar to the monatomic model of [5], the following expression is adopted for the drift coefficient

$$A_i^{\alpha} = b_i^{\alpha} + c_{ij}^{\alpha}V_j^{\alpha\prime} + \gamma_i^{\alpha}\left(V_j^{\alpha\prime}V_j^{\alpha\prime} - \frac{3kT^{\alpha}}{m^{\alpha}}\right) + \Lambda^{\alpha}\left(V_i^{\alpha\prime}V_j^{\alpha\prime}V_j^{\alpha\prime} - 2q_i^{\alpha}\right), \tag{8}$$

where $\boldsymbol{V}^{\alpha\prime} = \boldsymbol{V} - \boldsymbol{U}^{\alpha}$ is the fluctuating velocity of species α. Here the model constants b_i^{α}, c_{ij}^{α} and γ_i^{α} should be chosen such to enforce consistency between the model and Botlzmann. The constant Λ^{α} is responsible to keep the drift coefficient stable [5]. The constant term b_i^{α} can be found from the linear model to be

$$b_i^{\alpha} = -A^{\alpha\beta}\rho^{\beta}\left(U_i^{\alpha} - U_i^{\beta}\right), \tag{9}$$

while the constants c_{ij}^{α} and γ_i^{α} should be found from relaxations of the pressure tensor and the heat flux vector. Consider $p_{ij}^{\alpha,\beta}$ and $q_i^{\alpha,\beta}$ as pressure tensor and heat flux of gas species, thus the relaxation equation for the pressure tensor takes form

$$\begin{aligned} \frac{dp_{ij}^{\alpha}}{dt} &= p_{ij}^{\alpha}\rho^{\beta}\left(-2m_0^{\alpha}A^{\alpha\beta} - 3m_0^{\beta}B^{\alpha\beta}\right) + p_{ij}^{\beta}\rho^{\alpha}\left(2m_0^{\beta}A^{\alpha\beta} - 3m_0^{\beta}B^{\alpha\beta}\right) \\ &+ \frac{3k(T^{\alpha}+T^{\beta})}{m_{\alpha}+m_{\beta}}\rho_{\alpha}\rho_{\beta}B^{\alpha\beta}\delta_{ij} \\ &+ m_0^{\beta}\rho^{\alpha}\rho^{\beta}\left(|\boldsymbol{U}^{\alpha}-\boldsymbol{U}^{\beta}|^2B^{\alpha\beta}\delta_{ij} + (U_i^{\alpha}-U_i^{\beta})(U_j^{\alpha}-U_j^{\beta})(-3B^{\alpha\beta}+2A^{\alpha\beta})\right) \\ &- 3p_{ij}^{\alpha}\rho^{\alpha}B^{\alpha\alpha} + \frac{3\rho^{\alpha}kT^{\alpha}}{m^{\alpha}}\rho^{\alpha}B^{\alpha\alpha}\delta_{ij} \end{aligned} \tag{10}$$

and for the heat fluxes, the relaxation becomes

$$\begin{aligned}
\frac{dq_i^\alpha}{dt} &= -q_i^\alpha \rho^\beta \left((3m_0^{\alpha 2} + m_0^{\beta 2}) A^{\alpha\beta} + 4 m_0^\alpha m_0^\beta B^{\alpha\beta} \right) + 4 q_i^\beta \rho^\alpha m_0^{\beta 2} \left(A^{\alpha\beta} - B^{\alpha\beta} \right) \\
&\quad + 4 m_0^{\beta 2} \left(A^{\alpha\beta} - B^{\alpha\beta} \right) \left(U_j^\alpha - U_j^\beta \right) \left(\rho^\beta p_{ij}^\alpha + \rho^\alpha p_{ij}^\beta \right) \\
&\quad - \left(A^{\alpha\beta} (m_0^\beta - m_0^\alpha) - m_0^\beta B^{\alpha\beta} \right) \left(U_j^\alpha - U_j^\beta \right) \rho^\beta p_{ij}^\alpha \\
&\quad + 2 \rho^\alpha \rho^\beta m_0^{\beta 2} \left(A^{\alpha\beta} - B^{\alpha\beta} \right) \left(U_i^\alpha - U_i^\beta \right) \left(\left| \boldsymbol{U}^\alpha - \boldsymbol{U}^\beta \right|^2 + 3k \left(\frac{T^\alpha}{m^\alpha} + \frac{T^\beta}{m^\beta} \right) \right) \\
&\quad - \frac{1}{2} \rho^\alpha \rho^\beta \left(A^{\alpha\beta} (4 m_0^\beta - 1) - 6 m_0^\beta B^{\alpha\beta} \right) \left(U_i^\alpha - U_i^\beta \right) \frac{3kT^\alpha}{m^\alpha} \\
&\quad - 2 q_i^\alpha \rho^\alpha B^{\alpha\alpha}, \qquad (11)
\end{aligned}$$

where $B^{\alpha\beta} = (\mathscr{A}_2/\mathscr{A}_1) A^{\alpha\beta}$ and $\mathscr{A}_2 \approx 0.436$ as shown in [3]. Here the reduced masses are denoted by $m_0^\alpha = m^\alpha/(m^\alpha + m^\beta)$ and $m_0^\beta = m^\beta/(m^\alpha + m^\beta)$ and δ_{ij} is the Kronecker delta. In order to facilitate the references to equations (10) and (11), the abbreviations P_{ij}^α referring the right hand side (r.h.s.) of (10) and P_i^α referring the r.h.s. of (11) are used. The model coefficients c_{ij}^α and γ_i^α have to be evaluated from the following moment relations

$$-\rho^\alpha \int_{\mathbb{R}^3} \left(A_i^\alpha V_j^{\alpha\prime} + A_j^\alpha V_i^{\alpha\prime} + D^{\alpha 2} \delta_{ij} \right) f^\alpha d\boldsymbol{V} = P_{ij}^\alpha \qquad (12)$$

and

$$-\frac{\rho^\alpha}{2} \int_{\mathbb{R}^3} \left(A_i^\alpha V_j^{\alpha\prime} V_j^{\alpha\prime} + 2 A_j^\alpha V_j^{\alpha\prime} V_i^{\alpha\prime} \right) f^\alpha d\boldsymbol{V} = P_i^\alpha. \qquad (13)$$

Note that the density ρ^α can be calculated from the conservation of mass. At this point, the system of equations (1) and (2) becomes closed, where the transport properties of the mixture are enforced to be the same as those resulting from Boltzmann.

3. Numerical Scheme

Although the system of equations (1) and (2) provide a closed model for simulations of binary gas mixtures, the high dimensionality of the corresponding solution domain makes its direct solution very expensive. To cope with that the system of partial differential equations (1) and (2) is transformed into Itō type stochastic differential equations (SDEs) [4]. Let the velocity and position of representative particles be $\boldsymbol{M}(t)$ and $\boldsymbol{x}(t)$, respectively. Therefore the state of species α should evolve according to

$$dM_i^\alpha = A_i^\alpha dt + D^\alpha dW_i, \qquad (14)$$

where the increments dW_i are independent Gaussian numbers with zero mean and variance equal to dt. Consequently the position of the particles follows

$$dx_i^\alpha = M_i^\alpha dt. \qquad (15)$$

Note that the system of SDEs (14) and (15) is equivalent to the system of PDEs (1) and (2). Since the above SDEs are continuous in time, it is possible to devise very efficient particle based solution algorithms [8]. For that reason a Monte-Carlo particle scheme is adopted. However, unlike in DSMC [1], here the simulated particles do not collide with each other but follow the stochastic trajectories along $\boldsymbol{x}(t)$. Therefore the collisional scales do not need to be resolved.

For simplicity we assume that $U_i^\alpha = 0$ and $U_i^\beta = 0$ in the following. Let us decompose the r.h.s. of (14) into a linear isotropic part L_i and a remainder N_i

$$dM_i^\alpha = \overbrace{-A^{\alpha\beta}\rho^\beta M_i^\alpha dt - A^{\alpha\alpha}\rho^\alpha M_i^\alpha dt + D^\alpha dW_i}^{L_i} + N_i dt \tag{16}$$

Considering only the operator L_i, the time integration can be performed analytically (for frozen coefficients) [8, 4]. Using the corresponding analytical solution together with an appropriate time integration for the nonlinear part, M_i^α at the new time $t^{n+1} = t^n + \Delta t$ can be computed as

$$M_i^{\alpha,n+1} = \mathrm{e}^{-\nu\Delta t} M_i^{\alpha,n} + \sqrt{S^\alpha}\xi_i + N_i\ \Delta t\ \mathrm{e}^{-\nu\Delta t}, \tag{17}$$

where the frequency $\nu = A^{\alpha\beta}\rho^\beta + A^{\alpha\alpha}\rho^\alpha$ and the diffusion factor $S^\alpha = D^\alpha(1 - \mathrm{e}^{-2\nu\Delta t})/(2\nu)$ are employed. Note that here ξ_i are independent normal variables with zero mean and unity variance. It can easily be shown that in the limit of $\nu \to \infty$ or $\Delta t \to \infty$ a solution with the Gaussian distribution is obtained. However, because of the error resulting from the approximate time integration of the nonlinear term N_i, the energy of the system does not evolve correctly. Therefore the solution needs to be scaled by a correction factor f, i.e.

$$M_i^{\alpha,n+1} = f\left(\mathrm{e}^{-\nu\Delta t} M_i^{\alpha,n} + \sqrt{S^\alpha}\xi_i + N_i\Delta t\mathrm{e}^{-\nu\Delta t}\right). \tag{18}$$

In order to find the factor f, the energy of the ensemble of the whole mixture should remain unchanged after the time interval Δt. Therefore, one obtains

$$f = \sqrt{\frac{\mathrm{e}^{-2\nu\Delta t} kT^\alpha/m^\alpha + S^\alpha}{\mathrm{e}^{-2\nu\Delta t} kT^\alpha/m^\alpha + S^\alpha + \Delta t^2\ \mathrm{e}^{-2\nu\Delta t}\ \mathrm{Var}(\boldsymbol{N})/3}}, \tag{19}$$

where $\mathrm{Var}(\boldsymbol{N})$ denotes the variance of the nonlinear term N_i.
The position of a particle should evolve consistently with equation (15). An appropriate numerical scheme to update the positions of the particles was proposed in [8], and is also used for this study.

4. Results

For validation of the model, an Ar-He mixture was studied in the hydrodynamic limit. The problem is a planar supersonic Couette channel flow and the flow is driven by motion of two parallel walls with respect to each other. The distance between the walls is $L = 1\ m$, which are considered to be diffusive and isothermal with temperature $T_w = 273\ K$ and move with the velocities $U_{up} = 500\ m/s$ and $U_{down} = -500\ m/s$. For the simulations approximately 10000 particles have been used for each species, whereas the time weighted averaging was adopted for reducing statistical errors. The cross stream direction was discretized into 100 cells and flow in the direction parallel to the walls was assumed to be periodic. Therefore in streamwise direction the periodic boundary condition was employed. After reaching steady-state, the viscosity of the mixture was calculated from the shear stress and velocity gradient of the fluid. Three different molar fractions of helium were considered, i.e. $\eta = 0\%$, $\eta = 19.9\%$ and $\eta = 37.7\%$. The number density of argon molecules was set to $1.4 \times 10^{20}\ m^{-3}$ and for helium it was calculated from the mixture fraction. Thus the Knudsen number for all cases were around 0.01. The viscosity coefficient of the mixture was found to be 2.22×10^{-5} and 2.24×10^{-5} for $\eta = 0\%$ and $\eta = 19.9\%$, respectively, which are in a very good agreement with experimental results of 2.227×10^{-5} and 2.270×10^{-5} [3]. It can be seen that the model correctly predicts the viscosity increase of the mixture for a moderate helium-fraction. Different profiles for the mixture velocity and temperature are shown in figures 1-2.

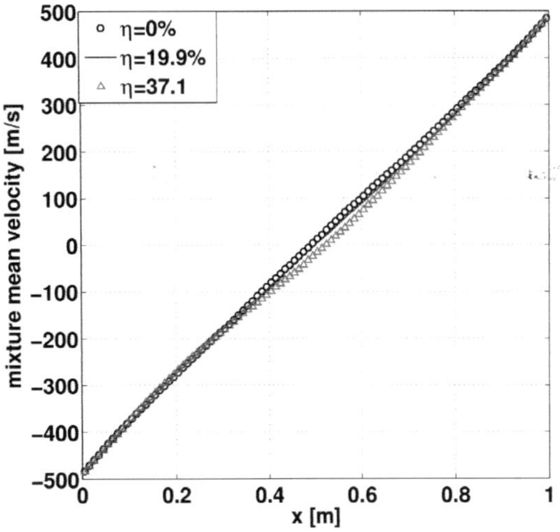

Figure 1: Supersonic Couette flow: depicted is mean velocity of the Ar-He mixture for different molar fractions, i.e. $\eta = 0\%$, $\eta = 19.9\%$ and $\eta = 37.1\%$ along the cross stream direction x.

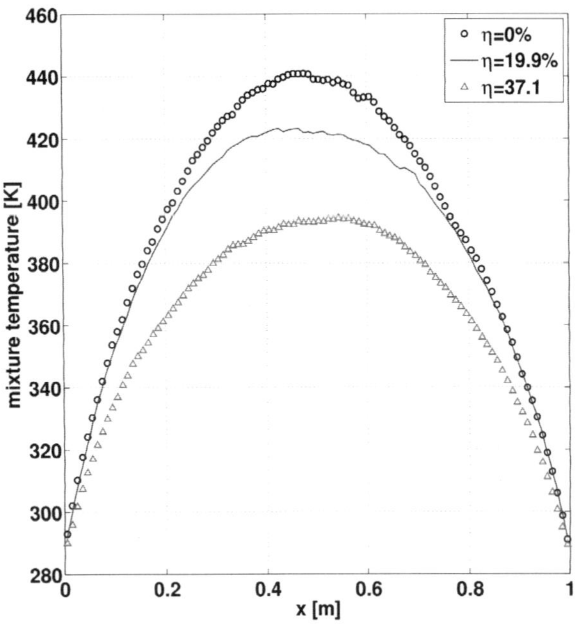

Figure 2: Supersonic Couette flow: depicted is temperature of the Ar-He mixture for different molar fractions, i.e. $\eta = 0\%$, $\eta = 19.9\%$ and $\eta = 37.1\%$ along the cross stream direction x.

5. Conclusion

In this study a kinetic model based on a Fokker–Planck approximation of the Boltzmann equation was proposed for simulations of rarefied gas mixtures. The model was constructed such that the evolution of moments up to heat fluxes becomes identical with that by Boltzmann. An accurate Monte–Carlo particle based numerical method was then devised, which conserves the total energy of the mixture. It was shown that the scheme does not need to resolve the collisional scales and therefore the computational cost of the method is independent of the Knudsen number. Results were shown for simple supersonic Couette flow of Ar-He mixtures in the hydrodynamic limit. The agreement with experiments was found to be good. More challenging test cases will be studied in the future.

References

[1] Bird, G. 1994 *Molecular Gas Dynamics and the Direct Simulation of Gas Flows.* Calerndon.

[2] Cercignani, C. 1988 *The Boltzmann Equation and Its Applications.* Springer-Verlag, New York.

[3] Chapman, S. & Cowling, T. G. 1970 *The Mathematical Theory of Non-uniform Gases: An Account of the Kinetic Theory of Viscosity, Thermal Conduction and Diffusion in Gases.* Cambridge University Press.

[4] Gardiner, C. W. 1985 *Handbook of Stochastic Methods.* Springer-Verlag.

[5] Gorji, M.H., Torrilhon, M. & Jenny, P. 2011 Fokker–Planck Model for Computational Studies of Monatomic Rarefied Gas Flows. *J. Fluid Mech.* **680**, 574-601.

[6] Gorji, M.H. & Jenny, P. 2011 Continuous Stochastic Model for Diatomic Rarefied Gas Flows. *J. Comput. Phys.* (submitted)

[7] Goldman, E. & Sirovich, L. 1967 Equations for Gas Mixtures *Phys. Fluids* **10**, 1928-1940

[8] Jenny, P., Torrilhon, M., & Heinz, S. 2010 A solution algorithm for the fluid dynamics equations based on a stochastic model for molecular motion. *J. Comput. Phys.* **229**, 1077-1098.

Thermal Lattice Boltzmann models derived by Gauss quadrature using the spherical coordinate system

Victor Eugen Ambrus[1,2] **and Victor Sofonea**[3]

[1] Department of Physics, West University of Timişoara, Bd. Vasile Pârvan 4, RO - 300233 Timişoara, Romania

[3] Center for Fundamental and Advanced Technical Research, Romanian Academy, Bd. Mihai Viteazul 24, RO - 300223 Timişoara, Romania

E-mail: victor.ambrus@gmail.com, sofonea@acad-tim.tm.edu.ro

Abstract. A hierarchy of thermal Lattice Boltzmann models is derived by separation of variables using the spherical coordinate system in the momentum space. The moments of the equilibrium distribution function are computed by means of Gauss-Legendre and Gauss-Laguerre quadratures. This procedure allows us to find the discrete momentum vectors for each model in the hierarchy. The Shakov collision term is used to get the right value of the Prandtl number. Computer simulation of Couette flow is used to illustrate the capability of these models to capture specific effects in microfluidics.

1. Introduction

Lattice Boltzmann (LB) models are derived from the Boltzmann equation using a simplified version of the collision operator. These models provide an alternative to the simulation techniques of Computational Fluid Dynamics (CFD) or Direct Simulation Monte Carlo (DSMC) and have been recognized as efficient tools for the investigation of single- or multi-component complex fluids with or without structure formation, as well as for the investigation of microfluidics problems [1, 2, 3].

At their beginning (more than two decades ago), the LB models were designed to retrieve the Navier-Stokes equation in the incompressible limit [1] by using a discrete set of velocities in the two- (2D) or three-dimensional (3D) space. Later, the LB models were rigorously derived by using the Chapman-Enskog method, as well as the Gauss-Hermite quadrature after separation of variables in the Cartesian coordinate system [4]. This way, the LB models approach the discrete velocity method (DVM) well known in kinetic theory of rarefied gases [5, 6].

As the Knudsen number increases and the fluid system goes farther away from equilibrium, higher order moments of the equilibrium distribution function need to be considered in order to account for the complexity of flow phenomena when using the Chapman-Enskog method. As pointed in the literature [4, 7, 8, 9, 10, 11, 12, 13, 14, 15, 16, 17], higher order moments of the equilibrium distribution function can be achieved in LB models by increasing the number of velocity vectors in the discrete set. Since the number of vectors in the discrete set increases monotonically with the order of the Gauss-Hermite quadrature, the corresponding LB models

[2] Present address: School of Mathematics and Statistics, The University of Sheffield, Hicks Building, Hounsfield Road, Sheffield, S3 7RH, United Kingdom

form a hierarchy. Higher order moments of the equilibrium distribution function are successively satisfied when increasing the position of a LB model in this hierarchy.

The LB models currently reported in the literature are mainly isothermal or of low order. High order LB models with variable temperature are computationally expensive and were not considered until recently [17, 18, 19, 20]. Application of the Gauss-Hermite LB models to investigate gas flow phenomena at non-negligible values of the Knudsen number is very encouraging [17, 20]. In these models, the Cartesian coordinate system is used in the momentum space. The purpose of this contribution is to introduce a new class of LB models in 3D, which are derived using the spherical coordinate system in the momentum space, as well as the Gauss-Laguerre and Gauss-Legendre quadratures.

All quantities of interest in the LB models introduced in this paper are made dimensionless by using three basic reference quantities [21]: a characteristic length l_R, a reference mass m_R and a reference energy $e_R = k_B T_R$, where k_B is Boltzmann's constant and T_R is a reference temperature. Other references are derived from the basic quantities: particle number density $n_R = 1/l_R^3$, speed $c_R = \sqrt{k_B T_R/m_R}$, time l_R/c_R, and so on.

2. Description of the model

We take advantage of the *spherical shell* models introduced by Watari and Tsutahara [22, 23, 24, 25, 26] and build a hierarchy of three-dimensional thermal Lattice Boltzmann models by using the spherical coordinate system in the momentum space. According to the discretization procedure [8], the continuum momentum space is now replaced by a discrete set $\{\boldsymbol{p}_\kappa\}$ having $\mathcal{N}$ elements. The corresponding density functions and equilibrium density functions are denoted f_κ and $f_\kappa^{(eq)}$, respectively.

The highest natural number N for which all tensors of order s $(0 \leq s \leq N)$

$$\widetilde{\mathcal{M}}^{(s)}_{\{\alpha_l\}} \equiv \widetilde{\mathcal{M}}^{(s)}_{\alpha_1\alpha_2\ldots\alpha_s} = \sum_\kappa f_\kappa^{(eq)} \prod_{l=1}^{s} p_{\kappa\alpha_l} \tag{1}$$

equal the corresponding ones in the continuum space [8]

$$\mathcal{M}^{(s)}_{\{\alpha_l\}} \equiv \mathcal{M}^{(s)}_{\alpha_1\alpha_2\ldots\alpha_s} = \int d^D p f^{(eq)} \prod_{l=1}^{s} p_{\alpha_l} \tag{2}$$

defines the order of the LB model in the hierarchy ($p_{\kappa\alpha_l}$ and p_{α_l}, $\alpha_l \in \{1,\, 2,\, 3\}$, are the Cartesian components of the vectors $\boldsymbol{p}_\kappa$ and $\boldsymbol{p}$, respectively). For given N, the momenta $\boldsymbol{p}_\kappa$, as well as the expression of the equilibrium density functions $f_\kappa^{(eq)}$ can be established by means of series expansion and Gauss quadrature.

The equilibrium distribution function in the continuum space, i.e., the Maxwell - Boltzmann distribution function, is given by [1, 27, 28]

$$f^{(eq)} \equiv f^{(eq)}(\boldsymbol{p}; n, \boldsymbol{u}, T) = n(\beta/\pi)^{D/2} e^{-\beta(\boldsymbol{p}-m\boldsymbol{u})^2} \tag{3}$$

where $n = \mathcal{M}^{(0)}$ is the particle number density, $u_\alpha = \mathcal{M}^{(1)}_\alpha/mn$ is the fluid velocity and $T = \delta_{\alpha_1\alpha_2}\mathcal{M}^{(2)}_{\alpha_1\alpha_2}/Dmn - mu^2/D$ is the fluid temperature, respectively ($\beta = 1/2mT$, $D = 3$, m is the mass of fluid particles and $\delta_{\alpha_1\alpha_2}$ is Kronecker's symbol; the sum rule over repeated α_l indices is understood). Since mass, momentum and energy are conserved during fluid particle collisions, the local quantities n, u_α and T are also recovered by computing the corresponding moments of f_κ and f [1, 8, 13, 27, 28]. Splitting Eq.(3) yields ($p \equiv \sqrt{\boldsymbol{p}^2}$)

$$f^{(eq)} \equiv f^{(eq)}(\boldsymbol{p}; n, \boldsymbol{u}, T) \quad = \quad nF(p^2; T) E(\boldsymbol{p}; \boldsymbol{u}, T) \tag{4}$$

$$F \equiv F(p^2;T) = (\beta/\pi)^{D/2} e^{-\beta p^2} \tag{5}$$
$$E \equiv E(\boldsymbol{p};\boldsymbol{u},T) = e^{-\beta(m^2\boldsymbol{u}^2 - 2m\boldsymbol{p}\boldsymbol{u})} \tag{6}$$

According to [8], we have

$$Ep_\alpha = (T\partial_{u_\alpha} + mu_\alpha)E \tag{7}$$

This means that the sth order moment

$$\mathcal{M}^{(s)}_{\alpha_1\alpha_2\ldots\alpha_s} = \left[\prod_{l=1}^{s}\left(T\partial_{u_{\alpha_l}} + mu_{\alpha_l}\right)\right] n \tag{8}$$

is a polynomial of order s in the fluid velocity $\boldsymbol{u}$. Hence Eq. (2) is preserved for $0 \leq s \leq N$ when replacing E in (4) with its series expansion with respect to $\boldsymbol{u}/\sqrt{T/m}$ up to order N:

$$E^{(N)} = \sum_{j=0}^{\lfloor N/2 \rfloor} \frac{1}{j!}\left(-\frac{m\boldsymbol{u}^2}{2T}\right)^j \sum_{r=0}^{N-2j} \frac{1}{r!}\left(\frac{\boldsymbol{p}\boldsymbol{u}}{T}\right)^r \tag{9}$$

($\lfloor x \rfloor$ is the largest integer less than or equal to x).

When using the spherical coordinate system (p,θ,φ) in the momentum space, such that $p_\alpha = pe_\alpha(\theta,\varphi)$, $e_1(\theta,\varphi) = \sin\theta\cos\varphi$, $e_2(\theta,\varphi) = \sin\theta\sin\varphi$, $e_3(\theta,\varphi) = \cos\theta$, we get $E^{(N)} \equiv E^{(N)}(p,\theta,\varphi;\boldsymbol{u},T)$ and

$$\mathcal{M}^{(s)}_{\{\alpha_l\}} = n\int_0^\infty dp\, p^{D-1} F \int_{-1}^{+1} d(\cos\theta) \int_0^{2\pi} d\varphi\, \mathcal{S}^{(s)}_{\{\alpha_l\}} \tag{10}$$

where $\mathcal{S}^{(s)}_{\{\alpha_l\}} \equiv \mathcal{S}^{(s)}_{\{\alpha_l\}}(\theta,\varphi) = p^s E^{(N)} \prod_{l=1}^{s} e_{\alpha_l}$ is a trigonometric polynomial [29]:

$$\mathcal{S}^{(s)}_{\{\alpha_l\}} = \sum_{0\leq\mu+\nu+\lambda\leq N+s} A_{\lambda,\mu,\nu}(p,\boldsymbol{u}) \times (\sin\theta)^{\mu+\nu}(\cos\theta)^\lambda(\sin\varphi)^\mu(\cos\varphi)^\nu \tag{11}$$

According to [30, 31], the following quadrature formula

$$\mathcal{Q}^{(s)}_{\{\alpha_l\}} - \int_0^{2\pi} d\varphi\, \mathcal{S}^{(s)}_{\{\alpha_l\}}(\theta,\varphi) = \frac{2\pi}{M}\sum_{i=1}^{M} \mathcal{S}^{(s)}_{\{\alpha_l\}}(\theta,\varphi_i) \tag{12}$$

with $\varphi_i = \phi + 2\pi(i-1)/M$ and ϕ an arbitrary angle, is exact for any integer $M \geq 2N+1$. Terms with odd $(\mu+\nu)$ in $\mathcal{S}^{(s)}_{\{\alpha_l\}}$ do not contribute to the integral with respect to φ. The even powers of $\sin\theta$ are easily changed in powers of $\cos\theta$ and $\mathcal{Q}^{(s)}_{\{\alpha_l\}}$ is eventually expressed as a polynomial of order at most $2N$ in $z = \cos\theta$ [29]. The *Gauss-Legendre* quadrature [31, 32] gives

$$\mathcal{E}^{(s)}_{\{\alpha_l\}} = \int_{-1}^{+1} dz\, \mathcal{Q}^{(s)}_{\{\alpha_l\}} = \sum_{j=1}^{L} w_j^{(P)} \mathcal{Q}^{(s)}_{\{\alpha_l\}}(z_j) \tag{13}$$
$$w_j^{(P)} = \frac{2(1-z_j^2)}{(L+1)^2[P_{L+1}(z_j)]^2} \tag{14}$$

where z_j, $1 \leq j \leq L$ are the roots of the Legendre polynomial $P_L(z)$ of order $L \geq N+1$. Because of the symmetry of the integration domain, odd powers of $z = \cos\theta$ do not contribute

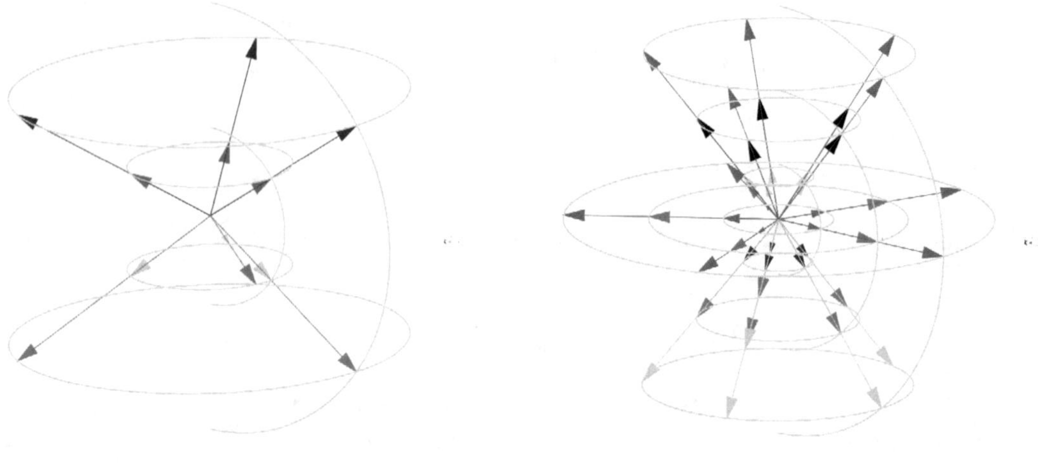

Figure 1. First two members of the SLB family: $SLB(1,2,2,3)$ and $SLB(2;3,3,5)$.

to $\mathcal{E}^{(s)}_{\{\alpha_l\}}$. Thus, the appropriate use of quadrature formulae during integration over the angular coordinates (θ,φ) gives us the polynomial $\mathcal{E}^{(s)}_{\{\alpha_l\}} \equiv \mathcal{E}^{(s)}_{\{\alpha_l\}}(p^2;\boldsymbol{u},T)$ containing only even powers of p [29]

$$\mathcal{E}^{(s)}_{\{\alpha_l\}} = \frac{2\pi}{M}\sum_{j=1}^{L}\sum_{i=1}^{M} w_j^{(P)} p^s E^{(N)}(p,\theta_j,\varphi_i;\boldsymbol{u},T)\prod_{l=1}^{s} e_{ji\alpha_l} \tag{15}$$

where $e_{ji\alpha_l} \equiv e_{\alpha_l}(\theta_j,\varphi_i)$, $\theta_j \equiv \arccos(z_j)$ and $\varphi_i = \phi + 2\pi(i-1)/M$. This allows us to use the *Gauss-Laguerre* quadrature [31, 32] after changing the integration variable from p to $x \equiv p^2$ in Eq. (10). Since

$$\mathcal{M}^{(s)}_{\{\alpha_l\}} = \frac{n}{2}\int_0^\infty dx\, x^{1/2} e^{-x}\mathcal{F}(x;T)\mathcal{E}^{(s)}_{\{\alpha_l\}}(x;\boldsymbol{u},T) \tag{16}$$

contains the term $x^{1/2}$, we expand $\mathcal{F}(x;T) \equiv e^x F(x;T)$ with respect to the generalized Laguerre polynomials $L_\ell^{(1/2)}(x)$:

$$\mathcal{F}(x;T) = \pi^{-3/2}\sum_{\ell=0}^{K-1}(1-2mT)^\ell L_\ell^{(1/2)}(x) \tag{17}$$

The order $K-1$ of this expansion must be greater than or equal to N because $\mathcal{E}^{(s)}_{\{\alpha_l\}}(x;\boldsymbol{u},T)$ is a polynomial of order at most N in x and hence is orthogonal to all $L_\ell^{(1/2)}(x)$ of order $\ell > N$. Application of the Gauss-Laguerre quadrature rule gives [29]

$$\mathcal{M}^{(s)}_{\{\alpha_l\}} = \frac{n}{2}\sum_{k=1}^{K} w_k^{(L)}\mathcal{F}(x_k;T)\mathcal{E}^{(s)}_{\{\alpha_l\}}(x_k;\boldsymbol{u},T) \tag{18}$$

$$w_k^{(L)} = \frac{x_k\Gamma(K+3/2)}{K!(K+1)^2[L_{K+1}^{(1/2)}(x_k)]^2} \tag{19}$$

where x_k, $1<k<K$, $K>N+1$ are the roots of the generalized Laguerre polynomial $L_K^{(1/2)}(x)$, $w_k^{(L)}$ are the quadrature weights and $\Gamma(x)$ is the factorial function [32].

Comparison of Eqs. (1) and (18) gives the expression of the equilibrium distribution functions $f_\kappa^{(eq)} \equiv f_{kji}^{(eq)}$ in the LB model of order N with Gauss quadrature in spherical coordinates [29], as well as the Cartesian components $\boldsymbol{p}_{\kappa\alpha} \equiv \boldsymbol{p}_{kji\alpha}$ ($\alpha = 1, 2, 3$) of the corresponding momentum vectors $\boldsymbol{p}_{kji}$ ($1 \leq k \leq N+1$, $1 \leq j \leq K$, $1 \leq i \leq M$):

$$\begin{aligned} f_{kji}^{(eq)} &= nF_k E_{kji} \\ F_k &= \frac{w_k^{(L)}}{M\sqrt{\pi}} \sum_{\ell=0}^{K-1} (1-2mT)^{\ell} L_{\ell}^{(1/2)}(p_k^2) \\ E_{kji} &= w_j^{(P)} E^{(N)}(p_k, \theta_j, \varphi_i; \boldsymbol{u}, T) \\ p_{kji\alpha} &= p_k e_{ji\alpha} \quad , \quad p_k = \sqrt{x_k} \end{aligned} \tag{20}$$

The members of this family (hierarchy) of *spherical shell* LB models will be denoted SLB($N; K, L, M$). More details on the derivation of these models can be found in [29]. The first two members of the SLB family are shown in Figure 1.

3. Preliminary results

To illustrate the capability of the SLB models introduced in this paper to capture the physics of flow phenomena at various values of Knudsen number, we consider the problem of Couette flow between parallel plates perpendicular to the z axis and located at $z_b = -0.5$ and $z_t = 0.5$, respectively. The plates have equal temperatures $T(x, y, z_b) = T(x, y, z_t) = T_w = 1.0$ and move in opposite directions along the x axis with speed $u_w = 0.63$ (as stated in Section 2, all quantities of interest are non-dimensionalized). To ensure the right value of Prandtl number for the ideal gas ($\mathrm{Pr} = 2/3$), we use the Shakov collision term [33, 34, 35, 36] in the evolution equation of the distribution functions, instead of the BGK collision term [1]:

$$\partial_t f_{kji} + \frac{p_{kji\alpha}}{m} \partial_\alpha f_{kji} = -\frac{1}{\tau}\left[f_{kji} - f_{kji}^{(eq)}(1 + S_{kji})\right] \tag{21}$$

where $\tau = \mathrm{Kn}/n$ [27] and

$$S_{kji} = \frac{1-\mathrm{Pr}}{nT^2} q_\alpha (p_{kji\alpha} - mu_\alpha)\left[\frac{(\boldsymbol{p}_{kji} - m\boldsymbol{u})^2}{(D+2)mT} - 1\right] \tag{22}$$

$$q_\alpha = \frac{1}{2m^2} \sum_{k,j,i} f_{kji}(p_{kji\alpha} - mu_\alpha)(\boldsymbol{p}_{kji} - m\boldsymbol{u})^2 \tag{23}$$

The evolution equations (21) are numerically solved by projecting the discrete moments $\boldsymbol{p}_{kji}$ on the Cartesian axes and using the MCD flux limiter scheme [17, 29, 37, 38, 39]. Diffuse reflection boundary conditions were implemented on the walls [3, 17, 21, 29, 39]. We used a cubic lattice with 100 nodes in the z direction and 2 nodes in the x and y directions where periodic conditions apply. The lattice spacing had the value $\delta s = 1/100$ and the time step was set to $\delta t = 10^{-4}$.

Simulation results recovered with SLB(6; 8, 8, 13) for two values of Kn (0.01 and 0.10) are compared to DSMC resuls for hard sphere molecules [40, 41, 42, 43] in Figures 2 and 3. The Knudsen layer, where the velocity profile is nonlinear and a longitudinal heat flux is generated, is well captured by the *SLB* models up to Kn=0.10, despite the large value of u_w. Differences between the SLB temperature profiles and the corresponding DSMC results were observed for larger values of Kn [29].

4. Conclusion

The spherical coordinate system may be used in the momentum space in order to derive a hierarchy of Lattice Boltzmann models $SLB(N; K, L, M)$ based on Gauss-Laguerre and Gauss-Legendre quadratures. These models ensure the recovery of all moments of the equilibrium

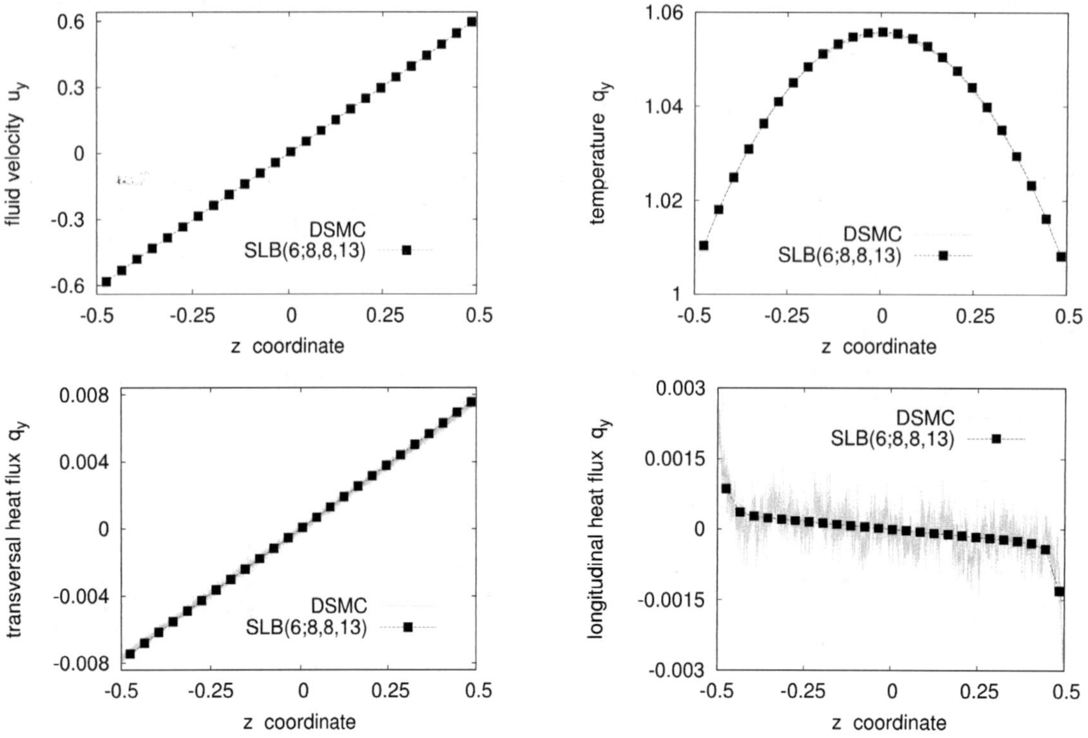

Figure 2. Stationary profiles (velocity, temperature and heat fluxes) across the channel in Couette flow at Kn=0.01 ($T_w = 1.0$; $u_w = 0.63$).

distribution function up to order N. Couette flow simulation was used to validate these models and the preliminary computer results fit well to DSMC simulations up to Kn=0.1. Further investigations need to be performed in the future in order to explore the capabilities of the SLB models, as well as to compare them with the existing LB models in the literature.

Acknowledgments

This work was supported by CNCSIS-UEFISCSU project number PNII-IDEI code ID_76/2010. The authors are grateful to Professor Henning Struchtrup (Department of Mechanical Engineering, University of Victoria, Canada) for kindly providing the DSMC simulation results used to validate the SLB models introduced in this paper. The computer simulations were done on the IBM Blue Gene / P system at the West University of Timişoara using the Portable Extensible Toolkit for Scientific Computation (PETSc 3.1) developed ar Argonne National Laboratory, Argonne, Illinois [44]. V.S. would like to express his gratitude to Professor Michihisa Tsutahara (Department of Mechanical Engineering, Kobe University, Japan) and to Professor Minoru Watari (LBM Fluid Dynamics Laboratory, Gifu, Japan) for fruitful discussions.

References

[1] Succi S 2001 *The Lattice Boltzmann Equation for Fluid Dynamics and Beyond* (Oxford: Clarendon Press)
[2] Aidun C K and Clausen J R 2010 *Annu. Rev. Fluid. Mech.* **42** 439
[3] Ansumali S and Karlin I V 2002 *Phys. Rev.* E **66** 026311
[4] Shan X W, Yuan X F and Chen H D 2006 *J. Fluid Mech.* **550** 413
[5] Broadwell J E 1964 *J. Fluid Mech.* **19** 401

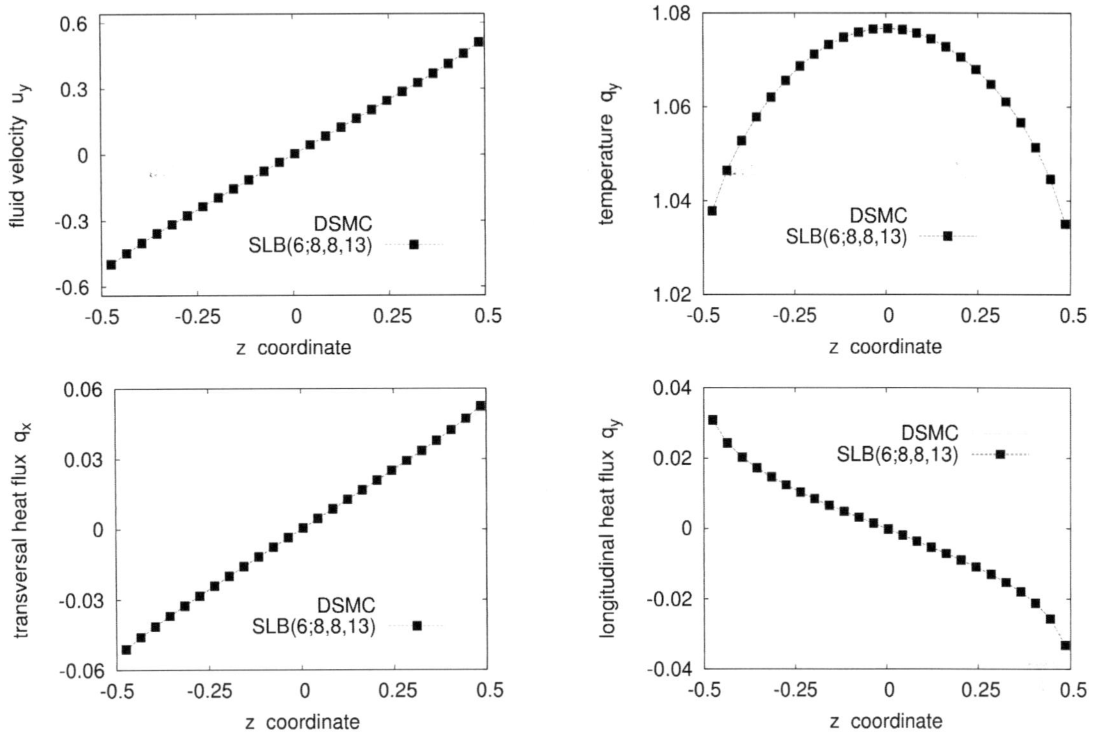

Figure 3. Stationary profiles (velocity, temperature and heat fluxes) across the channel in Couette flow at Kn=0.10 ($T_w = 1.0$; $u_w = 0.63$).

[6] Abe T 1997 *J. Comput. Phys.* **131** 241
[7] Kim S H, Pitsch H and Boyd I D 2008 *J. Comput. Phys.* **227** 8655
[8] Chen H D and Shan X W 2008 *Physica* D **237** 2003
[9] Philippi P C, Hegele Jr. L A, dos Santos L O E and Surmas R 2006 *Phys. Rev.* E **73** 056702
[10] Surmas R, Pico Ortiz C E and Philippi P C 2009 *Eur. Phys. J. Special Topics* **171** 81
[11] Piaud B, Blanco S, Fournier R and Clifton M J 2005 *J. Stat. Phys.* **121** 119
[12] He X Y and Luo L S 1997 *Phys. Rev.* E **56** 6811
[13] Shan X W and He X Y (1998) *Phys. Rev. Lett.* **80** 65
[14] Zhang R Y, Shan X W and Chen H D 2006 *Phys. Rev.* E **74** 046703
[15] Chen H D, Goldhirsch I and Orszag S A 2008 *J. Sci. Comput* **34** 87
[16] Shan X W 2010 *Phys. Rev.* E **81** 036702
[17] Meng J P and Zhang Y H 2011 *Phys Rev.* E **83** 036704
[18] Yudistiawan W P, Kwak S K, Patil D V and Ansumali S 2010 *Phys. Rev. E* **82** 046701
[19] Meng J P, Zhang Y H and Shan X W 2011 *Phys Rev.* E **83** 046701
[20] Meng J P and Zhang Y H 2011 *J. Comput. Phys.* **230** 835
[21] Sofonea V and Sekerka R F 2005 *Phys. Rev.* E **71** 066709
[22] Watari M and Tsutahara M 2003 *Phys. Rev.* E **67** 036306
[23] Watari M and Tsutahara M 2004 *Phys. Rev.* E **70** 016703
[24] Watari M and Tsutahara M 2006 *Physica* A **364** 129
[25] Watari M 2007 *Physica* A **382** 502
[26] Watari M 2009 *Phys. Rev.* E **79** 066706
[27] Karniadakis G, Beskok A and Aluru A 2005 *Microflows and Nanoflows: Fundamentals and Simulation* (Berlin: Springer)
[28] Struchtrup H 2005 *Macroscopic Transport Equations for Rarefied Gas Flows* (Berlin: Springer)
[29] Ambruş V E and Sofonea V 2012 submitted to *Phys. Rev.* E

[30] Mysovskikh I P 1988 *Soviet Math. Dokl.* **36** 229
[31] Zwillinger D 1992 *Handbook of integration* (Boston: Jones and Bartlett)
[32] Abramowitz M and Stegun I A 1972 *Handbook of Mathematical Functions, 10th printing* (Washington: National Bureau of Standards)
[33] Shakov E M 1968 *Fluid Dynamics* **3** 95
[34] Shakov E M 1968 *Fluid Dynamics* **3** 112
[35] Titarev V A 2007 *Computers and Fluids* **36** 1446
[36] Graur I A and Polikarpov A P 2009 *Heat Mass Transfer* **46** 237
[37] Shi W, Shyy W and Mei R 2001 *Numerical Heat Transfer* B **40** 1
[38] A.Cristea, V.Sofonea, Cent.Eur.J.Phys. **2**, 382 (2004); V.Sofonea, A.Lamura, G.Gonnella, A.Cristea, Phys.Rev.E **70**, 046702 (2004); V.Sofonea, R.F.Sekerka, Int.J.Mod.Phys. C **16**, 1075 (2005).
[39] V.Sofonea, J.Comput.Phys. **228**, 6107 (2009).
[40] H.Struchtrup, M.Torrilhon, Phys.Rev.Lett. **99**, 014502 (2007)
[41] M.Torrilhon, H.Struchtrup, J.Comput.Phys. **227**, 1982 (2008);
[42] H.Struchtrup, M.Torrilhon, Phys.Rev. E **78**, 046301 (2008);
[43] P.Taheri, M.Torrilhon, H.Struchtrup, Phys.Fluids **21**, 017102 (2009).
[44] Balay S, Buschelman K, Eijkhout V, Gropp W, Kaushik D, Knepley M, Curfman McInnes L, Smith B and Zhang H 2010 *PETSc Users Manual* Argonne National Laboratory Technical Report ANL - 95/11 – Revision 3.1, *http://www.mcs.anl.gov/petsc*

Efficient Time-Step Coupling For Hybrid Continuum/Molecular Modelling of Unsteady Micro-Scale Gas Flows

Duncan A. Lockerby
School of Engineering, University of Warwick, UK
E-mail: d.lockerby@warwick.ac.uk

Carlos A. Duque-Daza
School of Engineering, University of Warwick, UK;
Department of Mechanical Engineering, Universidad Nacional de Colombia, Bogota, Colombia

Matthew K. Borg, Jason M. Reese
Department of Mechanical Engineering, University of Strathclyde, UK

Abstract. In this paper we describe a numerical method for the efficient time-accurate coupling of hybrid continuum/molecular micro gas flow solvers. Hybrid approaches are commonly used when non-equilibrium effects in the flow field are spatially localized; in these regions a more accurate, but typically more expensive, solution procedure is adopted. Although this can greatly increase efficiency in steady flows, in unsteady flows the evolution of the solution as a whole is restricted by the maximum time step allowed by the molecular-based/kinetic model; numerically speaking, this is a stiff problem. In the method presented in this paper we exploit time-scale separation, when it exists, to partially decouple the temporal evolution of the two parts of the hybrid model. This affords major computational savings. The method is a modified/extended version of the seamless heterogeneous multiscale method (SHMM). Our approach allows multiple micro steps (molecular steps) before coupling with the macro (continuum) solver: we call this a multi-step SHMM. This maintains the main advantages of SHMM (computational speed-up and flexible application) while improving on accuracy and greatly reducing the number of continuum computations and instances of coupling required. The improved accuracy of the multi-step SHMM is demonstrated for two canonical one-dimensional transient flows (oscillatory Poiseuille and oscillatory Couette flow) and for rarefied-gas oscillatory Poiseuille flow.

1. Introduction

Advances in micro and nano technologies are presenting new challenges for engineering science. In fluid dynamics, the number of flow systems that need an appreciation of the multiscale physics involved is increasing substantially. Modelling flows with a low degree of scale separation requires accounting for microscopic effects in the macroscopic behaviour: micro- and millisecond effects are important in micro and nano flows, but depend on the outcome of pico- or nanosecond molecular processes. The design of future technologies that exploit micro and nano scale flow components will therefore require the ability to resolve phenomena across scales of at least 8 orders of magnitude in space, and 10 orders of magnitude in time — this presents a formidable multiscale problem.

Any fluid flow could in principle be simulated by employing a suitable microscopic model over the entire flow domain. While this is practicable when studying the flows in a carbon nanotube, a small group of proteins in solution, or other very small-scale systems, in engineering problems the simulation domain is often much larger and more complex, making this approach computationally impractical. In many cases, the degree of scale separation may vary across the flowfield, as well as with time, and using a model of microscopic interactions in scale-separated regions is unnecessary because perfectly adequate macroscale models exist, in particular, the traditional Navier-Stokes-Fourier equations of fluid dynamics.

A more rational physical way of modelling the flow is to couple a macroscale model to a microscale model. The most popular of these hybrid approaches is domain decomposition, in which micro models are used in part of the computational domain and macro models on the remainder. But a generalised hybrid framework has been proposed that exploits scale separation, both in time and space, wherever it occurs in the domain in order to make computational efficiencies (the reader is referred to [1], for an in-depth review). In this Heterogeneous Multiscale Method (HMM) the micro model is decoupled from the physical domain of the macro solver, and there is no direct communication between the different micro simulations. No a priori knowledge of the macro form of the fluid constitutive and boundary behaviour is needed: the micro models are used to provide this missing flow property data at different spatial locations. Each local micro simulation is in turn constrained by the local output of the macro solver.

The strategy for exploiting timescale separation in HMM is illustrated in Figure 1a. A macroscopic time step is set, Δt, which is small enough to adequately resolve macroscopic variations, but which is much larger than the microscopic time step, δt. For each macroscopic time step there are performed multiple microscopic time steps, such that the time evolved microscopically is roughly equal to the microscopic relaxation time, T_{micro}; more than this, and computational effort is wasted.

A major improvement on this method has recently been proposed by E et al. [2]: the Seamless Heterogeneous Multiscale Method (SHMM). Here, the micro simulation does not need to be continually re-initialised between macro timesteps, which can be costly: the interaction between the micro and macro solvers is continuous (seamless), as illustrated in Figure 1b. SHMM exploits temporal scale separation in the given problem by running the macro and micro solvers on different clocks (t and τ, respectively). The computational time steps, although coupled as if they were of equivalent times, are asynchronous. The micro model is not reset at every coupling, instead it dynamically adapts to changes in the macro environment (manifested through the constraints applied from the macro solver). Implicit time-averaging of the micro data therefore replaces the explicit time-averaging of most other hybrid methods. This asynchronous coupling has the effect of "fast-forwarding" the macro constraints (from the perspective of the micro-solver); in timescale-separated conditions this produces a good approximate response to the macro constraints being applied in real time, provided the SHMM macro time step is sufficiently small. A further advantage of the SHMM approach is that it allows control over how aggressively timescale separation is exploited for computational efficiency (at the expense of error). The ratio between the scales of the micro and macro clocks (which we refer to in this paper as the temporal gearing) directly determines the computational saving relative to a simulation that is performed entirely using the micro model; in SHMM, the gearing can be set explicitly.

A drawback of SHMM is that in regions of high timescale separation, many more macroscopic time step calculations, and transfers of data, are performed than are necessary in HMM; these additional macroscopic computations may not always be negligible in cost. Furthermore, in such conditions, the HMM approach is more accurate, because at individual macro instants the HMM micro solution is fully relaxed to the actual macroscopic constraint and not a time-averaged one, as in SHMM.

In this paper we propose a modified time-stepping strategy for exploiting timescale separation: a *multi-step* SHMM. This approach adopts the same asynchronous coupling of two simulation clocks, and the continuous solution of the micro-solver (i.e. no re-initialisation is required), but allows the number of micro time steps per coupling instance, N, to be greater than 1; see Figure 1(c). Figure 2 shows the sequence of solution for each coupling instance. The multi-step number, N, provides an additional control, which can be increased in regions of high timescale separation (where $t \gg \tau$) to obtain maximum accuracy and to lower the macroscopic/coupling overhead, and reduced in regions of low timescale separation ($t \to \tau$) to approach a full microscale simulation. The multi-step approach can thus be viewed as a hybrid of HMM and SHMM, combining desirable features of both: simple control of the computational gearing (saving over a full micro simulation); no requirement for reinitialisation of the micro simulation; highest accuracy over the whole range of timescale separation; and fewest macroscopic computations and coupling instances performed in fully scale-separated conditions.

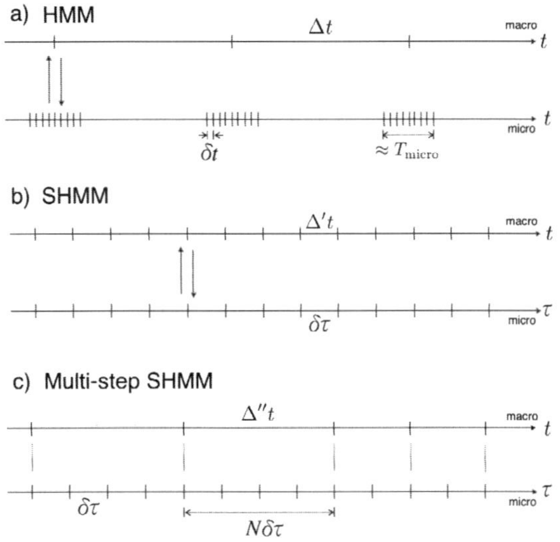

Figure 1. Alternative multi-scale time-stepping approaches: a) HMM; b) SHMM; and c) the Multi-Step SHMM

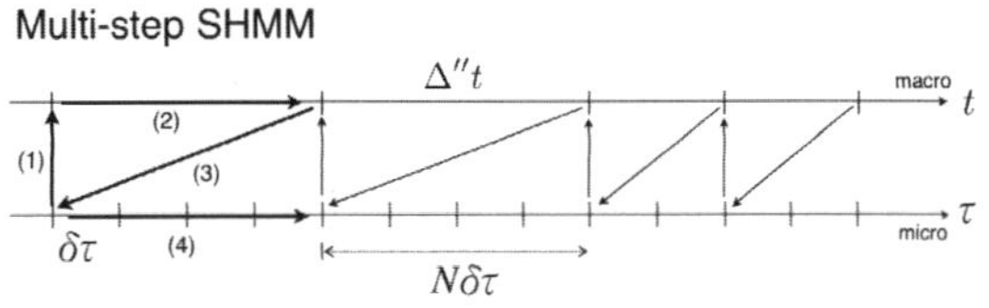

Figure 2. Sequence of solution for the multi-step SHHM, for a single coupling instance. (1) micro data compressed to macro model; (2) macro solver advancement; (3) constraints applied to micro model; (4) micro model advanced N steps.

2. Time-stepping with timescale separation

In this paper we define a dimensionless timescale-separation number, S, as:

$$S = \frac{T_{\text{macro}}}{T_{\text{micro}}}, \tag{1}$$

where T_{macro} is a macroscopic time scale and T_{micro} is a characteristic microscopic relaxation period. For $S \gg 1$, the micro and macro scales are purely separated; for $S = 1$ there is no separation (i.e. the micro and macro scales act over the same time frame); a "mixed" or transition regime between the two extremes is here defined as $1 < S < 10$. The exact division between the mixed and pure-separation regimes is arbitrary, in the same way as are the divisions of rarefaction regimes in terms of Knudsen number (e.g. slip, transition, free-molecular).

Here we define the *temporal gearing* of a hybrid approach, g, as the macroscopic time evolved per unit of microscopic time evolved. If the macro solver costs are negligible, this is equivalent to the computational saving compared to a full micro simulation. Clearly, to exploit scale separation fully, at higher timescale-separation numbers the temporal gearing should increase.

The temporal gearing in HMM is the ratio: $T_{\text{macro}}/T_{\text{micro}}$, i.e., $g = S$. Depending, though, on the error that is acceptable or the accuracy desired, this particular relationship between temporal gearing and timescale separation may not be appropriate. In this respect, the SHMM (and also the proposed multi-step SHMM) has an advantage. Here, the temporal gearing is simply the ratio of the two clock time scales, $g = t/\tau$. This means that any functional dependence of the temporal gearing on the timescale separation number can be implemented.

In this paper, we choose a straightforward functional relationship between temporal gearing and timescale separation, as follows: when there is no scale separation, the macro clock should approach the micro clock (i.e. $g = 1$ at $S \leq 1$); when there is pure scale separation ($S \geq 10$), the temporal gearing should increase proportionally with the degree of scale separation, (i.e., $g \propto S$):

$$g(S) = \begin{cases} 1, & \text{for } S \leq 1, \\ kS, & \text{for } S \geq 10, \end{cases} \tag{2}$$

where an increase in the constant of proportionality k increases computational saving, at the expense of accuracy. For values of S in the "mixed" regime ($1 < S < 10$) we choose the minimum-order polynomial that enforces continuity of value and slope of g at S=1 and S=10:

$$\begin{aligned} g(S) &= a_1 S^3 + a_2 S^2 + a_3 S + a_4 \\ &\text{for } 1 < S < 10, \end{aligned} \tag{3}$$

where the coefficients a_n depend on the specific value of k. Figure 3 shows a plot of temporal gearing g against S, for k=0.25 (note, this is a more conservative temporal gearing as compared to HMM, which implies $k = 1$).

2.1. Multi-step SHMM

For high degrees of temporal scale separation ($S \gg 1$), the microscopic state is relaxed to the macroscopic state in a quasi-steady manner. This means, the closer the micro solution is to its locally (constrained) relaxed state the more accurate the result. In this scenario, for a given macro time (T), it is more accurate to have M micro steps constrained to the one macro state at T (as it is in HMM), than it is to have a converged micro state over a range of M-previous macro states (as is the case in SHMM).

On the other hand, when S is in the mixed regime, having the micro state converged over a range of previous macro states (as in SHMM), reflects the true micro state more accurately, as it is not locally relaxed at each macro instant (as in HMM).

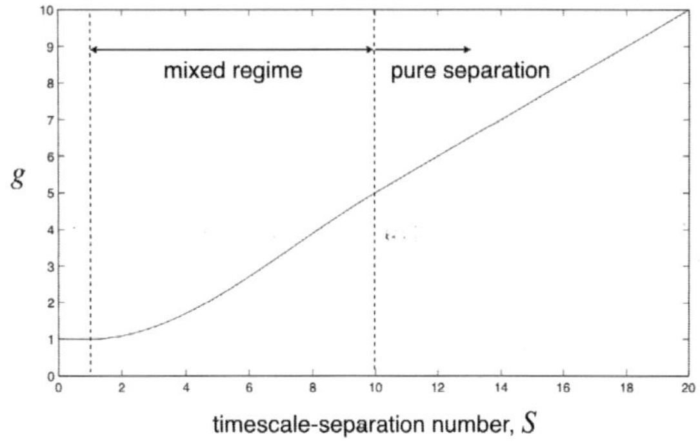

Figure 3. Temporal gearing, g, against the timescale-separation number, S; k=0.25.

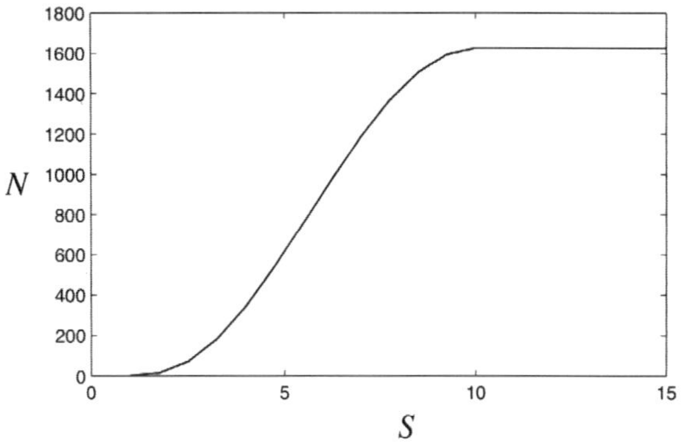

Figure 4. Number of micro timesteps per coupling instance, N, against the timescale separation number, S.

The multi-step SHMM we propose seeks to combine the best of both approaches over the full range of S. As per SHMM, the micro time step, $\Delta\tau$, is held constant, and is set by stability/accuracy considerations. In SHMM the macro time step can be set directly by the gearing desired (as determined by equations (2)-(3), for example):

$$\Delta' t = \Delta\tau \, g, \tag{4}$$

Here, though, differently to SHMM, we allow the number of micro steps per coupling instance, N, to be greater than 1 (as illustrated in Figure 1c). The macro time step is then set according to:

$$\Delta'' t = \Delta\tau \, g \, N, \tag{5}$$

under the proviso that it does not exceed the maximum macro time step allowable $\Delta t_{\max}$ (based on stability/accuracy considerations).

The most accurate/efficient number of multisteps, N, will depend upon the temporal scale separation, S. Given that for pure timescale separation, a larger N provides a more accurate solution (as discussed above), we set N to be maximum for $S \geq 10$. For non-timescale-separated conditions ($S \leq 1$), the macro and micro time steps should be synchronous, (i.e. N=1). For values of S in the "mixed" regime ($1 < S < 10$), we choose the minimum-order polynomial that enforces continuity of value and slope of N at S=1 and S=10:

$$N(S) = \begin{cases} 1, & \text{for } S \leq 1, \\ b_1 S^3 + b_2 S^2 + b_3 S + b_4, & \text{for } 1 < S < 10, \\ N_{\text{max}} & \text{for } S \geq 10, \end{cases} \tag{6}$$

where

$$N_{\text{max}} = \frac{\Delta t_{\text{max}}}{\Delta \tau\, g}, \tag{7}$$

and the coefficients b_n are dependent on N_{max}.

3. Test cases: transient one-dimensional shear flows

In timescale-separated simulations there are three sources of error: error due to the micro solution; error due to the macro solution; and error due to the time-step coupling scheme (e.g. SHMM). We restrict our attention to the latter.

The test cases in this paper are time-dependent flows between parallel plates. Here, the macro model calculates the pressure gradient or the channel-wall velocity, and the micro model calculates the spatial distribution of streamwise velocity (parallel to the walls). For these cases, the macro calculation is trivial, exact and known *a priori*. This is convenient, as it allows us to eliminate macro solver error in our evaluation of the alternative time-stepping schemes, and since there is no error in the constraints applied to the micro solver, all error is either from the micro model (which is deliberately kept negligibly small) or error due to the temporal coupling scheme. In all simulations we specify a maximum macro time step $\Delta t_{\text{max}} = T_{\text{macro}}/5$, which in practice would be set by macroscopic stability/accuracy requirements.

3.1. Oscillatory Poiseuille Flow

The flow between two parallel plates resulting from an oscillating pressure gradient is a canonical transient case in fluid dynamics. If this is a locally near-thermodynamic equilibrium flow [1] an appropriate micro model is the one-dimensional Navier-Stokes momentum equation:

$$\frac{\partial u}{\partial t} = \nu \frac{\partial^2 u}{\partial y^2} + \Pi(t), \tag{8}$$

where ν is the kinematic viscosity, y is a direction perpendicular to the walls, and u is the flow-velocity component parallel to the wall. The micro model is constrained through the time-dependent streamwise pressure gradient, $\Pi(t)$. The characteristic microscopic time scale, T_{micro} is taken as the startup time: the time taken, after a suddenly imposed and constantly applied pressure gradient, for the mean velocity (over y) to reach 95% of its steady-state value.

The macro model solver, in this example, is the model that calculates the pressure gradient, simply:

$$\Pi = \hat{\Pi} \sin(\omega t), \tag{9}$$

where ω is the radial frequency of oscillation and t is the macroscopic time. The characteristic macroscopic time scale is chosen to be the quarter-wave period:

$$T_{\text{macro}} = \frac{\pi}{2\omega}, \tag{10}$$

and thus the timescale separation number at a particular frequency is:

$$S = \frac{T_{\text{macro}}}{T_{\text{micro}}} = \frac{\pi}{2\omega T_{\text{micro}}}. \tag{11}$$

[1] note, timescale separation can be independent, as it is here, of the state of local thermodynamic equilibrium

We consider a range of oscillation frequencies, spanning timescale separation numbers from S=1 to 50. The temporal gearing for a given S is as described by equations (2) and (3), and as shown in Figure 3 (k=0.25). For the multi-step SHMM the number of micro time steps per coupling instance (N) is calculated using equation (6); N is plotted in Figure 4 as a function of S.

For evaluating the accuracy of SHMM and the multi-step SHMM, a classical analytical solution to the full problem (equations 8 and 9) is used:

$$u(y) = \Re\left\{ \mathbf{e}^{\mathrm{i}\omega t} \, \frac{\hat{\Pi}}{\omega} \, A_1(y,\eta) \right\}, \tag{12}$$

where

$$A_1(y,\eta) = \left(1 - \frac{\cosh\left[\eta(y-h/2)(i+1)\right]}{\cosh\left[\eta h(i+1)/2\right]} \right), \tag{13}$$

and η=$\sqrt{(\omega/2\nu)}$, with h the channel height (y=0 at the centreline of the channel).

Figure 5 compares the accuracy of the SHMM and the multi-step SHMM, showing the maximum error (in time) of mean velocity (over y) for a range of S. For all values of S considered, the multi-step method is more accurate (for the *same* gearing); for $S \gg 1$ the multi-step approach is over 8× more accurate. Furthermore, the multi-step HMM involves approximately 1600× fewer macroscopic calculations and instances of coupling (though, in this illustrative example, the macroscopic calculation has negligible cost). Note, direct comparison with the conventional HMM cannot be made, since the temporal gearing of HMM differs from that used in Figure 5. However, for indicative purposes, for HMM (with $g = S$), the error for $S \geq 5$ is approximately 5%; the error is greater for $S < 5$.

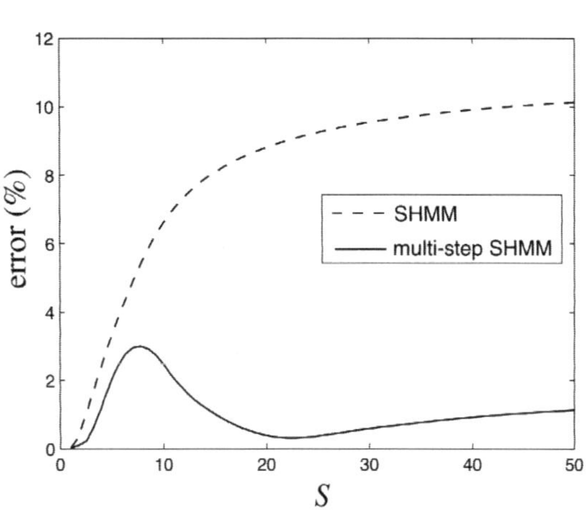

Figure 5. Maximum error (%) (over time) of mean velocity (over y) against S. SHMM (dashed line) against multi-step SHMM (solid line).

3.2. Oscillatory Couette Flow

Another canonical transient flow in fluid dynamics is that between two plates generated by the oscillatory transverse motion of one wall relative to the other. In this case, the macro model provides the constraint on the micro solution in the form of an imposed wall velocity, i.e.:

$$u(0) = \hat{U}_{\mathrm{wall}} \sin(\omega t), \tag{14}$$

where $\hat{U}_{\text{wall}}$ is the amplitude of the moving wall. For evaluating the accuracy of SHMM and the multi-step SHMM, a classical analytical solution to the full problem (equation 8 with the pressure-gradient term zero) is used:

$$u(y) = \Im\left\{\mathrm{e}^{\mathrm{i}\omega t}\, \hat{U}_{\text{wall}}\, A_2(y,\xi)\right\}, \tag{15}$$

where

$$A_2(y,\xi) = \left[\cosh(\xi y) - \coth(\xi h)\cosh(\xi y)\right], \tag{16}$$

and $\xi = \sqrt{(i\omega/\nu)}$ ($y{=}h/2$ at the centreline of the channel). Figure 6 compares the accuracy of the two methods. Again, for all values of S considered, the multi-step method is more accurate (for the same gearing); for $S \gg 1$ the multi-step approach is over 8× more accurate, and this is achieved using over a thousand times fewer macroscopic calculations and coupling instances than in the SHMM.

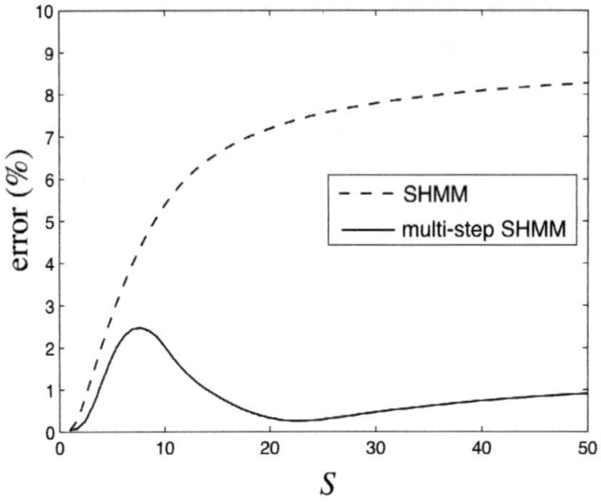

Figure 6. Maximum error (%) (over time) of mean velocity (over y) against S. SHMM (dashed line) against multi-step SHMM (solid line).

3.3. Non-equilibrium Oscillatory Poiseuille Flow

At atmospheric pressures, gas flow through micro/nano-channels can be highly non-equilibrial [3]. The final example considered here, is isothermal oscillatory Poiseuille flow (as in §3.1) of a monatomic gas in rarefied/non-equilibrium conditions.

The degree of departure from local thermodynamic equilibrium in a gas (the state of rarefaction), is expressed by the Knudsen number:

$$Kn = \frac{\lambda}{h}, \tag{17}$$

where λ is the mean distance between molecular collisions, and h is the characteristic macroscopic length scale; here, this is the channel height.

The micro model chosen here is the BGK Boltzmann equation:

$$\frac{\partial f}{\partial t} + \hat{v}\frac{\partial f}{\partial y} + \frac{\partial f}{\partial \hat{u}}F_x(t) = -\frac{(f - f_{eq})}{\xi} \tag{18}$$

where f is the molecular distribution function, f_{eq} is the equilibrium distribution, ξ is the molecular relaxation time, $F_x(t)$ is the time-dependent body force (equivalent to an oscillatory pressure gradient of opposite sign) and $\hat{u}$, $\hat{v}$ are the $x-$ and y-components of molecular velocity, respectively. The molecular relaxation time for the BGK model is given by $\xi = \mu/p$, where μ is the dynamic viscosity and p is the gas pressure.

We obtained solutions for the micro model (18) using a discrete velocity method (DVM) similar to that used by Valougeorgis (1988), [5]. This numerical scheme, which uses Gaussian quadrature to integrate in velocity space, has been tested extensively against a host of problems in rarefied gas dynamics and has proved to be both accurate and highly efficient (see, e.g., [4; 5; 6]).

As in previous sections, the macro model solver for the oscillatory driver is trivial. Simply:

$$F_x(t) = \hat{F}_x \sin(\omega t). \tag{19}$$

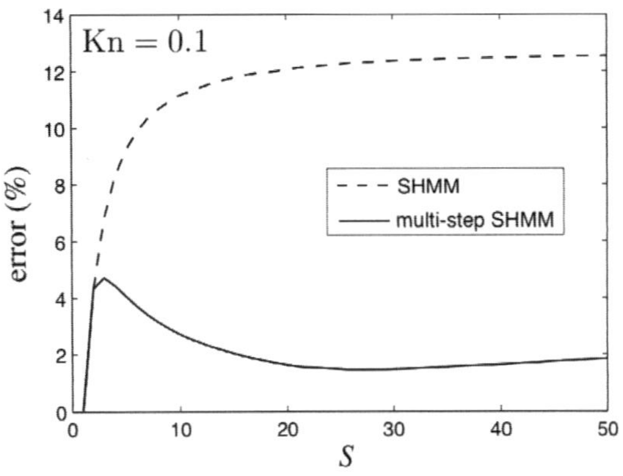

Figure 7. Maximum error (%) (over time) of mean velocity (over y) against S. SHMM (dashed line) against multi-step SHMM (solid line); Kn=0.1.

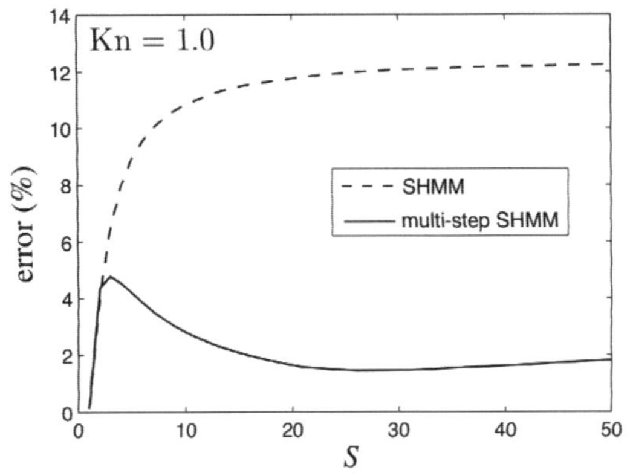

Figure 8. Maximum error (%) (over time) of mean velocity (over y) against S. SHMM (dashed line) against multi-step SHMM (solid line); Kn=1.0

Figures 7 and 8 show the maximum error over time, for varying degrees of scale separation, S (defined as in section §3.1), for rarefied oscillatory Poiseuille flow at Knudsen numbers of 0.1 and 1.0, respectively. Here, error is calculated with respect to a full BGK solution of the problem (i.e. with no gearing). The error of the multistep SHMM is lower than the SHMM for values of $S>3$, and no worse for $S<3$; this is achieved with far fewer macroscopic calculations and coupling instances.

4. Summary and Conclusions

A multi-step version of the seamless heterogenous multiscale method [2] has been proposed. It shares the main advantages of SHMM over HMM, e.g.:

(i) the micro-solver does not require re-initialisation at every macro time step;
(ii) the temporal gearing can be controlled in a straightforward manner by adjusting the ratio of the micro and macro clocks;
(iii) it can be used in both timescale-separated and non-timescale-separated conditions, thus appropriate for an adaptive scheme with time-varying scale separation;

while retaining the accuracy and macroscopic efficiency of HMM:

(i) when timescales are purely separated, multi-step SHMM provides a more accurate result;
(ii) and in such conditions, far fewer coupling instances and macroscopic calculations are required.

The multi-step scheme can thus be viewed as an amalgam of HMM and SHMM: a combination of the best features of both methods.

We have presented results from three transient shear flow problems. In each case, the macro model computation is trivial (i.e. the evaluation of a sine function). This has allowed us to focus on error arising from the particular time-stepping scheme. Two micro models have been considered: a Navier-Stokes micro model and a Boltzmann BGK micro model. In each of the three examples, the multi-step SHMM demonstrated significantly greater accuracy than the SHMM, and at a fraction of the number of coupling instances and macroscopic computations.

Acknowledgments This research is financially supported by the EPSRC Programme Grant EP/I011927/1. The authors would like to thank Prof. Dimitris Valougeorgis for providing a well-documented DVM code that helped to generate the BGK results presented in this paper.

References

[1] E W, Engquist B, Li X T, Ren W Q and Vanden-Eijnden E 2007 *Communications in Computational Physics* **2** 367–450
[2] E W, Ren W Q and Vanden-Eijnden E 2009 *Journal of Computational Physics* **228** 5437–5453
[3] Reese J M, Gallis M A and Lockerby D A 2003 *Philosophical Transactions of the Royal Society a-Mathematical Physical and Engineering Sciences* **361** 2967–2988
[4] Naris S and Valougeorgis D 2005 *Physics of Fluids* **17**
[5] Valougeorgis D 1988 *Physics of Fluids* **31** 521–524
[6] Valougeorgis D and Naris S 2003 *Siam Journal on Scientific Computing* **25** 534–552

Thermal-Pressure-Driven Gas Flows through Micro Channels

Hassan Akhlaghi Akhlaghi[1], **Ehsan Roohi**[2]

[1,2] Department of Mechanical Engineering, Ferdowsi University of Mashhad, P.O. Box 91775-1111, Mashhad, Iran

E-mail: akhlaghi@alum.sharif.edu

Abstract. In this paper we study mass flow rate of rarefied gas flow through micro/nanoscale channels under simultaneous thermal and pressure gradients using an improved direct simulation Monte Carlo (DSMC) method. Before targeting thermal/pressure driven flows, we first analyze pressure-driven flows and verify our DSMC solver. Next, we study micro-/nanochannels flows under simultaneous pressure-temperature gradients while our main objective is to predict mass flow rate increment due to thermal creep effects. The effects of thermal creep are studied over a wide range of flow rarefaction from the slip to free molecular regime. Our results showed that the non-dimensional thermal creep mass flow rate increments of the Poiseuille flow increases and approaches the value of 0.578 at free molecular limit.

1. Introduction

With the fast progress in micro/nanoscale devices and components, it is desired to obtain fundamental understanding of the transport phenomena in these devices. Additionally, a true understanding of the hydrodynamics and thermal physics of rarefied gas flows is crucial in design, fabrication, and operation of micro/nano-electro-mechanical-systems (MEMS/NEMS) [1-2]. Once the gas flow rarefies, the Knudsen number, defined as the ratio of gas mean free path to the characteristic length of the geometry, $\mathrm{Kn} = \lambda/\mathrm{H}$, increases and the flow analysis should be performed using accurate approaches based on the solution of the Boltzmann equation [3]. Discrete molecular modeling of direct simulation Monte Carlo (DSMC) is a key tool to model flow field in all degrees of rarefaction for both of mono/diatomic gases [4]. In DSMC, the gas is modeled as a collection of moving particles which interact through collisions. DSMC is known as one of the most successful particle simulation methods in analyzing the rarefied gaseous flows.

Literature survey shows that there are continuing attempts for mass flow prediction of pressure-driven and thermal-driven rarefied gas flows through micro-/nanoscale channels. In the case of pressure-driven flows through micro-/nanochannels, there are a wide set of analytical predictions for mass flow rate based on different wall velocity slip boundary conditions. In addition to pressure gradient, wall temperature gradient may generate mass flow rate at micro/nano scale. Thermal creep flows in gases occurs by non-uniform temperature distributions in the flow boundaries [5]. In this situation, fluid starts creeping from the cold region towards the hot one. Knudsen pumps, which operate based on thermal creep/transpiration flow due to non-uniform wall temperature distribution, are employed in MEMS/NEMS [6-9]. Thermal creep/transpiration problem is investigated by several authors using kinetic models of the Boltzmann equation. The flow between two parallel plates, where

[1] Corresponding author.

a constant unidirectional temperature gradient is applied to the plates, is a common flow configuration which is used in kinetic approaches. Loyalka *et al.* [10-14] presented kinetic theory analysis of thermal transpiration and mechanocaloric effects. They presented a general kinetic theory analysis and showed that the cross phenomena of thermal transpiration and mechanocaloric effect can be treated by the use of the variational technique for all rarefactions regimes. Kanki and Iuchi [15] investigated Poiseuille flow and thermal creep of a rarefied gas between two parallel plates using the (Bhatnagar-Gross-Krook) BGK model. They showed that the ratio of the flow rate of the thermal creep to that of the Poiseuille flow reduces to zero at the continuum flow limit and approaches to 0.5 as the density of the gas approaches to zero. Ohwada *et al.* [16] studied the Poiseuille and thermal transpiration flows of a rarefied gas between two parallel plates on the basis of the linearized Boltzmann equation for hard-sphere molecules. They predicted the mass fluxes of the Poiseuille and the thermal transpiration flows for the whole range of the Knudsen number with good accuracy compared to the existing numerical method and experimental results. They showed that in spite of Poiseuille flow, mass flux increase continuously in thermal transpiration flow versus the Knudsen number. Alexeenko *et al.* [7] numerically investigated the temperature driven gas flows in a two-dimensional finite length microchannel and a cylindrical tube using DSMC method, a discrete ordinate method for the ellipsoidal statistical model, and BGK model. They used a short channel with the length-to-height ratio of 5, pressure ratio equal to one and Knudsen number equal to 0.2. They generated creep effects with different wall temperature distributions. Their results showed that despite of the same inlet/outlet pressure values, there are pressure changes along the channel. They also examined the effects of various wall temperature distributions and compared different numerical methods. Taheri and Struchtrup [17] applied a streamwise constant temperature gradient over the walls of a microchannel and studied rarefied gas flow through the channel using fully linearized and semi-linearized forms of the regularized 13-moment (R13) equations.

Simultaneous thermally/pressure driven flows may be considered to control mass and energy flow rates in micro/nanochannels [17]. This physics is also employed in Knudsen pumps, as discussed before. Micro/nanoscale Knudsen pumps are essential for the realization of completely miniaturized devices such as gas sensors, fuel cells, and lab-on-a-chip systems [9]. Consequently, study of this class of micro/nano flows would be crucial for accurate design of MEMS/NEMS. However, there is not an archival study in the mass flow rate prediction of the Poiseuille flow under simultaneous thermal and pressure gradients. Therefore, the main objective of the current study is the mass flow rate prediction of the rarefied gas flow through micro/nanoscale channels under both thermal and pressure gradients over a wide range of Knudsen numbers. We use DSMC method as our direct simulation strategy.

2. DSMC Method

DSMC is a numerical tool to solve the Boltzmann equation based on the direct statistical simulation of the molecular processes described by the kinetic theory [4]. The primary principle of DSMC is decoupling of the motion and collision for all simulated particles. DSMC is considered as a particle method in which one particle represents a large bulk of real gas molecules. After fulfilling all molecular movements, the collisions between molecules are simulated in each cell separately. To simulate complex flow effectively, improvements in the basic DSMC algorithm has been recently suggested by Gallis *et al.* [18] and Stefanov [19, 20]. However, in the current work, we extend and optimize DSMC codes employed by Roohi and coworkers for simulation of different micro/nano flows [21-27]. The variable hard sphere (VHS) model is applied as our collision model in all simulations. We use the diffuse reflection model at the solid walls. In diffuse reflection, the emission of impinging molecules is not correlated with the pre-impingement state of the molecules. The outgoing velocity of molecules is randomly assigned according to a half-range Maxwellian distribution determined by the wall temperature. Full thermal and momentum accommodation coefficients are considered for the walls.

2.1. Inlet/Outlet Boundary Conditions

For a 2-D pressure driven flow, the values of pressure and temperature at the inlet and outlet and the value of pressure at the outlet should be specified. However, DSMC calculation does not use pressure magnitude directly. On the other hand, density and velocity at the inlet and outlet and temperature at the outlet are unknown while they are required for the DSMC simulation. For example, inlet and outlet density and velocity are required to calculate the number of particles injected into the flow field or the outlet temperature is required for calculation of the most probable speed of particles [4]. Benefiting from the characteristics theory, specified inlet/outlet pressure boundary conditions are implemented implicitly into the flow field through correcting density ρ and velocity U values [28, 29]. For a backward-running wave, we consider

$$dU/a = -d\rho/\rho \tag{1}$$

$$a^2 = dP/d\rho \tag{2}$$

where P and a are pressure and speed of sound, respectively. Applying the definition of speed of sound to a boundary cell, we find a relation between the density and pressure at the outlet as follows

$$\rho_{j,out}^{new} = \rho_{j,out}^{old} + \left(P_{out} - P_{j,out}\right)/a_{j,out}^2 \tag{3}$$

Subscript *out* denotes values at channel outlet boundary. The *old* and *new* superscripts are corresponding to the values at previous and present time steps. Once the density is calculated, outlet temperature $T_{j,out}$ at the cell j-th can be obtained from the equation of state

$$T_{j,out} = P_{out}/\left(\rho_{j,out}R\right) \tag{4}$$

where *R* is the gas constant. The velocity is also computed from the characteristic wave equation, obtained from multiplying Eq. (1) by speed of sound (a) and substituting definition of a^2 from Eq. (2) in the resulting equation. It yields

$$U_{j,out}^{new} = U_{j,out}^{old} + \left(P_{j,out} - P_{out}\right)/\left(\rho_{j,out}a_{j,out}\right) \tag{5}$$

Subscript *in* denotes values at channel inlet boundary. The inlet velocity is also calculated in the same way, as follows

$$U_{j,in}^{new} = U_{j,in}^{old} + \left(P_{in} - P_{j,in}\right)/\left(\rho_{j,in}a_{j,in}\right) \tag{6}$$

In addition, the density at the inlet is calculated from the equation of state

$$\rho_{j,in} = P_{in}/\left(T_{j,in}R\right) \tag{7}$$

Even though Eqs. (1) and (2) describe a wave in the continuum limit, their use in the DSMC simulation of rarefied flow is justified. In fact, DSMC algorithm assumes an equilibrium Maxwellian distribution for the particles which are injected into the flow field from inlet/outlet boundaries, as these particles come from free atmosphere [4, 28, 29]. From molecular gas dynamics, the Maxwellian distribution represents equilibrium condition, which could be described by Euler equations. As a result, application of wave Equations, Eqs. (1) and (2), is approved. Suitable accuracy of the above equations in the simulation of internal micro/nano rarefied flows also confirms the use of aforementioned equations [21-23, 25, 26, 28, 29]. In our DSMC solver, the above equations are called at the end of "Enter" function, which computes the properties of incoming particles to the computational domain. Consequently, the estimated values of density, velocity, temperature and number of inserted particles in the inlet/outlet cells are corrected through Eqs. (3)-(7). At the steady state condition, the corrections to the old values are zero in Eqs. (3), (5), and (6). Thus, there is no change in the hydrodynamics and thermal properties of gas flows at the inlet/outlet cells and correct values are applied.

3. Solver Verification

In this section we verify our DSMC solver. It should be reminded that our DSMC solver has been optimized by different features. The cell dimensions are chosen less than mean free path, i.e., smaller than $\lambda/3$ [30-33]. This criterion is implemented based on the local mean free path of cells. Each cell is subsequently divided into two sub-cells in each direction. The time step must be a fraction of the mean collision time and satisfy a Courant–Friedrichs–Lewy (CFL) number, based on the most probable molecular speed V_{mp}, less than one. Hadjiconstantinou [33] showed that the DSMC error exhibits quadratic dependence on the time step, and that for time steps of the order of one mean free time the error is of the order of 5%. According to mentioned criteria, we choose the time step equal to 0.08 of the minimum mean free time of all cells. Numbers of simulated particles are chosen such that there would be at least 20 particles in each cell [34]. Sampling of results starts as soon as the flow reaches to the steady state conditions. Flow sampling continues in sufficiently large time step intervals. In the beginning of the flow sampling period, flow properties set to zero and recalculated during the new sampling period. Using this process, the effects of initial unsteady results will be eliminated and convergence of the final flow properties could be traced.

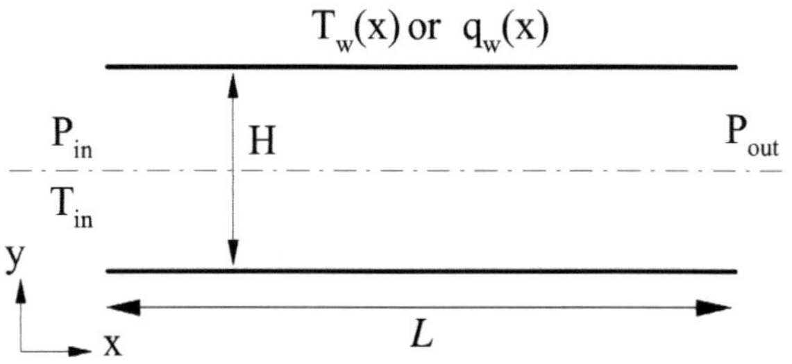

Figure 1. Micro/nanochannel geometry and imposed boundary conditions.

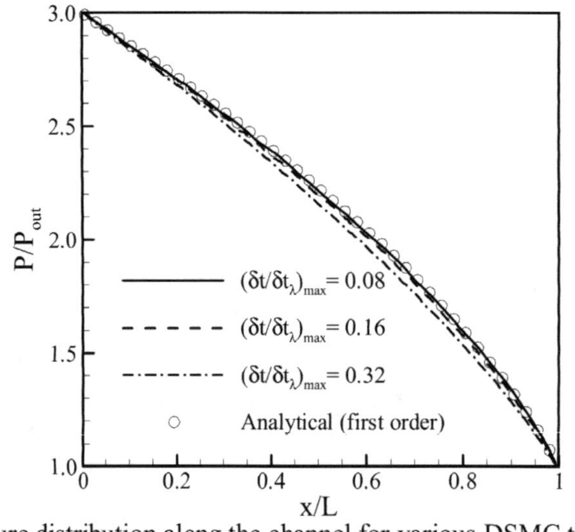

Figure 2. Pressure distribution along the channel for various DSMC time step values.

We evaluate our DSMC solver by solving subsonic pressure-driven nitrogen flow through microchannel imposing constant wall temperature condition. Channel length and height are 20 μm and 1 μm, respectively. Gas flow enters at the condition of $T_{in} = 300\,K, P_{in} = 4\,atm$. The channel pressure ratio is set to 3 and channel walls have a uniform temperature distribution equal to 300 K. Geometry of the channel and imposed boundary conditions are shown in Figure 1. Figure 2 shows the effect of the DSMC time step on the pressure distribution along the channel. The results are compared with the analytical solution based on the first-order wall slip boundary condition given by [35]:

$$\frac{P}{P_{out}} = -6Kn_{Out} + \sqrt{(6Kn_{Out})^2 + (1+12Kn_{Out})\frac{x}{L} + (\Pi^2 + 12Kn_{Out}\Pi)\left(1-\frac{x}{L}\right)} \tag{8}$$

Eq. (8) presents streamwise distribution of normalized pressure along the channel based on the outlet Knudsen number (Kn_{Out}), channel pressure ratio (Π), channel length (L), and streamwise location (x). Lines in Figure 2 correspond to DSMC times steps equal to 0.08, 0.16, and 0.32 mean free time, respectively. Knudsen number ranges from 0.013 to 0.040 along the channel. According to Fig. 2, expect the large disagreements in the case of $(\delta t/\delta t_\lambda)_{max} = 0.32$, other values of specified time step well predict pressure distribution along the channel.

4. Thermal-Pressure-Driven Gas Flows

Thermal creep flow is an induced flow from the cold region towards the hot one. Thermal creep effect occurs only in the transitional rarefied flows, primarily along walls with a superimposed wall surface temperature gradient, see Figure 3.

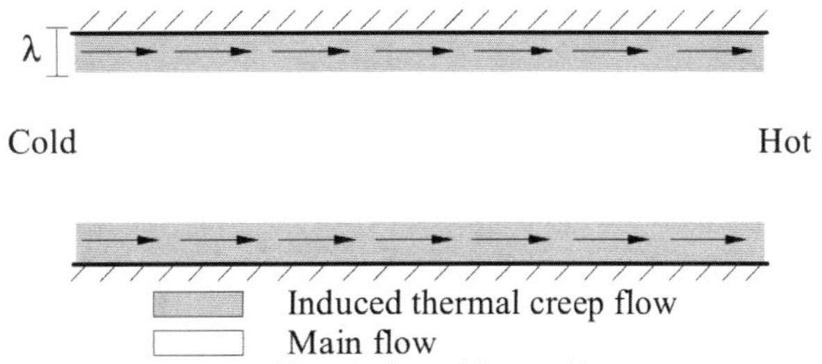

Figure 3. Thermal creep flow induced by wall temperature gradients.

For free molecular flows, thermal creep layer fills the entire channel, creating thermal transpiration flow. Meanwhile, in the case of transitional flow, the thermal creep effects are limited to the regions near the channel's wall [8]. At this stage, we consider the effect of thermal creep effect on the mass flow rate of Poiseuille flow. To achieve this, we simulate pressure-driven flows through channels with linear wall temperature distribution. To count the effects of flow rarefaction, we launch our study under different Knudsen numbers. Mass flow rate through micro/nano channels contains two parts as below:

$$\dot{m} = \dot{m}_p + \dot{m}_{th} \tag{9}$$

where subscript "p" corresponds to fully pressure-driven isothermal flow and subscripts "th" represents mass flow rate increment due to thermal creep effects. Thermal creep leads to an induced flow from the cold region to the hot one [36]. Therefore, streamwise wall temperature gradient increases mass flow rate and vice versa.

Figure 4 shows the effect of wall temperature gradient values on the thermal mass flow rate values. The results are obtained under different flow rarefaction conditions and streamwise wall temperature gradients. Figure 4(a) and (b) correspond to the bulk flow average temperature values of 300 K and 400 K, respectively. The x-component is wall temperature gradient in K, and X is the non-dimensional streamwise location, x/L, along the channel. According to the results, thermal mass flow rate increases linearly as wall temperature gradient increases for both average temperatures values.

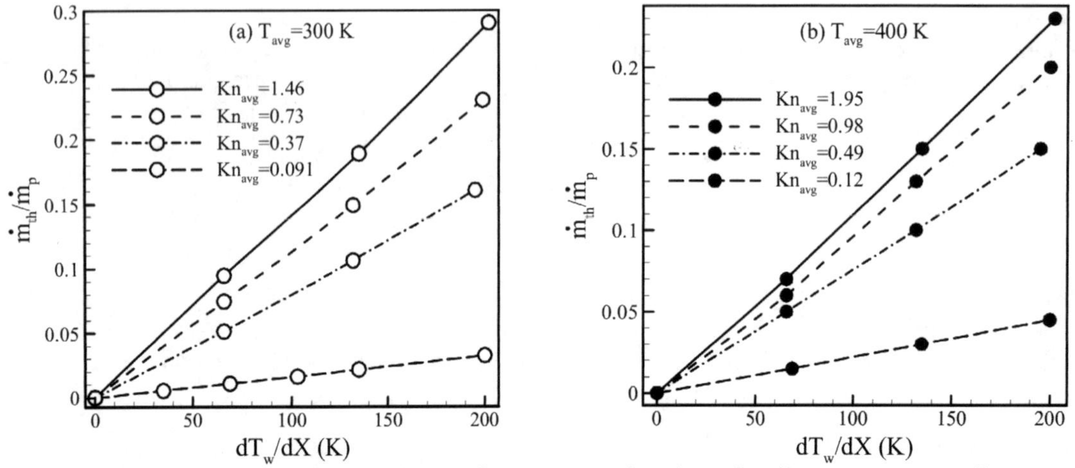

Figure 4. Non-dimensional thermal mass flow rate as a function of wall temperature gradient at different Knudsen numbers.

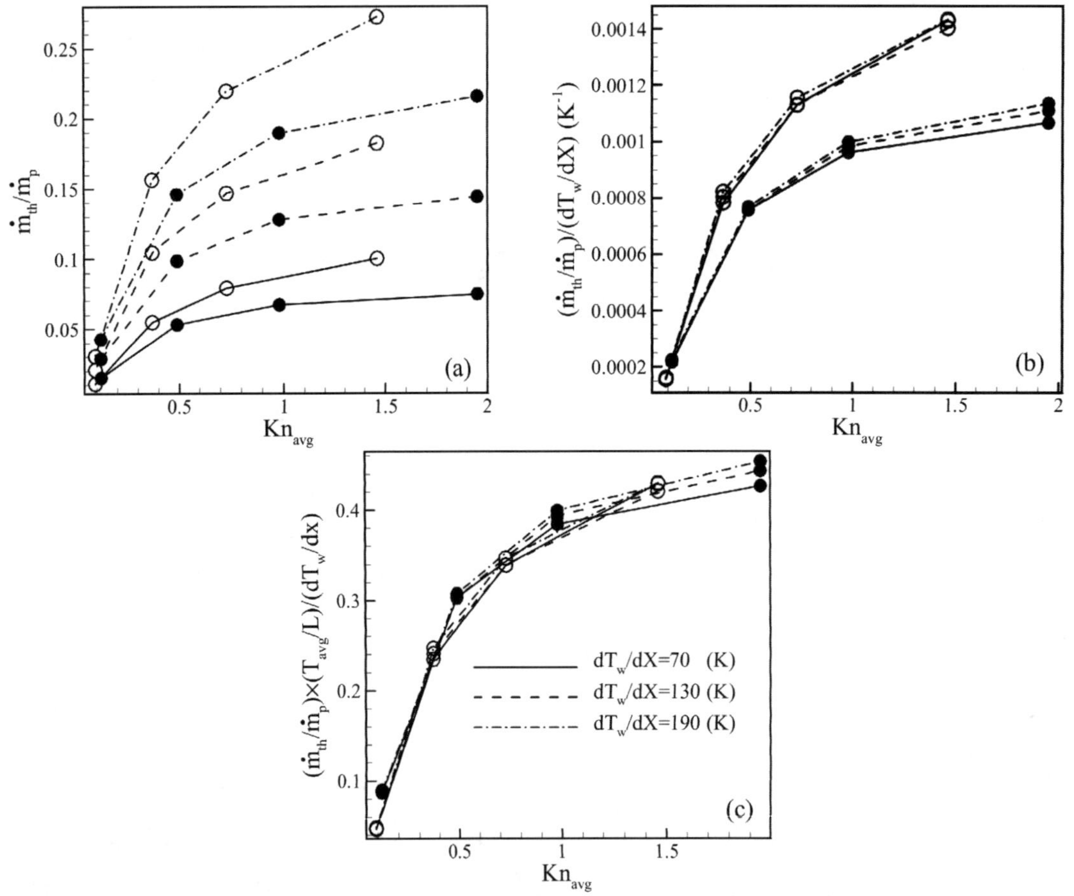

Figure 5. Thermal mass flow rate as a function of flow average Knudsen number, a) non-dimensional thermal mass flow rate, b) non-dimensional thermal mass flow rate independent from wall temperature gradient values, and c) non-dimensional thermal mass flow rate independent from wall temperature gradient and bulk flow temperature values. (White and black symbols denotes T_{avg} = 300 and 400 K, respectively)

The value of $\dot{m}_p$ is constant through variation of the wall temperature gradient and its value corresponds to the pure pressure-driven flow at zero wall temperature gradient condition. Figure 4 shows that increase in the thermal mass flow rate due to wall temperature gradient is faster for higher Knudsen numbers flows. It is also desired to study the thermal mass flow rate behavior with variation of the Knudsen number. For this aim, we have plotted the non-dimensional mass flux as a function of Knudsen number in Figure 5. Figure 5(a) shows the non-dimensional thermal mass flow rate as a function of the average Knudsen number at different wall temperature gradient values, i.e., 70 K, 130 K, and 190 K. White and black symbols correspond to the average flow temperature values of 300 K and 400 K, respectively. According to Figure 5(a), at a specified wall temperature gradient value, thermal mass flow rate increases as flow's average Knudsen number increases. The rate of increase is faster at lower Knudsen numbers and higher wall temperature gradient values. To make the results independent from the wall temperature gradient values, we normalize the results of Figure 5(a) by the wall temperature gradient values. This is shown in Figure 5(b). As it seen in Figure 5(b), the results for both average temperature values of 300 K (white) and 400 K (black) have the same trend versus the flow Knudsen umber. Therefore, we normalize graphs of Figure 5(b) to their average temperature values. It leads to Figure 5(c). It is well seen that both lines of Figure 5(b) have the same trend in Figure 5(c).

Actually the results given in Figure 5(c) are independent from the average flow temperature and wall temperature gradient values. According to Figure 5(c), at a specified wall temperature gradient and average flow temperature values, thermal mass flow rate increases as the flow rarefaction increases. This increase is slower at higher Knudsen number regimes. We extend the results given in Figure 5(c) to the free molecular regime to obtain a reliable correlation formula for prediction of the thermal mass flow rate at all ranges of the Knudsen number. As already stated, we produce higher values of flow rarefaction by reducing channel inlet/outlet pressure values. We have performed a large set of DSMC simulations in the third part of our investigations under different conditions of flow average temperature, rarefaction, and wall temperatures. An inlet pressure of 0.15 atm with a pressure ratio of 2 is considered.

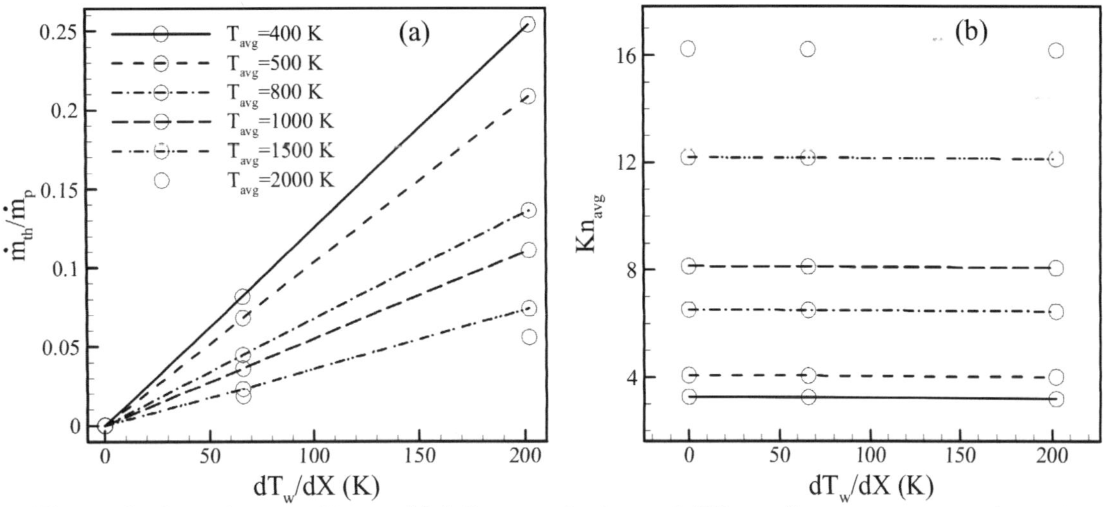

Figure 6. Thermal creep effects at high flow rarefaction and different flow temperature values; a) thermal mass flow rate and b) average Knudsen number.

Figure 6(a) shows the mass flow increments due to the wall temperature gradients under different values of the average flow temperature at higher values of Knudsen number. There is a linear correlation between the thermal mass flow rate and wall temperature gradient similar to that given of in Figure 4. Figure 6(b) plots the average flow Knudsen numbers for different values of T_{avg}. Each line corresponds to the same inlet/outlet boundary condition values, but symbols on each line are related to different linear wall temperature distributions. According to Figure 6(b), average Knudsen number is approximately constant on each line. It is because wall temperature was changed in such a way that average wall temperature is fixed to the desired T_{avg} value. This permitted us to study thermal creep effects only under different wall temperature gradient values at constant flow rarefaction.

We made a trend study for results of Figure 6 similar to what is performed in Figure 5. The results are plotted for all ranges of Knudsen number in Fig. 7. According to the Figure 7, we could cross a correlated line which contains the effects of flow rarefaction, average temperature, wall temperature gradient, and channel pressure ratio on the thermal mass flow rate increment of the Poiseuille flow. The resulting correlated line shows that the increase in the "non-dimensional thermal mass flow rate" slows down as the Knudsen number increases and reaches to a constant magnitude in the free molecular region. According to the Fig. 7, non-dimensional thermal mass flow rate increases from zero (in the case of continuum flow) to a limiting value of 0.578 at the free molecular regime. The results represents that thermal mass flow rate has a linear dependency on the wall temperature gradient value at the free molecular region. It also shows that the thermal mass flow rate increases as the flow rarefaction increases.

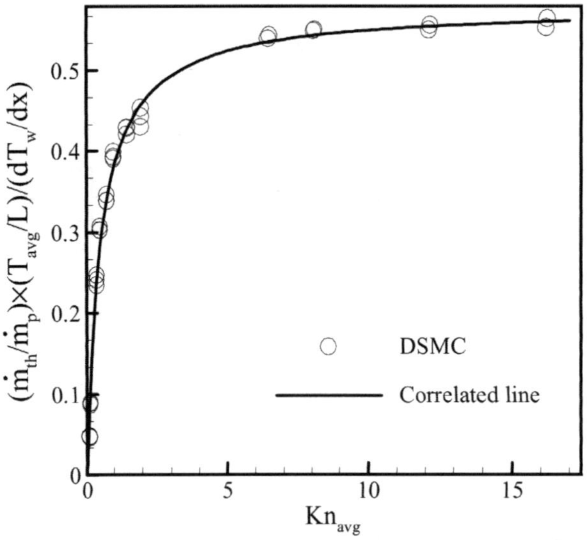

Figure 7. Thermal mass flow rate versus flow rarefaction from slip to free molecular regime.

5. Conclusion

We present a mass flow ate study for simultaneous thermal-pressure-driven flows through micro-/nanoscale channels using an improved DSMC solver. Thermal-driven forces were generated by non-uniform wall temperature distribution. The effects of thermal creep studied over a wide range of flow rarefaction from the slip to free molecular regime. We showed that thermal mass flow rate linearly increases as wall temperature gradient increases. The effect of wall temperature gradient is stronger at lower bulk temperature values. We presented a non-dimensional thermal mass flow rate which is just depends on the average flow Knudsen number. Our results showed that the non-dimensional thermal creep mass flow rate increment of the Poiseuille flow increases as flow rarefaction increases and it approaches to value of 0.578 at the free molecular limit. The obtained results would be useful in the mass flow rate prediction of Poiseuille flow under specific wall heat flux conditions.

Acknowledgement

The authors of the paper would like to acknowledge the financial supports of the Ferdowsi University of Mashhad under grant number 18888/2.

References

[1] J. Taylor and C. L. Ren, "Application continuum mechanics to fluid flow in nanochannels," Microfluid. Nanofluid. **1** (4), 356 (2005).

[2] C. M. Ho and Y. C. Tai, "Micro-electro-mechanical-system (MEMS) and fluid flows," Annu. Rev. Fluid Mech. **30**, 579 (1998).

[3] C. Cercignani, *The Boltzmann equation and its applications* (Springer, New York, 1987).

[4] G. A. Bird, *Molecular Gas Dynamics and the Direct Simulation of Gas Flows* (Oxford Science, Oxford, 1994).

[5] Y. Sone, *Kinetic Theory and Fluid Dynamics* (Birkhäuser, Boston, 2002).

[6] Y. Sone, Y. Waniguchi, and K. Aoki, "One-way flow of a rarefied gas induced in a channel with a periodic temperature distribution," Phys. Fluids **8** (8), 2227 (1996).

[7] A. A. Alexeenko, S. F. Gimelshein, E. P. Muntz, and A. D. Ketsdever, "Kinetic modeling of temperature driven flows in short microchannels," Int. J. Thermal Sci. **45**, 1045 (2006).

[8] Y.-L. Hana, E. P. Muntza, A. Alexeenkob, and M. Young, "Experimental and computational studies of temperature gradient-driven molecular transport in gas flows through nano/microscale Channels," Nanoscale Microscale Thermophys. Eng. **11**, 151 (2007).

[9] Y. L. Han, "Thermal-creep-driven flows in Knudsen compressors and related nano/microscale gas transport channels," J. Microelectromech. Syst. 17 (4), 984-997 (2008).

[10] S. K. Loyalka, "Kinetic theory of thermal transpiration and mechanocaloric effect. I," J. Chem. Phys. **55** (9), 4497 (1971).

[11] S. K. Loyalka, "Slip in the thermal creep flow," Phys. Fluids **14** (1), 21 (1971).

[12] S. K. Loyalka, "Kinetic theory of thermal transpiration and mechanocaloric effect. II," J. Chem. Phys. **63** (9), 4054 (1975).

[13] S. K. Loyalka and T. S. Storvick, "Kinetic theory of thermal transpiration and mechanocaloric effect. III. Flow of a polyatomic gas between parallel plates," J. Chem. Phys. **71** (1), 339 (1979).

[14] S. S. Lo, S. K. Loyalka, and T. S. Storvick, "Kinetic theory of thermal transpiration and mechanocalorlc effect. V. Flow of polyatomic gases in a cylindrical tube with arbitrary accommodation at the surface," J. Chem. Phys. **81** (5), 2439 (1984).

[15] T. Kanki and S. Iuchi, "Poiseuille fllow and thermal creep of a rarefied gas between parallel plates," Phys. Fluids **16** (5), 594 (1973).

[16] T. Ohwada, Y. Sone, and K. Aoki, "Numerical analysis of the Poiseuille and thermal transpiration flows between two parallel plates on the basis of the Boltzmann equation for hard-sphere molecules," Phys. Fluids A **1** (12), 2042 (1989).

[17] P. Taheri and H. Struchtrup, "Rarefaction effects in thermally-driven microflows," Physica A **389**, 3069 (2010).

[18] M. A. Gallis, J. R. Torczynski, D. J. Rader, and G. A. Bird, "Convergence behavior of a new DSMC algorithm," J. Comput. Phys. **228**, 4532 (2009).

[19] S. Stefanov, "Particle Monte Carlo algorithms with small number of particles in grid cells, in Numerical Methods and Applications," Lecture Notes in Comput. Sci. **6064**, 110 (2010).

[20] S. Stefanov, "On DSMC calculations of rarefied gas flows with small number of particles in cells," SIAM J. Sci. Comput. **33**, 677 (2011).

[21] E. Roohi and M. Darbandi, "Extending the Navier–Stokes solutions to transition regime

in two-dimensional micro- and nanochannel flows using information preservation scheme," Phys. Fluids **21** (2009).

[22] E. Roohi, M. Darbandi, and V. Mirjalili, "Direct Simulation Monte Carlo Solution of Subsonic Flow through Micro/Nanoscale Channels," Journal of Heat Transfer **131**, 92402 (2009).

[23] M. Darbandi and E. Roohi, "Study of Subsonic/supersonic Gas Flow through Micro/nanoscale Nozzles Using Unstructured DSMC Solver," Microfluidics and Nanofluidics **10**, 321 (2011).

[24] T. J. Scanlon, E. Roohi, C. White, M. Darbandi, and J. M. Reese, "An open source, parallel DSMC code for rarefied gas flows in arbitrary geometries," Computers & Fluids **39**, 2078 (2010).

[25] E. Roohi and M. Darbandi, "Recommendations on Performance of Parallel DSMC Algorithm in Solving Subsonic Nanoflows," Applied Mathematical Modeling **36**, pp. 2314-2321, (2012).

[26] M. Darbandi and E. Roohi, "DSMC Simulation of Subsonic Flow through Nanochannels and Micro/Nano Steps," International Communication in Heat and Mass Transfer, **38** (10), pp. 1444-1449, (2011).

[27] O. Ejtehadi, E. Roohi, and J. Abolfazli, "Investigation of Basic Molecular Gas Structure Effects on Hydrodynamics and Thermal Behaviors of Rarefied Shear Driven Flow Using DSMC", International Communication in Heat and Mass Transfer, **39** (3), pp. 439-448, (2012).

[28] W. Liou and Y. Fang, "Implicit Boundary Conditions for Direct Simulation Monte Carlo Method in MEMS Flow Predictions," Computer Modeling in Engineering & Science **1**, 119 (2000).

[29] M. Wang and Z. Li, "Simulations for gas flows in Microgeometries using the direct simulation Monte Carlo method," International Journal of Heat and Fluid Flow **25**, 975 (2004).

[30] F. J. Alexander, A. L. Garcia, and B. J. Alde, "Cell size dependence of transport coefficients in stochastic particle algorithms," Phys. Fluids **10** (6), 1540 (1998).

[31] F. J. Alexander, A. L. Garcia, and B. J. Alder, "Erratum: ``Cell size dependence of transport coefficients in stochastic particle algorithms" [Phys. Fluids [bold 10], 1540 (1998)]," Physics of Fluids **12** (3), 731 (2000).

[32] Z. X. Sun, Z. Tang, Y. L. He, and W. Q. Tao, "Proper cell dimension and number of particles per cell for DSMC," Computers & Fluids (2011).

[33] N. G. Hadjiconstantinou, "Analysis of discretization in the direct simulation Monte Carlo," Phys. Fluids **12** (10), 2634 (2000).

[34] C. Shu, X. H. Mao, and Y. T. Chew, "Particle number per cell and scaling factor effect on accuracy of DSMC simulation of micro flows," Int. J. Numer. Methods Fluids **15** (8), 827 (2005).

[35] E. B. Arkilic, M. A. Schmidt, and K. S. Breuer, "Gaseous slip flow in long microchannels," J. Microelectromech. Syst. **6** (2), 167 (1997).

[36] G. E. Karniadakis, A. Beskok, and N. Aluru, *Microflows and Nanoflows: Fundamentals and Simulation* (Springer-Verlag, New York, 2005).